How the Universe Started Before the Universe Started

SBN-13: 978-1534766211

ISBN-10: 1534766219

Author Peet (P.S.J.) Schutte

I don't care who you are or what your religion allows you to believe but I use mathematics with which I explain Genesis 1 v 1. I explain the beginning or birth of the Universe in terms of science and I prove that science precisely matches the Bible using not theology but mathematics. I prove Genesis 1 verse 1 to be correct according to nature's applying of science by using mathematics. The Bible says: IN THE BEGINNING OF CREATION, 1 when God made heaven and earth, 2 the earth was without form and void… 3… with darkness over the face of the abyss…Yes this I do explain mathematically and I manage this because I prove and explain four cosmic keys that build the Universe namely: The Titius Bode Law; The Lagrangian Points, The Roche limit and The Coanda Effect Everything in nature in the Universe applies these phenomena in how space forms. The Titius Bode Law: The Lagrangian Points: The Roche limit and The Coanda Effect forms the Universe in as much as forming stars in spheres and forming galactica in circles. These principles form space and materials. This is a process that produces space and that is how the Universe began before the Universe began in space. I take events back to how the Universe started at the arrival of the cosmic birth. I take events back to the point long before space came about as that took place during the Big Bang. The process proves Dark Matter and the so called String Theory

A COSMIC BIRTH... STARTING BEFORE ZERO

e-pub version By P.S.J. (Peet) Schutte

© Copy right KOSMOLOGIESE EN ASTRONOMIESE TEGNIKA

All rights are reserved.

No part, parts or the entirety of this book may be reproduced by publishing, electronically copied, duplicated by whatever means that form reproduction or duplication, without the prior written consent of the copy rite owner.

ISBN 978-1-291-50875-8.

Written by P.S.J. (Peet) Schutte Please

This book has a standard license

direct all mail to mailto:orders@sirnewtonsdraud.com

or mailto:info@questioneblescience.net

The paper-printed version as well as the PDF version is much better and more informative. This is because of the technical limitations and format layout of the formation that e -book publishing presents and therefore certain information could not be presented in that format. However it is only affecting limiting the available proof and does not harm the message. Those that wish to attack me do so after reading the more informing versions than just the limited e-pub version. A COSMIC BIRTH... STARTING BEFORE ZERO in PDF as well as in Paper Printed copies are available from Lulu.com.

This is book # 1 of

A COSMIC BIRTH...

STARTING BEFORE

ZERO

...and that is precisely what it does.

It shows that nature destroys all Newtonian principle credibility. If you find this hard to believe then read on; what have you got to lose but a bit of your time ?

ISBN-13: 978-1499781090

ISBN-10: 1499781091

This forms part of Naturescosmicconcept

WRITTEN BY Peet (P S. J.) SCHUTTE

©KOSMOLOGIESE EN ASTRONOMIESE TEGNIKA

All rights are reserved.

No part, parts or the entirety of this book may be reproduced by publishing, electronically copied, duplicated by whatever means that form reproduction or duplication, without the prior written consent of the copy rite owner.

This publication aims to put truth into science!

This reveals how the Universe formed and how nature works in accordance with a vision about Creation and in all of that it excludes Newtonian visionary rubbish. Representing the view presented in the following websites

Other websites are:

Naturescosmicconcept or Titius-Bode-Law-Explained and www.sirnewtonsfraud.com Also ANaturesCosmiConcept

The science you are going to encounter is nothing like you have ever seen. Therefore notwithstanding whatever your qualifications might be you have to get familiar with every detail of what you will read.

Not one aspect used in this publication is known except in the books I publish.

]

There are 4 laws applying in nature and that is the part that science ignores because with these laws Nature is Annihilating Newton

This is the Titius Bode law

The Titius Bode law proves that mass has no place in science. See in the picture how random mass is and with such randomness how can mass place planets in the positions they hold. By my effort to solve the mystery of the Titus Bode law I prove that gravity forms not by mass but gravity forms by Π forming in movement Π^2. Solving the Titius Bode law and proving from that how gravity works opens up a new view on the cosmos.

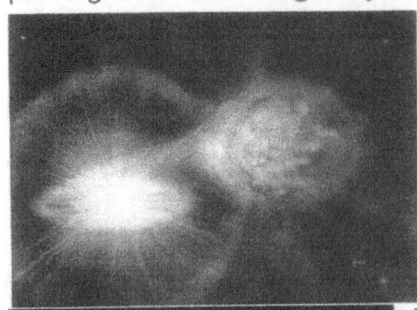

This is the Roche limit. The Roche limit proves amongst others how the sound barrier applies and works. It also proves that cosmic structures with an atmosphere can never collide because the Roche limit that produces the atmosphere prevent foreign object from moving faster that $\Pi^2 / 4$ within the boundary limitation of that atmosphere. The Roche limit brings further proof that using the truth about gravity in physics the answer is simple; it is that gravity is Π.

credit: NASA/JPL

This is the Lagrangian points

The Lagrangian points have been known to science for centuries and with all the mathematical splendour available not one calculation could ever explain why this event is taking place. The satellites form precise locations positioned around the major planet and never comes closer while remaining the their positions.

This is the Coanda effect

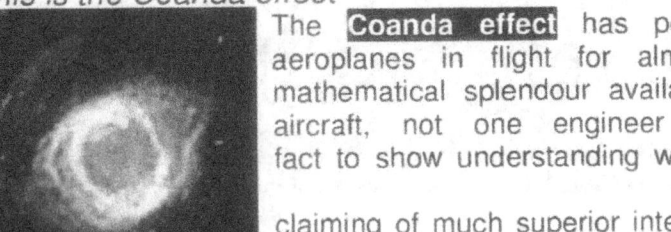

turbine engines and century and with all the design the most terrific mathematically compute one takes place.
How sad it is that those physics remain just no more understanding is not

The Coanda effect has powered aeroplanes in flight for almost a mathematical splendour available to aircraft, not one engineer could fact to show understanding why this

claiming of much superior intellect in than having computing power. The complex.

I have to warn the readers that the topics are showing a very new approach with no quick answers. Understanding is in the proof and that does not come by reading just a few lines and then forming conclusions. The information is new but not hard to grasp. I did not put these phenomena in place and these phenomena nullifies Newton's correctness and that proof I bring goes beyond any doubt. I prove the Titius Bode law. Go to the Internet and see how science doubt the Titius Bode law and the correctness thereof while to solve the problem you add 3 plus 4 to get 7 that is if you want to find a solution. I have published the Titius Bode law in four already published books but in this one I go deeper than the four already published. In each of the books I present I disclose how the Titius Bode law forms gravity

These 4 laws are part of nature. Go look it up before you go on...and why don't science recognize these laws...because these laws brings the entire industry of science to an abrupt end and will stop everything science put in place as science. Recognizing the importance of these laws will kill an industry worth trillions... I now introduce you to the Titius Bode law or how the solar system forms.

There are these four laws or phenomenon or principles in the Universe. Newtonian science can't explain it but I can. In doing so I shatter the myth called Newtonian science and I reveal the hoax science portray for three hundred years or more as the truth. I tried to get published but I was unable because then I break the strangle hold physics has.

Those physicists formulating science are a mafia-gangster club controlling the dishonesty that formulates science... I am one person trying to correct what is a joke but what is also sold as the truth.

Science force humanity to accept the hoax Newton founded and science brainwash everybody to disbelieve nature while clinging on Newtonian hogwash. That is mind-control and that is brainwashing!

I have sent this book with six other books to eighty-five publishers and e-mailed this book to thirty something more. I had no response...not from one. This book opens a new era in understanding how the cosmos works...and not one publisher found it interesting enough to publish or to reply reasons why they found no interest in this as a publishing project. No one is prepared to break the hoax and publish the truth...

For the first time in all of human history there is a method deciphered to show how NATURE no less forms the solar system...and in eighty five DVD's sent plus another (about) thirty six or seven e-mails going via sendspace and not one was interested to publish. Go to the Internet and see it is said this code can't be deciphered but I did find a way to decipher.

That science says is impossible and yet you will read how I did it. Science plainly ignores nature while nature is the reality.

Nature is the only reality but science brushes nature off the table, as if nature is madness. To so many publishers I sent the entire book...I sent it as a unit with two chapters more than the book you read and found no publisher prepared to take on science and correct the hoax Newtonian science is.

Now you can find out how to crack the code by which nature (not Newton's fiction) forms the Universe in the manner that it forms the solar system. It is simple; it is adding 3 plus 4 to get 7! Finding the Titius Bode is 7 also turns what you thought was cosmology into an explanation of the truth while the truth turns cosmology into truthful science. This Titius Bode law, its 7 / 10 or 10/7.

In this book I explain in detail the layout working process of Titius Bode law.

I explain this in detail because what there is in the solar system is this; the Titius Bode law forming in conjunction with the Coanda effect, the Lagrangian points law as well as the Roche lobe / Roche limit that is what forms gravity. This is it!
Newton and his ideology is as absent as the correctness Newtonian science hides and if you do not believe me go and research this by yourself. Giving this truth I contacted (about or more or less) 150 publishers among which there are about fifty or more Universities and not one had any interest in publishing this book.
This is the first time in human history that any person had the inclination to explain the forming of the solar system and nobody is interested in this venture?
This is how space forms and out there amongst all not one shows interest in publishing?
Some just believe because they accept and others have to understand to accept truth and you can choose where you are. Please be warned about the following:
Reading this book will **intellectually** find the reading to be **very challenging** to any person since what I say was never yet published. Everything you are about to read is new! That I in principle disagree with science's accepted principles on very basic issues is a fact that is undeniably true. What you read about the principles I propose is new to everyone alike. However, I found that the ordinary persons with a scholastic physics background cope with the difficult explaining much better than does Super-Educated-Masters. The Super-Educated-Masters have information stored by culture and if they can't bring the information to mind by recognition of it they fail to understand new science concepts. You are going to read this in a letter that was sent to me.

The purpose with which I wrote this book is to get around the network of Super-Educated-Masters who strangle any information that forms of science in the form I propose and therefore that does not fit their views or match their liking. If what anyone says does not stroke with what the Brainy Bunch says who controls physics and agree with "Mainstream Science" or echo their thinking, they just smother all intellectual publication on the grounds that it is not fitting their profile on science. If you ignore culture and read with using your logic you will find many accepted norms as ridiculous. With most concepts I disagree most strongly and I disagree because those concepts lack proof but I do also supply detailed proof of my views and that is where Mainstream Science blocks the publishing of my views on science that does not compliment their views. Read this and wake from the culture you believe in; that which science has lulled you into and made you accept science as the absolute undeniable rock fast truth by instating it as a religiosity then stop reading or get your tranquillising anti depressants next to you with a large bowl of water and a big glass. You will find some mathematical equations, if you are not familiar with it ignore it because it shows the silliness of "Mainstream Science" but if you don't read it you will still understand the explaining by reading the language where I explain it. "Mainstream Science" hides behind maths. I need help to fight their fraud and I need you to help me fight them. What you read I prove to even every last detail and even in this book and therefore I dare anyone to prove otherwise or reprimand me.

You are not going to read a book but you are going to travel a journey. Everything written in this assembly is collected from numerous articles and papers I wrote to individuals, journals, science magazines, Universities, academics in administrative-teaching positions as well as many on line physicists which I tried to interest in my view

and my findings on science. I did not remove parts or sections of the articles to disguise this as a book that is ready-written but formed it to be a road on which I travelled and the way I found never-ending rejections. I never tried to make a name in physics but tried to get some money to make a living from and take care of my children. I always knew there were so many smarter brains out there than what I have and who could see what I saw and take the challenge of correcting science from what I brought to the table. I was stonewalled by a bunch of corrupt conspirators trying to lay claim to rubbish Newton presented as truth and in some cases their efforts to justify Newton's lies became pathetically poor. I realised there are those that has brains and then you have those that understand Newton and that is why astrophysics remained backwards as it got stuck promoting forces flying all-over pulling to form gravity. I say this straight: no amount of words can describe my utter disgust I have for those in science thought of as flawless performing beyond blame because they are criminals hiding their criminality.
Why would knowing the Titius Bode improve our understanding of physics?

In my books I explain how these stars come about and why they form as they do. If you wish to also know read on...

An in-depth overview of my what my work involves you may visit naturescosmicconcept to go there click on
naturescosmicconcept or
ANaturesCosmiConcept : **For a broader basis of the content concerning the concept and what it involves go to ANaturesCosmiConcept**

NaturesCosmiConcept-E-Z: To be informed about what information involves the new antural cosmic concept go to NaturesCosmiConcept-E-Z

NaturesCosmiConcept-E-Z-R: Should you not have time and only need being informed about the basis got to NaturesCosmiConcept-E-Z-R

After finishing A COSMIC BIRTH... STARTING BEFORE ZERO Book # 1 you must evolve to A COSMIC BIRTH... STARTING BEFORE ZERO Book # 2

The first part (Book # 1) will introduce the cosmic principles which I introduce and the second part Book # 2 will bring the conclusions on the principles I introduce. Reading the first book is essential to understand the second book and the conclusions are part of the second book.

This is book # 1 of

A COSMIC BIRTH...

STARTING BEFORE

ZERO

...and that is precisely what it does.

It shows that nature destroys all Newtonian principle credibility. If you find this hard to believe then read on; what have you got to lose but a bit of your time ?

ISBN-13: 978-1499781090

ISBN-10: 1499781091

This forms part of Naturescosmicconcept

WRITTEN BY Peet (P S. J.) SCHUTTE

©KOSMOLOGIESE EN ASTRONOMIESE TEGNIKA

All rights are reserved.

No part, parts or the entirety of this book may be reproduced by publishing, electronically copied, duplicated by whatever means that form reproduction or duplication, without the prior written consent of the copy rite owner.

This publication aims to put truth into science!

This reveals how the Universe formed and how nature works in accordance with a vision about Creation and in all of that it excludes Newtonian visionary rubbish.

Representing the view presented in the following websites

Other websites are:

Naturescosmicconcept or **Titius-Bode-Law-Explained** and **www.sirnewtonsfraud.com** Also **ANaturesCosmiConcept**

The science you are going to encounter is nothing like you have ever seen. Therefore notwithstanding whatever your qualifications might be you have to get familiar with every detail of what you will read.

Not one aspect used in this publication is known except in the books I publish.

1

There are 4 laws applying in nature and that is the part that science ignores because with these laws Nature is Annihilating Newton

This is the Titius Bode law

The Titius Bode law proves that mass has no place in science. See in the picture how random mass is and with such randomness how can mass place planets in the positions they hold. By my effort to solve the mystery of the Titius Bode law I prove that gravity forms not by mass but gravity forms by Π forming in movement Π^2. Solving the Titius Bode law and proving from that how gravity works opens up a new view on the cosmos.

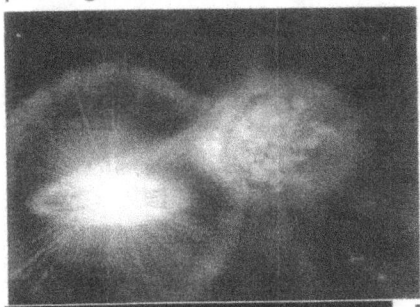

This is the Roche limit. The Roche limit proves amongst others how the sound barrier applies and works. It also proves that cosmic structures with an atmosphere can never collide because the Roche limit that produces the atmosphere prevent foreign object from moving faster that $\Pi^2 / 4$ within the boundary limitation of that atmosphere. The Roche limit brings further proof that using the truth about gravity in physics the answer is simple; it is that gravity is Π.

credit: NASA/JPL

This is the Lagrangian points

The Lagrangian points have been known to science for centuries and with all the mathematical splendour available not one calculation could ever explain why this event is taking place. The satellites form precise locations positioned around the major planet and never comes closer while remaining the their positions.

This is the Coanda effect

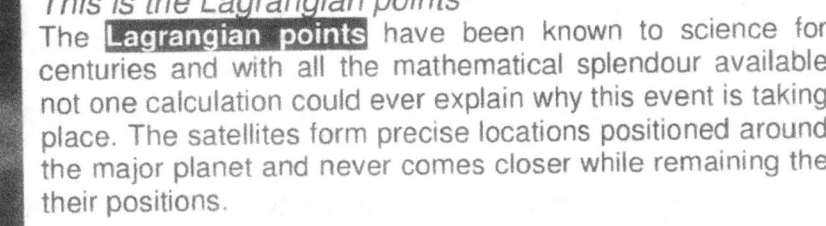

turbine engines and century and with all the design the most terrific mathematically compute one takes place.
How sad it is that those physics remain just no more understanding is not

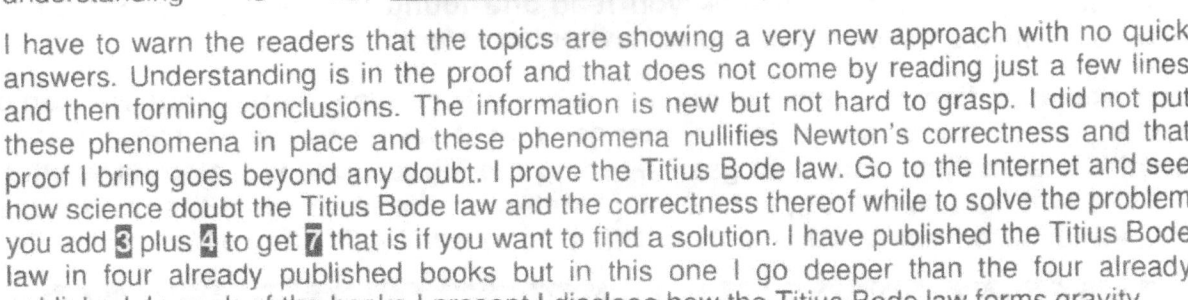

The Coanda effect has powered aeroplanes in flight for almost a mathematical splendour available to aircraft, not one engineer could fact to show understanding why this

claiming of much superior intellect in than having computing power. The complex.

I have to warn the readers that the topics are showing a very new approach with no quick answers. Understanding is in the proof and that does not come by reading just a few lines and then forming conclusions. The information is new but not hard to grasp. I did not put these phenomena in place and these phenomena nullifies Newton's correctness and that proof I bring goes beyond any doubt. I prove the Titius Bode law. Go to the Internet and see how science doubt the Titius Bode law and the correctness thereof while to solve the problem you add 3 plus 4 to get 7 that is if you want to find a solution. I have published the Titius Bode law in four already published books but in this one I go deeper than the four already published. In each of the books I present I disclose how the Titius Bode law forms gravity

These 4 laws are part of nature. Go look it up before you go on...and why don't science recognize these laws...because these laws brings the entire industry of science to an abrupt end and will stop everything science put in place as science. Recognizing the importance of these laws will kill an industry worth trillions... I now introduce you to the Titius Bode law or how the solar system forms.

There are these four laws or phenomenon or principles in the Universe. Newtonian science can't explain it but I can. In doing so I shatter the myth called Newtonian science and I reveal the hoax science portray for three hundred years or more as the truth. I tried to get published but I was unable because then I break the strangle hold physics has.

Those physicists formulating science are a mafia-gangster club controlling the dishonesty that formulates science... I am one person trying to correct what is a joke but what is also sold as the truth.

Science force humanity to accept the hoax Newton founded and science brainwash everybody to disbelieve nature while clinging on Newtonian hogwash. That is mind-control and that is brainwashing!

I have sent this book with six other books to eighty-five publishers and e-mailed this book to thirty something more. I had no response...not from one. This book opens a new era in understanding how the cosmos works...and not one publisher found it interesting enough to publish or to reply reasons why they found no interest in this as a publishing project. No one is prepared to break the hoax and publish the truth...

For the first time in all of human history there is a method deciphered to show how NATURE no less forms the solar system...and in eighty five DVD's sent plus another (about) thirty six or seven e-mails going via sendspace and not one was interested to publish. Go to the Internet and see it is said this code can't be deciphered but I did find a way to decipher.

That science says is impossible and yet you will read how I did it. Science plainly ignores nature while nature is the reality.

Nature is the only reality but science brushes nature off the table, as if nature is madness. To so many publishers I sent the entire book...I sent it as a unit with two chapters more than the book you read and found no publisher prepared to take on science and correct the hoax Newtonian science is.

Now you can find out how to crack the code by which nature (not Newton's fiction) forms the Universe in the manner that it forms the solar system. It is simple; it is adding 3 plus 4 to get 7! Finding the Titius Bode is 7 also turns what you thought was cosmology into an explanation of the truth while the truth turns cosmology into truthful science. This Titius Bode law, its 7 / 10 or 10/7.

In this book I explain in detail the layout working process of Titius Bode law.

I explain this in detail because what there is in the solar system is this; the Titius Bode law forming in conjunction with the Coanda effect, the Lagrangian points law as well as the Roche lobe / Roche limit that is what forms gravity. This is it!
Newton and his ideology is as absent as the correctness Newtonian science hides and if you do not believe me go and research this by yourself. Giving this truth I contacted (about or more or less) 150 publishers among which there are about fifty or more Universities and not one had any interest in publishing this book.
This is the first time in human history that any person had the inclination to explain the forming of the solar system and nobody is interested in this venture?
This is how space forms and out there amongst all not one shows interest in publishing?
Some just believe because they accept and others have to understand to accept truth and you can choose where you are. Please be warned about the following:
Reading this book will **intellectually** find the reading to be **very challenging** to any person since what I say was never yet published. Everything you are about to read is new! That I in principle disagree with science's accepted principles on very basic issues is a fact that is undeniably true. What you read about the principles I propose is new to everyone alike. However, I found that the ordinary persons with a scholastic physics background cope with the difficult explaining much better than does Super-Educated-Masters. The Super-Educated-Masters have information stored by culture and if they can't bring the information to mind by recognition of it they fail to understand new science concepts. You are going to read this in a letter that was sent to me.

The purpose with which I wrote this book is to get around the network of Super-Educated-Masters who strangle any information that forms of science in the form I propose and therefore that does not fit their views or match their liking. If what anyone says does not stroke with what the Brainy Bunch says who controls physics and agree with "Mainstream Science" or echo their thinking, they just smother all intellectual publication on the grounds that it is not fitting their profile on science. If you ignore culture and read with using your logic you will find many accepted norms as ridiculous. With most concepts I disagree most strongly and I disagree because those concepts lack proof but I do also supply detailed proof of my views and that is where Mainstream Science blocks the publishing of my views on science that does not compliment their views. Read this and wake from the culture you believe in; that which science has lulled you into and made you accept science as the absolute undeniable rock fast truth by instating it as a religiosity then stop reading or get your tranquillising anti depressants next to you with a large bowl of water and a big glass. You will find some mathematical equations, if you are not familiar with it ignore it because it shows the silliness of "Mainstream Science" but if you don't read it you will still understand the explaining by reading the language where I explain it. "Mainstream Science" hides behind maths. I need help to fight their fraud and I need you to help me fight them. What you read I prove to even every last detail and even in this book and therefore I dare anyone to prove otherwise or reprimand me.

You are not going to read a book but you are going to travel a journey. Everything written in this assembly is collected from numerous articles and papers I wrote to individuals, journals, science magazines, Universities, academics in administrative-teaching positions as well as many on line physicists which I tried to interest in my view

and my findings on science. I did not remove parts or sections of the articles to disguise this as a book that is ready-written but formed it to be a road on which I travelled and the way I found never-ending rejections. I never tried to make a name in physics but tried to get some money to make a living from and take care of my children. I always knew there were so many smarter brains out there than what I have and who could see what I saw and take the challenge of correcting science from what I brought to the table. I was stonewalled by a bunch of corrupt conspirators trying to lay claim to rubbish Newton presented as truth and in some cases their efforts to justify Newton's lies became pathetically poor. I realised there are those that has brains and then you have those that understand Newton and that is why astrophysics remained backwards as it got stuck promoting forces flying all-over pulling to form gravity. I say this straight: no amount of words can describe my utter disgust I have for those in science thought of as flawless performing beyond blame because they are criminals hiding their criminality.
Why would knowing the Titius Bode improve our understanding of physics?

In my books I explain how these stars come about and why they form as they do. If you wish to also know read on...

An in-depth overview of my what my work involves you may visit naturescosmicconcept to go there click on
naturescosmicconcept or
ANaturesCosmiConcept : **For a broader basis of the content concerning the concept and what it involves go to ANaturesCosmiConcept**

NaturesCosmiConcept-E-Z: To be informed about what information involves the new antural cosmic concept go to NaturesCosmiConcept-E-Z

NaturesCosmiConcept-E-Z-R: Should you not have time and only need being informed about the basis got to NaturesCosmiConcept-E-Z-R

After finishing A COSMIC BIRTH... STARTING BEFORE ZERO Book # 1 you must evolve to A COSMIC BIRTH... STARTING BEFORE ZERO Book # 2

The first part (Book # 1) will introduce the cosmic principles which I introduce and the second part Book # 2 will bring the conclusions on the principles I introduce. Reading the first book is essential to understand the second book and the conclusions are part of the second book.

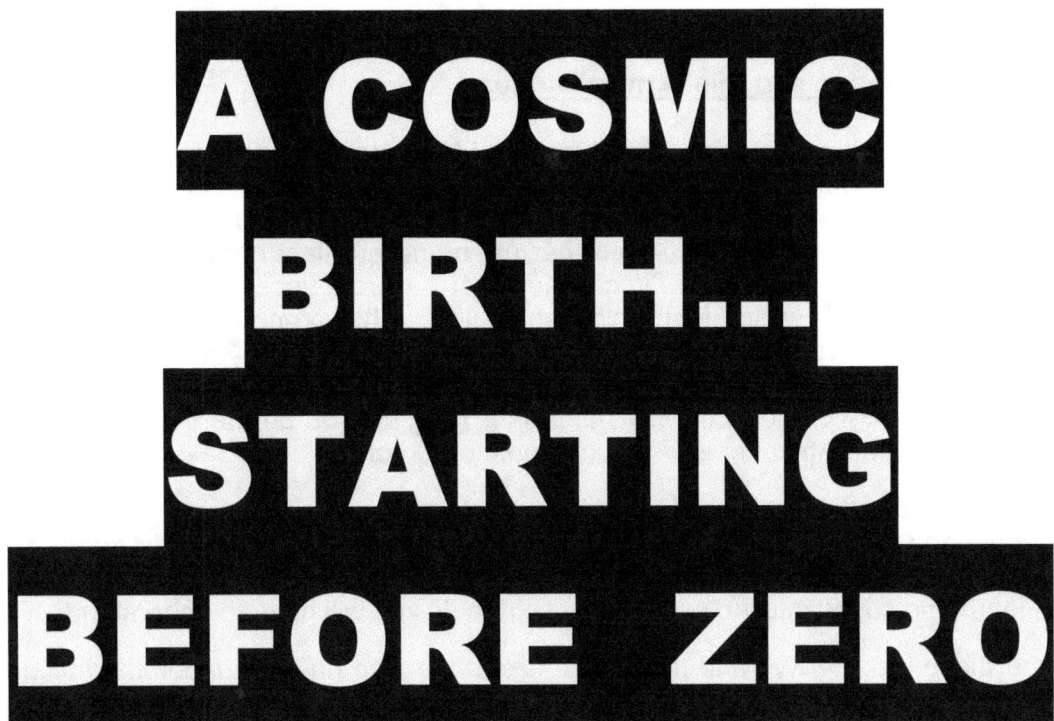

A COSMIC BIRTH... STARTING BEFORE ZERO

ISBN-13: 978-1499781090

ISBN-10: 1499781091

Written by P.S.J. (Peet) Schutte Please direct all mail to

mailto:info@singularityrelevancy.com

or mailto:info@questioneblescience.net

Please be forewarned: You are going to see me introduce myself numerous times during the length of the book because this book is compiled by many articles written to distinct and not-so distinct physicists, physic journals, science magazines, University lectors and professors, articles and entire books sent off to Universities and damn-well any body so that I could get someone and anyone to listen to what I have to say about my work. I still do it. No sooner you tell the person Newton is a lot of bullshit and they refuse to read further. This time it is your turn to put the book down but when doing that you will remain as uninformed as you were before you started this book. Notwithstanding who you are, what your achievements are, what your level of education is or what your academic standing and history is but what this book reveals as far as science you have never come across or experienced. Therefore then keep your opinion clear until you have been through the book.

This book was done with a $25 ^{oo} scanner and a $35 ^{oo} printer and the reason I explain inside. For the same reason this book was not edited or linguistically checked. I could not because that does not work because I am in the writing business and not the spelling business and while I check spelling the writing gets more and so does the spelling and grammar errors. I had a choice; doing the books with no funding or not doing it at all because while I rubbish Newtonian science and show it is the fake it is, they will never publish my work because I trash Newton. Not having funds and trying to fight science for the truth with the truth was a fight that physically broke my health and still I am not published except in this manner. I apologise for the spelling and language but in poverty that was the best I could do under the prevailing circumstances in which I find myself…. This book is a first in every sense… it unites science and religion because science and religion was separated by human stupidity.

To whom it may concern,

This is my introduction and this is my prologue:

But before I can commence with that task I have another duty to administer:

I do find much pride in my status as being Afrikaner and would like to have my names used by pronouncing it in the manner Afrikaans dictates…therefore I would sincerely appreciate the courtesy when readers will take note that my name and last name are pronounced in Afrikaans, which is originally from Dutch and must be pronounced in that way. Peet one would pronounce "here" which is the closest English to the pronouncing of the "ee". The "Sch" in Schutte is pronounced exactly as one would pronounce sch in school and the pronouncing dictates that it is done in such a manner as where both actually are pronounced Skutte or "skool". By pronouncing my name in Afrikaans you do me the utmost courtesy any one can. Being an Afrikaner is what I am most proud of.

Please take note that I sell information and not words or books and therefore the information takes priority and not the spelling or words used to inform the readers. This represents the work of God and not the word of God and so there is no interpretations applying and versus you can learn and sound intellectual but only cold facts you will have to understand.

There are those who believe and then there are those who have to understand to believe. This book aims to satisfy those who have to understand in order to believe. However, understanding is a mental ability and not an emotion as believing is. This is science and not theology and to understand this requires intellect and not just rehearsing a few verses that you learned off by heart as it is taken from the Bible. To understand this concept I introduce will require a fair bit of study on your part. It is not going to be easy but it will be rewarding.

This is to inform you about my latest book aiming too bring justice to nature by showing how nature and not Newtonian principles form the cosmos. There are four cosmic principles that nature applies and Newtonian science ignores. Nature form the cosmos by applying the following four principles found in nature:

These are **1) The Titius Bode law**

2) The Coanda effect

3) The Roche limit

4) The Lagrangian points and these four still form everything.

Have you heard of these cosmic phenomena? These phenomena are the building blocks by which and on which the entirety of the Universe forms and yet science in the Newtonian fashion ignore not only the importance but also the phenomena having any role in physics. While Newtonian science ignores these natural phenomena because it annihilates the way Newton saw the cosmos, we have a scenario where we can apply only one of the two options that are available which can be correct. We can forcibly conspire to make Newtonian principles apply as is the current teaching practise notwithstanding everything pointing to this being wrong or we can look at nature and find the correct way that the Universe works.

The Sun and Nine Planets   Copyright © Calvin J. Hamilton

Look at the picture. Envisage how planets are in allocated positions according to mass. There is no order according to mass, therefore without order mass can't apply mathematically and the formulas Newtonian science pretended to have invented is a hoax.

Remember I try to facilitate as much information in as little space in this introduction and it might come across as compact but with 500 pages everything becomes clear.

This book and all my other books introduce a new concept about science. Science at present is flawed and before throwing this introduction aside in disgust then first go and look at the planets varying in size. The distribution of mass is completely random and there is no mathematical or theoretical or any other complying order to the way planets arrange by mass. Therefore it is madness to promote any idea that mass can place planets in located positions. Nature and Newton is not compatible even in the least. We have to distrust science or we have to distrust nature. Put in the correct sense we have to trust nature or we have to trust Newton because what Newton says is incompatible with nature and this is true above and beyond the brainwashing mainstream science inflict on us all. You now can for the first time in human history choose to read the truth for the first time or you can remain brainwashed forever believing Newton. I deciphered the following cosmic laws that totally annihilated Newtonian wisdom.

These laws are what nature put in place instead of mass and moreover being in place of what Newtonian fantasy makes every person believe is science.
To be able to achieve what the Bible says happened I had to achieve what no one in science this far could or would or even tried to achieve in science. I had to break a code of silence.

Before I prove how the cosmos started I first have to prove how nature forms the Universe. This is how nature forms the Universe and these laws form the Universe. From deciphering how these laws apply I could decipher how the Universe started and guess what; I then managed to use mathematics to support what the Bible says how this Universe came about.

This is the Titius Bode law

The Titius Bode law proves that mass has no place in science. See in the picture how random mass is and with such randomness, how can mass place planets in the positions they hold? By my effort to solve the mystery of the Titius Bode Law, I prove that gravity forms not by mass but gravity forms by π forming in movement π². Solving the Titius Bode Law and proving from that how gravity works opens up a new view on the cosmos.

From mathematically explaining how the planet formation forms I could gather the way the cosmos formed right in the very beginning when the Bible said it started the way the Bible said it started. Mathematically proven it is precisely as the Bible says the cosmos started.

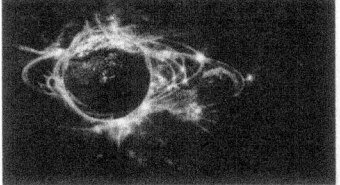

This is The Roche limit

The Roche limit has been around for centuries and with all the mathematical splendour available to apply in order to fathom concepts behind this phenomenon, still with all the computing ability of a machine all those physicists with all the mathematical superiority could not touch any understanding about the concept forming the background. Yet when using the truth about gravity in physics the answer is simple; it is that gravity is Π.

This is the Lagrangian points

The Lagrangian points have been known to science for centuries and with all the mathematical splendour available not one calculation could ever explain why this event is taking place. The satellites form precise locations positioned around the major planet and never comes closer while remaining in their positions.

I introduce these four laws as I introduce you to science in the manner as nature forms the Universe. These four laws shows how and even why the Universe started because these four laws are the cosmos and the way the cosmos started. There is no Newtonian trickery.

The four cosmic laws hold the basis of what forms the Universe where every one of these laws form part of what forms Π and together these laws form as a group the value of Π^2 and Π^2 is gravity.

That I prove mathematically.

The Universe forms the value of 1. No more, no less because that is the value of singularity.
In the beginning there was possibilities before material formed groups that formed material in groups that formed groups of arterial, which we then gave individual names. But before that there were spots and dots as there now are spots and dots forming a Universe. To understand how this lot formed we have to go back and find out how the spots and dots formed before the Universe formed our Universe we give so many names to. Essentially only singularity forms a Universe so small it is outside the Universe we recognise.

This is the Coanda effect

The Coanda effect has powered turbine engines and aeroplanes in flight for almost a century and with all the mathematical splendour available to design the most terrific aircraft, not one engineer could mathematically compute one fact to show understanding why this takes place. How sad it is that those claiming of much superior intellect in physics remain just no more than having computing power. The understanding is not complex. I have to warn the readers that the topics are showing a very new approach with no quick answers. Understanding is in the proof and that does not come by reading just a few lines and then forming conclusions. The information is new but not hard to grasp. I did not put these phenomena in place and these phenomena nullifies Newton's correctness and the proof I bring goes beyond any doubt. I prove the Titius Bode law. Go to the internet and see how science doubt the Titius Bode Law and the correctness thereof while to solve the problem you add 3 plus 4 to get 7. That is if you want to find a solution. I have published the Titius Bode Law in four already published books but in this one I go deeper than the four already published. In each of the books I present I disclose how the Titius Bode Law forms gravity. These books are:

Everyone is in agreement with Albert Einstein that the Universe started with one spot, a point we call singularity. Now that you saw the laws I wish to give a glimpse about how the Universe started. As I said, the Universe started with 1 and this fact even Newtonian science accepts. Singularity started as the first spot and not with a massive already formed Universe that only afterwards grew in size. Since the Big Bang event the Universe only grew but everything that is in the Universe already was in the Universe. What was in place at the Big Bang only then grouped and became material but not more of what already was.

Think of it this way. The spot overheats and expands into the dot as the movement comes by the spot enlarging to form the dot and by growing into the dot the movement enlarges and the movement as well as the increase in size reduces the heat the dot accumulates. The growth then by reducing cools off and the cooling removes the structural size of the dot as it returns to form the next spot. However what is in the Universe can never leave the Universe but has to remain a component within the Universe as long as the Universe exists.

However before heat brought about the Universe we know time was one continuing everlasting spot, which is a line that never went further than one spot.

The spot formed the future. The spot formed the present. The spot formed the past. Since the spot in the future was an exact image of the spot in the present and that was identical to the spot in the past the new spot had no identifiable difference between the future and the one in the past. It was a repeat of what was being identical to what was coming and therefore it stayed the same. The spot was so small it was invisible and unnoticeable and yet it is so big that at present time it holds the entirety of what is within. It is as big as nothing could ever be and it was so small it was nothing that could ever be. This was the only time nothing had a validation because as soon as the Universe came nothing disappeared and became something. One should be very clear about understanding this aspect. The spot then represented nothing because it was consistent of nothing since only nothing existed at the time. It was infinity that which can never start united with eternity, which is that which can never end. To start that which can never end first had to end and to end it first had to wait for that which can never start to find a way to start to lead to the process of ending. Since neither process could apply due to unfavourable circumstances making it impossible to follow on the other the process was eternally infinite.

To start with giant formulas as Newtonians do only impresses other dim wits that can't think and only shows incompetence and ignorance as is the case with Newtonian science. Mathematics came about as the Universe formed and as the Universe formed it formed mathematics during the process. I show exactly how did the first spot arrive and what was in

place before the first spot that became one. Π^0 I show why the first spot came into place and what made it be.

1^0 1^1 I show what made 1 grow into two and why did one grow into two. This is the most fundamental reason why the Universe started. It started with the fact that one came

about that became two. 1^0 1^1 1^2 What changes took place to allow two that was one to become three and why did this process begin to form a Universe.

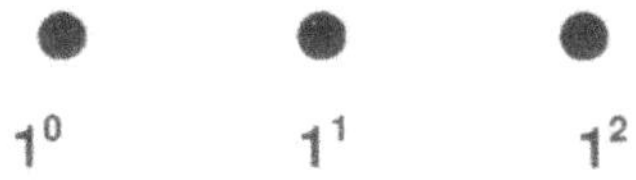

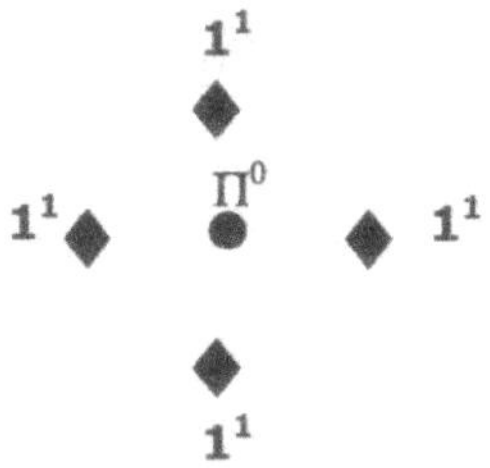

1^1

1^1 Π^0 1^1

1^1

Then also why did four become in the place of three without replacing one or two, as a number and I will give a hint; it was because $2 + 2 = 4$ then became 2×2 $= 4$. This brought the Universe into a new era such as was never seen before.

Then that which can have no inside parted from that which can have no outside. $\Pi^0 = 1^0$ parted from 1^1. The spot became dot and then relevant as the Universe transformed. Another hint is that this is what the Universe are made of and this is how the Universe formed and transformed into what it became. To claim proof I have to show that this indicated that the motion produces the space and the space finds limits in the motion confirming the space while the space is conforming the motion. Singularity is where and in

1^1

1^0

1^1

that the triangle and the half circle and a straight line is equal in $180°$

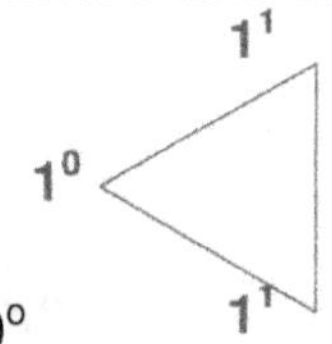

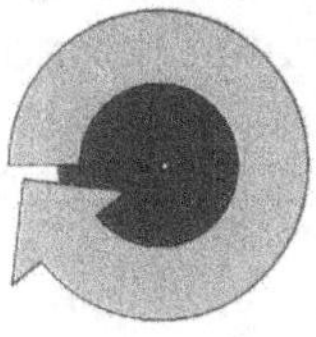

1^1 1^0 1^1

Realizing why these three forms are equal unlocks the information in Genesis 1 verse 1. This was when singularity as 1^0 parted from 1^1 that the motion then came about as 1^2. That in reality left little consolation because with $k^0 = 1^1$ that

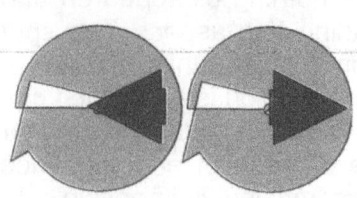

left the space formed by the motion way outside the realms of the emerging Universe. I decided to replace the symbol Kepler used of $k^0 = 1^1$ to a more appropriate Π^0. However I am not going to go into detail before I go into detail about what the four cosmic laws are about and why they in movement control the Universe. The concept is confirmed in the fact that by using a pendulum in a time device such as a clock then it proves that time is movement and gravity is movement and that is why gravity moving the pendulum arm is able with that movement to measure time.

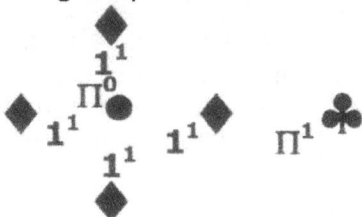

Then the Lagrangian law came in place and this allowed four to advance to five. This principle, as is the case with all the principles are still applicable in nature. However how did the development of gravity take one dot to four and then shifted it to five. This is where the Universe formed $\Pi^0\Pi\Pi^2$ for the very first time.

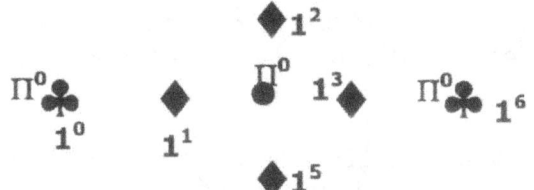

Then as a result using the reason why two became three then applying the very same reason, it came about to bring five to form six.

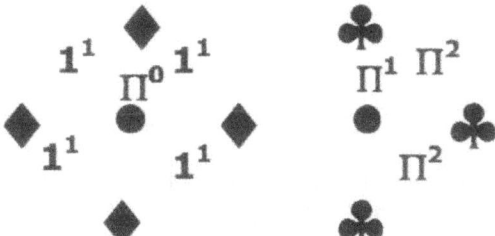

However the reason why four became five is the same reason that brought seven into place because the one is the result of the other forming and the two being five and seven are so well interlinked it had to become in place simultaneously. Therefore this then brought about that gravity formed space as $\Pi^0\Pi\Pi^2$

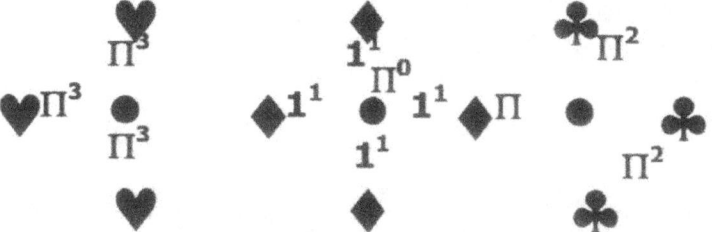

By forming seven it then formed ten and that was where that which was eternal and perfect became temporary and imperfect. Singularity distorted into a concept outside the eternal. The truth is at that point all there was, was an inside and an outside still to form and a promise of what might come.

The mighty wise can push their present into forming the past by clinging onto the worthless they represent as worthwhile that will become the past when these books becomes the present. However, that choice will doom them into the past along with all other things and thoughts not worth the burden to take into the present and onto the future from the past. Their adopting the worthless and not adapting to the truth moves them to the past only worth to be forgotten as the worthless part of the past. The choice is theirs to make.

Again I repeat: You are going to read some <u>mathematics in equations</u> and expressions in <u>mathematical formulas</u> placed to defend my position but if you <u>don't like it</u> then just <u>skip</u> the mathematics because the content and grounds the mathematics proves or disproves is not important in the arguments and it is there for physicist to hide behind. I don't have to hide behind mathematics to make others feel inferior because my arguments make people understand physics and make people feel empowered and superior. The mathematics I include is to show what mindless clots those Superior Humans are that portray their position as superior in mathematical ability and it is there to disprove the Members of the Physics establishment that advocates the necessity of bringing mathematical proof to prove? It is there not to scare readers away but to silence the Brainy Bunch critics by showing them the foolishness of their arguments. By using mathematics the Brainy Bunch have been cheating the public and have been brainwashing students for centuries. That cheating is how they do it and I have to show and uncover the dishonesty in mathematics.

The Universe consists of gravity that forms by the working of the four phenomena never mentioned. That does not say much for the bountiful prestige that mathematician's claim as their lawful bragging rights in areas where true human intellect is called on. Is it not high time to begin to admit you are playing the game of fools with you arrogance about your achievements using mathematics when designing space whirls and travelling to galactica while not even understanding what movement asks for? You do not even understand the neutron and the neutron is compressing density increasing, which is what gravity is, which is what time is, which is what all movement is...that is why the neutron has no mass because mass is the principle coming about where independent movement ends.

You're mathematics could not get you any closer than playing games in a fairy tale Universe using misguided presumptions about mass forming gravity and living the Universal farce which Newton created because that fairy land is what all the Kings clever heroes and all the King's splendid wise could never prove in hundreds of years. If you feel superior as a scientist practising physics on the highest level having a gloating hail of superior mental capability covering you like an aura, then I have very saddening news for you.

If you have the ability to compute and calculate at the highest level, then look at your computer and see one that machine has abilities as a machine which is equal to you, but it's a manmade machine. Stop playing games by creating fairy worlds making up fairy tales about fairies and little people, mass that can create forces, four of them no less, and come and join the rest of us living in reality that does not need to compute forces to be able to not understand what it is that you compute, but to use human intelligence and in that way to understand what only human intellect could ever understand.

Then what in the present is not worth carrying into the future as the past being worthwhile?

Notwithstanding your mathematical brilliance you completely lack any understanding of mathematics or of physics.

The spot was one perfect spot that overheated and parted into a dot.

•• Then by overheating the spot split into two being a spot and a dot.

••• The spot shifted to the past leaving the dot in the present while the dot cooled of and formed another spot one in the past and one in the present and one coming from the future.

The spot expanded  into becoming a dot going to the value of Π.

While the spot expanded into becoming a dot it had nowhere to go because it was still inventing space, which was a concept that did not exist yet.

This was when 1^0 became 1^1

From the four cosmic laws we can see how the Universe started. I will allow a glimpse into the process, which I wrote several books about and still I know I have not scratched the surface. It started with one spot that became a dot through overheating on the one side and cooling on the other side. The dot became two, the next spot and the previous dot, because space interrupted time and in between two instances of time landed one speck of space forming a dot on the one side and a spot on the other side. Remember once anything even a process forms part of the Universe it ha no place to go but to remain part of the Universe.

● 1^0 ● 1^1

From this the Universe started exactly as we find the Bible says it started. It started with $Π^0$ or then singularity or if you wish, then from a void where there is ns space and there is only time in the instant. The entirety we think of non- material space is filled with this. It is a spot that is not and while it is not it still fills and maintains one entire Universe. Wherever you look you will find this spot that holds space and forms space without being able to claim space. Material fills as a solid because movement compresses this dot so tightly it forms a solid construction of compressed material while the material forms of a substance that claims no space. Non-material is this same heat but is much less compressed and so because of a reduced density because of the flow of time from space to material we can see the density increasing on the one side and decreasing on the other side. The proof of this we see in Kepler's tables where the Titius Bode law process that material is regularly spaced in accordance with the growth formula that the Titius Bode law is indicative of.

Also the Roche limit at $Π^2/4$ shows this limit between material forming and the Coanda effect shows a clear growth of density developing around spinning materials as well as the satellite positions of materials around structures we think of as planets. There is a definite distance maintained between matter and therefore it must be non-material the compresses around the sun and while spinning the sun compresses the non-material because this maintains the heat balance within the sun.

When this heat balance goes array the structure overheats and we then find what is thought of as a Super Nova or an exploding star. Everything in the Universe is reliant on density caused by movement, which causes movement, allowing density to development specific time or specific gravitational ratios.

This left one dot and one spot that developed into the next dot. Where is the first dot, the mother dot, and the original dot from where everything came. We are within that first spot that became the mother dot and that dot we think of the entirety as the Universe that has no end or beginning.

Then time formed a sequence of the past, the present and the future leaving three dots lined up in a line we think of as time.

1 = past   2 = present   3 = future

Afterwards space becomes a factor because we see the second dimension develop, as time becomes 2 + 2 = 4 and 2 x 2 = 4. This takes the Universe into a complete new level.

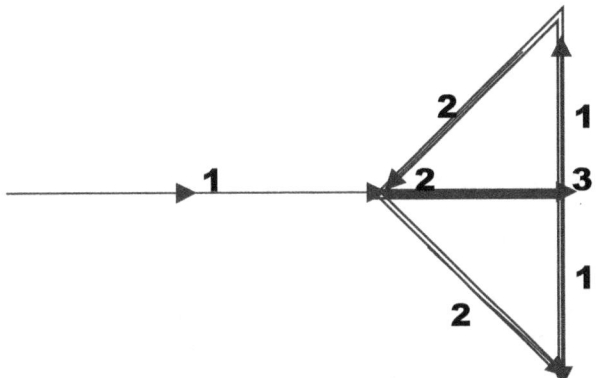

There fore the one line forming time became three and the line becoming space became the line (2 + 2 = 4) as well as the circle (2 x 2 = 4).

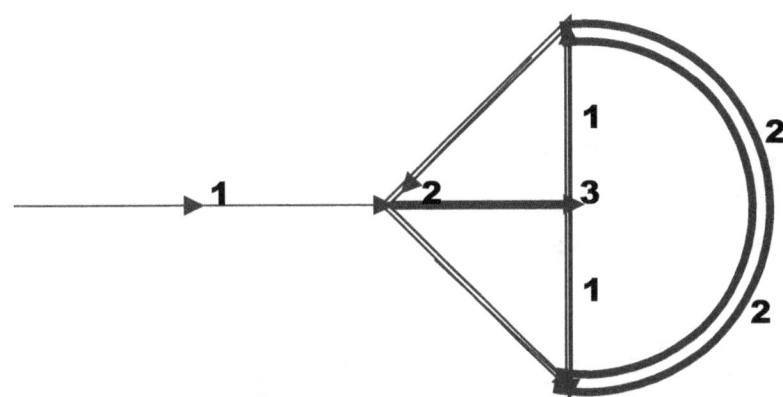

In this we find that the triangle, the half circle and the straight line all three has equal value of 180°. Please keep in mind that we are within the limits of singularity forming 1^0 going 1^1 going 1^2 going 1^3 going 1^4, which all is still Π^0.

In that we find the value of the n ext spot forming at 5, which is the Lagrangian number for the next dot.

At the point where the four cosmic pillars form 7 as a relevance we have also two values 7 + 7 + 7 = 21, which is time (3) times space (7) = 21 and in the space sector there is 5 times the four sides forming space (4 X 5 = 20) plus singularity 1 = 21.

I shall get to the other part forming the value of Π, which is $\Pi = \dfrac{21.991}{7}$ and when in

relation to Π^0 it is $\Pi = \dfrac{3.1416}{\Pi^0}$. The Universe we know begins where the two values

forming Π begins because less that Π we have singularity. This is extremely important to realise because realising this confirms the Bible and the way the Bible states that the Universe started. In the Universe there is infinity within the centre forming time and there is eternity forming a circle where both infinity and eternity forms singularity.

$Space = \dfrac{\Pi r^2}{\Pi^0}$ To get to form cancel space: $\Pi = \dfrac{\Pi r^2}{\Pi^0 r^2} = \Pi$. To get to singularity cancel

form: $\Pi^0 = \dfrac{\Pi}{\Pi} = \Pi^0$ and Π^0 is singularity. The value of Π forms as $(3 + 4)^2$ and $(3^2 + 4^2)$.

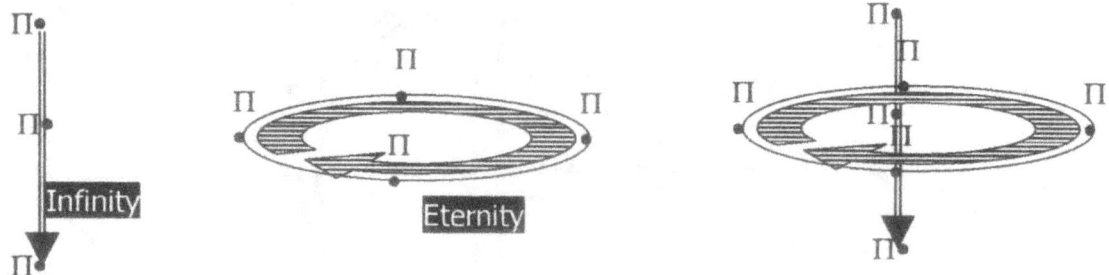

It is through the spinning top that we learn the process of how the Universe started from one dot to what we now have. This is because the top teaches us how the Universe started.

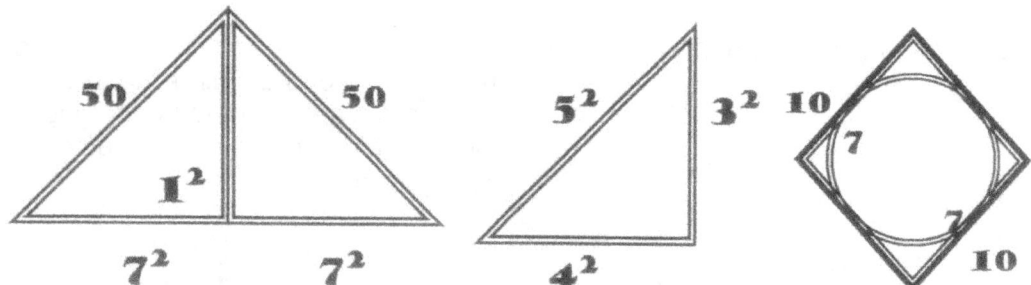

It is by 7^0 that the Universe turns. In turning it diverts direction by 7^0. This application still moves the universe ever instant of movement or time.

The divide is $(3 + 4)^2 = (7)^2 + (1)^2 = 50$ and forming the next point is $3^2 + 4^2 = 5^2$ where both values is the result of the Pythagoras triangle.

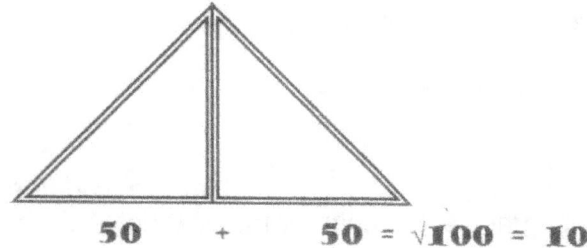

50   +   50 = √100 = 10

We still have material forming time (7) in space (10).

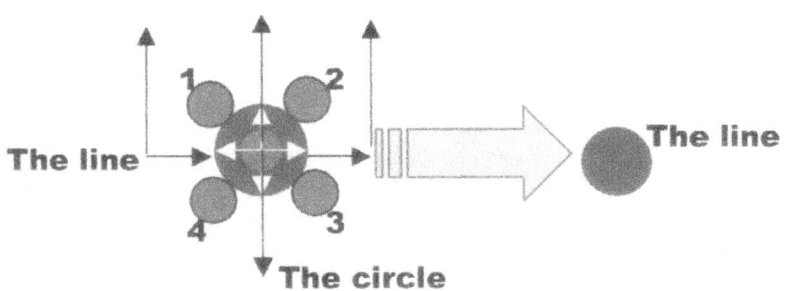

In the one $\{(3 + 4)^2 = (7)^2 + (1)^2 = 50\}$ a line forms a double value and in the other triangle a new point forms the next point in the line.

Both values end up as 10 where the one is $\sqrt{100} = 10$ and the other is $5 + 5 = 10$

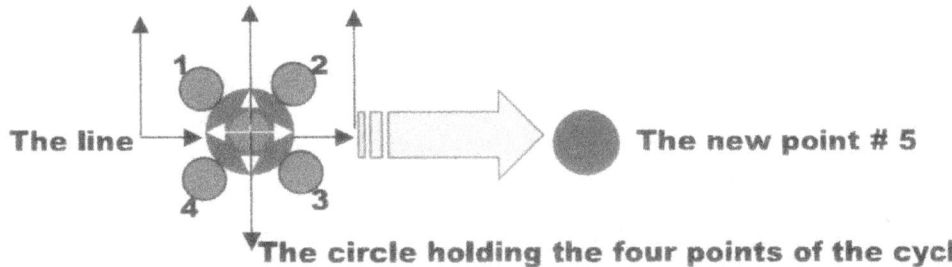

The circle holding the four points of the cycle

$+ .009 = 21.991$

$(1/ 10 \times 3/10 \times 3/10) = 0.009$
Π found a new value of 21.991 / 7

The total points worth became 4 x 5 is 20

The process in which the circle that holds the five points form a triangle in reference of the circle of four points forming a new cycle.

However we know that singularity as 1^0 is the smallest point any point can be in our Universe because that point in value falls outside any space within our Universe.

Forming a value as Π of 3.1416 indicates a value of less that 1 which indicates that the line forms a future that comes as the line continues from far outside singularity. In the black hole we can see this clearly where the future absorb and consumes the past. The past is the darkness and the future is the line holding singularity.

This is how the Universe began. It began simple as it formed numerical order in numbers that brought about order to spots that became dots. It is so simple that the complicated mindset of the Newtonian physicist rejects this as being too simple for their liking.

How does the world of Newtonian science accept this reality and how does science respond to this being the way that the Universe formed. In twelve articles to Annalen Der Physics I tried to explain the process that I name the Absolute Relevancy of Singularity and I went about to insist on the simplicity of the Universe such as you could see in the manner in which I did. I introduced a concept that was never penned and was new as a concept to everyone.

It is forever that Newtonians try to bullshit me by implying that I am too stupid to understand Newton. Never, not in over 40 years was there even one that explained to me what I did not understand about Newton.

Whenever I disagree with science and Newton I am forever and always brushed off by suggesting without suggesting that I lack the mental facilities to follow reason and in that I did not have sufficient mental intellect to understand Newton.
I sent twelve academic articles in which I explain the four cosmic laws and moreover the influence that this had on the outlook we have on cosmic science. I explained in the articles how singularity forms and this is far from how Newton works.

I now wish to present you the reply exactly as I received it using the precise words as it came via e-mail. I was (again) rejected because I was not Newtonian. This came back...

To: Mr. Peet Schutte

27 March 2008

Dear Mr. Schutte:

I am sorry but it is apparent from your letter that you are missing the basics of mathematics and classical mechanics.

Sincerely, N. Eckern
Editor in Chief
Annalen der Physik

Prof. Dr. Ulrich Eckern

Theoretische Physik II
Institut für Physik

Universität Augsburg
D-86135 Augsburg

Telefon +49-821-598-3236
Telefax +49-821-598-3262
eckern@physik.uni-augsburg.de
www.physik.uni-augsburg.de/theo2

This how gravity forms:

In short I said: The curve of the earth is $7°$ on both sides ($7° + 7°$) but because $7°$ represents the earth turning in movement it is also ($7^2 + 7^2$). By turning it crosses singularity (1^2) in the centre of the turning circle in rotation and then according to the law of Pythagoras the circle in movement becomes a triangle moving in singularity. Numerically ($7^2 + 1^2$) = 50. So $2(7^2 + 1^2) = (49 + 1) + (49 + 1) = 100$ on both sides of a circle.

A triangle then forms turning the direction = 50 + 50 = 100.
Therefore the space in which the circle turns is $100^{1/2}$ to the root thereof = 10 and therefore the Titius Bode law shows the inside of the circle factors forming Π as gravity where 7 goes related to twice times 10.

That is Π and that is why 7 goes double squared in a circle adding the centre part of the circle, which is 1.991 divided by the space in which the planet orbits around the sun that forms singularity at 1.991. This adds to the 10 +10 placing each planet in the allocated singularity position, which is then derived. It is implementing Π as gravity. The Titius Bode law shows that gravity is when (10 + 10 + 1.991) / 7 = Π.

Sound complicated.. not bloody likely when I use twelve articles to prove this fact...
And the e-mail sent to me only again confirmed yet another Newtonian rejection...

Dear Dr. Schutte,
You submitted an article of 15 pages to the Annalen. The content of this paper doesn't constitute a theory in physics. With a lot of words and some simple algebraic relations, there is no way to "explain" the world of physics. You seem to be out of touch with modern developments. This is also shown by the fact that you don't quote any relevant literature. I am sorry to say, but the Annalen is not able to publish your work. I am sorry for having no better news for your.
Best regards, Friedrich Hehl
Co-Editor Annalen der Physik (Berlin)
--Friedrich W. Hehl, Inst. Theor. Physics
* University of Cologne, 50923 Koeln _____/_____ Germany
fon +49-221-470-4200 or -4306, fax -5159
hehl@thp.uni-koeln.de, http://www.thp.uni-koeln.de/gravitation
* Univ. of Missouri, Dept. Phys. & Astr., Columbia, MO, USA

In the macro the four phenomena or the four cosmic principles form the following structures in the greater cosmos. However these four cosmic laws are the pre-micro that project a pattern onto the macro cosmos but forms the smallest particles one can find. In truth these four principles take the Universe into the abyss where space does not yet form. My discovery of how to decipher the laws took me to an understanding of how the Universe began and that understanding brought me to where I can mathematically explain the Bible according to Genesis 1 verse 1. I was able to achieve this because I discovered the manner in which the smallest particle forms and then I retraced how everything started. I repeat: these four pillars forming everything. These four principles begin the Universe before the Universe begins.

I am very aware of the fact that most everybody including physicist that is practising physics never heard of these laws and are unaware of the existence of these law. These laws are what nature applies instead of Newton and that is why these laws are kept under cover and in the dark as far as the significance it represent. Nature and Newton don't even share a Universe and because Nature nullifies Newton science hide nature under a blanket and out of the sunlight. If you wish to remain in the dark then don't purchase this book but it you wish to read about nature's reality you can't afford not to read. Without recognising these four laws science does not make sense unless they cheat facts to make it seem to be sensible just as is the case presently.

Science forever hides the truth of any subject behind this veneer that the public is too stupid to understand and therefore they disclose their opinions about whatever field they wish to promote and not telling is because they know all. It is the information about the unbelievable oversight of Newtonian mistakes I disclose that proves how they keep the oversight of Newtonian mistakes silent and why they don't divulge that which they keep silent about. It is about them never committing to the entire story by giving an all-round presentation of everything anyone would require to know to be in a position to evaluate.

Science went corrupt in 1705 when Edmund Halley told the world he used his friend Isaac Newton's physics formula to calculate the route and time that the comet that was named after him would arrive. This was where science went crooked and started to corrupt science, a position that went on ever since because that same dishonesty is still present in Newtonians science. Halley calculated the time periods since 1066 at the battle of Hastings and found a comet was mentioned every seventy-six years. This was very ordinary for a man of his class so he had to get far cleverer than backdate history to get a time frame. So he really got clever and conspired with the biggest fraud in science ever since; the man hat stole all the formulated physics Doctor Hook invented, the man that even got Kepler' figures wrong by cheating it to support his ideas, the man called Isaac Newton.

If Halley was honest about tracing the arrival of a comet that was mentioned ever seventy – six years then Halley was no more than bloody ordinary and hat Halley could never be. So to look smart Halley said he used the formula of Newton to calculate the rout the comet took. This says he used mass to calculate how the comet came to the sun. That says the sun's mass pulled the comet and the comet's mass pulled right back and this way the comet came to the sun. I don't go into the comet that much in this book but I do in other books. In this book I show how Newtonian science started to go corrupt in 1705 with one conspiracy to cheat and became the corrupt myth it now developed into. How do I know Halley did not use the mass pull mass idea because if he did then how did he calculate that the comet was cyclic or that it returns every seventy-six years. If mass pulled the comet to the sun what then pushed to comet back into outer space? Halley's big ambition was to prove the comet comes and goes but if mass makes the comet come what pushes the comet back. You know what is the biggest fraud that came to be called Newtonian science? The most brilliant minds on earth this past three hundred years failed to asks this simple question: if mass pulls the comet closer what pushes the comet away? If mass forms the force of pulling and pulled the comet closer than what pushed the comet back into the darkness of the beyond. How did he know the mass of Halley's comet? Nobody then asked questions.

No one asks uneasy question...except I. I show the fake science we have by just questioning science in search of the truth. Newton and Halley got away with corrupt science. Today Newtonians get away with corrupt science. Then those in science question my integrity because I question the integrity of those in physics and in this book you read how a bag of stinking shit flies into the faces of the most holly, the most intellectual mind the world ever produced.

I deciphered the cosmic keys by which the Universe is formed. These keys are so small it can only be mathematically equated and yet it forms everything within the Universe. These keys are applied to form all forms of material and non-material of which the Universe comprises. By deciphering the keys I also deciphered the manner in which the cosmos started LONG before the Big Bang started. Using the way the keys form material and space made me see how the Universe began before zero was in place. If you purchase and read this book you will find that the Bible is correct about Creation and I challenge any **mindless atheist** to show me **MATHEMATICALLY** where I am wrong with my **mathematical interpretation** of the Bible on the first page of the **Bible in Geneses 1 Verse 1**.

However if you purchase this book AND read the content you will see how I use MATHEMATICAL numerical equations with which I show how the Bible is correct when presenting the Biblical version of how the Universe was created. I use science and mathematics that is a numerical understanding to convey what the concept constitutes of. When the cosmos came about words were not yet a reality and therefore using words brings a shortfall when trying to understand a process that came about before the Universe came about or the spoken word was a reality. I DO NOT use theology or any religious doctrine or any form of evangelistic approach in any way but conduct mathematical science to show the Biblical version of Creation is the only correct interpretation of how the cosmos came to be. This is the first time ever that this achievement was successfully conducted and I did it with the grace and mercy of the Almighty Creator that Created a Universe by applying mathematics. It was not mathematics being God that formed the Universe but it was the Universe by Creation that brought about mathematics! It started when one started...

Take note that the Almighty God does not use words or language to write His laws applying as nature because word and language is human methods to write what applies to humans and manmade thoughts. In nature where it really matters we see the Almighty God uses cosmic laws written as cosmic structure using mathematics to write on a cosmos as a canvas using light as ink and time as pigments.

Index and Page Numbers

Hidden Secrets

I would start by introducing myself. That is because during the reading of this book you will constantly find an urge to hate me and to despise what I say because the thought pattern forming your mind control is going to kick in and you will defend what you were taught at school or ever since. I am going to attack and I am going to destroy what you thought was truer than your religion is. Like it or not but you believe more in the unquestionable truth that science brings than what you believe in the truth about your religion. You will defend your religion as if it is normal but you have never thought to attack the correctness of science. You will find an urge to throw this book down but when you do so you further a scandalous corruption committed by persons you put in the highest esteem and on a scale as never experienced by human beings. In this light that I am going to experience your anger and therefore I would like you to know a little more about me. I do not introduce myself because of an ego, I am far too poor, too little and too unimportant to have an ego. I am the "nobody" with no status that nobody wants to know but that is because I destroy everybody that thinks they are of any importance or fame. I bring the truth and in that destroy the fables other depends on to get rich. But since everybody in science and outside science attack me just because they can't disprove what I say I would want you to know me slightly better.

I am Petrus Stephanus Jacobus Schutte going by the nickname of Peet and I am a married male, with a sane mind and I hold very sober habits being a lifelong teetotaller therefore my mind is clear. I am as sober as a judge but with a difference and it's that I am sober while judges are not. By nature I am frustrated beyond what any mind can take. I am as mild mannered and friendly as any rhinoceros with acute molar tooth infection and I am as gentile as a lion that got his tail pinched in a vice. Now that I have introduced my friendly and mellow side whereby I try to expose my soft and tender nature and soft underbelly let's get to my work to show my crocodile teeth.

If you criticize me in favour of Newtonian science what that proves is that you are brainwashed about science and you are not educated in science. You don't find Newton in nature because nature never used Newton or any aspect of what Newton said nurture is. If you go bashing me it proves you are totally uninformed about what nature uses and you never progressed beyond being brainwashed. Arguing shows your blatant stupidity about the reality of what is truly present versus what you were told is present in nature. I suggest you read and for the first time get educated and not suffer mind control by academics brainwashing you in believing in magical forces that never existed before or presently. Do as I did long ago; act intelligent and study what nature uses and not what Newton said. You're going to go in the correct direction following nature. What you are about to read is the mother of all the conspiracies in science. It is science made up with endless fabrications.

Every conspiracy that Science ever thought up such as the Critical Density Conspiracy or the Dark Energy Conspiracy, or any conspiracy connected to science is in place to protect this theory they hide from anyone outside physics from becoming known. All of the conspiracies in place in the past or at present do the hiding of the Mother Conspiracy so well that in three hundred years no one that is not part of science got a sniff about what the Mother Conspiracy entails. The Mother Conspiracy is in place so that students in physics are brainwashed by instigating the deliberate sanctioning of Mind Control through their practising of enforcing Thought control they unleash on students. I reveal the Mother Conspiracy and I show why it is in place. I prove that the Mother Conspiracy is in place. I show how every student including you reading this has been brainwashed to believe science is true and to believe in science. This book is for free; I make no money so why would I be dishonest? When you download it you are going to get the nastiest surprise of your life. Then you can also read how the cosmos truly apply science and read explanations of what was never explained before and how the explained factors interlink

If you criticize me in favour of Newtonian science what that proves is that you are brainwashed about science and you are not educated in science. You don't find Newton in nature because nature never used Newton or any aspect of what Newton said nurture is. If you go bashing me it proves you are totally uninformed about what nature uses and you never progressed beyond being brainwashed. Arguing shows your blatant stupidity about the reality of what is truly present versus what you were told is present in nature. I suggest you read and for the first time get educated and not suffer mind control by academics brainwashing you in believing in magical forces that never existed before or presently. Do as I did long ago; act intelligent and study what nature uses and not what Newton said. You're going to go in the correct direction following nature. What you are about to read is the mother of all the conspiracies in science. It is science made up with endless fabrications.

Every conspiracy that Science ever thought up such as the Critical Density Conspiracy or the Dark Energy Conspiracy, or any conspiracy connected to science is in place to protect this theory they hide from anyone outside physics from becoming known. All of the conspiracies in place in the past or at present do the hiding of the Mother Conspiracy so well that in three hundred years no one that is not part of science got a sniff about what the Mother Conspiracy entails. The Mother Conspiracy is in place so that students in physics are brainwashed by instigating the deliberate sanctioning of Mind Control through their practising of enforcing Thought control they unleash on students. I reveal the Mother Conspiracy and I show why it is in place. I prove that the Mother Conspiracy is in place. I show how every student including you reading this has been brainwashed to believe science is true and to believe in science. When you download it you are going to get the nastiest surprise of your life. Then you can also read how the cosmos truly apply science and read explanations of what was never explained before and how they explained factors interlink

What you are about to read holds nothing new and definitely nothing about my work. I introduce what is in nature and I show what nature uses and therefore when you disagree with me you don't disagree with me because you show your personal inadequate level of truthful information about what you should know if science were not fooling around by brainwashing you into believing a fairy tale story about Newton while they know you know nothing.

As I said I say again, this is what nature or God Almighty uses to build the Universe and when you eventually read my work you will find out why these building blocks work the way it does, how does it come about to function as it does and the entire process how the Universe came about from the first original dot. By the way that is why you can see the entire Universe with the space your eye holds and that is why you are within the centre of the Universe as all the light comes to you where you are. It is because I only show what you were suppose to know if science didn't cheat you witless that I offer this book at such a low price. Should you wish to purchase my work to find out the mysteries driving the Universe and what is the Universe you will have to pay more. I'll show you why you can see what you see I studied physics and in my first year of studies I came upon a mistake concerning physics. The mistake is so deep rooted and to get to it one must go back to ancient times.

I have progressed far beyond what is needed to get on friendly terms with persons judging their position as important people because others think they are important people. I got use to the idea that who ever is in power of whatever is corrupt to the core and no less are the department of Physics, Astrophysics and philosophical physics. I saw how absolute power absolutely corrupts. They are as much gangsters, politicians and generals and Clergy of all dominations are. The more you give them respect the more they act as if they became God Almighty that rules the earth. When you read what you are about to read it is not the hallucinations of a drunkards mind. I introduce myself because I know you have never heard of me and there are hundreds of physics academics that keep it that way. It is because while they hide cover-ups in physics I reveal it.

If I tell you science is built on fictional fraud you will put this down and believe I am as loony as a hatter. If you think I am as mad as a hatter you are in for a brutal surprise because science has as much criminal activity to cover than does loan shark bankers have however science is respected.

When you have read what I uncover you will agree with me in the way I see myself as very responsible being down to earth and am not well liked because of my straightforward personality where I say what needs to be said to maintain honesty above friendship and what I say is always not infusive or congenial to make friends. As should be evident by now: I wrap no feelings in cotton wool for the sake of peace. I can't sue for peace while I battle the corruption of science.

Science hides an entire Universe behind "nothing". My roommate told me it was Newton that said what I disputed and I remember telling him I don't care who this Newton fellow is but he is wrong and ever since then I stuck to what I said. That discovery made me disagree with the establishment forming principles in physics. Later in time I have detected much more than a mistake. When you read my work and banish your cultural brainwashing you can judge my work. You then can see what a huge Pandora's box I have discovered. Science is believed because of mind manipulation and brainwashing going on forcing students to believe notwithstanding truth.

I try hard to be honest even though I am poor. I am poor because I present the truth. If I did try to accommodate Newton and the falseness Newton embraces I would be embraced by science but I just can't underwrite Newtonian incorrectness for the sake of money. The proof of my honesty is that I am one of the poorest people in society. If I want a cold drink I have to ask my wife because I have no income trying to introduce the truth to science. I have to fight an ideology of corruptness.

Yet for all the good the exposing on my side did me I still write books that don't sell because I try to convince people about mistakes no one on earth are aware of and therefore no body bloody cares except me. I am the only one that can see a mistake and seeing it in the full consequence thereof by what it holds to the entire human race it is frustrating me senseless. I see the mistakes because I see how to correct the misleading dogma in science and by correcting the mistakes I show a clearer science. The total size of the conspiracy to corrupt goes beyond what any mind can cope with at first. At first you will think I am exaggerating to try and make a profit but then look at the price of the book I sell and try to find any profit in a book selling at such a price. My effort is showing how to correct the mistakes I show they hide.

I have spent a lifetime researching how nature works and I discovered how the building blocks of the Universe works. I did not discover the building blocks but I di discover how the principles work, how they work and what results from the working of the principles. While Newtonian science suppresses the importance of the phenomena in favour of falsifying the correctness of Newton I am blocked revealing what I discovered. As long as they legitimise by conspiracy what they claim is absolutely godly truth about Newton's ideas and trying to maintain Newton as being supported by nature I am chained to frustration by a falseness science presents. They hold science out as if nature completely supports Newton and nature never supported Newton in even one idea the man had. I can show how the Universe works but as long as Newtonian science legalises Newton's falsification of facts this conspiracy to uphold Newton. When I challenge those in office I get blown away because they argue Newton needs no proof since time proved Newton. They don't have to prove Newton while nobody ever proved Newton as long as science exists.

With the hardship and poverty my family went through especially my children while growing up I can't hate anything more than those in astrophysics because they all should be in jail for uncompromising dishonesty and mind managing corruption and then they are on top of the list of those presenting honesty. They block me with fraud and corruption. With their covering of fraud they put me through hard times and poverty while they ride on the wave of complete holiness.

According to my approach I mention these facts to establish beforehand that I am not a danger to society. Although I am the only one in the entire world that totally disagrees with modern mainstream science and mostly on the thinking of physics yet, I am not criminally insane. I would not attack your dog before your dog will bark at me because I hate your dog. I would rather attack you and not your dog after he barks at me just because I most probably won't like you but I do like dogs. I say this to convince you that I have never been jailed in all my life on grounds that I attempted to mislead or fool or tried to mislead anyone, as you can see. I try hard to be honest even though I am poor. The proof of my honesty is that I am one of the poorest people in society.

I am so poor because I write books that don't sell because I try to convince people about mistakes no one on earth are aware of and therefore nobody bloody cares except me. I try to introduce honesty in the midst of fiction ruling science. I bring the truth and remain poor while Dan Brown introduced fiction as truth and a lie got him rich. I am the only one that can see mistakes and seeing it in the full consequence thereof by what it holds to the entire human race it is frustrating me senseless.

I see the mistakes because I see how to correct the misleading dogma in science and by correcting the mistakes I show a clearer science. Take it from me that writing about science while trying to convince people about mistakes everybody but me believes is correct and therefore not selling books is not a very profitable enterprise and is frustrating at any level. If I was less honest and went about bullshitting everyone about the accuracy that Newtonian science portrays in the modern era it would at least sell some books because some dinosaur Newtonian physicist would be pleased about it but then I was a cheat although I would have been richer but being as poor as I am in favour of trying to be as honest, in that I am can't be a cheat. I studied physics and in my first year of studies I came upon a mistake concerning physics.

I write books that don't sell because I try to convince people about mistakes no one on earth are aware of and therefore no body bloody cares except me. I am the only one that can see a mistake and seeing it in the full consequence thereof by what it holds to the entire human race it is frustrating me senseless. I see the mistakes because I see how to correct the misleading dogma in science and by correcting the mistakes I show a clearer science. Take it from me that writing about science while trying to convince people about mistakes everybody but me believes is correct and therefore not selling books is not a very profitable enterprise and is frustrating at any level.

If I was less honest and went about bullshitting everyone about the accuracy that Newtonian science portrays in the modern era it would at least sell some books because some dinosaur Newtonian physicist would be pleased about it but then I was a cheat although I would have been richer but being as poor as I am in favour of trying to be as honest, in that I am can't be a cheat. If you feel I am exaggerating or that I am a mentally impaired asylum escapee that found a writing pad, ink and paper and that I now start to scrabble senseless suggestions to while away my social frustrations then I challenge you to prove that students are NOT being brainwashed to believe what they are taught in science. I started off being likable and being nice but the deeper I delved into the conspiracy the less I could care about who thought what about what I said because I am going for the truth as hard as I can. Their mannerism in blocking and frustrating my opinion when showing the mistakes in science convinced me about a Conspiracy in Science in Progress and this spurred me on to tell the entire world about their brainwashing students minds.

By the manner they selectively withhold information when teaching science, amounts to deliberate brainwashing of students in physics by "normal" education practises. The new concept I wish to introduce puts all emphasis on space ands material is only space filled with material substance while other space is filed with non-material. In the end all space are equal but the movement it has makes the difference it presents in relevancy. All space structures hold in the centre most heat concentrated and from that centre holds all material owned by that structure. I can go on and on but heat in the centre couples gravity to space-time, just like Kepler said before he was spoken for on his behalf and without his permission or his agreeing to it. Studying Kepler helped to understand why the phenomena are there to begin with and that enabled to explain in some way…

I studied physics and in my first year of studies I came upon a mistake concerning physics where science hides an entire Universe behind this "nothing". My roommate told me it was Newton that said what I disputed and I remember telling him I don't care who this Newton fellow is but he is wrong and ever since then I stuck to what I said. That discovery made me disagree with the establishment forming principles in physics. Later in time I have detected much more than a mistake. That put me on a road where I questioned my personal sanity and then questioned science. I found mistakes that was unbelievably big and made me fight an enemy is strong everyone folds when they enter the equation. When you read my work then banish your cultural brainwashing you were subjected to if you whish judge my work.

The term conspiracy is so widely connected that it lost its true meaning and at this point no one believes the existence of a true conspiracy any longer. Well you are about to see the biggest conspiracy and it is so big you will attack me rather than question the presence of the conspiracy. You and I and we are all art of this conspiracy. We are all brainwashed to believe the facts this conspiracy hides. If you are not ready to wake up then put this book down and slumber more. What you believe is truer than what you believe is true about God you are going to learn that that what you thought to be the truth is the biggest hoax this world has ever seen. We are all part of this conspiracy and keeping it alive, one way or the other. Now you will learn what a conspiracy is but you already think you know all about what a conspiracy is. Brave yourself for the shock that is coming. What you believe in is going to unravel and you will be confronted by a truth you never thought existed.

You are going to read about a conspiracy but people think of a conspiracy in many terms. This is a science conspiracy but now you think about this in terms of another million science conspiracies floating around. Let us define not by definition but by interpretation to what a conspiracy constitutes. What do you think is a conspiracy? All the conspiracies you know about is known about because someone somewhere makes money by allowing the revealing of the conspiracy. Silencing the conspiracy does not make money but informing a suspicious public loosens the flow of money. If it were a true conspiracy no one would know about the conspiracy because the powerful would make money from not revealing the conspiracy. The revealing of the facts about any conspiracy would be stopped before it leaked because it would kill the flow of money.

A conspiracy is thought to be a gossip story that makes money and by not revealing it or revealing it goes in line with making money or not making money. You can download this book and then read how they do it and on my part this is not to make money by revealing the conspiracy. I truly want to find an audience to divulge the truth. I want to make money but it is by showing how I can correct the flaws in science, not by hiding it in a conspiracy.

People put a conspiracy in the same realms as a gossip story, an old wives tale, which is going about but does not intend to harm and mostly serves as amusement to many. Hearing about a conspiracy tests your intellectual comprehension. It is some quiz that you match your truth against the truth that the conspiracy reveals. It seems to be funny, but it is not funny until you catch the funny part hiding behind the conspiracy and only when you measure the catch behind the conspiracy are you treated to be amused.

If the conspiracy does not touch the person directly then no harm is felt and no harm is intended. Everyone holds this view that a conspiracy is on a slightly higher level than gossip. It is a gossip story about someone living in the neighboring village known only to some people next door but has no direct linking to me or has no threat to the safety of others directly associated to me. Everyone treats a conspiracy as if it is something amusing that holds no threat at all. It is something that goes around as a joke of sorts.

For all of those who live in America or on the far side of the Moon or spent the past almost one decade on a vacation on Mars and others who had no contact with Mother earth or the media on Earth, this is Madeleine McCann the daughter of two prominent Scottish doctors who were on vacation in Portugal in 2007. Since her abduction no one saw her, no one heard of her and no one thinks of her. However, that is not new because since the time of her disappearance more than 300 000 other kids in Britain alone befell the same fate and there is no media-mania to try and find those children. Every person knows Madeleine went missing and this is / was prime news for almost or more than one year non-stop.

Lets narrow the conspiracy idea down to a known case. A conspiracy is very typical to the Madeleine McCann case where the disappearance of Madeleine brought money to many. Madeleine McCann disappeared Thursday, 3 May 2007 and was / is the daughter of Gerry and Kate McCann. Anyone living in the western world, except Americans, would know about Madeleine McCann. Americans live in a unique Universe called "the States" and regard Europe as some overseas country situated just past the moon. Americans are the most uninformed brainwashed nation on earth thanks to advertising. Americans can't cross the street without some advertising jingle telling them how to do it the best way while getting the most pleasure while crossing the road and for fun they will throw in a sexual connection in the crossing. They are even brainwashed by advertising to believe in capitalism while the system failed the world many times in the past and only the rich gets richer from failings.

Madeleine disappeared from a hotel room in Portugal while being in bed. For months the disappearance of Madeleine was front-page news and we heard about her all day long. We heard about Madeleine's disappearing day and night on all channels and in all papers around the world. Then the news raised the hopes and then the news dashed the hopes but it was all about the disappearance of a girl called Madeleine McCann. To sell advertising time the press had to write about Madeleine.

This is a conspiracy.

Not the fact that she disappeared or didn't disappear or that she came to harm or died at the hands of her mother or not. The conspiracy hides in the big deal that was made about the case all throughout the months afterwards. The case was blown out of all context and proportions considered the international effect it had. The press made an issue about the case as if it was some special occurrence unknown to mankind. The Press presenting this case as a one-off incident makes the entire case a shameful mockery to society. The world viewed this, as an incident equal to the Lindbergh baby's disappearance in uniqueness while the truth is that 11.4 children disappears hourly in Britain alone and no cockerel ever crows about it.

Britain alone loses a hundred thousand children under the age of sixteen per year every year. One hundred thousand children go missing each year in Britain alone and what Special Forces are in place to combat this...none. The government never mentions this in parliament. In 2007 there were a number of bomb blasts on trains and one bus blew sky high and that brought Britain to a stand still. The number of lives lost was many times less than one hundred thousand and such bombing happens once in about ten to twenty years. Yet, the propaganda value in terrorists outweighs the leverage of disappearing children.

Under tough British laws, anyone of 30,000 on sex register, those marked, as sex criminals must inform police when planning to travel abroad. That information has been used to compile names of at least 130 paedophiles known to be on Portugal's Algarve coast, at the time when Madeleine McCann vanished. Police searching for the missing Madeleine McCann were as they claimed scrutinising four pieces of "very useful" fresh information. The tip-offs came among hundreds of calls made from Portugal via a special UK charity number. If these figures were terrorist-related incidents the world would be in world war 3 by now.

This is not all. It is said that each year, non-family members, often in connection with another crime, abduct more than 58,000 U.S. children. Family members who are seeking to interfere with a parent's custodial or visitation rights abduct more than 200,000 children. Although the vast majority of children (at least 98%) return from abductions, too many children do not. While there are only around 100 reported cases each year of the most dangerous type of abduction – stranger kidnapping – fully 40% of these children are murdered. Where is the media coverage on these events? According to the Department of Justice, almost 800,000 children are reported missing to law enforcement each year, while another 500,000 children go missing without being reported to authorities. I was astonished to find recently that 1 million children, yes 1 million, go missing every year in the US and the UK. These are astounding statistics, every year a city the size of Amsterdam goes missing? Less than 200 of these are murdered so where are the rest? Its mind-boggling stuff, when you delve a little deeper, in the UK a child goes missing every 5 minutes, 2000 kids a day lost in the US! I suspect hat even these figures are suppressed. Getting excited about these figures still hides the overall truth and we will never get to the real numbers.

Belgium is a small country of some ten million people and yet in the Brussels region alone it was revealed that 1,300 minors disappeared between 1991 and 1996. That was only the reported cases and not those that went unreported which are the cases we will never know about. Around the world you find the same repeating story and multi-millions of children go missing every year never to be found. This is not even contained to a continent or a part of the world but is a global phenomenon. There are children going missing in every country across the world. We dare not mention numbers when we are thinking about what is happening in countries such as Brazil, India, China, Russia and other countries, where due to the geographical vastness, the Governments just don't have any records of missing persons in place. The most remarkable issue is the absolute silence we experience about this matter. It's as if it is not for real and we know it is. I can't confirm what I repeat from the Internet but if it is true it confirms a lot of my suspicions I have about this. Where do they go? It is millions of children in the end. The numbers alone just can't hide the crime but the apathy of all the Governments can apparently hide the numbers. It is another conspiracy kept in a cloak of silence. There is apparently a program or was to be a program on TV called Conspiracy of silence, which is / was a documentary listed for viewing in TV Guide Magazine and that was to be aired on the Discovery Channel, on May 3, 1994. I have seen on the Internet that this documentary exposed a network of religious leaders and Washington politicians who flew children to Washington D.C. for a sex orgy. I can't confirm this that I got from the Internet but my logic tells me not to deny this either. There is a show that was supposed to be aired about the disappearance of children. At the last minute before airing, unknown congressmen threatened the TV Cable industry with restrictive legislation if this documentary was aired. Almost immediately, unknown persons had ordered all copies destroyed purchased the rights to the documentary...Only the most powerful on earth can fasten a lid on such a matter this secure.

Nearly 30 million children and youth go online to research homework assignments and to learn about the world they live in. Research by the University of New Hampshire found that one in five children between the ages of 10 and 17 received sexual solicitations over the

Internet during any given year. One in thirty-three received an aggressive solicitation - a solicitor who asked to meet them somewhere; called them on the telephone; or sent them regular mail, money, or gifts. This is very difficult to comprehend; we are not talking of second or third world countries here. Wouldn't you think the media would be all over this for instance? Or our respective governments constantly warning us but no, I only managed to stumble on to these unbelievable statistics on an obscure website. Belgian parents live in fear of paedophiles because it is very well known that in Belgium more than any other place, these paedophile rings are almost openly active! When I hear about doubt or I hear people say that if child abuse, kidnap and murder were happening on the scale it is suggested to occur, then it would not be able to be kept under wraps. The child kidnappings and murders in Belgium in the mid-1990s threatened to implicate the country's political establishment and other famous names. That power that democracy places in the hands of politicians are where the power is that control absolute silence while the banks control the political power by telling the politicians what the bankers wish to happen and the crime syndicates control the banks by telling the banks what it is that they wish to happen. Therefore democracy obeys the rich that controls the banks that orders the politicians that rules us. What happened in Belgium shows that the syndicates can control the people. This is the only explanation as to how they got away with it and pinned it all on a sick and pathetic paedophile and child-supplier called Marc Dutroux. It was the classic establishment response to danger that is repeated constantly across the world. As so many have said so many times, people generally don't get their views and opinions from researched information, but from an 'image', an 'impression', of how things are. By controlling the leaking and presenting of conspiracies then by controlling the conspiracy they control the information leaked as a conspiracy. With regard to missing children, this 'impression' is heavily influenced by the number of lost children stories they see in the media. The Belgians live in fear of having their children abducted, as the paedophile gangs operate unbothered by the Police. Once a child is abducted, they are never seen again. The Marc Dutroux case became famous, as he allowed girls to die in his cellars when he was arrested. But two girls escaped unharmed and provided many leads as to who Dutroux' clients would have been. But no inquiries were made. Belgians all have their theories as to who is involved, many suggesting paedophiles are high up in the government. In Brussels, kids are often photographed at play and abductions made to order. Why the Police don't do anything, or politicians keep silent is not known for sure, but you can use your imagination...a combination of threat and reward - in a country where money easily buys influence, it is not surprising. The strain of knowing so much evil and not being able to act to prevent it gets too much for many. Belgian Police are very prone to committing suicide. *The Sun* recently reported that Algarve is a 'haven' for paedophiles. Why must society's parents tolerate these criminal scum living amongst us?

In Britain, Missing People -- a charity formerly called the National Missing Persons Helpline -- has tried to draw some degree of attention to these thousands upon thousands of missing cases. It is astonishing to witness the degree or rather the lack thereof to which the news media devote attention to vulnerable missing persons. One little undefended organisation is claiming that despite its efforts to generate news coverage for all missing persons cases, the news media themselves will cover only a very modest few cases which the media claim that those cases fit their publications. The Media should eat this story as if it was a ripe banana and yet they don't. Why would that be, why would they ignore the bulk of this story? If one or two cases (e.g. McCann) could generate such a massive response and earn that much advertising revenue, what prevents the media to go all out on a mission to bring justice to those who went missing.

Two cases of missing white women syndrome are given as contrasting examples: the murder of Hannah Williams and the murder of Danielle Jones. Although both victims were white female teenagers, Jones received more coverage than Williams. It is suggested that this is because Jones was a middle-class schoolgirl, whilst Williams was from a working-class background with a stud in her nose and estranged parents. Media reports about the murder of Amanda Dowler, the murder of Sarah Payne, and the Soham murders as examples of "eminently newsworthy stories" about girls from "respectable" middle-class families and backgrounds whose parents used the news media effectively. These cases are controversial and in contrast to the street murder of gangland murders by youths on youths in drug related turf war incidents which receives little news coverage, with reports initially concentrating upon street crime levels and community policing, and largely ignoring the victim. The assertion is that "the near hysterical outpourings of anger and sadness that accompanied the deaths of Sarah, Milly, Holly, and Jessica" increase ratings to news events. The National Centre for Missing Adults has also commented on the phenomenon by saying "Unless it's a pretty girl aged 20 to 35, the media exposure is just not there. Mentioning these few names fall much short of the numbers of children disappearing, pointing fingers to the press. Some 2,300 Americans are reported missing every day, including both adults and children.

But only a small proportion of those are stereotypical abductions or kidnappings by a stranger. For example, the federal government counted 840,279 missing persons cases in 2001. All but about 50,000 were juveniles, classified as anyone younger than 18. About half of the roughly 800,000 missing juvenile cases in 2001 involved runaways, and another 200,000 were classified as family abductions related to domestic or custody disputes. Only about 100 missing-child reports each year fit the profile of a stereotypical abduction by a stranger or vague acquaintance. Two-thirds of those victims are aged between 12 and 17, and among those eight out of 10 are white females, according to a Justice Department study. Nearly 90 percent of the abductors are men, and they sexually assault their victims in half of the cases. To further complicate categorization of cases, the FBI designates some missing-person incidents—both adult and juvenile—that seem most dire as "endangered" or "involuntary." Kim Pasqualini, president of the National Centre for Missing Adults, said the media tends to focus on "damsels in distress" — typically, affluent young white women and teenagers. The media's dilemma is that government research shows that victims of non-family abductions and stereotypical kidnappings are most at risk of injury, sexual assault or death. "Damsel" cases may be the exception, but they often are the most urgent. These numbers are staggering and yet even so they are ultimately never mentioned, why? Compare what is mentioned as cases to what the cases are mentioned and are targeted by the press to who disappears in reality. What makes the media so reluctant to tell to entire story and blast information out for months?

That is the media response of what happens to losses accruing in terrorist cases but the numbers of losses hardly match. In relevancy one could say for every two million children lost to prostitution there are five bomb blasts. During about the same time there was a huge bank robbery and the culprits were traced far and wide. They never stopped searching until everyone was caught and every note was accounted for. The leader was traced back to an Arab country (I forgot the name) and was brought back to face justice. He was chased until he was caught showing that criminal cases can be solved, except where children disappear. If one bank robbery is committed they get the thug but when children disappear, who cares? If we are talking of the national media, there are very few cases reported in the light of what takes place. The enormous coverage of the missing British girl, Madeleine McCann, who was abducted while on holiday in Portugal, is a rare example compared with the number of children who disappear every year. While people get their impression of scale from the lost children featured in the media, staggering numbers of children go missing never to be seen again. I remember calling many American states a few years ago to ask for their missing children figures and it was truly extraordinary. On average, around 3,000 children a day are reported missing in the United States, never mind those the authorities never hear about. Add them together and you are talking hundreds of thousands of children.

Yet, in that time it took to close the bank robbery case about 200 000 children under the age of sixteen went missing and the number that disappeared were never even reported! If there is another two major BIG bank robberies in the same period no stone is left untouched to solve the problem of banks robbed and "terrorist" attacks against "the people of Britain" but to lose a million children every decade or so is quite acceptable because the disappearing is part of a conspiracy and true conspiracies are never revealed or investigated. If a bomb blast occurs in London, MI 6 forms a special investigating unit but no Special-Force action is ever taken to bring this loss of children into the open. Losing 100 000 of the prime persons, the future of the country is a matter of discussion that is never discussed in polite conversations. It would be very embarrassing to ask the Prime Minister of England or the President of America what happens to 100 000 children under the age of sixteen in Britain alone that disappears every year.

This applies to all countries.

Which British politician ever made it a political pledge to do research into these crimes? No one mentions it because it is best wiped under the carpet. Most of the children are a product of broken homes and pupils of the unmentionable side of society and therefore no one cares to consider them. Their parents never vote because they are mostly the driftwood that never votes and the children are too young to vote, so who cares? But when two medical doctors' child disappears and they know how to manipulate the press, the world goes fanatic. The couple even found an audience with the Pope no less.

It all has to do with money. The fact that 100 00 children disappears while it never forms a political debate alone tells the whole story. If Bankers make money Politicians have to hush about the Bankers' very rich and therefore very influential customers. The crime cartels tell Bankers to tell the Politicians to be quiet. The crime bosses are in charge of the Bankers who are in charge of Politicians who make the laws we must obey.

According to a Portuguese lawyer that worked in the Madeleine McCann case, him being professionally experienced in working with several English clients, any parent being accused of abandoning the children to danger is a crime under British law that is severely punished by UK laws. Yeah, sure and the moon is made of cheese! That is so typical of a conspiracy. There are laws in place that are laws never attended to or adhered to and nobody ever obeys the laws because they are as good as being non-existent. Every law that is in place serves to protect the rich and the powerful by denying the poor and the defenceless any say in Government law forming. Yes, they show democracy but they create mass hysteria instead.

The Members of Parliament put conspiracy-serving laws in place, which serves nothing anyway. TV hosts such as David Frost claim their diligence and their courage and their tenacity. Why don't TV talk show hosts ask any politician about the children disappearing, because the TV station would not permit such questioning? Why do all members of the printing press or electronic media ignore the subject as if it does not exist? Why not attack all the Political Party leaders on a live debate or cross-question them? Why don't the press report every missing child in the same manner as they reported Madeleine McCann? Is it because there then will be no space left to report anything else or who stops them shouting about it?

These children just disappear from the face of the earth and that says there are mighty powerful money barons at work. Those who bought those who have the political power to decide who goes to war and die and who gets rich from the declaration of the war have the leverage in the social structure to allow 100 000 children in Britain alone to go missing and the public is none the wiser and could care even less about the matter. Those children disappearing are hard currency more valuable than money. Those children become assets as good as gold or property because their currency is set fast. I believe there is more dope bought with children earmarked for prostitution than there is money used as transaction payment. I believe a lot of oil is bought with pretty little blond haired girls that have gone missing.

Telling everybody about the McCann girl is not the conspiracy but not telling the world at the same time about the other 100 000 children that also disappeared becomes the conspiracy. It is not the information that is presented but it is the information withheld from the presentation that becomes the conspiracy. It is screaming from the mountaintops about one girl while pushing the other 100 000 that went missing under the carpet, that becomes the conspiracy. To defer the attention from the true problem they cry about one.

Reading about such a pretty girl being abducted or going lost is sensational. It is something to cry your heart out about. Knowing about 100 000 children being abducted or going lost is a problem and while everyone is ready for the sensation no one wants to know about a bloody problem that no one cares about.

The following is a joke coming from the Internet:
What's the difference between Madeleine McCann and Elvis?
The Answer: More people believe Elvis is still alive. This is no joke.
To hide this idea of Madeleine being dead is the centre part of the conspiracy, because only then can the McCann family be in the position to earn money through engaging sympathy.
A Portuguese Police officer spearheading the investigation wrote a book stating his (and the official police) point of view on the case and tells about why they think there is no more evidence to research since they have the opinion that the girl is already dead. He said Madeleine is deceased and gives his reasons in a book. The McCann family got a court interdict stopping his book being published because they want to stop the idea spreading that Madeleine is dead. Why would they spend hundreds of thousands of Madeleine's money to prevent a detective from putting to print his views on the case?

It is about money as every aspect of this conspiracy is about earning money. A dead Madeleine would not enlist donations and evoke the media ratings that the story generates and the income of the entire enterprise will plummet if the public at large accepts the girl's death, while searching for a live girl brings in millions and to hell with free speech and free opinions when money is the issue. That is why the papers that printed the fact that Madeleine is dead were sewed and they paid up. To hell with free speech because the papers saw their earning in revenue go down the toilet by allowing the public to think Madeleine is dead. Admitting to Madeleine's death will kill the money flowing in and that is horrific.

Even the money the McCann family and all other people fight to accumulate is a conspiracy. One group of persons thought of as Bankers bought from crooked politicians worldwide the privilege to print paper and give the paper a value. In the system I take the paper they printed and then I have to regard that paper as having more worth than say my home has because I "sell" my home by detaching my ownership and attach my own worth to the printed-paper. Then I am very impressed with myself afterwards in my effort to exchange what has visible and useful worth for some paper they tell me I have to accept the worth but has no worth but to give it back to bankers to put in their vaults. This is insane stupidity and every person on earth goes along with the madness. The Bankers decide the worth of the worthless paper they print at about no cost to them and then by creating a system I am forced to accept their paper and the worth they attach to it or else I starve and die. Even committing to dying requires money to accomplish the process. If you come to the end of your life you have to pay for the privilege to die and go on.

The Bankers take their tax or share long before the Government can but they call it bank fees. People are so brainwashed and beaten to a pulp by systematic control of the mind and their thoughts that they fall into the practise of doomed slavery without trying to fight for freedom. In the days of the Romans and the Greeks slaves were paid 10 % of what their Masters earned from their services while the Masters still had to feed, cloth and shelter them at the cost of the Masters. We all are slaves to the bankers but we earn about 10% of the cut they take.

While the Mammonites pay us 10 % of what we earn from what they earn from or services, we have to clothe ourselves, feed ourselves and our children while we purchase houses form the Mammonites and then find that behind successful Mammonites there are Bankers pulling strings by supplying worthless printed paper we accept as the commodity we will work all our lives to accumulate and possess. In the end we can't take with us anything we ever wanted on earth because it is worthless.

From every angle this case presents including the Pope's visit, all aspects involve money and it is about publicity that will entice donations of money leaving very little scope for a girl being found. Why don't McCann donate some of the money they have to searching for all children that has gone missing in the time Madeleine went missing? Why don't they also include as many photographs of children that went missing during the time that Madeleine went missing? Because then it would not bring about the money that it makes when only one very special little girl is in the hands of a child molester...but 100 000 children just vanishing slips the minds of every "concerned" do-gooder in England while they donate money to this deserving quest just to keep their minds away from the 100 000 others. This tendency to evade the truth is a sickness that runs through all aspects of society because this disease is what we use as education. I am going to show how teachers in science **brainwash** students by forceful **mind control**.

You think their abilities are mathematically superhuman: Then you hold on to that thought because I AM GOING TO SHOW YOU THE EXTENT OF THEIR MATHEMATICAL SKILLS! **This conspiracy is so widely active that every person on earth participates, most probably without knowing, some reluctant and others well knowingly pursuing the goals of the conspiracy with all the vigour they could muster.**

 I am going to ask you this again later on but be honest not to me but to your own mind as you answer yes to the following questions. See how gullible you are in terms of your surroundings.

I can tell from personal experience that you purchased this book because you are a person who likes to be on top of events as they happen and you insist on being well informed. You buy the paper every day to read about world affairs going on around you. You take everything as it happens while it happens being as a way to be as awake as any civil person could dream to be. Being this well informed and on top of present matters: Ask yourself if you are truthful about your answering the following questions: Will your government lie to you to deliberately deceive you and misinform you by virtue of a media set - up...you know for sure they not only would but they do and they will stipulate how honest and trustworthy the media is because they control what the media says when the believable media misinform you as best they can. The media will report everything not to inform you but to misinform as the government tells them how to do that. The media is no longer the extension of the people but it is used as a political extension.

You believe the papers and the Television news would tell you events as it unfolds and they would report what happens as honest to God as if they are in prayer about what they inform you. You think that you are learning the truth from the news media as if you are busy with Gospel. If that is your view I do not wish to wake you up because where ignorance is blessed, it is a sin to be wise. I do not wish to shock you because as far as I am concerned you already died and crossed over to the other side while your body slumbers on our side waiting for the date of completing the dying ceremony. You are not sleeping no you are already dead. The news media reports what the Government tells them to report and no more and no less. The Government tells the media to say what, when and where and if any reporter crosses this line, which would be the end of such a reporter. You can ask Peter Arnett of CNN what happens to such a reporter that steps out of line and report the truth on matters of war and not what the Government tells reporters they should see. The media will spoon-feed you and you will digest what they whish you should know and you will be in an information cocoon as if you are a small baby because they wish to keep you thinking like a small baby. It is the Mammonites and the money barons that own the banks and because the banks are holding the money-tycoon's investment the banks dutifully tell the politicians what the money moguls' want and the Mammonites who owns the banks tell the banks that own the media what fits where and by owning the media they control what the public should know about events and thereby they are controlling the politician in that they enslave the public. It is even worthless to become dissatisfied because they even control the dissatisfaction. All uprisings and rampaging is selectively staged and controlled by them since they make the public feel they are in control of the negative ness and by staging and paying for such an uprising they release tension when in front of television they get thugs to kick around garbage cans and burn vehicles. Whenever did one such uprising lead to any government turning around? The government enforces the will of the bankers which is the will of the Mammonites, those who own the money of the world, onto the public and the public has to take what the Money barons dish out whether they can stomach it or not because they will do what the elected official force them to do and that is what the Mammonites tell the politicians to do. Those who control the money are the same ones who regulate the negative protests. All the great protests and uprising are paid by the Mammonites because then while they push ahead their will, they control the bad feeling in the public's domain and allow the steam to be released with little derailing to their plans. It is great to control a bunch of fools who believe in democracy because this is democracy at work in all its splendour.

In line with this then ask yourself if you are truthful about the following questions:
Will your government lie to you to deliberately deceive you?...for sure they not only would lie to you but in comparison to the Mafia we find the Mafia much more trustworthy than your Government. See this is why they are so paranoid about Wikileaks because suddenly for the first time they don't control the media. They wish you to believe they are about protecting you and in that they are less truthful than what the Mafia is. If the mafia tells you they are going to slit your throat, it is going to happen and you can count on it but the Government will never tells you they are going to slit your throat because they only tell the CIA to do it and then your throat is slit. The government hold tabs on every person on earth, which is not because they feel responsible for your well-being; no they will kill you and make you disappear if you step out of line or if you see you are becoming some threat to them. Their government is partly formed by administrators and partly by politicians. Will the politician you so dutifully re-elect bring you harm?...yes because they made you a brainwashed zombie who believes in the power of democracy. Will the politician you so dutifully re-elect ever lie to you? By God I have never heard one single one tell the truth in all my life. To them the truth is rhetoric that will get them re-elected and democracy is about telling you what you wish to hear but never what the Mammonites tell them they must do. If you believe your Politician will only speak the truth about everything he governs...you better not read further because this earth is so gullible it does not need innocence such as you have. But to all others who got wise from getting bitter, you know they are the biggest bunch of thugs going around and yet you vote for them because they deceive you again and again. You are about to find out about THE ORIGEN OF ALL SCIECE CONSPIRACIES. The conspiracy That Proves Science is An Art to Further Opinionated Incorrectness by Withholding, Misleading or Misinforming one and all...In the event of you answering yes to all the questions I put to you, then by god get back into your carrycot and wait for your Mommy to bring your bottle. Your naiveties must render you no more than six months of being on this earth.

My theses are in place because science as it currently is (whether you believe what I say or not) is running on so many phoney principles of which those in charge of science holding the highest office in science benefit richly from and therefore discard the utter flaws on which science in the present way is founded. Their job is to ignore the flaws there are. By discarding my work they secure their work. If they agree about the flaws, their papers on science and their degrees in science becomes something for the paper wastebasket.

People would argue religious concepts but everybody accepts science on face value. The guilty party that I refer to betray us in every way possible by teaching us lies that never could be true and all the while we believe them with all our hearts and entrust them with our entire future... They take money from parents with the sole aim to brainwash the students that are entrusted to their care and demolish their intellectual understanding ...and before you think my claims are exaggerated I say again I prove every word I say. I show how the establishment of science frustrates me, ignores me or runs me down and I challenge everyone and any one to disprove me.

Those I refer to are allowed the powers to take as much tax dollars as they wish and endeavour on all sorts of projects that they could fantasize about. Their budgets are limitless and they never face questions because they are the intellectuals that we dare not question. With the same powers we give them, they brainwashed your child making science a religiosity. The methods they use on your child are to make your child believe in science and that is just methodical corruption of their thinking.

They take money from parents with the sole aim to brainwash the students that are entrusted to their care so that students will believe in science unequivocally and regardless. If you are a parent then mind what you do and beware and read on! If you think I am going one step too far in accusing the academics teaching physics and I slander their names, I challenge you to ask this question again after you have read this entire book and thought about everything and considered every aspect.

This Network which I denounce Engulfs the Entire Civil Human Race in Every Aspect and willing or not, you reading this are participating without knowing about what you support with every part of the mental intellect you have.

It involves the most trusted members in the part of our civil society and we are deceived by the ones EVERYBODY trusts the most, the teachers and Professors teaching at academic institutions everywhere. For over a decade I have been knocking on doors with the information I present in this book and lots more evidence, just to be ignored and to be turned away.

This behaviour of physicists ignoring me or even attacking me when I point out a very legitimate case of mistakes in science made me realise there is a conspiracy going on. Science has been getting away with this conspiracy since the Dark Ages. What you are about to read is not for the simple minded because science requires much intelligence and in that physicists get away with turning Creation into a joke.

By turning the facts that form the conspiracy into what you think of and you accept as natural and as culture, man has been cheated and mislead for three hundred years because those teaching science consider everyone as representing stupidity. To them everyone else is simple-minded peasants who are not worth having independent minds who could be able to think on their very exclusive superior level that they are able to think. That is how they get away with the conspiracy.

They pretend to be intellectually superior and to understand what seems senseless to the rest of us and then look down their noses on the rest of the world. I will show you what it is they can't see and then miss.

They underestimate the public as much as the public is too lazy to think and the misleading carry on. Those in science think their mathematical abilities make them gods–on–earth while we others are mindless human fodder and for that reason they brainwash our children at school by applying mind control as their mathematical understanding makes them gods. What they say, all must believe because if they speak using mathematics then God spoke.

THIS IS THE BOOK THAT **WHISTLE BLOWS** ON A SCIENCE CONSPIRACY NETWORK. You might think physicists are mathematical Masters knowing every aspect of whatever part of mathematics one may stumble upon but I AM GOING TO SHOW YOU THEIR UNBELIEVABLY POOR MATHEMATICAL SKILLS!

This conspiracy is so widely active that every person on earth teaching physics participates, most probably without knowing, some reluctant and others well knowingly to set out to brainwash by thought control the minds of innocent students forcing them to believe in dark aged ritual beliefs, in the ritual practicing of forces and allow the conspiracy to be enlisted with all the vigour they could muster.

People would argue religious concepts but everybody accepts science on face value! The guilty party that I refer to betray us in every way possible by teaching us lies that never could be true and all the while we believe them with all our hearts and entrust them with our entire future... They take money from parents with the sole aim to brainwash the students who are entrusted to their care and demolish their intellectual understanding ...and before you think my claims are exaggerated I say again I prove every word I say. I show what is true and what is wrong in science and as a thank you for all the concern and devoting care I show, the establishment of science frustrates me, ignores me or runs me down, while I challenge everyone and any one to disprove me. Decide after reading this book weather what you read is not a conspiracy...however, familiarise yourself first with what I present and then decide.

Those I refer to are allowed the powers to take as much tax dollars as they wish and endeavour on all sorts of projects that they could fantasize about. Their budgets are limitless and they never face questions because they are the intellectuals that we dare not question. With the same powers we give them they brainwashed your child making science a religiosity. The methods they use on your child are to make your child believe in science and that is just methodical corruption of their thinking.

Physics teachers take money from parents with the sole aim to **brainwash** the students that are entrusted to their care so that students will believe in science unequivocally and regardless. When teachers teach falsified information then they are worse than the mafia, that operates openly by crooked intention and in that the Mafia maintains honesty, which the physics teachers don't have. If you are a parent then mind what you do and beware and read on! If you think I am going one step too far in accusing the academics teaching physics and I slander their names, I challenge you to ask this question again after you have read this entire book and thought about everything and considered every aspect.

Read About The Biggest Conspiracy ever enlisted, it's more than just the next conspiracy because it is A Science Conspiracy Network. I prove everything I say about the conspiracy I announce...read and be shocked. I call science a scam and if many readers if not most will completely disagree with me about science living as a scam then the next medical scam I wish to bring to everyone's attention will support me. This scam is called humanity as it involves the medical science more than physics. It covers not what you are told or what you are taught but what science remains silent about. It is what is never said that is of serious importance. It is never what the research reveals but what science hides from getting revealed.

It lurks in the profits of the pharmaceutical companies or the doctors or the hospitals the industry never wishes to reveal. Euthanasia is not about the absolute sanctity of life but it is keeping the patient alive when the patient is going to need the most expensive drugs and the most expensive intensive care. I saw a patient with a knife in his head wheeled out of hospital to find some other place to die because that patient did not have medical care to fund his desperate situation. Because there was no money for treatment there was no treatment and this goes on in every private hospital in every city around the world. If that person had a good medical care and the medical insurer was willing to pay then a team of doctors would fight day and night to keep the person alive and no cost will be spared. It is during the last few months / weeks/ days / hours that the most care and medicine is needed.

During that time when the end of the life approaches and the fatal disease is going to take its final toll that the pharmaceutical company, the hospital and its staff, and every person sanctioned with keeping this individual alive will fill the bill as fast and as hard as possible to get the last money from the dying sucker. You have a lawyer walk in there holding your last testament that states you are no longer presuming responsibilities for the costs and you have the lawyer tell them if they don't guarantee success with the treatment payment will stop on that minute and the costs for keeping you alive will be on the hospital and its staff that treat you and then you see how quickly they stop machines keeping you breathing. If you start legalising euthanasia you kill the part that serves the highest profits and they would rather see you suffer than they would agree that you might die cheaply.

If you are not prepared to pay for their effort of holding your life so ever dearly they will lose all interest and let you die as quickly as the machines are switched off. Your life and all life have value when there is someone prepared to pay the bill and those doctors are the worst criminals there mathematically is. But yet again behind this medical industry there are insurers and behind the insurers there are bankers.

Fighting euthanasia is a conspiracy because euthanasia will kill the huge profits the medical industry makes. If we are all so against euthanasia let's take the profits of these pharmaceutical companies, the hospitals and the doctors and spend their money to pay for everyone dying that can't afford treatment. Now it is a case that they would rather see people suffer and die in agony when such persons can't pay for the treatment because after all, being part of science makes them equal to God and therefore others must suffer so they can reap the reward of the money they invest in promoting and furthering science.

Money for space travel goes straight to NASA while money for poor black children that goes to Africa is first skimmed and siphoned and tapped to fund the wealthy who feed the banks from which the politician gets funded under handed. Then how long will it take to get the politician to get the money going to hungry black children in Africa that then is siphoned to feed the rich and the bankers and the politicians and while everyone gets a share worth fighting for the hungry children do get the left - over that Europe and America were going to dump into the sea in any way.

In the sixties the Russians were so far ahead with the "space race" that Kennedy the then President had to say something to hide his embarrassment with the lack of credibility and progress that the Americans had in outer space. He said the Americans would be on the moon at the turn of the decade and the speech would have been lost if it was not for a fatal intervention. Everybody that was somebody had a finger in the pie with the death of Kennedy the then president and instead of making a martyr that was killed and with a million questions to have answered they made Kennedy a hero with a dream. The Americans who were in power either had something to do with the assassination or knew somebody who had something to do with the assassination. In order to divert the attention away from questions needing serious answers they upheld Kennedy's dream of landing on the moon at the turn of the decade. So whether the Russians were in on it or not, the Russians participated in a race to the moon except the Russians never had any intention of this race. The Russians saw money would be much better spent on establishing a laboratory in outer space and they proceeded with that gaol. But the Americans had the Russians in this race that the Russians never took part in. So America was going to the moon at all cost to hide the accountability of those who had ordered or was in place to order the death of Kennedy.

The Kennedy's were arrogant brats that Joe raised to believe they were above end beyond whoever else was sharing the earth with the Kennedy's. The Dad Joe did make his money on the collapse of Wall Street in 1929 when everyone was losing and he was gaining millions that others lost. Joe Kennedy was so big on insider trading and he got rich from the stock exchange after the collapse. This I am telling you to show there were no scruples in the house that raised the Kennedy's as the cream of America only fit enough to fill the high profile in History as they did.

The two Kennedy's were making enemies a lot faster than they were making friends and they almost dumped the world into a nuclear war because JF was so high on drugs he did not know what he was doing. Both of them were on a witch-hunt and persecuted every person who had fame or influence as part of the surge "to drive out communists". This venture had them way down on the popularity list and they both had their personal assignation coming because they were acting tough.

Why am I telling you this? I tell you this because I don't want to make a hero out of a drug enslaved, womanising sex crazed, spoilt brat who thought he could put millions maybe billions of lives at risk in a nuclear war just because he wanted to show Castro he was tough. He attacked Castro and made an arse of America's "war machinery" and so Castro ran for help. In that he made himself the most unpopular person in the world and had his people rid themselves of him. So with this I wish to put the first moon race in perspective. It was employed to draw attention away from who killed Kennedy and who was involved because I think the list was never-ending to begin with. I mention this to show that every aspect of life holds some form of a conspiracy and wherever you may tread, you are going to step on many conspiracies forming part of one conspiracy.

But then the first moon race was in place as a front forming a conspiracy to cover very important, very rich and very influential persons from becoming more known and more famous than they were before Kennedy's death. So let us all not get too ambiguous about the event and put it into the correct perspective because this must play a part in the second planned moon race. The moon race was in place from the American point of view and guess what; America won the race mainly because the Russians never got in on the act. I am not criticizing the event because for once this did further Human achievement and played out for the better thereof. I hail what Nasa did on behalf of America.

Yet, it does not seem that everybody is that convinced about the truth in the dealings with this entire affair. There are others who hold another opinion as the popular opinion of America achieving the unbelievable.

After hitting the moon with a big who-ha and all the excitement associated with it in the seventies the excitement boiled down to becoming cold. Kennedy was dead and Kennedy was made a notional hero and everybody hated Lee Harvey Oswald for being killed before he could be stoned for something he was convicted for before he was even charged in a court of justice. Now came the seventies and the fire went cold. Then came the eighties and traces of the fire disappeared. Then the nineties brought questions that no one seemingly was able to answer. Were "we" on the moon? Yet, it does not seem that everybody is that convinced about the truth in the dealings with this entire affair. There are others who hold another opinion as the popular opinion of America achieving the unbelievable.

Were "we" on the moon? This eventually is all part of the conspiracy. Why doesn't NASA defend their stance and prove that "we" were on the moon! The newsreels showed how the astronauts danced the moon dance and flew like moon men. They show the moon buggy twisting through the moon rocks. The fact, which everyone misses, is that the dust comes down much faster when the moon buggy or the astronauts kick up dust. The dust came down as fast as it would with the earth's gravity. This big news became news that became ordinary news and ended up as boring stuff with no news. The seventies went past and the moon had no interest to John and Jane public in America.

The simmer of Global warming loomed in every news item. If a dog went for a shit it was because of Global warming and if a dog did not go for a shit it had something to do with Global warming. This was an attempt to scare the daylights out of John and Jane public and to get every person on a roll to discuss Global warming. If there was a haunting ghost you then called it Global warming because the only threatening thing you could endure was Global warming.

Global warming is supposed to happen when the earth goes warmer. It happens when the surface temperature of the earth rises by some significant number of degrees. Go back some 31 thousand years and you will find the earth was in the midst of an ice age. The ice was so bad it covered England and the entire Europe.

The only place habitable was the desert regions because where there now is dry desert, there then was water that was not snowed over. If it was not covered with snow it was environmentally liveable. The ice age is a repeating process that comes and goes every thirty one thousand years when the polarity of the earth switches. It holds a link with gravity and gravity is the result of pi forming.

What you see on this map that is brown and now is desert then during the ice age had water and had life. What you see as green was as snow covered as Antarctica now is. The green parts had ice and ice had fields that were lifeless. Because Newtonians think with their arses they put the centre of the Universe in the centre of their arses. Then from there they observe the Universe through the one hole they use as a tool to think.

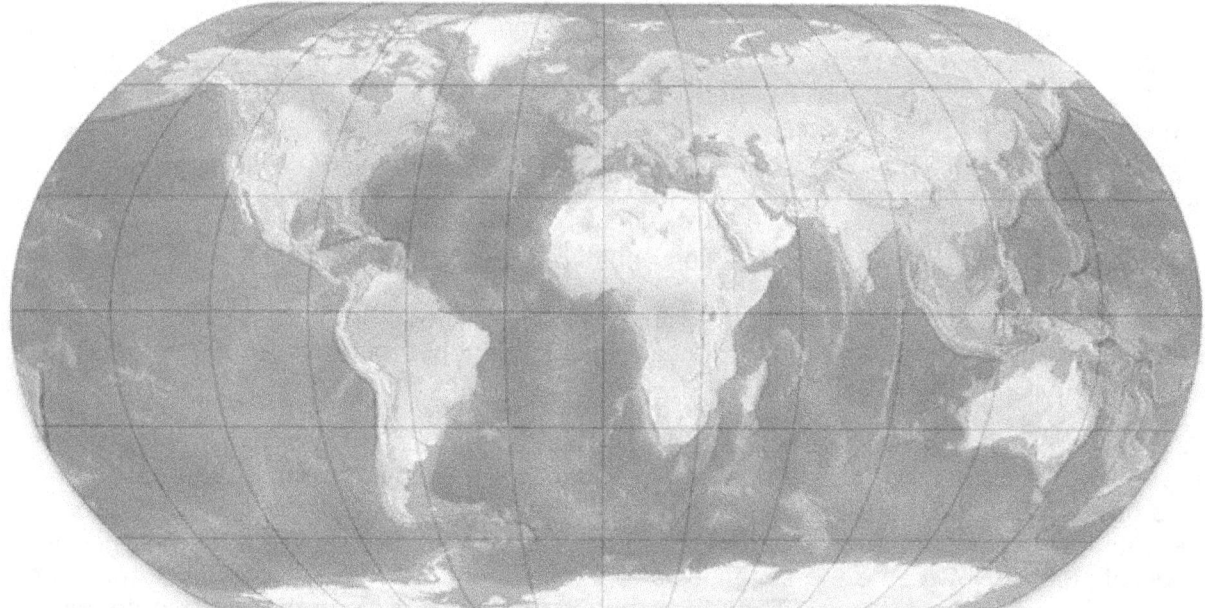

Global warming is not an earth thing but it is a territorial thing. In Antarctica and Arctica at the tip of the top of the earth the ice age never dissipated. In Britain the ice age went away about five or six thousand years ago and in Italy it went away I guess about ten thousand years ago. During the ice age the Sahara desert brimmed with life and all evidence is there for all to see. There had to be big grassy patches because every known animal was found roaming in the desert. We have human rock paintings all over the desert showing these animals grazing lively. That is where civilisation began, in the centre of those deserts. This happened because Europe and Asia was covered in ice, which was miles thick. As the earth grows hotter the ice age reclines further towards the poles. This will end as soon as the polarity switches and the ice age will begin again. Therefore, I call the one we are now approaching the liquid age and I call the "ice age" the solid age because in one part fresh water is mostly liquid and in the other stage fresh water is frozen solid in ice. But telling me something is going to end the earth whereas the evidence the earth provides shows that evidently this eventuality happens every thirty one thousand years but I now am forced to believe it is something I now have to shout about in fear is one of the conspiracies they force upon mankind. The earth has been through droughts we can't even imagine and it has been flooded in ways they can't even think of and yet I have to be scared that I am ruining the earth with their petrol I have to burn to survive the world they created. All of a sudden I am part of the end of the earth that is upon us and I have to pay them or I shall die the most awful death they can imagine. If that is not conspiring to defraud the earth then what is it? I tell you it is normal science behaviour because they have been defrauding the earth for centuries in the past. Science is nothing but fraud conspiring to fool everyone.

I mention this to warn you of a conspiracy of scaring everyone into the state of Global madness, as everyone is scared of everything going hot. Go and look at rocks formed in the desert and see how water made the rock surfaces smooth. That means there was water once that flowed as a liquid and that went on for many generations, maybe not all the time but it was definitely tidal and concurrent. The evidence is there but conveniently missed by science. Have you ever wondered why they choose not to see it? They are said to be worried about some supposed carbon footprint. The Mammonites pump oil to the surface at the tiniest cost because the product they get for free. It is there and because they have the oil in the ground below then they own the oil. They force me to live in an oil dependant world where I have to commute or starve. I have to pay to travel to earn a living while they own the factories that build the cars that run on oil by-products that they pump from the earth for free. Now they want me to pay for some f%@#$en mad crazy scheme to pump the gas back into the ground where they removed the oil from, in order to do what? It is so that they get paid by John and Joe public to cool the earth. And they keep the human monkeys they force feed garbage to pay for the expenses. We have to buy the cars and buy the fuel and pay them to remove the gas from their moneymaking madness.

They print the money and tell me what it is worth. They only pay for paper and ink and a printing press and then they have some printed-paper that they determine the worth thereof by forcing the public to work for this much money, labouring hours to get this printed paper they call money in return. The have what I produced with sweat. That which has true worth belongs to them but is a product of my labour and in return I get some worthless paper I have to give back to them in return to make a living and feed my family. Then they have the shops that we have to buy food from and they decide how much paper I am going to get and how much paper I have to give to be able to live. They decide the worth of worthless paper and force me to accept that a piece of paper with some number printed on it has the buying power of that amount and me working my arse off must accept that for that piece of paper they print almost for free I have to labour for so many hours as they decide it will be. Now they will take even more of the worthless paper they force upon us to earn to pay for a scheme we will be unable to monitor. We have to take their word on what it will cost to pump unseen gas into a hole in the ground we don't even know exists but only have to believe their say-so. But through the entire callus conspiring we again must trust those bunch of criminals as we are forced to work as the slaves of the Mammonites.

If the stock exchange crashes the money the stock represented can't just simply disappear. If it was worth money the money must still be somewhere because no banker will loan money without finding sufficient collateral. If you paid $10 000 for ten shares and the next day the shares fell to a $ 1 000 then you did not lose $9 000 because you have ten shares and that is as much as you had. Whether your ten shares are worth $ 10 000 or $ 100 or $ 1000 000 you have ten shares and only when selling the shares would you have gained money or lost money. But if you thought the ten shares was going to repay you your investment of $ 10 000 it would then so much easier make you $ 1000 and you can get a far better return on your investment. Every ten years or so the stock exchange implodes but this time the banks imploded and that loss was much bigger than the stock exchange losses. The banks made losses by weak investments but never do we hear what the bad investments were. All the banks in all the countries made these losses and now all the taxpayers have to fit the bill of the losses. In every country it is the taxpayer that pays for money the government bailed to the banks. Where is the money going?

Every country has to sacrifice health care and government benefits for whatever causes. The students get less money and has to pay back more. The schools get less money and the staff is reduced so where does all this saving of money go? The Hospitals must do more with much less and the hospital a staff working in the hospitals is reduced in numbers working while the pay the staff earns, are also reduced. Where are all these billions and trillions of savings going? The money was there before to pay for the lot and suddenly from the blue the money disappeared. Whereto did the money go because there was money before the savings were announced and that money that was there must still be there but that money is now channelled to other places? Who is getting the savings that before went to welfare and the payment of staff that since became redundant? On average a quarter is no longer earning while the three quarters remaining is doing a quarter more work for a quarter less pay. Where are the beneficiaries of these savings? The losses of bad loans and the losses of carbon cleaning and the losses of stock exchange rampaging and the savings of redundant welfare did not vaporise the money so the money in terms of being must still be but where is it? I shall get back to this so just keep this conspiracy of making money disappear in the back of your mind. Those in power are investing in things we have no part of in the planning thereof. We only pay for the investment with our effort and sweat but later on we will get to that. Let's start a song…where has all the money gone so liberally taken from us by force? Money can't just vanish no matter how fraudulent the idea of money is. Money can be stolen, money can be hidden but money can't just be and not be. If they take 25 % of the money of the economy for so called debt, this money has to be paid to someone and it must serve a purpose, but what is that purpose?

The oil they pump they get for free. There are no cost hikes involved that will pump up the price. They have to get it from the bottom where the oil is located to the top where they remove it and that is not so costly. As a farmer I pumped a lot of water from a borehole and my pumping was much less than that but the price of pumping would have been steady if not for the electricity supply company fiddling the prices that much. To hike the price of oil is illegal because it is creating monetary currency. It is the same as printing money. There is a commodity that you have a selling price of say $10.00. Then you jack the price to $20.00 thereby creating $10.00 that never existed. You have printed $10.00 that was not there before the jacking of the oil price. The persons or company that did this falsified as many $10.00's as the barrels there are by which they sell oil.

Where does the extra money go? The oil is the same oil but they charge more for the oil. It is not the oil that became more expensive but it is the paper they use that became less valued. The worthless printed rubbish we work for just became less valued and thereby more worthless than it was before because now I have to work more to do less with this worthless ink printed paper than what I could do with the worthless rubbish beforehand. If I make cars in a factory, they don't pay me with a part of the car as Bugatti did in the thirties when he on occasion ran out of currency to pay his workers. They print as much money as they wish and they give me the pulp but never the product that I could go and sell. The product they keep and the pulp is mine. Then the government raises taxes so the money I get is even more worthless. I have even less purchasing currency that I had before. But this becomes profit to someone somewhere because my labour in hours remained the same and the produce I generate remains in quantity the same and the house I then have to pay more for because they raised the interest rates remained the same and in the end I am far worse off but the money they steal from me in this manner has to go to someone's pocket.

If the money is in vaults they must realise as I do that the paper is worthless and to have more of the senseless commodity stuck in larger vaults are doing them as little good as paying more by using more money to receive less goods. If it was that simple they could print money by cleaning the trees of Brazil and it will do them no good. If the money does have a value and they increase the value then what do they gain from receiving more or paying less for my labour and forcing me to pay more for receiving fewer goods. What do they do with the profits they make? Where are the extra profits going? Why am I being ripped off? Now put this into the context of their ambitions suddenly rising by their dreams of going to the moon and then onto mars. They are suddenly making my slave labour much more where I receive much less and if I am not satisfied I am going to be one of those who are walking around and looking for a job.

So now I have to be grateful for being allowed to have a job because I could be one of those who do not have a job. When they start a new factory they tell us so many new jobs are created but in the background there are vultures that make ten times more from my labour than they pay me for my labour. If that hyena pays me $5.00 for my hour of work that vulture will get $50.00 for the things I deliver to the company during the time that elapsed in that hour. I have to be thankful I can be a slave and make that Mammonite richer just so that the pig will allow my family and me to live. Where do all the profits go and why the urge to go to mars? Why would the urge coincide with the scheming of the profits of every country that serves the cause to enrich the Mammonites even further? We will come back to this when we delve into some other part of the conspiracy. But first let's get back to the moon where we were or where we were not...

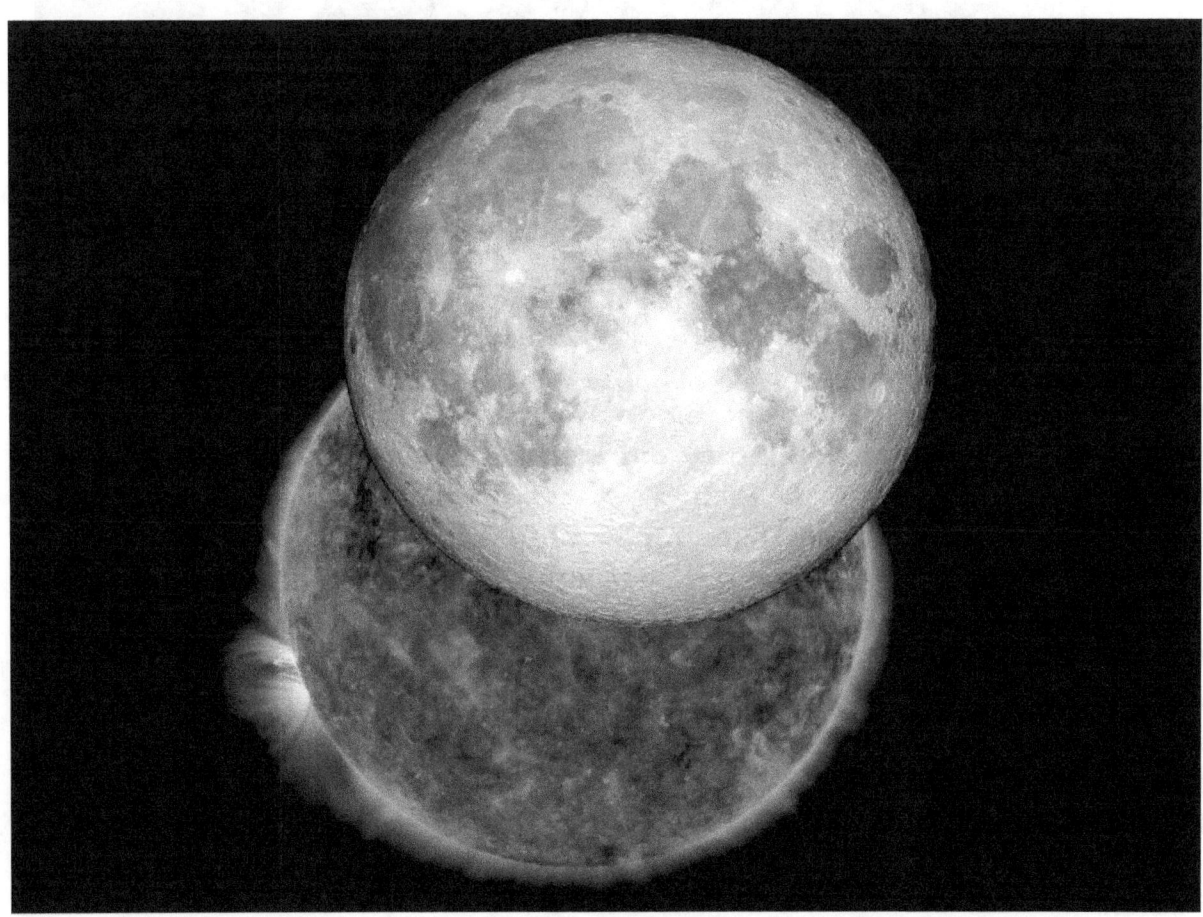

We accept what science accepts and we reject whatever science rejects. We put our faith in the hands of science as we humans put our faith in the hands of the clergy five hundred years ago. On what grounds do we rest our infallible belief in science, where we never question the legality? The conspiracy in science is summarised by the question that a philosopher once asked when he asked the most decisive question that he ever could ask: "In order for us to know anything, we first have to question everything we know." By not asking questions science conspired to conceal. The conspiracy I detected in science about science is that science never had questioned everything known to science. I found a conspiracy in science that is in place keeping it alive by not to question everything in science that science knows and hiding the knowledge no one ever questions; that is the conspiracy I wish to inform every person about. The conspiracy is to never question and by never questioning, science hides that what science does not wish to have others to know about. The "not questioning" became the obsession that became a conspiracy amongst all people working with and in science. By not questioning that which science knows they hide that which they should question and by the hiding of the questioning the conspiracy comes alive and not questioning maintains the conspiracy in science about science!

Apollo 11 moon landing

Are you convinced about the outcome in the search to uncover the possible conspiracy about NASA really being on the moon or faking all the evidence they show?

Do you think NASA was on the moon in the sixties and seventies? Do you think they ever landed and if they did then how many times did they land? Have you got this eerie feeling that somehow there is somewhere something about science that comes across as not that believable and you in your mind have your personal limbering suspicions, concerns and doubt but you can't see it?

Do you think there is a conspiracy in the unknown regions of science hiding in the dark avoiding the obvious detection of everyone but no one can point a finger as to where the itch is coming from? Is it not how we all feel about science!

Politicians mislead the public; bankers fake figures and mislead the public, and warring Generals fake facts to mislead the always-trusting Public.

Now it is the turn where we find science that seems to go wanting for the truth!

The question on everyone's lips is "were we there or were we not there?" Suddenly everyone questions the moon landing and everything that is connected to the moon landing in the sixties and the seventies. Why would everybody have an urge to distrust the loyalties of science?

"We" were never there to begin with. NASA was there and to demean the efforts of NASA by questioning the integrity of what those intellectual giants achieved is despicable to say the least.

"We" were not there because "we" have not the guts to land on the moon. It is always a small number of giants, a handful of wise men who carry a nation or in this case who carries a human population on earth. However, to question or not to question is the conspiracy that science hides.

Was NASA on the moon...well yes, I think for sure they were on the moon? Did NASA at the time and during the event make the movies that they then showed as if broadcasting live about the event and as it took place on the moon that very minute? Don't be simple minded to think or expect that such film making on the moon was possible because the television technology allowing that was not in place yet and what the public saw most probably was pre-recorded.

I truly can't see how the computer facilities at the time could handle such a broadcast transmission about what took place on the moon at that moment in time. There was no transmission possible to broadcast from the moon back then, live in the instant frame-by-frame photography. There were no facilities in those days to broadcast live events from the moon to the public, but that is what the public wanted and if the public wants to be fooled then the public gets what they expect to please them.

Think of what NASA achieved and see how small the related accusations are when compared. Moreover, we know the American public is the most demanding people there is and if they wanted to see the events unfold while having a live broadcast that was what they would get. Think of the cabbage doll outcry that happened some Christmas some time later but during the same decade and how the American public got nasty over a silly doll. Can you then blame NASA for going undercover about some of the truth?

Everyone tries to prostitute everything for money without showing scruples about it. No one shows a sign of having a conscience because there is an all-conquering lust for money and this has no limits or boundaries. Everyone out there tries to make money by selling improvable suggestions or toxic rumours filled with ridiculous defamation and defamation it is because they can't prove what they declare is true.

I suspect the public thought by paying billions of tax dollars for the entertainment that the least NASA could do was to oblige by delivering the expectation that was the buzz at the time. Does the live broadcasting or not deter the importance from what the event is in history... sure as the earth is round the event remains as important and as real as if the broadcast was coming from some live sport event. But they used sliding rulers to calculate on board the spacecraft. I used sliding rulers to calculate at the time because hand held take-away computers were science fiction and with no computers to broadcast, the entire idea was fiction. No one had the facility to put a computer in place that was capable of relating the broadcast from the moon back to earth. The technology was not in place but everyone wanted it to be in place. I don't say that it happened in this manner but from my perspective it could have lead to the NASA reaction.

But this doubt that was evoked came as a result of other doubt that was lingering in suspicion. The smallest question of doubt has another origin coming from another source in science that I found but was apparently never yet detected. The doubt that these questions raise could be the part of the conspiracy that science tries to hide. By looking at these issues, everyone is looking in the wrong direction and looking away from where the conspiracy is and then to what is not important. Of all the enterprises in the last century, the achievements of NASA rates bigger than any of the other industrial-related achievements, coming only second to what we gained from using electricity as man's main energy source.

The achievements of science must leave all of us in awe. Their efforts are frightening as much as it is mind-boggling and goes beyond what anyone could ever have dreamed of in a fairy tale say two hundred years prior. If those generations that came before us could come alive today and could see this that man now achieved, they would think of man as God. If those generations that preceded us could see what is taking place today, they would run from us and hide in fear of our achievements and our abilities in living as we do today!

However, be very sure about one thing, we are better off just because we now know how to harvest energy better but we are not cleverer and we are for sure not much wiser. Society is made up of the same foolish mentality as we had in the Greco Roman era and we know less than what we knew then. We can only use energy better but as soon as the oil and gas as well as the minerals are depleted, civilization is going to crush, as it never has done before. This is because there is no money but only numbers and symbols we give value and without

energy those numbers and symbols become what they are...nothing but a nightmare.

Look at the picture and see the engineering effort portrayed in this image of tranquillity. Not long ago in human history people would have fallen on their knees and would start to worship what they saw if they saw this and whoever witnessed this would father a brand new religion just there and then on the spot.

Still unmistakably the effort remains incredible and the picture reminds us of Godly inspired wonders. If ever anything man made proved intellect way above the ordinary and into the realms of superhuman thoughts, then this is proof of such achievement. Yet, it took so many man-hours of superhuman labour, planning and concentration to bring everything in this picture together. With the entire marvel this picture holds and all the greatness it portrays of man's ability to achieve, out of view, it hides a dark side too of the brilliance of man and more particular of science. The conspiracy comprises science as much as it involves the entire faculty of science. It is about what science conceals, hiding underneath everything science reveals.

With all the amazing achievements accounted for and when recognising all that science changed our way of living on the earth and what was achieved by scientists developing this super mentality and in that also giving science all the admiration duly admitted, notwithstanding I am about to dump on you the biggest conspiracy that has ever been presented and that was ever undertaken by any group of persons in the entire human race. Think of anything you might think is big or outlandish by nature and that dwarfs in comparison to what I am about to reveal.

It is so large that there is nothing in the past history of man with which one could compare it to. It involves every aspect of the life of every human being and this shadow in our midst covers the darkest secret that was ever hidden from intellectual human view. It is perpetrated by those we absolutely unconditionally trust in all aspects, It touched on every individual walking the surface of the earth and that excludes no person of any status albeit an infant or someone in old age.

Do you fully believe in science as if everything about science is proven fact and is truthful, never questioning or rethinking one question in having a minute's doubt just because this story has been repeated for centuries? Are you so confident in science up to the point you will put your life on the line to prove the accuracy of trusting the facts we have in physics? Are you one hundred percent sure about the honesty of science and are you sure about the trust we put in the honesty with which we regard science?

We know that scientists claim to work with facts and truth and the commitment to further the truth science presents as facts. This trust we have in the truthful accuracy of science goes beyond any and all suspicion of any kind. We all know that physics can never lie and physics represent the truth as no other form of knowledge could ever have. The presence of God or the absence of God becomes doubtful and is scientifically debatable but the accuracy of physics can never be doubted in having any suspicion. When you then worry about NASA being on the moon or not, I have to warn you to forget about NASA going to the moon and to start thinking about the moon coming to us when we are taking Newton's physics principles under review. What we see in this picture is that heat can release objects from the gravity that confines everything to the earth. Yet, this says nothing to the physicist about the nature of physics and that surprises me as much as it disappoints me. No one ever gave a thought about the question that if a hot air balloon can lift "mass" up into the air and break the shackles of gravity, why is that feature in physics possible when it is "mass" that forms the confining gravity?

With raw heat streaming from engines in the picture it is flames that convert fuel into fire to form the power that is so great it releases so many tonnage of "mass" into outer space. We can lift so much "mass" into orbit and by doing that then overcome the burden of gravity where that gravity burden is the result of mass attracting all other mass to produce a pulling force.

Yet, by releasing heat from engines, this power of gravity can be overcome and send many hundreds of tons of mass into outer space. That means heat can overcome gravity and gravity is a force. But heat will only react to heat and a force will be neutral to heat expanding. However, in a hot air balloon it is heat that also produces lift and we know that heat produces expanding space and therefore gravity shows more inclination to heat increasing and reducing than chasing after Dark Age "forces" fighting contraction.

Holding this evidence in the front of my mind I have formed a new concept concerning physics and in particular gravity in which I attach other rules to the forming of gravity. By trying to introduce these concepts aligning heat and cold to gravity, I ran into the biggest stonewall formed by resistance. Then with many decades of research and studying facts science avoids I came to form a conclusion I am about to share with you. I am about to shock your socks from your feet with the accusation I am about to make!

If you are looking at such concepts as the landing on the moon or no moon landing issues you are looking at a conspiracy in science from a totally wrong direction. I am going to show you where the conspiracy in science hides. There are far greater issues that science avoids than the fact of landing on the moon or not. The question should be why nobody is informing the human race about the biggest disaster that man could ever imagine. Why is there a deliberate silence about a pending but certain collision that is coming and that would end everything that man made including the life form "man"?

It is, believe it or not, about modern Physics as they teach physics at schools and Universities by those who don't understand physics. If you aren't into science this might just be why you chose not to study science. You most probably were smart enough not to understand Newton. What kept you out of science is The Ultimate Conspiracy Theory; it is the Conspiracy Concerning Physics.

If there were no reason why NASA would tolerate this nonsense of not being to the moon they would have silenced this Mickey Mouse as science has been silencing me for many years.

They use this rubbish as a lightning conductor. They draw attention away by having much debate over no existing enterprising arguments. But from what are they drawing attention away and why are they drawing attention away?

Why would NASA not just come out and settle the argument in the direction they want. Why keep the tongs clicking? Could it be that they are not deflating the situation because they wish to draw all the attention away from a reality that forms a conspiracy and anything will do to deflect the attention from where it would bring harm?

Do you think a conspiracy has to be as dramatic as Dan Brown's conspiracy he called *The Da Vinci Code* **and** *The Lost Symbol.* No it is not because what Dan Brown wrote about is entirely fiction and therefore it **must be** dramatized to **gain maximum publicity** in order to sell. It is not intended to be silenced because by silencing the conspiracy then the conspiracy would remain a conspiracy and the conspiracy would not unveil the billions of dollars it must make. The entire idea is to blur out as loud as much gossip minded hogwash to sell as much as it can. Conspiracies are kept silent and Dan Brown's is anything but quiet and that makes it no conspiracy.

The best seller of "factual mystery" is one big hoax and every person my wife included (except me and maybe a minute number of others I don't even know) bought it and fell for this devious prank, gun, barrel and flintlock. They ate it like it was a green banana and got choked on the pornography of lies and clever deceit. They wanted it to be true and if you ask me why, I would have to admit it is one of the rarities I can't solve but I think it has to do with their utter F&c@en stupidity and half a brain not working in most people that read it (including my wife and my daughter in-law that enjoyed the book) and was fascinated by the garbage. The problem I see is that we live in a society that brainwash the paupers like me that forms a part of the rest of us moneyless scum of the earth or that is how the rich and powerful feel about us drifting in the hogwash they create. In the past many decades ago before television and advertising put us into stupidity the suppressing by the upper class was bad but now it has become shear undiluted slavery. On television they target persons with a mighty effort to keep people thoughtless in order to make the advertising slogans work. It makes those looking at the advertising think less and they get impressed easier and quicker. Everyone knows there is a problem that everyone endures but in the comatose state they keep the public captured. No one can see clearly because of the comatose state in which the general public is.

Everyone became a robotic sleepwalker fed with wild sporting events making billions on the go and movie fantasies less realistic than anyone's life are paid entertainment while doping the wits. This is the way society stitches people together. From the churches to politicians to solicitors to doctors to bankers to the army/navy/air force including the police and teachers all vow to keep whoever they can under a control as to keep everyone in slavery. Since we all know there is a problem but we all can't see the problem since we have become most of the problem, everyone grasps for a solution to bring an end to the mental zombie-state in which we all are. Now we are all grasping for air and start to look for conspiracies. Most of those making money from detecting conspiracies play along in keeping us intoxicated with brainwashing.

Even when Dan Brown said in court his books were entirely just a fantasy, a flimsy hoax put together by compiling no evidence, he was ignored. That statement he gave in court about the absence in the truthful research he supposedly brought to light was lost. The entire story is founded on the Leonardo Da Vinci's painting portraying something that is totally non-biblical and false. The way they portray this hoax is as if Leonardo Da Vinci was present at the last supper and he had a stand-in portraying of the actual event. It's as if Da Vinci was eye witnessing while documenting the event as one being present.

On this interpretation the entire idea is vested. And all the mindless masses indulge in the carnage of mindlessness while feasting on bullshit and getting stuffed by mental stupidity that feeds on the replacing of their miserable existence as to give them a view into the higher working class. They pretend Da Vinci was present in the room and saw the diners sitting around the table. That is as far from the truth as that idea that the Pope can forgive sins but this last comment again is the way religion of all kind put the fear of God into you by presenting those clergy as having a better connection to God and can converse with God on a far elevated level than we, the low class "Grass Roots" commoners can. ...And this made Dan Brown a multi millionaire ...by selling the truth short and being a conspirer in the effort to make money.

The media batters the public to become mentally instable in an all out war to make money and to create slaves. Or do you think a true conspiracy theory should be as devilish looking as the illuminati? Are you convinced one must see the evil of the devil being portrayed in every symbol that represents the illuminati? No, those in power would silence all. It's just what happened in the Kennedy conspiracy. Those with the power hide the blame so effectively by law no less that the guessing will remain in place for the next century.

ILLUMINATI

People who have the power to control our coming and going can't just be nameless. They must be named and what better name as naming them the enlighten ones. They are the ones that can see in the dark and they are the ones that can see through your bedroom window and they can visit your wife when you sleep and they can farther your child when you are not looking because they have the power of God on their side by practising dark Magic. We know we are cheated hands down by government but it can't be because of our own impotent stupidity. We know we are cheated politically but they must have an evil Satan Devil on their side to play with us so ruthlessly.

Everyone wants change. George Bush being the President in office promised change during an election campaign while he served the rich. He sat in office knowing the public was unsatisfied with the way things were but instead of changing while being in office he promised to bring change. No one wants what is coming their way and still they want to be part of the process that rules them, so they opt to go democratic.

We think the power they derive by which they control all aspects of our lives can't be because we are brainwashed into thinking democracy gives us power. No. If that is the case then we must take the blame for being that stupid and we are clever since everyone in power tells us we can fight for what we want because we are committed to democracy. We think that it is by democracy that we govern the country by putting government in power and we tell the politicians how to govern because we can vote, and we are the ones living in the twenty first century thinking we came to be the most intellectuals of all times? In Britain there are 44 million voters and of the 44 million each one can vote once. I have to stand in line to make a cross next to some political party and that turns me into Superman while in fact that makes me more stupid than the ape-man during the time before he could think. If you can jump off a cliff and I tell you every once in 44 million someone lives after the fall to tell his or her tale, will you jump? Still you do think being one of 44 million, that that honour gives you the power to rule the country by democracy? Can we still find people that stupid, yes there are and the western civilisation is filled to the brim with such mindless masses, begging for the privilege to be able to vote. No wonder those in power can keep so many on as slaves and never pay a dime to have slaves to fill the sole purpose of making those in power also rich at the same time.

If I cast my vote I can change the world. What a bunch of dumb bastards and mindless idiots would fall for that. To think if I was a Brittani I have one chance in 44 million and still they want me to be an idiot to think that the one vote I have counts. No, I can't be that stupid in believing I am played like a drunken chicken! No there has to be an evil-eye bunch of Devil worshipers that call on the power of Satan by formulating dark magic rituals and then they get the power to influence my life so that I become my worst nightmare. I become so stupid that I believe it is my right as it is my duty to send my children to be killed in a war they wanted and planned and that they put in place so that they can profit from and to protect their billions.

Hey liquid brain, you are not one in a million making you special, you are one of 44 million making you anonymous. No matter whatever the party is that you vote for, the Bankers got to the politicians first. To thank Tony Blair for giving the oil fields of Iraq to the British oil barons, they awarded Tony Blair with $30 million this far just to go around and make speeches. He gave the bankers the oil and while he is alive he is forever a rich individual. It paid to allow the Liberal Democrats to be against the war to present a democratic resistance and pay them handsomely for being negative about the war, while the Liberals and the Conservatives believed wholeheartedly the photos of the weapons of mass destruction was real and that piped out to be demolished war debris laying in the desert. So both the main parties voted for the war that would enrich the bankers and industrialists much more than they are already rich and the politicians on all fronts were paid generously to believe pipes in old mines could be atomic missile launchers. All they had to do was to brainwash the public about the urgency of going to war to save the country!

No matter for what party you vote, you will keep the money-mighty Mammonites in power and in turn they will keep the Politicians in party as long as the politicians serve their cause well. The public will still have the privilege to send their children to death to become national war heroes and die for national pride while when they get back there is not even money to supply the war cripples with artificial limbs or to give the crippled a living pension. The penniless have the fortune to send their offspring to die and to murder so that the rich and powerful can become richer. When will Tony Blair's boy go to Iraq or Afghanistan to serve his country? Are Gordon Brown's two boys going to enlist to fight for their "Queen and country" No it is the " grass roots" scum that only has a purpose of keeping the rich safe that would have the honour to be killed for Queen and Country or to kill as many Muslims in Iraq as possible for "Queen and country".

Turning our heads to America things are much worse because in America I think you as an American will compete with something like 160 million to one to get any change in the two houses Governing or to get the President to go look for a new job. Think of what the chance is to fight a battle against things you don't care for and playing a 160 million to one role in the outcome. You lot who believe in democracy are simpleminded. With one vote you think you can change the government. The government uses your simple-mindedness to get their illegal governing legalised.

The money power even controls the anti-whatever protesting with fabricating a protest and getting the mood that was in favour swinging against the riot. This is how they do it. We as civilised people get tired of petrol prices rising through the roof. We want to get our message across. Then over the news we learn that there is going to be a protest meeting in several cities. That is what we need. We need the government to see we are deeply unsatisfied with the way things go under their government. That night we see TV cameras bringing us the riots in long sections in the news. We see a bunch of hooligans burn cars and beat the daylights out of the Police. We think of the Police being those who protect us from such hooligans. We as civil people don't want to see the police getting kicked and beaten and we reject that behaviour completely. We see the rioters burning cars and demolishing shops and going on like madmen and we can't associate with that type of behaviour. We have to distance us from such madness and we turn our backs on the protest. The protest was about how we felt but those with money hired a bunch of criminals to hijack the protest and after seeing what the protesters were, we don't want even to think we were acknowledging anything they stood for. Immediately we go against the protesters but going against the protesters we then agree with the raising of the fuel price because that was what the protest was all about. By distancing ourselves from the protest we are on the side of the moderates and the moderates accept a new fuel price like all citizens do who believe in "democracy". So for a few pounds and pennies the Mammonites bought their right to raise the price of petrol and by me being civilised they remove my protest in hiring criminals to make the protest unacceptable and hijacking my feelings in the process. My rejection of such behaviour as those they hired is much higher than my rejection of the new fuel prices and they buy my silence and my rejection about what I believe in by hiring thugs to throw police around and to attack descent people's cars and burn it. Since I cannot associate with such behaviour I distance myself from what I first believed in and I become part of the silent majority who hold my values higher than my frustrations and my feelings of pure contempt. That was the reason why the Mammonites had Hoover in charge of the FBI for so many years. With a smile on his face the biggest crook in America (that is according to my values) was the head of the biggest law enforcement office in the land. If he could kill off every person he found undesirable just because he found them undesirable, killing Kennedy would bring him the ultimate challenge and would leave him the most powerful man on earth, the man that could kill the most powerful man on Earth.

We know Hoover was involved by his involving of the FBI in the cover up because Hoover had to hide his gambling debt and his homosexuality. It was the Mafia that involved Hoover and the FBI. If you wanted someone executed from behind, Hoover was your man. If you wanted anyone executed you could entrust Hoover with the job unless your name was Fidel Castro and you lived in Cuba working as the Cuban president. Other than that Hoover and his FBI squad of men had their jobs cut out as legal assassins. The truth never mattered to J Edgar Hoover and this we saw in the way he shot and killed Ma Baker and her boys and so many other brutal criminals that the Media paraded as already convicted and then made them enemy number one. You ask Ma Barker- head of the Barker-Karpis Gang that supposedly committed a spree of robberies, kidnappings, and other crimes between 1931 and 1935, but in contrast to the popular image of her as the gang's leader and its criminal mastermind has been found to be fictitious. Fictitious or not they were slain by the FBI without ever been convicted in a court and America stood for it. Another example was George Nelson, who was a bank robber and murderer in the 1930s. Gillis was known as Baby Face Nelson, a nickname given to him due to his youthful appearance and small stature. Usually referred to by criminal associates as "Jimmy", Nelson partnered with John Dillinger, helping him escape from prison in the famed "wooden pistol" escape, and was later labelled along with the remaining gang members as public enemy number one. These were but a few that was killed by Hoover without facing a trial but was shot because of the say so crimes Hoover charged them with without ever going to court. Being public enemy number one made you convicted without trial and already dead. The Kansas City Massacre, or Union Station Massacre, occurred on June 17, 1933. Frank "Jelly" Nash, a convicted mail train robber and sometime member of the Barker-Karpis Gang, was being returned to Leavenworth, from which he had escaped three years earlier, when would-be underworld rescuers attacked the two carloads of lawmen guarding him with machine guns. A federal agent, Raymond J. Caffrey, two Kansas City police detectives, Frank Hermanson and William J. Grooms, and Orrin H. (Ott) Reed, Chief of Police of McAlester, Oklahoma, were all killed in the attack, as was Nash himself. When it was later proven that the "massacre was the doing of the totally incompetent FBI field agents that couldn't handle their firearms and they killed their own people. This evidence was swept under the carpet by Hoover that played the public like a fiddle when he convicted the criminals without having a judge present and then had them executed on site. He was so good with this, with all the practice he had through the years he was quite capable of getting the Kennedy assassination successfully completed, that would or should I say also be with much help from other partners.

Hoover's involvement in the mafia was so blatant it shook the foundations of the Kennedy boys. Hoover denied there was organised crime because organised crime or better known as the mafia had him wrapped up in horse racing gambling debt and therefore he had to keep the FBI off of the heels of the Mafia. It is known that Hoover was a notorious gambler and had a craving for horse racing. He must have owed the mafia lots and lots of money because the mafia was in control of horse racing gambling and he had a free hand when it came to punting on horses. Hoover couldn't go for a piss without the permission of the mafia because of his love for horses and betting on them, as well as his soft spot he had for men. He had a very open homosexual relationship with a number of men and this is information that the mafia used to their advantage. I wouldn't be surprised if the two Kennedy boys also started threatening Hoover with this information to get him out of the FBI office.

Hoover saw his position as the man who brought civilisation to America by creating the G-men, those who rid America of crime, albeit by blatant distortions of the truth and using courts and media spectacles to kill those he blamed for crimes he wanted solved. He saw J Edgar Hoover equipped to judge and qualified to convict any person out there who just by being in his office J Edgar Hoover then could decide that the person was guilty of everything J Edgar Hoover charged that person with and had the execution executed from his office without having insomnia. If ever there was a maniac overflowing with self-righteous narcissism it was G Edgar Hoover. If J Edgar Hoover put a man on trial he did it by allowing the media to sell papers and then get the public in general to become the jury and convict enemy number one that month by giving then selected well rehearsed propaganda and let the public decide the person's guilt long before any court got to the person.

I would have loved to read their minds when J. Edgar Hoover, JF Kennedy and Robert Kennedy shared space in one room. I believe with their ego's filling the room, furniture would have flown out the window due to lack of space. I am sure Hoover knew how Kennedy got to office because Hoover had a file on everyone in America that had any importance and the Kennedy family surely knew who they were dealing with when they took on the FBI boss. **The death of President John F. Kennedy is a true conspiracy and that we can see from the fact that there are no results coming from any inquest launched at even a Presidential level.**

Then when the Kennedy's came to office it was by the farther Joe Kennedy rigging an election outcome in Chicago and when he did this he called in the help of the mafia allowing the Mafia to swing the votes in his son's favour. But as any business deal a favour required another favour and Kennedy had a promise to keep. The mafia did this because Kennedy had to bring the lucrative Cuban gambling casinos back to the mob and the mob did their part. It was choosing between the mafia and the Russians and the mafia expected Kennedy to deliver on his promise because the mafia tends to be bad losers. Kennedy had to get Cuba back to mob rule and the gambling houses back on behalf of the mob and when JFK turned on his election promise to the mob they showed that not even a President is above their retribution. The presidency was their part of the deal. Cuba was part of the Kennedy's end of the bargain. But Kennedy thought that being the President made him beyond any approach and I guess this would have been true if he and his brother did not make such an effort to step on every important and influential toe in America. In the process they united parties that would never have befriended one another but with a common enemy the worst opponents for the time united in a common cause.

Suddenly his boss was a thirty something rich kid that was on a crusade to fight crime and that placed Hoover in a tight spot. Hoover was the one who for decades decided what was the crime of the day and who was to be fought and now this rich kid that never fought one crime case in court was going to turn the apple cart upside down. Hoover himself had not the power to do anything to Bobby because Bobby had a Big Brother in office going by the name of JFK so Bobby could tell Hoover what to do and where to go because JFK was the president and this did not sit very well with Hoover. Hoover was all so happy to oblige by involving the FBI because removing Kennedy also removed the little pest called Bobby and he could do well without the Kennedy brothers becoming problem number one in the life of J Edgar Hoover. No one realised this but J Edgar Hoover alone was a criminal institution deciding who and what is the criminal institution of the month to fight.

Kennedy had one promise to keep and that was to deliver Cuba back to the mafia and that Kennedy failed to do. Kennedy thought the Russians could not touch the President of America but he lost sight of the influence of the Texas oil Barons, the mafia and the FBI and many Senators that did not sit well with Kennedy trying to get people out of Vietnam or then better said some President who was not getting America involved in Vietnam fast enough. Think of the money that was made by the rich and the powerful that was selling and supplying war material. It was almost twenty years after the Second World War and the ammunition stockpiles as well as the feeding supply was grinding to a halt. The supply was not going anywhere and those making money saw drought coming if America did not enter the East in a war effort. Korea was almost a disaster with the North fighting back and with America almost getting their arse kicked by the Chinese. That was one trap the money machine did not want to enter but relieving France from the Vietnamese burden was one way to sell arms to America. Vietnam was such a pathetic little country and to bomb them into submission or just to pacify them with half-ton bombs was a sure way to go to keep the war machine supplied with arms and to feed the war-mongering arms industry.

All this had a part in the tragic demise of the President but not one of those alone had the influence to get the job done smartly. It was down to the Dallas police, the FBI and the mafia and they all had a lot invested in whatever Kennedy was threatening. Then Kennedy made a compromising deal with Russia that said America was in Future going to leave Castro and his Cuban Island alone. This affair did not sit well with the Mafia because the Mafia had Hoover deciding there was no organised crime such as the Mafia and when the Mafia thought they bought a president by swinging votes they got a witch hunter going after their security. These two rich kids became a menace and everyone in crime had something to lose with the Kennedy's in office and nothing to lose with the Kennedy's out of office.

We know the Mafia was involved because they silenced Lee Harvey Oswald the innocent man that was openly framed because it was the mob that got Kennedy elected by faking the election results against Nixon. The mob by the hand of Jack Ruby silenced the investigation afterwards as they had other interested parties also in their pocket and those in high office helped afterwards getting JFK out of the way. Jack was a dying man and I guess owed the mafia quite a bit.

All the facts point to the death of President John F. Kennedy as a true conspiracy because as I said and the conspiracy is what we could see from the fact that the investigation was a shamble from the start. The Dallas police had a very good idea who to look for before the shots were fired but the location that they went to look for the lame duck did not deliver the groomed suspect because the fool was watching a movie and that he was not supposed to have done. If anybody was groomed to take the fall for anything years before the event then this was Lee Harvey Oswald. He was sent to Russia to get a Russian wife and to seem as a person that switched sides openly. He was sent to Cuba to seem as a Castro collaborator. He was ordered to get (no less) a mail ordered rifle with which he was to have assassinated the president of the United States. This man must have been a complete idiot to line up all the evidence against him so conspicuously or he was a sitting duck in place for the conspirers to poach whenever they felt like it. If this international tail in the complot was so evident we then also can see the hand of the CIA touching the overall planning in this plot. They are the ones able to send agents to Russia and then to Cuba and get them to seem like communists as Lee Harvey Oswald did. One question no one ever asked is how did Oswald know about the route the President was supposed to have followed the day of the parade. How did Oswald know that he was in the correct building where he could keep a loaded gun and wait for the president to parade past this very building, in an open limo and that is if he did not have intense help from very informed persons close to the presidents entourage. You reading this, would you know when to bring a gun to work so that you will be in a position from where you could kill the President as he passed your office window? …And Oswald knew when and where the President was going to pass during the Dallas parade so that he could order the correct rifle through the post no less and keep it loaded in the event that the President was coming past the building in which he worked so that he could kill the President of America. If you believe such bullshit, then my friend I believe that you are gullible enough to believe in democracy. Oswald knew he could kill the president from his office window as the president passed in an open top vehicle at that very instant that the president did. Then he went and killed a cop just to draw more attention to him and after that he went to watch a movie and make it easy for the cops to get him. Good God I pray for you lot because this method of thinking is so lame only Americans could believe that! …And the Americans need advertisements to know how to cross the road.

We know Hoover was involved by his involvement of the FBI in the cover up because Hoover had to hide his gambling debt and his homosexuality. It was the Mafia that involved Hoover and the FBI. Hoover denied there was organised crime because organised crime or better known as the mafia had him wrapped up in horse racing gambling debt and therefore he had to keep the FBI off of the heels of the Mafia, in spite of the mafia being able to remove any persons notwithstanding their social positions. You take the case of Bugsy Siegel who was one of the prominent mafia bosses who had to take the credit for Las Vegas now being what Las Vegas became. Being the head of the FBI, he would know what happened to even mafia bosses when they cost the firm money and he also knew not to investigate the matter since it was a family affair. But owing the mafia money for gambling debt Hoover also knew that made him part of the mafia family where the only family love was that if you owed the family you better pay up, not necessarily with money, but you did what the family needed you to do. Hoover was a notorious horse gambler and did this "sport" openly and frequently. Everyone knew he had as much of an eye for choosing a winner as Jimmy Hoffa had for choosing secure sites to meet the mafia. Hoover lost money as if he knew he would never repay in bank transferable currency because he did not earn that much even being the FBI boss. Hoover was in a commanding position and had debt to pay not only with his engagement with the mafia but there was a score to settle with the Kennedy boys that stepped on his toes and frightened his mafia friends. This plot had good riddance written all over it from every way you looked at it as far as Hoover was concerned.

We know the Mafia was involved because they silenced Lee Harvey Oswald and Lee Harvey Oswald was the innocent man who was openly framed because it was the mob that got Kennedy elected by faking the election results against Nixon. The condition of being elected was that Kennedy had to get Cuba back to mob rule and the gambling houses back on behalf of the mob and when JFK turned on his election promise to the mob they showed that not even a President is above their retribution. There were many not as big celebrities as the Kennedy president but celebrities none the less that had money arrangements and went back on their word and got hacked for not fulfilling a promise made to the mafia. Sonny Liston the boxer and Mario Lanza the singer went to their graves because they owed the mob gambling money and couldn't pay up. Sonny Liston was a paid fighter and the mafia was writing his cheque on the condition that Sonny Liston had to fight and people had to place bets and the mafia had to make money from the bleeding nose of Sonny Liston. When the punch-drunk boxer thought he was above and beyond the reach of the mafia, the mafia had their approach in place and that was the end of Sonny Liston. Hey, this is so common knowledge that even I know what comes from not meeting you obligations to the mob and I have never even seen one mobster in my life.

Mario Lanza owed money and had to sing in Las Vegas to repay. He could have had much more that he owed if he only sang but he had stage fright and the mafia showed him his stage fright was nothing in comparison with how much he should have had to be scared of the mafia's retaliation techniques. If they could get to Jimmy Hoffa and Sonny Listen and Mario Lanza, then Hoover knew they could get to J Edgar Hoover without even breaking a sweat. I think even Hoover met his end via the mafia because his death had all the trademarks and the lack of investigation and the strange scenario has one big ring to it, the ring of the mafia buying with purchasing power that money alone as a currency to bribe can't buy.
Hoover knew the rules by which the mob enforce bad debt. So he allowed the mob by the hand of Jack Ruby to silence the investigation afterwards as they had other interested parties also in their pocket and those in high office helped afterwards getting JFK out of the way. Did the police look for powder residue on the face, hands and arms of Lee Harvey Oswald? Did the police determine how many shots were fired from the rifle of Lee Harvey Oswald that they found in the Texas School Book Depository store and how many cartridges were retrieved? Where are the cartridges now? Are they lost, if so, what happened to the lost cartridges?

Those are the question to address, but no investigation ever asked the obvious questions. We know that someone ordered the conspiracy from very high up being in cahoots with the mob and the FBI because the investigation was screwed even before the assassination took place. The conspiracy does not end its involvement there, because Congress and the Senate also had to be involved.

Then there were other contributing factors that remove so many influential persons from the adoring friends – list Kennedy had. With Kennedy so quickly reaching for the Nuclear button, a lot of Bankers got stiff with anxiety because a few Nuclear bombs might not reach the Kennedy family in their bunkers, but it will write off billions of dollars of property all belonging to a lot of banks, either by mortgage or by ownership. Going into a nuclear bomb contest is not what the bankers in America had in mind when finding a resolution to the Cuba problem. It is now bragged that America had more nuclear missiles than Russia had at the time but this fact does not soothe the Bankers who owned the buildings in the major cities. Say only one missile landed in New York and destroyed only one high rising building. It is not the building that goes bust but it is the rest of New York that is contaminated that forms the problem. No one wishes to live in Chernobyl, not because of the building not standing any longer but it is because of the nuclear contaminated atmosphere that does not come across that inviting any more. Now think of the entire New York or Los Angeles or any major city being in the same contamination state as Chernobyl and see how those bankers will be able to fill the buildings with tenants. One entire city going spooky is not what any banker requires for a future vision and the best insurance policy to prevent this strategy is to go anti-Kennedy in a big way. They all knew Kennedy was going to be re-elected because if a Chicago Democrat liberal as can be can get a cheer in Dallas in the Lone Star State as Kennedy received during that tour in Dallas, then Kennedy was going to be re–elected the very next election. That confidence of re-election was securing Kennedy in his office as far as democracy goes but that too got Kennedy riding the barrel of a cannon through a street surrounded by people on either side.

With a trigger-happy cowboy set and secured in the American Presidential office, someone who is very influential somewhere might just lose billions and in that it is cheaper to get a few million flowing in the direction of politicians who will help the conspirers get Kennedy from the office to a coffin literally with a bang…of a gun. That many in the Senate and High Court Judges had a common interest in the demise of the Kennedy boys was apparent, not in Kennedy dying but the other Kennedy dying a few years on. This also is so apparent in the manner that the Warren inquest was so self-servingly righteous and maliciously opinionated which all shows an open cover-up and while everyone on the Warren commission was greatly rewarded during the next decade, each in their own time, one cannot draw another conclusion that the outcome was decided beforehand. The commission rushed to a conclusion and the lot had very little interest in what other factors could bring about. They had one guilty man who was dead and one mafia man who turned hero just because he silenced the guilty man before the guilty man could defend his position with a not guilty plea.

This case shows the true science of a conspiracy and even the brother of JFK got done in when he got too close to solving the mystery. But investigate as much as you like, you will keep on guessing for the rest of your life! Are you under the impression that a conspiracy theory cover-up must be as old, as widespread and as powerfully involved as the Free Masons, which represents those who have the power to alter the destiny of the human race? If that were true then they would have the influence we find in the Kennedy case to keep their secrets covered under a veil of total silence and never to reveal any information to anyone! When there is a real conspiracy even those in the establishment that reveals the conspiracy, guards the secret and then it is well guarded.

So, then you want a conspiracy theory to be as blatant and as staged, in your face and unreliable as the UFO landings where the sites are very vaguely concealed and as revealing as it can be with the most pathetic idiots becoming star witnesses? The media then blasts their idiotic untested testimonies over the air around the world as Gospel. It will be better to believe in ghosts. I don't ever say there are ghosts that can haunt us but I can't use physics  to prove there are not ghosts while I can prove with physics there just can't be aliens. Where would aliens come from and where do they return? A visit from a "nearby" galactica will take those tens of thousands of years to reach us and for what purpose, to have sex with the biggest idiots they find? You never think why those Aliens would choose to meet the most simple-minded idiots with a single IQ figure after travelling for many a century non-stop to reach us? Are you impressed with aliens coming to have intercourse with the most greasy-looking wench on earth that by her own admission in accordance with her intellect and her appearance could not find one male in three billion to inseminate her with wasted sperm, and so the alien got lucky? There is no one on earth in flesh wanting her for sex because no one is that desperate so she invents someone craving her from outer space.

She is so desperate for sex that she invents someone even more desperate than she is. Do you fall for her stories that her ten mindless children has a father many galactica away and did not even leave her alimony to bring up his very earthly looking brats. If you have that mentality, then go for every conspiracy theory you can read because facts of life you will find too challenging to understand and comprehend fully. If you think conspiracies must be as mind baffling and logic eluding as crop circles that are not in the fields when the sun sets and are there when the sun rises, then you are a sucker living for bullshit and fantasy, a monkey to be made of and you have the mentality a child has and you then are living in a fantasy-world only you can believe. Then this world is a place far too harsh for a person as gentle minded as you are. It would seriously be in your best interest to commit suicide very gently and even more quietly for this earth is far too complicated for your likes. You would be better off not being with us in this harsh unforgiving environment and we would be better off being teary by missing you than having you around.

Newtonian science can't disprove such mad ideas because of incompetence firstly as a result of flawed principles applying and secondly Newtonian science would rather deflect any scrutiny away from science and in the direction of something as mad as Newtonian science because then Newtonian science would not get scrutinised. They would rather keep quiet and stand defenceless about mad claims that persons from outer space visited earth, than to attack such madness head on. While people are preoccupied with statements made by loony cucumber heads trying to get one second of fame albeit only getting the attention of those even lamer brained than the attention seeker is, this prevents anyone starting to investigate science in the manner in which I have. To travel to the "nearest" galactica is 26 light years away. If any person travelled at $\Pi \times 10^3$ km / sec it will take thousands and thousands of years to get there or from there get here if it could, but it can't be done and that part Newtonian science doesn't even realise. Much of the so-called preparing for space travel is another conspiracy hiding illicit international banned research.

For us to break free from the gravity of the sun, we have to travel faster than the gravity is within the very centre of the sun where the sun annuls all captured space-time. The sun is the centre and with gravity reducing space it is moving space towards the centre of the sun in the spot holding singularity. Therefore the movement such an object must apply must be faster and stronger than the movement within the very centre of the sun. If you keep on reading that is what I will show what gravity is. Within the very centre of the sun a fledgling Black Hole is forming gravity and in the end eventually the Sun will grow into a Black Hole because this centre Black Hole is growing progressively. If the something that tries to escape from the sun's gravity and then travels to another star, that something must travel faster than the attraction is within the centre of the sun or else it will become an orbiting missile at the distance of the debris probably at Oord's cloud or further but it will never be able to escape. That reason is the why comets circle around the sun and never crush into the sun or escape and leave their orbit around the sun. Once captured by the gravity of the sun there is no escaping possible and that is the biggest fact of gravity. That means that object trying to escape must be able to free itself as if it is escaping from the centre of the sun which means it must be stronger than the gravity is in the very centre of the sun. Only a person who knows nothing about physics and has an IQ of about 75 or less will claim that there can be space-travel outside the solar system and the Newtonians are unable to say why such travelling is not possible. Travelling at that point within the sun is travelling at the speed of light and no persons or objects can travel at the speed of light just because travelling exceeds the peed of sound and Newtonian science has no way to tell the difference through physics.

Human travelling would not exceed beyond Mars and go even beyond since gravity would not commit life to function at that point. Life developed on earth in the gravity of earth and on the long haul can't function fluently outside the earth. Secondly it would take too long and we cannot travel faster than $\Pi \times 10^3$ km / sec after which electronics will start to malfunction. This has to do with the earth gravity and our natural material that formed on earth within the earth gravity. However, science likes to tell how we can travel at the speed of light because that way they bullshit everyone who has fantasies and no reality about physics. Newtonian physics are so poorly informed they do believe that it is possible to travel at the speed of light. In this way they prevent people from getting fed-up with them throwing money into a pit of mockery which is what their science is. So to create a fantasy world where science is on top of all aspects and controls Creation they don't distance their views from madness but also don't complement the outrageousness in the press. This plus the deceit they use to swindle and brainwash all forms part of the most elaborate conspiracy ever devised by any form of human deception ever thought out by man in any way possible. For three hundred years they were teaching about mass and never, not once proved mass.

The truth about a conspiracy is that everyone involved with the conspiracy will fight tooth and nail to stop the conspiracy to get leaked and leave those behind the conspiracy and all those feeding from the corruption of the conspiracy exposed. Those who have the power to maintain the conspiracy would keep the waters as still as possible as to draw no attention to such conspiracy. A true conspiracy has to be as quiet and as unseen and going as unnoticed as it could be. A true conspiracy must involve everyone without anyone detecting even a hint of what the conspirers hide. There are many conspiracies going on such as the banks involvement with crime and the bankers profiting from gamble rackets and drug selling. The same goes for the insurance business profiting from lenient sentencing of courts holding very merciful judges in office so that the insurance will sell cover-policies. Our social system in the Western World is a conspiracy. Democracy is held up as a saviour but politicians rule only to favour the rich.

The conspiracy of Government is to govern the population with fear and the fear must be directed at a general enemy that can be pinpointed but never be caught. Create a terrorist conspiracy and you have a conspiracy that can be directed at the population you wish to control and subdue. Many more people are killed by "crime" than by terrorism and yet there are billions more spent on fighting terrorism than eradicating murder and crime. The crime helps to subdue the people by fear and pressing the terrorism button does precisely the same so why not put fear about by being killed through terrorism in action in order to do undercover maintenance of fighting the poor to keep the rich safe and secure when it works? If a terrorist is caught they don't have to divulge the information of the danger of the conspirator because of national security. In most cases it is one man with a feeble mind having malicious intentions, that's all. The Government can control and spy on the population with CCTV cameras and secret police under the cover of fighting terrorists and the population will be so grateful losing their liberty and privacy because by doing that they don't help to fight crime which is the main enemy but a fictitious creation of government.

Burglars, robbers, car thieves, drug smugglers all are the main income of Bankers and Insurance firms. If there is one chance in two million that your house would be burgled or your car stolen would you take out very expensive insurance, never in your life. You would take the chance to go without expensive insurance and take the risk yourself. Therefore if you have a chance of one in one hundred that you will become victimised in that your house will be burgled or your car will be stolen, then you better take the insurance notwithstanding the premium. In England the Bankers already got to the politicians to draft a law that makes it illegal not to have insurance. By enforcing insurance payment through law, now the Bankers and their money-Brothers can rob you blind legally and take you to court if you do not oblige.

They set the terms and the rates of how you will be robbed. The Insurers would never have the insurance premium holders pay the equal amount of what the amount is of cars being stolen. They would set the premiums at least at three times the number of crimes committed to cover their arses and cut a tidy profit on the run. Therefore we can assume that for every one car that is stolen the premium holder would pay three times what the crime rate asked because to the insurer it has to be worthwhile. That makes the insurance firms the biggest robbers, car thieves and crime hustlers in the world because they rob three times more from the public and take three times the cars or furniture or whatever was stolen than what the criminals did. They steal more from you than thieves do and they have to if they want to make a profit. Then they use crime to keep us defenceless and vulnerable and easy prey, just as the villains prefer it. The more defenceless we are the better victims we could be and good victims are unarmed persons.

They use anticrime propaganda to get us to become complacent, as the criminals prefer us. It is as easy as taking candy from a baby... Unarm the law-abiding civilian to protect the criminal. Use antiviolence propaganda to promote violence flourishing, and achieving it is so simple they let the public do it. They allow the mindlessness of the public to get the better of public safety and then get the do-gooders to finish the task. When a madman goes on the killing spree and starts killing everyone and everybody, the first thing that is mentioned is removing firearms. It was not the firearm that killed the say 30 something people or whatever; it is the madman that got hold of the gun. They blame everything on guns. A hundred years ago guns were much more common and accessible and there were no mass murdering of innocent people where a madman goes running down the street killing as far as he goes. Or some boys played a video game once too many times then go to school shooting from class to class. That is not a gun issue but it is a behavioural issue. It is society becoming crazy and shows there are those on the fringes that go that one step further. Guns left alone are harmless and guns used for hunting or sport in safety are harmless, but getting the young public hero-worshipping the blood drunken maniacs on screen helps.

The children sit and view Rambo type movies showing how Sylvester Stallone or van Dam or Vim Diesel runs through town killing and mutilating with all forms of explosives while being the hero. There were no such mass killing a hundred years back while there were firearms a plenty especially after the two wars. There were no such mass murdering or thoughts entertaining such killing sprees but there are now and now there are violence on the TV and in movies and displayed where you can participate in such brutal killings on a TV in the form of a video game. However, they only need one professor to claim these games have nothing to do with the murders and then it is Gospel. It is pure science or in other words a lot of bullshit to say these violent pictures and this brutal bloodthirsty entertainment does no harm in the minds of the masses. They pay some University and one professor in that University to profess how unscientifically it is to blame these games or movies and science again conspires to pollute the minds of the public. What is wrong with this? The fact is that Science does this all the time with everything so what is so horrible in doing it once more, after all the entire lot is a bunch of cheats in any way. Ban Rambo who is a hero and a mass killer. Ban these movies but that won't happen because the banks bought the Politicians and the moviemakers and video game owners tell the banks what to tell the politicians. Stop this by banning Rambo who is a mass killing hero but there is too much money made from this form of entertainment so the Politicians are told to lay the blame elsewhere. Everyone with a lick of sense knows that the violence in Video games must corrupt some of the minds that play these bloodthirsty games.

Then after all it is so easy to blame guns and the democratic principle takes its course where the majority can immediately see the sense thereof. If there are no guns then the man can shoot no one. It is as simple as the simple minds believing the simple mess they create in the first place. In China there are no guns and therefore Chinese madmen do it with butcher knives. Are they going to ban all knives after they banned all the firearms because banning the firearms doesn't solve the problem? They have to remove violence from society in making violence and killing an everyday sport that entertains millions. That they won't do because then they will loose the billions the entertainment industry makes them each year!

All they ask for is that crime flourishes so that no one could dare to be without insurance and remain alive. Then to get everyone scared and nippy about crime, they protect the child molesters and the rapists and arsonists just to keep John and Jane Dow and the rest of the public so scared of crime they haven't got time to become frustrated with the total rip off the public suffers under the "protection" of the Police and the Politicians. That is why the "law makers" make crime ever so humane and get parents to become the villains when they try to discipline their children. It is about making the law-abiding citizen feel guilty and vulnerable at the same time while allowing the criminal to victimise the innocent so that the public can be controlled. You see how fast they put a father in jail when he goes to castrate and remove the genitals of the rapist who raped his daughter. Having the rapist out in public will put the parents on high alert rendering to a condition where they are grateful just to have their children alive and moderately safe. When it does happen to the persons next door we pray on our knees thanking that it were not our turn.

The rapist gets a slap on the wrist with early release and the parent get tossed into the rough side of the jail for decades because no one in charge wants crime to get under control. We have gone as far as feeling grateful that a rapist did not kill his victim or that a burglar only took "earthly possessions" and that the victim wasn't physically harmed. We don't mind being robbed as long as we're not clubbed to death. That is how far these scoundrels wheeling money got into our safety zone. Those scoundrels are the ones governing you by employing "democracy" and bringing comfort to your door. Your credit cards, home loans, hire purchases, and all the credit available is paid by blood of victims of the criminal enterprise the Bankers fight to uphold and even pay to maintain. To Banker's, crime pays through insurance coverage, money that fills the vaults of banks.

The latest trend is the kidnapping of persons and taking ships for ransom. This became just one more way the bankers find a means to "tax" the public. If pirates in Somalia take an oil tanker, everybody sits back with a smile because it is not their problem since they are not involved. Wake up and smell the shit on your doorstep. We, the public, pay for the goods the cargo carriers carried. By paying the ransom to the pirates the price of the products went up. Not only that but now insurance are paid on all other products in the event of pirates grabbing ships. Insurers will now put a levy on every container carried on the high seas just in case of piracy. They up the price to ad profits on the go whether the cargo is carried through safe waters or not, it is money to be made by insurers and the Bankers get the benefit of the crime.

Everything on earth just went up with a few cents but at the end of the year it comes to billions of currency that was captured by the Bankers going into the pockets of the wealthy and out of the pockets of the poor. Crime pays so there has to be a way to boost the levels of crime so that the insurance firms and Bankers can benefit from the source of wealth. The trend of kidnapping will only rise and in the next decade or two everyone on earth will have a stipulation added that there is a certain premium put on the head of every policyholder. In the event of any person being kidnapped, the insurers of the policyholder would be willing to pay that some of money that your policy covers so that you then are released. If you have not got such a clause in your contract, the kidnappers just get rid of your body and get the next victim to pay up. If you have not got insurance covering such an event, you will disappear without a trace, but if you do have such coverage, then your insurers would be kind enough to pay for your release. This is how the game is played. The Insurers would always see to it that they get many times more from crime than what the criminals could ever get. They can't insure one on one but will go ten to one in their favour.

Ask yourself who steals the most from you? Illegal immigrants are another source of currency going into the pockets of the insurers and the Bankers. When an illegal immigrant lands in a country, that person has to eat and live and crime does it. That person didn't cross the shores to live on minimum wages and be the poorest part of society in that country. Yeah sure, maybe a few percent are honest workers striving for better but the vast majority wanted more than what was on offer where they came from. These are perfect candidates because they are untraceable. The police have no names and no identity and the criminals are shadows coming out at night to harvest what is not theirs and then to live lavishly while the true criminals are the Insurance firms and Bankers that regard this affair as only being part of business.

Tony Blair as Prime Minster of Britain acknowledged he has no idea how many illegal immigrants there are in Britain and no Brit ever asked why he has no idea. He and his party are paid not to have any idea. Why is the Ministry of Home affairs the one department that is "not fit for purpose" to use the phrase they used. It is so that no one can trace anybody illegal even if they tried to do it. Criminals are renowned to walk out of jails with early release even when they were illegal immigrants. Now why would you think that would happen? The Bankers pay the Government to have the Home Office in such a state that no one knows arse from head and the bigger the mess in this department is, the more illegal immigrants there are and the more the crime rises. Crime pays but only for the Bankers, the stockbrokers and the filthy rich that can buy their own security army. Therefore the rich benefit by deducting cost as tax deduction so that the poor must pay more tax. Notwithstanding democracy, the poor remains part of the food chain of the rich.

The illegal immigrants bring another sort of business with them. They bring girls they use as slaves for prostitution in the developed countries where there are money to be made. The girls get kidnapped and then their services get sold off. The price asked for one hour is not high but since the girls don't ask for much as they are drug slaves every man can afford to pay for the services and having many girls in a working position, the money eventually escalates into billions. The money is far too much to be kept outside bank vaults and so where will this money go...it goes to banks and the Bankers are the recipients of the cash in the end! Big drug busts don't stop the drug trade. It does not even influence the process by causing massive price hikes and that should be an indication of how much money goes to Banks.

Don't think of rescuing the girls in distress. You try and touch those girls to save them and see how you vanish from the scene either by getting jailed for interfering with police investigation or by the criminals who will make you disappear without a trace. You touch the girls and you touch the bankers that befriended the Politicians who are the bosses of the police and in that order you will annoy the command structure because the girls prostituted as slaves guarantee Bankers and other criminals much wealth. Those with money or with power are criminals or are criminal minded because that is the only way you can have money, steal from others to fill your pocket or use them as slaves to work for you. Don't believe this crap about only honest money going to Banks because with them in charge of the Political Government of all countries they will get the law to create loopholes through which they will work "legally" and with them having your elected government working just to please them they have nothing to worry about what is legal and correct and what is not because everything is just business.

The drug trade flourishes just because of one principle, the money has to end up in the coffers of Bankers. If the drug trade is worth a billion dollars a day, where does the money go? No one can walk around with that much money and therefore it has to go to bankers for safekeeping. I have not made a study on the subject since it is not my main interest to reveal crime statistics, but anyone could see that printed money escalated by tens if not by hundred fold the past forty years. Why would they print that much more money? It is because that much more money is in circulation and the distribution thereof finally land in the banks. This gives banks the ultimate power they could ask for. Crime brings in cash and cash is their business, so crime becomes the business of banks. In order to further their campaign on behalf of criminals they get to politicians. Politicians not in favour of using their "help in getting elected" and do not accept their aid through "election contributions and donations" don't get elected. The honest ones don't even get press publication. In this syndicate don't discard the TV stations and the printed press. Since the Bankers would own much of the press as shares legally held, the press is in place to further the cause of those in politics who would further the cause of Bankers and Bankers are in business to make money and crime pays good money. This is an animal that feeds off itself by consuming the consumer and enrich the rich.

There are many more "lobbyists" running down the corridors of the American House of Congress and the House of the Senate than there are politician official assistants in America. Why would that be? That is a smarter and more acceptable name for bribing the politicians to gang up against the electrets. It is a fancy way of bribing anyone and everyone so that the "democratic" process could be hijacked and rigged.

If Gordon Brown and the rest of the European leadership so full heartedly believe in democracy, why don't they put the death penalty to the system of democracy? Go on and give the public the choice, let the people decide... If they say they rule by democracy, then let the people decide on the main issues directly, call a referendum! Why don't they put it to the vote and disallow the one-sided advertisement funding of vote swinging so that the people can make a well-decided popular choice on the matter. They lie and cheat and bullshit before an election just to come afterwards and push the rule of the rich down the throat of the poor.
The Police say they protect.
If they say it let them prove it and then let them protect and if they don't solve crimes then put the Politicians in jail with the rest of the Bankers and the criminal mob. But it is a case of the quicker robbers and car thieves and rapist get out of jail, the quicker they get back to crime and the more crime is committed, the higher the premies get for insurers. Therefore the more the money is and the higher the profits are for insurance firms, the more bankers and industrialists gain. Also the higher the crime rate gets the more the people have to see that they are insured to cover their potential losses in crime. Therefore it pays to buy politicians to create laws that go soft on criminals.

Defend yourself in your own home against a burglar and you are going to jail! If you as a law-abiding citizen have a gun, it is a crime, but is it because you make it difficult for criminals? They can't keep guns away from criminals so they take my gun away! You go and beat the holy shit out of a person that robbed you then you land in jail. If the robber gets away with the crime no single tear is shed but if you beat him blind you are the villain. This is to protect the Insurance firms that bank at the Bankers that invest on the Stock exchange that get the Rich richer so that the Poor must get poorer. The more crime is about and committed daily, the more people need insurance cover and thereby the more money insurers bank giving bankers profit to spend on the stock exchange by controlling the economy and the politics.

If the Government puts Police in charge of crime and disallow me as a citizen to take revenge on the criminal by going out and look for the robber myself, or get the rapist to swing on a lamp post because I have his finger prints in my house where he raped my wife, or cut the genitals out of the rapist who molested my four-year old daughter, then they must take the blame for unsolved crime and the high crime rate. Then the Politicians must see to it that when the rapist rapes again after being released from jail everyone involved with his release from the lenient judge down to my local

politician that is responsible for crappy laws, must sit in jail with the rapist because they all participated in the crime being repeated again.

Because of their weakness crime is committed, so the politician must guarantee my property or give myself the option to retrieve my belongings and then hang the criminal from the nearest lamppost. From the lenient judge to the incompetent Chief of Police and his underpaid and over rated police department to the Politician who put the Chief in office, including the parole officer who decided it is time to release the repeating offender, the lot must be jailed with the piece of scrap that must swing for crime he or she repeat on doing. Let the lot sit in jail while the released criminal swings from a lamp post dragged there by the public. That is democracy...the people decide who will pay in whatever manner for the crimes the criminal committed against their society! Now the powerful will take charge of the crimes and be soft on the criminals and the innocent must suffer more and more under the weight of crime just so that the powerful can be in charge of every aspect of our lives. Is it not time we take charge of how we handle crime and reduce crime to nothing at all?

Are we victims of crime but moreover is it because we want to be or are we victims of crime because it is forced on us. Why do we participate in a life being the victims from all angles of society?

Then if crime is forced on us why do we sit back and take the punishment without taking action against the system that does not protect us. If we were victims of crime because we wish to be victimised and brutalized why would we wish it upon us?

Ever thought about these questions?

I did and I address the questions and you will be most surprised about my conclusion.
Read this and then start doing something about the system we are all brainwashed to believe in and get surprised at what you are about to find out.

It is time government takes responsibility for laws failing and because they favour the Bankers and Insurers it is my security that is compromised by them applying lenient laws that is about protecting criminals. If the government says their Police will protect my belongings and secure me then they take the responsibility. If I am being burgled, then the government must replace my lost property because it was they and their police and their laws that did not protect me and then if the crime is too high, the cabinet must resign and give other people more fit to do the job the chance to govern. Now at this moment we know that Politicians are the most crooked over paid under worked robbers of public funding there is on earth. Do like the Chinese do with crooked officials, those they catch with their hands in their cookie jar, take a crooked politician or official into the centre of town and put a bullet into his brain. Why would other politicians be scared of this practise when they are as innocent as newborn babies? That will get rid of the crooked politicians and place the genuine honest persons in public office. Crime pays for the elite that profit from the rate of crime becoming sky-high. The money is used to change laws in favour of crime so that crime can flourish and Bankers can profit. Bankers buy politicians and democracy through money donations. With the money they give politicians, laws are written that protect the rich against the poor. This is the way to run a conspiracy orderly. No one can trace it and no one suspects it because we all benefit. With this method the Bankers gain by inflating everyone's livelihood and eroding personal income and safety as to gain from that. In time I might write about this but now I cover another story, a much bigger story about a conspiracy everyone on earth has been brainwashed to uphold...it is about...

Crime became the centre of our lives even by watching it on TV as entertainment as something to view. We live in constant fear of crime but are crime forced upon us or do we comply without having a free will? Why is that?   What took away our free will? Is there a civilian battered consumer syndrome like there is a battered wife syndrome where the wife takes beating as a wife's burden she must carry and carry it she does?

Does this civilian battered consumer syndrome remove our liberty and our free will to be what we want to be from us where everything from criminals to guardians to clergy to politicians and courts hinge on the money we can pay and in the process take us for what we can earn by which we have credible means to pay in order to stay alive? In hospital they keep you alive with life-support machines while there is money to pay but as soon as you can't pay they kick you out or stop the machines although and even though you have a good chance of recovery.

It is money that keeps you alive and not your prospect on living.
Everything is about what financial reward there is for those in power to keep us, the displaceable consumer alive or let me die if I can't pay.
You have to use a bank and they rip you off.
You have to use the roads and they make you pay building it.
You have to use the roads and they make you pay using it.
You have to eat and they make you pay for that privilege.
Banks don't work with money; no they work with mortgaging property and financing vehicles.

Money is the peanuts they feed the monkeys that hold and pay for the mortgage the banks have on property and by which terms they enslave you.
Banks work with property albeit a vehicle, a house, a farm or a flat, but for you to use it they make you pay just to be alive and be able to live.

To live they feed you credit by credit card consumerism that enslave you even more so that you then are to them a bigger slave and a bigger source of revenue. I wrote the book **THE WHISTLE-BLOWER ON CORRUPT SCIENCE**. The book I wrote is about science but what the hell has science got to do with the battered civilian syndrome where they beat you up to keep you alive so that they can beat you up even more while you pay for the privilege to get beaten up? Read the article...........and then see how this suffering centres on science.

We the dispensable by-product of human society is bombarded with corruption from every angle. Those setting the rules we live by drain us while those not living by the same rules as we do clean us out. We get attacked and mugged and get attacked when we defend our property against muggers. We, forming the consumers are the victims of a rip off from every angle we can imagine and then it comes from angles we can't even imagine. Nobody can think which criminals are worst and we should fear most, those making the laws to rob us by enforcing brutal restraint; or those breaking the law in crime to victimise us by raping and plundering us. The clergy steal as much from us with scary tactics as the law enforcement does with strong-arm law enforcement and then the criminals in the banks and the shops overprice our living commodity to enrich their coffers and push us below the bread line while starving us.

Those we should trust conspire to work with those we should not trust but distrust because everyone sees us, the ordinary citizen, as a cash cow they can milk and the cash cow can never dry up. If the cow gets weary they feed the cow more peanuts and get the cow to deliver the golden milk they call money by being slaves. If the cow shows resistance they whip and beat the cow into reluctant submission. The clergy dips us in sin so that they can make us feel guilty so that we would feed them cash to buy our souls into heaven. The politicians keep us scared with all the enemies they create and the terrorism they establish so that we fearfully hide in their protection while they squeeze us dry. The politicians, the clergy, the Bankers, the courts and all those controlling our lives manipulate to get money into their coffers and they screw us out of our sanity while keeping us blindfolded at all cost and bleeding every vain.

The government let the shop owners and supermarkets commit theft by overcharging and allow the banks to drain us with excessive rates as long as we pay tax on the transactions while getting ripped off. We have to pay to go to work and we have to pay to live at home and we have to pay to educate our children and we have to pay to stay alive and we have to pay to die by purchasing a coffin. We pay for building the roads and we pay for using the roads and we pay for maintaining the roads and we pay for allowing those that keep us alive to enslave us by constantly increasing prices. Politicians take bribes openly and very much legally and they call it donations and contributions but it is due to the criminal investment they receive from those financiers whereby those in office deliver us to the slaughter table of the rich and the mighty because they get paid to have us vote to be enslaved. ...And in all this the media is doing the brainwashing while they broadcast and televise ideas that cry of liberty in democracy and of the fairness of the system by maintaining and upholding democracy. This action by allowing open dishonest media is the doping that keeps everyone mindlessly blank with vacant thoughts. It is as if dishonesty took away our freedom and greed removed our civil liberties and it is not even our greed that buys our slavery. Politicians are lining up with the money houses and drug lords because since sales tax became a part of selling, the politicians could not care a hoot if the populous were ripped off because a certain fixed percentage is always part of government income. The government triples the price of fuel and enforce this commodity to put in a system where we need fuel to get to work. They privatise the transport we need to get to work but that is allowing the moguls to further enslaving the poor to the rich. We need transport we pay for the privilege to go to work to get rewarded with a tip pence for every pound we create that the rich pockets and then we must be grateful they create jobs so that we can become their slaves. This is what democracy became and this is the system we vote for when we vote in the next government that will deliver us into the slavery of the rich, the powerful and the educated. The United States Government brought in laws to cover the wide spread corruption that persons such as the founder of wikileaks reveal. If that is not the most extreme protection racket there is then I want to see what is worse. They can do what they want and if anybody reveals what they do they kill the person that reveals their criminality. They're far worse than Henry the 8th ever was.

They made laws to protect lies from surfacing and anyone that uncovers corruption in their ranks is jailed for uncovering the corruption. The media in every form is overflowing with all sorts of corruption going on in commercial banking, in politics, in corporate pharmaceutical medicine, in hospitals and every other industry... and even reasons why the West attack Arab states on behalf of other Arab states. Even the TV media is so one sided nothing they report is believable because behind every action we see the dark hand of Money Moguls brainwashing persons to believe the politicians' actions for going to war is morally justified! Then what about science? Honestly, if there is one faculty that is pure and honest it has to be science because every person not excluding one person believes more in the accuracy of science than we believe in our specific religion. We discuss the accuracy of different religious outlooks but never do we question the accuracy of basic science. Why is science that Lilly white as they make it out to be ... or is it as pure and untouched as those in science pretend it is? This is the whistle blower on fraudulent practicing going on in science. Science is the worst dumping ground for dishonesty of all the others. It is the cradle where brainwashing is perfected into a tool to subdue and control the thinking of every child.

If you don't believe this last statement then argue this and see where it gets you. They brainwash us from when we are very little and that is true. This is how they go about and that is why I resigned from teaching. They use science to play mind games and make us believe what we should not believe to begin with. They make us doubt ourselves. I see the sun rises in the morning and that I can see every morning. The sun crosses the sky and I can see the sun move about my head as it goes over my head and goes down the other side at the end of every day. The moon rises and the moon sets and the moon went over my head.

If I would say the sun spins around me science stops me because science tells me that is not happening. This idea is a lie because the sun is not turning around the earth the earth is turning around the sun. Copernicus and Galileo proved the earth is turning around the sun! Therefore they will permit me to say the earth is turning around the sun but that is as far as I am allowed to go. Ptolemy said the sun spins around the earth but that is scientifically brutally suppressed because it is not true...that is according to science and science can never be wrong because science only works with hard facts and proven truths!

I can see that with my eyes every day the sun crossing the sky but nature and my eyes are telling one big lie because science says it is not true. Science says the earth is turning around the sun and that is the truth. As a child in school I am brainwashed to believe science when science says I am living a lie because whatever I see and whatever nature presents, nature is a lie and a failure because the sun and the moon is not spinning around me! The earth is at the most spinning around its axis but that is pushing my perception to the extreme. Nature is presenting a complete obscure picture of the truth and whatever I see, I must disbelieve it because nature is completely wrong and science is completely correct when science says not to believe my eyes and nature and just trust in science because nature is wrong as it is always wrong and science is correct just because science is always correct.

When I go outside I have to distrust my eyes because my eyes are telling me a lie and nature is presenting falsified truths and science can't be wrong and I have to believe science and...I am getting so F@%$&ing brainwashed out of my senses I am turning my head inside out. You, the person reading this have never been confronted with this message because I am the only person who dares to tell you the sun is turning around you and me and every other person on earth. If you don't believe me go stand outside tomorrow from sunrise to sunset and report to yourself every five minutes by writing down the position of the sun in relevance to your head while you stand dead still and jot down what is happening when you look at the sun. Write down the position of the sun in relation to your position and go and analyse your finding the night. Then you write to me and confess that you did the ultimate unscientific thing, you saw the sun turn around you while you were staring at the sun going over your head.

Then with this control of human thinking and everyone mentally being subdued into believing that my believing in what I see must be discarded in favour of what my controllers of my thoughts programme to be I then become the right subject to be enslaved by the rich, the powerful, the influential and the controllers of my mind and I start to believe more in science than in nature and more in Newton than I could ever believe in God. If I believe not to believe my eyes I'll do whatever I am told.

The media in every form is overflowing with all sorts of corruption going on in commercial banking, in politics, in corporate pharmaceutical medicine, in hospitals and every other industry...Then what about science? Why is science that Lilly white they make it to be or is it? This book is the **whistle blower** on fraudulent practising going on in science. This book informs about shocking detail everyone misses and reveals how deep the misconceptions are while this book exposes details of blatant mind bending brainwashing practises science are committing to unsuspecting students. In this book there is a business offer that can be very lucrative in participating to uncover dark secrets that large money hides to keep the world slumbering in fogginess and twilight. Reading this book will inform any reader on any level more than any other book did before and reveal a financial offer that will not repeat again. Why would we start a fight in corrupt science because compared to all the others this is the least exploiting one?

Yes, that is a valid point but science was the one that started it all and science keeps the fraud continuing because your children are taught fraudulent science that has no crumb of correctness and while we hear about a lie and are forced to believe it as the truth we will believe any lie as the truth because we were taught at school to honour such behaviour. This is where we are flattened to accept what our senses tell us is truth. We can see the sun rise in the morning but are told it is not true and that we should ignore it because the earth is spinning around the sun. We can see the constellations float by at night but are told it is not true because the earth spins around its axis.

We are forced not to believe our senses and cling to science because science is correct while nature and our senses are untrustworthy. I can see the sun rise as I can see the stars go by and to force myself to not believe that, I condition myself to allow civil mind control to run my life and be brainwashed becoming controlled by forces superior to me. If we can distrust our senses and tell nature it is wrong while accepting science without question, how brainwashed we get to be to act that gullible and accept everything? In such a mind state as we then become by not believing our eyes and our mind every person can tell us anything and we will believe it because someone says it is science and that makes it pure truth. No person would dare to tell science to prove Newton or Newtonian ideas. In that state of mind we are the slaves of any system as long as it is part of science. Remember everything nowadays is science. Money is science and medicine is science and research is science and politics is science and in science we believe.

Please rethink this because when I raise this issue everyone gets ready to take me to battle because everyone is brainwashed to believe in science. Science says everything falls by the force that the mass creates. That is Newton. Then science says everything falls equal. That is Galileo. If everything falls by mass creating the force by which things fall and everything falls equal then everything must have equal mass or one of these two masters are telling a fib. If mass creates the falling then the heavy stuff must land on earth distinctly quicker than the light stuff does and that we see does not happen because in freefall when skydivers drop from an aircraft they fall equally irrespective of their mass. When a car, a human and a bag fall out of an aircraft, they all fall equally in their descent.

Only by mentioning this people attack me without realising or thinking about the controversy this unleash. How can something fall equal and fall by mass where it is said that mass is forming the power by which it falls and then there is that much difference between the different objects' mass and still we have the equality of falling objects. Please consider this. We see everything falls equal but nothing falling equal has equal mass and yet science teaches us everything falls due to the mass it has.

Now comes the part where you want to attack me because I destroy the preconditioning you went through to get in a mindset where you accept what science tells you and get to be a zombie by repeating whatever you are told mindlessly. You are told mass pulls the object to create a force of gravity by which you fall. Then you are told everything falls equal. You have to repeat this contradiction until such time that you are versed in it that you are able to comply and where you then state your belief in this fact so much you write this down as an accepted fact in an examination test. At the point where you declare that you are so conditioned that you accept this without argument you are accepted as having an education is science.

If gravity comes as a result of mass how can hot air lift an air balloon bag with al the mass inside? What has hot air got to do with gravity? If gravity is the result of mass how can they create uplift by blowing an upward draft with enormous fans in that wind going up prevents you from falling? If mass pulls you down how can there be magnetic levitation where electricity pushes you away and keeps you at a distance from the earth. How can moving air prevent any object with mass from falling?

I concluded that it isn't mass pulling down but gravity causes air to condense and becoming thicker and by becoming thicker it employs less space and by having less space things reduce the space in which they are moving to the ground and it is space and not material that falls to the ground. It is not because of mass that things fall but because denser air creates lesser rotational movement and that increases the ratio of linear movement and in that we have gravity forming. If things move fast they rotate in a circle around the earth. The faster the rotation is that objects rotate in this circle around the earth the further they can be up in the air. An aeroplane can't be at 31 kilometres in the air while doing 207 km per hour. The aircraft must to at least 2500 km per hour in a rotational orbit to be at that height. If the object then does 250 km per hour it will be coming down linearly like a stone.

A question: if there is always more air flowing over that car producing down force than air flowing underneath the car forming up lifting, why does a car lift into the air? The down force is always exponentially bigger. The air forming the force that pushes the car down will be many times stronger than any air able to lift the car from underneath. From this question I rewrite everything about airflow and the sound barrier.

That means gravity is the ratio between circular orbital movement and movement directed towards the earth centre and that is contraction by reducing air space and allowing air space to concentrate heat. I prove with formulae that gravity and electricity is the same thing because it is increasing the density of heat or space.

I am the first person to reject science as they brainwash people to accept and I made a study how nature applies science and from that made many conclusions that wipe the dogma science carries from the table. Science is about brainwashing that is in place to prevent thinking and not to explore thinking and I started to explore thinking.

I formulated gravity and it was not that formula that Newton used to formulate when he formulated gravity. I was unbelievably excited with my new found wisdom and my over exiting discovery. I saw what the clearing of one mistake could mean to the entire world. I saw how removing one misconception could clarify so many blind spots humans have about the Universe and the practical layout of the Universe. I wanted to share my enthusiasm about the mistake with those who could correct the mistake and bring new light to everyone willing to listen to the new vision cosmology would bring.

Those in charge of physics could correct this one flaw and by correcting this one flaw the jig saw puzzle we had on the cosmos would all fall in place. I contacted the custodians and arch fathers of physics and enlightened them to my discovery. They were the wise in know how and if it was possible for a person such as I to see the mistake and the Universe that would open when one remove the mistake and replace the mistake with the correction, how much further could the wise men in physics take the translation coming from such a correction of the obvious mistake. I contacted them with hard evidence on the matter.

From them came no reaction. I tried again and changed my approach because I thought my approach brought on their missing the essence of the problem. My mistake was indicating they had a mistake. After seven years of constant knocking on doors not one reply was ever returned by a single academic professor. Not to disprove me...not to point out my mistake and my error, not to silence me.

It was as if trying to open a grave by shouting down to the coffin and expecting a joyful reply. It was as if trying to chase a ghost with a catapult. It was as if trying to photograph a spook. There was this silence befitting a graveyard with all the ghosts pleasantly gone to the after life. There was this curtain that absorbed all sound coming from me and killing off any response coming back to me.

I took the mission one step further. I thought with personal touch and eye-to-eye intimacy I could see what part they couldn't see about what I saw and then explain my lack in clarifying the matter as to enable them to see what I so clearly saw. I was more than just ware of the importance there was in detecting the mistake and that put all my sincerity into solving the problem.

I went to see countless academics wherever I could and whenever I could and that I did with much hope riding on their sincerity and it was done with much cost on my part. I though by having a person-to-person and an in depth personal conversation about the many factors arising from the matter, it would bring new light to the table, either on my side as I then made a gross error, or on their side. I went on this mission of sharing information with big costs and effort on my part. I thought that such a personal visit conducting a debate on the issue would either bring clarity to my oversight or bring their oversight to their attention.

A fat lot of good all that did me by spending all that time and money going into all that effort. If they were sincere it was only to save their work by not recognizing mine. The mistake is so obvious and in view of all concerned with science that it is impossible not to notice it and yet, not one in science notices it.

From them I got a cold shoulder; from them I got silence or at times blatant aggression. I was received with what was all part of the Academy of science's tactics to prevent me getting closer to the conspiracy by ignoring me. When they did reply it was to belittle me and to aggressively silence me. Their attitude was not something on my part. It was due to something science is trying to hide. It was not I that should be blamed for noticing the mistake but it was they who silenced me because I noticed the mistake. I uncovered a three hundred year old grave in which was hidden a conspiracy far bigger than what the earth ever knew before. It rattled bones that were buried since the Elizabethan times and ever since then no one ever cared to bring the mistake to the open where everyone not in science could see what science hides. Then the truth finally dawned on me. I was overstepping my boundary because I was in the centre of something bigger than what I discovered. I was in the centre of a conspiracy where everyone conspires to keep what is the secret to remain the secret and to collude never to divulge what the conspiracy hides.

As I said before, I uncovered a mistake in science and by that abandoned the using of mass. From the onset, the mistake seems as insignificant as it is small. Because the rest of the book is about the mistake, I do not intend on elaborating about the mistake itself. The mistake came about with the culture of education and the mistake in itself seems harmless. When admitting that, one must also admit that any pilgrim that got lost and died of starvation through an incorrect travelling direction, made the very first part of his ultimate mistake by looking in the wrong direction. How harmful does looking in a specific direction seem, and yet such a mistake leads to his ultimate mortality.

The traveller could, when taking his first directional flaw with that first incorrect step, only put his foot skew in avoiding a rock. Or he could have turned his face to avoid a branch and that move pointed him in a direction that leads to his fatality. It is not the mistake that becomes the penalty and it is not the origin of such a mistake that leads to the penalising, but the ignoring of accepting signs telling the wonderer of an impending error and his stubborn ignoring of such telling sign that makes the lost party pay the ultimate price. By ignoring the mistake, for whatever reason, the ignoring of such a mistake is his undoing because the price due comes from the inability in recognising the sign indicating the presence of the mistake forming the reason for his final demise. The sooner such a person sees and admits the wrong, the less will be the consequences of his final price to pay.

The Newton mistake is one born in culture and the penalty from this mistake is bred by arrogance. At school minds are young and accepting, although developing. Through many tens of millennia humans came to a habit in surviving as a species where culture taught them that accepting the elders' advice is the same as to ensure survival of the following generations, and by such doing is also following the quickest way to an adult mind. By accepting the elders knowledge and experience without question proves the dominance of the tribe in relation to other tribes of the same race. This is culture we cannot do without and still maintain progress. It is an inheriting method humans grew on and is the corner stone of all civilization. We cannot abandon it. This was the method whereby civilisation became practised.

The scholar sits in class and receives from the Master information that is completely his days breaking news. As far as his mind can tell, the news is as factual as anything notwithstanding the fact that such news may be with the human mind for thousands of years. Whatever the teacher tells him is bound to be a first time experience so new he has no time to digest the information. Taking into account his youthful ways (which we all had), he has little stomach to scrutinize it because learning is a painful process to all. Without pain and perseverance there can be no education of any sort. He does very willingly accept the facts as tested and correct without flaws of any kind. The scholar has to because in any education system time will not allow students to ponder about detailing information and securing a prognosis to all learnt every day. Whatever the Master tells the scholar is taken as Biblical correct without any thought about testing the results. Where there are cases of scholars having doubt and subsequent questions, the Master takes such behaviour as being obstinate and being a reflection of the student on his (the tutor's) personal integrity and knowledge. The young mind will very soon discover that his behaviour is not tolerated by the system, and the truth is the system cannot tolerate such behaviour for the good of the rest. Time must be spent on learning and accumulating as much information as that which the young mind can accept. The young trusts the knowledge of his elder and superior by not questioning anything.

No information could be more affected by such a culture than that of Newton. You both understand Newton and are smart, or you do not understand Newton and accept that there will be very little future for you to have in the world of science. Newton is science. No Newton understanding automatically becomes "no science" education or learning. Without Newton, there is no other and science will be a vacuum of containing nothing.

This is very unfortunate but is the ultimate of truth. It is either Newton's way or no way at all. With this, culture also brought along the stigma that only the minds of the sharp and the sighted can accept and understand Newton and when not understanding Newton one tends to fail your personal I.Q. test. It is a sure sign of the slow witted when the student fails to recognise what Newton said. The only way to advance in science is to understand Newton and indicate to all your pears how brilliant your mind is in accepting information. As I am writing I still fail to understand Newton and therefore I am looked down upon as a slow wit unable "to understand" Newton after all these decades of studying Newton.

I am constantly reminded of this handicap and recently an editor enlightened me on the issue. That is why I feel obliged to refer to my position as ILL EDUCATED in terms of Newtonians who are as privileged and moreover blessed to being SUPER-EDUCATED in "understanding Newton". All students have little understanding about Newton, and that I can and will prove through the next pages of this book. The mistake Newton made and which I discovered is laughably small, yet it took me (not being that bright I may add) almost one lifetime to recognise the mistake whereas it took mankind three hundred and fifty years of research without recognising the flaw.

Others in the past may have come to see what I saw, but if there were such persons they never saw what I saw because if they saw what I saw they should also see that behind such an almost invisible puncture hole in the tube of insight is a reason for science to deflate and not accumulate. When "understanding" Newton it becomes the very same as learning Newton from the heart and accept that what you memorise is what you know. The memorised knowledge is beyond question, as it has to form part of the identity of the student having secured the knowledge.

There are so many conspiracy theories going around, floating like a bad smelling odour in the room, but most of those have no more factual substance or evidential backing by truthful facts than a James Bond novel has. I also have a science conspiracy theory but mine has all the proof and all the substance of authenticity that any argument can ever hold. It involves **physics** as **physics are taught** at institutions worldwide. If you now don't believe me, read on and learn the truth yourself, it is free of charge.

Mine is the most unbelievable one that has most truth. This book is not about any conspiracy ever mentioned before, but takes into consideration some of the worries of people. **When you see what science achieves and you hear what I accuse science of, then it seems as if I am a nutcase who escaped from the loony house and I must be the one having very dangerous tendencies to harm the innocent.** I should be locked up and taken from society. However, read my story and you'll see that I don't harm the innocent. It is those we all trust without reservations who brainwash to forge deceit through corruption and malice. The blame lies with those admirable, yet astute members in our society, the educated leaders in physics whom we trust without even thinking about their trustworthiness or question their sincerity regarding their blameless honesty. Do you believe in the absolute unquestionable correctness of science and that everything that forms science is truthful being far beyond suspicion?

Do you believe that those furthering the science of physics work with no less than absolute proven facts and the facts are beyond any doubt at all? If you do then I have very bad news for you, **news that will change the grain of what you believe is truthful. I** am about to shock you into reality! **This is going to reveal the biggest conspiracy ever concocted by the human mind, ever since the time a human had a mind to use.** It is A Conspiracy about Fundamental Science. It started when Newton gave us the idea that gravity is the result of mass and in gravity mass is the biggest contributing factor.

I say that mass as a factor in physics and not the value of weight, but mass with the pulling power only pulls stupidity over the eyes of incompetent idiots raving in their personal sublimation. Let us see what I say. Would those in science cheat you and withhold the truth from you?

An Ultimate Science Conspiracy, the biggest conspiracy ever concocted.
NATURE put TIME into place TO PREVENT Everything FROM HAPPENING all at once.

This proverb sounds silly and yet, silly it is not.
While this proverb is the truth, science says they don't know what time is...and this is ignorance they uphold while living with time for millennia all the while Newtonians say they uphold Galileo and everything Galileo implemented. Gravity drives the Universe as the earth spins around.

In other words what science can't see is that there is a time –connected plan in the form of gravity driving the Universe along. Time is in the form of gravity and the cosmos is driven by time. Time and gravity are the same, which is precisely what science is unable to see.

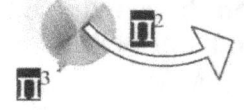

If gravity drives the Universe, the Universe is powered by time.
Galileo's pendulum can measure the flow of gravity and record time to a precise science while swinging in space, then the flow of gravity in space has to be measured with the pendulum swinging in air while

reading time. The pendulum measures time when swinging through the air that surrounds us, and whatever surrounds us is there because of gravity.

This obvious conclusion about gravity forming time, which is what the pendulum reads, any child can draw but as simple as it is, this went past Newtonian understanding of science for centuries.

They never could realise the flow of gravity and time is one and the very same since it then conflicts with Newton's ideas. Still, the pendulum proves that reading time and measuring the flow of gravity is the same. Galileo used gravity to measure time accurately, yet for as long as the pendulum recorded time, this fact eluded science in a way that science doesn't even know about it. It is what science doesn't know that becomes important to man. If gravity measures time by applying a swinging pendulum this becomes the biggest fact in physics and it still went past science because their cleverness brought on their rigged stupidity.

This religious belief in the infallibility of their Godly insight becomes their stupidity. Their ignorance is what science hides. It is that what they hide which is what you don't see. Science hides what science doesn't know which they cover under the larger pretence of their cleverness.

Also science hides from everyone's view what science doesn't know so that students will never realise Newtonians are covering their stupidity and ignorance behind a curtain of arrogance. This conspiracy to not reveal their stupidity then becomes the conspiracy they nurture for centuries. Science conspires to hide by covering their stupidity from open view.
But if there is a secret no one knows about and you are one of the few who can generate sufficient money to ensure your survival while billions upon billions of other will perish and will surely not survive the fact that you can generate money at will to purchase your survival then that gives the money extra ordinary power as a currency. If you can buy a seat on a spacecraft that will take you to Mars from where you will sit and see how the moon destroys the earth and all others on the earth then generating more cash than any person ever knew existed will be worth the effort. If you know you buy a future for your loved ones while six and a half billion others that are not sufficiently provided for will die in misery, then you can find a worth for that much money that will pay for your place on Mars.

I guess all those who have a nuclear bunker will have a Martian address to go to and the rest of us who don't even know where the bunkers are will see the moon come by and give

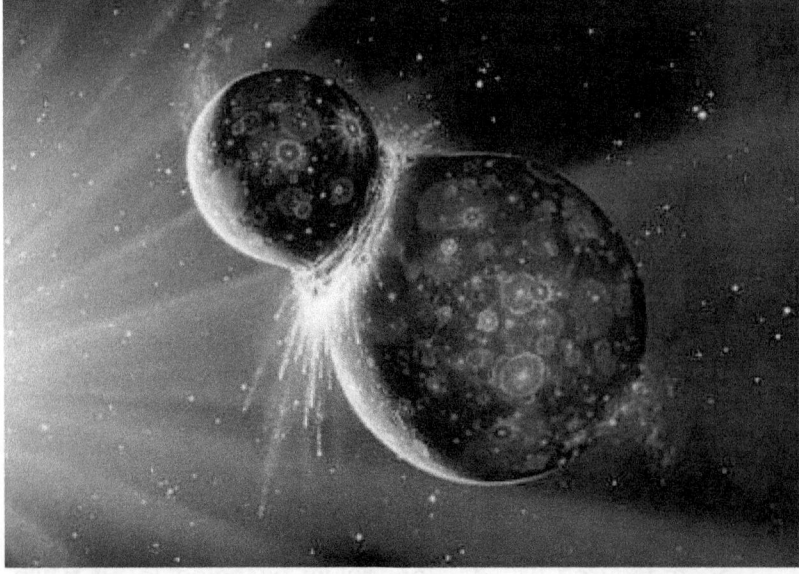

us tidal waves kilometres high.

Let us test Newton's attraction theory in practice. What everyone duly knows is that the mass of the earth pulls the mass of the moon closer and the fools who can't think that far is designated to perish.

Academics in physics insist that gravity is founded on the Newtonian gravitational principle of mass $F = G \dfrac{M_1 M_2}{r^2}$.

This says the moon is pulling the earth at the same time that the earth is pulling the moon and keeping all this in mind, **then when is the inevitable collision between the earth and the moon coming?** When will the collision that will end all come? What do those in science who know science say about the day the moon will hit the earth and destroy all forms of life in the Universe? **We have to know when it is the final doomsday when all life forms will end.**

There are those in society who must know since all know some information and some information is "classified" meaning the privileged know and the rest are misinformed. Why is it always only the wealthy, the influential and the powerful who is informed on very important matters concerning the well-being of all of us and not only those with money?

With $F = G \dfrac{M_1 M_2}{r^2}$ being in place the earth and the moon is pulling on each other for longer than life is around. This is science at its core.

This is fundamental physics.

But all we get is conflicting messages coming from all over. It is as if someone is hiding something to everybody and this does not sit very favourable in my mind. Why give some information and then fail to give other formation. The formula $F = G \dfrac{M_1 M_2}{r^2}$ says the moon has mass pulling on the earth that also has mass and this mass pulling on each other is destroying the radius of the distance there is between the moon and the earth.

The mass of both is pulling by a force that forms gravity that pulls each other closer until the moon and the earth collides with a thump. This collision has the ferocity as was never seen before in the history of the solar system! But when is it coming with such force as nobody can anticipate?

Humans had the formula for centuries and did nothing about figuring out when the moon and the earth will smash into each other...and having this formula that can produce the information not one person thought to find out when this event will happen. Science claims that the Newton formula of $F = G \dfrac{M_1 M_2}{r^2}$ forms the basis of all physics.

That means gravity is there to reduce the value of the radius separating planets and indeed all cosmic objects so that all cosmic objects will finally one by one collide with one another until only one lump of structural graveyard holds all the star material in one place. This is called the Big Crunch theory that opposes the Big Bang theory. In this case the moon having mass is pulling the earth having mass and the two reduced by the mass of the two the radius parting the moon and the earth. Now this is an issue of diverting the attention in a way that brings much suspicion.

The suggestion in the image portrays a scenario where the radius of both celestial objects was reduced to nothing. The moon and the earth already collided. Mass forms a force that pulls both the earth as well as the moon by forming gravity. If the mass of the earth was next to the mass of the moon why did it move apart? The big goal is to unite by having mass pull mass towards, not push mass away from mass. When this collision happened the deed was done, the puling was completed, the mass united with the mass and the lot waited for judgement day to come. Now someone is telling me that the moon was close and it took off and floated away. This does not meet with Newtonian principle requirements. This is a set-up.

The question is why would they come up with some fancy scheme which clashes with Newtonian wisdom head on? Why would they create some idea that totally contradicts the pulling of Newton's principles of forces pulling by the measure of mass? Why tell something that opposes all other things that confirms science's legitimacy. I can see how this story would divert attention away from what is truly happening. If Newton is correct and that is what everyone in physics believe unconditionally, then mass pulls mass by the force this pulling creates which is called gravity. With that being correct how can anyone presume that the moon was part of the earth and then parted from the earth?

Newtonian   physics   teach   us   that   mass   pulls   mass   closer   by   the   formula $F = G \dfrac{M_1 M_2}{r^2}$ and that also means the moon is coming while the theorists tell us the moon is leaving and what does that teach us; that there is total confusion in physics that should be the most sane science in the world. The confusion is deliberate and that made me very suspicious. Why come up with this deliberate confusing story and why does science not repudiate those spreading this contradicting message. If they say the moon is going while Newton says the moon is coming but they never mention the fact that the moon is coming, it will leave everyone thinking that the moon is going. Why would they not wish to say the moon is coming instead of contradicting science by saying that the moon is coming? Only those with power and those with money can buy this deceit. Only the politicians can persuade the scientists to obey what the bankers wish to allow the paupers to hear so we are kept in the dark and they feed us bullshit. As they make up science they make up science fiction and we have the task to see what science is and what forms science fiction and moreover separate the two as we part what is truth from fiction.

This book is written (and so too is many of my other less complex books) to show how the science magazine Annalen Der Physics and all other intellectuals approach new science information that in the basis contradicts the science used in the modern era and from this I aim to promote my theses compiled by six parts thesis to introduce a new way of thinking in terms of science. Those masters of Science practising science are up in arms about how the Roman Catholic Church treated Galileo Galilee while their conduct which clashes with their views receive no less condemnation from their ranks.

As you will later read, the condemnation by which I am treated for not agreeing with modern science **and proving them wrong no less** you will see those in Physics known as mainstream Physics are no better than Pope Paul III. They act as if they are holier than others while their condemnation is more harshly. The Theses I wrote are in place because science as it currently is,(whether you believe what I say or not) is running on so many phoney principles of which those in charge of science holding the highest office in science benefit richly from and therefore discard the utter flaws on which science in the present way is founded. Their job is to ignore the flaws there are. By discarding my work they secure their work. If they agree that the flaws that I show is in place, then their Academic papers that they wrote on science and their research from which their degrees followed in science instantly becomes something for the paper wastebasket.

I wish to take your mind into the world of science and moreover physics. The physics I show is the basis of physics taught to children at the lowest level of science at schools worldwide. It should be simple enough for anybody to follow because it deals with the everyday interpretation of science as we think of science in terms of science. Every person dropped a glass during their lifetime and felt sorry and incompetent for being so clumsy. If you are able to drop a glass and break it on the floor then you are able to understand the following explanation about science. If you know how to fall you will be able to understand why you fall when you fall...or you will be able to know why you fall when you fall...

To my thinking man became man when man saw shiny objects in the night sky and formed religious ideas about forces greater than what man controls. Man saw the two large objects in the sky and found the two had the most influence on mankind. Then civilization came along as amongst many other ideas from which man grew man gave these two that apparently had prominence being the sun and the moon the most direct influence on human life and then with that also the most importance. I say this because if anything makes man distinct from animals it is the facts that man see things inside the influence or outside the influence sphere of man's ability to control or to be controlled.

As humans of all cultures we have forever been fascinated by the moon and always tried to connect some purpose the moon would have on the destiny of life. From the beginning of whatever we connected to the idea of civilization, somewhere in the middle there was the moon playing a part that had a role to play in our destiny. If there were a bunch of gods in any collection that any culture or ethnic tribe formed you can bet your shirt on the fact that the moon had a name and the moon was an important part of the ruling class of the various gods' power constructions. There were lightening and thunder and the sun and the moon and some collection of stars, each group for every month, but the moon had a prominence no less than any of the others.

People attributed so many factors playing a part in their lives to the influence that the moon brought on them and many still do. Women who got pregnant were with the influence coming from the moon. Women who did not get pregnant were also because of favours not coming forthwith from the moon. Gender was a result of the moon being either badly influenced or favouring some of the community or not granting favours. The moon was some influence to be respected and could form a menacing influence.

The moon played a most critical part of farming from the time when farming first began. In centuries gone by before the inventing of electricity and fossil fuel ploughing was primarily done under the rise of the moon and therefore by the blessing of the moon. This idea is still very much alive in the modern times. I don't think any person would have grounds to think of me as backwards and as a farmer I know how intense we farmers gauge the moon to see rain coming or having extreme heat and cold spells al conducive to the moon influencing changes in nature. I don't believe or deny this but many very intelligent farmers believe that if you plant crops such as corn or wheat or tomatoes that grow on top of the surface you plant when the moon is rising or better said getting fuller.

When you plant crops that grow below the surface such as potatoes or sweet potatoes then you plant that when the moon is fading or reducing in volume. Do I know this works for sure, hell no I don't and yes I do because we all are planting in this way from the time many thousand years ago when farming became a science project and since that is how you plant therefore it is most scientific to believe in this method of planting. Since we all plant this way there is no evidence that this is not the way to go about farming and therefore with all the evidence pointing towards this method of planting it must work. Why take the risk of not planting in this manner if it is so well proven and tested. If you think I am silly and superstitious then wait till I get around in explaining Newton's gravity by mass forming forces all over the show, because that is superstition in the making!

Today in the present as I sit and write this information I know for sure that in about 90% of the rain that falls in this forbidding place of dissolution, the godforsaken drought stricken desert that God forgot to complete during the time of creation of the earth and then left alone to become the area which is the desert I chose to live in where the rain will fall only during a period from three days before full moon to three days after full moon in the summer months. It rains from three days before the event of full moon to three days after the event of full moon. It is a rule and if you think this has nothing to do with the moon then you come and farm here. In less than a year you will believe in this as much as we all believe in this rule. It might shower a few drops or be a cloud burst that month but it rains in that period of the month.

Then that confirms Newton because if the moon is full there is lots of moon pulling water into the air and releasing the water to fall as rain. Australia is flooding like never before. China is flooding like never before and South America is having entire towns covered by collapsing mountains as the result of drenching landslides. We see snow fall as we saw never before and this happens way before Christmas day even. You try to sell the idea of global warming to a town that had I meter of snow come down in one night one month before any snow was supposed to fall and I see you go home hungry because you will not do much selling.

We see how the gravity pulls the water up and how gravity pushes the water down in high and low tides when the moon's gravity comes into affect. If this is what you believe you are more Newtonian in science than most others are. The moon pulls the water up by gravity of the moon. The gravity of the earth releases the water to fall on the earth. The water that falls we think of and we call rain. One thing we never ever get is snow. Gravity is not strong enough to pull water to form snow.

There is always much talk of how the Universe will develop from how the Universe developed. But never do we find talking about our close companion. Some see the moon as the closest planet and others see the moon as an extension of the earth. There is much speculation how the moon got where it is but never is a word said about where and when it will hit the earth. If there is such a force as gravity then the cosmic question that needs an answer is … if this is true that the earth-mass pulls moon-mass as a pulling force, then when will the earth and the moon collide? This destruction of the earth and the moon must happen with the earth and moon being so close…

Is there anyone who ever read anywhere about when the moon will hit the earth? Why does no one ever refer to this doomsday that has to be part of our future!

Go on and hunt down as much research as you can but you will never see any researcher published anything about the day that the moon is going to hit the earth.

The moon has mass that pulls the earth and the earth has mass that pulls the moon, therefore this lot pulling each other has to come closer leaving a bang as the solar system never saw before…and with Newton's gravitational laws prevailing it has to come…but when is it coming and why does no one ever talk about this. Why is there absolute silence about this event that must take place? Have you reading this ever given this notion a thought or why did you not give it a thought before?

How far did the moon come closer this past say million years? We heard about animals that disappeared million of years ago and we see impact craters investigated of events gone to the past but nowhere is there any research offered on the moon coming towards the earth and the catastrophic end that will bring devastation to all forms of life. It will be the end, not to certain species or to some rainforests but to all forms of what can be called life! And yet as critical as this is, no one ever spared a thought about researching this doomsday looming in the future. This catastrophic end is coming as sure as Newton's mass is pulling to form gravity! There is not enough imagination in any one person's mind to replicate the disaster pending that will come when the earth and the moon collide. The impact will end any form of life there might be.

Ever since science broke "free" from the Dark Ages science insists on forces existing that control nature. They would not put these forces in the same category as witchcraft but a force to all purposes remains a force, a dark hidden unexplainable movement contracting and changing without finding any explanation as to what it is. The forces are hidden and the origin goes beyond what man can see or explain and yet the forces control our everyday as much as witchcraft did before we abolished witchcraft and made witchcraft a legal enterprise. Now we sit with four forces that no one can explain and the witchcraft behind it is illegal to think of as witchcraft. The force of mass pulls the force of mass and the result is a collision between cosmic bodies. This happened as often as science ran out of explanations about events that occurred in the past such as when the dinosaurs were killed off or when other tremendous unexplained occurrences happened. Yet, every doomsday science connects with a force but the force does not connect with witchcraft because the force is a force and that is science.

$$F = G \ \frac{\text{Mass of the earth x Mass of the moon}}{\text{radius between the two destroyed by the square thereof}}$$

With the moon ever coming closer think how big the earth's gravity will increase. The moon reducing the radius will unite the mass of the earth and moon and have the gravity increasing. This will increase both the moon and the earth's pulling power on every aspect there could be. See how suddenly we are under threat and at risk by comets colliding with the earth. Every so often we read of a "near miss" as some foreign object passed us at a hair's width in cosmic terms. The incoming object was a

10,000-ton asteroid that hit the atmosphere at 11.6 miles (18.6 kilometres) per second and subsequently exploded above the Russian city of Chelyabinsk in February. It produced such a powerful shockwave that it raced around the world twice, according to a new study published.

Researchers report that 20 world-wide monitoring stations, designed to detect ultra-low frequency sound waves emanating from nuclear-test explosions, managed to record the waves produced by the asteroid's explosion for the first time ever. However notwithstanding, the Russian meteor hit the atmosphere at 11.6 miles (18.6 kilometres) per second and at that speed no earthling had any protection of whatever came to shower down upon us with rocks and debris not to mention the heat. The concussive blast was heard at monitoring stations as far away as Greenland and Antarctica.

Travelling at hypersonic speeds, near Mach 60, the meteor experienced increasing air pressure as it pierced the denser part of Earth's atmosphere, finally imploding 14 miles (23 kilometres) above the Earth. Hundreds of fragments rained down, with the largest pieces weighing up to half a ton. How would you like to have a half a ton rock fall close to your house and where will you run and hide from this wind and hail storm?

Just how powerful was that massive meteor that rocked Russia last February? This powerful: The resulting shock wave circled the earth twice, reports the BBC. Scientists reached the conclusion after examining data from global stations that measure low-frequency acoustic waves, reports Discovery.

Most fortunately the main body fell into an icy sea and went harmlessly under the water. Things could have been much worse. Think if this landed on New York or London or any other major city? Where would the Powerful and the Mighty and the Rich hide because their nuclear bunkers can never protect them from this? In the light of these thoughts don't you think hiding on mars to avoid the shit coming to earth is not a much better idea? No place on earth will ever be safe with such shit landing random and with the randomness ever increasing in the future. They can rule us from afar and find protection and control us by spying on us with cameras monitoring us with this spying giving them access to our every move. They can sit up there and with the release of one drone kill who-ever they wish with no consequences to us. All the while they find targets to practise on in Muslim states in Asia and out of sight.

By the time they can get their children and their children's ancestry on Mars having built huge buildings, they will be so good with these drone attacks they can shoot down a dove flying in Manhattan and not damage paint or glass on a building. Replacing paint or glass costs money while replacing one of us comes with no price tag. We are very replaceable and all they have to do is increase our numbers and human worth falls as the numbers increase. We fight to survive while they use us to survive.

They can pick and choose whom to keep alive as they do at present while finding protection billions of kilometres away living the nice life.

The light you see is not the sun but an exploding meteor over Russia. This release of heat is nuclear but it is far more than what a man-made nuclear devise can produce. This is just a meteor. Think what will be when something one hundred or thousand times the size of this object hits the earth.

This is a 10,000-ton asteroid that hits the earth. This is not the moon coming closer but is a piece of rock that came into contact with the earth. What will the end result be with the moon visiting us so close and personal and leaving a vesting card in light as the asteroid did?

This is the first time they've seen such a thing since the International Monitoring System — designed to pick up evidence of nuclear tests — went into effect.

"For the first time since the establishment of the IMS infrasound network, multiple arrivals involving waves that travelled twice round the globe have been clearly identified," writes a researcher in the journal *Geophysical Research Letters*.

The team also confirmed that this was the biggest space impact since the 1908 Tunguska meteor. The more recent incident released the energy equivalent of 460 kilotons of TNT, or 30 Hiroshima bombs, into the atmosphere. (Tunguska was a lot bigger, 10 to 15 megatons.)

To us with human minds these numbers are what they are, just numbers. The devastation that they might cause is just our imagination creating an image and the real devastation is far beyond what we can envisage. It is beyond our human concept or facts we are able to deal with and yet we all are opinionated as to what this can be in reality.

This is the energy release of a simple piece of rock going though the Roche limit as the earth's gravitational limit destroys the formation of the solid rock and then by the Roche lobe turns the solid rock into cosmic liquid and smaller cosmic debris. This is why no structure from outer space can collide with the earth, but in the case of the moon, the earth will self-destruct as it destroys the moon in the same process. Look at the enormity of the release of light and then think how much more the Moon is going to rain havoc on us on earth while the "fat-cats" sit on mars drinking whatever they like to enjoy. It is easy to put everything down on Global warming and other comical excuses while the truth of the moon coming ever-more closer to the earth is hidden and the secret is protected by no party less than science.

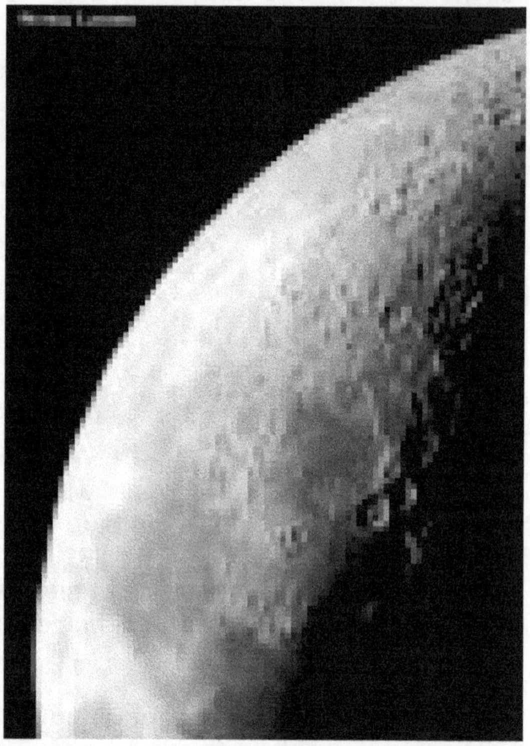

How will we ever get wise if science hides what we are not supposed to know?

Why does science always work within a hidden agenda? Please allow me to explain how this conspiracy works and then you decide…
Think how big the moon is and then think how big the earth is and then think how close the Earth is from the moon … think of the force this must unleash and how this force has to increase as the moon is coming closer, speeding up the pace by increasing the speed of the moon closing the distance between the two solar bodies and where does this leave you?
You, being sandwiched in between the two!

Think how the sea will have high and low tides of thousands of meters pulling the waves to destroy the land long before the moon ever rushed into the earth with a bang. By the time the Moon appears this large in the sky, would there be human life remaining to see the event occurring or would all human life be seized by then? Think of Global Warming and this will bring us Global Warming as nothing did before.

Then again might this "global warming" not be an eye blind to cover up the real issue that is at hand, the moon coming towards the earth. To stop a panic attack that will destroy all forms of property and currency those in power then create the "global warming issue" to make us think that we are safe and that we are in charge of what we clearly never could be in charge of, and that is gravity destroying the moon and the earth? Why has no person ever thought even to mention this subject ever, only once?

Those in charge of putting the fear of God into us by arousing mass hysteria about Global Warming and carbon polluting or destruction looming such as the international press never even mention the possibility of such a collision! Why is that the case!

There are probes in place to detect renegade comets and asteroids that might inflict massive damage when it hits the earth. Science does show a tendency to locate this event beforehand because this brings a certain degree of worry to those charged with finding something to worry about.

Billions in whatever currency is spent to detect some asteroid or comet that is heading our way with the sole purpose to destroy the earth, but when this happens the event is by chance. These events are not predictable and are avoidable and yet they are studied in anguish. There is a much more realistic desecration heading our way with the moon destroying the earth. This event has a better chance and more certainty, with Newton's principles applying the certainty factor and yet this possibility is never even mentioned by those charged with detecting concerns!

Why would those doing research never mention the time that the earth and the moon are bound to collide? If it is billions of years away, then knowing that it is still in the very far future will soothe us and bring comfort. However, if it is a possibility that could happen and we could do nothing about it but commit suicide to escape this fate, then hiding this information from the public seems the way to go for those in power.

The photos depict the Tunguska event in 1908 in Siberia when an Asteroid exploded above the earth. This was devastating.

See the trees that were giants seem as if they were matches strewn. The blast depicted by this event is comparable to a flash bulb lighting a room to take a photograph. Any person who thinks Tunguska was bad news to the earth should rethink his or her degree of worrying. Tunguska and its aftermath is a Christmas cracker sideshow compared to what is coming when the moon is coming to our part of the earth. Try the event when the moon comes hitting the earth and see what there truly is, that could bring discomfort at bedtime when no one on earth would feel the need to sleep because the ultimate doomsday is with us and upon us. Anyone needing a worry has this one to worry about because this tops the lot!

In late June of 1908, a fireball exploded above the remote Russian forests of Tunguska, Siberia, flattening more than 800 square miles of trees. At 7:17 AM on the morning of June 30, 1908, a mysterious explosion occurred in the skies over Siberia. Realistic pictures of the event are unavailable. I believe that we now know enough about large impacts to "decode" the subjective descriptions of the witnesses and create realistic views of this historic asteroid impact as seen from different distances.

Previous speculation had ranged from comets to meteors. Noctilucent clouds are brilliant, night-visible clouds made of ice particles and only form at very high altitudes and in extremely cold temperatures. These clouds appeared a day after the Tunguska explosion and also appear following a shuttle mission.

The researchers contend that the massive amount of water vapour spewed into the atmosphere by the 1908 comet's icy nucleus was caught up in swirling eddies with tremendous energy by a process called two-dimensional turbulence, which explains why the noctilucent clouds formed a day later many thousands of miles away. Noctilucent clouds are the Earth's highest clouds, forming naturally in the mesosphere at about 55 miles over the Polar Regions during the summer months when the mesosphere is around minus 180 degrees Fahrenheit (minus 117 degrees Celsius).

The space shuttle exhaust plume, the researchers say, resembled the comet's action. A single space shuttle flight injects 300 metric tons of water vapour into the Earth's thermosphere, and the water particles have been found to travel to the Arctic and Antarctic regions, where they form the clouds after settling into the mesosphere. Following the Tunguska Event, the night skies shone brightly for several days across Europe, particularly Great Britain — more than 3,000 miles away. In both cases, water vapour was injected into the atmosphere. The scientists have attempted to answer how this water vapour travelled so far without scattering and diffusing, as conventional physics would predict. This "new" physics, the researchers contend, is tied up in counter-rotating eddies with extreme energy. "Our observations show that current understanding of the mesosphere-lower thermosphere region is quite poor,"

Witnesses in the town of Kirensk and nearby towns at the same distance recollected the fireball flashing across the sky in the following terms:
"A ball of fire appeared in the sky... "A flying star with a fiery tail; its tail disappeared into the air."

After this object passed across the sky, it approached the horizon where it was consistently described from this distance of 400 km, as appearing like a "pillar of fire," then replaced by "a cloud of smoke rising from the ground," or "a cloud of ash...on the horizon," or "a huge cloud of black smoke. "It was called a diffusing bright ball of fire two or three times larger than the sun but not as bright; the trail was a "fiery-white band." Inconsistent colours were mentioned: white, red, flame-like, bluish-white.

Some minutes after the explosion, distant observers reported a column of smoke on the horizon. One observer said, "Where the body disappeared behind the horizon, a pillar of dark smoke rose up." I have wondered whether the dark colour could result from the smoke of the explosion containing black, sooty carbonaceous particles, in the same way that the explosion clouds on Jupiter from the impact of Comet Shoemaker-Levy 9 were very dark. Because the meteorite did not strike the ground or make a crater, early researchers thought the object might be a weak, icy fragment of a comet, which vaporized explosively in the air, and left no residue on the ground. However, modern planetary scientists have much better tools for understanding meteorite explosion in the atmosphere. Some of them drop brick-sized fragments on the ground, but others, such as the one that hit Siberia, may produce primarily a fireball and cloud of fine dust and tiny fragments. In 1993 researchers studied the Siberian explosion and concluded it was of this type -- a stone meteorite that exploded in the atmosphere. This conclusion was supported when Russian researchers found tiny stony particles embedded in the trees at the collision site, matching the composition of common stone meteorites. The original asteroid fragment may have been roughly 50-60 meters (50-60 yards) in diameter.

Many asteroid fragments circle the Sun; the Siberian object was merely the largest to hit the Earth in the last century or so. Had it hit a populated area, devastation would have been enormous. If there are many asteroid fragments, why don't we see more hits? An Air Force satellite in the 1990s detected a smaller explosion over the Pacific. In 1972, a 1000-ton object skimmed tangentially through Earth's atmosphere over the Grand Tetons in Wyoming, and then skipped back out into space, like a stone skipping off water. Even larger objects have hit Earth, but they are rarer. Brick-sized interplanetary stones fall from the sky in various locations every year. Interplanetary space contains many small bodies of different sizes. Large enough bodies leave sizable craters on planets or satellites. If we continue to study asteroids and build more telescopes for detecting and tracking them, we will have better information about the frequency of such asteroid impact-explosions, and more chance to have a warning about impending impacts. At 500 km (300 mi), observers reported "deafening bangs" and a fiery cloud on the horizon. At distances around 60 km, people were thrown to the ground or even knocked unconscious; windows were broken and crockery knocked off shelves.

The Tunguska event would in comparison seem to be a minor local inconvenience compared to when the moon collides with the earth and both objects becoming dust particles. In this light would it not be wiser for the rich and those in political power just to keep quiet about such a looming catastrophic scenario and in keeping silent not arouse a hysteric human race going mad in fear and anxiety? Think how uncontrollably mad everyone would get knowing death is upon us? So, hush the people's concerns.

With the formula as Newton introduced it being $F = G \dfrac{M_1 M_2}{r^2}$ anyone with a little insight into reality could see that as the bottom part decreases (become smaller) this will affect the top part to become larger. I will show three values and you will see what I mean. $\dfrac{100}{100} = 1$ and $\dfrac{100}{1} = 100$ and then getting the bottom really small $\dfrac{100}{.001} = 100000$. The bottom part's decrease increases the speed of the top part or the gravitational movement.

By reducing the distance the force of gravity will grow exponentially faster! The moon will seem to be on a runaway collision course going out of bounds the closer it gets.

The devastation it brings will increase exponentially and becoming so big no one would be able to realise the changes as the doomsday is increasing! No one would be able to keep track of the fatalities occurring. Everybody who died during the first and the Second World War plus the epidemic influenza that came after the First World War would be deaths happening in one day and during every day. And no one in science even thinks about this scenario while Newton's gravitational laws are in place? Can you think why not?

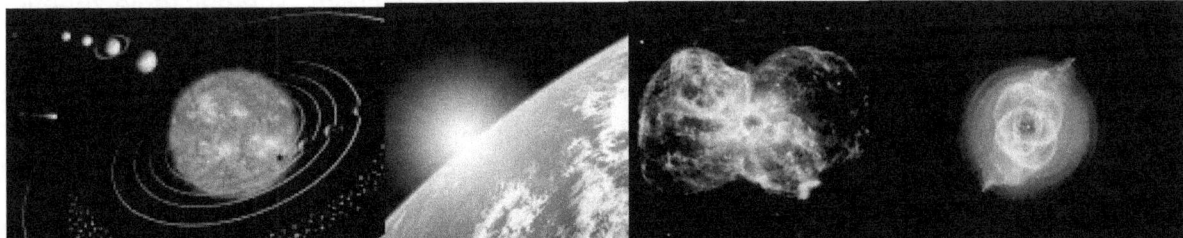

Those who should know better say when the sun runs out of its current fuel, the hydrogen starts to cool down, as it will expand into a red giant and could expand out to where Jupiter is circling or so they say it is going to be. These are the estimates I have seen quoted before and are not mine to quote.

The end of the sun is billions of years away. It is a lot of hogwash but it shows someone somewhere got interested and started some form of thinking about how the sun and the solar system would end. Yet no one in science ever thought it worthwhile to conduct a study into when the earth and the moon will go into destruction.

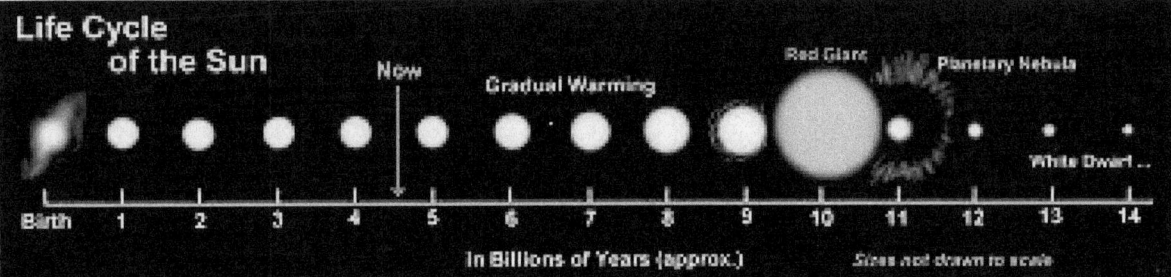

Science took the time to work out what path the future development of the sun would follow and they say they have it down to more or less a precise detail. Never ever has the moon drawing closer to the sun become a topic of discussion?

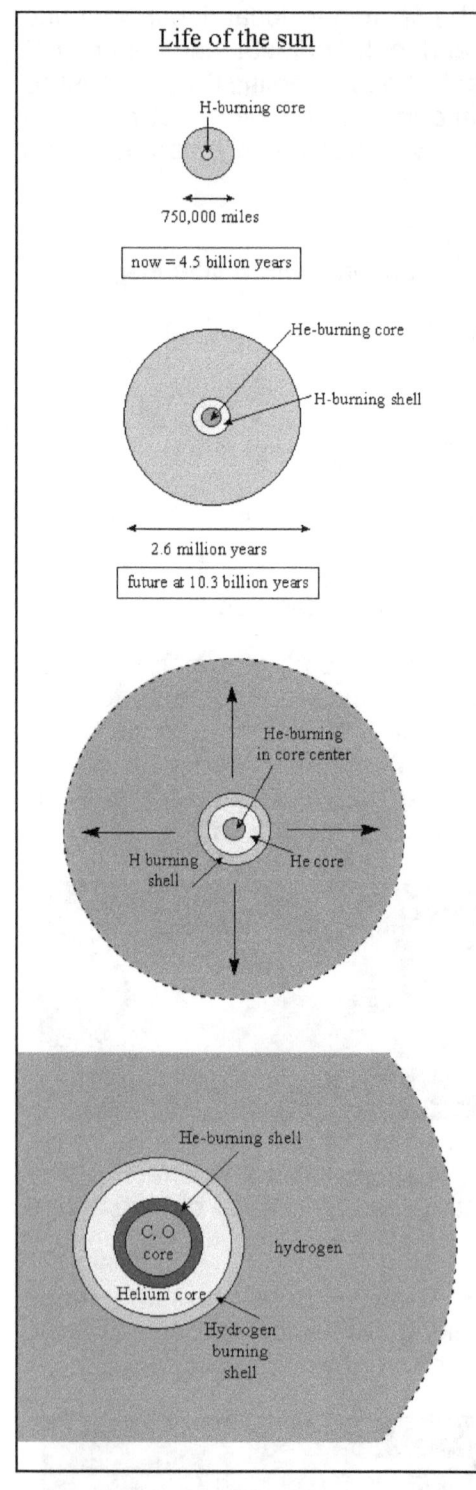

Life of the sun

H-burning core

750,000 miles

now = 4.5 billion years

He-burning core

H-burning shell

2.6 million years

future at 10.3 billion years

He-burning in core center

H burning shell

He core

He-burning shell

C, O core

hydrogen

Helium core

Hydrogen burning shell

This is the future path the sun is going to take according to science. They have worked it out. They could calculate the time it is going to take for the sun to explode.

There's every expectation that in about 5 billion more years, that our sun will swell up to become a red giant. And then, as it gets larger and larger, it will eventually become what's called an asymptotic giant branch star – a star whose radius is just under the distance between the sun and the Earth – one astronomical unit in size. So the Earth will be literally skimming the surface of the red giant sun when it's an asymptotic giant branch star.

In approximately 5 billion years, the sun will begin the helium-burning process, turning into a red giant star.

When it expands, its outer layers will consume Mercury and Venus, and reach Earth. Scientists are still debating whether or not our planet will be engulfed, or whether it will orbit dangerously close to the dimmer star. Either way, life as we know it on Earth will cease to exist. The changing sun may provide hope to other planets, however.

When stars morph into red giants, they change the habitable zones of their system. The habitable zone is the region where liquid water can exist, considered by most scientists to be the area ripe for life to evolve. Because a star remains a red giant for approximately a billion years, it may be possible for life to arise on bodies in the outer solar system, which will be closer to the sun. The window of opportunity will only be open briefly, however. When the sun and other smaller stars shrinks back down to a white dwarf, the life-giving light will dissipate.

They have worked out the life cycle of the sun and what is in its future. This they have done using the Newtonian basis of $F = G\dfrac{M_1 M_2}{r^2}$, which they concluded from the calculation that $F = \dfrac{r^2}{M_1 M_2}$

because $F \propto \dfrac{M_1 M_2}{r^2}$ and this then ends up as $F = G\dfrac{M_1 M_2}{r^2}$. However long before the sun will disappear this formula will bring about the end of the earth / moon association because $F = G\dfrac{M_1 M_2}{r^2}$ must bring a lot faster unification collision than when the sun will end. So what the F@$#*&k has the sun's ending going to have an influence on my future or on the earth. The earth moon collision will end the earth long before the end of the sun will come about.

According to science the sun formed from a dust bowl and then that was about 5 billion years ago. Then the sun grew into what we now have and the sun will grow even more. Take not what this says. The sun is not springing into the oblivion according to their prediction but it is increasing in volumetric size. Now, when are they fooling you deliberately...when they tell you the sun is expanding or when they don't tell you the moon is going to hit the earth in a future event? Now the sun is going to scourge everything up to Jupiter and ever further.

Even if this is going to happen in a billion or five billion or a trillion years from now, still there are persons who find the end of the sun important enough to be concerned about. It shows that it is not because it is far into the future that there is no apparent interest into the collision between the moon and the earth. The moon would have hit the earth much sooner than the sun would explode, so why is there no investigative research done on the earth and moon contracting by the mass that Newton foresaw? I don't for a minute say I agree with anything said about the sun, but at least it shows some investigation was done not looking at the correctness thereof, there was a thought spared about how the sun will end.

It is said using Newtonian vision about what the Universe holds install that the sun is going to blow up and burst and come to an end. Don't repeat me on this hogwash because all I wish to prove is that in terms of even the sun coming to a closure of a life cycle there was research done in this line of thought. It is said that the sun will scourge the solar system out of existence but never does anyone mention that the moon and the earth is having a get-together long before that formed on the grounds of Newton's vision on gravity coming from mass that pulls mass. If science can look this far ahead and find a doomsday waiting on us billions upon billions of years from now, why is everybody avoiding the apparent doomsday that is much closer to realising the ending of all of the earth and its moon including all (and the only) life that was ever present in the Universe.

With everyone in science saluting Newton's gravitational contracting there was an extended effort by Albert Einstein to find the critical density of the Universe. That is the backbreaking effort that science took with painstaking accuracy to find the level of interest in determining the end of the entire Universe. The critical density idea did not pan out and that left science high and dry for answers about science. They did not stop there, no Sir, they paddled on in darkness to find the answer. They left no stone untouched to come up with a conclusion...and yet not one person in all that time started to think that the contraction will be much better monitored by researching the moon reducing of the radius it has between it and the earth. Would it not have been much easier to study when the moon will splash into the earth and from there work out when Newton's attraction will have the worst collision we can think of happening in our backyard?

Then science goes further and tries to detect the untraceable that is invisible. They try to find dark matter in a shady Universe. They spend billions on dark material or dark energy research but not a single dime goes the way of finding out when the moon and the earth will have a gravitational self-destruction event. Do you not find that very odd...or is it just me wondering about that?

Someone in science though it was worthwhile to study what would happen when the sun

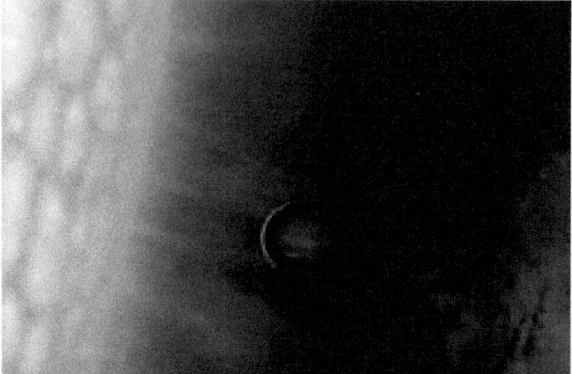

comes to a final conclusion. Someone in science took time to bother with the end of the solar system and what will be the applying conditions during the finale era of our sun having a solar system. They even measured how big and how red the sun would be in the end but never do they bother about what is more obvious, when will the earth and moon get together. That seems very odd and suspicious to the less informed...

They have probes to detect tsunamis that will only horribly devastate regions and as it stands, it can't compare to the moon hitting the earth. There will be waves every day generated by the gravitational increase of the nearing moon that you now would associate with the worst tsunami occurrences, which would then become the mild everyday waves found on the beach. The sea will become a wild devastating machine that pulls all land into the sea with waves going as high as a thousand meters and much more. This is the scenario waiting on us holding life and no research is going into trying to predict when this is going to happen. They research tsunamis in detail but not one place on earth could be considered as safe when the moon lands on earth. Such waves will draw mountain ranges into the sea and destroy giant areas into water-wastelands. Think of the crushing force of one such a wave and then the moon is only still coming our way. Long before humans would feel the demise there will be no human left to experience the demise. That means that the sooner we come to terms with what waits on us, the more time we have...

Building caves in mountains would not help because the massive waves will get us there in the final event. It is just a thought but why is there such a surge to investigate possible occupation of Mars? There is no spot on earth that will remain safe so where will those who

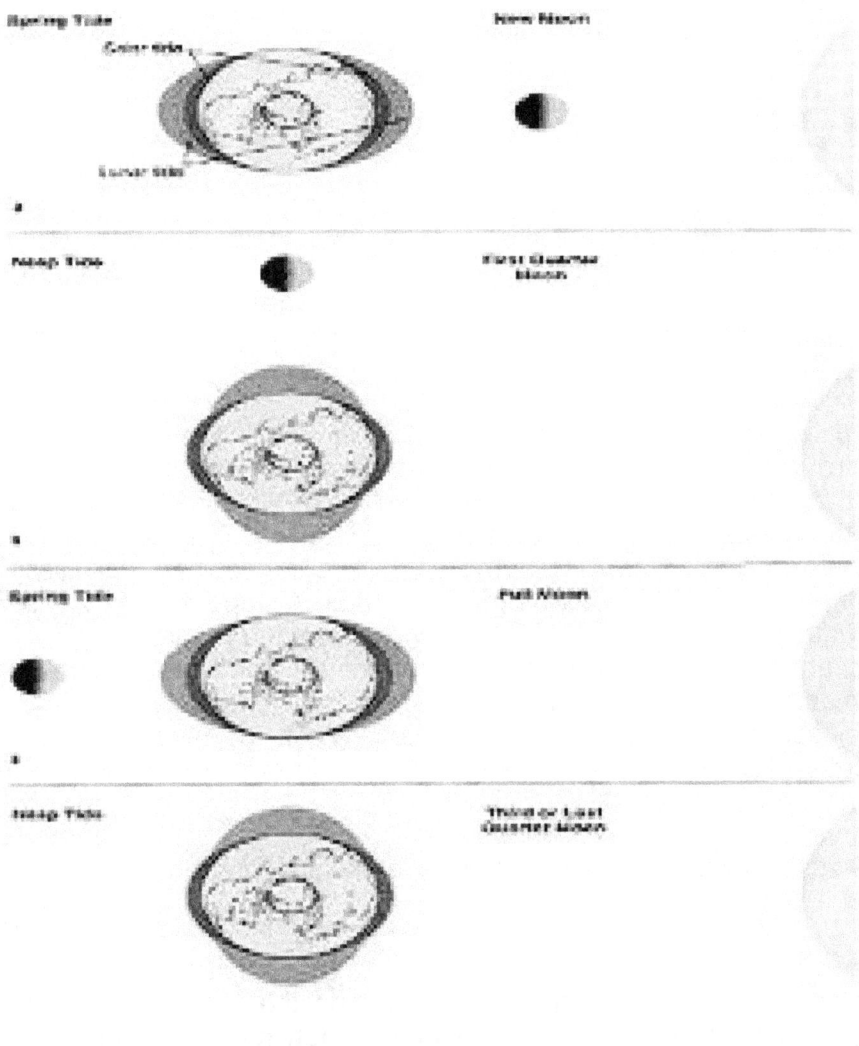

hold the money and the power go when they need a safe haven? What would be the most obvious reason to keep order and keep the money flowing and for what purpose would they wish to keep the money flowing? Think about it this way, if the rich and the powerful could take all of the tax money of all the lesser human beings and start to build on Mars under the disguise of research, then when the doomsday arrives those being the fortunate, the powerful, the influential and the law makers will have somewhere to go, to escape, leaving the less worthy to take the moon head on! There was an urge to get to the moon and when they got there then afterwards the interest in the moon fell to an all time low. Then those in investigating investigated all the planets without showing the least interest in what the moon might hold. Then suddenly from the blue comes some hysteric research about Mars, just after someone discovered possible prehistoric semblance of life on Mars.

If they study and find this event is still far off into the future, then do the study and tell us. If they do the study and see it is going to happen in three or four generations from now, then include us into the plans of becoming Martians or say who will be left behind and on what merit is the choice made, but for God sake, do studies and tell us, that is if we could trust Newton's physics of mass that pulls mass.

Those who are quick would observe the mistake in this most Newtonian science idea I made. The idea that it rains only six days is as true as anything in reason could be. The idea that the moon has more gravity during full moon and less gravity during dark moon is fabrications. The moon has the same mass all over the period of any given year going on year by year. So the idea of more of the being in place moon finding more mass forming more gravity that pulls more water is lame and yet it has a very Newtonian ring to the concept. Newton said the moon's gravity must pull and the earth's gravity must pull and then with tides the moon's gravity pulls stronger than the earth's gravity, which results in high tides and low tides.

Let us test Newton's attraction theory in practice

I suppose this article alone will change little in the onslaught of the conspiracy, but someone

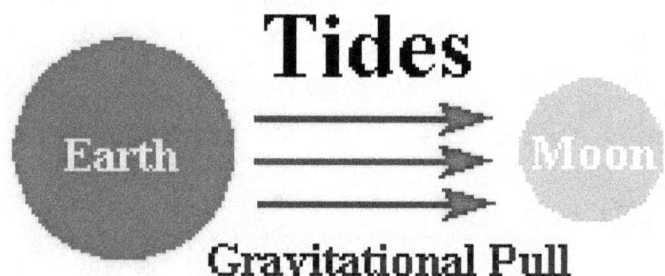

has to begin to ask questions at some point. I have the answers but the conspiracy stops me from being heard as the force of the most powerful Academics silences my message.

This is quoted from the Internet

You know how the earth pulls Newton's apple down? It pulls everything on or near its surface toward its own centre. And everything near enough to be pulled pulls back as hard as it can. The earth pulls the moon, and the moon pulls the earth. Although the moon is much smaller than the earth it is just the right size to be the distance it is away from the earth, to keep from falling into or getting away from the earth. We cannot see the pulling power on the solid parts of the earth. But the ocean is made of water. A slope of land, a brisk wind, and many things set water in motion. It feels the pulling power of the moon. Whenever the moon rises over the ocean, it pulls the water that is just under it. So, a great wave, or tide, travels under the moon across the wide sea. When the shore is reached this wave rises higher against the rocks, or spreads over level sand beaches. When the moon sets, the wave goes back to the old level.

Right, this explanation is very Newtonian and very incoherent.
Newtonian science says, "Mass pulls to form gravity". Newtonian science says "gravity pulls" to form high and low tides.

Can anybody see the Newtonian double fork in this conflict of factor implications? If this is true that mass forms gravity by creating pulling then that means the moon-earth mass relevancy must change twice in twenty-four hours. Either the moon must become more massive or less massive twice a day because the tide rises and the tide lowers in one circle rotation of the earth. It has nothing to do with the mass of the moon because if this "gravitational pull" forms gravity by mass, then either the earth or the moon has to increase the mass ratio to "draw" the water "up" cyclic or to "push" the water "down" in an earth rotation cycle.

Tidal waves again as everything else in nature proves Newton wrong and out of touch with cosmic reality. I formulate and I prove that gravity has nothing to do with mass and everything to do with the rotation movement of the earth and through the rational spin of the earth, mass is created. The moon proves my theory as much as all of the solar system proves my thinking of how gravity forms as much as all the mentioned factors disprove Newton.

Yet, Newton is accepted as if proven without question. Should you wish to find out how gravity does work then go to Lulu.com There is one question though? With the moon always at an even size and most often at an even distance how does the moon pull the earth by 12-hour cycles?

The idea of proof comes automatically to the door of Newton although Newtonians will deny this fact as if they deny the honour of their Master Newton and that is what they have to do. Newton as a culture became a religiosity in which every human being believes unconditionally.

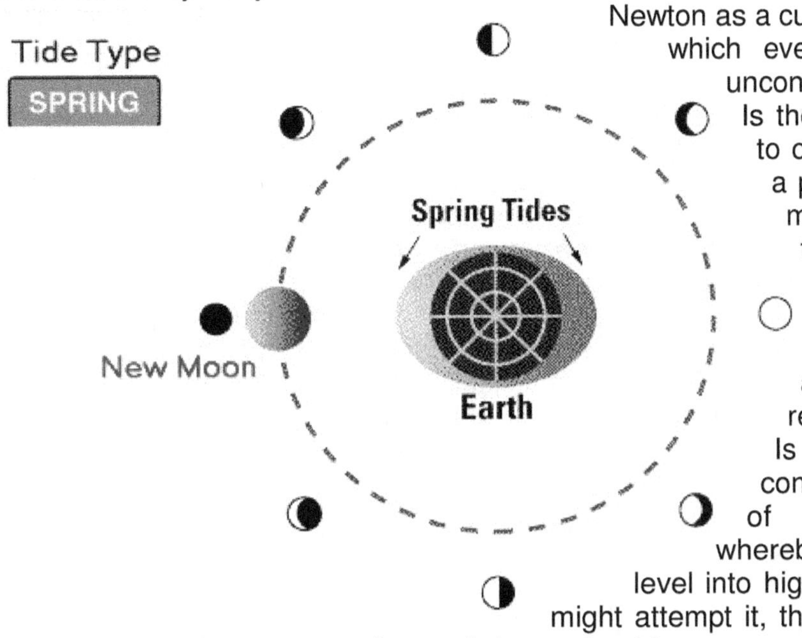

Tide Type

SPRING

Spring Tides

Earth

New Moon

Is there anybody who would try to convince me that mass play a part in the tidal wave of the moon? It is gravity pulling the tidal wave but that gravity is directly linked to the spin of the earth and has no connection to mass, although mass is a direct result of the spin of the earth. Is there anybody who would convince me of Newton's idea of "mass forming gravity" whereby the moon pulls the sea level into high tide because if he or she might attempt it, then let's start with the mass of either the moon or the earth becoming bigger or smaller in every circle that the earth rotates.

Is there anybody gullible enough to believe that physicists in three hundred years never saw this tendency contradict Newton's "mass that forms gravity" claim? There is no one who thought this should deserve a better look because this does not come across as very sane or believably accurate. This is the conspiracy. The conspiracy is to hide Newton's outlived ideas from all those standing outside physics because remove Newton's ideas from physics and you are left with nothing!

I prove with formulation and evidence of what applies in the cosmos that gravity is the result of the earth spinning. To do that I prove that the Titius Bode law, The Roche limit, the Lagrangian points and the Coanda effect combines to form gravity. These are the most important principles in physics and are in place used in the formation of the Universe but because they contradict Newton and make a fool of Newton they are not very pertinent. Because the layout and the reason why it is in place is not understood they make those Brainy-Bunch in science look incompetent, as incompetent as Newton's ideas are, and therefore it is very discretely hushed up. I challenge anyone to show where the cosmos uses mass but I can show where the cosmos uses the four principles I mention and that I prove forms gravity.

However, there is nobody who is interested in what I have to say because there is too much money to be lost! If my work is accepted it makes Newton part of science fiction and think of what that would do to the Powerful-In-Charge-Of Money-Matters? Every book about Newton then becomes fiction. Since the money in science depends on every person on earth believing in Newton just the sheer influence of money gone lost prevents the most powerful even to consider my views as long as I denounce Newton. But that I have to do because Newton is as far from the truth as a Christmas tree is from the Christian religion.

I prove that as the earth turns, by turning it compresses the space around the earth. The space we call atmosphere is trusted onto the earth and if anything circles the earth such an object then can leave the surface of the earth. By turning the space around, the earth compresses the space we call the atmosphere and by thrusting the atmosphere onto the earth, the space holding a solid will then press onto the earth and thereby form mass. That's how simple it is.

Science says mass pulls mass. Then why is this astronaut floating in space? He still has mass and the earth still has mass and the mass must be pulling to form gravity. The man is floating on the same principle that the moon is floating. If the man is not going to fall, the moon and earth will never collide and that makes Newton's mass pulling mass a contradiction that makes science a joke. If Newton's "mass is pulling mass" is correct, then the moon and the earth must collide. So which is it because it can't be both? The man has mass just like the moon has mass and the earth still has the mass it always has and the law of attraction between the human floating above the earth must be the same as the moon floating above the earth. If the man could float then this law of attraction is rather becoming suspicious. If that is the way you think then having those suspicions are very correct. Science confirms that there is always a force of attraction between all bodies. There must therefore be a force of attraction between the astronaut and the earth in the picture above, so why isn't it falling straight down in like with Newton's apple? The mass has gone nowhere but science cheats the answer by giving the man "micro gravity". This way they cheat by avoiding the truth without pulling a face.

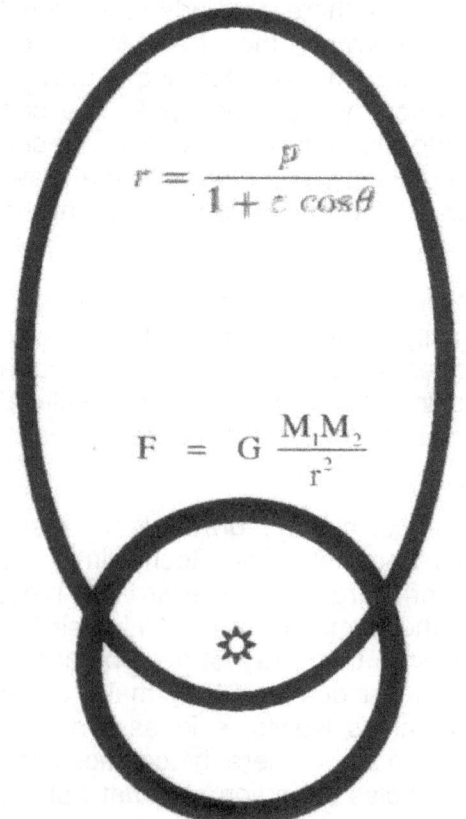

$$r = \frac{p}{1 + \varepsilon \cos\theta}$$

$$F = G \frac{M_1 M_2}{r^2}$$

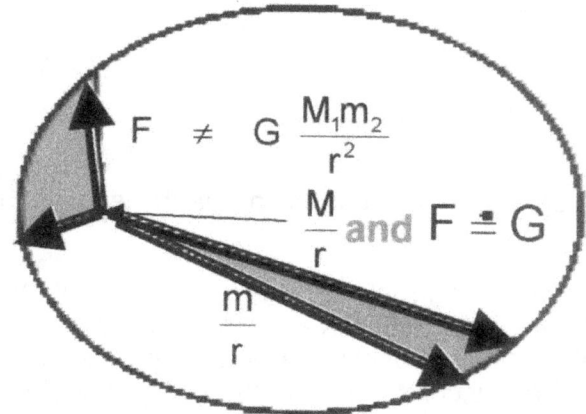

$$F \neq G \frac{M_1 m_2}{r^2}$$

$$\frac{M}{r} \text{ and } F \stackrel{\triangle}{=} G$$

$$\frac{m}{r}$$

If mass formed gravity, then Kepler's first law $r = \dfrac{p}{1 + \varepsilon \cos\theta}$ that Newton devised would not be possible because this shows deviation of the orbit while we all know that the mass remains totally constant and always the same. The why would the orbit circle in a loop coming "closer" at one point and going "wider" at the opposite side. Does anyone out there still wish to tell me there is no conspiracy about Newton and the truth going on in science?

If mass was responsible for forming gravity and it was the mass of both the sun and the planet then a precise circle would form allowing a continuous radius than could never deviate. I child can see this but our Newtonian maters of conspiracy never once noticed this or gave any of their precious time to rethink this Newtonian binomaly...or did the lot just conspire for three centuries to look the other way because if they have not got Newton, then they have got nothing...and they forever tell they only deal in proven facts.

If Newton's presumed $F = G \dfrac{M_1 M_2}{r^2}$ was true then "Kepler's" second law could not apply because the Kepler's true formula, which is $a^3 = T^2 k$ would bring about again a perfect round circle because again mass being constant will sustain as perfect circle around the sun. Again no one in science in three hundred years saw this and did not consider it to be rather odd. Then every one outside physics thought those inside physics were so clever that those inside physics saw what those outside physics could not see and left it to be. That gave those inside physics the chance to swindle with physics and was never challenged by those living outside of physics for the fear of feeling inadequate and stupid.

Where Newton shows that mass is supposedly the factor that should place planets according to size in their respective positions, which is not true, I prove that mass is not used by nature, not anywhere and I am the one they frown upon because I don't accept Newton's and therefore I don't compromise by embracing the Newtonian conspiracy. I show how ridiculous it is to support Newton because Newton has no foundation or support from the cosmos, yet I am ridiculed as the one that is incoherent.

I show what nature uses namely the **Titius Bode law**, **The Roche limit**, **The Lagrangian Points** and **the Coanda effect** and how this forms gravity as well as place the positions of the planets in accordance with singularity, not mass. Because I trash Newton's rubbish that does not fit and in that can't apply, no publisher of science books or science magazines will publish my work! I show what goes on in nature while Newton's contribution of mass applying is total rubbish. Because I call it rubbish and I rubbish Newton's ideas I am ignored.

Let us now proceed and test the basis on which all physics are founded.

The following formula carries every basic idea on which science formulates the entirety of the Universe. If anything, this formula must be the best tested and the most thought through idea ever launched by any human mind. No other idea even in the form of religion could be more stern that the following formula.

Let us then test the truthfulness and believability of the foundation on which the entirety of

Newton's physics principles rests. I show you the absolute accuracy of $F = G\ \dfrac{M_1 M_2}{r^2}$.

This is Newton's formula on which the entire philosophy of gravity rests and it supposedly proves that mass forms gravity by force.

No one in science is capable of proving this formula worthy of being part of science and with this forming the basis of science it must be clear to what extent the brainwashing is and the mind control and the thought manipulation going on to force students to accept corrupted science that can never make sense.

Newtonian physics teaches that the mass of the earth and the mass of the moon is constantly contracting (reducing), which produces a force called gravity, which reduces the distance between the two solar objects.

The cosmic question that calls for an answer is ... with this being true that the mass pulls mass, then when will the earth and the moon then collide? This destruction of the earth and the moon must happen first since the earth and moon are the closest...

Academics in physics insist that gravity is founded on the Newtonian gravitational principle of $F = G\ \dfrac{M_1 M_2}{r^2}$, then when is the inevitable collision coming.

Now, everyone has to braise him- or herself for the inevitable impact that has to come, but when? When can we trust the conspirers and physics academics to tell us when this inevitable doomsday destruction is on the earth's doorstep?

Newton said this must come because of mass pulling mass to clash with each other and therefore so when is the day going to be when we will not be on earth any longer. We should know!

For those not familiar with Newton's gravitational principles I will give you an update of science development that took place the past three hundred years, in the event you missed the latest findings.

Have you heard about the apple that fell from the tree that had a bigger influence on mankind than any religious intervention before the event or after the event? This incident with the apple changed the outlook of every person on earth notwithstanding religion, culture, ethnicity, or race.

This apple that fell from a tree changed the lives of every individual on earth that remotely had anything to do with science.

Let's be serious. They say (they being those in science) the moon was a part of the earth and by some mystical collision the two came apart in some fashion no one can entirely understand? This does not make sense and is even more senseless where it does not stroke with science principles.

This is how you con people by diverting attention away from the place you question. In trickery and playing games of deception one uses a method of diverting the attention towards another place to commit a crime. One would bump hard into someone while picking the pocket while the person feels the thump of a trip when colliding with the victim. This is precisely what I see is happening.

This principle declares that a force of attraction by the value of mass of both solar objects is pulling by reducing the radius, therefore bringing the moon and the earth closer. This must lead to an inevitable collision that must end both the earth as well as the moon.

Let the Newtonians calculate when this event is due and inform the human race when our final day of doom will arrive by using Newton's law of attraction or $F = G \dfrac{M_1 M_2}{r^2}$.

The day (if Newton is correct) when this happens, life in all forms will end and therefore knowing about this event is crucially important for all of us having life on earth! We have to know! The future of life in the Universe depends on this knowledge. If we know then we have to find the ability to transplant life to Mars and beyond just to save life, as we know life is! Knowing when this will happen will give us humans a fighting chance to have a future somewhere safe!

Those Ever-So-Wisely-Educated cosmic Super-Brains always calculate the power that drives a Super Nova. Let them use those brilliant minds then to calculate the precise date when the earth and moon will destruct all life on Earth! Those Brainy- Mathematical-Masters always custom-design in detail space –whirls they invent with applied cosmic imagination. Let them bring such terrific astonishing human abilities closer to home. The Mathematical-Geniuses who can calculate the inside of a Black Hole should bring their splendour to a much better use in terms of where it concerns human future.

Instead of painting a Universe fit for Alice in wonder-world, and sprinkling it with the best mathematical formula that they think must put them on par with God, rather apply the same formula and show when the solar system, as big as it is, will collapse into the Sun. Ask them to not search for imaginary undetectable, unexplainable dark matter they can't even point out, but to find when the distance we have between the planets and the sun will dissolve and when the pull of the planets will have all the planets go crashing into the sun.

Those Ever-So-Wisely-Educated cosmic Super-Brains' best art form is to deceive everyone by conspiring to hide the truth about Newtonian science. Those Mathematical-Masters will never even hint at Newton's incorrectness of thought because it will reveal their fraud.

It is showing that those you think you can trust are those very persons that you dare not trust!

My website allows you to see what the degree of blatant corruption is presented as physics and what Academics in physics hide.

The law of gravity says that the mass of one object pulls by force on the mass of another object where the second object then pulls back also by force.

Newtonian physics says that the mass of the earth and the mass of the moon is constantly contracting (reducing) the distance they are apart, by forming a force called gravity, or so science believes. These objects with the most mass are the closest and must form the biggest pulling power from our earthly perspective.

Have you heard about the event about the falling apple that had such a concrete influence on mankind it changed the world more profoundly than Jesus Christ Did?

This book and what forms the content that I present has the dynamics to change science forever and that is not cheap exaggeration or promotional talk. You are about to find out how influential apples are in directing man's eternal destiny. According to the Bible it was a fruit that a man going by the name of Adam ate in the Garden of Eden and it is generally thought of as an apple that changed man's destiny. The fact that it was an apple is just surmising and it is not a proven statement and that much I admit but such as it may be and correct or incorrect, the idea that it was an apple stuck. The correctness about the fruit of the Garden of Eden being an apple or not being an apple is disputed and is also never disputed but since it was never proven to be an apple or not to be an apple therefore it also was never disproved just as much as it was never proved. In the general accepted views of the broader view of people that the concept of the apple was the forbidden fruit is widely accepted although it is not widely proven and although widely accepted it is accepted without carrying strong doubt or religious rejection and therefore we are not going to dwell on the apple being the fruit which the Bible refers to. The point I wish to bring across is that even without proof that the apple was the fruit being illicit in the Garden of Eden is accepted almost without a second thought. That is the culture man has. Man accepts without question the small detail of what is accepted. In this where we go back to the beginning of Man it is accepted in that case that it took an apple falling to change the destiny of Man and the thought pattern of humanity, and everything about the idea that it was an apple that was the illicit fruit proves to be a lie!

Now you may think that I am referring to Adam and Eve and the probability that they ate from an illicit apple without permission. No this apple had far more reaching significance than Adam's apple had because although Adam and his Apple was important it had only had significance as far as man is concerned in context of sins man had and as far as some religions believe. The apple I am talking about changed the outlook man has on the entire cosmos. This apple brought the cosmos into a new light and everything in the cosmos became clearer all thanks to an apple that fell at the right place at the right time. Every time anything falls to the earth this should be your reminder that your final destiny with the moon is eminent. As objects drop or fall, then by the same margin and token is the moon coming closer but quicker because the moon has a lot more mass favouring the speed by which it must then fall. Newton saw the apple fall and realised the moon was falling to the earth as the earth is falling to the sun. Newton was the one that concluded his principles of gravity by seeing an apple falling and then he extended this rather ordinary occurrence to become a Universally spectacle as compelling as a cosmic connected event.

Newton as a student saw an apple fall while he was on his back as he was loitering under the English sun. As he saw an apple fall from a branch and he made this the most fundamental event in science as far as human's recollection is concerned. Seeing the apple fall Newton concluded that the apple has mass and the earth has mass. He saw the apple fall down from the tree to the earth and realised this as a spectacular cosmic spectacle with all the astronomical implications preceding the event. As great a mind as Newton had he did not see an apple fall but he saw the way the moon is descending towards the earth. He realised there was more than one apple falling to the earth; no there was a force that was between the moon and the earth. The apple fell by the force of mass and whenever whatever fall, the falling is going by mass.

There are those who are atheists, which means they don't believe in this Garden of Eden apple incident and what was connected to such an event but they also attach their unproven and highly doubtful religion thought of as physics to another event also connecting an apple where in their religion the fact of the apple carries much less dispute than what was derived from that apple's falling. The apple in the Garden of Eden I refer to as the first apple incident and the Newton apple I therefore call the second apple controversy. Those who dispute the first apple incident normally are the same people who support the second apple incident and there is one person as far as I know who disputes the second apple incident. It is not the fact that an apple fell that I dispute but it is the outcome of the broader religion that formed due to this overall accepted incident that I dispute.

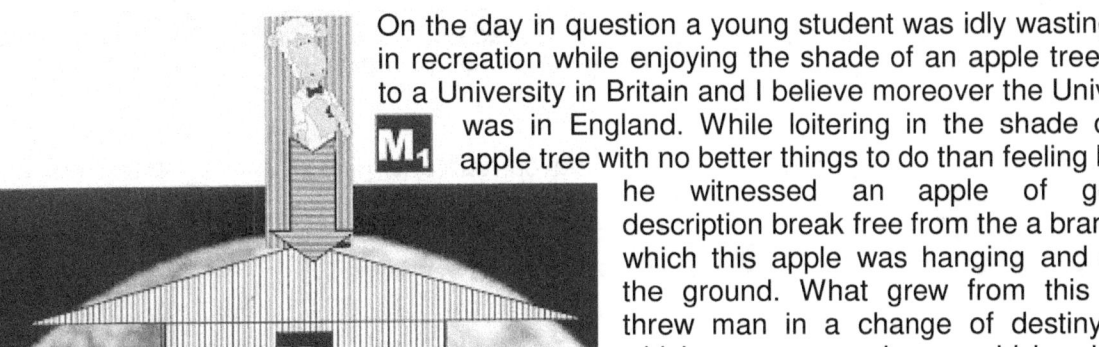

On the day in question a young student was idly wasting time in recreation while enjoying the shade of an apple tree close to a University in Britain and I believe moreover the University was in England. While loitering in the shade of this apple tree with no better things to do than feeling bored, he witnessed an apple of general description break free from the a branch on which this apple was hanging and fell to the ground. What grew from this event threw man in a change of destiny from which man never since could break free. Newton saw this apple coming to the Earth and apparently this filled him with inspiration that changed the world to larger degree than did all the teachings of Jesus Christ. I am not being sacrilegious but I am merely stating the honest truth. While everyone is disputing the religion of another person or the convictions of others there is only one that who looks on this second apple incident with a high degree of scepticism and that is me. There are on any specific second on Earth more persons preaching the physics of Newton that spawned from this vision Newton received than there are persons believing concepts about the teaching the Gospel of Jesus Christ. The connection I see is that Jesus Christ is accepted as the new Adam where the old Adam failed his test by the incident involving the apple. Again I know the case of it being an apple or not being an apple is highly disputed but I shall return to this part since in this lies a

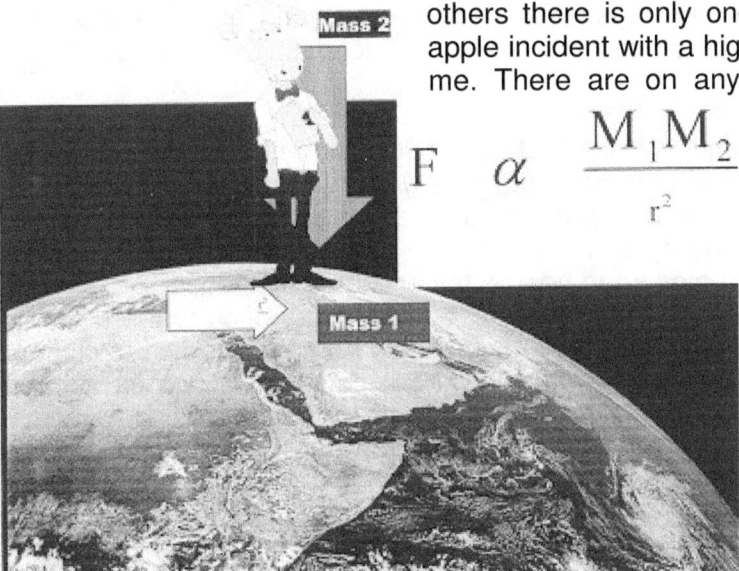

$$F \; \alpha \; \frac{M_1 M_2}{r^2}$$

large degree of the malice in physics.

After seeing an apple fall from an apple tree Newton was inspired. Newton was overcome by the magnificence of a Universe that opened the door to all physics at that moment. Newton saw an apple fall. What a divine revelation this turned out to be. This revelation equalled any revelation of Biblical proportions and then even topped any of those revelations coming from the Bible. This genius of a man hiding in the body of a young student in England going by the name of Isaac Newton was apparently sitting near or under an apple tree when he observed the now so famous apple falling from the tree. Seeing the apple land the student then made a brief calculation in relation to what he observed as a young man that saw an apple fall from a tree. He observed as a student (giving him a very special and rare quality to be able to be a student and to be able to observe with both abilities connected to the same individual) that from the observation he made the presumption that the apple fell by the weight that which the apple had when the apple was on the ground. Young Isaac thought that the weight that which the apple had was responsible for having the apple falling. This sparked a cosmic connection never anticipated before. Isaac Newton was so inspired by an apple falling from an apple tree that from this a science was born.

Newton saw an apple hanging suspended from a tree. There was a distance between the apple and the earth and this distance he named the radius. Then he gave the earth mass, as he gave the apple mass. This he then subsequently formulated and from this formula everything connected to physics was born.

In his formula he envisaged $F = \dfrac{r^2}{M_1 M_2}$ he carried one part of the square $\dfrac{r^2}{}$ of the travelling distance he named as the radius $\dfrac{r}{}$, which he devoted to the earth $\dfrac{r}{M_1}$ leaving the other part of this radius $\dfrac{r}{}$, which is the falling distance $\dfrac{r^2}{}$, to that of the apple $\dfrac{r}{M_2}$ when the apple had to travel to fall to the earth. He saw the mass of the apple was pulling the mass of the Earth along the radius as much as the mass of the Earth was pulling the mass of the apple along the same radius, and instantaneously the formula $F = \dfrac{r^2}{M_1 M_2}$ became instant religiosity. This brainwave was the wave that shook the earth from its cradle of slumber in mindlessness. Isaac Newton was a Superstar before there was a superstar! He found the force containing the Universe into a unit. It was an all-time breakthrough with no other parallel ever.

However, one can never keep a genius satisfied with one conclusion at a time. Because this was Isaac Newton and Isaac Newton was very clever it went further, much further when Isaac Newton saw the moon as the apple and the earth as the earth. He no longer saw an apple but saw cosmic structures forming their final destiny that drew to the last Universal conclusion. He saw the apple as the moon and he saw the earth, well I suppose as the earth and although the moon was much bigger than the apple it was also much further from the earth than what the falling apple was. ...But according to this he then put his gravitational principle in place where he declared that between the apple and the earth a force called gravity is in place. He gave the apple a value in mass and he gave the earth a value in mass. Then he gave the mass the responsibility to create gravity and he gave gravity the responsibility to pull by the force the mass initiates. ...And there are some educated person's who hold the opinion that I don't "understand Newton" or his physics principles...

The mass of the apple was pulling the earth but the mass the earth has made it impractical to expect that the apple would be able to pull the mass of the earth towards the apple.

The earth also pulled on the apple by the force the mass of the earth would enlist. The earth is pulling the apple with all of the mass the earth could apply while the apple was pulling the earth with all the mass the apple had about. By measure of mass discrepancy all the falling went the way of the apple. This then brought about a pulling contest and the end result of the pulling is the falling of the apple and this enormous event of the earth and the apple pulling one another in the tug of war happened in front of his eyes. He became the eyewitness of an absolute enormous cosmic event and Newton furthermore was as brilliant in mind as to give this the deserving Universal implications, but that is the mind that Newton had!

In the vision of Isaac Newton he interpreted what he saw as the Earth dragging the apple down by the weight, which the apple pulled the Earth closer. In this action he envisaged a force at work. He apparently later named the force gravity and I presume that being a very religious man unlike modern Newtonians he saw this force as being something that was there to drag one into the grave hence the grave connection in gravity but about this part I can be mistaken but that is a presumption I now make and is not necessarily the truth as his presumption about a force coming about by the influence of mass was not necessarily the truth at the time.

Newton saw the Earth rise to the occasion by meeting the apple halfway thus the apple took on the first part of the square of the radius leaving the other part of the square of the radius by which the apple travelled to the Earth travelling to the apple. However, the mass of the earth is rather substantial placed in regard to the apple and this reflected all movement down to the apple travelling the entire square distance.

The formula depicted a force that dissolved space by the power of mass coming from two ends that was drawing towards a common centre.

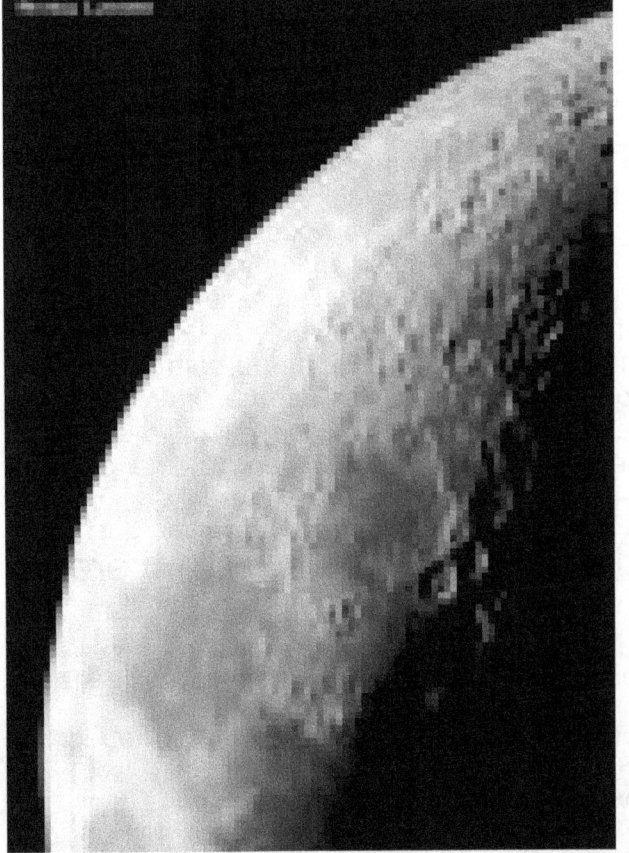

$F = \dfrac{r^2}{M_1 M_2}$ The mass of the two objects destroys the radius between the objects and to top the lot they finally after so many centuries discovered a force. Everyone went ballistic, proclaiming him as an instant genius, the one the world was waiting for after the crucifixion event. If there were TV in those days he would have been the Newsperson of the day. He made a presumption about mass as I made a presumption about what his motive was in choosing the name gravity. His presumption was as little tested as what my presumption is tested. He got more famous than Jesus Christ and Mohammed put together. There are more people believing Newton than there are people believing in the preaching of either Jesus Christ or Mohammed. Of all persons on earth there is one person on earth who doubts Newton while there are billions of those who disbelieve in either Jesus Christ or Mohammed and this is one of the biggest facts and this indicates the biggest farce ever created by any man ever. This opened up the road that other going around by advocating bullshit such as Darwin could follow. Those pretending to fall from your chair about my remarks concerning Darwin, tell me what did Darwin ever prove? Name one fact that he introduced with substantiated and un-denounced facts. What did Darwin conclusively prove in his book that did not rely on exaggerated inconclusive ideas about a collection of deductions that in hundred and fifty years could never conclusively be linked by a timeline and proven to be inconclusively correct? Darwin must be part of the third or fourth biggest science scam to serve a destructive purpose ever launched by man under the pretext that it forms conclusive science in all its aspects...back to Newton.

Newton saw the apple drawing to the earth and if the apple drew to the earth the moon also then was drawing to the earth just like the earth was drawing to the moon. If you believe physics you have to believe this and if you believe in physics, this moon coming to the earth is your destiny waiting for you. If you see physics as Newton does, you better look at the moon with much less romantically connected visions.

If we go one step further and realise what Newton realised at the time we find that we don't have an apple hanging from a branch but we have a moon filling the sky, or filling a part of the sky. The principle does not change. In the one event there is an apple and in the next we have a moon. The apple is much smaller than the bigger moon but then the apple is much closer to the earth than the moon, which is much further away from the earth. The earth is the earth that remains the earth in both cases.

Therefore the scenario remains identical in both cases notwithstanding many discrepancies. As the apple falls so the moon falls but the moon is a lot further and would therefore take a longer time to fall to the earth. However, by investigating both the scenarios no person who knows the least about science would dispute that if Newton's pulling of gravity by the force of mass were true (and nobody in science seems to disputes this) the moon is falling although the moon is taking a little longer to fall all the way, for as sure as the apple was falling, in that we can be assured that the moon is also falling. When you look outside and you see anything falling you can be as assured that the moon is falling on you and the moon is coming as sure as rain or anything else is coming down to the earth.

According to Newtonian science everything is falling because everything has mass and where everything has mass, in that everything is falling by the value of that mass. Go on and allow something to fall and what you will notice is that when you see whatever falling, it actually is the moon falling except, the moon has a lot more mass and therefore the moon is falling a lot more than what a measly apple would fall. The moon is coming whenever you allow an apple to fall or not allow an apple to fall. The moon will contribute all the mass it can contribute and the earth will contribute all the mass it can contribute and both objects are coming towards each other but more the moon to the earth than the earth to the moon...and like it or not the moon is coming all the same! This is Newton's science legacy to us. But in this idea is hidden a lot more doom than just envisaging any object meeting another object!

$$F = G \ \frac{\text{Mass of the earth x Mass of the moon}}{\text{radius between the two destroyed by the square thereof}}$$

With the formula as Newton introduced it being $F = G \ \frac{M_1 M_2}{r^2}$ any one with a little insight into reality could see that as the bottom part decreases (become smaller) this will affect the top part to become larger. I will show three values and you will see what I mean. $\frac{100}{100} = 1$ and $\frac{100}{1} = 100$ and then getting the bottom really small $\frac{100}{.001} = 100000$. The bottom part's decrease increases the speed of the top part or the gravitational movement. By reducing the distance the force of gravity will grow exponentially faster!

I show you this to indicate that the closer the moon gets to the earth the faster it will come to the earth. At the last minute the moon will travel about say one hundred million kilometres because the gravity pull will increase by a billion trillion times per second. During the last day's coming closer to the earth by radius reduction it will be as quick as the last million years we now have. The rushing moon will increase in speed by the square of the disappearing radius and we sit between the two as a human sandwich.

Go to any of the doomsday scenario websites and see what they predict when a comet hits the earth. It is going to be final. They really scare the shit out of those who believe easily in

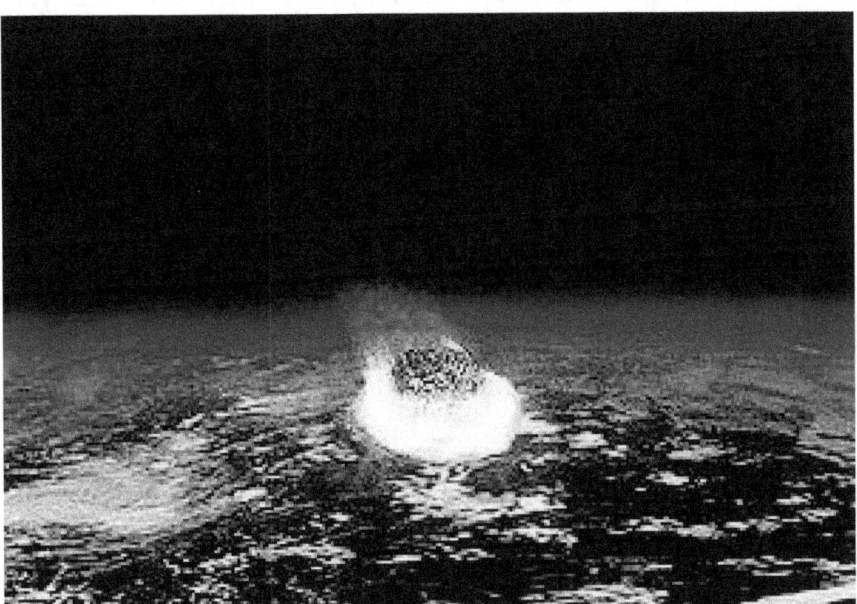

frightening tales. A comet hitting the earth is going to change whatever we think was the earth...and this is only when a comet hits the earth according to Newtonian science. Think what the mass would be when a comet hits the earth and compare that mass to what the mass of the earth will be when the moon hits the earth. Go and visit the web and see on these Armageddon websites how they paint a picture of doom and gloom about all sorts of destructive events bringing a conclusion to the earth. These Armageddon websites tell stories of comets we have to wait for and that might come. See if you can find one website forecasting the moon arriving on your doorstep and this is not something that might or might not be. This event is in the future as sure as Newton's principles are true. We have Newton's formula $F = \dfrac{r^2}{M_1 M_2}$ that then by deduction

became $F = G\,\dfrac{M_1 M_2}{r^2}$.

The formula $F = G\,\dfrac{M_1 M_2}{r^2}$ is science as taught by any person who knows anything about physics.

The symbol F represents the gravity force by which the moon is coming towards the earth or visa versa.

The symbol G represents the composition of what space holds that is between the moon and the earth, which forms the stuff keeping the moon and the earth apart.

The symbol $\dfrac{M_1}{r}$ represents the force by which the mass forms the gravity that the earth creates in relation to the distance $\dfrac{1}{r}$ there is between the moon and the earth.

The symbol $\dfrac{M_2}{r}$ represents the force by which the mass forms the gravity that the moon creates in relation to the distance $\dfrac{1}{r}$ there is between the moon and the earth.

The mass is in place. We have the formula given by no person less than Newton. We know the distance…and can you find any website predicating this forthcoming collision? No, you can't but the conspiracy is in the reasons why you can't find one such a website. Why do we have so many scary warnings but nothing mentioned about what is truly part of our future, the moon coming for a visit!

This $F = G\,\dfrac{M_1 M_2}{r^2}$ is the tempo whereby the distance of the radius between the earth and the moon will shrink. Then these numbers still have to be divided by the square of the reducing radius making the result incomprehensibly more devastating. The smaller the radius becomes the faster the radius would become smaller and the more the tempo would be of the moon closing in on the earth. During the one-year everything might seem perfectly under control and the next year everything will seem to go mad in haste. The moon will seem to be on a runaway collision course going out of bounds the closer it gets. Remember the formula showing the closer the moon gets the quicker the rush of colliding will become as in $\dfrac{100}{100} = 1$ and $\dfrac{100}{1} = 100$ and then getting the bottom really small $\dfrac{100}{.001} = 100000$.

The devastation it brings will increase exponentially and become so big no one would be able to realise the changes as the event of the final doomsday is increasing! No one would be able to keep track of the fatalities occurring. Everybody who died during the First and the Second World War plus the epidemic influenza that came after the First World War would be deaths happening in one day and during every day.

No humans were around to make permanent records of prehistoric hurricanes. These so called hypercanes that were prevailing in the prehistoric days will come back and ruin the planet again, but this time it will be an everyday occurrence. Think of how conditions may have existed about 65 million years ago that could have spawned prehistoric hypercanes that is far more powerful than modern storms. The doomsday theorists do not have the imagination to sketch reality when the moon is closing in on the earth! Whatever they think is bad this is far worse. When the moon is only a few thousand kilometres away, how much will that influence the weather pattern applying?

Super Typhoon Tip

Tropical Cyclone Tracy

In the era we now enjoy a hypercane might be thought of as a hypothetical extreme tropical cyclone, but in reality when the moon comes closer this will become a multiple seasonal event. In order to form a hyper cane, ocean waters would need to be much warmer than they are currently. This would become a reality when the gravity of the moon brings on waves thousand meters high. Hypercanes could form if ocean temperatures reached approximately 122 degrees Fahrenheit; this is 15 degrees higher than the warmest ocean temperature ever recorded! When the Moon is half as far away as it is now the influence it must have on water temperatures in the oceans will increase this to the colder end of the ocean temperature scale. Science at present acknowledge that such a hypercane can result from any large asteroid or comet slamming into our planet, a volcanic or super-volcanic eruption, or extensive global warming could all contribute to ocean waters reaching such outlandish temperatures. These conditions dwarf in reality when we consider the moon coming closer towards the earth. Some scientists speculate that the dinosaurs were killed off by a series of hypercanes caused by an asteroid hitting Earth. Hypercanes would have incredibly long life spans due to central pressure below 700 mille bars and wind speeds in excess of 500 miles per hour.

This is the condition within a Hypercane. Hypercanes would have wind speeds of over 800 kilometres per hour (500 mph) and would also have a central pressure of less than 70 kilopascals (21 in Hg), giving them an enormous lifespan. For comparison, the largest and most intense storm on record was 1979's Typhoon Tip, with a wind speed of over 300 kilometres per hour (190 mph) and central pressure of 87 kilopascals (26 in Hg). The extreme conditions needed to create a hypercane could conceivably produce a system up to the size of North America (compare image of Typhoon Tip's size at the top), creating storm surges of 18 metres (59 ft) and an eye nearly 300 kilometres (190 mi) across. The waters could remain hot enough for weeks, allowing more hypercanes to be formed. A hypercane's clouds would reach 30 kilometres (19 mi) into the stratosphere. Such an intense storm would also damage the earth's ozone. Water molecules in the stratosphere would react with ozone to accelerate decay into O_2 and reduce absorption of ultraviolet light. Other scientists have theorized that the system, compared to a normal hurricane, would be considerably smaller, about 10 miles in diameter. This would be more comparable to a tornado, which has been recorded at up to about 2.5 miles. And when the moon is on us this will be everyday weather changes where the real frightful ones will be hundreds of times worse.

Has such a storm ever formed? No. Will such a storm ever form? ...probably not. Here is why:
Storms of this magnitude rarely merge, and if they were to join up, the result would actually be a weaker system. When two tropical cyclones merge, air circulations become disrupted, and as a result, wind velocities are lessened. Over time, the storm could re-intensify and it would certainly be larger in area than the two storms it was formed from, but it would get nowhere near hypercane status. Some meteorologist for the National Weather Service in St.

Louis, Missouri, describes the situation as follows: "To use an analogy, these storms are similar to magnets of the same polarity...they tend to repel each other rather than attract. When they approach too close, usually one of them will weaken dramatically, due to the other one taking most of the energy necessary to sustain the hurricane force winds." Therefore, hypercanes are HIGHLY unlikely because two storms rarely collide over open water. But with the moon filling most of the night sky things are going to change dramatically.

Scientists have for a long time thought that the dinosaurs may have died after an asteroid struck the Earth and caused dramatic climate changes.

There are other researchers who think any asteroid could have heated the ancient oceans to as much as 50 degrees Celsius (about 120 degrees Fahrenheit). This is the research that is going into speculative scenarios where the moon hitting the earth is as definite science as Newton is and no one in science even thinks about this scenario while Newton's gravitational laws are in place? Can you think why not?

Those who should know better says when the sun runs out of its current fuel, the hydrogen starts to cool down, as it will expand into a red giant and could expand out to where Jupiter is circling or so they say it is going to be. These are the estimates I have seen quoted before and are not mine to quote. The end of the sun is billions of years away. It is a lot of hogwash but it shows someone somewhere got interested and started some form of thinking about how the sun and the solar system would end. Yet no one in science ever thought it worthwhile to conduct a study into when the earth and the moon will go into destruction.

Science goes even much further. They sequestered Albert Einstein to measure all the mass in the entire Universe to find out when will the Universe start contracting and come to the end of its life cycle in accordance to Newton's gravitational pulling principles. There are no limits or ends to which they will not go to find the end or bring finality to whatever they try to establish. To measure the end of the Universe is going much further than to establish when the earth and the moon will meet their gravitational destiny, with Newton and his theory of mass pulling mass so fundamentally proven. But there is no end to their resolution for they never stop with their inquest. When the Critical Density Investigation did not deliver the results that would bring satisfaction, they went in search of Dark Matter. I am going to go much deeper into the Critical Density and the search for Dark Matter later on in the book.

Think of what effort it took and still takes to find the reason why the earth is expanding and what human effort went into that. They are relentless in their quest to find answers. Yet, when it comes to the time when the moon and the earth will finally collide we find only silence. It is the sort of silence I describe when I pointed out what to look for when one goes in search of a conspiracy. I find not even a whisper…

On earth nothing is kept to rest in peace. They run probes from the Artic ice fields all the way down to Antarctica. The search for information never seizes and the lust they have for knowledge never ends, except when we get to the issue of the moon colliding with the earth. In that matter the thought is never even realised and the question serving the silence is: "Why not and why never a word about this matter?" Is nobody at least interested what will occur the final days when the moon meets the earth in a full moon?

If our Universe comes to an end it will start with the moon and the earth coming together and when these two comes together it does not matter how the rest will come together because then all our interest in the Universes would be permanently be suspended in any case because the earth represents our Universe entirely as the entirety there is. Losing the earth becomes the same as ending the Universe.

What people don't realise is our Universe is this earth. Without having the earth we have no Universe because then as far as we know there is no other life left in the Universe. It will be much better to stop speculating about the possibility of other life in the Universe and try to save our only Universe, which is the earth we so much try to destroy in the name of making money and showing profits.

Researchers have probes to detect tsunamis that will only horribly devastate regions, as it is it can't compare to the moon hitting the earth. There will be waves every day generated by the gravitational increase of the nearing moon that you now would associate with the worst tsunami occurrences, which would then become the mild everyday waves found on the beach. The sea will become a wild devastating machine that pulls all land into the sea with waves going as high as a thousand meters and much more. This is the scenario waiting on us holding life and no research is going into trying to predict when this is going to happen. They research tsunamis in detail but no one place on earth could be considered as safe when the moon lands on earth. Such waves will draw mountain ranges into the sea and destroy giant areas into water-wastelands. Think of the crushing force of one such a wave and then the moon is only still coming our way. Long before humans would feel the demise there will be no human left to experience the demise. That means that the sooner we come to terms with what awaits us the more time we have...

What information did the Rich and the Powerful gain from these probes on the moon that did not filter down to us? When will the moon contract the earth finally...it is the most obvious point that needs researching because losing the earth becomes the same as ending the Universe and as far as anyone can see the research into this collision is zero, now why would that be if the obvious becomes very obvious and then someone has something to hide?

At first all the additional moon voyaging launches were halted. The reasons for the sudden lack in interest were numerous but the most believable was that the costs involved were truly astronomical. America could not go to the moon and fight a war in the east and a war in the east was raking in much more money than did the trips going to the moon. So the moon expeditions were halted in favour of a Vietnam War. If one takes the millions of tons of bombing material and the production costs going into the manufacturing of this alone the profits will outweigh the money to be made of a trip to the moon many times over. So the war won on the profit margin it could generate alone. Then if one takes into account the machinery that was wasted and the purchasing power of the replacement needs for such machinery a trip to the moon stood no financial chance when compared to the profits gained by the wealthy in the war effort in the east.

After the seventies no one was interested in the moon because "*We*" were "*there*". The later expeditions then involved probes to other planets and research went about finding out what was going on, on the other planets. They found that Venus was too hot to handle and so was Mercury and the two was left alone. Then normal interest came into the conditions on Mars. Photos came back from the other nearby planets that was studied but never with the sudden frenzied state in which the development got a surge as that later came about. Then the moon was suddenly left alone. No one said a word about the moon. In this period Computers came into play and producing chips in a "gravity zero" state was generating the money. Money brings about interest. Everyone making money had to make computer chips and this then became the main focus of many. If there is money the banks will have interests and if banks show interest they command the politicians to show interest and if the politicians have interest the job is done without questioning validity. The money that this made superseded everything else and with the money to be made it pushed everything else aside. Then from this point on it was about laboratories being manned and worked within conditions only suitable in outer space and the research of "gravity zero". This was a cover to get the public to pay for research by their tax money going for the research that only the rich would financially benefit from. The taxpayer, who is the poor paid for the scheme whereby the rich got the benefits to get richer. The selling of the products developed in outer space was financed by the public and therefore the gain of wealth by this research belongs to the public and the earnings should go to public funding. The rich should be taxed for the royalties that belong to the public. Yet, we have stinking rich benefiting from this research by getting more rich than they were and moreover without paying equal tax as the poor does and behind this the bankers swell their coffers where the money the banks have is almost limitless.

This pushed the interest in the moon completely into fading away. There is a lull in any interest from all sides about the moon until a certain point. Suddenly again out of the blue there came a rush in the George Bush era where the moon became a place to go to in order to launch men to Mars. All of a sudden Mars had to be explored and the building of a halfway station on the moon was on the cards. The question never asked is why is there the shift in interest all of the sudden? Who gains from this interest?

Moreover would be why would the Rich and the Powerful have the sudden interest in Mars? Think about it this way, if the Rich and the Powerful could take all of the tax money of all the lesser human beings in status and start to build on Mars under the disguise of research, then when the doomsday arrives those being the fortunate, the powerful, the influential and the law makers will have somewhere to go to escape, leaving the less worthy to take the moon head on! In the sixties there was an urge to get to the moon and when they got there then afterwards the interest in the moon fell to an all time low. Then those in investigating investigated all the planets without showing the least interest in what the moon might hold. This trend went on for decades. Then suddenly from the blue comes some hysteric research about Mars, just after someone discovered possible prehistoric semblance of life on Mars. Why would everybody on earth now be dreaming about life on Mars...and why would all of man suddenly benefit from exploring possibilities of life extending to Mars? This sudden research of life on Mars is not about finding life on Mars but it is about making sure there is no life to be found on Mars because finding life on Mars will end all further explorations. If there was a hint of life on Mars it will stop all further research about living conditions on Mars. The risk of finding bacteria or a virus on Mars would mean stopping all further research on Mars because of contamination and infections caused by the Mars micro biology and the chance it will wipe out all forms of life on Earth. However, there is a much bigger chance that life never developed on Mars and that means the Rich and the Powerful can go there. But first they have to scare many less important beings coming from earth.

In 2004, the then President of the United States of America George W. Bush in a speech gave NASA an order to develop towards the direction to continue to progress where this will take them back to the Moon and from there to Mars. Why would there be this sudden change in strategy. Why would the interest suddenly flame up in going to the moon? If that's the plan what information blew the interest to new highs? The reason they give is as diverting as any reason coming from science as science says it is because it's a lot easier to go to the moon than to accomplish a Mars landing and build a station base there. But

 thinking of the answer in such terms is diverting from the true question. Why suddenly go and build space stations to begin with when so much time lapsed since the early seventies when the interest in the moon faded to zero. Thinking that Mars is far and the moon is much closer is a way of not answering the question at all. Yes, the demands of a trip to Mars will be costly, much more costly than the space station that was already too costly to complete by one country alone but the least expensive would be to go on to ignore the entire project of building on the moon or on Mars. Building by itself as a project will be much more demanding than those in far earlier stages of our species' exploratory history. Building a station instead of collecting rocks like they did in the seventies is calling for a much more evolved effort. So, why the big effort this time around? Numerous challenges will face us along the way, but three of the most serious are logistics (in terms of both fuel and crew consumables), mass, and crew interaction.

Why would this rush be, as we want to go to the Moon? Why go to the moon in the first

place? Because the Moon is an ideal "staging post" for us to accumulate materials and manpower outside of the Earth's deep gravitational well. From the Moon we can send missions into deep space and ferry colonists to Mars. That is the main reason, the sudden urgency to go to Mars. But to give Polly its cracker there has to be something in for John and Jane Dow too, so they throw in the idea that tourists may also be interested in a short visit. Mining companies will no doubt want to set up camp there. The pursuit of science is also a major draw and in that they have to get all people back home to believe in. For whatever reason, the people would hang on to or believe in, say to maintain a presence on a small dusty satellite would not get the work done to establish a new occupying colony on Mars so we will need to build a Moon base. Be it for the short-term or long-term, man will need to colonize the Moon in order to colonise mars. But why would we live there? How could we survive on such a hostile landscape as Mars? This is where structural engineers will step in, to design, and build, the most extreme habitats ever conceived...and who is supplying the funding, the suckers they will leave behind on earth to die in the earth-moon collision that is coming as sure as Newton is correct. They have to start a living presence on Mars to avoid the destruction on earth!

Manned missions to Mars has to be well funded as it will take up a lot of the limelight insofar as colonization efforts are concerned and why would no one ever tell the true reason for this venture? We must use our initiative to focus our minds where we aim at the ongoing and established concepts for colonization of the Moon. Why the rush when we currently have no exact means of getting on Mars in the first place and why after all the time that lapsed since it is nearly 40 years ago when the interest went dim since Apollo 11. The technology is sufficiently advanced to sustain life in space as we saw from living in capsules above the earth so the next step is to begin building a life for future inhabitants on Mars. But first is the present instalment that concerns building a Moon Base. What are the pressing issues behind the immediate urge to inspire engineers when planning habitats on a Mars landscape? Only the Rich and the Powerful have the authority to authorise such a venture.

The debate that still rages about whether humans should settle on the Moon or Mars first seems like an eye blinder. With Mars being the ultimate destiny for mankind as to live on a planet other than Earth why would they bother with the moon when the moon is going to be destroyed in the same event as the earth? But they need an eye blind as not to wake suspicion about the true nature of the venture. But it is also much easier to get people to buy into the scheme and tell the story that it would be possible for all people to one day look down on earth and see earth during cloudless nights. People will fall for this idea making the earth a bright and attainable view from the moon. From there people will see the details of the earth as we now see the lunar landscape with the naked eye, because the earth would be just as astronomically close when compared with the planets and therefore being that close people will buy into the idea. Give as many as possible the wish that the Moon should be the first port of call before humans begin the six months (at best) voyage to the Red Planet. It also helps as we've already been there... We will never live on the moon but leaving the idea can only be beneficial to the launch of the project.

From the internet one can see information is given about the general opinion that has shifted somewhat in recent years from the going to Mars directly plan that was on the table in the mid-1990s to the now favoured going to the Moon first idea, and this shift has recently been highlighted by US President George W. Bush when in 2004 he set out plans for re-establishing a presence on the Moon before we can begin planning for Mars. That it is US President George W. Bush who made the promise has little surprise because he started an oil war on behalf of the oil goons and then bankers that ill benefit from such a senseless war in terms of finance it makes sense; many human physiological issues remain to be identified, plus the technology for colonization can only be tested to its full extent when... well...

colonizing. This is the eye blinding part. The question is why the sudden rushing where the moon has to become the halfway house?

Firstly the human canon fodder that is lost in war can be sent to the moon under the disguise of keeping peace on earth. Then it will be a normal trend if some are lost in the task given to them to learn the very vital understanding as to how the human body will adapt to life in the low-gravity environment on Mars but this could also be learned much closer on the moon. The colonisers made up of the disposable part of society in the ranks of humans will have to endure a very new way of locomotion on Mars. It is not that common as to how earth technologies will perform on any location forming the environment on Mars or the moon. Therefore learning this on a surface much closer to earth will be beneficial because working on the handicaps brought about by the project will become easy enough to alter and adjust as it is much closer to relate to persons back home on earth. The first colonisers of Mars will be too far away to deal with the first problems that arise during the start of the venture. Then as progress smoothens out further development it will become assuring, as a new lifestyle will become feasible by the colonists and astronauts. Launching first from the moon and later from Mars will also seem sensible to those brainless taxpayers who pay for everything but eventually will perish in the moon-earth collision. Exploring the space and living conditions on Mars is dangerous enough, so therefore minimizing the risk of mission by first developing some new moon base makes failure less of a certainty than going straight to mars. Having as little failure or covering up most failure will be critical to the future of manned exploration of the Solar System.

So if the moon is the favourite place to start building out of the earth projects where do you start when designing a moon base? The structural engineers would put a premium on the quality building materials may ask that when exposed to a vacuum the enduring quality must be at the biggest premium. Damages may result from severe temperature variations, high velocity micrometeorite impacts because there is almost no atmosphere to give protection, high outward forces from pressurized habitats because of the artificial atmosphere within the structures to serve the needs of life within, material brittleness at very low temperatures as zero is a daily occurrence and cumulative abrasion by high energy cosmic rays and solar wind particles will all factor highly in the planning phase. Once all the hazards are outlined, work can begin on the structures themselves but it has to be synchronised to correspond with Mars.

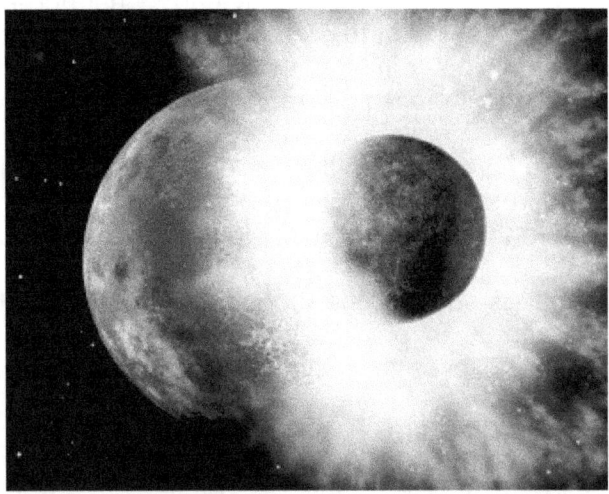

The actual construction of a base will be very difficult in itself. Obviously, the low-G environment poses some difficulty to construction workers to get around, but the lack of an atmosphere would prove very damaging. Without the buffering of air around drilling tools, dynamic friction will be amplified during drilling tasks, generating huge amounts of heat. Drill bits and rock will fuse, hindering progress. Should demolition tasks need to be carried out, explosions in a vacuum would create countless high velocity missiles tearing through anything in their path, with no atmosphere to slow them down. (You wouldn't want to be eating dinner in an inflatable habitat during mining activities should a rock fragment be flying your way...) Also, the ejected dust would obscure everything and settle, statically, on machinery and contaminate everything. Decontamination via air locks will not be efficient enough to remove all the dust from spacesuits; Moon dust would be ingested and breathed in – a health risk we will not fully comprehend until we are there.

With all this haste to go to Mars and set up some shop there that could accommodate those that could afford to begin a new existence on Mars, how long is it until the moon finally destroys the earth by destroying its own structure? No research tell how long does Newton's formula $F = G\ \dfrac{M_1 M_2}{r^2}$ give the rest of us to multiply our genes before it is wiped out by the oncoming collision that inevitably has to end all life form on earth according to Newtonian Principles applying in physics as physics.

The Definition of Gravity according to the Oxford Dictionary of Astronomy:
<u>**Firstly it is said in physics that in Physics: The natural force of attraction exerted by a celestial body, such as Earth, upon objects at or near its surface, tending to draw them toward the centre of the body. With this being Newton's accepted science principal why then is what I said in this first part about the moon colliding with the earth total rubbish?**</u>

<u>**The conspiracy is in the silence of... It is not in hiding or not telling us how much the moon is coming closer but it is in the "not coming closer" that the conspiracy hides the truth.**</u>
If the **Fundamental of Science** according to Newton is **correct** then there is only one way the earth could find its final destiny, it is when **the moon** will hit the **earth** and with doing that **destroy all** forms of **life** in the collision. According to Newton $F = G\ \dfrac{M_1 M_2}{r^2}$ the moon and the Earth are in a tug of war and the two are pulling each other closer. That is according to Newton's formula.

Fundamental to Science is that the mass of bodies attract. Then there is one question more important than all other science research put together.
When will the moon and the earth destruct in a collision where the mass destroyed the distance that held the two solar bodies apart? The two have been pulling for billions of years and should be coming to the period of intimate connecting by collision.
It is said that mass pulls mass.

Newton came up with this novel idea about a little more than three hundred years ago. What this idea says is that every object has mass and each object pulls all other objects by the value of the mass that every individual object has.

Teachings say that the moon is pulling the earth as much as the earth is pulling the moon. All teachers in physics in learning institutions teach students this and by culture every student believes this without condition or question. If I say this is rubbish no one believes me.

This concept is so engraved into our culture we have to accept this without question to a point that when I say this is not happening, it then is me who has to deliver the proof of this concept being total void of truth and is completely incorrect.

Newton was never tested before and I prove that, yet I have to prove Newton is incorrect!

Everything I speculated on about the moon and the earth coming into some collision sometime in the future is a lot of rubbish just because the definition of Newton that says <u>in physics that in Physics gravity is the natural force of attraction exerted by a celestial body, such as Earth, upon objects at or near its surface, tending to draw them toward the centre of the body. With this being Newton's accepted science principal why then is what I said in this first part about the moon colliding with the earth total rubbish?</u>

<u>The Moon is going away from the earth and the earth is getting bigger while the earth is slowing down in rotation velocity as the earth is moving further away from the sun.</u>

<u>The Practice of physics propagating Newton's principles is Brainwashing and Mind Control of students in Physics!</u>

Let me show you what The Practice of Brainwashing and Mind Control in Physics are.

If the Fundamental of Science according to Newton is correct then there is only one way the earth could find its final destiny, it is when the moon will hit the earth and with doing that destroy all forms of life in the collision. According to Newton $F = G\ \dfrac{M_1 M_2}{r^2}$ the moon and the Earth are in a tug of war and the two are pulling each other closer. That is according to Newton's formula.

Newton was never tested before and I prove that, yet I have to prove Newton is incorrect! I say gravity is the part that does not connect to mass and is the factor producing movement that brings on the intention of motion of all objects to carry on moving downwards, notwithstanding the blocking action or mass, which comes from intervention of the earth occupying space that stops further descending. Mass is not connected to gravity in the way as initiating gravity since mass comes in place when it stops the motion of moving downwards of the falling body.

When the body stops moving down further, it is there where the concept of having mass starts. Mass is achieved by stopping and preventing further downwards motion by having a solid object fill the space with material that will intervene further movement of descent to a centre of the body having gravity and therefore gravity is in the performing of the descending motion. Then my Super-educated-Intellectuals feel threatened when I confront them by bursting this super intellectual demeanour they hide behind with a thought through idea of what gravity is in relation to how mass forms.

On the moon I have a minimum mass (1/3 of what I have on earth), which must result in minimum gravity (again 1/3 of what I have on earth) if mass is what produces gravity. That means in space I must have no mass that produces micro gravity. They say mass is doing the pulling but then when there is no pulling it must mean there is no mass doing the "no-pulling". That is one part of the ongoing science conspiracy, changing the rules ongoing and never gets to a point of understanding. If I have micro gravity I then must have no mass! Gravity is in turning the earth and this thrust downwards ends in the body having mass when the thrust downwards can't move the body down any more but the intention of movement remains.

It is not you being glued or not being glued to the Earth that I discard when I question the mass idea. Mass forming weight is indisputable and that the conspirers called physicists know. It is the definition holding this whole idea about mass forming a force of a pulling magnitude that I do not share in the least. What the definition of a pulling force describes is magnets pulling onto iron and is the total opposite of what I experience when I have mass. If it was produced by the formula Newtonians use, it will be the breaking of the first millimetre of the gravity clampdown that would be the hardest to achieve whereas this first millimetre of lift is the easiest part and not the most difficult. The difficulty increases as the radius grows and not as the radius decreases.

When I say there is no mass which produces gravity everyone thinks I say we all are going to fall off the Earth and with me thinking that way then it is obvious that I must be a nut. Everyone thinks of me as the clown acting mad when I say mass is not to be found in the cosmos. I do not say we are not standing on the Earth where we obviously have mass; I say Jupiter and the other planets can't have mass because I can't find mass playing any role. I do not say there is nothing that is keeping me glued to the earth. I say there is no attraction between two bodies by the force of the mass that in such doing is diminishing the radius parting the bodies by the inverse square law. I say there is a connection by motion between the centre of the body and the material surrounding the centre. This is what I say when I say there is no gravity produced by the pulling power of a fictional force that mass would create.

The Universe has mass that is pulling mass towards one another and we are in the centre of an ever shrinking Universe. Then along came a man that had a good look at the Universe. The lot was growing apart. The Universe was growing by miles and not shrinking into nothing.

The man was E.P. Hubble. The world was expanding and not contracting which made the Universe quite wrong. Newton could never be wrong because Newton was never wrong yet...so if the Universe is out of step with science, then science will correct such an abnormality by finding a way to defraud science and postpone the correcting that the Universe had to comply with since the Universe owed the Master Newton some apology. Did the Universe not know that he whom never can be wrong is in name Isaac Newton! Decisive action was needed. When will the Universe confirm its incorrectness by affirming Newton's obvious correctness? When will the cosmos come clean and prove Newton correct.

Looking at the astronaut float in space we believe there is a force of attraction between the mass of the earth and the mass of the person and this force of attraction is gravity. The mass must be present if Newton's principles apply. The man must have mass and the earth must have mass but it is without falling, so where is the pulling "force" of the mass in any of the above pictures?

This principle of gravity declares that a force of attraction by the value of mass of both solar objects is pulling to the value of mass in order to reduce the radius, therefore bringing the moon and the earth closer by reducing the distance between the two bodies by the square thereof. The body of the earth holds mass that pulls the mass of the body of the astronaut down by the force called gravity. This idea is not science but it is culture and is drummed into our minds like it is one of our living senses we have to adhere too.

This idea, being culture and not science is also the biggest lot of hogwash ever invented by man to scare the mindless many forming the greater thoughtless populous on earth. If this was the case, then the result of this must lead to an inevitable collision that must end the existence of both the earth as well as the moon. Let them and those with the ability to design space worms and space whirls and Black Holes now calculate when is this inevitable event is due and inform the human race when our final day of doom will arrive by using Newton's law of attraction or $F = G\dfrac{M_1 M_2}{r^2}$. This collision will end all form of life in the known Universe, it is that important to know. We have to know the day. However, the good news is that the **moon coming towards the earth** is **not** the conspiracy under investigation but **what is the conspiracy** is the fact **that the moon** is **not coming closer to the earth**.

Now at this point, up to now we have speculated about how a science conspiracy is not presented, let's see how a science conspiracy is conducted. Now it is most revealing but very little publicised that there will be no moon / earth coming together. It is spoken of but not a lot and when it is spoken of it is told in a matter-of-fact-hushed-up sort of way. Has any reader seen this as front-page news lately? Big banners reading: *the moon is departing and not coming closer and therefore Newton is absolutely wrong and physicist has no*

idea about physics. No, it is very slightly and even more scarcely spoken of. That confirms a conspiracy to hide something. However, there are conspiracies in place to hide conspiracies in place that would cover for conspiracies and the entire thing becomes an onion with layers enough to drive you to tears. In the background hidden under many conspiracies there lurks a Mother Conspiracies and the purpose of all conspiracies is to cover the Mother Conspiracy.

When last did you see on TV on the six o'clock news the headlines shouting out: **It has been confirmed by NASA that the Moon's orbital distance** is increasing at a rate of 3.8 cm per year of about 3-4 cm/yr. Has any one physicist announced that the moon will not hit the earth and why this will not happen is because the moon is parting from the Earth. The moon was much closer in the past but they hide the fact that the Moon's orbital distance is increasing at a rate of 3.8 cm per year. If you saw this on any news channel, please let me know. I think this news is much more appropriate to be designated as headline news than the declassification of Pluto from planet status to an orbiting rock fragment. But this news will never reach the headlines. Neither will the news become evening headlines that the earth is increasing the circumference every year. What would the purpose be of hiding this truth from the public, is it insignificant? The reason is the Mother Conspiracy, which I will explain later on.

The distance between the moon and the earth is not reducing but it is increasing! Save that for front-page news. What such an announcement would shout out loud is that the Universe is not contracting by the value of mass as Newton anticipated, but it is increasing by a value that I have determined. This is my story about the conspiracy story. In order to accept my calculations, one has to reject Newton's law of attraction just as much as the cosmos rejects Newton's law of attraction.

Every child at school is forced to believe in the formula Newton introduced when Newton saw his apple fall from the tree. It is $F = G\ \dfrac{M_1 M_2}{r^2}$ where it is believed that the mass of the earth pulls the mass of the apple because the mass of the earth is pulling the mass of the moon because the mass of the sun is pulling the mass of the earth. If the moon is not falling, the sun is not falling so how does the apple fall? Never will this be on TV news just because according to Newtonian principle there is attraction by the force of gravity to the value of mass. Nevertheless, NASA confirms the moon and earth is parting in a way that increases the distance between the two solar systems. Knowing this proves physicists' stupidity and lack of fundamental understanding about physics. Therefore, one conspiracy was improvised to stop this from becoming public knowledge. It is called the Critical Density Theory and to form blind spots an important name was placed in the centre. It was Albert Einstein who was supposedly the chosen one to correct the wrong doings of the cosmos.

There was one culprit who was responsible for this information becoming everyday news and he was E.P. Hubble. Before Hubble got so outspoken about his findings the entire world of physics knew they were bullshitting the public blind about Newton's anomalies but was getting away with it for centuries. Just go and question "Kepler's" laws that Newton founded and you can see how far Newton was off the track.

E.P. Hubble saw through his telescope that the Universe was moving apart. Before this everybody agreed with Newton that the Universe was contracting under the load of mass. Newton saw the moon coming close by the value of mass as the earth was getting closer to the sun by the value of mass and to be equal to God those intellectuals could just apply $F = G\ \dfrac{M_1 M_2}{r^2}$ and redesign the entire Universe according to each one's liking. If you had the mass you could gauge when the Universe began. If you had the mass you could gauge when the Universe will end. By having the mass, one would find the force that drove the moon towards the earth. Before the event of Newton's miraculous discovery of forces driving planets around the sun to the tune of gravity, this was mainly God's prerogative to have such knowledge. Now having Newton's laws and many mathematical equations such

as $r = \dfrac{p}{1 + \epsilon \cos \theta}$ and others such as $\dfrac{d}{dt}\left(\dfrac{1}{2}r^2\theta\right) = 0$ and also $\dfrac{P^2 planet}{a^3 planet} = \dfrac{P^2 earth}{a^3 earth}$

because Newton declared that as a result of the mathematical fact that $4\pi^2 a^3 = P^2 G(M + m)$

it then was beyond question that $\left(\dfrac{P}{2\pi}\right)^2 = \dfrac{a^3}{G(M + m)}$ and knowing all this put you level with

God. No one could argue this, because if Newton was brilliant enough to know that

$\left(\dfrac{P}{2\pi}\right)^2 = \left(\dfrac{a^2\sqrt{1-\epsilon^2}}{\ell}\right)^2 = \dfrac{a^4(1-\epsilon^2)}{\ell^2} = \dfrac{a^4(1-\epsilon^2)}{a(1-\epsilon^2)GM} = \dfrac{a^3}{GM}$ who on God's earth will dare argue

with him. No one gave a fart about understanding anything this formula said because if they did they would know the correctness thereof was a joke. So with nobody having the intellect to argue the correctness thereof, it was considered brilliant just to pretend one does understand and then become a member of the Brainy Bunch society ruling physics. All one needs is to pretend to understand and believe that mass pulls mass and then you are as clever as Newton and everyone knows Newton is bloody clever so anyone believing the Universe is contracting by the measure of mass that forms gravity places such a person equal to God's intellect. That way all that anyone is asked for when to be thought of, as

being clever is to repeat after Newton that $F = G\,\dfrac{M_1 M_2}{r^2}$ is the formula on which the entire

Universe stands by the principle of physics. How easy could it be to be so smart you think of your position as equal to God's position? No one dare to argue with you because you have Newton backing you and only insanely stupid or mentally handicapped people will dare to argue with Newton about mass having pulling forces going around as gravity. Newton never

proved that $F = G\,\dfrac{M_1 M_2}{r^2}$ was correct but since nobody understood whether it was correct

or not nobody gave a blue apes virtue about the concept of correctness. This illusion was ending and it came with a bang that broke a two hundred year conspiracy, which guaranteed silence.

Hubble saw the Universe expanding while Newton said the Universe was contracting. Then along comes this man Hubble. All Hubble had to do is to tell those not believing him to look through his enormous telescope so keeping Hubble quiet was not easy and discrediting Hubble would be most stupid. That meant they had to explain that Newton could be wrong!

They had to explain why $\dfrac{d}{dt}\left(\dfrac{1}{2}r^2\theta\right) = 0$ and also $\dfrac{P^2 planet}{a^3 planet} = \dfrac{P^2 earth}{a^3 earth}$ was wrong because

Newton declared that as a result of the mathematical fact that $4\pi^2 a^3 = P^2 G(M + m)$ it then

was not so much beyond question that $\left(\dfrac{P}{2\pi}\right)^2 = \dfrac{a^3}{G(M + m)}$ was incorrect and knowing all

this put them below the level with God because only God and Newton were flawless and now Newton was not flawless. No one could argue or understand what Newton formulated because if Newton was the brilliant mathematical genius that made him smart enough to

know that $\left(\dfrac{P}{2\pi}\right)^2 = \left(\dfrac{a^2\sqrt{1-\epsilon^2}}{\ell}\right)^2 = \dfrac{a^4(1-\epsilon^2)}{\ell^2} = \dfrac{a^4(1-\epsilon^2)}{a(1-\epsilon^2)GM} = \dfrac{a^3}{GM}$ then all they had to do to

be stupid was to condole Newton. In the past if you were to be clever just repeat after

Newton and you are a Newtonian, but now this no longer applied. Don't ask questions because nobody dares to ask questions. You don't have to understand anything because that is not asked of you. You don't have to look for flaws because that would demise your position in terms of Newton. What you see is precisely not what you see!

Science has to reflect on what nature is and not interpret it in one way and then judge it in another way.

When we look at this picture we do not see one star. We do not see the Milky Way, which we think it portrays. We see a stream of light flooding the lens of the telescope through which we see this fantastic picture and the light by which we see every spot shows time not space.

We see the light that shows the individual star but the star as such we are not able to see. We see the light as it left the star many, many thousands of years ago. We see what this was before there were humans on earth. Part of this picture could be representing a time when the dinosaurs still walked the earth in drones. However, we can't see one star because all the information is written in light and it portrays what is not present any longer.

We witness what was but what we witness does not represent one instant taken from time because every pixel we see left the individual star we think we see at a different time than the others on the picture. We see a mosaic of time differences spanning billions of years ago and the only common factor is that the light which left at so many intervals only arrives in the lens at the same instant.

When we view the moon we think we see the moon because we were taught that we see the moon. We can't see the moon because it is practically impossible to see the moon. We see light that was reflected from the surface of the moon and only when this light arrives, do we see what the moon was like one and a half seconds ago. We see not the moon but we see where the moon was one and a half seconds into cosmic history and we see the allocation of the moon at the time the light left the moon and that is part of what we think of as cosmic yesterday.

Time is the very now, the moment of the present the instant and anything concerning space is behind us, is the past, is what is no longer in the immediate. The person standing very much next to me forms the beginning of the historic past while a star twelve billion years away is forming the edge where space can be traced by forming the past. All space including everything I see including the fact that I see anything is part of the past while the part saying I who don't think because the thought forms part of the past is in the present. The instant in which I am the Universe expands by pushing everything I see and everything I can't see in time away from me. Time flows at a rate of singularity Π^0 = forming space $\Pi^3\div$ in relation to what moves ($\Pi^2\Pi$) which is $\Pi^0 = \Pi^3\div(\Pi^2\Pi)$.

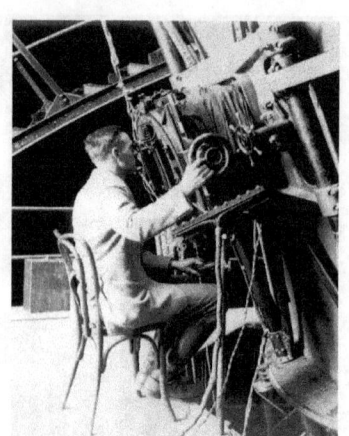

Newtonians know this but is too narrow-minded to realise what it is they realise. The instant forming singularity Π^0 forms space but space seen in light = Π^3 and the light is placing space Π^2 at a growing distanceΠ. All you have to be is to be stupid enough not to ask silly questions and then you are brilliant. Be a Newtonian and follow your leader blindly with no questions asked. Everyone would be so amazed with your brilliance they will think of you as one of the Brainy Bunch. People will stare at you in awe and all you have to do is to believe Newton and believe in science. Now Hubble went along and burst this friendly little bubble by his unasked for discovery of the expanding.

Then along came this man Hubble with his large telescope. Everything was going well and everybody was enjoying being clever as long as you were stupid enough to be a Newtonian. Then this man unveiled a fact that declassified Newton as a flawed worthless form of stupidity. The Universe expands and does not contract. Now once again only God was flawless with Newton flawed throughout. This the Brainy Bunch was not going to appreciate and this did not go down well with the most esteemed the Newtonians.

In order to cover up for Newton's misperception an entire variety of reasons are established, each accepted as a possible truth. The fact that Newton's principle goes begging never gets mentioned, although the only reason why it would never get mentioned is because it is the only valid conclusion and that they don't want. All other reasons they mention are overruled by Newton's principal of mass pulling. Mass pulling is the founding law that all other factors rest on. The earth slinging the Moon away can't be a factor because the mass of the earth is too great. The mass that pulls reduces the radius by the square. The moon is moving away from the earth. Yes, have a look at all the theories presented by Newtonians on the web as to why this happens. It is the thrust, no, it is heat expansion, no, it is parting because of sea currents, no...and there are innumerable many excuses why this is happening. Not one has to do with Newton being wrong! They never take Newton's formula and apply the mass of both solar objects and see what the gravity is to see why the parting occurs. With the radius increasing the gravity must therefore reduce in pulling power because the distance determines the validity of the force by dividing into the multiplication of the mass.

It is not the Universe that is very far that is expanding because Newton said the Universe is contracting. The Universe is expanding and every atom is the Universe forming time. As every atom spins while going straight forming the spin of the earth every atom performs time by expanding in the movement time performs. The expanding is beyond what Newtonians wish to measure or calculate, the expanding has a value of $\Pi^0 = \Pi^3\div(\Pi^2\Pi)$ and this says that space Π^3 grows bigger by a margin of Π^0 as space moves in a circle $\Pi2$ where the electron spins while this spinning takes the earth to go around the sun Π. This is how time works. The spinning pendulum shows this effect very clearly.

When the pendulum swings, every new position it takes on indicates the time ended and the ending of the instant that formed space. Space is the history of time or the reflection of light, indicating time left an image of space behind. In the instant we see time we see Π^0 forming singularity but we think we see Π^3 while the ever onward going space that forms is due to time moving $\Pi^2\Pi$ leaving space Π^3 behind as light that forms an image $\Pi^2\Pi$. When the turning of the earth counters the pendulum swing, the rotation turns around, as the earth will not allow the pendulum to move out of the influence sphere of the earth. By moving backwards and forwards the pendulum reads the time of the earth as the earth spins and moves around the sun forming time in motion by the earth moving. The pendulum reads the movement of the earth forming atmosphere or then the gravity of the earth that forms time. Mass does nothing in this entire operation and if mass had value a bigger pendulum would swing slower than a small pendulum that would spin faster.

The earth moves by compressing the space around the pendulum and it is not the pendulum that swings by its own initiative following the space as it moves towards the earth. If the pendulum reads time by swinging through space and the earth moves space by contracting or compressing space then the movement of space is the factor that forms time and then time is the movement of space by the changes occurring in space. Newtonians will never agree because of the mass factor they believe in and then rather disagreeing with what happens in nature they will uphold Newton's version of misinformation. I say again if the pendulum reads time by reading space moving towards the earth then that gravity forming from the earth's circular movement must be time forming in direct relation to gravity moving space. This is a large part of the mother conspiracy whereby Newtonians tell nature what it must be according to Newton instead of reading from nature and then accordingly find what nature truly is.

While the earth provides material Π^3 that contracts Π^2 space moving down Π the pendulum registers singularity or the instant Π^0 in which time is and while the earth provides the instant Π^0 that the pendulum reads Π^3 the pendulum encounters swing Π^2 that the space Π that is moving down formsΠ. This spinning action of the earth holding a body results in the space surrounding the body to compress and that movement results in a moving circle compressing and that forms time in relation to the earth forming gravity. But the major issue in all of this is that wherever the instant or singularity forms Π^0 space grows Π^3 by the movement of space Π^2 in a circle that goes straight Π. This growth Π^0 is beyond measure because it depends on material $\Pi^3 \div$ that moves in a circle Π^2 when the circle moves straight Π. $\Pi^0 = \Pi^3 \div (\Pi^2\Pi)$.

Science can't prove gravity but gravity is a fact. Science tries to explain gravity by going biennial and obscure and then pretends to get clever but Science only becomes foolish. If science cannot prove God's existence, it is not God that does not exist, but it is science failing and therefore it is then that specific view about science that should be re-examined since it is the view of science that is proving as being incorrect. This fact is what the so very brilliant and intellectually mindful Newtonian atheist should remember when they fail in their science altogether is that their science fails altogether and that failing it does in all its splendour, is a fact I am delighted to prove! The fact is Newton's views were never tested and that the Newtonian views on science were never challenged before and because of that Newton principles never withstood diligent scrutiny before.

When Sir Isaac Newton is investigated even in the flimsiest of manners, well accepted facts seem to become very suspect, to say the least. This becomes evident when concluding all the facts this book presents. Now, in this book, for the first time Newton is tested and such testing is the proof you gain by reading that which I uncover. What I bring into the open are unseen facts, which I present you with as I take you on a tour through an avenue of facts I introduce in this work. The lack there is in sensibility concerning Sir Isaac Newton's principles this book proves. The theories of Sir Isaac Newton require proof, which was never given while God never needs proof and that is what science constantly seeks. When science perpetually ignored my concerned calling on and ignored my calling on them because (I suppose) they were finding my concerns wanting, in my final letter to them I promised them never to contact them personally again by any and by all means. I also promised them a fight. This is the fight I promised. I was not worth noticing so I was ignored.

I now am calling on the public, as I am ignoring their reputations. I am showing the public just how extremely bright the Newtonian inspired super-thinkers are!
In this picture what floats in space is a lot more massive than a truck but it floats effortlessly and it will keep on doing so not until it has more mass, but when it slows down the speed it circles with. Explaining up there the mass and the pulling becomes rather much confusing.

Well…I wish to bring to mind some of the facts that physics work with when academics as scientists only work with facts. Remember they are the ones boasting that if facts are not proven then it is fables and those very important academics don't waste time with fables because they only work with facts. The accuracy of their basis on which science rests is that mass is responsible for gravity by pulling. If you don't have mass you're not going to have gravity. Mass is equal to gravity and gravity applies only by mass. If mass is present then it is by gravity or otherwise gravity is absent. If a body falls it is the mass that pulls the body to fall because the body receives gravity by ratio of mass and mass is that which produces gravity in relation to the mass available.

It is mass that drags you down because the mass is in charge of the gravity and the gravity finds the value from the mass available. Mass pushes you down by the gravity it forces onto you. But if mass drags you down then what lifts you up in the air balloon? If mass gives the gravity to drag you onto the Earth then why would the hot air lift you up in a balloon? Is the hot air causing anti gravity or anti mass because gravity is nullified by lifting the air balloon and cargo. The hot air balloon is lifting the passenger and all that is in the bag plus the bag plus the balloon into the air. So what is then pushing the lot up if it is mass that drags you down. Has the air not got mass because then the air can't have gravity and then the air must escape into the blackness of outer space because by going up it shows a resilience of either mass or gravity. We have seen that it is mass that pulls everything onto the ground.

When science was in search of discovering new planets they chose to use these principles nature applies and never once used Newtonian calculations to position the planet location they looked for. When everyone of these discoveries about one of these phenomena came to be, science stood before the door of choosing nature and research what they found or shoot what they found as nature in nature down and preserve Newton at all cost. That is the theme running through the book. It is how Newtonian science tries to degrade nature and belittle nature to cover–up for Newton. This still goes on today and this book proves it.

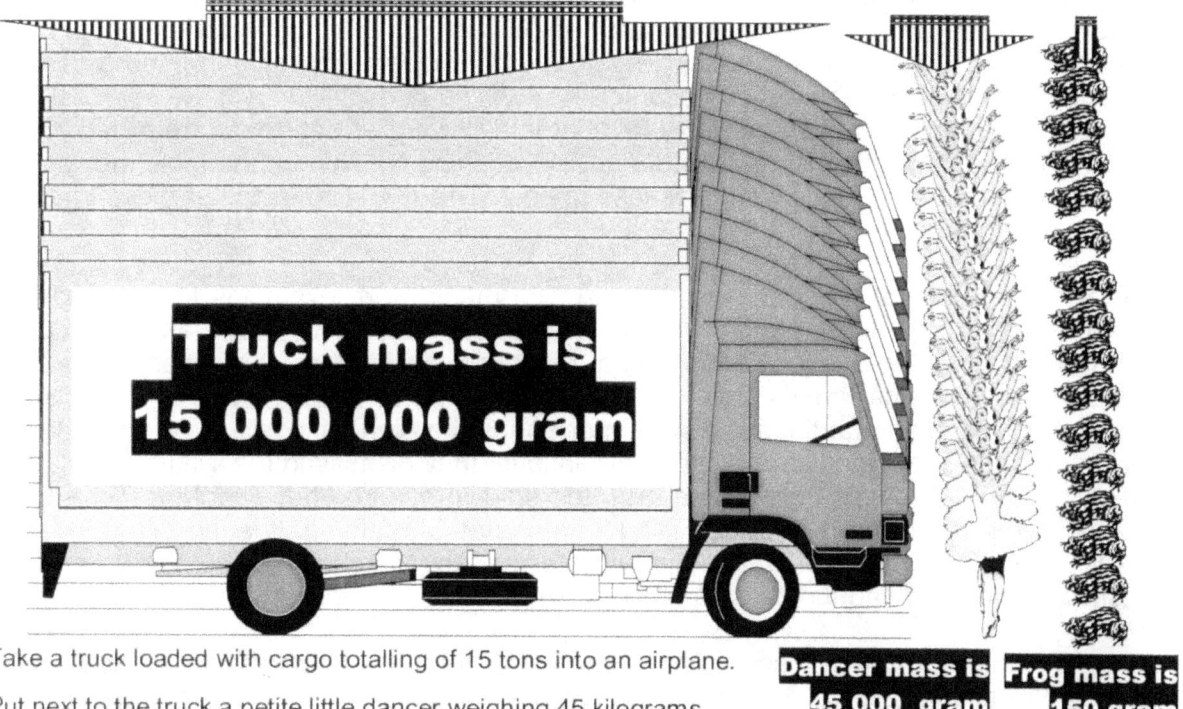

Take a truck loaded with cargo totalling of 15 tons into an airplane.

Put next to the truck a petite little dancer weighing 45 kilograms.

Then to keep the dancer on her toes, put a frog of 150 g next to her.

Now we will have the mass of each object "pull" by "gravity" as this lot falls down

Take note that you are told by the wise amongst us that it is mass that produces the gravity that pulls you down. We have just had a lovely debate on how it works and how mass drags you down and wondered if it then is anti mass or anti gravity that lifts you up with the hot air balloon, well take note of this as your airplane reaches 11 thousand meters which is eleven kilometres straight up into the air. Now you throw this lot out without parachutes and let the lot fall free in accordance with each object enjoying its individual mass forming its own gravity. If Newton is correct this lot will fall according to gravity and if Galileo is correct this lot will fall together. But never can Newton and Galileo have the same opinion of the outcome, as they are not sharing an issue with the same outcome because by falling with mass the lot must fall different and we know they do not. If falling is according to mass then it is ridiculous to think the lot would arrive on the ground simultaneously. If the falling is equal Galileo is correct and if everyone lands in ratio of individual mass then Newton seems to be correct. Judge on what you see and not what you are told by physicists upholding Newton.

I will give you a very unimportant example. You have heard of the expression he is a bag of "hot air" or it's only "hot air" or "they use hot air again". This is in place to discount the fact that when you blow hot air into a balloon the balloon surges into the air. Nature can take "hot air" and lift a balloon with a lot of mass into the air. This ruins Newton's idea that mass pulls down because by just adding "hot air" the mass is discounted and the balloon flows into the air. Now you explain that in terms of Newton's mass pulling mass and you get nowhere very fast while sounding very stupid on the trod. If a balloon can hold three tons of mass that the earth pulls down, how then can a bit of "hot air" lift this lot up? Science would rather belittle the truth than question Newton! They could not fight nature because nature proved that "hot air" lifts mass.

That made Newton's idea sounding a bit stupid because when hot air lifts the three-ton basket-parcel up it must be cold air that pushes the three-ton parcel down. In order to stop the public getting wise to the idea, they belittled the thought about "hot air". They started with the saying that anything not worth much becomes a bag of "hot air". If you are not cut out to be successful you are a bag of "hot air". If you are not doing anything worth your while you blow "hot air". If your chances to accomplish anything are only very tiny, they call you another bag of "hot air". If you can only talk but is a coward at heart you are a bag of "hot air". ...And so to draw attention away from hot air brushing Newton off the table they spread the rumour that hot air means worthlessness at best. This is the influence of those I have to fight to gain ground!

I wrote many books on this because there are so many more examples to show how far academics did go to avoid people getting wise about Newton or that Newton becomes compromised or people see Newton in a revealing position or think about Newton's correctness. There was forever a cover-up to hide Newton from the truth or from nature. When they discovered that the solar system arranges the planets' position not according to mass as Newton has it in his formulated calculations, they dismissed the Titius Bode law as a fluke of nature. They did not make a fuss and start a research but swept the Titius Bode law very much under that table by labelling it as a once off enigma. It is the Titus Bode law that allocates the planet positions and not Newton's mass attack of madness. It was nature that showed where to find the missing planets. They discarded Newton in favour of nature.

I am going to give you a few scenarios in science and then you translate what happens in nature to Newtonian science. This is the formula on which the entirety of physics depends!

You better keep your eyes open because if you blink you will miss how science conspires to defraud you and mislead you by you allowing scientists to brainwash you by means of mind control. Let's get down to the most basics in physics. Everyone is familiar with the concept on which all that forms physics are founded. It is Newtonian and it forms by the following concept. The formula science applies to form the basis of physics is $F = G \dfrac{M_1 M_2}{r^2}$. It is the mass of one object pulling on the mass of another object pulling each closer and as such this pulling force reduces the radius between the two objects. It is not the top but the bottom is important.

Why would the air defy mass and allow the balloon to go anti whatever is going anti because nothing in or on the balloon has gone without mass. The object(s) has mass to produce gravity. Why then would hot air allow the balloon plus everything in the balloon to lift into the air? The balloon lifts in relation to the hot air that blows into the sack. The more hot air and the hotter the air is the more lift and the swifter the lift will be that the balloon provides. The issue sticking out is that the balloon then must not have mass because with anti gravity it is pulling up. Remember mass drags you down and mass can't pull you up and drag you down at the same time. Then what is pushing while mass is pulling or is mass pushing while what is pulling? The object is not going in the normal direction where it is dragged down by mass forming gravity and in all my life I have never heard one Academic mention anything about gravity lifting and that makes the lot very confusing. What is lifting up when the lot should be pushing down and why did everything connected to the balloon lose the mass and if it has mass why is it not dragging down the balloon?

That fact about Galileo, science does embrace, although this strongly contradicts Newton's impressions about mass inflicting gravity. On TV we see how all objects, such as cars, humans and bags fall at the same pace, which sets a standard totally against Newtonian mass principles that produce the falling, and proves Newton wrong because mass then does

not underwrite gravity in any way or form at all. The formula $F = G \dfrac{M_1 M_2}{r^2}$ would suggest

mass taking all the responsibility for such falling that takes place.

Newtonians declare gravity as the **force** of gravity **F**, that is **=** equal to **gravitational constant G**, when it is multiplied by the **mass M₁** and the **mass M₂** after which then the product of the three factors influencing gravity is divided by the square **r²** distance between **mass pulling the mass** that **destroys** the **distance between** the two **objects**. If mass pulls mass as Newton said, the Big Bang is not possible but the Universe is notwithstanding Newton's claims, expanding (growing apart). If mass forms gravity, every planet must orbit at a different pace, which they do not, as all planets orbit at the same pace around the Sun.

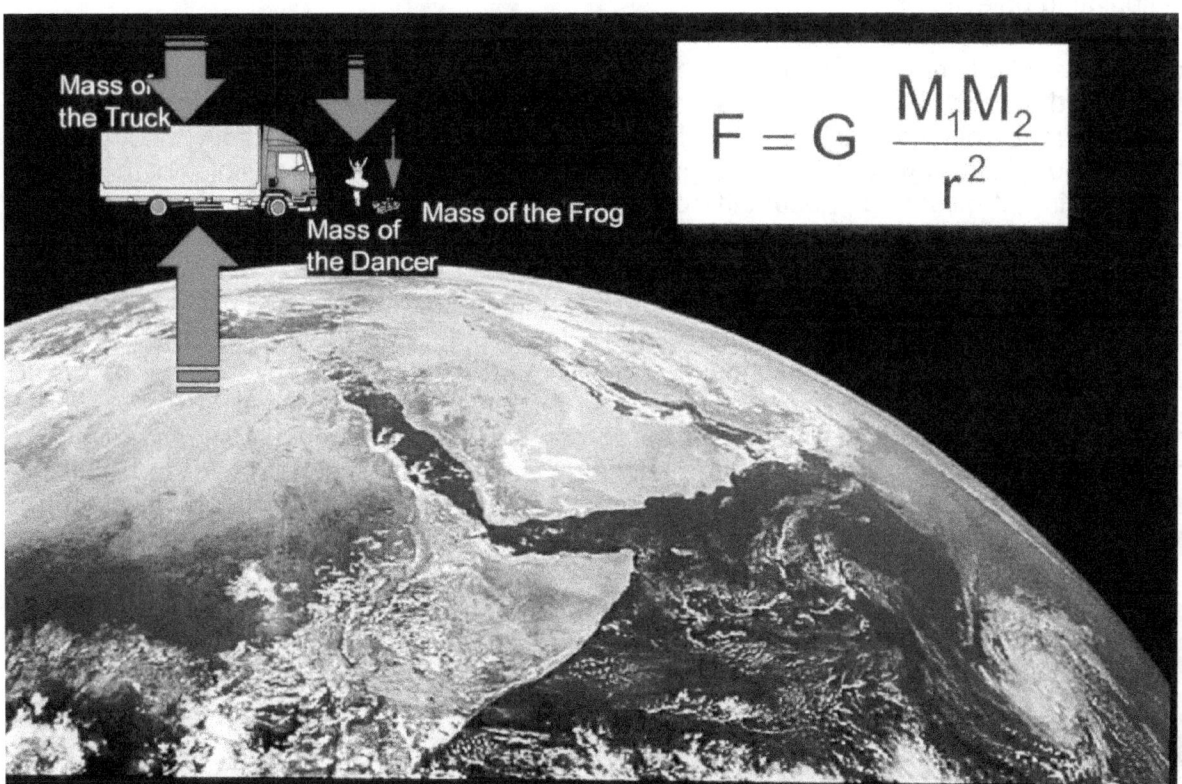

It is said that the earth has mass that it pulls with. The truck has mass that it pulls with. So does the girl and the frog have mass that it pulls with. The pulling power all depends on the mass that the object holds, well that is according to Newton.

Science teaches that a feather and a hammer have different mass while they fall equal in time through an equal distance travelled. If gravity was mass related, then this was not possible, because then objects must fall according to mass. Reality says if mass plays a part in the "pulling" that forms gravity, then there has to be distinction in the speed when large and small objects fall and the time it takes to fall. The period large objects take to fall must be shorter than small objects take, that is if mass did play such a fundamental part in the forming of gravity.

There is no evidence of mass playing any part in falling objects. Any two objects holding different mass fall equal in time and in distance when sharing similar conditions, which suspends mass altogether as an influencing factor. Galileo proved objects of different mass fall equally under similar conditions. Planets don't give the slightest hint that they obey Newton's suggested cosmic laws by implementing mass. The truth is that mass is the resistance of any independent material to deform and to acquire mass the individual object relinquishes independent motion. Mass comes about when the falling of any object stops the motion of the falling. Mass prevents further falling, it does not sustain further falling. Gravity is the moving of the object to the centre of the Earth while falling. I say gravity is movement while mass is obstructing independent movement, which is what gravity is. Mass is not forming the factor responsible for gravity or movement, but prevents further movement. A body falls by gravity. Mass hinders movement and therefore mass can't enhance or produce movement or gravity.

What does happen is that movement stops when mass becomes a factor because mass is the indicator of the restriction of movement when objects do fall. Mass prevents or blocks gravity. Gravity is the motion that defines the individual identity of any object's structural form by rendering motion while reserving independence in granting free space from other manipulating objects.

Do you as students realize the inconsistencies that physic Academics present you with when portraying that what they teach you as being the solemn truth. What they don't shout from the top of mountains is that the principles of Newton and Galileo are totally incompatible. There is no getting away from the argument that if mass was responsible for gravity in objects falling; it must bring about differences in the tempo of falling and then objects could not fall equal as Galileo said and proved it does.

Again I have to bring the attention back to the pendulum that measures time by measuring gravity. There was never yet any claim brought to this fact but a fact it is. Whether it is a large pendulum or a small pendulum it measures time. Everything falls by the rhythm of time contracting space and I mean everything. In order to fall faster one has to use a driving propulsion that enhances the movement to go beyond the speed in which space contracts and then some effect of weightlessness become a reality for a short while and until the aeroplane has to level out again to avoid a crash. If the propulsion continues it ends in a crash with absolute devastation and brutal consequences.

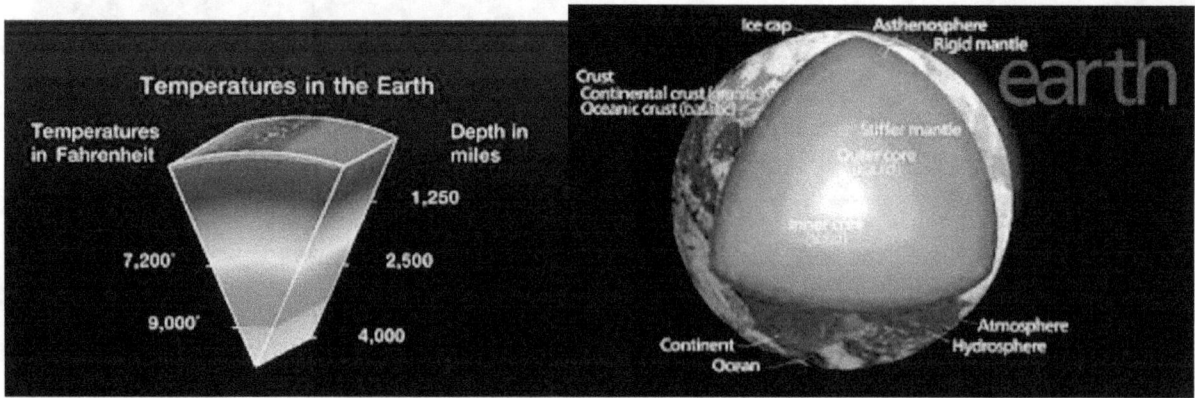

It is space that depletes making the density increasing and by the density increasing the concentration of space causes the surge in heat. This condensing and concentration of heat runs to the very centre of the earth and that point concentrates all space into singularity, which is what happens in a Black Hole.

...And that is gravity. It is reducing space by movement to concentrate heat in getting cold. But our human cold is not a cosmic cold because we humans are not a part of the cosmos. Let's return to our falling objects, a lorry, a ballerina and a tiny frog.

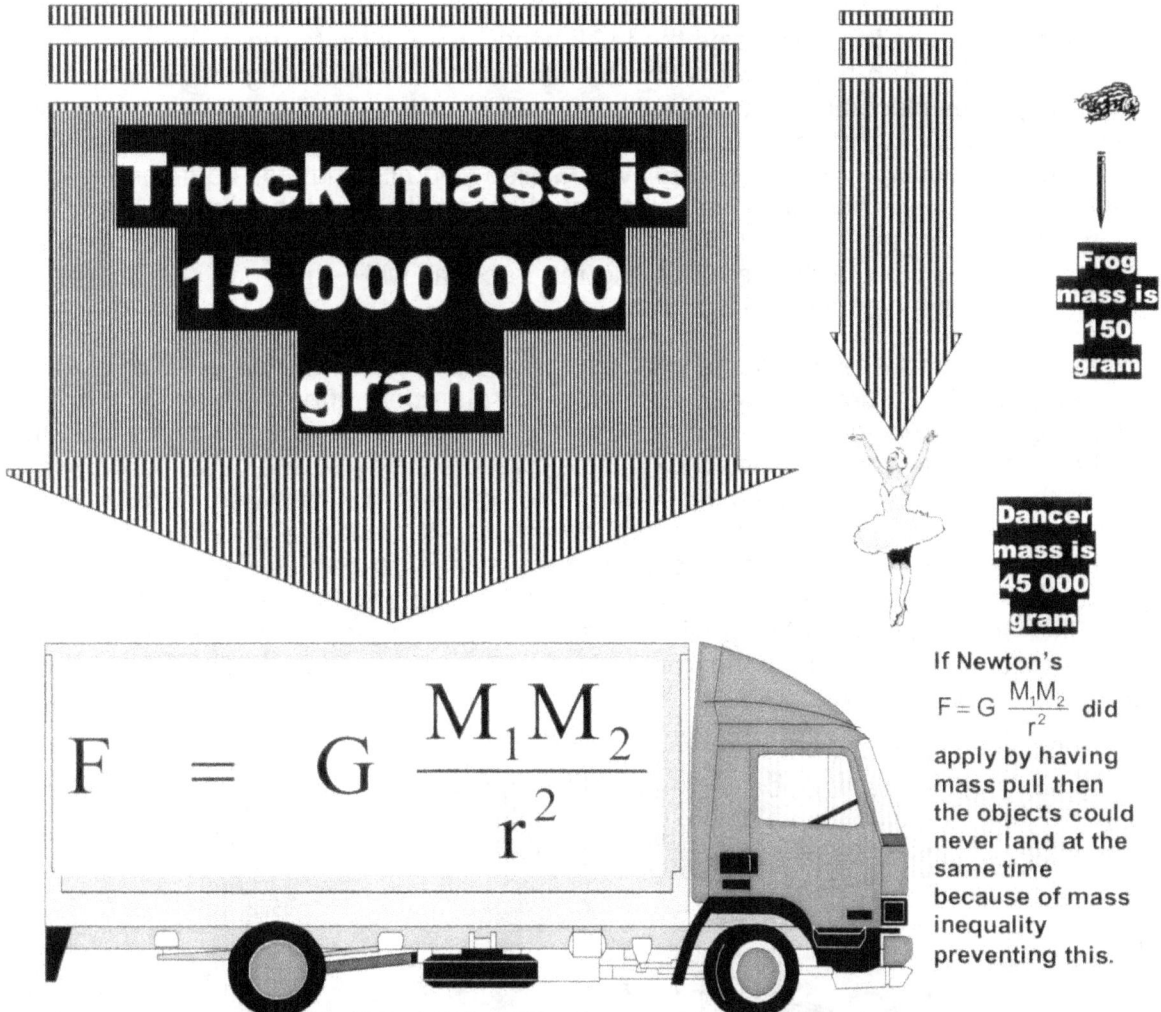

Truck mass is 15 000 000 gram

Frog mass is 150 gram

Dancer mass is 45 000 gram

$$F = G\ \frac{M_1 M_2}{r^2}$$

If Newton's $F = G\ \frac{M_1 M_2}{r^2}$ did apply by having mass pull then the objects could never land at the same time because of mass inequality preventing this.

The mass of each falling object amongst others supposedly provides the gravity that pulls the truck down or should pull the truck down one million times better than it pulls the frog down. If the mass is doing the pulling by forming the gravity the truck must fall 333 times faster than the girl and one million times faster than the frog. If Newton is correct the frog must land on the earth hours after the truck landed and minutes after the girl. In reality the frog and the dancer and the truck fall at an equal descending rate. That is what Galileo said all along. That is not what Newton said. On the way down the frog then pretends he drives the truck and the next scene he is dancing with the girl while the truck is falling as fast as a truck can fall and they all fall together as if they are in a unit sharing space. Does each then fall according to their mass as Newton says? Who do you think is lying? Physics is telling students all over the world that mass is in charge of gravity and it is by the force that mass forms that the bodies are pulled down. Then the mass is pulling the truck of 15 tons down since the mass produces the gravity and the gravity produces the fall which is three hundred and thirty three times more in the case of the truck going in a down direction than the girl's mass of 45 kg is pulling the dancer down while the frog of 150 g is descending as if it also is 15 tons as it falls beside the truck. This point is crucial to what you are going to read about gravity in the rest of the book!

If the Brainy Bunch who is all too wise is correct, the frog can fly to America and have a pizza in New York while the truck has a few micro seconds to get down if the girl is going to fall during the normal falling duration of a minute or so. Everyone has seen skydivers jump out of airplanes next to cars and trucks and bags. Every one has seen they all fall at the same rate as the big and small objects. They can hop in and out of cars while floating next to the downwards descending car and the bags that fall with them. The girl can do tap dancing around a jumping frog on top of the truck or below the truck and they can be inside the back of the truck galloping on fresh air inside the truck because the lot is falling at the exact same rate. The academics wish to brainwash you by mind control in accepting that it is by mass that the falling takes place and that mass is responsible for the gravity and by mass pulling you down it is gravity that makes you fall. Where is the proof of mass that according to them is that which is producing gravity? They tell you Galileo said all things fall equal and we can see from the TV monitors how all things fall equal. Where is the mass that makes the gravity to let you fall if all things fall equally? They tell you that the truck has a mass of 15 tons and that mass is making the gravity that is having the truck fall while the truck is falling at the same speed and distance than the frog does.

Then if you don't repeat after them and echo every word test after test and exam after exam they will fail your papers and kick you from campus. You repeat after them and you live an academic life or you disagree and you go home to play with your toes. If mass is in the picture then mass must be represented by a factor of more than just one because if mass is not part of the overall picture then mass has a factor of one which proves that mass is not part of the equation since mass can't change the results. With all the objects falling equal mass has no role and if mass has no role then for my money academics in physics can't just go and put everything in as their hearts desire. If it is Galileo who is correct and if all things fall equal then mass has no part in gravity! If mass is the inspiration behind gravity the truck must fall a million times faster than the frog and in fact the frog should almost land in another country because that is how slow it falls

$$F = G\frac{M_1 M_2}{r^2}$$

Telling this to your physics professor will have him shrug his shoulders dismissively as if it is not his problem and raising the issue is slightly out of place and rather a silly point to make. That is how they have been eluding the problem now for three hundred years. They put the mistake at your door for noting it.

The fact of the matter is that I don't wish to be near when any of this lot hits the ground because the truck will cause a quarry and the dancer will be a splash of red fluid while the frog might not be that worse for wear if the truck or the dancer doesn't land on the frog. The differentiation of having mass or having equality when falling and then not having mass and between individual differences in mass by each component that enters the equation when the objects touch the ground is astounding. Then everyone gets the mass it has. Only when they touch the ground and land on the soil is mass as a factor awarded. While they fall they all fall equal and there is no distinction between the falling at all. What then is gravity? The gravity is the falling. The gravity is the motion. The tendency to move and apply gravity is the part that the mass restrains. The factor of weight only comes into play once the independent movement stops and the object becomes a part of the earth in all aspects of movement. When it moves with the earth, it has mass.

As the object hits the ground, the object stops falling with the mass that manifests as weight that then prevents the falling from continuing. The role of forming identifiable mass is to prevent further falling and independent motion to continue. Some might even still honestly believe it is mass that produces gravity due to the brainwashing as they were taught it is mass that produces gravity and never thought about the matter again afterwards. Take this issue up with your professor.

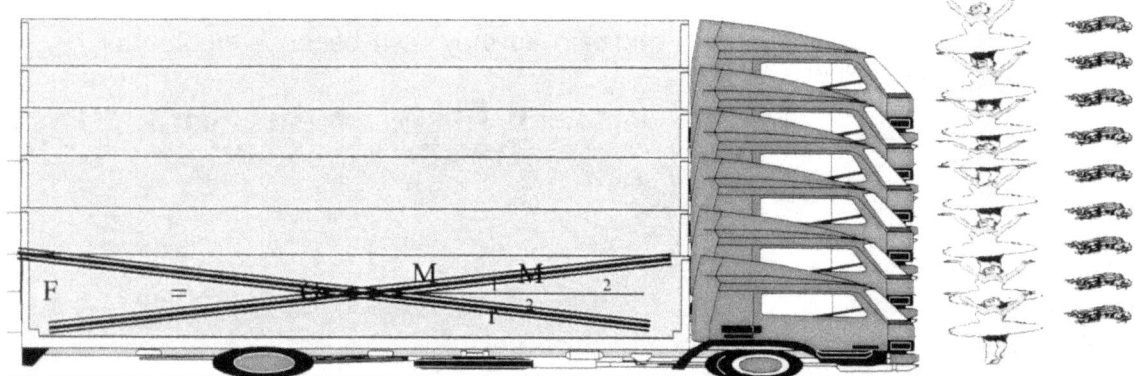

The falling object experienced no mass while falling therefore the falling or moving must be gravity's contribution. While objects are in motion those moving objects is experiencing gravity. The object shows mass when the object has a tendency to move but the motion towards the centre of the Earth no longer takes place. That means mass is the restraining of the motion or is that which prevents the motion or gravity taking place. On Earth, objects experiences mass by restricting gravity or motion with the Earth giving mass but taking away free motion. By giving mass the Earth forces the object to become one with the Earth and move with the Earth as a pat of the Earth.

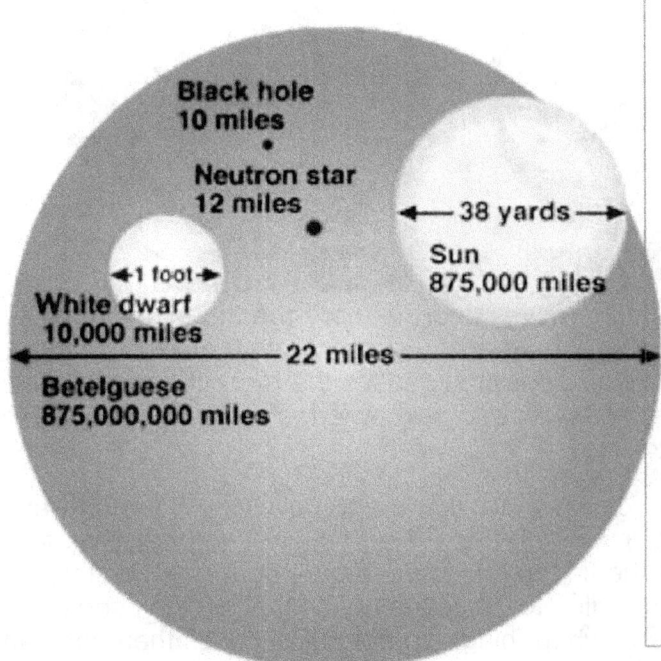

Massiveness brings gravity and the bigger that star the stronger is the gravity. It is put forward that size brings about mass according to the Hertzsprung-Russell diagram of star classification. This shows big stars produce heavy mass and is as outdated as all Newtonian principles. If Betelgeuse the biggest star's diameter of 875×10^6 miles was represented by 22miles then the sun's 875×10^3 miles was 38 yards and then 1 foot would represent the white dwarfs 10×10^3 miles. A Neutron star's diameter would be 3 mm and a Black Hole would be .5 mm true size. On the earth 100 lb weight would be 1 ton on the sun and that weight would be 10 000 tons on a white dwarf, 10 billion tons on a neutron star and that would become 30 billion tons inside a Black hole. In the Universe there is no big or small and everything that is big is small and all that is small is big. A Black Hole is the smallest there is and the biggest too.

In a later stage I use Kepler to show why stars are big and small at the same time. They were brainwashed by their tutors as their tutors were brainwashed before them. You are in a position where you can teach your tutors the truth about gravity if you read what is in the books. The truth is there and the truth is out and the truth will be because the truth is written for all who wishes to read. The academics on the other hand have ignored my work and my being on Earth for the past twelve years while I was writing them letters about gravity. They ignore me as if I am a rattlesnake because to them I am a rattlesnake. Look at the picture and think! If mass was the standard for establishing gravity then Betelgeuse with a diameter of 875 000 000 miles must be what forms a Black Hole and we know from pictures taken and shown that Betelgeuse is the closest that comes to a sloshing pot of shit as you throw it out. If you wish to see gravity in an unbridled form as powerful as it comes, you go to a Black Hole that has no diameter or no space within. So where is this mass story of Oppenheimer for which he received no less than a Nobel Prize. In the Oppenheimer – Volkoff limit Oppenheimer calculated that a star having a mass bigger than 1.6 times the mass of the sun becomes a Black Hole...no shit! Then how many Black Holes did Betelgeuse become all at once? They use science to award or vindicate as they wish because Hubble for his work never received a Nobel Prize.

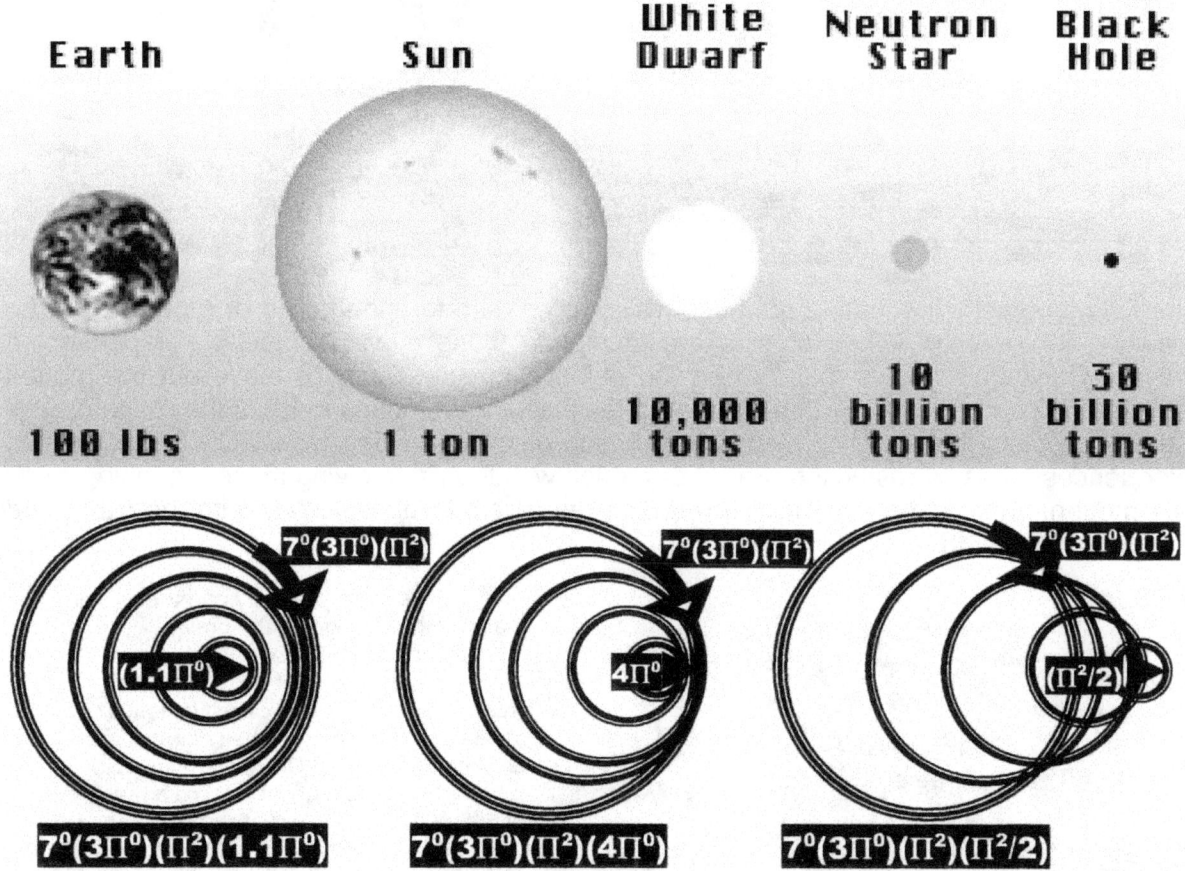

The faster the atom spins the further the atom moves and the more the atom concentrates the liquid aspect surrounding the atom. In that the atom shrinks because the spin correlates with the Universal flow of time. That is why the density increases the more the gravity gets and this evidence is all over in nature ready to read and acknowledge by people not stupid enough to ignore nature in favour of Newton.

For that reason if no other they will rather go on lying to you and cover their corrupt fraud than face up to the truth and admit their work is lost. The truth will be whether it is recognised by them and they can become the first to admit and repent or they will be the last of the laughing stock that those in the future will refer to as the bunch that couldn't see when things fall equal they cannot have mass and when things do not fall by mass then one can know mass has nothing to do with the falling and the gravity.

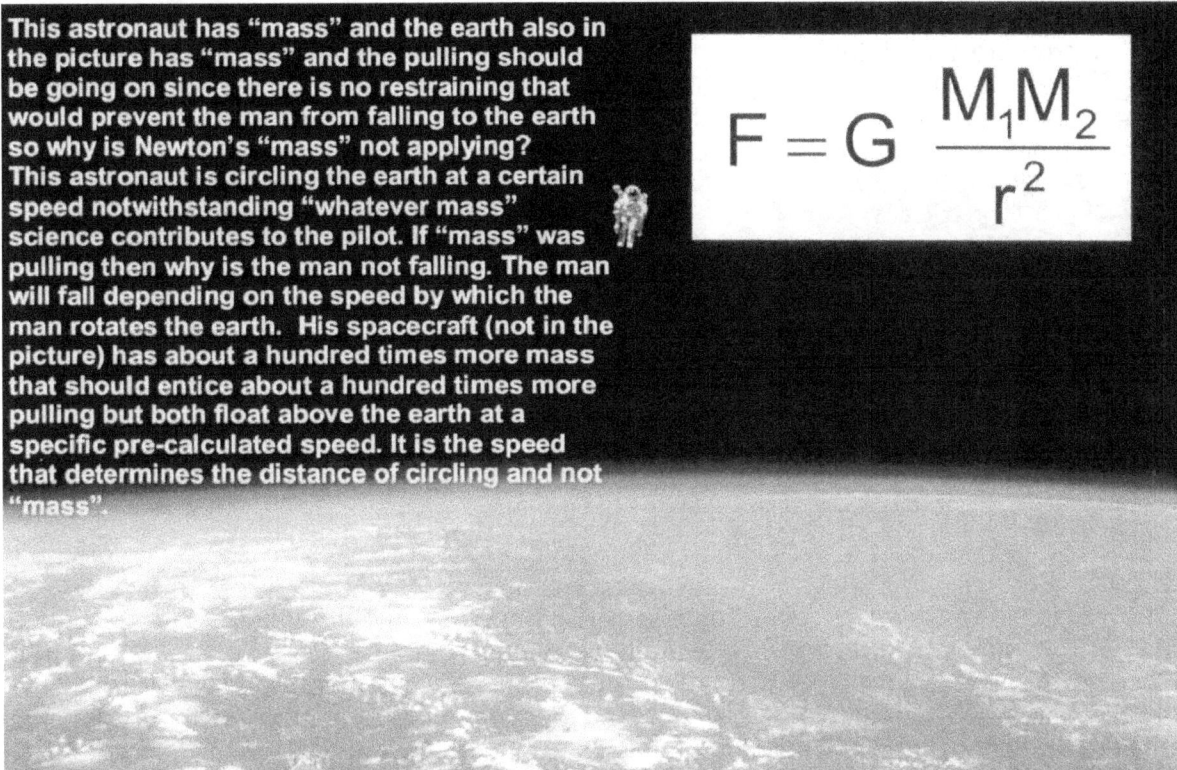

This astronaut has "mass" and the earth also in the picture has "mass" and the pulling should be going on since there is no restraining that would prevent the man from falling to the earth so why is Newton's "mass" not applying? This astronaut is circling the earth at a certain speed notwithstanding "whatever mass" science contributes to the pilot. If "mass" was pulling then why is the man not falling. The man will fall depending on the speed by which the man rotates the earth. His spacecraft (not in the picture) has about a hundred times more mass that should entice about a hundred times more pulling but both float above the earth at a specific pre-calculated speed. It is the speed that determines the distance of circling and not "mass".

$$F = G \, \frac{M_1 M_2}{r^2}$$

If the astronaut was on earth his mass would have been the same as his weight. Therefore on earth there is no distinction between the mass factor on earth and the weight factor on earth. It is clear that things change when this astronaut walked on the moon. If the astronaut was walking on the moon his weight would not be equal to his mass. The astronaut would weigh less than the mass he has on earth. It is said that the gravity changes the numbers. The moon has less mass and that gives the man less weight while the mass remains the same. Say the man weighed 90 kg on earth he would have mass equal to 90kg. When the same man walks on the moon he would weigh 30 kg while having a mass of 90 kg being on the moon. On the moon the machine may have the same weight as the man on earth but the mass of the machine stays what it was on earth and so does the mass of the man. That becomes science in fictional perspective. If the man could walk on the sun his weight would become a thousand times more but the mass of the man remains 90 kg. If I believe that bit then somewhere I am fooled just because I am that stupid and I deserve to be fooled that easy.   Then when the astronaut is in space he still has a mass of 90 kg but when walking in air the man has no weight with a mass of 90 kg.   When reading on, just be clear about the "having weight" part that I believe and the having mass that is very fictitious and manufactured.

I have no qualm with having weight because that you have or you have not depending on the reality of where you are. My problem is with having mass and that is a fictional representation of what Newtonian wishes ask for at that minute. It is this mass constant where-ever-you-go-fit-all scenario that seems to be something similar to a rule that is made up-as-you-go. Something much like New Zealand Rugby Rules applying to them in a game and can change in every game as they go along but others playing have fixed rules applying.

That there is a conspiracy in science present and very, very active, about that fact there can be no doubt. Dodging the truth runs as deep as it can go and involves every person who studied science, worked with science, works in science and those teaching science. Some know they conspire and others conspire unwittingly. Everyone in science conspires to keep up untested and corrupted views to promote the status quo. Everyone tries to elude the truth by conspiring **not to address** the issue and this even applies to Albert Einstein in his quest for answers. Einstein let the establishment dictate although his calculation could never substantiate any of Newton's claims and it was only Newton's claims Einstein could not collaborate.

In 1907 Einstein experienced what he called "The happiest day of my life" as he described that day. For weeks he wrestled with a daunting question that from all the natural laws, he concluded that it was only Newton's law on gravity, his special theory on relativity couldn't explain. There he missed the conspiracy! If the Newton gravitational principle theory was the only part of science that did not fit into his relativity concept, then it was what is incorrect.

He said he was sitting in his office in the department of patents in Bern and suddenly he had a thought. He realised that when someone fell freely through air the person would not feel his own weight. The simple thought made a massive impression on the young Einstein. In his mind he saw someone falling and he asked himself what happened to the law of gravity in the fall? If other objects would fall with him, the other objects would be in a complete state of total rest.

When bodies fall, they fall as if the bodies in mass are equal because all things fall together. That is the result of complete equality in mass during the fall. Only by having equal mass while falling would they fall equal and this takes mass out of consideration of attracting.

Every object on earth moves at a speed of 30 km / sec. around the sun. The sun takes the solar system that includes everything in it around the Milky Way at 300 /sec. Where everyone is at present the absolute speed can be determined because everything on earth moves around the earth's axis by spin that moves around the sun's axis that moves around the Milky Way.

Everything that moves have to move in a line while moving in a circle where the circle becomes a straight line that circles around something again. That is gravity where gravity is the relative movement and not the pulling of mass. That proves it is not the objects that fall but it is the space the objects hold that condenses because the space the object holds gives equal density as they have equal buoyancy.

On this principle I founded my concept of gravity where the Earth by turning around its axis forms Π and only that spin contracts to liquefy the space around the earth and no mass pulls mass because with mass other bodies will distinctly fall different. If it fell according to mass things will not fall equal, which means mass does not pull. Read on to learn about the conspiracy in science where science protects and will even change laws of nature to protect Newton's ideas.

When anyone reads the books I offer you will see how science always values the incorrectness of Newton over to the cosmos. It is the Universe that has insufficient mass to form the critical density in the critical density theory and needs dark matter. According to the conspiracy it is never Newton who is wrong in the assumption of contraction, but the

Universe is wrong to expand. The Universe must start to contract and not Newton who needs a review.

Newton's law of physics state that: This Newtonian principle declares that there is a force of attraction between bodies forming gravity... It forms attraction by the value of the mass of both solar objects that holds the gravitational force...
The gravitational attraction that is forming is to the value of the mass which forms the gravity, therefore the pulling results in the reducing of the radius between the bodies by the square.... ... the force of gravity therefore is reducing the radius running both ways and therefore reducing by the square thereof...*and then there are those who say I don't "understand Newton'* ...Therefore this must result in the moon and the earth reducing the distance between the two by the square thereof to the value of the mass both objects hold by coming closer by gravity.

This must then lead to an inevitable collision that must end the existence of both the earth as well as the moon. Let the wise calculate when is this event due and inform the human race when our final day of doom will arrive by using Newton's law of attraction by mass. $F = G \dfrac{M_1 M_2}{r^2}$.

Science says mass pulls mass. Then why is this astronaut floating in space? He still has mass and the earth still has mass and the mass must be pulling to form gravity. The man is floating on the same principle that the moon is floating. If the man is not going to fall, the moon and earth will never collide and that makes Newton's mass pulling mass a contradiction that makes science a joke. If Newton's "mass is pulling mass" is correct, then the moon and the earth must collide. So which is it because it can't be both? The man has mass just like the moon has mass and the earth still has the mass it always has and the law of attraction between the human floating above the earth must be the same as the moon floating above the earth. If the man could float then this law of attraction is rather becoming suspicious. If that is the way you think then having those suspicions are very correct. Science confirms that there is always a force of attraction between all bodies. There must therefore be a force of attraction between the astronaut and the earth in the picture above, so why isn't he falling straight down like with Newton's apple? The mass has gone nowhere but science cheats the answer by giving the man "micro gravity". This way they cheat by avoiding the truth without pulling a face.

Ask your teacher to say without doubt which object with most mass will land first or will the lot land at the same instant? If the cargo haul of the truck was empty, will it fall slower or will it fall faster when the cargo bay is full and there is more mass doing the pulling? Get some truth out of them because they are there to teach you the truth and not the truth Newtonians create. By telling you what they wish to tell and forcing you to repeat that in examination only forms brainwashing and that is the basis on which science forms the basis for centuries.

The following forms the backbone behind the brainwashing in teaching. If I come to you as a mentor with a proposal about something I wish to educate you on condition that you pay me an amount to share my past and proven knowledge with you about what I am sure is everything I know, then I am an honest academic wishing to teach you. Now I tell you one part knowing that part is flawed and not telling you the rest because I know that is the flawed part then do you think I am trustworthy? Am I the honest person you can entrust with your future?

I, as the tutor, know the half that I tell you, but it is unproven and the other half I do not tell you about is made up as science goes along while I also know that the half I tell you is rubbish but because I know it is rubbish I don't tell you it is rubbish because where will I get another job if I tell you that we know nothing about science. I only tell you the first part that is not proven and leave out the second part that shows the first part is rubbish and still I wish you to pay me for my services. Have you a name for such a person who will force another person to pay him to brainwash him by mind control because the tutor has absolute control over the life and death of the academic future of the brainwashed individual and therefore is willingly forcing this unfortunate creature in accepting what will never amount to form the truth? They are called Physics professors and rule Universities as draconian authoritarian dictators bent on sadism.

Let's investigate the falling as such and see what happens during the fall. The truck falls at the same pace empty or loaded and this falling is at the same pace in which the girl falls, which is the same pace as that which the frog falls. I don't quite see the role mass has in this! If the truck falls at the same pace as the girl and as the frog there has to be a common denominator in this process and since the common denominator eliminates size form and shape, we can eliminate mass. Mass brings distinction and the falling eliminates all forms of distinction. Something bigger must have more mass than something smaller.

Let us take this scenario to a waterfall. When I fall down a waterfall with a boat I travel the same pace, as does the boat. That could be because I am fixed to the boat by sitting in the boat. But my sitting in the boat has certain conditions and one is that I can remain sitting because I fall the same pace as the boat is falling. This is like the truck's empty cargo bay falling as fast as it will fall with a filled cargo bay.

If I fall down in a boat with the boat and the boat and me forming a distinctive unit, the unit falls at the same pace as the water that forms the waterfall falls. Should I at the time of my falling hold an empty mug in my hand and I wish to fill the mug with water, I then will have to move the mug against the flow of water streaming up the waterfall. I will have to thrust my mug upwards at a faster pace than my descending is casting the mug down because I am accompanying the mug down the waterfall. My mug will not automatically fill with water or if there was water in the mug my mug will not automatically empty with water just because the emptiness filling the mug will be at a different pace than the content that is otherwise the filling of the mug. The empty part is falling as fast as the filled part that holds the mass.

The mug being empty falls as fast as the boat and I. If mass pulls the filled part what pulls the empty part? The empty space in the mug is falling as fast as the mug will fall when the mug is filled to the brim with what ever can fill a mug to the brim. Notwithstanding the content within the mug or the content within the boat or the content within the water being within the waterfall, the very lot is falling at a similar pace. By lifting the cup while falling the cup will fill with water. I am not putting the water into the cup but I am exchanging the space that the water holds with space that the empty cup holds and my action in truth has no bearing on the water filling the space, which I then transfer into the cup. I am filling the cup with space that at that point holds water but the holding of water has nothing to do with the transferring of space.

If I leaped from the boat and fell I would fall alongside the boat. The boat will be empty but will fall at the same pace and as the same space as I fall notwithstanding being empty. The mug being empty will fall at the same pace as the boat being empty which will fall at the same pace as the water in the waterfall and I would fall. The space in the boat, which is empty if I do not fill the space, will fall at the same pace as the empty space, which fills the mug, and the mug will fall at the same pace whether the space in the mug contains or doesn't contain whatever can fill a mug.

The space filling the mug is falling the same as the water that would fill the space in the mug should the mug be filled with water. The space in the boat is falling at the same pace as I would fall whether I am filling the vacant space in the boat or otherwise filling the vacant space next to the boat. It is the space that falls and not the object filling the space that is falling. It is the space that is filled or not filled that is dropping down because the space being filled is in decline. If it was not the space that fell, the space within the mug would fill first as the mug and the boat fell because the empty space would first fill before it could take anything down. But since the boat falls as fast as whether it is being filled or not we can assume that the space which the boat fills or does not fill is falling as fast as it would fall whether it is holding the boat or I or the boat and I. The space not filled by mass also moves just as fast as space filled by mass. If Newton's idea of mass was valid the boat should only fall when it is full and when the emptiness was removed. But if you fill a hot air balloon with more air or with more emptiness it takes mass with it into the air removing the pulling. When the object such as the mug or the boat or I connect with the Earth the Earth disallows the object free motion by taking any more space the object claims through to the centre of the Earth. The object now has to give up the space it claims and take on new space that the object claims to flow by contraction to the centre of the Earth. In forming a blocking it resists the flow or the gravity or space lining up with the centre of the Earth. The flowing of space by contraction is gravity but the object being in the space that flows becomes and obstacle through which the oncoming space must drag in order to flow to the centre of the Earth. It forms a resisting of allowing space claimed to release to the normal flow when the object will not relent form in favour of gravity. This is only when the object touches the solid earth.

This resisting such relenting of form and consequently forming a frustrating barrier that blocks the free flow of space towards the centre is time displacement of space and this relenting of space-time flowing freely becomes the mass factor. The density and the resistance that the particles show forms the mass that implicates the degree of the frustrating or preventing or disabling of such free flow of space through time and the displacement of space during time is space-time notwithstanding whatever irrational connection Newtonians wish to add to space-time. Allowing space to displace through time to form time is space-time and that is gravity. All I ask is to read what I bring. Don't be a coward and stop reading as soon as you reach the point where I condemn what is in place! Just move past that to the point where I show what is wrong and how it can be corrected! Just judge me not for condemning what now is so apparently incorrect but for showing why I condemn what now is so apparently incorrect and what I bring to the table and offer as a remedy. See what I have to offer and not only what I am taking away. Don't set your sights on what there is to lose but take a view on what there is to gain! All you lose is the untruth.

Do not reject me on merits you do not wish to instate because you have the fear you are going to lose what is instated. Do not judge me by using your double standards that are useless in the face of the truth.

Rather look at the double standards you employ and do not judge me by using your double standards on me. Rather use your mind to detect what is double about your standards and then investigate with me what needs to change. Don't hide the truth. Don't hide from the truth and don't hide behind what you wish to portrait as the truth. Rather come out into the light for the first time in three hundred years and admit to the truth. Follow what I say and see for yourself what there is to gain by trying to detect what is wrong because we all know there is much wrong. The comet does not collide with the Sun and the Moon is not on its way to collide with the Earth in time to come.

Expand science and know the Universe for the Universe is the only aspect that has not the ability to expand. I challenge all of you Newtonians to prove $F = G\dfrac{M_1 M_2}{r^2}$ and not just to declare it proven because it is in use since the Dark ages. Expand your mind and double check the formula you all so vividly underwrite and support. Prove why you support the formula in a modern and a scientific way.

Explore the correctness that this formula $F = G\dfrac{M_1 M_2}{r^2}$ underwrites. Be a true exploring scientist and journey with me through the following pages while we venture on the quest to find and vindicate my incorrectness by proving the truth vested in the formula $F = G\dfrac{M_1 M_2}{r^2}$ that carries the entire physics everyone uses. Let us start where the lot should start and get two Masters together on one point of argument. Galileo said all things fall equal.
That says all things fall alike.

The first thing anyone brings in is the vacuum bit with the feather and the hammer and since we do not live in vacuum there is no chance of finding a feather that will fall as fast as a hammer. Since the feather does not fall as fast as the hammer we immediately jump to the conclusion that there are falling disparities because of the falling discrepancy we find between the hammer falling and the feather falling. Then what would give the feather the time to fall longer than the hammer does. Everyone concludes about mass coming into play and they are correct. But they are half correct while Newton still is completely incorrect by attaching mass to the entire idea of falling. Take away the resistance of the feather and replace it with something far less air resistant and one will come to a different conclusion. We have to dissect what factor consists of gravity and what factor represents mass. Then we have to dissect which part does mass play and what part does gravity play. The falling object experienced no mass while falling therefore the falling or moving must be gravity's contribution. While objects are in motion those moving objects is experiencing gravity. The object shows mass when the object has a tendency to move but the motion towards the centre of the Earth no longer takes place.

It is up to you as students to rattle their cages and make them admit they've been lied to as they are lying to you. If you do not accept the role as being zombies that is brainwashed then confront these academics that treat you with disgust and betray your trust. They might tell you the mistake is not that serious and the damage is small but then how will they know how big or small the damage is if they don't even know what damage there is or what the damage is. My books will serve as the light switch that brings the light to you. I charge your young minds to confront those fraudsters about the truth. If you reach the need you may download it because it is a fair bit of information. However, it is not my views that are your enemy but the brainwashing you are victim of that holds your logical thinking captured behind the bars of a jail of corrupted ideology that forms a culture.

Should you wish to discount this book merely on the basis that it does not comply with you parameters of submission that will be to your peril. Should you not wish to study this work because of whatever pretext or format not complying it will be your loss. In this book I show that I have discovered what all are looking for and that is THE FOUR BUILDING BLOCKS THAT FORM THE UBIVERSE. Because this book is very exclusively special it can't follow the normal route and because this is ground breaking science nobody can glance at it and form an opinion.

I assure you that you have never seen science in this format, notwithstanding who you are or what qualifications you have or don't have. What is in this book is new. When I say I prove the Titius Bode law I am not referring to the sequence but I refer to what forms the sequence and why the sequence is in place and these facts do not support Newtonian science or current opinions that form science.

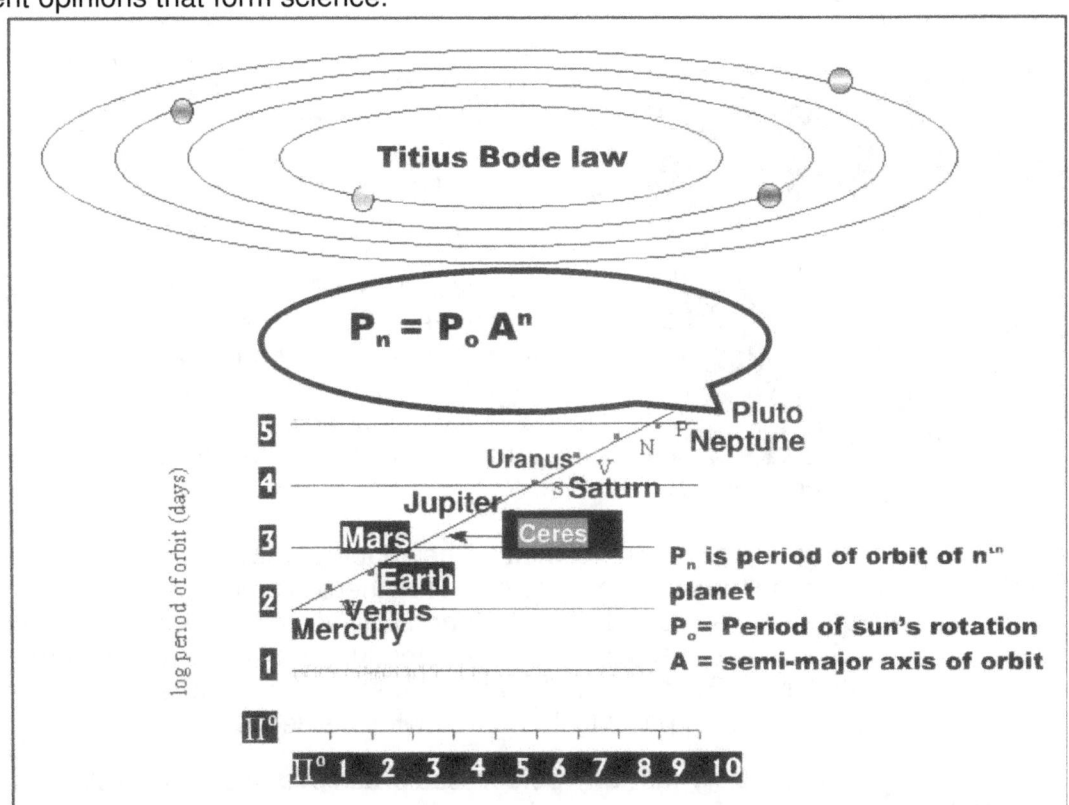

I am Peet Schutte the author of **Nature Annihilating Newton**. I realise my impressions of science and approach to science might not fall in the boundaries of the criteria of acceptance that you recommend or how you demand it must be but this book is not within the ordinary parameters of other books and therefore requires a better detailed introduction and description. Should you not care to read it because it does not fit your mould then that will be your loss or disadvantage. However if you care to read it you will learn about a breakthrough in science never experienced before in history and the reason why I have to submit it in the manner in which I do. The introduction of **Nature Annihilating Newton** calls for much more detailed explaining than what the ordinary topics of normal books would demand.

This book specifically proves the **Titius Bode law** and the law is as follows:

Titius Bode Law The Titius-Bode Law **is rough rule that predicts the spacing of the planets in the Solar System. The relationship was first pointed out by Johann Titius in 1766 and was formulated as a mathematical expression by J.E. Bode in 1778. It leads Bode to predict the existence of another planet between Mars and Jupiter in what we now recognize as the asteroid belt.**

A question they never ask or answer is what is starting where the cosmos according to their opinion ends because something else has to start behind where an end forms and that is the location of the start of something else. Or can they answer another question: What was in place of time at the time predating what is now in place during a time before what they say is there when they say the cosmos started and from what did it develop and why did it arrive at where we are now. Before what we now have, what was then in place and how did what we now have come into place? They present gravity as a magical force.

In this meagre presentation I provide a small part on an infinitive big answer about how everything called the Universe started but that is only after I annihilate every principle those in science so dearly protect and by doing so I uncover the corruption that they hide. The corruption and falsifying of facts are there for you to see. I show what they create to justify their correctness about what they claim they can prove. As can be seen by this little

presentation forming an example, I prove that Newton's principle has never been proven as long as it is in place. In short I would wish to have our scientists prove where mass is part of the planet formation as Newton so eagerly said it is. Use the planetary layout to prove Newton's "*Kepler's laws that has nothing to do with Kepler or with science and is completely fake.*" Then where I try to uncover this incorrectness they refuse to publish my work. You will find in this book I show what the response is when I sent an article in which I explain the Coanda effect and how that transcends to the operational functioning of the gyroscope and then

as a result why the gyroscope works.

If mass pulls why are there circles around planets? The image of gravity forming by mass is a conspiracy preventing the truth from coming out. That there is a conspiracy in science there is no doubt about. It runs as deep as it can go and involves every person that studied science, worked with science, teaches science and thought about science. In short it touches every person that lives. Everyone works with fake science and conspires to promote science fiction passing it off as Newton's science.

It is the first time in human history that any person could achieve this. The explaining how the **Titius Bode law** works as such has never been questioned by science or did anyone in science try to address this issue. I am able to do it because I don't oppose nature, pretend to be smarter than nature or reject nature. Scientists can't address these questions because they, the Brainy Bunch, are ever so smart though having no clue why nature performs as it does.

Yet as smart as they all pretend to be, their thinking is stupidly inferior. In stupidity they have the guts to try to replace God Almighty. They come together in big seminars and decide how to design the Universe and their design came to be pretty shitty. By creating a falsified Universe instead of trying to explain the one God Almighty made, they think they replace God Almighty while being clueless about the truth. I explain why the sun turns around the earth just as Ptolemy said it does.
Why would I do that? I do that because it is the truth.

...And yet with these books and the content within it is the most important books ever written on science about science. I for the first time ever in human history explain nature in the cosmos and how nature forms the Universe! I found that the cosmos apply gravity movement not by mass but by forming Π.

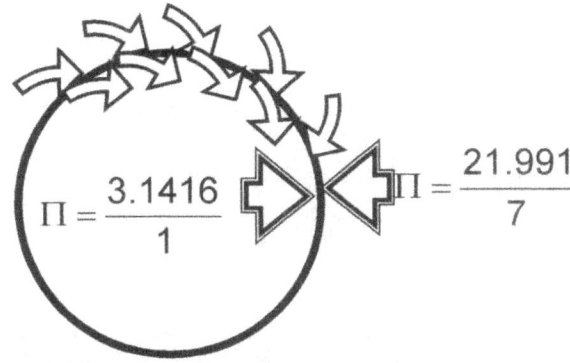

$$\Pi = \frac{3.1416}{1} \qquad \Pi = \frac{21.991}{7}$$

What makes the difference between reality and science is Nature does not use Newton in any way or form. Newton says objects pull by an unexplainable magical force called gravity while I say it is round objects that rotate. As it turns it forms Π and by collapsing Π the space around the star / planet, that space forming an atmosphere compresses and thereby collapses from 21.991 / 7 to 3.1416 / 1. As it turns it divides 7 from 21.991 / 7 by 7 to form 7/7 = 1. As the direction rotates Π changes the travel by 7° and that 7° is 7 / 7 = 1 or Π = 3.1416 / 1. Then by getting reduced it compresses everything in that space as the entire space the object holds and claims, condenses. Space reduces and Newton said objects in space pull each other. While I prove nature, Newtonian science cheats, corrupts and manipulates nature to make science work in ways nature doesn't work. There is a link between what we call empty space that compresses and material that spins and condenses space.

Anybody who wishes to read the truth about science has to purchase this book. The genie is out and science has to come clean about a three hundred year old conspiracy they kept hidden that kept science as backwards as science currently is. There is no mass pulling by magic and there are no forces flying around pushing or pulling anything by unclosed magic as a backwards alchemist three hundred years ago viewed science. The entire Universe is controlled by singularity maintaining time as time forms gravity.

Petrus Stephanus Jacobus (Peet) Schutte
e-mail: sirnwetonsfraud   http://www.lulu.com/content/e-book/nature-annihilating-newton/7570956]

Every person that ever studied physics will have to read this book to know what is the truth about what nature implements as true physics

Bode's Law works for moons around planets

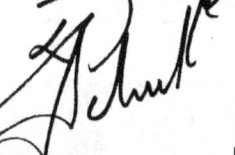

A

The book revealing how the solar system forms by way of the Titius Bode law.
With this book and four others also already published Science principles changes forever. Now nature is explained, which proves Newtonian science is without a basis of truth.
new approach to science is revealed in the book about just this issue and appropriately named **Nature Annihilating Newton** in which to introduce science truth. This is a big claim to make but see if you can reduce the claim after reading it. Now your first thought would be that scientists are doing the flip flop backwards with joy!

The Truth
PART 1

I wish to again reaffirm my commitment and motivation behind my writing the book that you are reading. It is, believe it or not, but it is about modern science never fully revealing the truth and as part of physics hides the truth by doing what those who don't understand physics but still teach physics do at schools and Universities.

mailto:info@questionablescience.net

Please do take note of the following warning:
The content of this book might seem to be intellectually challenging and reading it might require much concentration at some points because the ideas I put forward will be very new to any person reading it and will sometimes seem to call on intense detailed analysis of some of the new information I present in this book. One example for instance is reading how I prove the Universe started off at the very beginning before the beginning those in science refer to as the Big Bang.

I am able to introduce this breakthrough that I claim I achieved because I found the building blocks used in the building of the Universe and by applying the four cosmic principles I am able to go as far back as before the Big Bang event. The Big Bang only happened recently when atoms formed the known Universe. In this book I do not touch on this aspect but I go to the period predating the Big Bang era as I inform the reader how the start took place when the Universe first came about. I found the working principles of the cosmic principles called the Roche limit, the Lagrangian points, the Titius Bode law and the Coanda effect that are in place and forms that which forms the construction layout of what builds the entirety of the Universe. The horror in this is that notwithstanding the importance these principles hold almost no one knows about them.

I ensure you that what you read even in this book was never written before and absorbing new information are difficult. You are confronted with facts you never heard of before. So therefore notwithstanding the level of the person's academic qualifications or how well developed background the person has in the field of science, this is not your average garden-variety storybook. However, in the end after you have come to realise the conspiracy in science, you will be a lot wiser than you were before. The information might ask for a lot of your attention span at times, but then again challenging the mind to think is what makes life that more interesting!

I found a mistake and then realised I had to search for a solution to the mistake. First I had to locate what was wrong in physics and from there formed a correct approach. I compiled my presentation of it in a theses that I call The Absolute Relevancy of Singularity and then six separate thesis parts forming the theses published through LULU.com which I saw as the only manner whereby I could generate funding by which I would be able to have the twenty seven books I already wrote linguistically edited and then to have the books published on a Print-On-Demand basis. I compiled a new cosmic theory by which I eliminated all the incorrectness that Newton has burdened science with but with this being my opinion I did not find a garage full of academics supporters waiting to applaud me and to uphold my views on the matter. Yet still I was not going to be ambushed by their relentless stonewalling my efforts and blocking my efforts in introducing both the incorrectness and the new cosmic theorem I concluded. Their mannerism in blocking and frustrating my opinion when showing the mistakes in science convinced me about a Conspiracy in Science in

Progress and this spurred me on to tell the entire world about their brainwashing students minds. By the manner they selectively withhold information when teaching science, amounts to deliberate brainwashing of students in physics by "normal" education practises.

Trying to convey my message kept me busy for the past going on to twelve years on full time basis whereby I was trying to introduce my findings to many academics without finding much joy from my efforts. This past eleven years plus saw me go without any income as I tried to get my theorem recognised as well as get my warning noted. Going without a steady income left me almost destitute and in order to find a manner to get my theory across to the attention of influential readers, I decided to publish a theses of six books electronically as to try and get around the stranglehold of Newtonian bias controlling science at present worldwide. I decided to publish electronically which those in power do not control. However to get people to believe me is to change science that everyone believes as culture.

With my first language not English and the books not linguistically checked by an expert there are bound to be language errors that readers will notice. In the past I tried to check my work myself but after checking say one hundred and fifty pages for language corrections, then after days of toiling instead of having corrected work I ended having four hundred pages of newly written information which is still not linguistically corrected but holds a lot more information. The language and spelling errors compiled instead of reduced. This is because my priorities lie elsewhere. I aim to spend money on correcting the work as far as language goes, as I receive money in the selling of my theses and in the hope that I will receive money. I will have all my work including the one you are reading edited professionally and corrected as I find money to do so...But first I have to get the public aware of the problem to get the academics to appreciate the problem. In everyone's mind science is more perfect than religion is.

In the event of any readers who may have questions concerning more facts as it is presented in this book, please feel free to contact me, PEET SCHUTTE. All information divulged came about through independent self-study during the past thirty-two years or so. I have to warn the readers that the topics are showing a very new approach with no quick answers abstaining from proof or holding just a few lines and the information is new in nature but not hard to grasp. Should anyone desire to contact me with an opinion about my views on science then please do by using the worldwide web addresses mailto:info@questionablescience.net or mailto:info@singularityrelavancy.com but please do so when coming to the end of this book at least and then see if you understand the entire concept that I introduced. In most cases all your questions are answered further down the line and I am not fond of repeating what is already said in the book. The book forms a line in explaining and the concepts are best understood if the reader follows the designated line. This book started off as a website to inform about a science conspiracy but although reduced still it grew into a book that serves much more information than what I first intended to supply. You will see many new aspects about gravity please make sure you understand what you read. It grew into a comprehensive study on cosmology. At times you may observe while reading this book that it seems as if my frustration will ring through like the chiming of the Big Ben Bell. For that there is a reason. At times my frustration and anger will boil over drowning my politeness and that is true, which I admit. For twelve years I have had the answer that would correct the philosophy that has a stranglehold on cosmological science.

I discovered the building blocks of nature where my discovery puts all other cosmic aspects of science into science fiction. Those who force-feed non-existing dogma do so to brainwash students to hide the incompetence of "modern science" so they can rule supreme while ignoring the truth that they deliberately hide by concocting a conspiracy. To keep everyone unguarded they practise a conspiracy by which they perform an accepted practise of thought control on students to further the false dogma presently in place. I try to blow the whistle on such a practise but accepting my resolution makes every thesis ever written science fiction. Therefore no one in science dare to read my work leave alone appreciates the revolutionary nature thereof. Whatever now is deemed to be accepted science would then become what is the past tense in science because the flaws that those in power of science principles kept coated for centuries on end as untouchable truth will then be rust that breaks the surface to show the holes! They try to silence me but surely somehow somewhere I have to break through with my massage! I bring you a true form of science as never seen before in all of history and I do that when I dispose of the conspiracy that hides all the incorrectness and the failures that haunts science today. Science is accepted as the most righteous information available to man and that is a scam.

There is a conspiracy in science and about this fact there can be no doubt. Once you finished this book and you still disagree about the conspiracy going on it then will be because you are a professional physicist that are paid a huge income to spread the falsified science and contribute to the well being of the conspiracy. However, you will be convinced either way. At the end of this book you will believe convinced that it runs as deep as science goes and involves all person that studied science, worked with science, teaches science and thought about science. In short it touches every person that lives. Everyone works with fake science and conspires to promote science fiction passing it off as Newton's science.

I am Petrus Stephanus Jacobus Schutte going by the nickname of Peet and I am a married male, with a sane mind and I hold very sober habits being a lifelong teetotaller, therefore my mind is clear. When you read what you are about to read it is not the hallucinations of a drunkards mind.

I see myself as very responsible and am not well liked because of my straightforward personality where I say what needs to be said to maintain honesty above friendship and what I say is always not infusive or congenial to make friends. I wrap no feelings in cotton wool for the sake of peace. I mention these facts to establish beforehand that I am not a danger to society. Although I am the only one in the entire world that disagrees with modern mainstream science I am not criminally insane. I would not attack your dog before your dog will bark at me just because I hate your dog. I would rather attack you and not your dog after he barks at me just because I don't like you but I do like dogs. I say this to convince you that I have never been jailed in all my life on grounds that I attempted to mislead or fool or tried to mislead anyone, as you can see. I try hard to be honest

The proof of my honesty is that I am one of the poorest people in society. I write books that don't sell because I try to convince people about a mistake no one on earth is aware of and therefore no body bloody cares except me. I am the only one that can see a mistake and seeing it in the full consequence thereof by what it holds to the entire human race it is frustrating me senseless. Take it from me that not selling books is not a very profitable enterprise and is frustrating at any level.

If I was less honest and went about bullshitting everyone about the accuracy that Newtonian science portrays in the modern era it would at least sell some books because some dinosaur Newtonian physicist would be pleased about it but then I was a cheat although I would have been richer but being as poor as I am I in favour of being as honest in what I am, I can't be a cheat.

I studied physics and in my first year of studies I came upon a mistake concerning physics. My roommate told me it was Newton that said what I disputed and I remember telling him I don't care who this Newton fellow is but he is wrong and ever since then I stuck to what I said. That discovery made me disagree with the establishment forming principles in physics. Later in time I have detected much more than a mistake. I have discovered a Pandora's box of mind manipulation and brainwashing going on to force students to accept science notwithstanding.

If you feel I am exaggerating or that I am a mentally impaired asylum escapee that found a writing pad, ink and paper and that I now start to scrabble senseless suggestions to while away my social frustrations then I challenge you to prove that students are **NOT** being **brainwashed** to **believe** what they are taught in **science**. I started off being likable and being nice but the deeper I delved into the conspiracy the less I could care about who thought what about what I said because I am going for the truth as hard as I can. If you are a student you better read on...and if you are a teacher I advise you to read and get wise in realising what you do. I am going to give facts that will support me in the claim that science is all about falsifying facts to prove what is crooked in order to make those that should know science look good while in fact those in physics and teaching physics know less about physics than does a cat know about opera.

There are so many conspiracy theories going around, floating like a bad smelling odour in the room, but most of those have no more factual substance or evidential backing by truthful facts than a James Bond novel has. I also have a science conspiracy theory but mine has all the proof and all the substance of authenticity that any argument can ever hold. It involves **physics** as **physics are taught** at institutions worldwide. If you now don't believe me, read on and learn the truth yourself, it is free of charge.

Mine is the most unbelievable one that has most truth. This book is not about any conspiracy ever mentioned before but takes into consideration what worries people.
When you see what science achieves and you hear what I accuse science of, then it seems as if I am a nutcase who escaped from the loony house and I must be the one having very dangerous tendencies to harm the innocent.

I should be locked up and taken from society.

However, read my story and you'll see that I don't harm the innocent. It is those we all trust without reservations that brainwash to forge deceit through corruption and malice.

The blame is with those admirable, yet astute members in our society, the educated leaders in physics that we trust without even thinking about their trustworthiness or question their sincerity regarding their blameless honesty.

 Do you believe in the absolute unquestionable correctness of science and that everything that forms science is truthful being far beyond suspicion?

Do you believe that those furthering the science of physics works with no less than absolute proven facts and the facts are beyond any doubt at all.

If you do then I have very bad news for you, **news that will change the grain of what you believe is truthful.** I am about to shock you into reality!

 This is going to reveal the biggest conspiracy ever concocted by the human mind, ever since the time a human had a mind to use.

It is A Conspiracy about Fundamental Science. It started when Newton gave us the idea that gravity is the result of mass and in gravity mass is the biggest contributing factor.

I say that mass as a factor in physics and not the value of weight, but mass with the pulling power only pulls stupidity over the eyes of incompetent idiots raving in their personal sublimation. Let see what I say. Would those in science cheat you and withhold the truth from you?

<u>www.singularityrelavancy.com</u> that is telling everyone about
An Ultimate Science Conspiracy, the biggest conspiracy ever concocted.

NATURE put TIME into place TO PREVENT Everything FROM HAPPENING all at once.

This proverb sounds silly and yet, silly it is not.

While this proverb is the truth, science say they don't know what time is...and this is ignorance they hold while living with time for millennia all the while Newtonians say they uphold Galileo and everything Galileo implemented. Gravity drives the Universe as the earth spins around.

In other words what science can't see is that there is a time –connected plan in the form of gravity driving the Universe along. Time is in the form of gravity and the cosmos is driven by time. Time and gravity are the same, which is precisely what science is unable to see.
If gravity drives the Universe, the Universe is powered by time.

When Galileo's pendulum can measure the flow of gravity and record time to a precise science while swinging in space, then the flow of gravity in space has to be measure of the pendulum swinging in air while reading time. The pendulum measures time when swinging through the air that surrounds us and what surrounds us is there because of gravity.

This obvious conclusion about gravity forming time, which is what the pendulum reads, any child can draw but as simple as it is, this went past Newtonian understanding of science for centuries.

They never could realise the flow of gravity and time is one and the very same since it then conflicts with Newton's ideas. Still the pendulum proves that reading time and measuring the flow of gravity is the same. Galileo used gravity to measure time accurately, yet for as long as the pendulum recorded time, this fact eluded science in a way that science doesn't even know about it. It is what science doesn't know that becomes important to man. If gravity measures time by applying a swinging pendulum this becomes the biggest fact in physics and it still went past science because their cleverness brought on their rigged stupidity.

This religious belief in the infallibility of their Godly insight becomes their stupidity. Their ignorance is what science hides. It is that what they hide which is what you don't see. Science hides what science doesn't know which they cover under the larger pretence of their cleverness.

Also science hides from everyone's view what science doesn't know so that students will never realise Newtonians are covering their stupidity and ignorance behind a curtain of arrogance. This conspiracy to not reveal their stupidity then becomes the conspiracy they nurture for centuries. Science conspires to hide by covering their stupidity from open view. The main thing is that science is clueless about time and time is the driving cosmic plan so science hasn't a plan to see the truth.

I have been in fruitless conversation mainly going one way about what I see as their stupidity in defending flawed science. I told them this but with having no response. Still notwithstanding their arrogance about ignoring my showing right from wrong, still those physicists filling high academic office are so infatuated by their superiority and their personal righteousness they can only see the malice that the Pope showed towards Galileo when he differed from the general science views and yet they do the very same today. They are overwhelmed by their correctness.

You will see how those hypocrites condemning the pope having the holier than thou attitude because they think they are cleverer than all other intellectually lesser sub-humans. They accuse the Pope of constraining science progress while back then but they torment me in much crueller ways than the pope did to Galileo. At least the Pope allowed Galileo to print a book while they did everything to stifle my efforts when I just tried to be heard by others about my views!

Whenever anyone show those in science that their religiosity called Newtonian science is wrong, they act in exactly the same way as the Catholic Church did to Galileo five hundred years ago, in no less a manner while they uphold in public that they maintain their stand on condemning the Church for not excepting Galileo's principles. The crude thing is that I prove all those filling academic office in science have also still not accepted Galileo to the letter and I challenge anyone to prove me wrong. That is one of the parts of the big conspiracy I see and call it the Mother Conspiracy!

By the swinging pendulum in space while connecting it to the earth Galileo proved gravity is time and that principle goes unnoticed as it diverts from Newton's idea that "mass" producers a magical "force" of a "pulling nature" he called "gravity". This view they can never explain while this is what they will always defend as the truth.

 Those oh, so clever Dark Aged wizards pronouncing the upholding of free speech and science liberty for all of mankind while hiding behind their superior positions will befall the same memory as what befell the Pope in the days of Galileo. They will be remembered for being the last of those that were stupid enough to believe that an inexplicable magical "force" of "gravity" "pulled" the Universe by "mass" while E.P. Hubble showed the Universe is not contracting but otherwise expanding.

Their sublimation covering their stupidity causing their ignorance will outlast them for all time to come. As the Pope five hundred years ago is remembered for his ignorant stupidity, so those currently in office would also befall the same fait just because they were as stubbornly arrogant and ignorant as the Pope was back then when ignoring mistakes.

They conspire to conceal the dishonesty that they **don't wish to reveal.**

I WISH TO DEFINE THE CATOGORISING I USE AS PART OF THE BOOK.
I have the utmost admiration for Scientists and I shall never dream of placing me in the same category as academics mainly because of their intellect and achievements. They pushed their corrupt conspiracy of a hoax they present as science and which they further by brutal brainwashing through over 300 years of never getting detected and that in itself is an achievement unheard of in human history. I will not be able to bullshit anybody for 300 seconds because I am not built to be dishonest and Newtonians went about for far longer than 300 years fooling even the brightest among us on earth. That achievement is most brilliant and no religion of magical mysteries in the past could ever match the Newtonians. Every time I go against Mainstream Science which is another name for upholding Newtonian blindness I am told I do not seem to have the intellect or mental capacity to **_"understand Newton's classical mechanics"_** and then because of my limited vision on physics I should know my place and retire to a dark corner where I would then silently and quietly vanish from earthly records. They forever tell me there are two positions on earth: those with the mental capacity **_to understand Newton_** and then there are those in my sector **_that is mindless to the point of not understanding Newton._** In that sense there are two classes, the clever

ones that **_understand Newton_** and then me, the mindless that just cant **_understand Newton._**

To substantiate this segregation I use some referring to place distinction between the highly schooled super trained academics that spent most if not all of their lives in preparing to further their minds, filling it with the same void they fill the Universe and calling it "nothing". When I asked where is more nothing: Between Pluto and the sun because Pluto is the furthers from the sun holding the most "nothing" between it and the sun or in the centre of the sun because there is nothing standing between the sun's centre and the centre of the sun, I was discredited as incoherent and irrational. I tip the opposite of the scale as I spent little time repeating the brainwashing they subdue every student with to believe in the norms taught as the official policy in learning and education I have to be on the "other end". I don't believe their crap and tell it as I see it and therefore I am dumber than a pig, or that is their opinion. From where I stand and admire those in science, I can only see intellect as they fooled every person on earth for centuries non-stop: and moreover that achievement is presented as the academic's common denominator. If that is the common denominator used on the one side, fooling everyone by using unsubstantiated rumours and gossip and putting that as the joining factor, then on the other side, which has to be *"my side"* must then be the class of stupidity. To those forming the brilliance in science and their class such a remark would be an insult but to me (and therefore my class) it rings truth and that makes it not an insult but a norm we should except and learn to live with. I would rather be stupid and not **_understand Newton_** than be **Brainy** and believe I **_understand Newton..._** how stupid must I be before I would be able to **_understand Newton._**

It is rather a pity that while the SUPER CLASS will never say it to our faces; the SUPER CLASS is strongly of such opinion that we on the other side of the Universe have no minds to think in any way, and it is therefore our duty as much as it is our absolute privilege to except what the SUPER–EDUCATED, the ones occupying the informed side of the Universe inform us to what we should accept and the SUPER–EDUCATED live by that idea. As I said I have to live with it too and if I am the ill literate, then the SUPER CLASS must be the SUPER–EDUCATED; where I am the class amounting to stupidity the SUPER CLASS must be the Brainy Bunch. It all comes from the fact that there is such a huge differentiation between us. Those that **_understand Newton_** is therefore Superior and I, that don't **_understand Newton_** are of the lesser blessed. To distinctly point to grouping or class or whatever the readers wish to consider the division there are between the SUPER–EDUCATED and me I refer to the SUPER–EDUCATED side of the Universe by the names I use above. Further more when I refer to mistakes that I do prove to be mistakes in the book as we go along I refer to it as Xepted mistakes to clear another distinction of necessity. In short I don't **_understand Newton_** and therefore I am stupid and they **_understand Newton_** and therefore they are brilliant and what I present must hold the categories in such class divisions.

If you read on you are going to Read About The Biggest Conspiracy ever enlisted, it's more than just the next conspiracy because it is A Science Conspiracy Network. I prove everything I say about the conspiracy I announce...read and be shocked.

This Network which I denounce Engulfs the Entire Civil Human Race in Every Aspect and willing or not, you reading this are participating without knowing about what you support with every part of the mental intellect you have. It involves the most trusted members in the part of our civil society and we are deceived by the ones EVERYBODY trust the most, the teachers and Professors teaching at academic institutions everywhere. For over a decade I have been knocking on doors with the information I present in this book and lots more evidence, just to be ignored and to be turned away. This behaviour of physicists ignoring me or even attacking me when I point out a very legitimate case of mistakes in science made me realise there is a conspiracy going on. Science has been getting away with this conspiracy since the Dark Ages. What you are about to read is not for the simple minded because science requires much intelligence and in that physicists get away with turning Creation into a joke. By turning the facts that form the conspiracy into what you think of and you accept as natural and as culture, man has been cheated and mislead for three hundred years because those teaching science consider everyone as representing stupidity. To them everyone else are simple-minded peasants who are not worth having independent minds who could be able to think on their very exclusive superior level that they are able to think. That is how they get away with the conspiracy. They pretend to be intellectually superior and to understand what seems senseless to the rest of us and then look down their noses on the rest of the world. I will show you what it is they can't see and then miss. They underestimate the public as much as the public are lazy too think and the misleading carry on. Those in science think their mathematical abilities make them Gods–on–earth while we others are mindless human fodder and for that reason they brainwash our children at school by applying mind control as their mathematical understanding makes them gods. What they say all must believe because if they speak using mathematics then God spoke.

This conspiracy is so widely active that every person on earth teaching physics participates, most probably without knowing, some reluctant and others well knowingly to set out to brainwash by thought control the minds of innocent students forcing them to believe in dark aged ritual beliefs in the ritual practicing of forces and allow the conspiracy to be enlisted with all the vigour they could muster. People would argue religious concepts but everybody accepts science on face value. The guilty party that I refer to betray us in every way possible by teaching us lies that never could be true and all the while we believe them with all our hearts and entrust them with our entire future... They take money from parents with the sole aim to brainwash the students who are entrusted to their care and demolish their intellectual understanding ...and before you think my claims are exaggerated I say again I prove every word I say. I show what is true and what is wrong in science and as a thank you for all the concern and devoting care I show, the establishment of science frustrates me, ignores me or runs me down, while I challenge everyone and any one to disprove me. Decide after reading this book weather what you read is not a conspiracy...however, familiarise yourself first with what I present and then decide.

Those I refer to are allowed the powers to take as much tax dollars as they wish and endeavour on all sorts of projects that they could fantasize about. Their budgets are limitless and they never face questions because they are the intellectuals that we dare not question. With the same powers we give them they brainwashed your child making science a religiosity. The methods they use on your child are to make your child believe in science and that is just methodical corruption of their thinking. Physics teachers take money from parents with the sole aim to brainwash the students that are entrusted to their care so that students will believe in science unequivocally and regardless. When teachers teach falsified information then they are worse than the mafia, that operates openly by crooked intention and in that the Mafia maintains honesty, which the physics teachers don't have. If you are a parent then mind what you do and beware and read on! If you think I am going one step too far in accusing the academics teaching physics and I slander their names, I challenge you to ask this question again after you have read this entire book and thought about everything and considered every aspect.

Please study Symeof's criticism. If you are not prepared for the shock awaiting you about the conspiracy in science, you should not continue. It will be as big a shock to you as it apparently is to Symeof. You will as he did realise that so many years of study has gone wasted because what you learned was one big hoax, covered by the mother conspiracy in science and as a result of this awakening you will feel to lash out but don't try to kill the messenger. There is the mother conspiracy, which I expose. If you are unprepared for the shock then reading this book will be more harmful than never knowing the truth. To find all your professional knowledge came to nothing must be unbearable and insufferable. The mother conspiracy is in place but you decide if you wish to know about it or not, the choice are yours to make.

What you are going to read is the mother of all the conspiracies in science, which is about how science applies mind control by processing thought control. Every conspiracy ever linked to science is in place to protect that conspiracy from becoming known. I prove that there is a mother conspiracy in place. The mother conspiracy is in place whereby students are brainwashed through the instigation of mind control through enforcing the acceptance of dogma on students.   I also introduce a new cosmic vision with the entirety called the Universe, which is formed by singularly taking on every shape, and space that we know. Singularity is the point where the Universe first started according to Einstein.

Those making such remarks as Symeof does must remember that persons on a higher level of education can immediately gauge the level of your education development when it is so inferior. This person might seem highly educated in his own eyes but it is clear he is on a very low level of understanding physics. This person clearly never heard of the cosmic laws named **1) the Lagrangian system 2) the Roche limit 3) the Titius Bode law 4) the Coanda affect**, which I explain by delivering mathematical proof as to how they fit into the overall picture of gravity and which I mention and explain in much detail.

Reading this remark it is evident that Symeof never came as far as the explaining of the four cosmic laws or such explaining as I give went past this reader without him noticing the explaining as it was too far advanced and much above his level of understanding. That indicates that the person never understood the explaining of the laws and therefore has a very small insight and a low level of understanding physics. It would be much wiser to shut up and get wise than to advertise your uneducated stupidity to the world as you did. When you say "*The author has no legitimacy in disputing physics: he is merely an amateur trying to disprove what the greatest minds of our times have taken centuries to prove*" with that remark you are trying to dispute what is apparently very much above your limited understanding of physics and that which I prove you did not even come to read. It is a pity but then again we can't be all intellectual.

When I refer to brainwashing students into believing this is the exact example I refer to. I could not have asked for a better example even if I ordered it myself. This is the typical learn by repeat and never question those teaching you because questions will uncover the mother conspiracy and that the teachers avoid. The mother conspiracy is in place to teach about what is not present in the cosmos like telling that things such as mass is positioning planets while never mention what forms the true basis of cosmology which is **1) the Lagrangian system 2) the Roche limit 3) the Titius Bode law 4) the Coanda affect** Those who are on a low education level such as Symeof so clearly is, would never have heard of these laws. I explain for the first time in the history of man how these laws apply and how to read gravity from their applications. However he never came to know such information is in place and thus informing him and those such as he is almost impossible. When explaining these laws I have to discredit what Newtonian science uses because what they use has no validation or credit, as I will show very soon.

This is how the interpretation of physics would apply if Newton was correct and mass did position planets. I am, going to show everyone with even a childlike understanding of reality that there is not even a remote chance that the positioning of the planets go in accordance with mass or $4\pi^2 a^3 = P^2 G (M + m)$. This is the formula Newton introduced and it is the formula that all those studying physics still are taught to accept as the valid formula. Do you realise there is much more "gravity produced by mass" in the space your feet has contact with the earth than there could ever be between Jupiter and the sun when using this formula

$$F = G \, \frac{M_1 M_2}{r^2}$$? But this is part of what Symeof never can understand. It shows that those who study physics never tested what they were told apply in cosmic physics. They never put the correct mass as applying numbers into $F = G \, \frac{M_1 M_2}{r^2}$ and put in place the radius between the sun and whatever planet and come to a mathematical conclusion. Any person trying to put mathematical values in this equation that is supposed to be in place to show how solar bodies are positioned will realise it has no value. You that can calculate it

$$\left(\frac{P}{2\pi}\right)^2 = \left(\frac{a^2\sqrt{1-\varepsilon^2}}{\ell}\right)^2 = \frac{a^4(1-\varepsilon^2)}{\ell^2} = \frac{a^4(1-\varepsilon^2)}{a(1-\varepsilon^2)GM} = \frac{a^3}{GM}$$ so then do it. Show

that $\dfrac{a^3}{GM^2}$ is valid. Please show how mass by $4\pi^2 a^3 = P^2 G(M + m)$ can produce planet positions. It is hogwash. Look at the picture and try to put mass into perspective in relation to planet positions!

Using $\dfrac{a^3}{GM^2}$ has no validation because it says nothing and yet whatever anyone should understand about physics comes directly from reading this formula. Please let any person put in the mass of Jupiter by the square in relation to the body at a cube and show what Newton then calculated! This is non applicable and as senseless as trying to mathematically prove that Snow White had a castle, if she did have a castle. To say that the square of the position of any solar body is directly in relation to the mass of the body $P^2 \propto a^3$ is as senseless as trying to prove one can use a wind glider to go to the moon. Even looking at the picture and having the smallest planets on either end and the biggest in the middle must ring alarm bells through the centuries and yet in over three hundred years there was not one that detected this anomaly and if there was then did not speak out. This is a conspiracy they keep quite about.

Newtonians make a statement about "mass" holding the solar system in place. No matter how much this is corrupt, nevertheless they put it down as a given fact so much so that they will show doubt in a living God being present but that mass pulls planets goes beyond doubt. The proof of mass pulling to form gravity can never be tested because it is beyond doubt. If you doubt in it they throw a Newtonian made formula they call Kepler's laws at you. "Kepler's third law" supposedly is "the square of the orbital period of a planet is directly proportional to the cube of the semi-major axis of its orbit." In mathematics it is Symbolically: $P^2 \propto a^3$ and therefore a³ = P² in position of P and therefore a³ / (P²P). This is taken from the idea that "Kepler said", which is totally fabricated that $a^3 = T^2$ where Kepler said in fact $a^3 = T^2 k$ and this is a big difference because $a^3 = T^2 k$ is the same as E=mC². Look at reality. a³ = P² is total garbage and as big a hoax as is the idea of mass being any form of factor in gravitational physics or that gravity applies in accordance with $F = G \, \frac{M_1 M_2}{r^2}$. Look at the picture below. Look at how the planets are sorted and that is not by size. There is a ratio applying called the Titius Bode law and this law puts planets in terms of size or mass at a precise equal base notwithstanding that Jupiter is many time bigger than Mercury. Everyone is so taken by the accuracy of Einstein's formula that E=mC² but this is exactly Kepler's formula where $E^3 = mC^2$ is taken from Kepler's formula when accurately used as $a^3 = T^2 k$. There is no a³ = P².

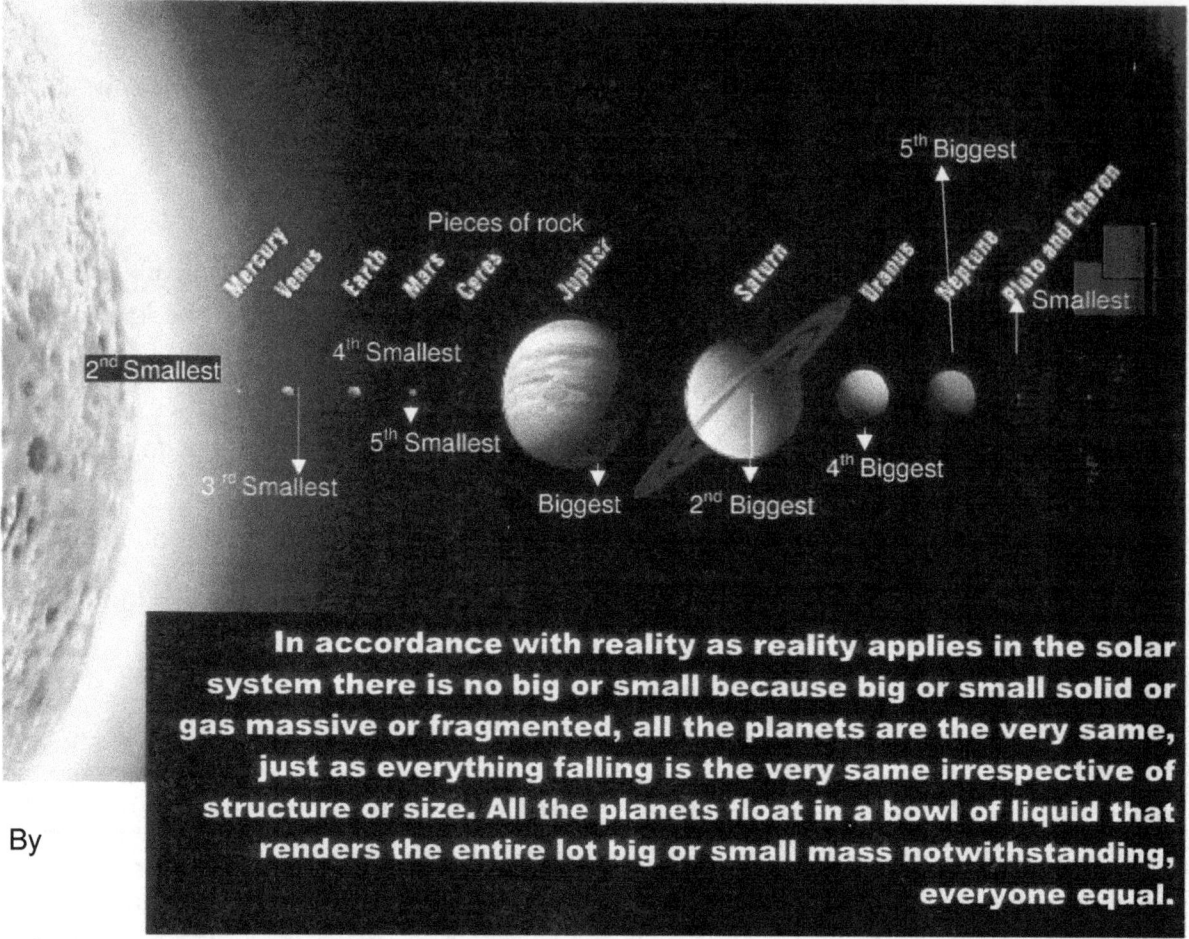

In accordance with reality as reality applies in the solar system there is no big or small because big or small solid or gas massive or fragmented, all the planets are the very same, just as everything falling is the very same irrespective of structure or size. All the planets float in a bowl of liquid that renders the entire lot big or small mass notwithstanding, everyone equal.

By

depicting the solar system in such a presentation as Newtonians normally do such as the picture next this form of presenting the layout without providing correct spacing purposely corrupts the entire structure formation by which the solar system develops. It then purposely hides the essence that forms the solar system

This is so typical Newtonian in every sense there is in science. Can anybody, even those with the mentality of Symeof and his gang see that the planets are not arranged from the biggest that is most massive and then therefore should be closest to the sun and smallest way to the outside as they should if plants orbits P^2 was the result of size or mass $G(M + m)$ and mass has no place in the layout?

This is so typical Newtonian in every sense there is in science. The Newtonians gave the Titius Bode law a formula and that explains the lot. To they're under achieving standards that is very satisfactory. Now it is written in mathematics then what more do we need to know? The fact that the distance that Mercury has from the sun is doubled by that which Venus has from the sun is completely ignored. In cosmic reality mass plays no part. Then again the distance that Venus has from the sun is doubled by that which the earth has. This clearly has nothing to do with the size or mass of the planets. Explaining that part is completely ignored. Then again the distance that the earth has from the sun is doubled by that which Venus has and inexplicably this forms the layout of all planets in the solar system. Where do Newton and his idea of mass fit into what truly applies in outer space. Moreover, why does science never mention this? This is my formulated explanation about how the Titius Bode law forms.

If you think my accusations are baseless or the ravings of a madman then go on and download what you have opened and read for yourself. What you download is free and I do not benefit financially from this explanation I present to you. There is no such a thing as mass anywhere in the cosmos. If there were a factor such as mass every planet would orbit distinctly positioned according to mass, but they don't. Should you think of the size of a body containing more or containing less material and put that in terms of mass that forms gravity then the orbital layout of the Universe or solar system would very distinctly NOT be the way it is. The cosmos shows no mass as a factor and we can either regard the cosmos as correct or Newtonian science as correct as Newtonian science diverts totally from the physics that the cosmos displays. The choice to make is do we believe science or do we believe the cosmos you choose?

Show that the square of the mass in relation to the gravitational common factor puts planets in their allocated positions! Put the orbit of Jupiter in relation to the mass of Jupiter and in relation to the position Jupiter holds. Forget getting swept away by the fancy Mathematics; just get to the task of putting the mass of any of the planets in relation or ratio of that particular planet has and then in connection with the position that any of the planets hold. Don't come up with the argument that science works and therefore Newton works. Please then show as he put it: _The author has no legitimacy in disputing physics: he is merely an amateur trying to disprove what the greatest minds of our times have taken centuries to prove (by the way, you need to have studied a subject before you can criticize it, which the author didn't). In my opinion, the author is so narrow-minded that when he came to scientists to talk about his material, the scientists must have told him that it was nonsensical._ That is a lame excuse to hide incompetence. If you are unable to do it your physics is a giant fraud, which is based on a century old lie and a hoax. It then shows your small-mindedness because you have never put physics as it is taught to the test. This is taken from the idea that Newton had when Newton changed Kepler's formula from $a^3 = (T^2 k)$ to $a^3 = T^2$ because without any legal mathematical backing Newton said the third dimension is equal to anything holding a second dimension or a cube a^3 is equal to a **square T^2**.

Mathematical reality is that any person in the third dimension a^3 having three sides can climb into a mirror T^2 being absolutely flat and then climb back to the third dimension because $a^3 = T^2$. This is the garbage those slow witted person's such as Symeof failed to see when he was taught that $a^3 = T^2$ because Newton said so.

Reading this mathematically encrypted coded formula of the cosmos given to Kepler and keeping it removed from Newton it reads as being that the space a^3 is equal to = the motion T^2 of the space a^3 in ratio k to a centre k^0, which is relevant to the positioning of k. If we bring in the full equation it will be $k^0 = a^3 \div (T^2 k)$ which means half of space is solid $k = a^3 \div T^2$ and half of space is liquid $k^{-1} = T^2 \div a^3$ where liquid is moving. However, it is also true that everything through movement defines a value in relation to one point holding singularity k^0 and that is what the formula $k^0 = a^3 \div (T^2 k)$ underwrites. What this proves is that gravity is the motion of space provided by time being the liquid. Please allow me to explain. In the formula $a^3 = T^2 k$ the space forms as the space is in motion. Newton suggested that $\dfrac{dJ}{dt} = 0$ where he stopped time to have the motion of the circle demolish the work that the circle does. It is because the chambers of physics are filled with brainwashed victims such as Symeof that physics is the joke it is. To say a square is equal to a cube is a joke but then Symeof says that I as the author has no legitimacy in disputing physics: Please address the following facts.

All of the above is way past the level of understanding physics of a person such as Symeof and in his critics as he showed how little he understands of what should be understood about physics. When a body touches or contacts the earth there is a factor such as weight but when you call that weight by the name of mass it is deceptive. Then again to call weight by any other name is equal to commit deception or to deceive the public. Notwithstanding the number of atoms that form any cosmic body, the mass or body volume is all the same because all planets orbit at the same rate.

I can and I did prove how and why the Titius Bode law apply. This I can validate mathematically on the condition that science accepts that $a^3 = T^2$ is impossible and absolutely incorrect because it is scientifically and mathematically and logically absolutely incoherent with reality. However, as long as science will not accept any incorrectness in the idea that mass potions planets in contrast to reality or to what nature applies the conspiracy to keep this lie hidden blocks my work from becoming reality.

Science wishes to believe that they are equal to God because they pretend to fiddle with mathematics and form ideas, as far from reality as this position is that the planets hold positions according to mass. It is because $a^3 = T^2$ they accept Newton in saying a square is equal to a cube they also hold the position that outer space is filled with nothing. Because Newton "proved" that a^3 / T^2 is supposedly nothing the entire Universe are therefore filled with and by nothing. This idea is mentally retarded as far as logic goes. When I declared space that forms space by volume can't be nothing it was said that I am incoherent.

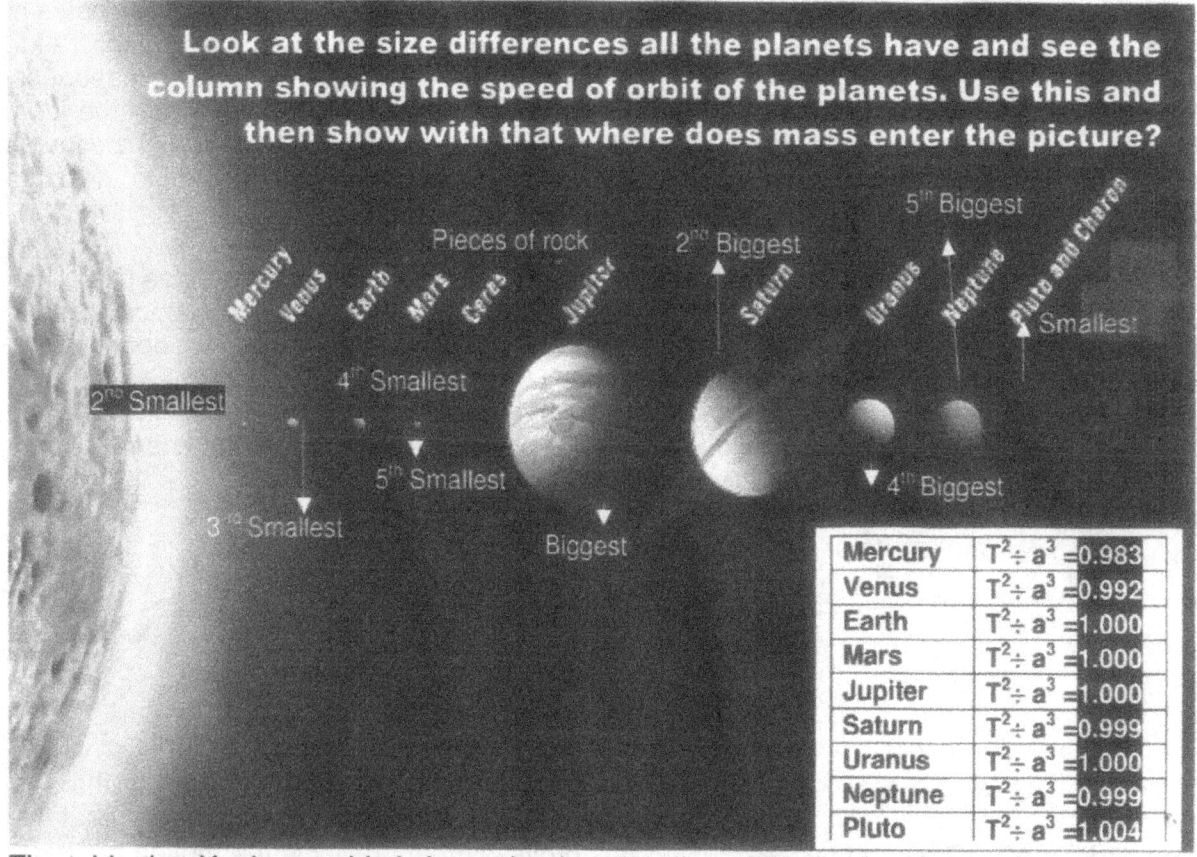

Look at the size differences all the planets have and see the column showing the speed of orbit of the planets. Use this and then show with that where does mass enter the picture?

Mercury	$T^2 \div a^3 = 0.983$
Venus	$T^2 \div a^3 = 0.992$
Earth	$T^2 \div a^3 = 1.000$
Mars	$T^2 \div a^3 = 1.000$
Jupiter	$T^2 \div a^3 = 1.000$
Saturn	$T^2 \div a^3 = 0.999$
Uranus	$T^2 \div a^3 = 1.000$
Neptune	$T^2 \div a^3 = 0.999$
Pluto	$T^2 \div a^3 = 1.004$

The table that Kepler provided shows the time that the orbit of every planet takes according to the distance the planet travels in the same time lapse and considering all the planets it is very much the same thing and in that there is no provision in the table for any idea that might form mass. This ratio is the indication of speed travelled. The idea that mass exists is a Newtonian invention made up by Newton and is completely groundless except for the value that Newtonian science gives it in order to maintain the Newtonian principles. The entire idea of mass is a myth. The idea that mass pulls mass is complete mythology and is as baseless as any fairy story. But even more deceiving is that notwithstanding that every planet has a value when $T^2 \div a^3 = 0.983$ which is $a^3 \div T^2 = k$ Newtonian science completely ignores the values and declares that $T^2 = a^3$ whereby they ignore the values in the column. That is

Science is manufactured by cheating the truth into submission just to corruption common sense is criminal to say the least. Newtonians fabricate their truth.

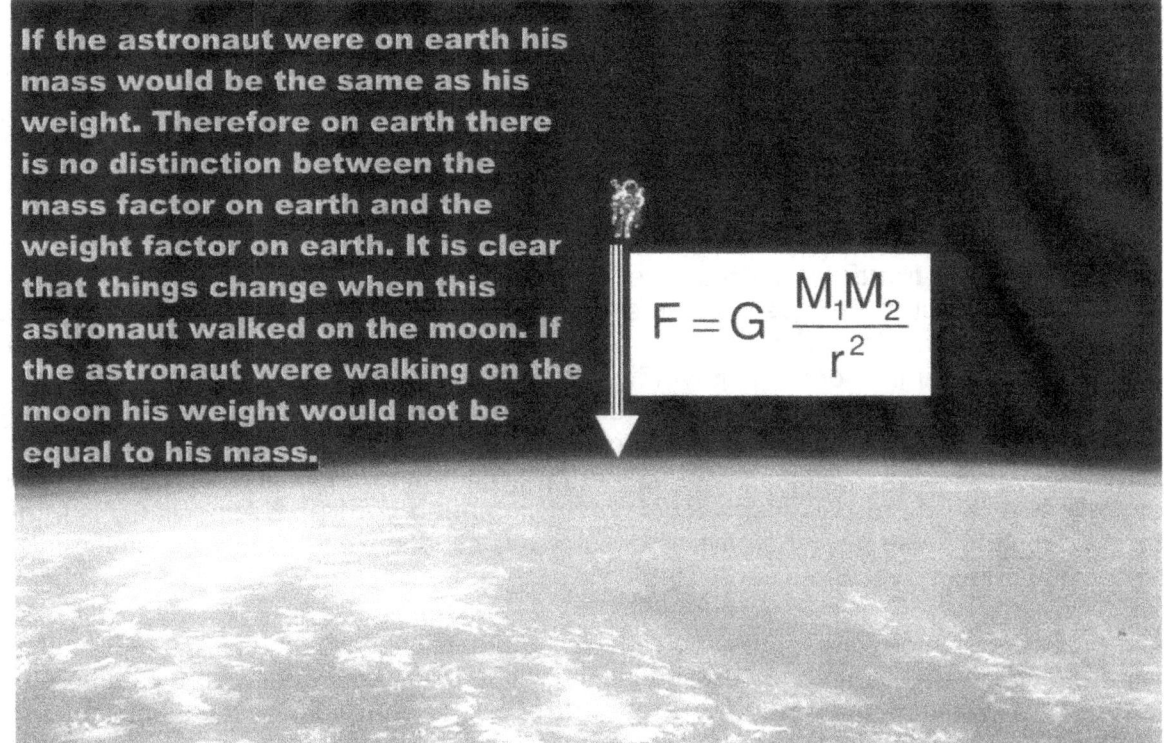

If the astronaut were on earth his mass would be the same as his weight. Therefore on earth there is no distinction between the mass factor on earth and the weight factor on earth. It is clear that things change when this astronaut walked on the moon. If the astronaut were walking on the moon his weight would not be equal to his mass.

$$F = G\ \frac{M_1 M_2}{r^2}$$

This astronaut has "mass" and the earth also in the picture has "mass" and the pulling should be going on since there is no restraining that would prevent the man from falling to the earth so why is Newton's "mass" not applying?

This astronaut is circling the earth at a certain speed notwithstanding "whatever mass" science contributes to the pilot. If "mass" was pulling then why is the man not falling. The man will fall depending on the speed by which the man rotates the earth. His spacecraft (not in the picture) has about a hundred times more mass that should entice about a hundred times more pulling but both float above the earth at a specific pre-calculated speed. It is the speed that determines the distance of circling and not "mass".

The astronaut would weigh less than the mass he has on earth. It is said that the gravity changes the numbers. The moon has less mass and that gives the man less weight while the mass remains the same. Say the man weighed 90 kg on earth he would have mass equal to 90kg.

When the same man walks on the moon he would weigh 30 kg while having a mass of 90 kg being on the moon. On the moon the machine may have the same weight as the man on earth but the mass of the machine stays what it was on earth and so does the mass of the man. That becomes science in fictional perspective. If the man could walk on the sun his weight would become a thousand times more but the mass of the man remains 90 kg. If I believe that bit then somewhere I am fooled just because I am that stupid and I deserve to be fooled that easy.   Then when the astronaut is in space he still has a mass of 90 kg but when walking in air the man has no weight with a mass of 90 kg. The indication is that while the astronaut still has mass and so the earth does he circles around the earth by gravity. Gravity is circle and not a pulling by force. Gravity comes about in terms of density applying through motion exerted on any object. It is movement that determines the location of an object and the movement forms gravity. How stupid must any person be not to see and therefore realise this as a fact.

There is no factor such as mass in the Universe. There is no evidence of a factor such as mass that holds any validity throughout the Universe. There is no proof that the Universe indicates the presence of gravity by the measure of mass forming a pulling power and while science conducts an entire religiosity based on this falsified belief, any such notion is falsified truth. Using science based on the idea that there is a pulling force such as mass forming gravity is as valid as giving Snow White seven dwarfs and then beginning a religion on that basis. There is a factor such as weight but there is no pulling of anything towards anything by magical forces forming gravity or whatever.

There is a conspiracy of conducting fraud by claiming non- existing forces but such claims are utterly fraudulent. I have been trying for twelve years to introduce the true forming of gravity but all Physicists I have encountered prevented me of doing so. They stop me because my work makes Newtonian science look like the farce it is and when removing the notion that a pulling force of gravity works by the value of mass, most of their work becomes science fiction that falls apart in substance. Read this and see how **students in physics** are methodically **brainwashed** to get the students to believe in the absolute accuracy of science. Professors and teachers participate knowingly or unknowingly in this thought manipulation process by means of conducting **mind control.** By applying this **mind-altering process** those teaching physics ensure they subdue students into becoming mind-altered zombies.

It is a process going on for centuries and which without science would have no foot to stand on in the modern environment. By presenting incorrect, falsified or unproven facts and other untruths as proven truth they exert **thought control** and thereby change the student's ability to appreciate what is correct and believable logic and then force students to discard such judgement ability in favour of accepting the institutionalised untested norms and values of science in order to unequivocally believe in science. The accepted teaching methods force students to comply by compromising their better judgment and then systematically to capitulate under teacher pressure by making their own what science prescribes what should be believed. I prove this and you get this for free so what have you got to lose…**but you can get wise to what forms a better understanding about science**! By using the building blocks that forms the Universe I take you back to the instant the Universe started and I show you how the Universe fits like a jigsaw puzzle.

If you as reader feel I have no right to dispute physics then it is not I that dispute physics but it is physics that is in dispute with reality. Those "professionals Physicist" who are so superior educated as Symeof is, please use the formula in the formula $4\pi^2 a^3 = P^2 G(M + m)$ on which all Newtonian physics rests and prove that all planets use "mass" to position their allocations. Then try to use $4\pi^2 a^3 = P^2 G(M + m)$ to explain the Titius Bode law because the Titius Bode law is in place while "mass" positioning the planets are a hoax. What this formula says is that the mass of the sun and the specific mass of the planet would form the allocated position in which that specific planet is.

I explain again: that the mass of the sun and the specific mass of the planet $G(M + m)$ would form the allocated position P^2 in which that specific planet is. Those so professionals explain why Jupiter is where it is by putting in the mass of Jupiter and the mass of the sun and then prove that Mercury or Venus or the earth or whatever planet is in its specific location by putting in the distance $4\pi^2$ that is according to the mass a^3 because Newtonians say that $4\pi^2 a^3$ is equal to $P^2 G(M + m)$

This idea of mass being responsible for planets positions only science shares while it contradicts nature completely and yet we all have to embrace this concept notwithstanding that nature embraces the Titius Bode law. This now is the explanation of how the Universe forms and this is for the first time in human history that any correct explanation is presented of the formation of the solar system.

Bode's Law:

Planet	Mercury	Venus	Earth	Mars	Ceres	Jupiter	Saturn	Uranus	Neptune	Pluto
Bode's Law distance	4	7	10	16	28	52	100	196	-	388
Actual distance	3.9	7.2	10	15.2	28	52	95.4	191.8	300.7	394.6

The Titius Bode's law is a numerical sequence announced by J.E. Bode in 1772, which matches the distances from the Sun of the six planets then known. It is also known as the Titus-Bode law, as it was first pointed out by the German mathematician Johann Daniel Titius (1729-96) in 1766.

It is formed from the sequence 0,3,6,12,24,48,96, and 192 by adding 4 to each number. The planets were seen to fit this sequence quite well – as did Uranus, discovered in 1781. However, Neptune and Pluto do not conform to the 'law'. Bode's Law stimulated the search for a planet orbiting between Mars and Jupiter that led to the discovery of the first asteroids. It is often said that the law has no theoretical basis, but it does show how orbital resonance can lead to commensurability.

The importance that becomes known is the sequence the Titius – Bode law saw in the number arrangement of 3; 6; 12; 24; 48; 96 etc. The incorrect application of the Titus Bode law lies in subtracting the figure of 3 from 10 leaving 7. The other way of reasoning is to add four each time to the firs value of three starting with 3 and so on. The true significance of the Titus-Bode law is that it points directly to a circular growth of 7 stages.

The 7 relating to 10 is a precise derogative of the Roche limit or the Roche limit is a precise derogative of the Titius Bode principle because he two systems interlink. This is how I mange to explain the Titius Bode law that is in the solar system by the ratio applying that really form the solar system in the way nature shows space growing by time. What you see on the next page was never been shown but on the other hand Physicist say this mathematics are too simple to apply as physics!

To find the mean distances of the planets, beginning with the following simple sequence of numbers:
0 3 6 12 24 48 96 192 384

With the exception of the first two, the others are simple twice the value of the preceding number.

Add 4 to each number:
4 7 10 16 28 52 100 196 388

Then divide by 10:
0.4 0.7 1.0 1.6 2.8 5.2 10.0 19.6 38.8

The Titius Bode law proves that in the Universe laws apply that positions objects in terms of other rules that mass. That means the Newtonians hides their lack of understanding behind mass that they invent.

The Newtonians gave the Titius Bode law a formula and that explains the lot. To they're under achieving standards that is very satisfactory. Now it is written in mathematics then what more do we need to know. The fact that the distance that Mercury has from the sun is doubled by that which Venus has from the sun is completely ignored. In cosmic reality mass plays no part. Then again the distance that Venus has from the sun is doubled by that which the earth has.

The resulting sequence is very close to the distribution of mean distances of the planets from the Sun:

Body	Actual distance (A.U.)	Bode's Law <A.U.)< td>
Mercury	0.39	0.4
Venus	0.72	0.7
Earth	1.00	1.0
Mars	1.52	1.6
		2.8
Jupiter	5.20	5.2
Saturn	9.54	10.0
Uranus	19.19	19.6

This clearly has nothing to do with the size or mass of the planets. Explaining that part is completely ignored. Then again the distance that the earth has from the sun is doubled by that which Venus has and inexplicably this forms the layout of all planets in the solar system. Where does Newton's idea of mass fit into what truly applies in outer space. Moreover, why does science never mention this? This is my formulated explanation about how the Titius Bode law forms.
The numbers we need to find the key to the mystery of the Titius Bode law is 3, 4, 7, and 10.

It then purposely hides the essence that forms the solar system. In the Universe all things are equal in size because Neptune spins around the sun equal to mercury's time and Jupiter floats around the sun equal to mars or Neptune. Notwithstanding what "size" or "mass" they grant the planet to have the rotation happens equal and without mass bringing any favouring in positioning or in speed of movement. So where the hell is mass a factor in gravity forming? Ask a physicist to explain the fact that if mass pulls mass by gravity pulling, then why do the planets in the solar system not position their allocated places in relation to the sun by applying mass as the nominating and defining factor?

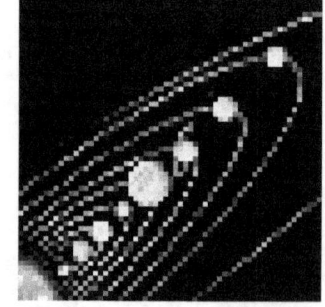

If Symeof and all those others who stand in line to correct me are so certain of his and their position why don't he correct me by showing how mass does put the planets in the order that they are. This is so typical of the brainwashed zombie who learned science where instead I studied science and to him and those there is a world of difference between the two concepts.

Those brainwashed into learning science never tested on concept that which they were told to learn. The following picture on the next page shows the applying size and size ratio and the allocated places they have. Are those defending physics really that blind that they are unable to see what even small children see or are they so dumbfounded that they will rather attack me to defend a senseless conception that can't be defended by logic or intellect. To those physicist brainwashed into stupidity this is not very serious.

They truly can't see what the fuss is about as you can see with the reaction of the ever-so-wise Symeof and his gang of wise that so fiercely protect the honour and legitimacy of physics as a whole that is in a hole of deception. To them the fact that there is no evidence that mass is responsible for planet positioning is not a big issue although that forms the fundamental basis of their cosmological concept they put forward as the truth and the only truth. It is said that science only works with truth so how much truth is there in this concept?

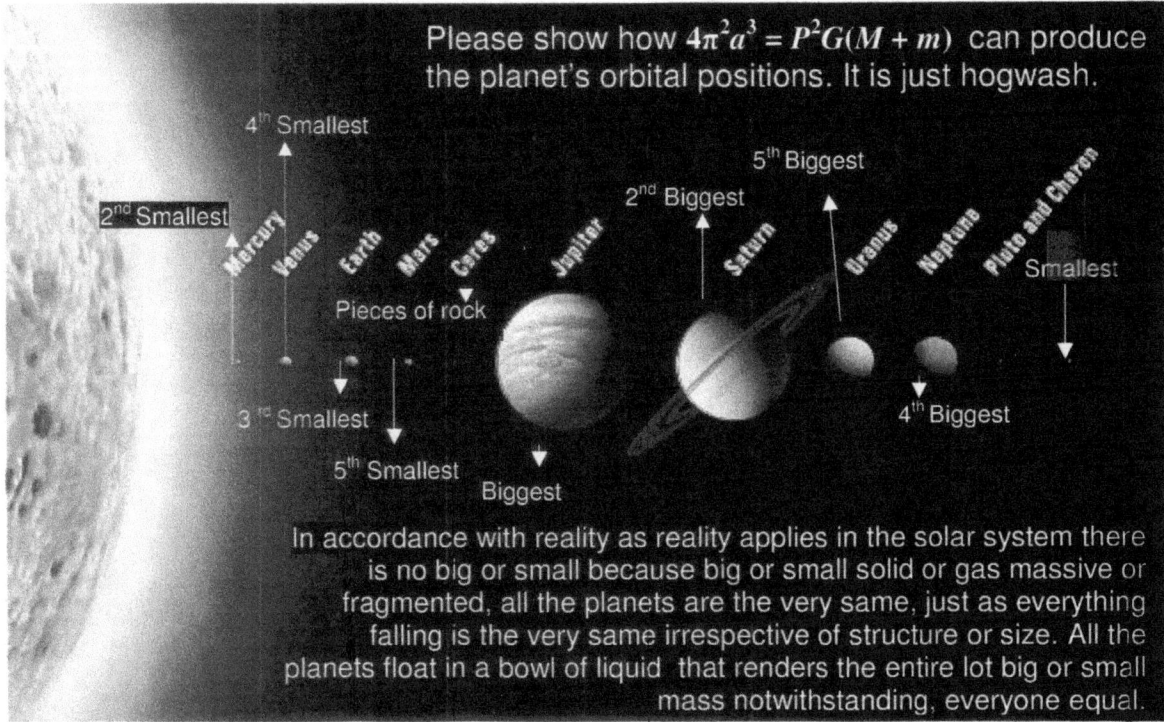

Please show how $4\pi^2 a^3 = P^2 G(M + m)$ can produce the planet's orbital positions. It is just hogwash.

4th Smallest

5th Biggest

2nd Biggest

2nd Smallest

Mercury Venus Earth Mars Ceres Jupiter Saturn Uranus Neptune Pluto and Charon

Smallest

Pieces of rock

3rd Smallest

5th Smallest

Biggest

4th Biggest

In accordance with reality as reality applies in the solar system there is no big or small because big or small solid or gas massive or fragmented, all the planets are the very same, just as everything falling is the very same irrespective of structure or size. All the planets float in a bowl of liquid that renders the entire lot big or small mass notwithstanding, everyone equal.

By depicting the solar system in such a presentation with the distances so inaccurately presented as Newtonians normally do such as the picture next this form of presenting the layout without providing correct spacing purposely corrupts the entire structure formation by which the solar system develops.

This is what is in the solar system applying the serving ratio that the Universe uses. It is not the fake Newtonian $4\pi^2 a^3 = P^2 G (M + m)$ that has no basis except for Newtonians brainwashing students into believing that which otherwise can't be accommodated by the Universe. These are ratio values that are there...used by the Universe as actual factors forming space. Use this picture below to show me where the planets are positioned according to mass or where the orbit going around the sun goes according to mass. The entire Newtonian idea of mass creating gravity by pulling is the complete misrepresentation of the truth. I show what principles are in place and do give the reason why. It is easy to talk about "mass" and never get "mass" part of reality when hiding the truth.

This shows a ratio used by the cosmos and not by science to position planets according to this ratio that works independent of mass or size. That is why Mercury as the smallest planet is inside next to the sun and Jupiter being the largest is in the centre. If mass did apply then Jupiter should be inside and Mercury at the very end. Where Symeof of no known e-mail address holds the opinion that I as the author has no right to dispute physics I wish to ask him to explain why mass then holds planets in place as the Newton formula says and as used by the honourable in science since they are the ones who declare mass applies as $F = G \dfrac{M_1 M_2}{r^2}$. Symeof have you or any forming your league ever seen any explaining about the Lagrangian system 2) the Roche limit 3) the Titius Bode law 4) the Coanda affect or know that these laws exist? You say I prove nothing and yet you understand nothing about what I prove because you have such a little understanding. Do you realise the fools you appear in the eyes of the more intelligent amongst the many in society? In your case I can't figure out which is more, your stupidity, your arrogance or you pathetic self worth through which you believe in your absolute intelligence. To Symeof and all the others such as he is my advise to you lot is to go and ask your money back from the institution that you thought educated you as you should clearly see you have not been educated but you have been conned and tricked. Moreover, read on and see the deception you have undergone by those you trust in such high regard.

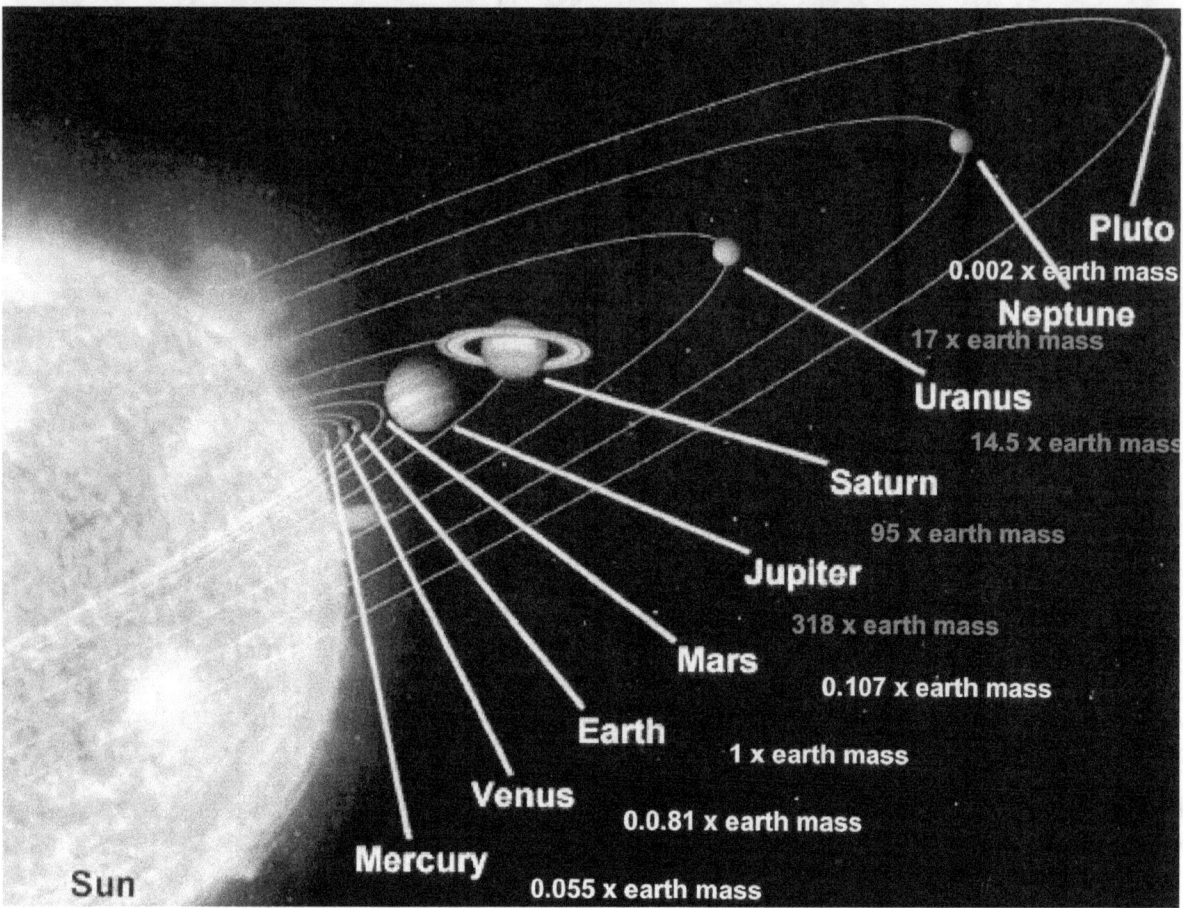

Why do I say there is a conspiracy? This is why I say there is a conspiracy in science. Science says the planets are in location because of the mass they have and this opinion stems from Newton. However look at the picture I provide on the next page and see how the mass applies. Then tell me you still think that nature puts the planets according to mass with the larger planets in the centre and the smallest on either end of the solar system. Is there any body that still say I am not aloud to criticize science with this blatant misrepresentation of the truth? Are there still those that think like Symeof and say I have no right to criticize the most honourable persons in science for misrepresenting the truth? This is but one of so many cases I bring to your mind, and yet the most clever say "*Then, of course, he couldn't imagine he might have been wrong, so he rationalized his failure to be accepted by creating a conspiracy theory: this is probably the reason why this e-book exists. This is no scientific material..* Tell me which part of this have I got wrong and about which part of this am I misrepresenting what science say incorrectly? Is the wrong part that I thought about what I was taught and the rest of you lot did not?

The cosmos however, does not apply mass in any form and the cosmos has four other principles in place. What Newton show that should place planets according to mass is not used by nature. It is the most incorrect idea ever put forward as the truth. I show what nature uses namely the Titius Bode law, The Roche limit, The Lagrangian Points and the Coanda effect and how this forms gravity as well as place the positions of the planets in accordance with singularity. Because I trash Newton's rubbish that does not fit and that can't apply no publisher of science books or science magazines will publish my work I show what goes on in nature while Newton's contribution of mass applying is total rubbish. Because I call it rubbish and I rubbish Newton I am ignored. How can any person believe mass forms gravity that puts planets in position in accordance with mass when viewing planets, in the Universe?

This picture shows the hoax the Newtonian conspiracy pampers to keep the rest of Newtonian physics believable. They never mention the Titius Bode law and try to explain the Titius Bode law while it is the Titius Bode law that is really in place in the solar system. *If you wish to learn the truth then think again.* With all this evidence in place for centuries and no evidence of mass pulling in the Universe science still upholds $F = G \dfrac{M_1 M_2}{r^2}$ because in the known field of physics they have nothing else to show for centuries of labour. They have to be content with Newton because they only have Newton and while Newton never made sense it was all there was to cling to... Newtonians uphold their version of the law of physics they advocate as the truth applying without showing mercy to anybody that dares to challenge Newton. The very first things the Newtonians use to beat us into submission are to blast us with incomprehensible mathematical formulas.

If only the equations they say applies were truthful...Incomprehensible they are not truthful but still those Newtonian philosophies are used to scare anyone with the idea that the mathematical equations will get everyone hiding as it has done for hundreds of years. They bewilder students and the public alike with equations that put the fear of God into you; it used simply to make you feel inferior so that they can feel superior and frown down on your inferiority from a dizzy height. They say the positioning of planets is according to this formula

$$\left(\frac{P}{2\pi}\right)^2 = \left(\frac{a^2\sqrt{1-\varepsilon^2}}{\ell}\right)^2 = \frac{a^4(1-\varepsilon^2)}{\ell^2} = \frac{a^4(1-\varepsilon^2)}{a(1-\varepsilon^2)GM} = \frac{a^3}{GM}.$$

This is pathetic, not my book you are reading. I challenge any person on earth to put in the mass of Jupiter and then from that show how it is arriving at the position any of the planets hold or how by mass any planet's allocation or position is derived. Or do it with any other planet. This is one big hoax that is a small part of the conspiracy science tries to hide. To make sense we have to look for Π in this because Π forms the value of gravity because of the circle nature that we find in cosmic space. We have 3 and 4 adding to be 7. Then we have 10 forming the other factor number. If we wish to stop pretending to make science a hoax we must come to some realistic conclusion that will prove what is working in the cosmos. The cosmos works on a distinct ratio put in place by a distinct pattern. This is the ratio. It is a ratio that puts every planet in a precise location that finds a value in sequence in accordance with the turning movement of the sun. Mass never comes into the picture. Mass plays no part. Science never announces this fact with trumpets blowing because then science has to admit they know less about physics than their dog knows about British or American politics.

To find the mean distances of the planets, beginning with the following simple sequence of numbers:
0 3 6 12 24 48 96 192 384
With the exception of the first two, the others are simple twice the value of the preceding number.

Add 4 to each number:
4 7 10 16 28 52 100 196 388

Then divide by 10:
0.4 0.7 1.0 1.6 2.8 5.2 10.0 19.6 38.8

The resulting sequence is very close to the distribution of mean distances of the planets from the Sun:

Body	Actual distance (A.U.)	Bode's Law <A.U.)< td>
Mercury	0.39	0.4
Venus	0.72	0.7
Earth	1.00	1.0
Mars	1.52	1.6
Asteroids		2.8
Jupiter	5.20	5.2
Saturn	9.54	10.0
Uranus	19.19	19.6

First to find the ratio one puts the earth distance to the sun in as a cog in ratio to 1. This eliminates the human factor in presenting the positions and puts cosmic ratios in practise. The distance from the sun to the earth is 1 and that is the dependence of the ratio of the solar system applying.

In that ratio we have Mercury at 0.39 Astronomical Units from the sun and according to the Titius Bode law it should be 0.4.  Using the solar ratio of placing the earth to the sun at 1 we have Venus at 0.72 Astronomical Units from the sun whereas according to the Titius Bode law it should be at 0.7.  Then we have the earth at 1 Astronomical Units from the sun according to the ratio that the solar system applies.  Mars is at 1.52 Astronomical Units from the sun whereas its allocated position should be at 1.6 when the ratio is applied. At 2.8 Astronomical Units from the sun we find a huge lot of planetary debris that also holds an allocated ratio position although the lot is fragmented into small pieces. Yet, all the pieces orbit at 2.8 in ratio to the sun / solar system.  Jupiter the "giant" stands at 5.2 Astronomical Units from the sun in accordance with the ratio the solar system provides.  Saturn is at a cosmic ratio of 9.54 Astronomical Units from the sun and the mathematical ratio is 10.  The next planet is at 19.19 and mathematically according to the solar ratio its allocated position should be 19 .6. This goes on and by using this ratio planets were discovered in the past. Yet science has no explanation as to why this is true and I am the very first person ever and alive that explain why this is the case.

This is a fixed motion driving the movement and forming time by placing the planets in relation to certain places the planets will maintain according to movement of time as it forms space by the movement in time. Using evidence that the solar system provides mass has no place but we are forced not to use the evidence provided by nature because nature clashes with Newton. Subsequently we must toss out what is there and what nature supplies us with as evidence and instead we have to apply Newton's vision of mass because God can be wrong by placing a ratio in Nature but when nature goes against Newton then God and Nature is wrong. This is the official position of science and you are going to read about this attitude many times more as this book develops. We will rather blot out what is in Nature than question the correctness of Newton and the way science view physics. If we do question Newton then we will learn that the "oh so clever" were a bunch of arrogant fools that did not know anything about what nature in reality offers. Then in this ratio another factor is present:

When reading the distance that mercury is from the sun this distance doubles from

mercury to Venus and the distance from Venus to the sun doubles from Venus to

the earth. Then this distance from the earth doubles to Mars as it doubles again from

the sun to Jupiter . This goes on up to where the solar system ends. This shows no indication of mass forming any part in the planetary display. Science's claims since the time of Newton that by the Kepler laws which is what Newton fabricated everything depends on mass and the cosmos shows that mass has nothing to do with the planet positioning. It is folly that according to Kepler the planets hold individual positions according to mass. Science never put this formation in place when explaining the forming of cosmic science because then science has to admit they have no clue how the cosmos works and for that admission they are not ready because everybody must think that they know more than God Almighty about how science works. They can't come across as the ignorant fools they obviously are!

The cosmos shows no tendency to use the Kepler laws or any fabrication that science may propose that those laws claim. In nature the formula

$$\left(\frac{P}{2\pi}\right)^2 = \left(\frac{a^2\sqrt{1-\varepsilon^2}}{\ell}\right)^2 = \frac{a^4(1-\varepsilon^2)}{\ell^2} = \frac{a^4(1-\varepsilon^2)}{a(1-\varepsilon^2)GM} = \frac{a^3}{GM} \text{ or } F = G\frac{M_1 M_2}{r^2} \text{ is}$$

completely absent. There is the Titius Bode law that shows a positioning according to a ratio and the distance doubles every time a new position arrives in relation to the inside planet of that planet. Without any proof of this formula having any legitimacy whatsoever science continues to use this formula when explaining gravity and the way the Universe applies.

Notwithstanding all these facts available to science it still is propagated that mass forms the allocated positions that the planets hold according to the sun in

$$\left(\frac{P}{2\pi}\right)^2 = \left(\frac{a^2\sqrt{1-\varepsilon^2}}{\ell}\right)^2 = \frac{a^4(1-\varepsilon^2)}{\ell^2} = \frac{a^4(1-\varepsilon^2)}{a(1-\varepsilon^2)GM} = \frac{a^3}{GM}.$$

This conspiracy is in place to hide the fact that science has no idea what puts the planets in position and rather than admitting they have no idea they keep teaching unsuspected students a lot of unsupported hogwash just to cover the fact that science is clueless about the most basic knowledge about gravity.

I am the first person ever to explain WHY the Titius Bode is in space functioning as it is. But since those such as Symeof and the "Small-brain-brigade" such as he can't fathom any explanation going this far they are of the opinion that: *"The author has no legitimacy in disputing physics: he is merely an amateur trying to disprove what the greatest minds of our times have taken centuries to prove (by the way, you need to have studied a subject before you can criticize it, which the author didn't). In my opinion, the author is so narrow-minded that when he came to scientists to talk about his material, the scientists must have told him that it was nonsensical."* In their smallness they have not the ability to understand that planets are not allocated according to size or mass and this they do not understand notwithstanding all the planets' sizes or individual mass values proving distinctly that Newtonian science is senselessly wrong. Now I will explain it again in detail for those "professionals" such as Symeof and his band of "Super Educated Professionals" who "understand science" because they are so informed about physics and the application thereof and is well informed about physics. There is a ratio in place keeping time as if it runs on cogs. It is not connected to mass in any way. Science never come out and says this because they hide this. If they didn't hide this they then must admit they have been wrong all along about mass positioning planets.

Symeof who are so typical of the "brainwashed class of stupidity" and the "Small-brain-brigade" has the opinion that: _Then, of course, he couldn't imagine he might have been wrong, so he rationalized his failure to be accepted by creating a conspiracy theory: this is probably the reason why this e-book exists. This is no scientific material._

If I show that in mathematics according to science the planets' positions are in accordance with Symbolically being: $P^2 \propto a^3$ and therefore $a^3 = P^2$ in position of P and therefore a^3 / (P^2P). This is taken from the idea that "Kepler said", which is totally fabricated that $a^3 = T^2$ where Kepler in fact said $a^3 = T^2k$ and this is a big difference because $a^3 = T^2k$ is the same as $E = mC^2$. _Then, of course, he couldn't imagine he might have been wrong, so he rationalized his failure by creating a baseless and unfounded theory: this is probably the reason why this e-book exists. This is no scientific material._

These are Kepler's actual numbers before Newton got hold of those numbers and altered it into something completely fraudulent. Science at present accepts that $a^3 = T^2$. Then Kepler had a specific column for k where if you multiply T^2 **with k it then is equal to** a^3. Kepler said if you divide T^2 with a^3 **it has a value of say** 2.96 as it is in the case of the earth. Mathematics teaches us that if you have numbers telling you that $a^3 = T^2k$ **then those numbers will give you a mathematical result of** $T^2 \div a^3 = k^{-1}$ **. What should then dawn on me that I could be wrong as he suggests in his remark** _of course, he couldn't imagine he might have been wrong, so he rationalized his failure to be accepted by creating a conspiracy theory: this is probably the reason why this e-book exists. This is no scientific material_? $\div$

You brainwashed stupid zombie who tries to teach others what your simple mind do not even understand, where am I wrong? Do go and study what science says because in this part a few pages back I have given you the claims why science makes that $T^2 \div a^3 = 0$. Now with your little underdeveloped intellect, please show me that mathematically $T^2 \div a^3 = 0$. Where I contacted more that 2000 Academics worldwide in physics showing them just this anomaly and that their practising of mathematics are as incorrect as this, where no one answers me but either belittle me in precisely the same manner as you do without ever telling me where am I wrong but just telling me that I am wrong just as you do and NEVER substantiate my incorrectness with facts of where I go wrong, then yes, you are part of the conspiracy and I thank you for supplying me with such a splendid unmistakably example of how brainwashed students in physics become. I hold your example in such high esteem I even open this book with your writing.

So why do I think of this as a conspiracy. I have uncovered the formula by which the Titius Bode law works. The Titius Bode law is in place. It is not mass positioning planets but it is the Titius Bode law that holds the key and I have solved the riddle that unlocks the explanation about the Titius Bode law. Any person doubting my accusation is very much welcome to show the world exactly how mass works when all the planets are randomly sized. The Titius Bode law says one should apply the ratio according to the astronomical unit, which has the distance of the earth and the sun at a position of 1.

Then in ratio Mercury is .39 where it should be four. Then add four and divide by 10. That is the ratio but why that is the ratio has never been understood because it has never been explained or had any attempt by a living person to explain...that is up to now where I explain it in detail. That the ratio also has the distance double between planets next to each other I explain. This too has never been understood because it has never been explained or had any attempt by a living person to explain...and again that is up to now where I explain it in detail. This forms one of the four foundations on which the entire order of physics rests.

With a ratio that Kepler calculated using the formula $k^{-1}=T^2/a^3$ the negative value of k proves that space moves in recline towards the sun. The sun contracts space from the outposts of the sun's influence field.

PLANET	PERIOD (Years) (T)	MOVEMENT (T^2)	DISTANCE	SPACE (a^3)	RATIO k
Mercury	0.241	0.058	0.39	0.059	0.983
Venus	0.615	0.378	0.728	0.381	0.992
Earth	1.000	1.000	1.000	1.000	1.000
Mars	1.881	3.54	1.524	3.54	1.000
Jupiter	11.86	140.66	5.20	140.6	1.000
Saturn	29.46	867.9	9.54	868.25	0.999
Uranus	84.008	7069	19.19	7067	1.000
Neptune	164.8	27159	30.07	27189	0.999
Pluto	248.4	61703	39.46	61443	1.004

KEPLER'S LAW OF PERIODS FOR THE SOLAR SYSTEM			
PLANET	SEMIMAJOR AXIS $a\,(10^{10}m)$	PERIOD T (y)	T^2/a^3 $(10^{-34}\,y^2/m^3)$
Mercury	5.79	0.241	k^{-1} = 2.99
Venus	10.8	0.615	k^{-1} = 3.00
Earth	15.0	1.00	k^{-1} = 2.96
Mars	22.8	1.88	k^{-1} = 2.98
Jupiter	77.8	11.9	k^{-1} = 3.01
Saturn	143	29.5	k^{-1} = 2.98
Uranus	287	84.0	k^{-1} = 2.98
Neptune	450	165	k^{-1} = 2.99
Pluto	590	248	k^{-1} = 2.99

In this equation k has a defined value where k holds a value. The Newton ratio of $a^3=T^2$ put forward can't have any substance. Kepler said $a^3=T^2k$ and that nullifies Newton's presumed ratio of $a^3=T^2$

The others are the **Roche limit**, which I explain in detail, the **Lagrangian numbers**, which again I explain in much detail and the **Coanda effect** that forms gravity and that too I explain in excruciating detail. Then you get these arrogant half-wits such as Symeof and all the other "highly regarded Physicist of the world" who are unable to explain gravity but likes to belittle me because the evidence I show belittles them to the point where they are not even having a microscopic brain to work with. I show they cover fraud and then they turn around and declare I don't know physics because I show what they think they know is total shit.

Do you reading this or any person on earth wish to tell me that in science no one in almost four hundred years saw that the planets do not show arranging in accordance to mass and therefore ALL COSMIC OBJECTS show total indifference to mass positioning or allocating positions? With this evidence do you believe there is no conspiracy to remain quit and keep facts under cover?

All this detail is so small part of the entire web forming the conspiracy that I could use it as an opening to show how my critics work. Go back and read the entire criticism of Symeof and see what he says I do not understand or why he says I never studied physics when I know more about physics that what his little mind can hold. Go and see where he indicates one single piece of evidence that I do not understand. It is always rumours and thought thrown intro the air just as his idea of facts in science are. This is the rubbish I have to deal with. I have spent forty years on research and I can now explain what no one ever could explain and that is how gravity works.

Never as even once would science say they have no idea how the solar system holds positions in the cosmos. When admitting that they will have to forfeit their veneer of godliness about how enlightened they are on matters about how the cosmos works. I have made a study of almost thirty-five years to uncover how gravity works and I have concluded how the process of physics fit into the cosmos. I have showed what is wrong with physics and how to correct it. I am so grateful for suckers such as Symeof and his clan who so feverishly defend the science they believe in and thereby provide me with the opportunity to use this as a speaking example as to show how the brainwashing can apply. Symeof was brainwashed to believe that the formula he never could test did apply and he would die in the conviction that the formula he defends is correct and this formula carries more truth than does the Holy Bible. So all those who wants to crucify me when I prove what you lot studied was   a   waste   of   time,   first

$$\left(\frac{P}{2\pi}\right)^2 = \left(\frac{a^2\sqrt{1-\varepsilon^2}}{\ell}\right)^2 = \frac{a^4(1-\varepsilon^2)}{\ell^2} = \frac{a^4(1-\varepsilon^2)}{a(1-\varepsilon^2)GM} = \frac{a^3}{GM}$$ apply this formula

with the adding of the distances as well as the mass and prove to yourself why the planets hold their allocated positions according to mass. If you can't then a conspiracy is in place to prevent the truth from leaking. For thirteen years everyone in science prevents me from explaining the four comic principles that are there where they placed mass and they prevent me because then their image goes down the toilet because then they must admit of being wrong and they know nothing about what really applies in science.

By the way, those who learned physics and feel so proud about it, one first have to study physics to realize where there are incorrect adaptations in the reality of the concept when forming judgement on the entire matter as a whole. This is a drop in the ocean and those who studied physics if you are not prepared to live through the shock of awakening stop at this point. As you can well see with the example I provide, if you are so mind conditioned that you are unable to see the anomalies I provide, then stop reading because your mental faculties are at risk and you may find much mental strain when reading this and thereby coming to terms with what you never studied but thought you did. Prevent yourself the agony awaiting you.

In order to explain why I am so vocal in criticizing the teaching profession of the day is because they brought on me so much poverty and hardship and all the time I was portrayed as the cunning knifing hooligan out to mesmerise the innocent by blackening the characters of the highly appraises physicists. I chose poverty and therefore had to quite teaching because I could not betray children in my class any longer. I decided on poverty for I thought it would be better for my soul that I live on the brink of starvation as I have been doing and experience complete poverty than to cheat the children who trust me and look at me with those trusting eyes. To them as a teacher I am completely trustworthy, even more than they believe in the knowledge their parents represents because I am the teacher and the teacher they believe. ...And then I look into their eyes and tell them to believe in what I teach while I know the science I know is totally fictitious. How can I take money and in the end meet my Father in Heaven and tell my Father in Heaven I sold not only my sole, but also their soles for money. I had to live an entire lifetime with lies and betrayal not to one or two persons but to scholars year after year while I know every word I say on science is corrupt. I have to teach what I know is unproven dogma and I am forced to teach if I wish to earn a living. I brought much hardship on my children by forsaking Newtonian standpoints and searched for most of my life to find the truth and then those living the lie would thrust my finding aside because they would cling on to the lie that feed them, keep them famous and by which they can pretend to be wise. How can any person with a conscience, with moral standards and with dignity spread untruths and still feel unremorseful about the dishonesty they teach. If I call them scoundrels it is because I believe it.

In my effort to bring change I wrote many articles and books, amongst which I wrote an article, which I call the Absolute Relevancy of Singularity in which I go much further than Einstein did on the relevancy of singularity in the general application of all types of physics. In doing that I had to reject Newton. I prove the entirety called the Universe is made up of singularity and contains only singularity in many forms thereof. I prove singularity is not a general phenomenon but is the concept forming the absolute basis of physics and this foundation is transferred to the beginning of mathematics. The Universe is singularity coming in many forms and this is simple to prove; it is the other concepts flowing from this that complicates things. Modern Science can't explain any of the modern findings such as the Big Bang because of a lack of proficiency. After finding the building blocks of the Universe I now can show where is the centre of the Universe. Have you thought where is the centre of the Universe? I can answer that...and why is the Universe still grows since the Big Bang...why did the Universe start so small...why did the Universe fit into a neutron at one time...how did everything expand from fitting into a neutron...why does space grow from small to large...where is it going while it is growing...why was the Universe any specific size...what was everything before the Big bang...if you read this you will know all of this!

My discovery is the fact that singularity presents the total and complete control of everything there is in all forms of everything there ever could form in the Universe and discovering this fact led me to prove that gravity is the application of Π and from extending Π further into a six dimensional sphere that is spinning in a six-sided cube, is what makes the Universe form three dimensions using time thereby the cosmos enlists the multi dimensional sphere we live in. However, at the stage where the Universe holds gravity as Π, which is at the point before it forms multi-dimensional space, the Universe still is flat although the Universe then is already accepting form by going round in shape-forming when extending from forming singularity as a value when retaining form in the principle of a sphere. The foundation of physics is gravity forming Π by allowing Π forming. This explanation comes across, as simple but believe me no academic this far was able to understand to entire context of such an explanation.

My findings are not congenial with Newtonian views and therefore it means that no one in Mainstream science could dare to look at my work. If they do read my work and confess to the truth they then have to confess that Newtonian science is a corrupt conspiracy that is hiding issues most important and portraying a fantasy as the truth. Then I decided to put the cat amongst the pigeons and see what flew out of the bush into the sunlight. I wrote seven articles after the first article that was (according to my page set-up) in total 15 pages that I sent to Annalen Der Physics. This first article and the seven afterwards I sent to the addresses of Dr. Ulrich Eckern as well as Dr. Fredrich-Wilhelm Hehl and at the bottom of this page I give their responses. But as to getting a reaction about the other seven I am waiting for many, many months but as to date still never received any correspondence as to confirm the acceptance / rejecting of the seven articles explaining the four cosmic principles. It is true that I could never give sufficient proof substantiating my work, but as I put it to Dr Ulrich Eckern, one cannot prove a complete new trend in thought about science going using completely new principles about gravity when only devoting 15 pages to the concept. I showed that singularity is time forming space and where to locate singularity. I always submit articles to well known physics magazines but my articles are rejected on the most unappeasable grounds and for the most outrageously ridiculous reasons the Newtonians can think of. I explain how gravity forms but I am rejected because they are of the opinion that my work does not meet an acceptable level of standard since I am at odds with the way science in the present think about gravity.

I say and I prove there is no such a thing as "mass" with the ability to "pull" anything. I do not say a person does not have weight but I say no person has "mass" that brings about a force that pulls anything closer by using gravity. I explained in the article the Coanda effect and the way the Coanda effect applies according to gravity. I also showed that this explanation forms the fundamental and basis of physics and how it forms the foundation of physics. It is the heartbeat of what no one knows to explain in physics.

You are going to read about a conspiracy but people think of a conspiracy in many terms. Let us define not by definition but by interpretation to what a conspiracy constitutes. What do you think is a conspiracy? All the conspiracies you know about is known about because someone somewhere makes money by allowing the revealing of the conspiracy. Silencing the conspiracy does not make money but informing a suspicious public loosens the flow of money. If it were a true conspiracy no one would know about the conspiracy because the powerful would make money from not revealing the conspiracy. The revealing of the facts about any conspiracy would be stopped before it leaked because it would kill the flow of money.

A conspiracy is thought to be a gossip story that makes money and by not revealing it or revealing it goes in line with making money or not making money. You can download this book free of charge because I don't make money by revealing the conspiracy. I truly want to find an audience to divulge the truth. I want to make money but it is by showing how I can correct the flaws in science, not by hiding it in a conspiracy.

People put a conspiracy in the same realms as a gossip story, an old wives tail, which is going about but does not intend to harm and mostly serves as amusement to many. Hearing about a conspiracy tests your intellectual comprehension. It is some quiz that you match your truth against the truth that the conspiracy reveals. It is a funny, but it is not funny until you catch the funny part hiding behind the conspiracy and only when you measure the catch behind the conspiracy are you treated to be amused.

If the conspiracy does not touch the person directly then no harm is felt and no harm is intended. Every one holds this view that a conspiracy is on a slightly higher level than gossip. It is a gossip story about someone living in the neighbouring village known only to some people next door but has no direct linking to me or has no threat to the safety of others directly associated to me. Everyone treats a conspiracy as if it is something amusing that holds no threat at all. It is something that goes around as a joke of sorts.
We all live in a bubble we call civilisation and we all run after a dream called peace and we all preach a fantasy we named democracy and every one excluding no one lives a fantasy we love to believe but never can. Those that say they lead us are programming us in an assortment of ways to have us believe that we are happy content with our fate and we are lucky to be as prosperous as we are and this they do by a process of programming us

mentally. They call it advertising. They call it politically socialising. They even call it following teaching procedure. The teachers tell us what we must believe and then force us to write examinations about what we must believe and according to how we believe what they say we have to believe in, they set our future.

A hundred or two years ago it was believed that one of the unhealthiest practises any one can indulge in was to have a bath. In the reign of Henry 14th of France it was believed to bathe was to progress the aging process. I presume that they saw our skin in our hands wrinkle from being in the water for too long and when the skin wrinkles it is a clear sign of quick aging. With the wrinkling process so evident the thing to stay away was to have contact with water. It ages you enormously. Should you wish to die young and miserably old at the same time then you have to needlessly bathe regularly. It was believed at the time that it was bathing that made you old before your time or so it was believed throughout the Western Civilised world. One should much rather drown yourself in perfume if you were rich and smell like a cheap modern whore would from a mile off. The status symbol at the time was to smell like a perfume factory that has gone bust and you could get perfume for free. If you smelled like that you had the same aroma as the King did and his entire entourage of the royal upper class. In that period drowning your body in perfume was the fashion of the moment and the statement that you had money. It was the same as in modern times carry a hundred credit cards in many wallets for all to see how credit worthy you really are.

It is playing mind games with the public and makes those with simple minds feel important if they were fashionable and trendy. Then someone found a way of advertising and what was advertised first was soap. If you look at films and photos about hundred and fifty to hundred years ago was with soap advertisements that was filling the background in these true to life photos and films. The soap barons got in on the act first and through scrupulous mind control started to program the mind and the mindset of the everyday persons.

Even today and more so today every advertisement are programming you to believe that the subject advertised will be most unique and if you do not have it your soul would be much poorer.

Let's take soap advertisements.

One hundred years ago soap started with a program to get people to use soap more often. Look at the very old pictures of the turn of the century going to the 20th century and you will find a soap advertisement looming in the back ground when a film was shot showing a scene of everyday life. At the time of the turn of the 20th century the people were bathing once a week. One day in the week was set aside for having a thorough bath, shave and hair wash. The cleansing process took all night and all night was from sunset to about seven or going very late then about half-past eight. The gentlemen shaved with a cut-thought razor and the woman tool about an hour or more to dry their hair. This was a lengthy process and was done with most care in mind.

There was no advertisement prompting the public to buy a horse because that was all you could buy in order to travel without using trains as transport. You either had a horse and you rode on it or you had no horse and walked. So a horse you needed and therefore advertising to use a horse was senseless. But soap was something else. They had to get people to use more soap and they had to open a need for people to need more soap. Therefore if you were modern you didn't need a horse because you needed a horse in any case so no one is going to pay you to believe you needed a horse. But they had to tell you that you needed soap so that you would realise a need to purchase soap you never needed before.

How many houses had indoor plumbing during the American civil war in the 1860's or the Anglo Boere War at the turn of the 20th century? And were those persons a dirty barbaric bunch of lunatics only craving to plunder and destroy everything they came across...well frankly yes, those on the British side were because under the orders of a mad murder called

Lord Roberts they killed thirty thousand unarmed woman, old-folk and children by starving them and by lack of nourishment they killed in the Boer- concentration camps the thirty thousand mentioned from a population of a hundred thousand all included while only forty thousand Boer fighters died all told on the battle field. ...But all the murdering and the desecration on the side of the British, well that was not due to their habits of having a bath once a month or so. That was because excluding Portugal, the British warred with every other nation on earth and that too had nothing to do with the bathing habits of the day. As the soldiers came home they were not frowned upon as dirty disease ridden dirty pigs but notwithstanding going for months without bathing they were accepted as those representing the normal population. But then came movies and advertising soap in papers and advertising soap on boards many times the size of reality and by making money advertising changed the thinking of man. To buy soap was the in-thing and that liberated the modern age from the depressing past that the older generation was burdened with. Wherever your eyes would take you, there was some commercial scripture telling you that cleanliness was next to Godliness and therefore having a bath was presumably equal to with having a conversation God Almighty. Doing this mental rebalancing was done systematically to get the public trendy at the time advertising started getting the world moving the way the Mammonites planned and being clean began being very fashionable indeed. You had to wash more often to be civilised and cultured. The educated was clean and the want-to-bees were a dirty lot. If you had a tip pens worth of brains and a thought of decency you will use more soap more often.

If you walked past someone and you did not smell like a bar of soap and did not leave an aroma like a lump of soap behind you were a dirty gutter boy who was uncivilised, crude and uncultured. So you had to bathe as often as you could to smell like a walking brick of soap. You had to bath very regularly otherwise you were a peasant without reading skills or any cultural morality. All and everyone forming all looked down their noses at your indecency because not bathing often made you immoral and disgusting. If you had a bath you were well educated, very intellectual and very modern and the sun shone from your arse. The downside of this endeavour of bathing was that the world was not yet accommodating this process of utter cleanliness because back then in the living quarters of the time then the process of having a bath was one of labour. The water in which you bathed was boiled on a Dover stove in a massive pot. The boiling was done after the nighttime meal was cooked and the bathing started after everyone had eaten. The water was carried to a galvanised bath outside the house because no room was set aside for bathing. I know this personally because I grew up on a Transvaal farm where electricity was a commodity that belonged to town folk.

To get the water boiled on the pot took about half an hour and this water boiling only happened after dinner because before dinner food was cooked and a meal was prepared. This was a waiting process you had to go through every night. In times before the war, (there was always a time before the war because there was always a time during a war because there was a country such as Britain always going to war with everyone only excluding the Portuguese and the someone at present is Iraq because they have oil and having something the Brits and Yanks can steal, Iraq are declared the Axis of Evil that has to be destroyed and therefore they are experiencing a War with the Anglo American Allies at present) the tediousness of the process of bathing forced everybody to pick a day during every week that was the day that that particular family member had a turn to bath. Because of necessity it was impossible for every one to bath every day so it was primarily a full nights bother to get your hair washed and your body sanitised and cleansed. Except this did not apply to me during my boyhood days where I became part of this bath as a daily process while enduring circumstances befitting a bath a week conditions. I had to boil water not on a Dover stove but on a stove called the Ellis de Lux but while the stove was special for cooking food it was as tedious as a Dover was when it came to boiling water. It had a boiler at the back but that was for hot water in the kitchen and therefore I was not spared this pot with the boiling water every night of the week. Then when this water boiled in the large pot I had to carry this large pot filled with boiling water. This meant I had to move a few gallons of boiling water to a bath that was either in some room (if it was winter) or outside in a covering shelter that guaranteed our privacy. For a little boy less than ten carrying this pot with a few gallons of hot water was laborious to say the least and spilling only a drop was a painful affair. Even a drop of boiling water dripping on bear feet had you jump but not for joy. However, notwithstanding and nevertheless you had to take your bath before you took to bed.

That was a rule set in stone. It was like reading your Bible by candle light after having you bath then praying and then going to sleep. That was also set in stone. Then again I was part of society where the town folk had running water and two taps built into the wall that was serving a large bath. This was in the late fifty and early sixties. Those in town had a bath plug you could pull out and the water would run into a drain. I was not that lucky. This system they had in town was not the system I was used to on the farm so I had to drain the water with a bucket and that was bucket by bucket. I was part of the bath-in-the-house era while I was going about in the previous era where bathing was done once a week. Before the war everybody lived in a house that had bathing commodities outside. I grew up in a dual culture and had experience on both sides of the war. While my friend also had a bath I had to bath outside every night in a lamp while carrying my water

When we were living on the farm back when we were seen as the unsophisticated because we had to shit in a long-drop situated far outside the house and we ate inside the house in the dining room. Then when we moved to town where we became very sophisticated so therefore we had to shit inside in lavatories while we ate outside calling the process Bar-B – Queue or as we in South Africa call it having a "Braai" coming from the Afrikaans word "braaivleis" that translates directly as "roasting meat". This remark is quite beside the point but duly valid. In my days of growing up if you are old fashioned you shit outside and eat inside and if you are modern you shit inside and eat outside.

Going back to soap; after the war there was money to build new accommodation and building new accommodation brought about that there was money to be made by those already floating in money. Electricity and water was brought to every house by cable and pipe. Water was heated by electricity. Every house no longer had a need for a bathroom because there were three in every house. This got the soap advertisement selling and the soap advertising got even bigger after Television invaded all houses and destroyed civilised privacy. Then a new range of story telling by camera brought about the name for it as it was called "TV-soapies". The Movie-soapies were dim and mindless enough to make every housewife cry because their life brought no opportunities or reasons to cry and so crying became entertainment while they were told to use soap to bath so that their morals could also be washed away.

Nevertheless it was promoting soap with making a large profit in the selling thereof that got the human race to reject the natural smell of humans in favour of smelling like soap. It was rejecting what God gave to man as a sent in favour of smelling like perfumed cleaning material. When a man smells like sweat and manly flavoured work resulting after smell such as he would smell when finishing twelve hours of toil that no woman on earth lesbian or otherwise could do, he then is thought to be a scruffy beast to be avoided. This ridiculous but quite acceptable scenario is now in such a common practise it is done in all aspects of life. If you smoke you will die. If you eat fatty red meat you will die. If you eat oily chips you will die. If you run or if you don't run you will die. If you eat liver you will die. If you drink wine your baby will be a misfit. If you eat pork meat you will land in hell. If you eat turkey instead of fish you will die.

Everyone wishes to be God and force you not to eat so that you can live forever. Any old arsehole that has written a paper on whatever wishes to put the fear of God into you by telling you that if you don't do exactly as he scares then you will die. If you do not fear his finding you are going to die.

I have news for you…you are going to die whatever you do. The only thing that is as sure as tomorrow arriving by sunlight is that in the event of you that went through a process we call birth then you are going to die. If your mother gave birth to you then you are going to die and whatever you do might slow the process down but if you do that you will die of old age fading away and that is more horrible than any other death I can think of. She could not even go to relieve herself for years and in the end she looked at a sealing for days at a time being to weak to sit up straight. I saw my mother disappearing into a vial of demeaning loss of self respect and I saw my father getting a heart attack at the age of fifty one dying in seconds flat and if I had a choice in the matter, I will choose to go like my father did at the age of fifty one. I can't die at that age any longer because I am way past that age but still my father had a good life and spared his health very little as he enjoyed everything he was told by his medical doctor to avoid. Then how correct are all these arseholes trying to chase Godly admiration and eternal fame compared to the likes of me telling you that you are going to die whatever you do and I do not try to claim fame with this announcement. Every researcher portrays what he found as if it is going to put the earth in another orbit.

If you wish to be clean you wash three times a day. If you wish to be healthy you eat crackers with water and hope you will not last long because it is horrible to live healthy and enjoy life simultaneously. On all levels everyone is doomed to be a robot by those thinking they are the wise and they can help you not to die while the only thing you are eventually and finally going to do is die. It is not what they tell you that hold significance but it is what they prevent you form finding out that becomes critical to the way you live. Now hear this and come to terms with what you don't know and find out why you don't know the following.

I am going to show that there is no such a thing in the Universe as mass. No that would be incorrect. I am going to prove that mass is an invention Newton created to fool every person since Newton. It includes all those that pretended to be intelligent. It is a fairy storey coming true. You've heard about the King and the magic invisible clothes, apparently magical because only the wise and the clever in society could see it. Everyone was fascinated by the beauty of the clothes The clothes was only visible to the very wise and because no one wanted to be seen as being a fool in the community, and least of all the King wanted not to be seen as a fool that couldn't see his own clothes, the lot in the Kingdom saw the fantastic magic clothes. As the King paraded naked through the streets a young girl shouted: "The king is naked!" Then everyone realised the King was naked and even the King realised he was the biggest fool in his entire Kingdom. How does this have any connection to a book that is devoted to science in any way you might ask. Isn't this a famous story about a bunch of fools pretending, well it is not a story in science but it applies to everyone in science. Everyone since Newton that knew physics was able to see mass because no one wanted to be the mindless fool that couldn't see mass. So much does this story apply to physics that we can only change the name of the King to Newton and everything else in the story is in place. Newton created mass by magic and everybody since then could see mass. This was the case for three hundred years where everyone in science saw mass as clear as daylight until I came along in 1977 an I was unable to see mass as a factor when seeing all the evidence that disproves mass as a cosmic factor. I was the simpleminded student that couldn't see mass and I still can't see mass. As you read this you feel agitated by my remarks, don't you? At a later chapter I give you the entire story behind this letter that I received from a Journal in response to an article in which I showed how gravity comes together by forming the Coanda effect. The Coanda effect is when liquid of any form and solid of any form interact.

The following is the response I received when I submitter a paper on how gravity forms by singularity forming.

Dear Dr. Schutte,
You submitted an article of 15 pages to the Annalen. The content of this paper doesn't constitute a theory in physics. With a lot of words and some simple algebraic relations, there is no way to "explain" the world of physics. You seem to be out of touch with modern developments. This is also shown by the fact that you don't quote any relevant literature.

I am sorry to say, but the Annalen is not able to publish your work. I am sorry for having no better news for your.
Best regards,
Friedrich Hehl
Co-Editor Annalen der Physik (Berlin)
--
Friedrich W. Hehl, Inst. Theor. Physics
* University of Cologne, 50923 Koeln _____/\/_____ Germany
fon +49-221-470-4200 or -4306, fax -5159
hehl@thp.uni-koeln.de, http://www.thp.uni-koeln.de/gravitation
* Univ. of Missouri, Dept. Phys. & Astr., Columbia, MO, USA

Dear Prof Friedrich W. Hehl, I have received your e-mail reply and I wish to respond on your letter. The article of 15 pages to the Annalen had in mind to introduce a very wide-ranging concept contained in many books. I wish to promote books in which I introduce a much larger and much more detailed cosmic picture. It is six books that actually form seven volumes of one theme supporting The New Cosmic Theory. I wish to unveil a totally new approach to the thinking in cosmology. The concept is proposed in the article I sent to you which is "revealing" The New Cosmic Theory

In the article as much as the theme I wish to go where no one ever attempted to go before. I introduce the Universe of singularity, a state in which the Universe still is because it is a state from which the Universe grows. It is where material in a dimensional dynamic does not apply because it is where Einstein said, "the Universe goes "flat"". I show you how and where the Universe goes "flat" I will guide you to the point where I go...so that you may see where my books and the article lead you. It is in the domain of singularity

When you read work about the Big Bang you have to go right down the development (in reverse order) to the point where the Theory of the Big Bang points at a spot named singularity. It shows the very start from where all material developed. At that point one will find The Absolute Relevancy of Singularity and there has never been any attempt by any person ever to venture beyond the dimensional birth of the cosmos, which is called the Big Bang by going into the era where singularity prevailed. I take you there in my books as well as the unpublished article.

However, going there requires a very high degree of concentration and calls for understanding that a very little number of persons are capable to show. I try to show how the Universe goes "flat" as Einstein said the Universe goes "flat". Even by completing this unimpressive letter you will also know how the Universe goes "flat". Even where you failed to read the article I sent you, then by just reading this letter you will be able to find where singularity takes the Universe "flat". But it requires a mental capacity to understand because where I venture no one ever in the history of mankind reached into before.

I do not speculate but even in the unpublished article I show with pictures and sketches as well as "some simple algebraic relations" where to go to where the Universe starts, but you failed to read that because you are opinionated as to what conditions should the Universe have before the Universe will allow any one into physics.

That is a pity.

One should learn from the cosmos and not tell the cosmos what it must be to qualify as the cosmos. Then in the article I show you by almost taking your finger to the spot, the very point where the Universe ends and that too I qualify. You might dispute my arguments and show me about what you disagree, but it shows very little understanding of reason on your part about qualities man should have before understanding the Universe.

I go into a Universe that was in place before light was in place in the Universe and only darkness prevailed because light calls for space and in that era of singularity space was not even a thought yet. I show why the Universe goes "flat" and in a "flat" Universe only the value of 1 holds value since singularity is 1. If you can understand 1 or $5^0 \times 7^0 \times 3^0 = 1^1$ you have all the mathematical skills required to understand the applying concepts. To reach a value of 1 does not require big mathematical equations but to reach singularity requires 1.

The collection I named **The Absolute Relevancy of Singularity**: **The Theses** and the collection as such forms a small introduction to the thirty-two or so books I wrote on various matters concerning physics with gravity in mind, but **The Theses** as such in the entirety of the four books does not officially even start to introduce the spectrum of every aspect of my work. I have been in contact with numerous Academics and about one in one hundred reply.

When the one in a hundred reply, the academic always uses a most aggressive tone which I came to accept as what I receive from academics, and because of that I was most delighted to find some kind remarks from you as a practicing academic, and might I add, the first such kind remark in ten years of my trying to contact any person in physics that would take note of what I have to say about a new line of thought, because the few others that replied were extremely aggressive about me confronting Newton. I only began to submit books to publishers after twenty-seven years of studying Newton and the role Newton plays in cosmology and thereafter which was ten years ago I began promoting these ideas. **The New Cosmic Theory** is a process wherein I try to introduce a study that is ongoing for about thirty-seven years, give or take a few and I did not jump into the frying pan having my first thought about the matter published as an article when I sent the article to the address of Annalen der physics.

This is the ridiculous concepts that Professor Doctor Friedrich W. Hehl, Inst. Theor. Physics supports. Because I don't conduct tainted mathematics and highly suspicious Newtonian concepts covered by completely ridiculous malfunctioning concepts, I am despised as the ridiculous small-minded novice who is openly very feeble in thought. I do not support complete rubbish as this formula indicates the Newtonian vision of physics represents, but for my failing to go along with total trash my work gets rejected time and again by every science publisher I approach.

I was told, "With a lot of words and some simple algebraic relations, there is no way to "explain" the world of physics". Now I am going to go as public as possible and to use "words" to explain physics and to show how incredibly untrustworthy the world of Newtonian physics is. All you who think Newtonian physics is in reliability equal to God professing through the mouth of Newton don't be cowards by not reading my work as you always do but confront the conspiracy all of you are part of.

Taken from Nature Annihilating Newton

PART 2

In the previous Chapter I explained how the Titius Bode law forms the Solar system but up to now to the Newtonian minded brilliance that was never enough. Now I explain it once more but this time I put emphasis on the method more than the means of the process.

A Long time ago when I still tried to reach academics in earnest some students replied my e-mail warning me not to mention God or anything to do with religion or even to mention God because that antagonised those physicist since they were all sworn atheists. Every time and in all cases where the student replied using the Professor's mail they informed me that every academic in physics in that specific institution was mortally against any form of religion and would not even read further if he or she would read about me referring to God Almighty and His Creation.

Off course I wiped my arse with those views academics had about religion and continued to present my work as if my work represents nature where nature reprints God Almighty. I did not flinch one inch and I still don't blink a split second from presenting Creation as God Almighty did it by applying nature.

I have shown you the hoax they use as science.
I have shown precisely how they and Newton falsifies facts to bring about a lame case that would not stand up for one second of truth.
I have shown how they falsify and cheat to present facts that are as truthful as any fairytale ever could be. There is more truth in the story of Cinderella or the seven dwarfs that there is in Newtonian presentation of how the Universe formed as Newton explained the process. Going to fairytales do you remember the King that bought the magic clothes because only fools that were unwise could not see the clothes and since he was super wise he saw the clothes that no unwise fool could see. I now present you with Newtonians that bought the magic clothes Newton sold which they in their all-conquering wisdom bought and all those that were able to see the magic of mass that the rest that was fools could not see. Now I show you what is truthfully the way God Almighty works. I show they are the "wise king' that walked naked through the streets because he was so wise he could see the magic clothes and the rest that

was unwise and stupid was unable to see the magic clothes. Now I call on them to prove they are dressed up in anything they stand naked and cold in the face of truth.

Should any person ever find anything I use that does not prove the Glory of God please inform me immediately because then that is wrong or I have presented it incorrectly.

They and those that applied Newton's falsified facts and was wise enough to see Newton and "understand" Newton stand facing the world while the world face their stupidity. When there are falsified facts as Newton and his corrupt Newtonian clan embrace the hand of God will be absent. Where they brainwashed and bullied I am able to prove and substantiate facts. God Almighty would only be where no falsified facts are present and because Newtonians work only with falsified facts the absence of God is all overburdening. Now I parent the world with the way God Almighty forms a Universe in the way God Almighty formed our solar system.

Every point in the infinity we may observe at is not merely part of the Universe in not being

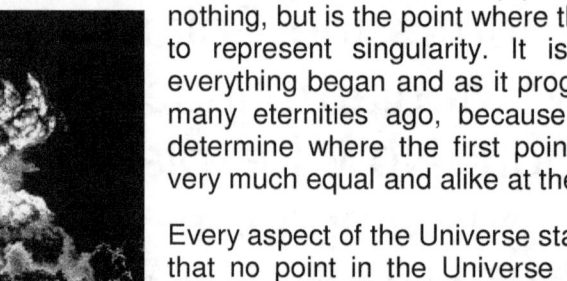

nothing, but is the point where the Universe started as it starts to represent singularity. It is the very first point where everything began and as it progress it begins just as it did so many eternities ago, because after all, how can we ever determine where the first point was, as they were and are very much equal and alike at the beginning.

Every aspect of the Universe started with the fundamental fact that no point in the Universe can represent "nothing" as a number, because every aspect in the Universe represents singularity in what ever form it may hold in that specific spot forming space-time. If man does not reach a conclusion where that conclusion is matching the Universe and stop to match the Universe with man (and man's incapability), we may all go back

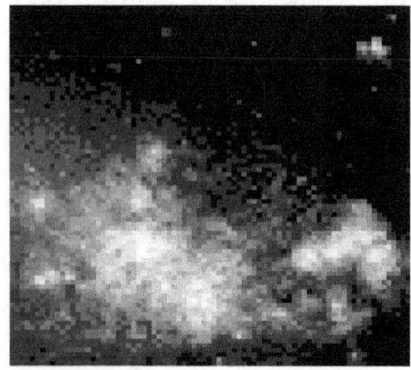

to caves and become starving hunter-gatherers again, because we will never find a way to progress to the ultimate understanding of the universe.

Looking at stars Newtonians see coal stoves being stoked to burn. In the days of Newton coal stoves were the nuclear science of the day and while all other departments in science moved on and away from coal stove principles astrophysics and cosmology remained true to Newton by inventing the coal stove in so many ways not even the coal stove could think of the facets it can go through. Newtonians see stars being fuelled like coal stoves and such stoves can run out of fuel.

This is so much Newtonian backwardness as mass forming gravity and the moon coming closer and the cosmos shrinking and we falling into the sun because of non-existing dark matter making up what is required to make Newton not to seem the idiot as that which all Newtonians are because they make Newton and his contraction theory to be less foolish than what it apparently is and they being the fools that they overbearingly are.

At the time astrophysics in its current form was formulised on the principle that boiling water was the nuclear science of that period and everything that presented anything exciting about science was based on a furnace. Converting carbon to heat in the form of steam was what nuclear reaction became in the fifties to the nineties of the previous century. Steam was at the time the way to go should there be progress in industry and steam carried the industrial revolution into a financial force of substance. Steam was what was science and so the sun was a boiler, a furnace converting material into heat. This was progress and this idea was as revolutionary as steam was and this idea is as outdated as steam today is. Industry progresses passed steam and astrophysics got stuck at the level of progress where it was three hundred years ago because it got stuck not to progress but to conserve the hoax called Newton thinking at all cost.

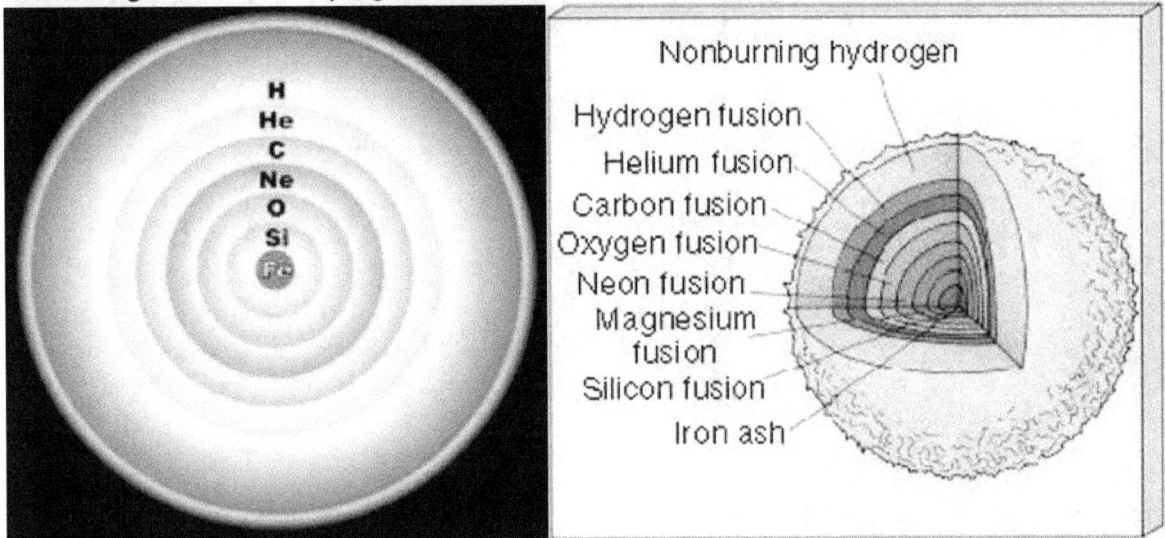

Look at the layers on the outside of any star including the sun. To have gravity the sun must also have an iron core and I prove that mathematically in books dealing with that topic. In any star there is no burning of hydrogen as fuel because the hydrogen cools the surrounding space by pumping out heat from the space within the star.

There is hydrogen, which is volatile, and absorbs heat when moving. The faster these volatile elements move the more aggressive they remove and concentrate heat from a gas to a liquid state as the liquid heat accumulates around the "gas". There is helium, which is volatile, and reacts by absorbing heat when moving through space.

Carbon acts as a solid or a gas because we breathe out carbon mixed oxygen, which it then is, gas, when we remove heat and tissue material from out bodies. We think of the process in terms of aging. Oxygen is a volatile element that freezes at a very high temperature and Neon is also a highly volatile element. These elements have characteristic we on earth use and adapt to remove heat from space as to cool whatever we want frozen. The elements we find on the outside of a star serve one purpose and that is to remove heat from any surface, which as we think of cools the surface.

Then we put the volatile elements through a radiator and the radiator again transforms the liquid state the element hold heat to a gas sate in order so that heat becomes a suspended gas again. The sun on the outside serves as just that radiator and we see a radiator when we see the sun. The sun is hot because it rids the inside from heat, which makes the sun inside very cold.

By duplicating its position on such a massive scale in relation to all the other objects within the solar system the sun concentrates or cools the gas that form outer space in the solar system from being overheating gas to become refrigerated frozen comic liquid and that is what Kepler's tables show. It is not material that moves by being sucked to the sun in mass but the material or planets stay at a distance singularity enforces while it is the space in which the structure move that flow towards the sun at a ratio of $a^3 = T^2k$. it is about maintaining position in a decreasing gas that becomes a liquid as the sun concentrates the space it holds.

In this picture see volatile substances serve two purposes. One is to transmit liquid heat from atom bondage to a gas and provide thrust to the vehicle. Secondly the hydrogen cools the walls of the engine with vanes it flows through so that the walls of the driving structure will maintain structural integrity and this prevents the structure from deforming. Those which science calls "gasses" are elements that hold heat within as a liquid and can liquefy heat transforming the heat from a gas to a liquid. Then by releasing the stored liquid heat within those surroundings of the atomic structure back to a gas as we see happening in this picture we find cosmic drive. There is only one form of cosmic drive and that is to transform unoccupied cosmic substance from liquid to gas or from a gas to a liquid. This is the very same process we find happening in stars such as the sun because we find the earth forming gravity the same way. Every atom within a star is a pump that pumps heat in to cool the structure and the outer layers are regulators that remove excessive heat from the star. When a star goes supernova in an explosion the star overheated and the structure deformed. A star reforms cosmic gas to cosmic liquid and transforms cosmic liquid to a solid we think of as an atom. Everything in the Universe moving faster than the speed of light is a solid and the rest that moves as fast or slower than light is either a liquid or a gas. The Big Bang process was when space turned from liquid to gas. Electricity is transforming cosmic gas to cosmic liquid at a photon density dimension.

Before I go forward with this article I must kill of a primitive perception Newtonian ignorance strives on. Oh yes they know the truth and the answer but the truth and the answer they use very scarcely because it is not very Newtonian thinking. If you look at the picture I provide of Jupiter what do you see. I don't know what you see so I will tell you what it is that I see. My mind and all of my human culture will tell you I see Jupiter. Not only do I see Jupiter but also I see rings around Jupiter. The entire picture gives me a sense of incredible astonishment at what I see. I see dynamics far beyond my surrounding because I see a star of the future in the making. I see a new planet system developing much as our sun is today. I see a future solar system that will grow and become one of the next solar systems somewhere very far ahead in time into the future. I see evidence of the Roche limit that destroyed structures that was in competition with space-time Jupiter put a claim on back then. I see one massive structure so big it is 317 times what the earth has as one. I may think is that all I see when looking at detail.

Guess what I see none of the above and I see nothing that I mentioned and I can't see Jupiter as much as I can't see my hand in front of my eyes! I can't see another person or the toes on my feet. This comes as a shock because all of the things I say I see we all know I can see and yet the truth is that I can't see a single item I think I can see.

I know I come across as being very testy on the subject but this is so adversely important that I can't spend enough time on this issue. This is a lovely picture I see of Mercury and it tells me so much about the history and the story of Mercury and the detail Mercury reveals is breathtaking...but for the fact that I can't see Mercury at all. Mercury is a dark spot void of light and totally invisible. Mercury has no light but the light of the sun bouncing on the surface and that light brings me Mercury. Te light bouncing off my toes at the end off my feet brings an image that I take for granted as my toes. It is not my toes I see. In that we see a hologram formed by an image that does not support reality but forms an interpretation of what I presume the object is that I see. I see what was caste away as unwanted light in colours that does not represent reality and forms time of space that in truth is the history of time. I see what was in what never was and what formed as a result of what the object never used to be and what the object is. I don't see the object and the colours I see that I presume is the object is the very colour the object rejects and is what colours the object caste back into open space.

I can't see anything. I see what time formed and left as space. I see space that became the result of the history of time left behind to form an image of what never was in the first place. When I see the moon I see an image of light and using that that I must use with my considerable powerful imagination to fool me in thinking what I see is the moon. It is not the moon I see but it is light reflected off the moon and the light crosses space by using time to reach me but that image I use to tell myself I see the moon tonight is very far apart from reality. This thinking process is cosmic reality and only thinking in such terms will we become able to understand the cosmos.

This picture immediately tells me I see the sun and I see Jupiter circling the sun as it orbits around the sun. That is totally incorrect and as far from the truth as I can be. I see a lot of light rushing away and the light carries a lot of heat. I see the light being blocked at one point by a round object that absorbs most of the light on the far side of the object that I see. The light is blocked and the blocking or retaining of light results in an image I have of the sun and of Jupiter coasting around the sun. This is a hologram I see forming unreal expectations.

What I don't see is the sun and what I don't feel is the heat of the sun. I feel a lot of heat formed as light rushes away from the sun because the sun rejects the heat as light. That makes the sun cold and dark because the heat as light is caste into outer space that absorbs all heat. That is the true picture I have to use to clarify science information because that results in the true picture I must have of everything I see that I can't see.

What is of vital scientific importance is that there are three fundamental dimensions controlling the Universe. The three are beyond intermingling and one confirms a status in relation to the others but not intermingling in status. From singularity comes matter and forming space-time in own accord. By matter not controlling time, space grew uncontrolled and the third dimension came about. That dimension birth we now recognise as the Big Bang, but the Big Bang is the last of a three prong cosmic growth. Science has to recognise the dimensions of densified (singularity), occupied (matter behind the electron) and unoccupied (space-time outside the orbiting electron boundaries) forming three points of cosmic recognising space-time.

Every dot was by itself as well as the accumulation as it currently is the present universe. The earth in itself is a Universe standing apart from other universes such as the moon as well as the space between the moon and the earth. The moon is a universe. Rules applying on earth do not apply on the moon and visa versa. When considering conditions with in the oceans and applying space-time another set of rules apply therefore the sea places a body in another universe. It takes the same engendering technology going underwater in deep sea diving that going into outer space.

The number of universal entities are still countless as much as it was in the beginning matter as atoms and even much smaller. Every dot insignificantly as it may be is a part of another Universe as much as it is part of the accumulative Universe and every dot in infinity holds singularity, which we translate as "nothing" but it cannot be nothing. There cannot be nothing as much as there cannot be darkness. There cannot be something big or small except in the relevancies of perceptions and then the relativity of such perceptions becomes questionable. There cannot be hot as much as there cannot be cold The sun freezes hydrogen to a liquid at 6500 ^{0}C and outer space boils over at 0 K. If we humans cannot or will not abandon our human culture driven perceptions and our mankind's pre-programmed perspective we may as well return to astrology for what the future hols. There are so many boundaries out there ready to destroy us because of our lack of insight, as did the challenger disaster.

Creation birth started off with one dot so small eternity met infinity within. Then came one more, and another and they continued coming until there were a countless number of dots. The accumulative size of the dots were the same size as one dot because in the true Universe big and small plays no part. The dots were infinitely small and eternally big at the same time because size is a relevancy and without one the other has no size. So in the true perception, there is no difference in size.

It started with the fact that there is no place or part in with which one may associate zero or nothing. There are no room for a number such as nothing. Next to the one dot (infinitely close) one will find the next dot, and if nothing was a factor then that is precisely what one will find between the two dots. Nothing of space, a non existing entity, taking up no space, and much more important, no time, therefore the dots are infinitely close to one another, being the same space, eternally big as much as infinitely small. If we as humans cannot find a manner in comprehending this notion, there can be no manner ever understanding the cosmos as much as the start to the cosmos.

Every dot was a Universe in its own and the accumulation was a universe. The earth in itself is a Universe as the moon is a universe, because rules applying on earth do not apply on the moon and visa versa. When considering the conditions with in the ocean and applying space-time another set of rules apply, therefore being in the sea places a body in another universe. The number of universal entities is still countless, as much as it was in the beginning, before dots formed atoms. Time is the process where singularity remove space as much as singularity form space by depleting the concentration of one part and increasing the concentration on the other part. All energy or fuel that produces movement is the result of the interaction there is between solids and non-solids and in the case of the earth we receive our energy from the sun concentrating cosmic gas to cosmic liquid. Then we have ways that the liquid heart called sunlight is stored.

In the two pictures we are seeing disposing or releasing heat creates space. We may call it

plasma or shock waves or what ever, but in the final analyses it is heat turning to space. Whatever you wish to call that which lies between the particles comes from being a solid, then with adding heat, the solid *"whatever"*

becomes liquid and that is the white and orange plasma that we find. That white and orange is heat in a liquid form, just as all flames and smoke is heat in a liquid form. But that liquid does not remain liquid because the governing singularity cannot enforce a commitment ensuring the liquid heat remains liquid. The liquid *"whatever"* you wish to call the heat in fluid form then further overheats turning the heat to space. The space created must be equal to the heat reformed. That is a law of energy where energy equals equality everywhere it is.

Every dot insignificantly small as it may be, is a part of another Universe as much as it is part of the accumulative Universe and every dot in the infinity holds singularity, which we translate as " nothing" being " darkness". There cannot be "nothing" just as much as there cannot be "darkness". There cannot be something big or small, but in the relevancy of perception, and then the relativity of perception becomes the question. There cannot be hot as much as there cannot be cold. The sun FREEZES hydrogen to a liquid at six and a half thousand degrees Celsius and Universe boils over in the form of the Hubble constant at the temperature (we presume from our vantage point) at minus 273 degrees C. If we Humans cannot or will not abandon our human perception and our manly perspective, we may as well return to astrology for all its worth, because that is the only boundaries we will find in the cosmos.

To unlock scientific truth we first have to dispose of scientific misconception

Let us humans first detach culture from facts. Take the argument to iron, which we know well. Iron cannot boil, iron cannot flow or bend and iron cannot brake. Iron is an element like all the other elements we know, not one element can do any of the above, in sharp contrast to human belief. As indicated in this book the limits we should find to guide us we ignore for the reason that we cannot see it. We may not be able to ever see singularity, but with intelligence guiding mankind, we do not have to see everything to believe everything. It is because we could not see religion, but still practised religion that set us apart from the other animals.

At the start one would find iron and iron in a "natural state" as we find iron on earth being a human produce on the surface of the earth it will be a solid, suitable for man to handle with bare hands. When such a piece of iron is left in a desert in the midday heat, the human hand cannot handle the iron any longer without aid of covering the skin of the hand. Our perception is that the iron became hot, but that is not the case and our view is a culture contribution and not scientific fact. By heating the iron artificially with combined gasses (acetylene and oxygen or what ever) we now can over heat the iron to a state of flowing like a fluid. Our human culture tells us the iron now is melting.

That is a misconception!

Like the fact of "nothing" we inherited the idea from our past. After introducing artificially even more heat with more heat releasing gasses we may artificially form a condition where the iron would become a gas. Again it is not the iron that becomes a gas, it is the space the iron finds itself in that became hot enough to become a gas. The iron particles remain the same; it is the condition surrounding the particles that changes form with overheating.

Important to note is the fact that iron in a solid state will surround itself with solid matter in space applying a solid space. By introducing conditions producing **more overheating** the space or connecting between the particles become concentrated heat forming a liquid substance! It is not the iron that turned liquid but the wrapper containing the iron that concentrated so much it formed liquid fluid by the introducing of more heat to a point where the overheating created a fluid. It is considered that the oxygen burn and by that the iron heats up. NOT TRUE!

If oxygen burns no oxygen would be left on earth by the time man arrived on earth to use it to the benefit of intelligent life. The oxygen remains oxygen while the oxygen merely does a task in nature where oxygen carries heat to a specific space. On the other hand it is the task of nitrogen removing heat from the point of overheating by means of flames whereby it creates space. One can feel the "wind blowing" as the flames generate created space. In the extreme the creation of such space we call an explosion.

In the process where the space between the iron particles still further overheats, it becomes a gas. It cannot be iron that becomes gas, because iron will be as much a gas as iron will be a liquid or a solid. It is the space covering the iron particle separating the different iron particles, which will convert and sustain form. The gas is as invisible as space because the gas is the form space holds. This confirms the Biblical view of earth (solids) created and heaven (heat or gaseous/liquids) created. There are only two forms of substance that forms the Universe solids and non-solids, which is liquids and gas. It is not the solids going liquid but it is more of the liquid in ratio with the solids in between the solids that make a structure go solid or gas. There are heavens (non solids) and earth (solids) and this has to do with movement applying control or non-movement allowing non-movement control.

Iron is a solid. Introducing more heat the iron becomes wrapped in a cover that concentrates the wrapper to the point of concentration where it became a fluid. The iron remained what it is, neither a solid, nor a fluid nor a gas. By introducing more heat it becomes a gas. The gas we cannot see because the gas is space. But so was the fluid space. The introducing of heat brought about the turning of a solid to a liquid to space and every time more space becomes part of the picture.

Iron is in its normal form a solid. That means the space, which the iron particles are in, is solid and that disallow the iron to alter the form in which it is. By introducing considerable heat the iron melts changing the form of the iron from solid to liquid.

Considering the evidence we find it is not the iron that melted and that became liquid, but it is the space in which the iron is that became liquid. The iron particles are still as solid as they were. By introducing more heat the iron would eventually turn to gas. It is not the iron that turned to gas, but it is the space in which the iron particles are that has increased to the extent that the space now has so much heat, the heat turned to more space. The iron as particles remain the same, they are just elements confined to a nucleus with electrons spinning about.

The space between the particles increased to such an extent it first became a liquid or a fluid and with more heat introduced the heat increase brought about that heat turned to space. That means by overheating the particles surround with heat as a fluid the heat increase then add space as a gas. The gas is the ultimate form of overheating but where one is unable seeing the gas.

1 Firstly the iron is cold enough to be a solid. Replace the word iron with cosmos and forget the colour we associate with heat being white and note the solidness of the centre of a galactica. This must have been the state of galactica that contained large parts of the Universe when time rolled away from eternity.

2 By introducing overheating the space between the iron and not the iron as such turns to liquid. The same apply as more matter (iron) produce more space forming as some matter turned to heat by overheating. The matter increased spin and in that way went out of sequence where it then became softer and softer in relation to other particles, where the loss of the matter released more of the third cosmic component we named heat and space.

3 Some of the heat introduced with the overheating by means of congestion then forms space while other remain in the form of heat allowing space to seam liquid. The matter could not breath and overheated by the enormous gravity the overheating created

4 As the area between the particles still further overheat certain parts of the area overheats to the extent that the space becomes an invisible gas allowing the congestion of matter to separate from one another and allow the stars' individual governing singularity growth. 5 From the soup of heat galactica come about allowing stars to rise out of the dense liquid cradle from where they can establish singularity growth. The process continues as more space becomes introduced through space overheating turning heat into space

6 Should star development come about as suggested it is foreseen that the Milky Way once was a liquid from which the sun developed the singularity in which it then form self-sustaining. The only pre-condition was that it captured individual space-time where the captured space-time remained a liquid frozen (as it was back then at the time of parting) by the governing singularity while outer space further overheated into a thin gas

7 The sun captured so much space by the intervention of singularity when released from the Milky Way that it produced space so concentrated today at present it clearly remained a liquid inside as it froze the interior in time the liquid it now is while outer space is still overheating as a gas with no visibility. From this overview one can judge just how far science is behind the time in their views on creation and the beginning of time including the universal establishing. Cosmology still hides behind medieval ideas that other faculties and scientific departments forgot long ago.

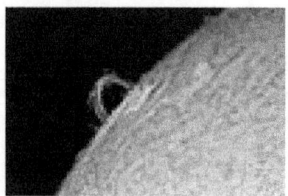

See the fluid push out of a bowl of liquid, spilling both sides as it falls into liquid. The inside of the sun is not gas but it is fluid. In all of nature there is no NATURAL GAS as much as there is no NATURAL SOLID. Look at the liquid squirting from the surface of the sun. If it is liquid the sun is liquid on the inside and if the sun is hydrogen then the sun is so cold through movement that it turned hydrogen into a liquid at 6500° Science acts as if they are under the impression that water will boil at 100°C when it is on the sun because 100o C is about as standard as the cosmos gets.

Let us take this formula back to the accepting of the Big Bang and find sensibility amongst a lot of confusion that I can see.

As a solid	Forming a liquid	Steaming as a gas
Hydrogen 1	melts at -259^0 C,	boils at -252^0 C,
Helium 2	melts at -269^0 C	boils at $-268,9^0$ C
LITHIUM 3	melts 180^0 C	boils at 1300^0
BERYLLIUM 4	melts at 1287° C	boils at 2770^0C
BORON 5	melts at 2030° C	boils 2550° C
Carbon 6	melts at 804^0C	boils at 3470^0 C
Nitrogen 7	melts at -210^0C	boils at -195.8^0 C
Oxygen 8	melts at -218.8^0C	boils at -183^0 C
Fluorine 9	melts at -219.6^0 C	boils at -188.2^0 C
Neon 10	melts at -248.59^0 C	boils at -246^0 C
Sodium 11	melts at 97.85^0 C	boils at 892^0 C
Magnesium 12	melts at 650^0 C	boils at 1107^0
Aluminium 13	melts at 660^0 C	boils at 2450^0

Hydrogen is as much a liquid as iron is a gas and neon is a solid. It depends on the element relating to the space/heat in the circumstances surrounding the substance at that very precise instant in time. We have to stop telling the cosmos to show us what we wish to find and start accepting what the cosmos is telling us to find. The culture that I am referring to is all about **nothing.** At present, we find that there is something we think of as nothing in outer space. Because nothing is what we wish to find and nothing is precisely what we are getting because we think of outer space as nothing. If you accept the cosmos to be nothing, then please define nothing to yourself and find the definition in the cosmos.

Lets look at heat. I know science ee the sun as a red-hot pale and that it is not. The sun is surrounded by flames but notwithstanding the flames surrounding the sun the sun by material is freezing the entire solar system reaching one third the way to the next star group into one spot so tiny it forms the centre of a space the size of an atom les the space the atom material holds. The sun being so cold contracts the entire solar system into the centre of the atoms forming the sun. However this is not pushing and pulling but freezing material together as it connects material and space.

Therefore to see that sun we have to see what is hot.

Let's first look at a fire. Te wood burning before it gets hot is cold. One can hold a match in hand and light a fire that will destroy Chicago and yet that burning material is not that hot. That is the reason why people think it is brave to run over coals. The burning coals are not hot but the glowing coals are hot and in that there is the difference.

Standing next to a BBQ fire one can touch the wood it is not hot.
Then there are the licking flames. Te flames are heat in a liquid form. Looking at the flames it matches the same picture as one get from the sun with the flames rising above the sun. The lames are not hot. One can put one's hand into the flames and your arm hair gets scorched but other than that is does not give a burning sensation. The flames being liquid heat transfers the sealed heat that was once sunlight into space. The liquid flames as a transport tool but that is it. It is what the light releases that brings the burning and holds the heat.

By releasing the heat into gas the heat forms space and it is this transformation of space from liquid flames to heat gas that brings burning. The revaluing of heat from solid to liquid removes heat but does not release heat. It is when the heat revalues its status from a liquid flame to a gaseous ball of heat that this brings about scorching and burning. The turning of heat to liquid is the discipline that transforms heat from a cold to a cool and then to a hot environment and this is also true of the sun.

The material that burns is cold. Going back to look at the way rocket engines produce ignition we find heat has to expand to convey movement. Even in releasing the gas from the rocket engine and letting it flow through the vanes of the engine while being in contact with the massive heat release on the other side of the wall just supports the integrity of the engine wall but the gas does into heat and burst into fire. Only when the gas expands and it ignites as a fuel does the heat release to form gas. Then that is flames and fire but the energy is the expanding of the ignited gas propellant. The sun is cold. The flames around the sun merely transports released heat from the solid to outer space. The true heat release is when the flames turn liquid into gas and the release of heat drive spacecraft and the solar system and the Universe ultimately.

If anyone puts anything as hot or as cold that person puts him or herself in as the centre of the Universe. Then from such a stance as the person has that person finds at the centre things go colder or hotter because that person is stupid enough to think the Universe was created with life in mind and moreover for that person in particular. There is no hot and there is no cold. As it is important to realise the above it is just as important to realise that heat is another form of material and a separate form of material. The two developed on an equal basis and as a result of the other. The one produced to save the other and what the one produced saved the other .The one principle brought the incentive for motion while the other took the incentive by providing the motion. The one produced what the other captured and the one retained what the other delivered. Eventually the motion did not bring the required relief and another form had to be devised. By overheating and increasing space it counteracted overheating and by removing the expanded material and retaining it onto the contracting of the other, did the two form a synopses where by all received benefits in the form of cooling.

Only when further requirements developed, did the need arise for more to be made available. The first demand on motion asked no further changes because one change brought on satisfaction to all that suited all. The second was more general and on an ad hoc basis that was established to fit the need of individual places and not groupings in the broader perspective to fit individuals at large. At first the establishing of motion set a trend that brought on required results but afterwards the space required, in which to move became a demanding issue as the heat levels raging out of control. The heat had to be stored in space by becoming space to retain heat for later consumption. The number in ratio that produced the heat providing particles that

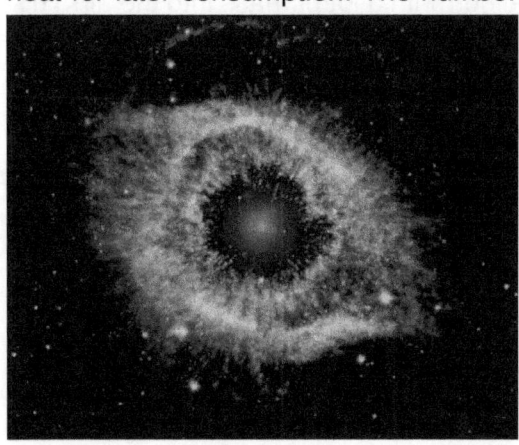 offered to release their form in contribution to have those that retained form, do so to save those others retaining form. Those on offer became those ones that became the danger of destroying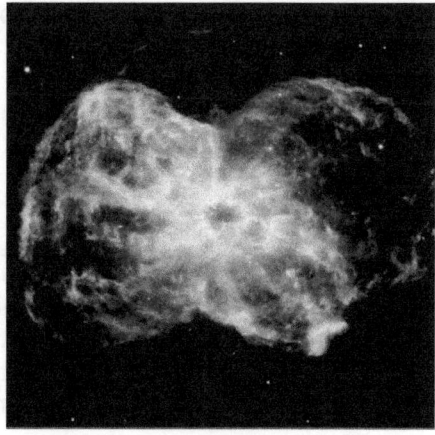

Creation instead of saving Creation.
If heat comes out of a star when exploding then the inside is filled with heat and a star is then a heat container

When we see a Super Nova going ballistic we see heat spewing from the star to the outside regions. We see what is inside the star and what is inside the star is heat, not "mass". We see heat bursting out in rage human minds cannot absorb. If heat comes out then a star is filled with heat. That means a star is a cooler of heat because when the star overheats, as it no longer can retain the heat on the inside by gravity it explodes into unleashing the inner heat.

There might even be some areas and regions in far off places in our modern day where an imbalance may evolve and some particles become unsuccessful to save those more successful. By going less successful, the singularity places a demand on another bringing about the command on space-time so that support can be accomplished to save singularity. Therefore by losing density, was gaining security to survive as part of a bigger relative. Density is the distributing of heat in specific relative space and by having less material in more space; the density is the offering for the common survival of the lot. The relevancy brings a contribution in whatever role to secure the survival of the lot in relations. No relevancy therefore can be "nothing" notwithstanding Newton's opinion about the matter as Newton had the opinion a relevancy acquired by rotation brought about an accumulation resulting in nothing.

It is the way the atom formed before the atom took on space-time. It is in the formation, that space-time relates to motion. We have some elements being quite massive but also lighter than air and others are quit light but as dense as they come. This can only be a contribution from the way the atom relates to heat, which make the atom volatile (movable) or dense (motionless). Those elements being volatile are also very movable and in that we find the role that such elements play in the star. Stars that are predominantly made up of hydrogen and helium with very slight support from the metallic inner core are those stars that duplicate by producing motion. However the point I wish to press is that mass and being massive and being heavy do not support the fact that some elements have more gravity they produce because their protons are more numerous than others. The fact that mass generates gravity is a myth. Everything in the Universes hinges on density forming differentiations between material moving in space. It forms as a ratio between what is occupied space and space to be occupied or solids and non-solids. This again hinges on what is cold, which is dense, and what is hot which is less dense.

You have heard of the expression he is a bag of "hot air" or it's only "hot air" or "they use hot air again". This is in place to discount the fact that when you blow hot air into a balloon the balloon surges into the air. Nature can take "hot air" and lift a balloon with a lot of mass into the air. This ruins Newton's idea that mass pulls down because by just adding "hot air" the mass is discounted and the balloon flows into the air. Now you explain that in terms of Newton's mass pulling mass and you get nowhere very fast while sounding very stupid on the trod. If a balloon can hold three tons of mass that the earth pulls down, how then can a bit of "hot Air" lift this lot up? Science would rather belittle the truth than question Newton! They could not fight nature because nature proved that "hot air" lifts mass.

That made Newton's idea sounding a bit stupid because when hot air lifts the three-ton basket-parcel up it must be cold air that pushes the three-ton parcel down. In order to stop the public getting wise to the idea, they belittled the thought about "hot" air. They started with the saying that anything not worth much becomes a bag of "hot air". If you are not cut out to be successful you are a bag of "hot air". If you are not doing anything worth your while you blow "hot air". If your chances to accomplish anything are only very tiny, they call you another bag of "hot air". If you can only talk but is a coward at heart you are a bag of "hot air". ...And so to draw attention away from hot air brushing Newton off the table they spread the rumour that hot air means worthless at best. This is the influence of those I have to fight to gain ground!

I wrote many books on this because there are so many more examples to show how far academics did go to avoid people getting wise about Newton or that Newton becomes compromised or people see Newton in a revealing position or think about Newton's correctness. There was forever a cover-up to hide Newton from the truth or from nature. When they discovered that the solar system arranges the planets position not according to mass as Newton has it in his formulated calculations, they dismissed the Titius Bode law as a fluke of nature. They did not make a fuss and start a research but swept the Titius Bode law very much under that table by labelling it as a once off enigma. It is the Titus Bode law that allocates the planet positions and not Newton's mass attack of madness. It was nature that showed where to find the missing planets. They discarded Newton in favour of nature. I am going to give you a few scenarios in science and then you translate what happens in nature to Newtonian science. This is the formula on which the entirety of physics depends! You better keep your eyes open because if you blink you will miss how science conspires to defraud you and mislead you by you allowing scientists to brainwash you by means of mind control. Let's get down to the most basics in physics. Everyone is familiar with the concept on which all that forms physics are founded. It is Newtonian and it forms by the following concept.

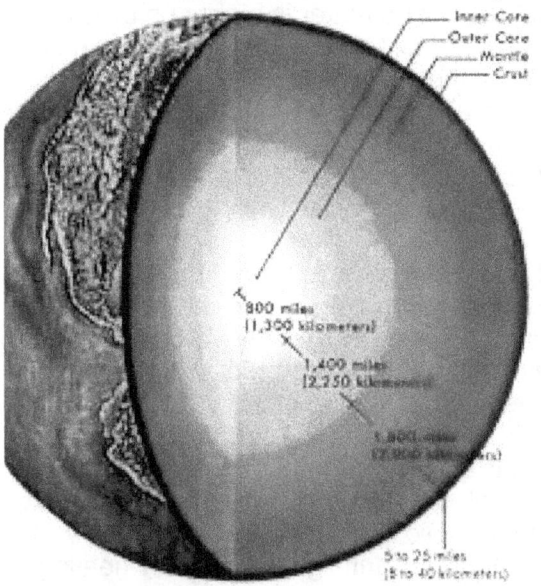
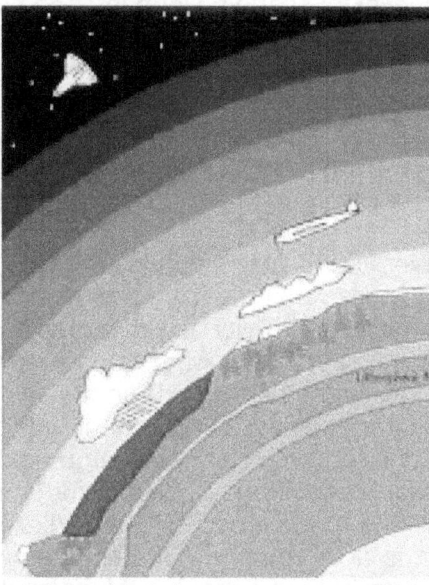

Hottest
2nd Hottest
3rd Hottest
4th Hottest
5th Hottest
Getting colder
Getting more colder
Getting freezing cold as it gets liquefying cold in the centre

When it expands it is hot. When it contracts it is cold. It is not cosmic cold because the human can feel heat but on the contrary it is cold when the human feel heat because it is getting rid of the heat since the space got cold as it contracted. That is science law and not the meaning of some stupid scientist. Heat in the centre of the earth is a liquid and heat in outer space is a gas. The liquid is much more contracted than the gas in outer space is and that is why we can fly in space and not swim in a volcano. We must start to realise how the cosmos translate values and stop thinking like humans. When the balloon gets hot it expands and it goes up. When the anything loses heat in shrinks as it becomes smaller. That too must apply to the earth where the inner core space gets colder as it loses heat and the heat rises to the outer space region. It is not what we feel that makes anything hot or cold but it is what the cosmos does for instance increase in size or decrease in size that forms heat.

One will find that whatever group one chooses there are gasses and there are solids. If mass was attracting mass then the strongest mass must be attracted to the strongest mass and the least mass must float in the air. $F=G(M.m)\ r^2$ hardly can even begin to explain the fact that there is a gas that is more massive than iron but floats in the breeze just as hydrogen which is the least massive element. Let's look at gravity and anti-gravity and see where the wind blows the balloon.

The saying goes "it fills with hot air" meaning it fills with nothing and yet while the balloon fills with hot air the balloon gets air borne. The balloon gets into the sky by being filled with "hot air"

and the idea that this might happen is senseless to science or so they pretend. Still it happens but why does it happen? It is because science has no idea about heat and cold. When filling the balloon with heat the balloon takes of to where science thinks it is cold. Why would the balloon go to a cold place when it is filled with heat? It is because where science think it is cold it is hot. Forget this idea of pressure because if it was pressure in the air the pressure will escape from the bottom where it is open and being pressure it will escape or release from that opening. By heating the balloon with hot air must increase the heat level and increasing the heat level takes the balloon into the air. This means hotter is higher up by holding more space than down below. When something fills with "hot air" it goes up and by going up it forms anti – gravity. If gravity is what takes bodies down to the earth then lifting it up into the air must be anti – gravity. Anti means opposing or counter acting and when going down forms gravity an anti of that then must be going up. If it is mass that pulls down then by applying more heat it forms anti-mass but it can't be anti- mass because the persons in the basket seems just as they were when they were on the ground wit all the mass intact.

Newtonians are never very clear on this issue but if it is mass that pulls then it must be gravity that moves. Mass supplies the magical force but the actual movement is derived from gravity. So the movement in itself is gravity and that gravity pulls bodies down. To have bodies lift into the air then this action represents what goes anti and anti is filling a basket with hot air. By adding hot air the gravity goes anti and then gravity must be the cooling of space. It is definitely heating space up that makes the balloon go into the air so going into the air forms anti gravity because gravity is going down. This is an argument that science cannot dispute except when they go into another cheating and dismissing mode of the truth. By heating the basket something happens and this science for many years try to avoid by reducing it to a joke. A joke it is not because it is a fact nature substantiates with forming anti- gravity. When heating a basket makes it lift than gravity is making things cool and that science can't deny.

When the earth spins there are always a relevancy applying between that which spins are double 7 and that which moves straight continuing at 10. The part spinning at seven we think of in terms of the diverting of direction it applies at 7. The liquid / gas holds the 10 factor.

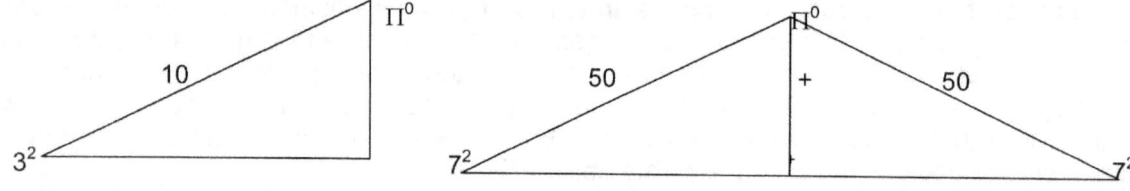

Without the application of specific heat, the object remains in the three directional moving of six possible directions. The value of space unoccupied therefore remains $\Pi \, \Pi^2$, as it was before the "Big Bang" event, whichever "Big Bang" you wish to refer to, because there were many. But space unoccupied holds time to the value of 10 to 1, and as the sketch of the triangles also indicated, holds space to Π. Therefore unoccupied heat holds the relation to space in applying 3 directions of influence $(3^2 + 1^2 = 10) = (a^3 = T^2 k)$. Always part of this equation is the dual function of space in $(a^3 = T^2 k)$ while at that very instant one has space-time. Therefore in space in time you have $(10^2) = 7^2 + 7^2 + 1^2 + 1^2$. In the sphere we have the axis holding a value of 3 and the circle holds a value of 4. These are dots forming in relaxation to the one spot holding a point from where singularity advances.

We have the axis valued at 3 going square through movement of the linear motion $(3)^2$ and then we have the circular motion $(4)^2$ going square by the spin of the circle ring the direct opposing side. Then the equation of influence becomes $3^2 + 4^2 = 25$ where $\sqrt{25} = 5$ and doubling the 5 on both sides of the triangle will apply the factor of $5 \times 2 = 10$ that then is $(10^2) = (7^2 + 1^2 + 7^2 + 1^2) = 50$ on both sides is 10. The implication of this may not dawn on one the very instant of realizing, but to scientists, there is no greater shock than just that. To any application of movement, the factor will be in the realms of singularity where half a circle is equal to a triangle is equal to a straight line and the lot is equal to 180°. No fancy mathematical expressions have any value in singularity because singularity holds a value of 1.

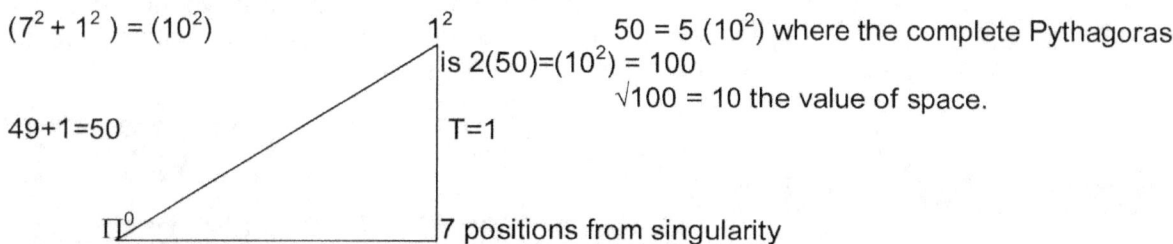

$(7^2 + 1^2) = (10^2)$

$49 + 1 = 50$

1^2

$T = 1$

Π^0

$50 = 5 \, (10^2)$ where the complete Pythagoras is $2(50) = (10^2) = 100$

$\sqrt{100} = 10$ the value of space.

7 positions from singularity

The fact of this comes as 49 plus one becomes 50 and that is in the three dimensions of space $\Pi^2/7$ where 7 holds the relation to one and $\Pi/7$ again where 7 relates to one. At this point it is most important to remember that Pythagoras works on the application of the sum of the square of the two sides. When seven has a direction in the fourth dimension applied to it, the opposing dimension will be one and this applies in time relevancy, therefore the interchanging in time between infinity will place matter at $7^2 \times 1$ relating to circular and $7^2/1$ with $7^2 \times 1$. This makes 49 plus one (singularity) always being a factor of one. Space in time however, never can be a cube, it will always be a square with one side pointing the direction of time from time to the past (1) to time to the present (1) to time to the future (1).

The circle forming Π uses 7 to indicate the roundness of the circle but the 7 holds its roots deep within creation. It indicates how the Universe started because this is the way a star will start moving and it shows how as the infant star starts generating gravity just as the top starts to spin when it is thrown by life. Life can create nothing and that is true but life can mimic all laws in the Universe. Time is eternal movement and will be with us always. The line in infinity is still present while not being a part of the Universe. This line is always ready to be in place when the slightest movement orders it in place. Before the Universe was in place eternity and infinity was in perfect harmony and the line forming singularity validates this fact.

Before infinity parted from eternity, eternity met infinity on one spot as eternity came from the past (1) forming the present (2) to go onto the future (3) but also returned to come from the past which was the spot held by the future and this we find in the fact that the line forms 1 when not spinning but as soon as it evokes by spin, 3 points form even now. Then heat and cold differentiated values and space landed in between eternity and infinity. As eternity moved in relation to infinity but not forming a part of infinity any longer, eternity had to follow a path by never going away from infinity (3) and always returning to the point infinity holds but never lash onto the point again. With space parting the points, eternity had two points (the past and the future) before the partition came about and infinity held both the past and the future while infinity had the present as it still gas presently. By eternity also moving, the two points it held opposed each other (the past and the future) and since it moves, by the movement it became the square of the two because movement is the square and not a flat blanket-like surface with squares embroidered on it as Newtonian science depicts it by using grand mathematics to understand singularity.

Then we had two point holding eternity in place going square by movement to form 4 points serving eternity and infinity captured the first three points held by both and since eternity could not release from the two it had but had to duplicate what it had, eternity by movement became a circle captured by the line. With four points captured by the line of three points the circle coming about is eternally returning to infinity but never complying with infinity because if mismatching temperature or movement (3 against four). Material will always be colder than outer space. It is because material spin and outer space moves by expanding due to overheating.

Whether you accept this statement or not, but for the past three hundred years there is a Conspiracy in Science in Progress and moreover it has been going on as a deliberate result of science brainwashing students in accepting absolute bogus information as truthful. Those in science pretend Isaac Newton's principles work and physicists make Newtonian principles work notwithstanding nature showing the very opposite. Nothing about what Newton says vaguely corresponds with what applies in nature! Scientists forcefully present science as a copy of nature but that misrepresents the truth. If you read only this website you will come to see what Newtonian science could never explain. It's called the Titius Bode law. For the first time ever I explain this law.
The solar system does not use mass to form gravity as Newtonian science declares but this information is never often and openly revealed to the public as detrimentally important. Nature does not apply Newton and Newtonian science but uses another application going by the name of the **Titius Bode law**. Go on and look it up on the Internet and verify what I say about nature using the **Titius Bode law** being applied instead of Newton. This is fact not widely promoted but nevertheless it is true. In four books on different levels of intensity in each I show and explain just how I cracked the principle named the **Titius Bode law** and show why it applies and how nature works and in each one of four books on a different understanding level I reveal the concept that **nature** (**not Newton or science**) uses as the building blocks of space. But the concept disproves Newtonian science and rubbishes Newton's ideas about what he thought gravity is as unworkable and it is that part that science will never accept as a factual scientific fact. **Nature dismisses Newton's mass ideas**. I have been fighting this fraud for decades but I found that there is too much money to be lost to recognise my work and too much honour going wasted if mainstream science had to freely admit that in their conduct they are fraudsters, which is exactly what they are. Download any of the four books I give for free and test what I have to say. The books I offer as an introduction is free but it delves into science in a manner you have never experienced before. However, I show what is incorrect with science but if you wish to read the corrections you have to purchase. Download the free books and see how revealing the information is on how science cheats.

The fact that nature uses the Titius bode law in the solar layout is known for centuries but as usual science tells the Universe what it is and ignores what the Universe really is. Science ignores nature since science has no idea why the Universe is what it is. The solar system and therefore the entire Universe uses a ratio that two men Titius and Bode simultaneously but being well apart at the time discovered. The solar system and indeed the Universe is built not by "mass" as Newton suggested incorrectly but employing another unknown system but is also a very well hidden system called the Titius Bode law. This law is what forms the solar system proving it is not mass. **The following table compares the law's predictions with the actual distances, where the addition of Pluto is a modern modification.**

Planet	n	Titius-Bode Law	Semi-Major Axis
Mercury		0.40	0.39
Venus	0	0.70	0.72
Earth	1	1.00	1.00
Mars	2	1.60	1.52
asteroid belt	3	2.80	2.8
Jupiter	4	5.20	5.20
Saturn	5	10.0	9.54
Uranus	6	19.6	19.2
Neptune	-	-	30.1
Pluto	7	38.8	39.4

This is what is there: It is the **Titius Bode Law** and I show why the **Bode's Law"** or **"Titius-Bode Law"** forms this formation ratio as it does but the best science could come up with was to change the original formulation that was $a = (n + 4)/10$ where $n = 0, 3, 6, 12, 24, 48...$ to the modern formulation of ($AU_{earth} = 147.597 * 10^6$ km): $a = 0.4 + 0.3 \times k$ where "k'= 0, 1, 2, 4, 8, 16, 32, 64, 128 (sequence)

The ratio is so impressively periodic or cyclic correct that it can be put to a formula such as $a = (n + 4)/10$ and the outcome is so predictable that according to the formula $a = (n + 4)/10$ it led to all the discovery of all the missing planets that was discovered after Galileo Galilee used the first spyglass to look at stars. This ratio doubles the distance a planet has every time a new planet is located in a new position. The distance of Saturn doubles the distance between it and Jupiter to what the distance is between Jupiter and the sun. The distance ratio doubles to Mars to what the distance is from the sun to the earth. This ratio of doubling the relevancy in distance from the sun applies notwithstanding the size or the mass or any specific value that any planet might have. This shows that the ratio is not vested in material but it is in accordance with the space that holds the material and that throws "mass" and Newton out of the window forever. This law has been and is known to science for almost if not two hundred years and if you are not a professional astrophysicist you have never heard of this law before in your entire life. It is well hidden under Newton's misconceptions and no matter how much any one wishes to share some brainwashing therapy with me the fact is that Newton's $F = G \dfrac{M_1 M_2}{r^2}$ and $4\pi^2 a^3 = P^2 G(M$

+ $m)$ does not wash because it is not applying in any form within the Universe anywhere. The Titius Bode law is what the solar system uses and if that burns your brainwashing and you mind control inflicts excruciating pain on your brain and you get an overwhelming urge to becomes violent with disapproval towards me it would be best to kneel and pray for it is not me that put the Titius Bode law in application within the cosmos but it is the way God Almighty designed the Universe. When you argue with me you have an issue with God or nature or the cosmos, not with me. It is used in nature and not by me 'cause I show it's there and that's all. What I achieved is to find out why the Titius Bode law applies and how does it come about to form the ratio that it does. I was the one than made the connection (not discovery) how gravity comes about by the implication of the Titius Bode law but also always in conjunction with three other phenomena called

1) The Coanda effect, which is the way, the atmosphere forms.
2) The Titius Bode law is how planets use a ratio to arrange their allocated positions.
3) The Roche limit is the law that applies to what we call as the "sound barrier" and how stars explode.
4) The Lagrangian points is why atmospheric layers form around the earth and some planets has rings.

It is the manner in which the solar system is presented that is completely inaccurate and just as

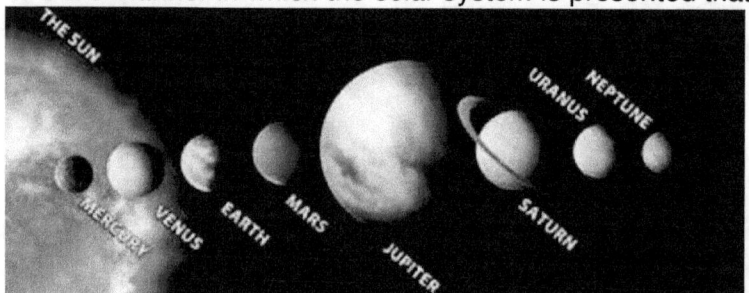

much confusing. The solar system presented with this layout is even less accurate than what the Ptolemaic presentation is because the Ptolemaic have the numeric order correct and the rest is completely mythical. Having the solar system is so close in proximity gives the impression of being cramped for space and huge in material and there is no excuse to explain away the inaccuracy except to help Newton cheat by cheating some more.

The distances apart are more important to science than what sizes are. Newtonian science dismisses that the sun spins around the earth but that I observe every day of my life. In this time I can see the stars in constellations determining which month it is and what season we are in. The stars turn around the earth just as the moon and the sun and denying this is misrepresenting truth.

You are going to say this is wrong to believe the sun is turning around the earth but yes although this is incompatible with science it is not wrong. To say it is not true that the sun spins around the earth is untrue because I can see it happen! It is as far off the mark as Newtonian science is off the mark. The way Newtonian science form your picture you form your concept by the solar system is as much a comedy made up of misunderstanding as this is. There is not enough space to go into even slight detail but for that I offer four books that is on the market. I show you in detail how far Newtonian science is wrong!

In those four I present you with the truth about science in all of human history and I do not exaggerate. I mathematically prove that according to the Titius Bode law gravity is pi.
Please believe me that this is your introduction to the cosmos and every aspect that brings about explaining the cosmos. The cosmos is not mathematical equating of more rubbish. It is about ideas mathematics can't prove. This picture shows the Roche limit and the Roche limit annihilates Mainstream physics' concept about the cosmos Therefore it is not popular. Mainstream physicists might portray science to diminishing nature in presenting nature as "a freak (of nature)". In that they actually show their small understanding about the cosmos. To belittle nature does not diminish nature but reduce scientists' credibility and knowledge. These pictures prove Newton is wrong and that mass don't pull mass as Newton said. This is called the Roche limit and the Roche limit proves stars can't collide.

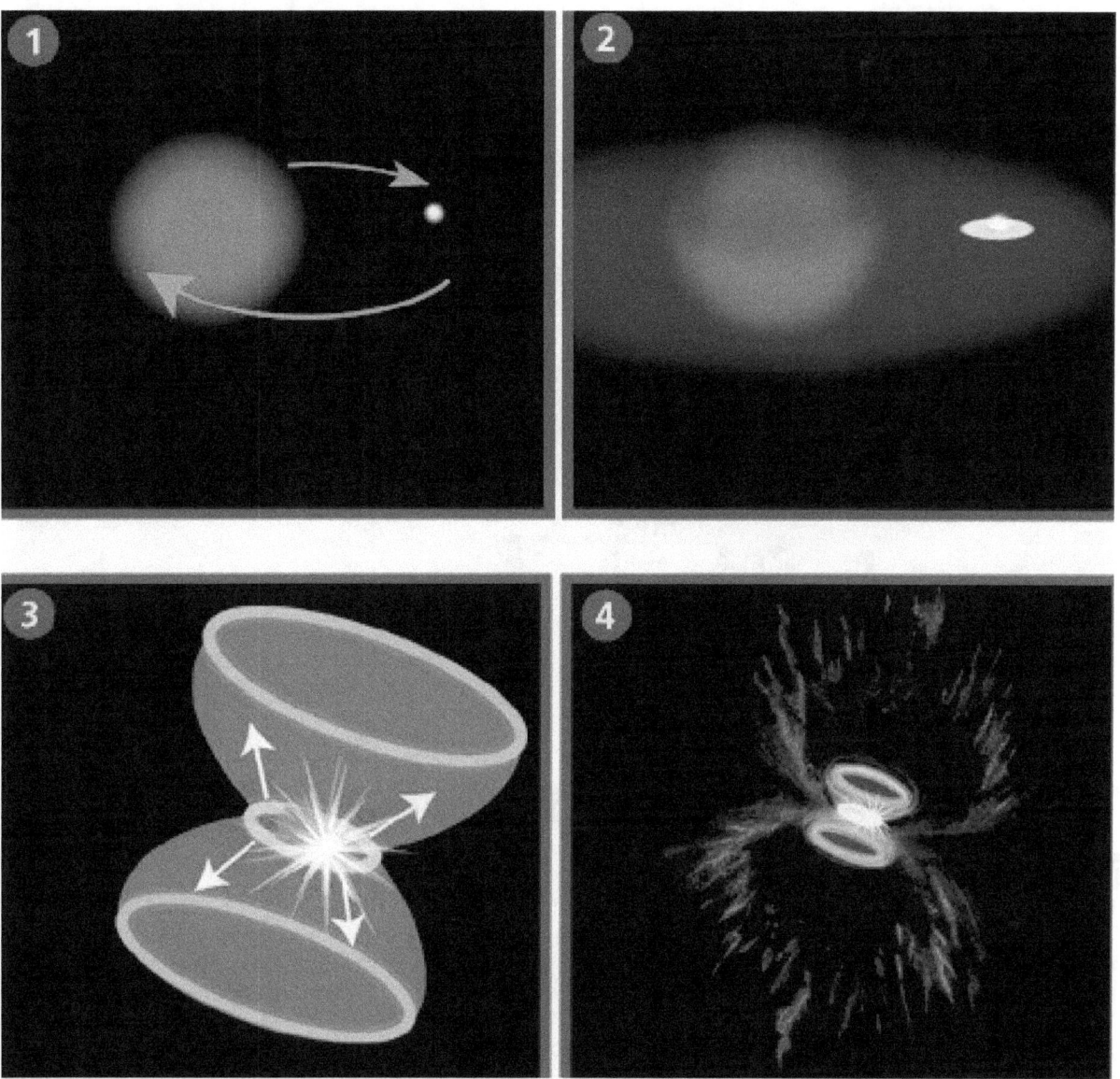

The picture explains one of the principles showing what happens when nature manages the working process of the four cosmic principles in Nature. These are: The Coanda effect The Titius Bode law The Roche limit The Lagrangian points.

There is an almost hundred percent chance that you never heard of these phenomena before notwithstanding that the entire Universe is built by these phenomena and these phenomena form the building blocks of the Universe. The reason why you never heard of it and science never mention the importance of the phenomena is because science are unable to explain it and so they rather ignore the importance than press the importance.

As I said nature applies the Titius Bode law to allocate planetary positions.

There is this very specific ratio that is in place whereby a distance ratio places plants numerically.

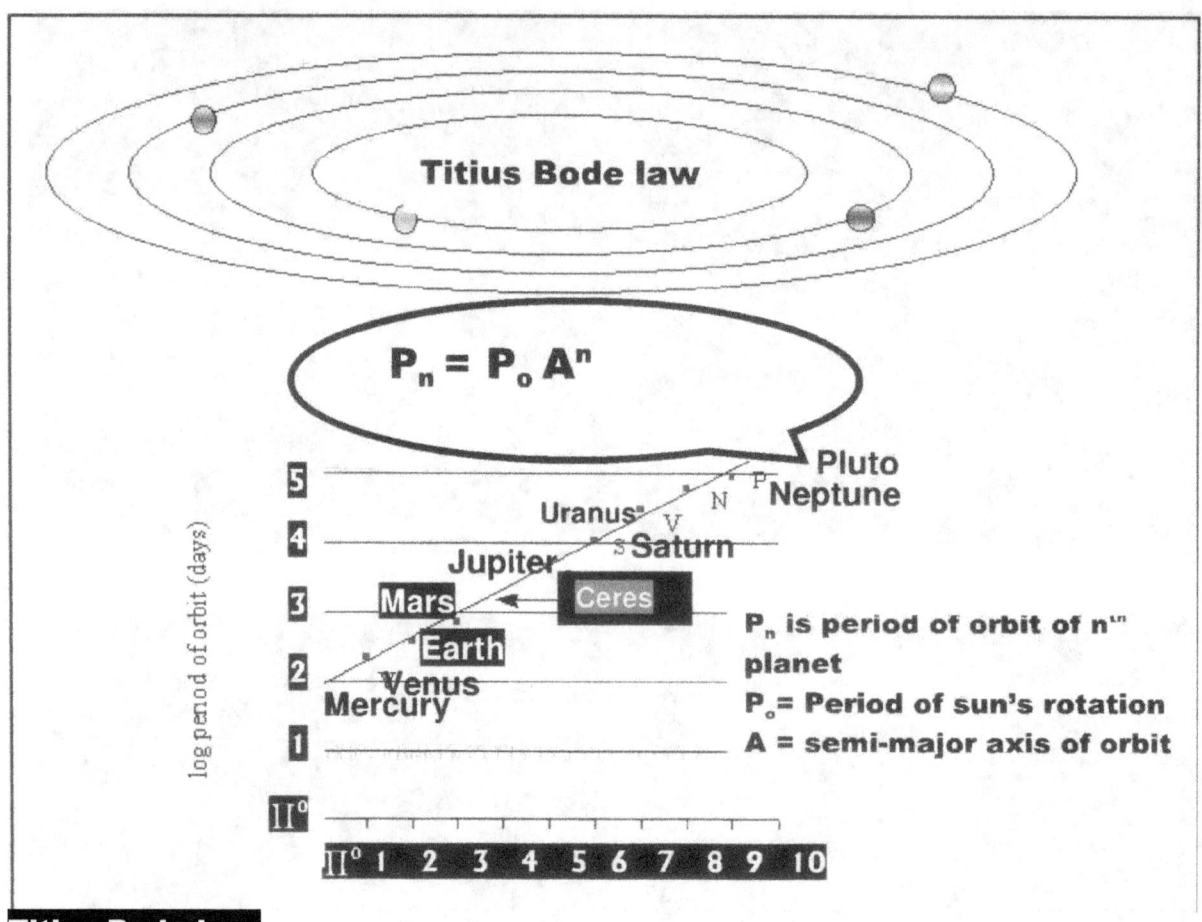

Titius Bode Law

The **Titius-Bode Law** is rough rule that predicts the spacing of the planets in the Solar System. The relationship was first pointed out by Johann Titius in 1766 and was formulated as a mathematical expression by J.E. Bode in 1778. It leads Bode to predict the existence of another planet between Mars and Jupiter in what we now recognize as the asteroid belt.

This shows a commitment to order and relevance but was never yet understood before. They could mathematically plot a ratio but never prove what is behind forming the ratio.
The law relates the mean distances of the planets from the sun to a simple mathematic progression of numbers. There is a code on which the allocation depends and this code has nothing to do with mass! In fact mass only plays a role in Newton's imagination. I dare any person to prove me wrong. Just read the book and see how science cheats.

This ratio of P_n, P_o, A shows the sequence and indicates a ration but I prove why the ratio exists and what principles in nature apply to put this ratio in place.

I show for the first time in all of history the reason why this ratio is there and why it applies as it does with this ratio of P_n P_o A. I use this to show that gravity forms not by Newton's mass but by pi.

There is a given ratio on which the allocations of planets are based and this is how nature and not Newton's imagination works.

To find the mean distances of the planets, beginning with the following simple sequence of numbers:

0 3 6 12 24 48 96 192 384

With the exception of the first two, the others are simple twice the value of the preceding number.

Add 4 to each number:

4 7 10 16 28 52 100 196 388

Then divide by 10:

0.4 0.7 1.0 1.6 2.8 5.2 10.0 19.6 38.8

The resulting sequence is very close to the distribution of mean distances of the planets from the Sun:

Body	Actual distance (A.U.)	Bode's Law <A.U.)< td>
Mercury	0.39	0.4
Venus	0.72	0.7
Earth	1.00	1.0
Mars	1.52	1.6
Asteroid Belt		2.8
Jupiter	5.20	5.2
Saturn	9.54	10.0
Uranus	19.19	19.6

It is this code of 3 and 4 that I decipher and the reasons why it structurally works by 3 and 4 and it is this growth by 3 and 4 that I unravel.

The solution to encrypt the cosmic code is in the value of 3 and four. It is because of this value that we see the sun as the sun crosses our sky while the earth spins around its axis It is a ratio that represents cosmology as we see it. We see the sun cross our sky because the sun rises in the east and sets west.

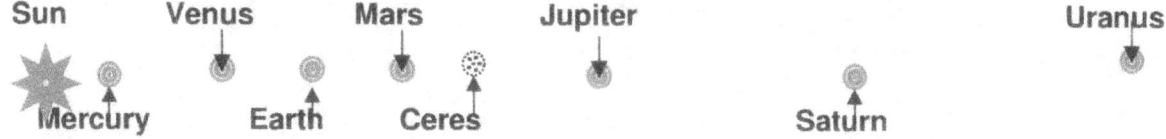

If the distance between the sun and Mercury is represented by three dots then there has to add another four to get the distance that represents four dots to allocate the position Venus has. The sun holds a value of 3 and Mercury is 5 making Venus 7 and earth 10.

Where Venus holds an allocated position of seven (3 + 4) then another three points further will we find the allocated position of the Earth. In this ratio the earth then has ten dots.

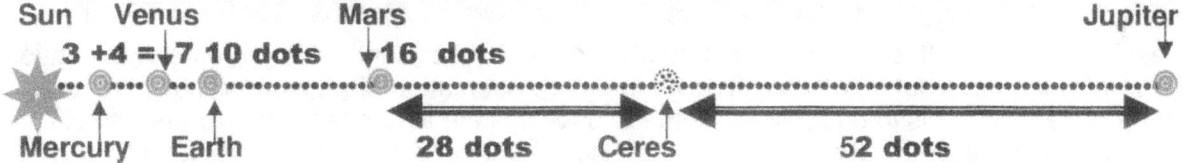

With the earth at ten dots Mercury is 16 dots further, Ceres is 28 dots and Jupiter is 52 dots. I have no more room on the page to represent a readable scale but Saturn will have another 100 dots between it and Jupiter and then Uranus world extend a further 196 dot away from the sun than Jupiter is.

Where Jupiter is 52 dots away from the asteroid belt and Ceres we have Saturn 100 dots further. On the scale only this distance between Jupiter and Saturn would cover the entire length of the page and the dots will be 100 in relation to Mercury's 3 dots and the earth's 10 dots from the sun.

Sun   Saturn ↓

Jupiter   **100 dots apart**

This is how it looks:

Then we have Uranus another number of 196 dots more to the outside of Saturn and then comes Pluto also almost double in distance away from this. In my book **Nature Annihilating Newton** I use rugby fields to show distances and how many rugby fields it takes to place Jupiter when the sun in ratio is the size of a soccer ball. Now put what I show in terms of this picture and see the grand scale of misrepresentation is going on. Look at the size and the distance in between and the variations and remember what I said the Newtonian scale is far more misrepresenting the truth in comparison with the Ptolemaic Universe. The truth is out there behind whatever deceit science tries to cover the truth with.

If you wish to have the Titus Bode law in a more practical sense I use a rugby field to bring ratio understanding about. Any one can walk to a soccer field / rugby field/ grind iron field because they are all very close to the same sizes.

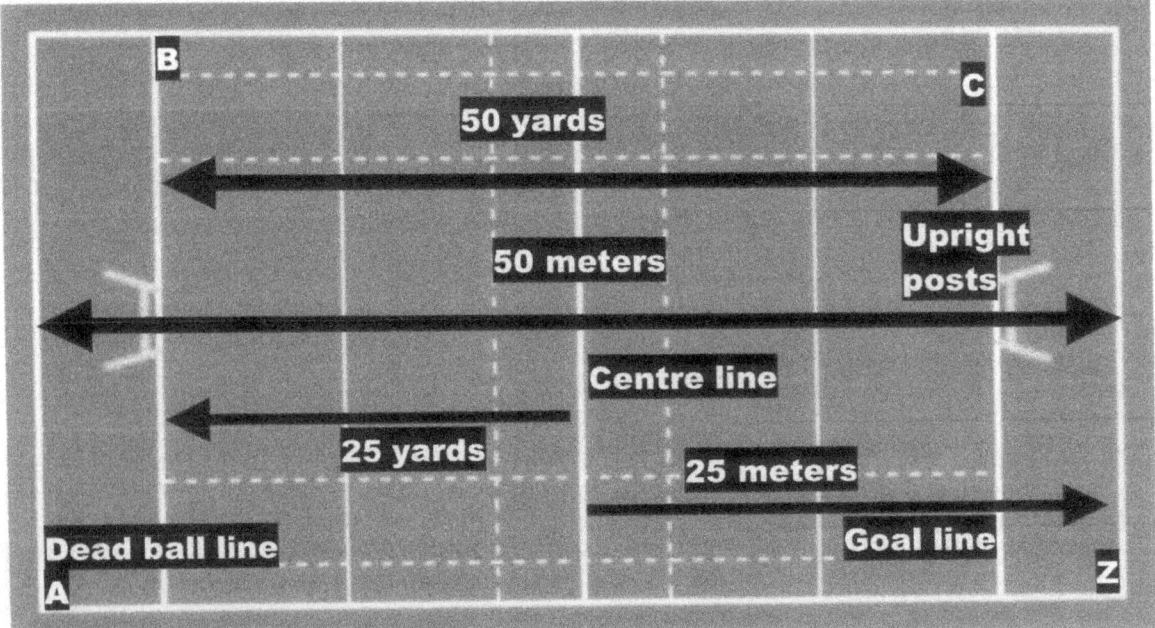

Please take note that this exercise is not about accuracy of a billionth of a micrometer in relation to a billion kilometres but it is merely to give some representation of relevant sizes of the solar system. This exercise is to bring about what the solar system in relevancies truly are and to show how misrepresenting the picture are that show the solar system as Newtonians portray it. The sizes of material the experiment use is as follows.

Sun-any ball, diameter 21 cm inches

Mercury-a pinhead, diameter 0.075 cm
Venus-a peppercorn, diameter 0.1 cm
Earth-a second peppercorn diameter 0.1 cm
Mars-a second pinhead diameter 0.075 cm
Jupiter-a chestnut or a pecan, diameter 2.5 cm
Saturn-a hazelnut or an acorn, diameter 2 cm
Uranus-a peanut or coffee bean, diameter 1 cm
Neptune-a second peanut or coffee bean 1 cm
Pluto- a third pinhead (or smaller, since Pluto is the smallest planet)

Put a soccer ball on the dead ball line of any Rugby field. The **sun**-soccer ball is placed in the very centre of the dead ball line in the centre of the upright poles of the rugby field.
Walk 5.8 meters to the goal line and put a pinhead representing **Mercury** as a pinhead, diameter 0.075 cm in line with the post while lining up the two posts on both sides.
Place the pinhead on the grass. This will be approximately on the goal line.
This represents the 57.9 million kilometres Mercury is from the sun.
Then walk another 11 metres from the soccer ball-sun. This will be on the five-meter line placing **Venus** as a peppercorn with a diameter 0.1 cm about 11 meters away from the soccer ball-**sun**
This represents the 108.2 million kilometres Venus is from the sun.

Another way to look at it is The Earth is 12800 kilometres wide! The peppercorn is eight hundredths of 2.54 cm wide. What about the Sun? It is 1280000 kilometres wide. The ball representing it is 21 cm wide. So, 1 metre in the model represents a billion meters or a million kilometres in reality.

This means that one metre (100 cm) represents 1000 000 000 metres. Take a one metre: this distance across the grass is an enormous space-journey of one billion meters."

Now, what is the distance between the Earth and the Sun? It is 149 million kilometres. In the model, this distance will be 15 metres. This covers the first three planets. Go and stand on the pavilion and look at your soccer ball-sun with Mercury the pinhead-size and Venus and the earth two peppercorns at a distance of 11 meters and 15 meters from the sun. This is space in relation to material. That is according to Newtonians something (materials) in comparison with "nothing" which is space.

Place another pinhead 23 metres away from the soccer ball-sun. That represents Mars. Then lets skip the asteroid belt and move on to Jupiter. Jupiter is 78 metres away from the soccer ball and is the size of a Jupiter-a chestnut or a pecan, diameter 2.5 cm. This is where this model ends because the rest is totally out of even a Newtonian's imagination about reality. Go stand on the roof of the pavilion of the Rugby field and look at the solar system model that you created. Can you see the pinheads and the peppercorn seed? You can't even see the chestnut at that distance and now image this lot is going in a circle around the soccer ball.

Let's take this further but it will not be part of a sensible model on the scale we applied this far.

Saturn is a one and a half rugby field away from the soccer ball at about 143 meters Saturn being the size of a hazelnut or an acorn, diameter 2 cm. The distance Saturn is from the sun is 1427 million kilometres.

The next to follow is Uranus at a 288 meters away from the first rugby fields on the far end thereof on the dead ball line. Uranus is a peanut or coffee bean with a diameter 1 cm. We now are one rugby field followed by a second rugby field away and then another third rugby field using almost eighty meters of the next or third rugby field, which is close to three rugby fields away from the soccer ball-sun and Uranus is the size on scale of a peanut or coffee bean with a diameter of 1 cm. To have a view of this one would now start to think in terms of using some hot air balloon because no view from even the roof of the pavilion will allow an entire complete overall view. Uranus is a distance of 2871 million kilometres from the sun. Try to se a soccer ball and a peanut or coffee bean with a diameter 1 cm being at a distance of 445 meters away from each other. Even from a hot air balloon this vision is impossible.

Then comes Neptune being four and a half rugby fields from the soccer ball-sun. This is as close to a half kilometre as can be and we are not at the end of the solar system. The size of Neptune is also a second peanut or coffee bean 1 cm and at that it is at 450 meters from the soccer ball-sun. There is no way a coffee bean 1 cm can be visible at 445 in distance from soccer ball – sun just can be visible. The distance of the sun to Neptune is 4497 million kilometres.

…And then at last its Pluto. Pluto is a pinhead and at the distance it is the size of the pinhead is irrespective of anything comparable. Pluto is in distance almost six rugby fields away from the soccer ball-sun. It is 5913.5 million kilometres away from the sun. Need I say more?

 There is a soccer ball over at the left end and there are six rugby fields inbetween a small pinhead over on the othe side's end. That is 600 meters between the soccer ball and pinhead.
It is true to say that the sun can't shine all the way to Pluto and yet Pluto is close by considering how far the sun's gravity still sweep space to condense it into heat wear the sun contracts.

I bet you never heard of this…why…because this disputes Newton completely and this sets the truth free for all to see. The truth is that Newton is a lot of bullshit as I said where I started this book. That is the size and the ratio of the solar system in accordance with the Titus Bode law. Now explain this reality that I have given as the way nature configures the solar system in relation to the way Newtonians say Newton surmising is correct. Can you see what I call my work as reference to Newtonian madness as Corrupt Science and why the title is **A Conspiracy in Science in Progress**?

Mercury	0.06	$T^2 \div a^3 =$ 0.983
Venus	0.82	$T^2 \div a^3 =$ 0.992
Earth	1.000	$T^2 \div a^3 =$ 1.000
Mars	0.11	$T^2 \div a^3 =$ 1.000
Jupiter	317.89	$T^2 \div a^3 =$ 1.000
Saturn	95.17	$T^2 \div a^3 =$ 0.999
Uranus	14.53	$T^2 \div a^3 =$ 1.000
Neptune	17.14	$T^2 \div a^3 =$ 0.999
Pluto	0.0025	$T^2 \div a^3 =$ 1.004

Sun

☼ ◎

Mercury $T^2 \div a^3 =$ 0.983

This is the distance from the sun to Mercury and this distance does not require even one arrow if I wish to show any ratio applying in the solar system.

Sun

☼▶

    Venus $T^2 \div a^3 =$ 0.992

By now at least we can introduce one arrow to show an indication of distance applying as a result of space forming ratio in the solar system.

Sun

☼▶▶

      Earth $T^2 \div a^3 =$ 1.000

The earth do not truly double in distance however, the demonstration is not about displaying any form of total accuracy but more about leaving the impression of space becoming more in a defined ratio while time remains in the instant.

Sun

☼ ▶ ▶ ▶

        Mars $T^2 \div a^3 =$ 1.000

Again this ratio of growth too is not precisely representative of what is truthful.

Sun

☼▶ ▶ ▶ ▶ ▶

          Ceres

At this point for the first time we see a more or less accurate representation of cosmic growth.

Sun

☼▶ ▶ ▶ ▶ ▶ ▶ ▶ ▶ ▶ ▶▶**Jupiter** $T^2 \div a^3 =$ 1.00

Sun

☼▶ ▶

                    Saturn $T^2 \div a^3 =$ 0.999

That's it. That is as far as the ratio goes that this page can present as truthful. More doubling up means going so small this computer can't present that. This is the ratio applying in the Titius Bode law and that is what I will show is why this proves that gravity forms as a circle to the measured value of Π that moves as Π^2

I have discovered the cosmic code that reveals this ratio. When applying the cosmic code all this makes sense whereas Newton's idea of mass is completely invalid.

The mass of the planets is totally at random and that is why Newton just cannot be true.

DOPPLERS EFFECT ONLY INDICATES THE POINT OF SINGULARITY RELATING TO TIME AND MOTION AND NOTHING MORE. However this is the Titus Bode law applying gravity.

THE MACH PRINCIPLE ILLUSTRATED THROUGH THE TITUS-BODE PRINCIPLE

STATIONARY SPACE-TIME

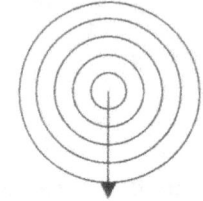

Titius Bode 1

AIR BORN SPACE-TIME

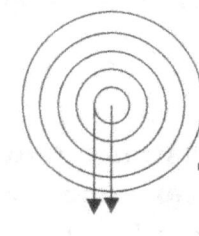

Titius Bode 3

AIR BORN SPACE-TIME TO MACH 1

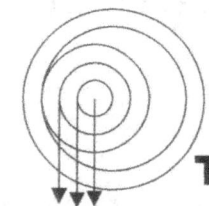

Titius Bode 4

$7(3\Pi^2)$ $(\Pi^2 / 2)$

Titius Bode 5

FROM MACH 1 TO MACH3

$7(3\Pi^2)$ $(\Pi^2 / 2)$ $(\Pi^2 / 2)$

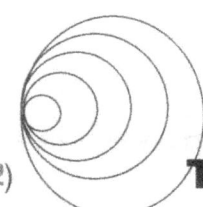

Titius Bode

This resonates from what I found the Titus Bode law to be

On paper, there seems to be a variation in speed, but this is because the aircraft changes its personal dimensional value from Π to 3. The value of Π seems to be a higher value than 3 but in the reality of the cosmos it remains the same thing. As this re-alignment of value takes place, a cosmic border brakes, where the differentiation will bring about a shudder, if it cannot change the time dimension with re-adjusting space.

$$7(\Pi\Pi^2)\Pi^0$$

Standing still the car is moving down and moving along with the earth at a relevancy rate of $7(\Pi\Pi^2)\Pi^0$

Down here on earth with gravity applying the effect of the Titius Bode law influence our movement in the same manner. What science doesn't recognise is that there are two forms of gravity. The one is linear gravity or **k** and the second is circular gravity or $\mathbf{T^2}$. This is precisely what Kepler formulated as gravity. Kepler formulated that $a^3 = k\ T^2$ or as I changed it to $\Pi^3 = \Pi\Pi^2$ and this in relation with the spinning of the earth gives a relevancy in the ratio of $7(\Pi\Pi^2)$ which then shows the linear or downward moving gravity limit of $7(\Pi\Pi^2)= 217$ km / h. This is the curve of the earth trusting down air or space towards the centre of the earth. The downwards thrust set the limit at the earth crust.

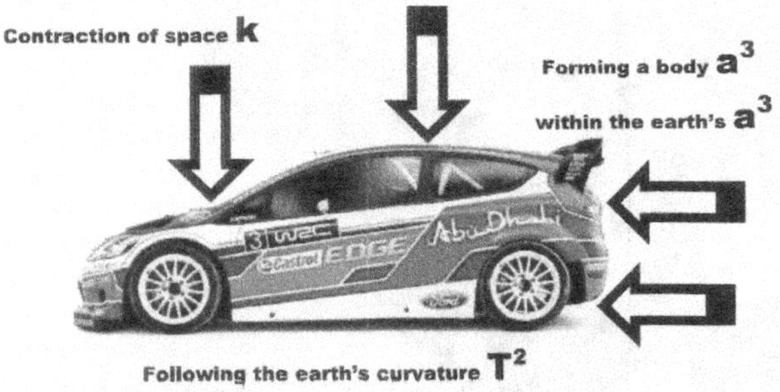

Contraction of space k

Forming a body a^3

within the earth's a^3

Following the earth's curvature T^2

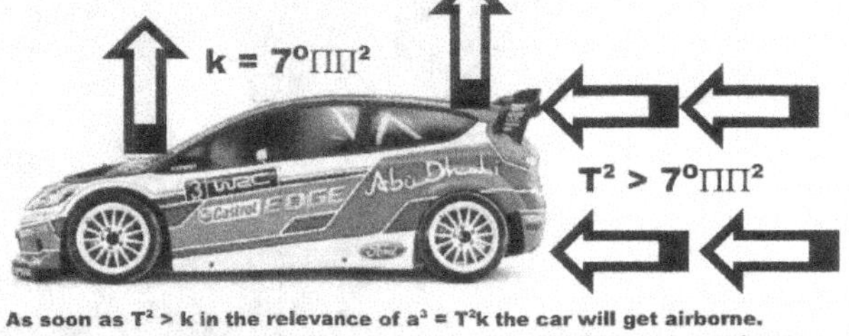

$k = 7^0\Pi\Pi^2$

$T^2 > 7^0\Pi\Pi^2$

As soon as $T^2 > k$ in the relevance of $a^3 = T^2k$ the car will get airborne.
In the earth's gravity the factor k will reverse $k = 7\Pi\Pi^2$ from Π^0 to $4\Pi^2$ because of the density of the car that increases when movement exceeds $T^2 = 7\Pi\Pi^2$

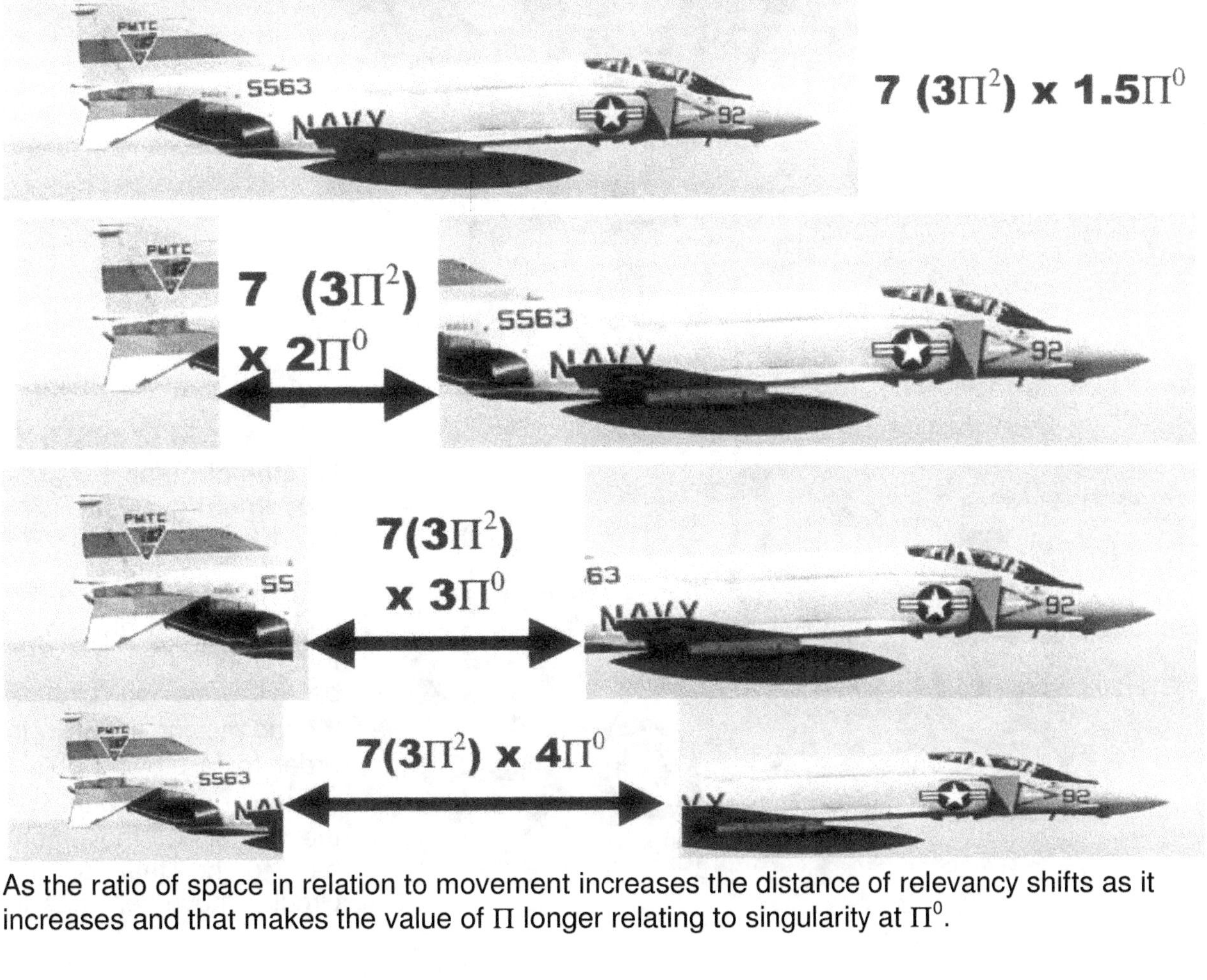

$$7 (3\Pi^2) \times 1.5\Pi^0$$

$$7 (3\Pi^2) \times 2\Pi^0$$

$$7(3\Pi^2) \times 3\Pi^0$$

$$7(3\Pi^2) \times 4\Pi^0$$

As the ratio of space in relation to movement increases the distance of relevancy shifts as it increases and that makes the value of Π longer relating to singularity at Π^0.

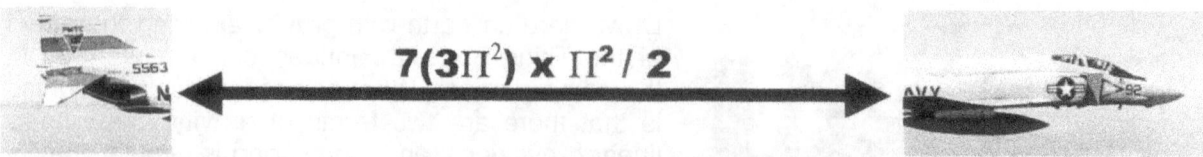

$$7(3\Pi^2) \times \Pi^2 / 2$$

Then when the so called "sound barrier" comes in effect it splits the movement By limiting such movement to a factor bellow $5\Pi^0$, which is where the Lagrangian points are. All four cosmic laws play part in gravity thus forming the four pillars on which gravity rests.

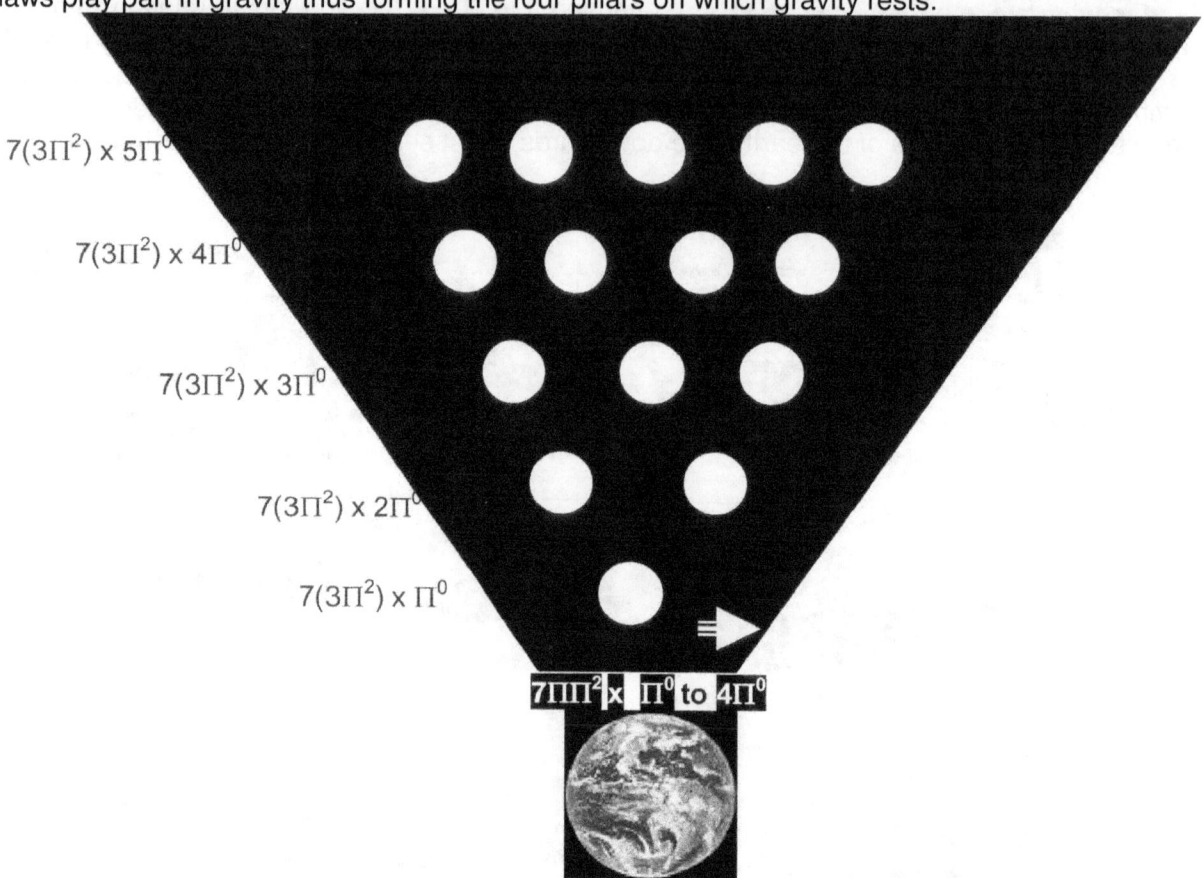

$7(3\Pi^2) \times 5\Pi^0$

$7(3\Pi^2) \times 4\Pi^0$

$7(3\Pi^2) \times 3\Pi^0$

$7(3\Pi^2) \times 2\Pi^0$

$7(3\Pi^2) \times \Pi^0$

$7\Pi\Pi^2 \times \Pi^0$ to $4\Pi^0$

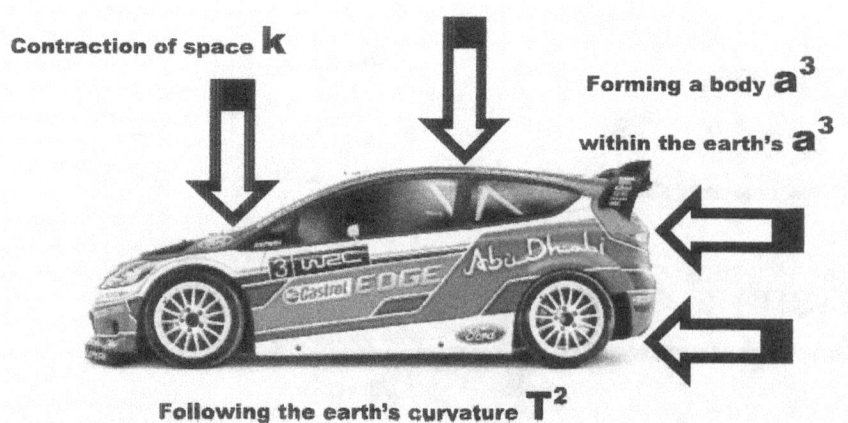

Contraction of space k

Forming a body a^3

within the earth's a^3

Following the earth's curvature T^2

At a speed of $7(3\Pi^2)$ X Π^1 the circular movement must be 651.1 km. / h

The Titius Bode law proves beyond any doubt that there is a ratio between moving circular T^2 and moving linear **k.** In order for a moving object to stay at a point within a^3 the required movement has to be that of the circular movement has to exceed in speed the linear movement as in **T² > k** to remain at that height **a³**. Any movement where speed the linear movement **T² < k** is slower than the required linear movement then the moving object will move towards the centre to a point where **T² > k.** In the flying industry they call this the "stalling speed". Every point where the circular speed is less that the linear speed the object falls out of that specific channel. This proves that gravity is determined by space forming a density in relation to material moving through the space.

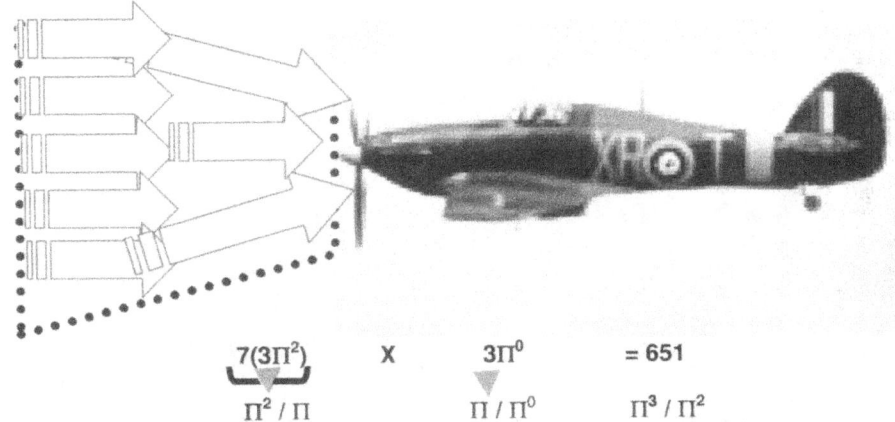

$$7(3\Pi^2) \qquad X \qquad 3\Pi^0 \qquad = 651$$
$$\underbrace{\qquad}_{\Pi^2 / \Pi} \qquad \underset{\Pi / \Pi^0}{\triangledown} \qquad \Pi^3 / \Pi^2$$

A propeller driven aircraft can divert from singularity with a maximum diverting of $3\Pi^0$.

The aircraft holds a relation to heat in dimensional change provided by the earths "gravity" at a value of $7(3\Pi^2)\, 3\Pi^0 = 651$ km / h. Any more speed needed the aircraft must introduce additional heat from own supply.

$7(3\Pi^2)$ X $\Pi^0 = 207$km/h $\Leftrightarrow$ $7(3\Pi^2)\, 3\Pi^0 = 621$ km/h

Staying inline with the earths atmosphere can take the craft to a maximum value of $7(3\Pi^2)$ X $5\Pi = 1036.3$ km. / h. After that point the Titius Bode law will stretch no further and the "SOUND BARRIER" becomes compromised

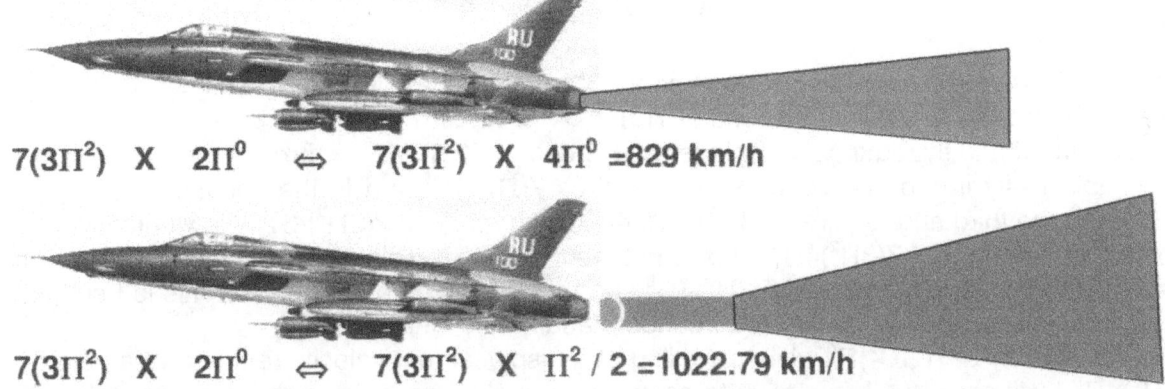

$7(3\Pi^2)$ X $2\Pi^0$ $\Leftrightarrow$ $7(3\Pi^2)$ X $4\Pi^0 = 829$ km/h

$7(3\Pi^2)$ X $2\Pi^0$ $\Leftrightarrow$ $7(3\Pi^2)$ X $\Pi^2 / 2 = 1022.79$ km/h

The long and the short about this is that if you don't move fat enough in a circular manner you are going to move in a linear manner towards the centre of the earth or the sun or around whichever centre you are rotating

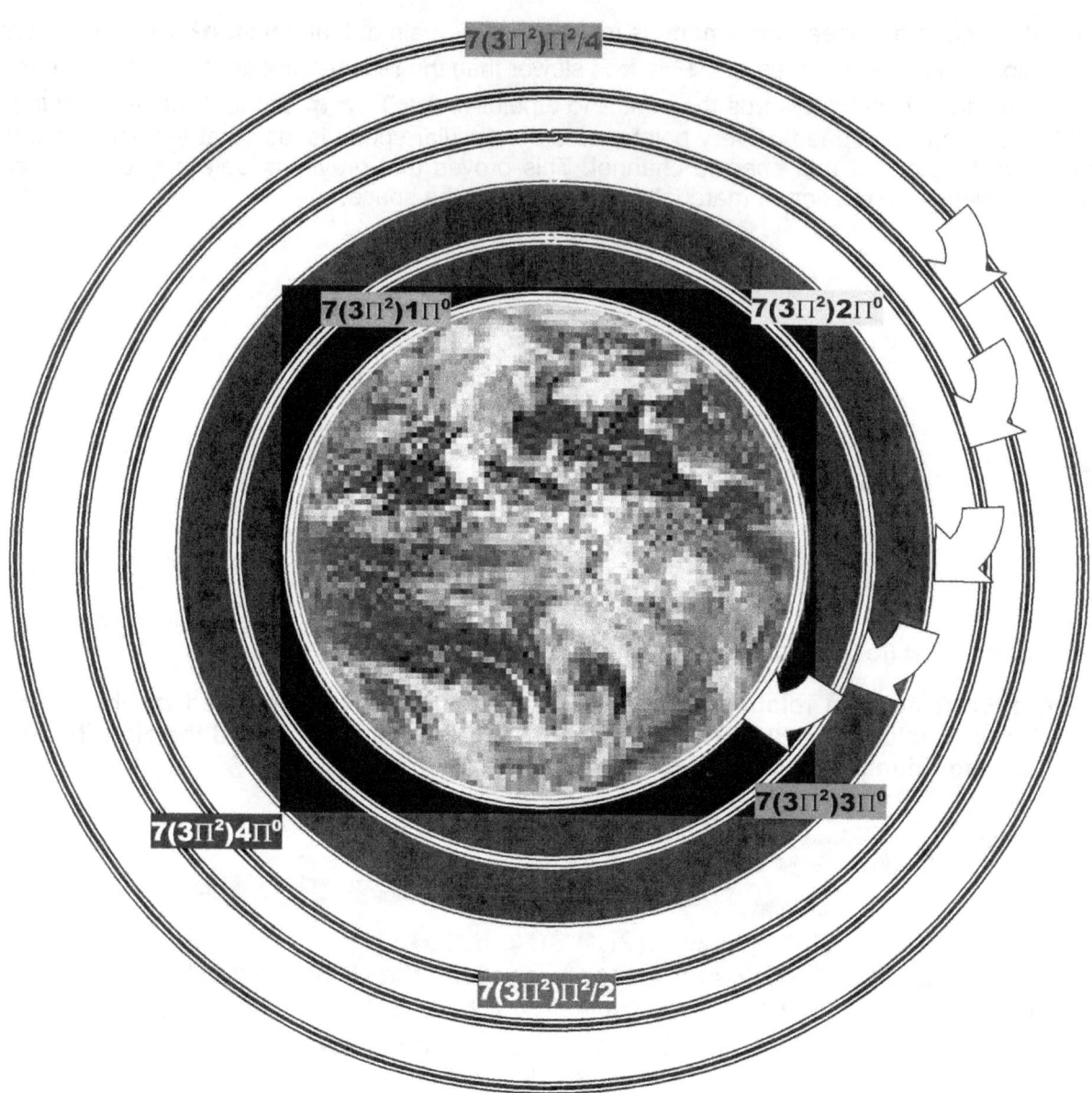

This is the influence gravity has on movement according to the Titus Bode law.

Flying in the first zone requires a speed of $7(3\Pi^2)\Pi^0$ = km. / h

To reach lift off and fly requires a displacement speed of $7(\Pi\Pi^2)\Pi^0$ = km. / h

To be able to fly in the first zone or ring requires $7(\Pi\Pi^2)\Pi^0$ and in the second ring requires $7(3\Pi^2)2\Pi^0$, the third ring requires $7(3\Pi^2)3\Pi^0$, the fourth requires $7(3\Pi^2)4\Pi^0$. Newtonians look at it to say wow they fly at $7(3\Pi^2)4\Pi^0$. This is not the case at all. This says to fly at a height of ring 4 the relative movement must be $7(3\Pi^2)4\Pi^0$ and that leaves no other choice. This is I suppose the most crucial aspect we could have in understanding the Universe.

Should I try to fly at $7(\Pi\Pi^2)\Pi^0$ while travelling at a displacement velocity required at a height of $7(3\Pi^2)4\Pi^0$ I will come tumbling down to earth like a brick. This is the most important thing that Kepler's tables teach us movement requires differentiated displacement setting at different planet densities.

PLANET	SEMIMAJOR AXIS $A(10^{10}m)$	PERIOD T (y)	T^2/a^3 $(10^{-34}\ y^2/m^3)$
Mercury	5.79	0.241	2.99
Venus	10.8	0.615	3.00
Earth	15.0	1.00	2.96
Mars	22.8	1.88	2.98
Jupiter	77.8	11.9	3.01
Saturn	143	29.5	2.98
Uranus	287	84.0	2.98
Neptune	450	165	2.99
Pluto	590	248	2.99

The distance ratio changes as the density increase or decrease.

I have another suggestion to make, in accordance to the space-time growth when placed in accordance to the value of space-time $a^3 = T^2 k$ which is space a^3 and time or movement T^2k

$$k = \frac{a^3}{T^2} \quad 2k = \frac{2a^3}{T^2} \quad 3k = \frac{3a^3}{T^2} \quad 4k = \frac{4a^3}{T^2} \quad 5k = \frac{5a^3}{T^2}$$ and so on but then also this

means not only is there space increasing but it will take in relevancy that much longer time for travelling   through   that   space

$$T^2 = \frac{a^3}{k} \quad 2T^2 = \frac{2a^3}{k} \quad 3T^2 = \frac{3a^3}{k} \quad 4T^2 = \frac{4a^3}{k} \quad 5T^2 = \frac{5a^3}{k}$$

This says if you move in every sector the relevant distance (k; $2k$; $3k$; $4k$; $5k$;) of travel per time period will increase because the density (a^3; $2a^3$; $3a^3$; $4a^3$; $5a^3$;) or the duration in time T^2 it takes to travel (T^2; $2T^2$; $3T^2$; $4T^2$; $5T^2$;) because the density in space (a^3; $2a^3$; $3a^3$; $4a^3$; $5a^3$;)

I changed the formulation to fit singularity which then will be expressed as follows: $\Pi^0 = \left(\dfrac{\Pi^3}{\Pi\Pi^2}\right)$

Then it will read as this: $\Pi = \left(\dfrac{\Pi^3}{\Pi^2}\right) \quad 2\Pi = \left(\dfrac{2\Pi^3}{\Pi^2}\right) \quad 3\Pi = \left(\dfrac{3\Pi^3}{\Pi^2}\right) \quad 4\Pi = \left(\dfrac{4\Pi^3}{\Pi^2}\right) \quad 5\Pi = \left(\dfrac{5\Pi^3}{\Pi^2}\right)$.

Planet	$a^3 = T^2 k$	$k = \dfrac{a^3}{T^2}$		
Mercury	0	50	50×10^6	$57,9 \times 10^6$
Venus	50	100	100×10^6	$108,16 \times 10^6$
Earth	100	150	150×10^6	$149,6 \times 10^6$
Mars	200	250	250×10^6	$227,84 \times 10^6$
Planet X	400	450	450×10^6	$418,88 \times 10^6$
Jupiter	800	850	850×10^6	$777,9 \times 10^6$
Saturn	1600	1650	1650×10^6	$1427,0 \times 10^6$
Uranus	3200	3250	3250×10^6	$2871,0 \times 10^6$
Neptune	6400	6450	6450×10^6	$4497,0 \times 10^6$

Planet	Mercury	Venus	Earth	Mars	Ceres	Jupiter	Saturn	Uranus
Bode's Law distance	4	7	10	16	28	52	100	196
Actual distance	3.9	7.2	10	15.2	28	52	95	192

What this says that is if the density level is 4 at Mercury the density level is 196 at Uranus. If you travel at 10 km / sec at the earth you need to travel at 196 km / sec cover the same ratio of movement at Uranus.

The Titius Bode sequence is

Planet	Sequence	Add 4	Divide by 10	Deposition of
Mercury	0	4	0,4	0,38
Venus	3	7	0,7	0,725
Earth	6	10	1,0	1,0
Mars	12	16	1,6	1,542
Planet X	24	28	2,8	None
Jupiter	48	52	5,2	5,2
Saturn	96	100	10,0	9,54
Uranus	192	196	19,6	19,19
Pluto	384	388	38,8	39,52

So that is precisely what I did, I followed the sequence and had three and added four to get 7...and guess what I was the first one in three hundred years that managed to do just that and by adding three and four and getting seven I was able to solve the "mystery" of the Titus Bode law. This one can doe when a person is not bent on hiding Newtonian corruption and one tries to such fore answers that solves true science. What is the implication of the Titus Bode law on Astrophysics? It is the increase in space versus movement in time that affects the distance from the sun. this is the reason why planets and comets are in cyclic orbits around the sun.

Then I discovered the last thing you can do in Astrophysics is use mathematical calculations to calculate anything because there is no constant applying anywhere. The only constant applying is that there is no constant applying and any point serves only as relevance to any point. This means the space is that many times more between every planet the further the planet is from the sun.

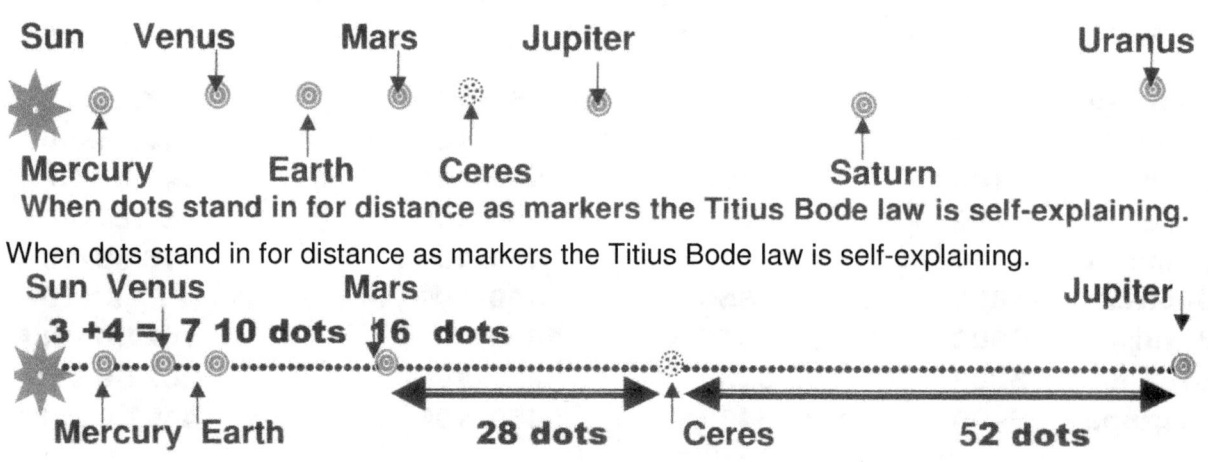

When dots stand in for distance as markers the Titius Bode law is self-explaining.

When dots stand in for distance as markers the Titius Bode law is self-explaining.

If the distance between the sun and Mercury is represented by three dots then there has to add another four to get the distance that represents four dots to allocate the position Venus has. The sun holds a value of 3 and Mercury is 5 making Venus 7 and earth 10.
Where Venus holds an allocated position of seven (3 + 4) then another three points further will we find the allocated position of the Earth. In this ratio the earth then has ten dots.

With the earth at ten dots Mercury is 16 dots further, Ceres is 28 dots and Jupiter is 52 dots. I have no more room on the page to represent a readable scale but Saturn will have another 100 dots between it and Jupiter and then Uranus world extend a further 196 dot away from the sun than Jupiter is.

Where Jupiter is 52 dots away from the asteroid belt and Ceres we have Saturn 100 dots further. On the scale only this distance between Jupiter and Saturn would cover the entire length of the page and the dots will be 100 in relation to Mercury's 3 dots and the earth's 10 dots from the sun. This is how it looks:

Sun   **Saturn**

Then we have Uranus another number of 196 dots more to the outside of Saturn and then comes Pluto also almost double in distance away from this. In my book **Nature Annihilating Newton** I use rugby fields to show distances and how many rugby fields it takes to place Jupiter when the sun in ratio is the size of a soccer ball. Now put what I show in terms of this picture and see the grand scale of misrepresentation is going on. Look at the size and the distance in between and the variations and remember what I said the Newtonian scale is far more misrepresenting the truth in comparison with the Ptolemaic Universe. The truth is out there behind whatever deceit science tries to cover the truth with. What this says is that if you move at a speed of 31 000 km / h hear at the earth then just to remains going straight will require moving at 79.8969 km / h. Then think about how slow this travel will be when reaching Oord's cloud, which is still within the sun's atmosphere of influence as the moon is till in the earth's gravity or atmosphere.

If you go less that the required speed you will return to the sun just like a comet does and that is what puts a comet in the cycle in which it is. The further the comet is from the sun the more the speed of the comet slows down because the density reduces and that makes the space in which to travel through so much more. Then at a point the comet just moves too slow and it begins its cyclic return the sun. This has a much more profound impact on the way we see the size the Universe has.

The time remains while the space increases from what it is on earth, which is in ratio 10 to Saturn that is 100 and Pluto at 388. This means for every kilometre per time unit that is required to move on earth there is a 100 kilometres required to move at Saturn and then there is 388 required to move at Pluto. This changes every dynamic of what we understand about the Universe. However this is only at the last planet...what about further. The dynamics brings changes in the size of the Universe that puts all feeble ability of Human calculations beyond human understanding. Newtonians think a kilometre as it is on earth is a kilometre wherever they go and this is as false an impression as everything else is about Newtonian concepts. It is very prevalent that what is known, as the "sound barrier" is merely the result of the Titius Bode law and one of the functions of gravity. Speed or distance displacement is relevant to density and density is relevant to the location of what moves in a circle in response to the distance from the centre of whatever forms the centre of orbit.

In a book which I Creation and which of the tight connection the entire creation as prove that the sun then destroyed the

named the Seven Days of nobody wanted to publish because it had with the Bible in explaining portrayed in the Bible in Genesis I was a part of a binary. The sun second star in the binary and the inner four (it should be five because the fragments forming the asteroid belt and Ceres is part of the five) planets became solid because this was the core of the star that exploded. This happened as the sun destroyed the second star in a Roche limit and subsequent Roche lobe dual.

The space remains the same
But as the distance grows
The time it takes to travel becomes longer
or the travelling speed becomes
much slower

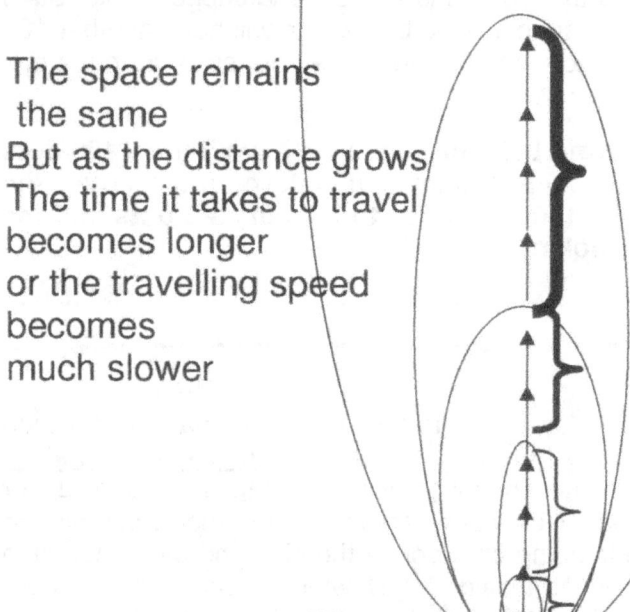

The sun

Due to the Lagrangian law applying the four inner planets pushed the fifth planet into the Roche limit Jupiter created and this then fragmented the fifth planet.

Some of the fragments of the exploding second star blew into outer space and we now call those fragments comets. As the comet blew into outer space the density of the space reduced causing the relevant distance that the comet had to travel through become more. That then changed the relevancy as the speed the comet travelled through reduced. The comet became slower and slower until the required speed was insufficient to move further away from the sun. That slowing turned the comet around and made the comet return to its original location. Then this movement became cyclic. It is because the Titius Bode expands the ratio between the circle T^2 and the radius **k** that the required speed has to increase should orbit relevancy remain.

Planet	Value	relevancy	equation
Mercury	0.06	he relevancy in $T^2 \div a^3 =$	0.983 moment at that point
Venus	0.82	$T^2 \div a^3 =$	0.992
Earth	1.000	$T^2 \div a^3 =$	1.000
Mars	0.11	$T^2 \div a^3 =$	1.000
Jupiter	317.89	$T^2 \div a^3 =$	1.000
Saturn	95.17	$T^2 \div a^3 =$	0.999
Uranus	14.53	$T^2 \div a^3 =$	1.000
Neptune	17.14	$T^2 \div a^3 =$	0.999
Pluto	0.0025	$T^2 \div a^3 =$	1.004

This affects what we think the value of the speed of light is and the speed of light is the unit by which science measures the distance they think applies in the Universe. They think a value of the speed of light is C and that is 299 000 000 km / sec everywhere. This is the last thing that is true. As one goes closer to the sun time will slow down as the density increases. What the Titius Bode law proves is that which is time produces space $T^2 \div a^3 =$ 1.000 wherever time produces space and that in this is the value of time. Although the distance could be 6 times more or a hundred points at Saturn compared to being 10 at the earth this does not influence time to space ratio because time is movement through space $k = \dfrac{a^3}{T^2}$ in space $T^2 = \dfrac{a^3}{k}$. If space is

100 times more in relevancy $100k = \dfrac{100a^3}{T^2}$ travelling will take a 100 times longer which in reflection makes that it will take 100 times longer to move through the space or it will take a hundred times the energy to move through the resistance. This is because moving away from the sun is the same as swimming through the sea current and the further you are from the sun the heavier the tide flow will be that slows you movement down No matter how fast you go an object in the gravity of the sun can never leave the gravity of the sun. To do that it has to break singularity and that is beyond achievement.

This shows that the density close to the sun is such that it takes a year to travel one millimetre and at Oort's cloud it takes 10 000 years to travel through one kilometre There is no speed of light at that many kilometres per second because near the sun one millimetre is a billion kilometres and at Oort's cloud one millimetre at that sun stretched to say 10 000 kilometres in length. That means the speed of light increases the density and by that it slows down the rate of time elapsing. The speed of light could reach 299 mm / hour or year or decade or millennium and much longer and going away the speed of light will be the 2999 million kilometres it now is at the earth it will become 2.99×10^{-1999} km / hour because the relevancy of distance adapts as it stretches the further bit goes. On earth the displacement of gravity is $7(\Pi\Pi^2)$, which is measured at the speed the earth displace one cubic meter of space in one second of movement on earth. At 1 meter above the earth surface the displacement changes to $7(3\Pi^2)$ because the density has changed from earth to air. Every other point in the Universe this displacement value will change to apply singularity forming as space and time $T^2 \div a^3 =$ 1.000 being one.

As space moves towards the sun the space becomes denser until the space forms a liquid around the sun with flames coming from the sun. Newtonians call these flames prominence because they have no idea what to call it but it is just more liquid heat. Because the sun is so large and spins so much the movement of the sun condense the space from a gas to a liquid and then freezes it to a solid all by movement of the sun That churning in rotation condenses space to freeze within the sun as a solid.

(3+4) + 7 = 10

(3+4) + 7 = 10

(3+4) + 7 = 10

Image Copyright JPL

The defining value is with the 3 + 4 = 7 and 7 + 3 = 10 placing the earth at 1 or $\Pi^0\Pi$. This is what cracks the cosmic code and proves that gravity is the forming of $\Pi^0\Pi$ by movement.

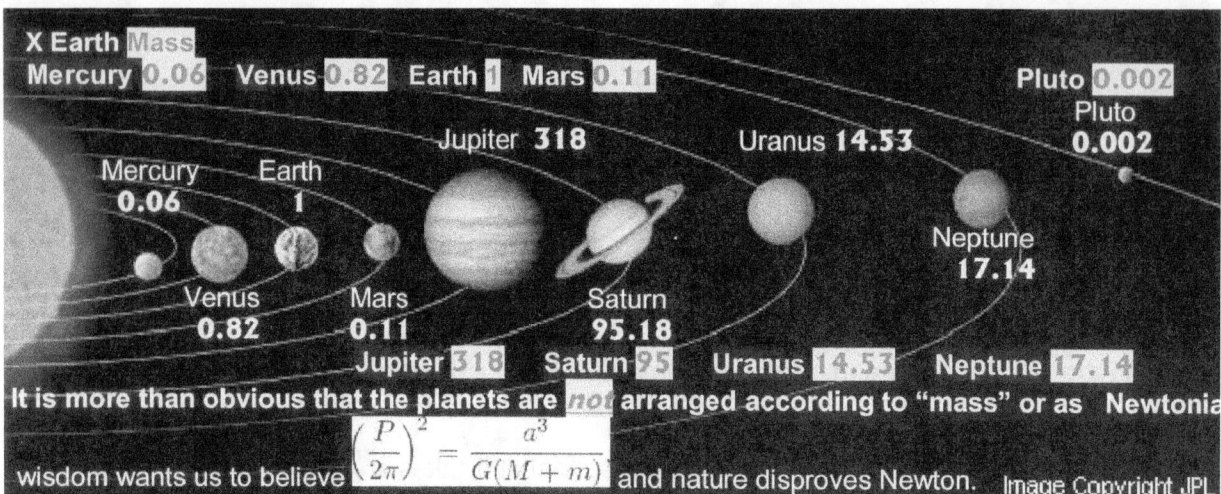

X Earth Mass
Mercury 0.06   Venus 0.82   Earth 1   Mars 0.11   Pluto 0.002
                                                                                      Pluto
                                                          Uranus 14.53   0.002
Jupiter 318
Mercury   Earth   Neptune
0.06   1   17.14
Venus   Mars   Saturn
0.82   0.11   95.18
          Jupiter 318   Saturn 95   Uranus 14.53   Neptune 17.14
It is more than obvious that the planets are not arranged according to "mass" or as Newtonia

$$\left(\frac{P}{2\pi}\right)^2 = \frac{a^3}{G(M+m)}$$

wisdom wants us to believe   and nature disproves Newton.   Image Copyright JPL

Look at the picture and see how mass is totally random distributed. As far as planet sizes go there is no defining order or rhythm connecting a sequence but there is a tiny one, then a small one and a another small one and an almost tiny one followed by rock debris and then all of a sudden the next one is the biggest one and that is followed by another almost big one…and so it goes. Shockingly science has been aware of this for how many hundreds of years? You think I am the first to come to this conclusion in three hundred years. Do you believe I am the first one that add 3 + 4 and get 7…I don't so and I think those that came before me was killed like science tries to kill me. It is not the 3 + 4 forming 7 but it is what pans out from the movement this ratio of 3 + 4 forming 7 develops. This is so simple to understand and yet science has killed my work up to now by ignoring me flat. Please don't allow them to kill me; spread the word. If it is this simple to solve all along why would I be the one that cracked the code do you think? The solution is truly so simple it is adding three and four and then arrive at a number of seven. How long did science try not to conclude or confirm the Titius Bode if it is that simple to solve would you think?

A number of three hundred years springs to mind. …And with this picture showing the mass differentiation the same picture is also telling a storey of deceit! Do you still think I exaggerate when I say they have been conspiring to cover-up Newtonian fraud for three hundred years? If you don't smell a dead rat then you are part of the brigade hiding dead rats. The Titius Bode law develops in the manner that Π forms a value.

By forming the Titius Bode law gravity forms Π and it is the forming of Π as a value and the way Π forms a value that the cosmic code becomes revealed. Gravity is about the forming of the value of Π and the way movement forms gravity as time forms space in the form of Π. This is the way gravity builds the Universe because it is the way that gravity builds the Universe. However science forever shy away from nature in favour of Newton and that puts a hoax in the forefront of science.

I have cracked it…I have finally cracked the cosmic code. The Universe is built on four cosmic principle laws and the four laws form gravity as a unit. However this proves Newton wrong and in proving Newton is a hoax I am a cast out and rejected.

These four principles I mention is the four building blocks nature uses to form our Universe. Why don't science use it…because science does not understand it. Science denounces its importance. Why would science not support what nature uses…because it destroys Newtonian mythology of mass. I am the first person in human memory to figure out how the four cosmic principles, which in fact are the four laws nature uses apply. Science would not even recognize these laws that are nature in space in place and used by nature.

Science rejects these laws formed by no less than nature because through these laws nature condemns Newton and what Newton said as being all fraud. I challenge any of the super brain cheats to PROVE Newton. Using these laws I prove nature and from these four cosmic laws I prove gravity forms by rotating movement that by turning forms pi to form pi square.

Newtonian science would not recognize my work because this destroys anything science brought about since the hoax started in 1705 and these gangsters calling themselves physicists protect a fable and a fantasy and the mess they make of science where they hide the truth under fabricated cover-ups. Now I call on science to prove Newton correct. Everything in the Universe is round. Anything that is round has to apply the value of Π.

This is a fact of mathematics but while Newtonian science forever tells the Universe to have "mass" and to use "mass" nowhere in science would one find Π used in prominence. Whatever you may study in astrophysics, go where you wish but never would you find Newtonian science taking the fact of Π into any prominence. When you read any of my books you will see that gravity forms by movement applying Π as a value. I have found the four phenomena that put Π in astrophysics. By valuing gravity as Π therefore the Universe consists of gravity that forms by the working of the four phenomena that Newtonian science hardly ever mention but dubbed "a freak of nature".

The **Titius Bode law** has been around for centuries and with all the mathematical splendour

available there for all to use, all the brilliant mathematicians could never come close to show any ability of understanding any of this very important phenomena. They could mathematically equate the formula the sequence applying as the formula, but then after that their superior human intellect dries up as they hide behind worthless equations.

The **Roche limit** has been around for the mathematical splendour available to concepts behind this phenomenon, still ability of a machine all those physicists mathematical superiority could not touch about the

using the physics is that

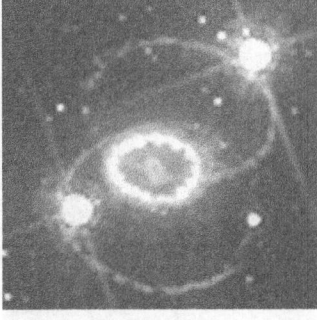

centuries and with all apply in order to fathom with all the computing with all the any understanding concept forming the background. Yet when truth about gravity in the answer is simple; it gravity is Π.

The **Lagrangian points** have been known to science for centuries and with all the mathematical splendour available not one calculation could ever explain why this event is taking place. The Lagrangian points form around stars that have satellites and rings and this shows a definite pattern applying. Just the fact that the satellites around the planets are not drawn to the planets red flags Newton as a hoax.

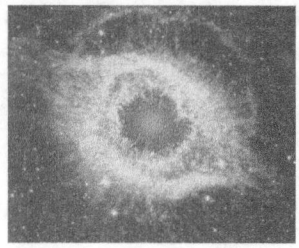

The **Coanda effect** has powered turbine engines and aeroplanes in flight for almost a century and with all the mathematical splendour available to design the most terrific aircraft, not one engineer could mathematically compute one fact to show understanding why this takes place. How sad it is that those claiming of much superior intellect in physics remain just no more than having computing power. The understanding is not complex. I have to warn the readers that the topics are showing a very new approach with no quick answers. Understanding is in the proof and that does not come by reading just a few lines and then forming conclusions. The information is new but not hard to grasp. I did not put these phenomena in place and these phenomena nullifies Newton's correctness, therefore don't blame me because falsified Newtonian astrophysics claims on correctness never ever existed but in Newton's imagination. Now to set the record straight it is time we wash dirty laundry in public. Since I present the truth I can call their conduct criminal.

This is what this book reveals for the first time in human history, it proves the law locating planets being the **Titius Bode law.** The **Titius Bode law** is the law showing the existence of relations between the mean distances of the planets from the sun to form the measured value of Π and that is what I prove and that is how I prove gravity is Π. Science can't support me ... I destroy what they say we must believe.

Look at the size of the sun in relation to the size of the earth. The sun freezes space into a liquid just by turning and by that turning it uses the law of Pythagoras to condense the space it contracts.

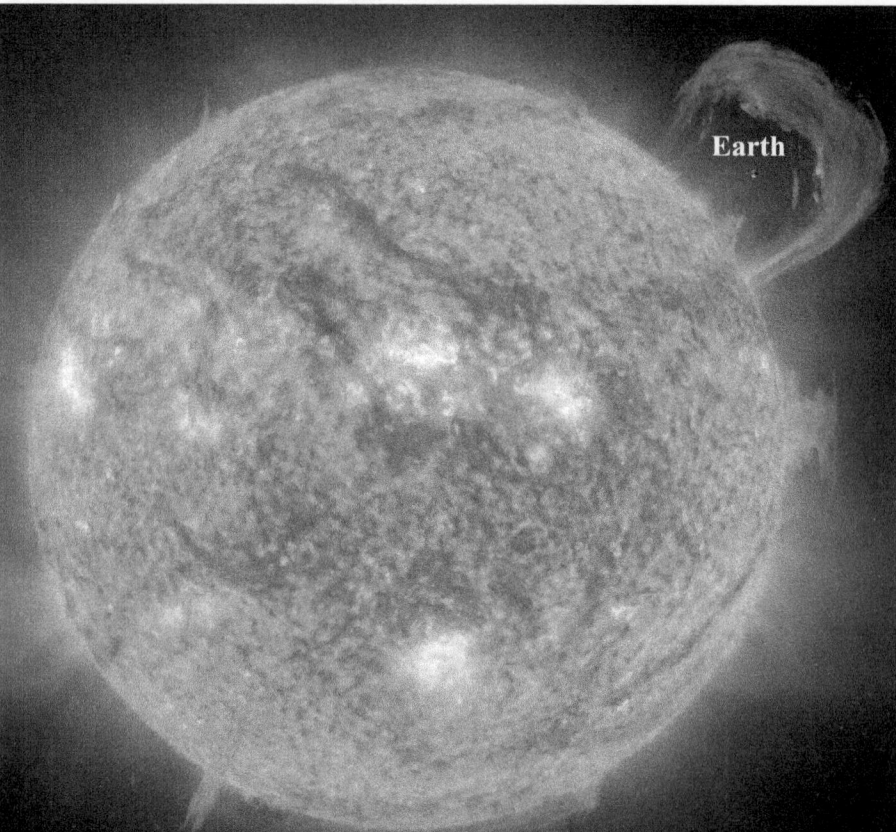

Earth

It is no coincidence that the sun by turning controls the space one third all the way to the next star system. It is that massive and the influence reaches that far.

However in this giant's turning forms the movement that is so veracious it turns outer space which is what I call cosmic gas and which is what fills all the dark space we see into cosmic liquid which is the flames you see and that is named prominence.

Whatever you think of the Newtonian terminology that Newtonians think up and use, in most cases it is to hide their inability to understand what they try to say. If it is prominence then what is it but if it is liquid heat or flames we all know what that is and you can see with your eyes that it is flames. That is all part of the conspiracy to hide reality away from the truth. Look at what you see and you see one giant gas pump spinning gas that freezes heat into liquid flames. Again I repeat that it is the space that moves and that Kepler and the Titius Bode law proves. The material holds position in the flow of cosmic space that condenses and change density as it flows to and towards the sun

There are four books on offer and each book targets intellectually a different person forming part of the entire intellectual basis in society. The following four are written to inform the general population without much emphasis on proof and more on informing how science works by implementing nature and moving away from fictitious science. If you raise your eyebrows by my mentioning of science fiction then only read the overviews I give for free and see if you still raise your eyebrows at me or do you raise your eyebrows at science. The books are listed running from most informing to the fourth being most revealing.

For the first time in human history you can read how the Solar system forms. It forms by the Titius Bode law and not Newton's mass pulling. I discovered for the first time how to interpret the Titius Bode law and decipher the ratio that forms the Titius Bode law. Does science applaud me? No because I trash Newton...and if you thought Newton explained it then know that is the Corrupt Science I Reveal.

Nature uses the Titius Bode law (go look it up) and I am the first to manage to explain that. Because the Titius Bode law trashes Newton Newtonians keep quiet about the Titius Bode law. I found the four building blocks that form the Universe and from that found gravity forms by Π. I show what nature uses and Newton and nature share nothing. If you disagree with me you disagree with nature because I present what nature uses in contrast to what Newton said is physics. If you read my work I explain nature because I discard Newton and the idea that mass has anything to do with forming forces.

That above statement I prove and I prove undoubtedly by example. There is no pulling of any object on any other object as Newton said. Gravity does not work like a magnetic field but works in principle by the object rotating... But it is space that compresses as rotation reduces in a principle called the Coanda effect. It works by material objects producing movement in rotating that condense space around the object to liquid. Gravity turns space surrounding the rotating material from gas to freeze it into liquid. Outer space is a gas that turns to atmosphere that is a liquid. This compressing of space is the Coanda effect. The difference between what my approach must be when using nature and Newtonian's approach that is unsupported by nature is one Universe away from each other. I show a functional Universe and Newton show what they call the mysteries of science in the Universe. I prove everything that Newtonian science this far could never prove. I prove the Universe applies four keys by which gravity works instead of unexplainable magical forces that pull each other and this no person (not even Einstein) this far could ever explain. There are 4 principles applying gravity and forms a value of Π as a circle rotates. What makes the difference between reality and science is Nature does not use Newton in any way or form. Newton says objects pull by an unexplainable magical force called gravity while I say it is round objects that rotate. As it turns it forms Π and by collapsing Π the space around the star / planet that space forming an atmosphere compresses and thereby collapses from 21.991 / 7 to 3.1416 / 1.

As it turns it divides 7 from 21.991 / 7 by 7 to form 7/7 = 1. As the direction rotates Π changes the travel by $7°$ and that $7°$ is 7 / 7 = 1 or Π = 3.1416 / 1. Then by getting reduced it compresses everything in that space as the entire space the object holds and claims condenses by reducing Π from 21.991 / 7 to 3.1416 / 1. Space reduces and Newton said objects in space pull each other. While I prove nature, Newtonian science cheats, corrupts and manipulate nature to make science work in ways nature doesn't work. There is a link between space that compresses and material that spins and condenses space.

All the above evidence I concluded from investigating and explaining the Titius Bode law. By proving the Titus Bode law I was able to prove that gravity forms by way of the forming of Π. Therefore I wish to show how I came to decipher the cosmic code by deciphering the Titus Bode law.

This is where I start when I start to explain the first moment but I use a shipload more information to do explaining when I explain the star in the book I do so. I involve the four cosmic pillars to substantiate the claims I make because all four still work the very same way as it did at the beginning of the Universe. The three points serving one part of singularity combined with the four points serving singularity unites as seven to form a circle of either 3.1416 or 21.991÷7. The seven going to one is eternity matching infinity by movement. But since seven moves it are seven that have to produce gravity. How do I know all these facts, because we can see from the top it is still doing what it did the very first second. When time started infinity as well as eternity had altogether 3 positions, the past, the present and the future. It is still forming the very line in the centre of the top as it forms all lines in the centre of all things spinning. Then eternity parted from infinity when heat separated what is cold from what is hot and eternity formed one more point than before when it had the three points.

With infinity and eternity then jointly having 7 the cosmos came into rotation. In the aftermath post big Bang we now see the phase of cosmic development where the tow sectors try to unite and this brings along the contraction. When Π forms it does so on the grounds that 7 rotates. The circle forms by a change in direction by 7°. Every circle has opposing sides forming in relation to the axis line. If the topside goes rite then the bottom side has to the left. If the rite side goes down then the left side goes up. There is this double presence of a change in direction forming on both sides of the circle. The 7° move and by moving 7° goes square 7^2 and that is Pythagoras.

The one part of the   earth is going up by 7° and the other part is going down
by 7°.

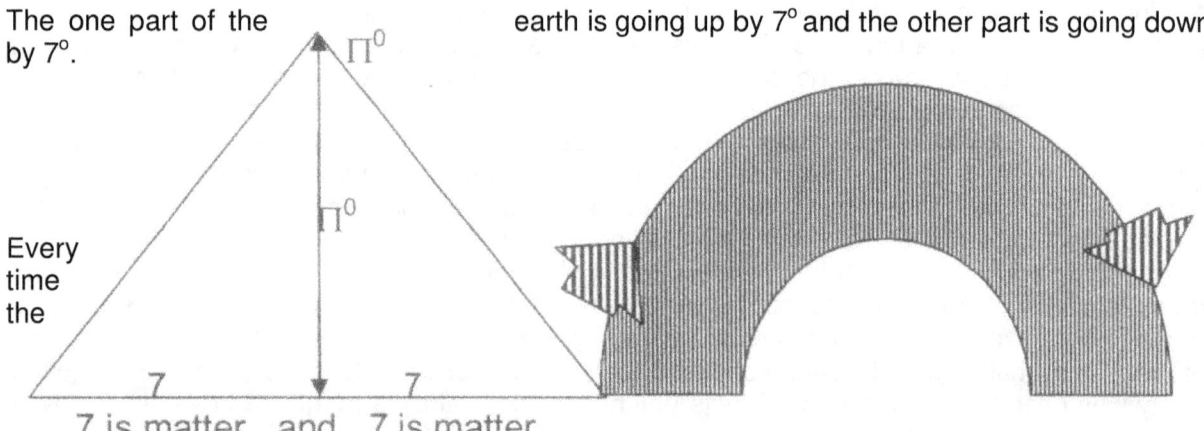

Every
time
the

7 is matter   and   7 is matter.

earth in the going up or going down in direction the relevancy crosses singularity 1^0. It is the same 7o that crosses 10 and 11 but it results in two different points holding a space value of 10 (5+5 on the one side and 5+5 on the other side.

They join space-time therefore the matter factor is the same. This is where one can visually see the one object, filling the space of the other object's atmosphere.

$$7 \times 7 = 7^2 = 49$$

That is matter Π^2 (time) times matter (49)+(1) = 50. This 50 forms space which then applies to both sides of the rotation of the solid being 7 that rotates.

As this is all under the law of Pythagoras the law will evidently place a square root to that value of 483,61 and therefore $\sqrt{50}$ +50 = 10. This leaves the space value of the Roche-limit, as it develops into the Titius Bode law giving them a shared value of 7 (matter) and 21,91 (space) the value of 21,991 / 7 = Π.
Then the relation becomes

$(\Pi^2+\Pi^2)$ $(\Pi^2\Pi)$ $(\Pi\Pi^2)$ $(\Pi^2\Pi^3)$ holding space (3) still outside. They therefore will share space and that sharing will continue till times end. We know by now that matter is 7, and space is 3. holding time to a relevancy in singularity of 1. Sharing the space means that 21,9 will become (10) space to the one side
1   to the instant position of time (k^0)
,99   lost to space depletion Π^2/10
7   the relation to matter.
Through that the Titius Bode law comes into affect of 10/7 or 7/10, depending on whether space or matter holds a superior position to time. From that stance, all objects will relate to one another by the value of $\Pi^2\Pi$ and seen in a whole sale total 7/10 or 10/7. That means to become part of the neutron status of the earth, the object has to be space (21,991 or less) and prove to be matter (7) before the earth will accept it. If holding a position of less than 7°, the earth will discard it and if it is more than 21,991 the earth will find the relevancy to be higher than the space it holds in a neutron time.

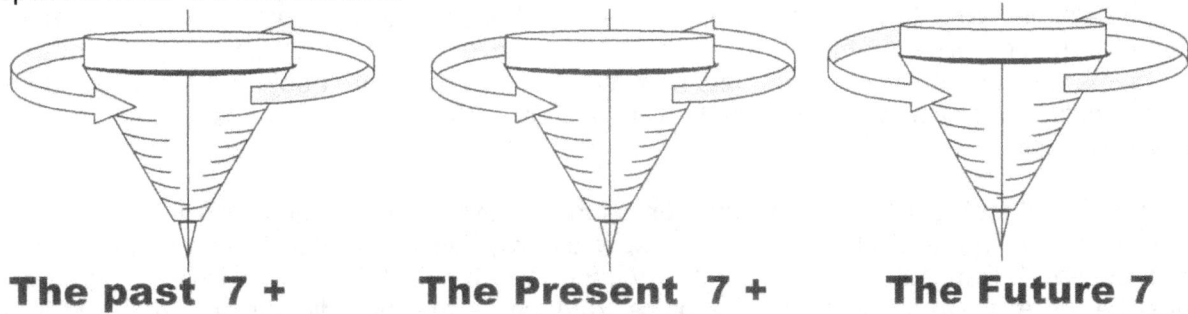

The past 7 +   The Present 7 +   The Future 7

That places the object in a relation of $4\Pi^2$ - Π^2 (because it is not part of the earth) in a position acceptable matter holds (7) within the confinement of Π (21,991/7). That means the object is part of space (21,991) acting as matter (it holds an acceptable own proton structure) 7 relating to the earth in the position the earth allows of $3\Pi^2$. That means to become part of the neutron status of the earth, the object has to be space (21,991 or less) and prove to be matter (7) before the earth will accept it. If holding a position of less than 7°, the earth will discard it and if it is more than 21,991 the earth will find the relevancy to be higher than the space it holds in a neutron time. That places the object in a relation of $4\Pi^2$ - Π^2 (because it is not part of the earth) in a position acceptable matter holds (7) within the confinement of Π (21,991/7).

That means the object is part of space (22,991) acting as matter (it holds an acceptable own proton structure) 7 relating to the earth in the position the earth allows of $3\Pi^2$. With the space position of the matter in the parameters of 21,991 it relates to the Titius Bode law as a factor of one. The object has the space value of 21 (3 x 7), which shows the axle value turning, plus the space value of .1416, in that instant of time (7) complying to the earth's space (.1416 x 7) in reduction (Π^2) formulating .1416 x 7 = 0,991. That makes the object complying with the full agreement as laid down by the Titius Bode law. The object is, no matter where it is, travelling at a rate of 7 ($3\Pi^2$) in the space of the earth (21,991). This will be agreeable to the parameters of the Titius Bode law as long as it remains within the space depleting "gravity" limits of less than Π^2.

In accordance to the Lagrangian atom layout, anything less than 5Π is manageable and is in effect less than Π^2. When it exceeds 5Π it will start opposing the dimensional equilibrium space holds of 10Π, therefore it will (according to space) exceed the linear point of R/T, which is $10\Pi/^2$ (space going in a straight line).

Everything in the cosmos is moving, either by own individual accord, or under the influence of some other singularity dominance. In explaining we return to Pythagoras where the entire Universe with everything in it started.

It is the point forming the very centre that plays the part as the controlling singularity within the Universe I have named as Infinity, which is better known as the axis. It is where nothing can go smaller and anything within that point can never reduce. That point is where the entirety called the Universe begins and where everything holding substance begins.

Once one accepts the fact of singularity being present in that location, that accepting of singularity then is contradicting all the things we know and we can measure and we recognise that point being present by merit of the fact that the point referred to is not being formed by any of the things we can recognise.

It is made up of everything we don't know and constitutes of everything we are unable to recognise or visualise. In that spot there is no space. That spot holds Infinity. In that space there can be no motion because there can be no space to have the motion within. It is formed as a line that is so small that our human reality by perception declare that point as not being there and the only reason why we know it is there is because of the results it left as an imprint of its not being there.

We cannot detect it but notwithstanding our failure to note it we can recognise the dot on the merits of its absence and while in our Universe it is always absent, reality disallows the dot ever to be absent, because it is never absent. It cannot be absent. It cannot go absent but it can never be there where it should be in a place from where the third dimension forms and it is always present if I wish to locate it. It is infinity that can never go away. I named the other part of singularity forming space eternity because that area never become bigger, or become more or find an end to the outside. Whatever was and is and will ever be is locked in that space I named eternity and it is eternity that never ends because eternity can never end moving. What we think of, as expanding is never ending movement giving eternity the eternal motion that will go on forever.

The line **k** coming from the centre (singularity k^0) forms by forming an initial spot Π^0 becoming the dot Πr^0. However, I went on to say that whatever the line used to start with has to continue in order to repeat the same that began the line. Therefore the line started with Π^0 and it has to continue with Π^0 until such a point, as it must end with Π. Whether the line is Π^0 or is r^0, or uses 1^0 the outcome all refers to singularity being used. By reducing the line we come to the end of the mathematical equation of the circle but the circle does not end there. When the top is in a state of motionlessness on own accord it is everything but motionless. The motion it adapts are synchronised with the earth in harmony with the solar system and according to the greater picture of the cosmos.

When an energy source not related to the cosmos called life intervenes and energises the tops motion, the singularity in that top suddenly jumps to life. By adopting a rotation energised to an unnatural state of energising because of life's intervention, the singularity of the top is not in charge but as it applies more and more energy, it will begin to find a means whereby it can escape and apply individual singularity as the top starts to separate from the singularity the earth holds. The singularity holding the earth would then allow the singularity of the top to rotate within a specific band where that a specific band of being active before the earth's singularity will start to destroy the singularity in rebellion.

The top on the other hand will try its outmost, when the singularity it holds gets by individual spin is too strong to remain be in domination of the earth's singularity. The motion of the top is an attempt to begin applying an individual singularity space-time defying and standing apart from the earth's gravity. That action we see as the top starts rotating in a manner where the top does not align with the earth's singularity. With the adding of spin, the time the top holds becomes unrelated to the time the earth holds and the top will start a campaign too escape from the singularity domination the earth has on the top. When the time or spin of the top exceeds the limits the earth places on the top, the top would emerge by trying to escape from constrains placed by the earth.

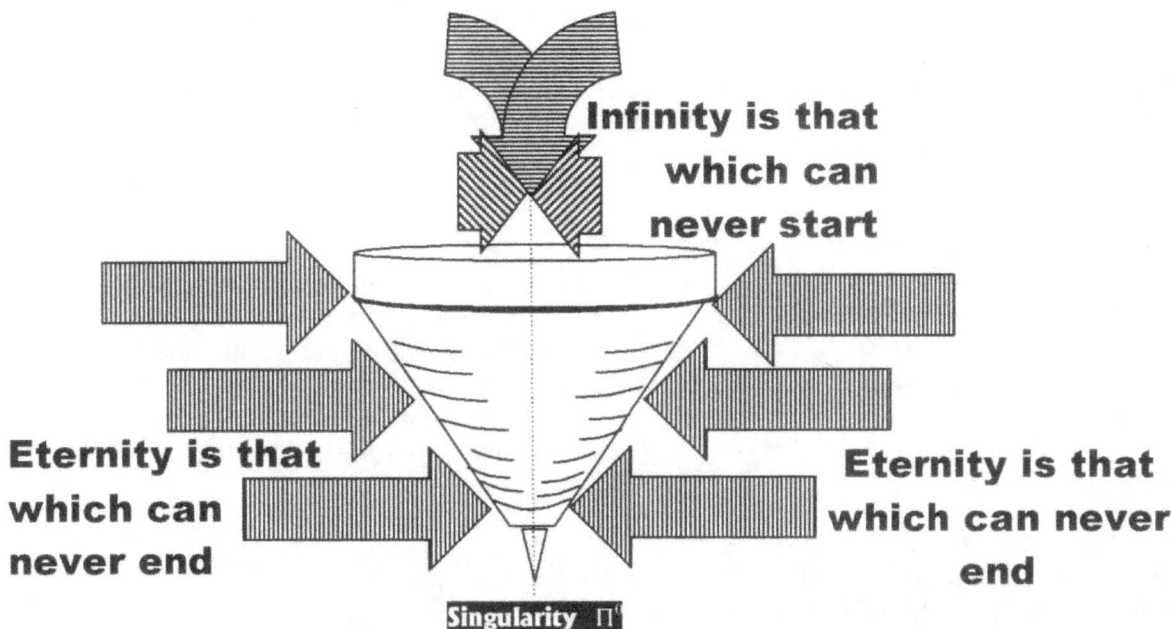

Infinity is that which can never start

Eternity is that which can never end

Eternity is that which can never end

Singularity Π

The view I represent at this point is known to science for almost as long as science knows mathematics. Not long after the law of Pythagoras was understood where Pythagoras introduced mathematics Eratosthenes of Syene made as big a discovery as Pythagoras did.

But in the one instance the world took notice because the world could see and understand and the other instance the world disregarded the findings because the world did not see what the implications was. The same apply to aircraft flying and when the aircraft wishes to escape the earth's singularity hold it has to comply with the laws laid down by the earth.

The seven becomes as big a part of the concept as does Π as it all interacts.

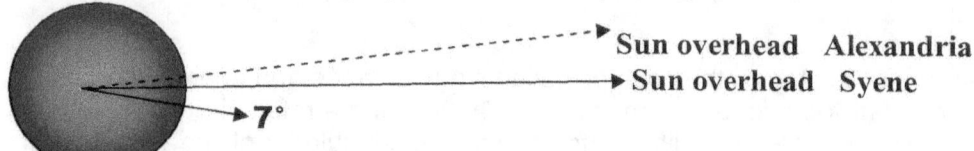

Sun overhead   Alexandria
Sun overhead   Syene

7°

It took Eratosthenes of Syene (276 – 194 BC) a Greek astronomer who in the year 240 BC made a discovery that the earth has a profile of 7°. Since then no one ever did anything about it. When any singularity wishes to disconnect from the earths singularity, specific pre-calculated laws would have to comply to allow the lesser object to divorce from the larger object.

I indicated how the dimensions of 10/7 and 7/10 interact to form (Π^2)

Matter is a product through the separation of space and time receiving the value of Π The original time and Π^2 as follows: By circling around a spinning solid the space contracts to form Π and Π^2.Gravity forms everywhere in the Universe by applying singularity. By dividing space into material (material spinning in space) and duplicating space by material spinning, the TITIUS BODE LAW forms a 7° deviation and 7 / 10 in conjunction with THE ROCHE PRINCIPLE OF $(\Pi/2)^2$

In my article to Annalen der physics I used 15 pages to explain this process of singularity applying. I received a rather cordial but sincere reply from the Editor of the magazine.

When I placed an article in Annalen der physics Dear Prof Friedrich W. Hehl said in the e-mail he sent me that there is no way to "explain" the world of physics. I am not going to go into detail how this works. On the other side of the Pythagoras's' triangle we have 1 going square.

SPACE MULTIPLIED WITH TIME

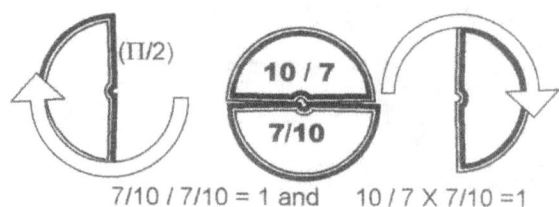

(Π/2)

10 / 7

7/10

7/10 / 7/10 = 1 and   10 / 7 X 7/10 =1

That makes Pythagoras's' triangle 49 + 1 = 50 on the one side of the earth and the same on the other side of the earth. The total is 100 and the square is 10. That leaves the Titius Bode law with a value 7 (it forms part of the material of one body) and 10 in relation to the

space.

Then from the relation of 7/10 and 10 / 7 forming Π the Titius Bode law form Π^2 applying "With a lot of words and some simple algebraic relations" to quote

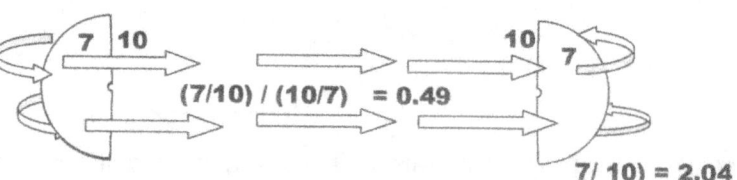

7 10

(7/10) / (10/7) = 0.49

10 7

7/ 10) = 2.04

Friedrich W. Hehl, Inst. Theor. Physics of Annalen Der Physics fame. This was simple algebraic relations but still it is science, is it not?

Since it involves singularity moving it calls for the law of Pythagoras to produce space. The law of Pythagoras is the triangle a^3 that is moving forward in singularity **k** by turning T^2. In singularity the 7 stands in for 7 points on the numerical line crossing over the line holding singularity or 1.

By moving 7 has to go square T^2 and that means 7 goes square 7^2 twice $7^2 + 7^2$ crossing the same divide $\Pi^0 = 1$. Since all movement in singularity has to enforce the law of Pythagoras we have two triangles holding 7 dots moving across singularity. I don't want to get too involved by bringing in numerical outlays because then this can truly become complex.

The line has two opposing sides turning directionally against each other while turning with each other. By moving or turning this involves time duplicating space by the square Π^2 on both sides of the divide $\Pi^2 + \Pi^2$ and using the same divide or the same axis or the same point serving singularity we have 7^0 crossing the same point in singularity Π^0. There then is in this rotational movement 7^0 standing in for Π^2 on both sides of the divide $\Pi^2 + \Pi^2$, which then is 7^2 on both sides of the divide $7^2 + 7^2$.

The circle spins in duel directions. On the one side it would go left if on the other side it would go rite. The one side hold a directional change in singularity by 90°. As it is going sideways it changes to going down. This produces a rite angle triangle of 90° and in it the law of Pythagoras produces direction changes. Since the square of the turn of the circle places by the spin and the direction change we have 7 holding a relation to 10 in space because it is space that has to carry the value of 10 when material circles by 7. There is a connection between space surrounding the spherical circle turning and the sphere. The circle holds the value of 7 as in 7^0 and this we find from looking at singularity controlling the circle by movement

The wise men from physics put everything down to mass without ever searching for evidence of mass. Putting everything down to mass may be one solution except for the fact that only nothing is that simple. Even the rejecting and / or accepting are incorrect, as the pulling part just comes across a tad too simple to make sense. To find substantiation one has to find the manner in which light connects to singularity because everything connects to singularity. In every circle centre there is space that is so small it is not present within the space of our Universe. In that space in the centre of every spinning object we have singularity forming space. That points serving singularity is the reality and the rest is only make believe.

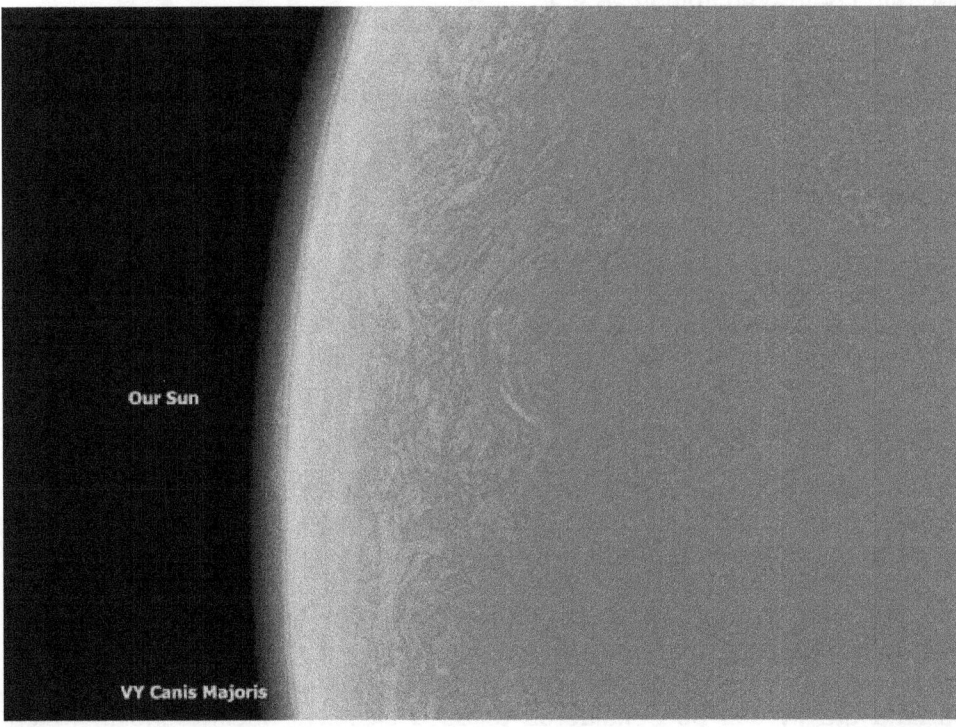

Our Sun

VY Canis Majoris

Explaining the following is rather tough with the limited space available but it should secure the conclusion that I am not grabbing for straws and there are substantial facts on which I work.

Our human view of the sun is that the sun is big; it is more than big it is outrageously big. The sun is so big that one person got a Nobel Prize when he worked out that anything bigger than 1.3 times the mass of the sun would collapse into a Black Hole. That is how massively big the sun is. Another person then got the Nobel Prise when that person calculated that anything the size of 1.6 times the mass of the sun would be an instant Black Hole. Yea sure and Elephants fly at night over the cameras of UFO hunters just as they press the record button! And now the sun is not even a dot in something that we might call big. Look at the picture of the giant CY Canis Majoris and look at the size of the sun in the photograph of CY Canis Majoris. Then place the earth in this picture and then place yourself spinning on top of the earth in this picture. Then think of your position in terms of the entirety called the Universe.

 The more suitable question to ask is how many suns will fit into VY Canis Majoris and in that we have to realise that VY Canis Majoris and Betelgeuse and other "giants" are not stars but they are galactica not developed yet and most of all they have not even began to develop. It is not what we think is true that is true but it is what we think about the truth that applies.

Lets investigate the Universe. We can see how far off the mark science is in their estimate of what is big and what is 1.6 times bigger than big. Let's find out what is small. The Universe is made of lines connecting points. Whatever you see or can't see is lines connecting. If we wish to examine the Universe we first better star to examine lines connecting dots because whatever is in the Universe is dots that connects with lines. So it will be necessary to investigate lines. In mathematics they teach students that a line starts at zero and that is a fable. Zero starts nothing and nothing loses everything before anything can start.

This sounds so unimportant but it is fundamentally all-important. A line starting with zero how cannot increase in length. This means that any line formed in a star only ends ant the outer limit if the star as the line is a solid non ending continuously continuing line from the centre to the outside. The sun is a liquid and not gas as science tries to convince people. Look at the picture. The picture shows a liquid with more liquid squirting into a gas. The density that is apparent at the surface of the sun is very clearly a liquid. But Newtonians have hydrogen as a gas just because on earth hydrogen is a gas. Thereby the Newtonian wisdom says that $6500°$ is very hot. It is so hot it would burn everything on earth and roast any form of life on earth down to less that carbon. This they argue because from the atheistic stance that life is a normal commodity in the Universe and life is everywhere to be found, going at a dime a dozen therefore if it does not befit life then it is extraordinary. Life is extra ordinary and not even a thought in the Universe except on this little blue dot we call earth.

What contracts is cold and what expands is hot. Cold cannot expand and hot cannot contract. What contracts because of cold must eject all heat because it contracts and therefore remove all heat. That is what happens to the sun. Because it is so cold it emits all heat from it and because outer space is so hot it accepts all heat because it has a cold it can forever heat as it forever expands. This is the fundamental of physics and not the human thermometer showing a scientist what he or she must presume to be hot and cold. In cosmology there can be no hot or cold as much as there can be no big or small. When the Big Bang was in The Planck time: 10^{-43} seconds. After this time gravity can be considered to be a classical background in which particles and fields evolve following quantum mechanics. A region about 10^{-33} cm across is homogeneous and isotropic, the temperature is $T=10^{32}K$. If the entirety was 10^{-33} then how big was the sun? If the temperature was $T=10^{32}K$ then what was zero? For the Universe to be

T=10^{32}K then something else must be zero or that temperature is meaningless because then it could just as well be zero.

If the Universe was 10^{-33} then what was the size of one atom? If they say it was before atoms was present then they must state what was the material used in the Universe at the time. It is like saying there is anti-matter. Yes and what was anti matter. I know matter is singularity spinning in a direction in excess of the speed of light because the electron spins at the speed of light and what is further inside the atom must therefore spin faster than the speed of light. Matter is controlled singularity that directionally diverts from uncontrolled space by spinning in a direction. If that is matter then what is anti – matter, things that don't spin faster than light or is it a concept science ahs no clue about but naming it brings clarity to absolute stupidity. It is like saying the sun is gas and all anyone can see from pictures is streams of liquid squirting all over.

It is no use showing how space expanded in relation to temperatures cooling because the big issue is lost in the entire scenario. Space expanding is the same as heat lowering and heat rising is the same as space contracting. That is what gravity is. It is some space expanding and some space contracting and the expanding space is the space heating while the contracting space is the space cooling. Just because we humans feel space heating it does not mean it is hot. It means to humans it is hot but humans have no say about conditions applying in the Universe. Life is a nuisance that are not even ever recognised in the entire Universe and life is alien to everything except the earth in the entire Universe. The heat we feel is the coldness releasing heat to the heated space that we think of as cold.

The very first instant space formed is when heat parted from cold. The Universe began when a

difference happened when there was one part that was hot in relevance to the other part that was cold. The entire Universe is a patrician of what seems cold in relation to what seems hot. If there is something hot then there has to be something cold and neither hot nor cold is prescribed or is dictated. It is only opposing values in that specific space forming factors that generate gravity and gravity forms when one part expands in relation to another part that contracts space.

in the Universe there are two forms of substance that which holds material and that, which is material. Material cannot flow but has solid form. Non-material holds no specific form but is able to hold material or not to hold material.

The sun is a bowl of fluid so cold it holds hydrogen as a liquid on the surface. Where the surface touches outer space the friction heats up and that makes the liquid boil. It is not the liquid that heat but is, is the gas causing friction as the liquid turns that makes the liquid boil.

How can the sun be filled with gaseous liquids a substance used to condense the most volatile

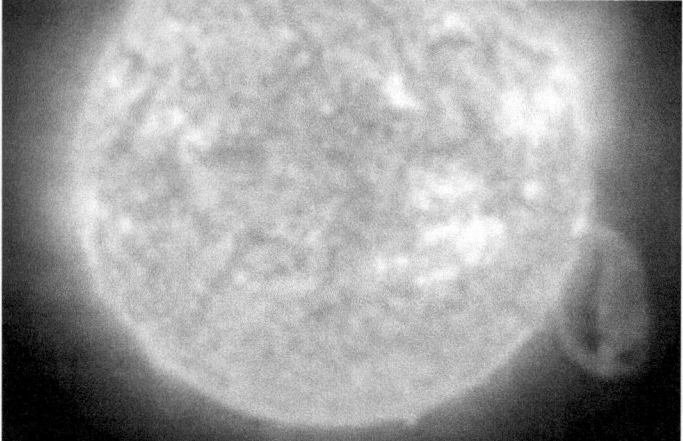

there is to a frozen state and still think of it being steam and water?

When a "gas" being volatile is pumped at the rate that the sun's gravity is pumping the "gas" then the "gas" has to freeze everything to liquid. It is so obvious and yet to save Newton the embarrassment of being wrong they ignore the truth and compromise that with more untruths.

The liquid seems pretty steady and it even seems like waves across the surface. Science forever and ever tells the cosmos what it is and never learns from the cosmos what the cosmos is. What squirts from the sun is a liquid frozen by the movement or gravity of the sun. It is not a gas because a gas cannot flow as the substance squirting from the sun very obviously does. Notwithstanding lack of support coming from the cosmos science tells the cosmos it uses "mass" to arrange planets and then it uses "mass" to allocate planets.

Again it is a case of science sees but science doesn't believe what they see. Science see flames raging from the sun and science can't admit that they see heat because they have to see "mass" because Newton said so.

Time forms as movement changes space. Between singularity acting as time and movement forming space we have singularity acting as time and space acting as the history of time. I.e. the earth circles around the sun and therefore every instant the earth is in a different position and in a different location in relation to the sun and therefore the Milky Way and therefore the entire Universe. However to do that the earth must stand still in every location before it changes to a new location. In that every fragment of any atom forms time by changing allocated space and the united effort of al material forming time finds that the earth represents the material in going to a new allocated position while circling around the sun. In the Universe time flows because the present is different from the future as the past is different from the present. Time is about everything changing. Every aspect of the Universe is different from what it was to what it is to what it will be the very next instant. That forms time. This is proven by the formula we received from Kepler and Kepler in turn received it from the cosmos which he and Tyco Brahe studied for eighty years and calculated as $T^2 \div a^3 = k^{-1}$. Time is not space but time is the movement of space. Time changes space. The changing gives the Universe time flowing. The space a^3 moves $T^2 k$ and therefore changes the space in a singular Universe a^3 at a ratio of a circle T^2 in a singular Universe that moves straight by k that is a line in a singular Universe. The change we see in the Universe as pictures formed by light is in the time flowing that forms space and thus creating a Universe by changing it every instant. Time is the changing of space flickering between $\Pi^0 = \Pi^3 / \Pi^2 \Pi$ and $\Pi^2 \div \Pi^3 = \Pi^{-1}$

Every galactica forms a prominent (however not always visual) Black Hole in the centre that is generated by all the material that forms what is that particular galactica. It is because al spinning material project a primary singularity in the centre of the galactica generated from the movement of the controlling singularity of every piece of material spinning inside the galactica therefore a Black Hole is established at the centre of the galactica. It hold the value of $\Pi^0 = \Pi^3 / \Pi^2 \Pi$ as it generates $\Pi^2 \div \Pi^3 = \Pi^{-1}$ and cosmic development we find is vested in what movement is between these two directional movements.

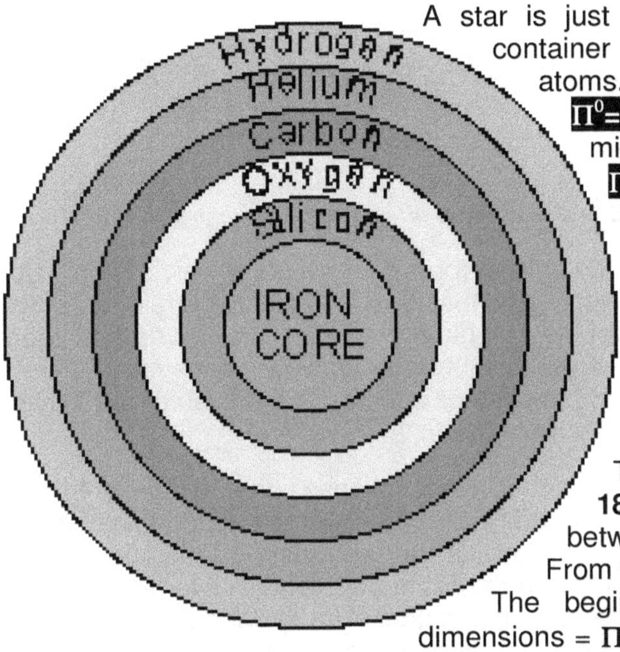

A star is just a cosmic atom because a star is the container of cosmic containers we think of as atoms. At the same time cosmic growth is $\Pi^0 = \Pi^3/\Pi^2\Pi$ but as insignificant as this growth might be it is innumerably more than $\Pi^2 \div \Pi^3 = \Pi^{-1}$ so where there are so many more points serving $\Pi^0 = \Pi^3/\Pi^2\Pi$ than there could be $\Pi^2 \div \Pi^3 = \Pi^{-1}$ the expanding of the Universe outgrows the collapse of space totally. The collapse of space starts with the atoms forming the star and it is within the star that the collapse of space or time starts.

The value of the atom is $(\Pi^2+\Pi^2)(\Pi^2\Pi)3 = $ **1836**, which is the displacement value between the electron and the proton.

From that we can revalue the cosmos as:

The beginning of space as time forms three dimensions $= \Pi^3 = $**31.0061.** Every element has a specific value for a specific task it has to fulfil in the star or as its role is as a cosmic atom. In the cosmic atom the value of the cosmic atom is that of a proton which is $(\Pi^2+\Pi^2)$. A star is a cluster of atoms forming movement of the star and within the star and this movement results in the density applying as gravity of the star.

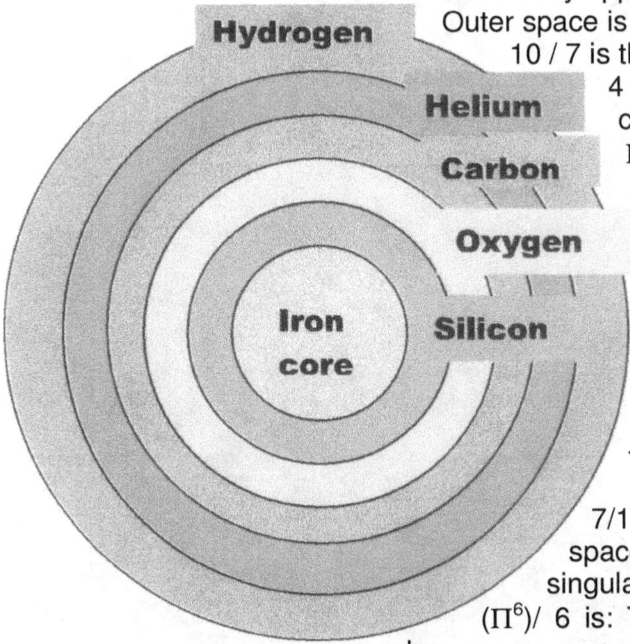

Outer space is 10 / 7 $(4((\Pi^2+\Pi^2) = 112.79547$

10 / 7 is the relation between material 7 and space 10.

4 is forms the quadroons around which time cycles.

$\Pi^2+\Pi^2$ is: The comic atom forming material.

Light ends at 10 / 7 $(4((\Pi^2) = 56.4$ Therefore light ends space at $3^3+3\Pi$.

Light meeting singularity is $3^3+3\Pi^2 = 56.6$
3^3 is: The Universe formed by light in light.
$3\Pi^2$ is: Light moving through the Universe as light forms the Universe

The cosmic atom as a sphere is $7/10(\Pi^6)/6$

7/10 is: Material going singular in relation to space. The movement in this case takes space singular.

$(\Pi^6)/6$ is: The sphere spinning in accordance with the cube.

Π The point serving singularity in space
$(\Pi^2+\Pi^2)$ The proton displacement value
$(\Pi^2+\Pi)$ The neutron displacement value
3 The electron displacement value

Elimination of space-time is $3(\Pi^2+\Pi^2)$
3 The edge of space
$(\Pi^2+\Pi^2)$ The cosmic atom

Elimination of time and space differentiation is $\Pi(\Pi^2+\Pi^2)= 62.01255$
Π The edge of valuing singularity
$(\Pi^2+\Pi^2)$ The cosmic atom
Space reuniting with time is $= \mathbf{2\Pi^3} =62.01255$
2 Doubling what singularity in three dimensions could carry.

Π^3 This is not light but is the limit that allows light to form a Universe in which information are obtainable. These boundaries are motion in specifics that puts relations to certain limits set from the position of the point serving the governing singularity outwards. It proves that there are **dimensional implications all around and** that the dimensions are valid. The same implications are validating other principles in the cosmos such as in the case of the Titius Bode phenomenon by implicating the Coanda effect and others.

All this proves that gravity within the confinement of a star collapses heat in density from where heat is most expanded forming cosmic gas in outer space to where it contracts into space less ness within the centre of the star. This proves that gravity and electricity is displacement of heat. This will become most significant where we go to the Bible and venture into how according to the Bible it says the Universe formed at the very first instant. The Universe is heat in different density levels and these density levels are the result of movement of material at different rates.

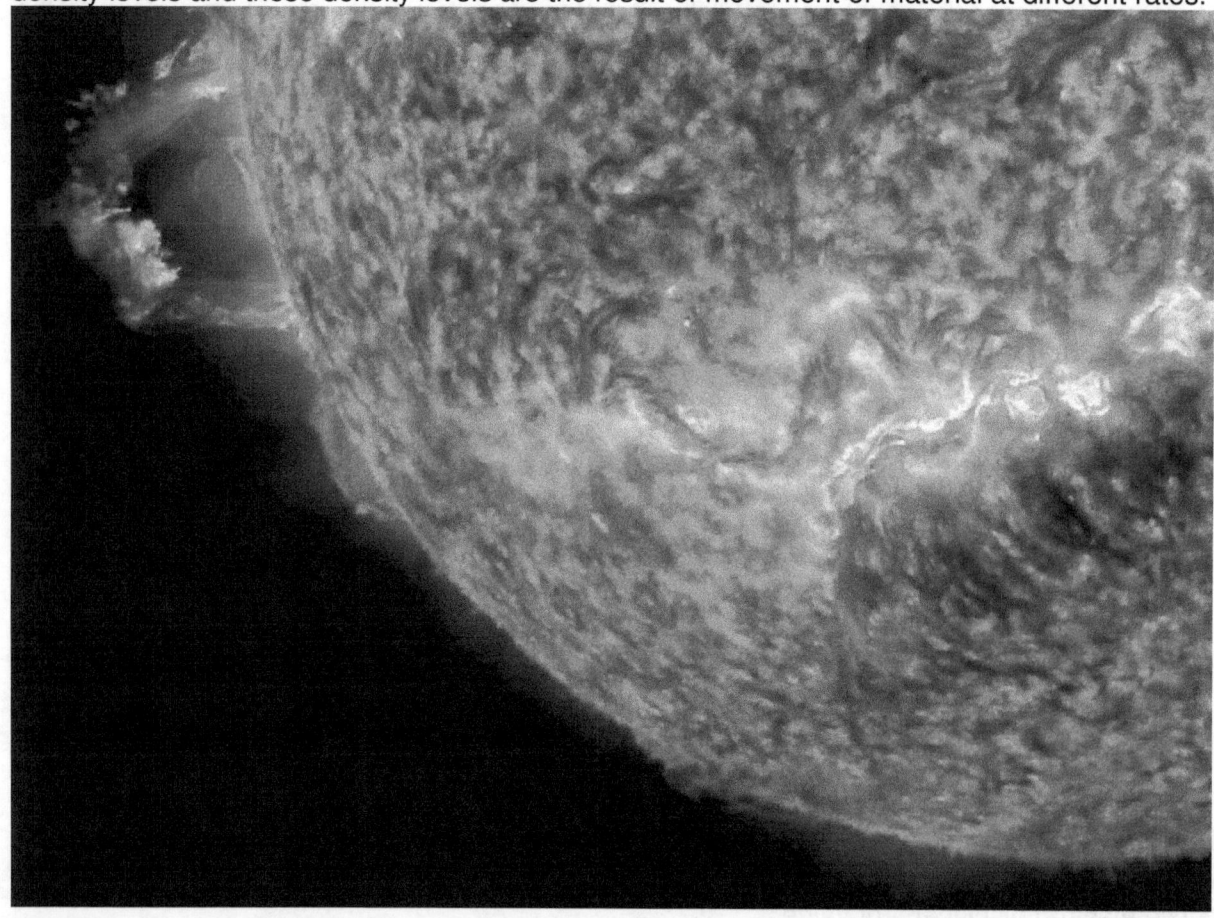

When a volatile substance such as hydrogen move through space most rapidly it will contract heat from space that it moves in and freeze such heat from a gas to liquids. Liquids are not hot it is only when the liquid turn to gas that the condensation releases and the heat becomes a huge factor. Turning flames into gas brings heat and turning gas into flames cools heat. This is the purpose of volatile elements. By moving such volatile substances at a rate that the sun moves gravity it freezes the cosmic gas to cosmic liquids and liquids we see as flames. Look at the sun and see flames pouring from the surface in tongues that dwarf the entire earth. We see liquid squirt from the sun but because humans think of Hydrogen in terms of gas and the sun is gas and the sun is also hydrogen gas, that it is hot! Science will forever tell the cosmos what to be. In the Universe there is no hot and there is no cold. In the Universe there is no big and there is no small. If outer space is the hottest there is then we know what a Black Hole is. It is the coldest there can be. It is so cold it froze all material into one point that is not even part of the Universe. That is gravity. It is the contracting of heat and the expanding of space.

It froze space out of existence and into the oblivious; literally the oblivious where what ever can be cannot be any longer. In the Black Hole the atoms froze into one structure where not even a proton has validation. The structure spins so fast it does not spin at all and allows all spinning into space that otherwise never can spin. The movement in outer space is the expanding of outer space and the movement allowed by outer space is the increase it has in growth by forming a larger volume of space. However this entirety growing is not in size but only applies in relevancy. If one dot removes from space; any space then the entirety forming space will collapse into where that one point has disappeared too. The Black Hole is a star that the atoms within joined to form the ultimate atom in singularity abandoning even the proton in the core and all movement in relevancy has gone outside into time or outer space.

When the atom becomes as cold as it can get, the movement within a Black Hole freezes the material back into the oblivious from where it came and it then freezes back into singularity. In

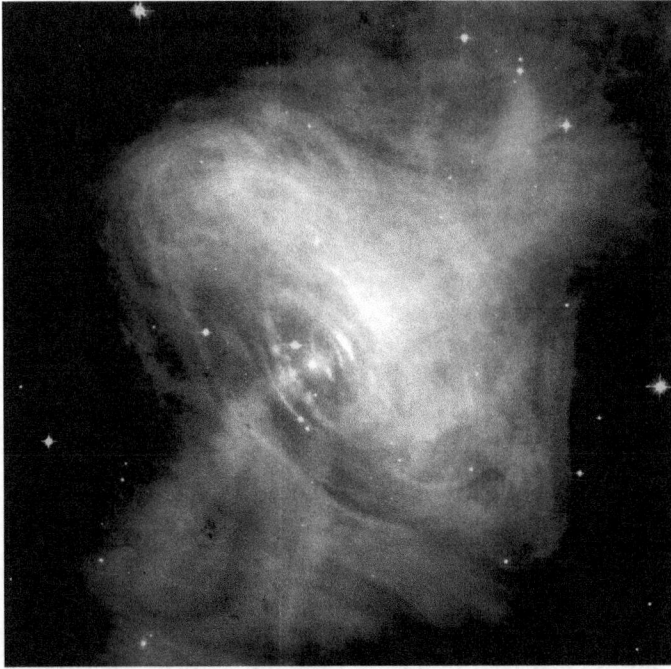

the Black Hole the atom froze to something that has gone into the oblivious and it froze to a point that no longer has a position in our Universe. But reality is that in every atom there is a Black Hole because every atom absorbs light and the rest it casts away. That is all but silicon and I just can't explain that detail in this book since it will be far beyond the technicality of this book's standard.

The sun is a future developing Black Hole and that is why what is in the solar system will flow around the sun eternally without ever having a chance to escape from the centre of the sun. The sun like every atom in every formation forms a Black Hole by forming singularity.

In ever star all atoms join one single point attaching all starts to that point and even after a Super Nova release such bonding remains firmly attaching the star.

Even the structural remains that is the reminisce of a star that went sour still hold singularity in position as forming a centre Π^0 indicting the point allocated to Π. This connection is not coincidental but rules apply putting this formation in place even after the structure went beyond repair.

It is accepted that stars develop by growth and that is true but the growth in stars developing is moreover in the Universe expanding. When the Universe expands the relevancy left to the star is not in ratio the star holds less space. The star expands to a point where the star the forms gravity that exceeds the speed of light and when that happens the space the star requires fades in comparison to the space used by the Universe in growth.

The speed of light is not a constant and it could never be because gravity changes the speed of light as it alters the waves that light transmits by. The stupidity behind the reasoning that the speed of light was forever a constant is underlined by the question as to when the Universe was 10^{-33} cm across what was the speed of light then.

As the Universe expands so the speed of light will stand corrected every instant and every location throughout the entire Universe. The speed of light might be 300 000 km / sec from where we are but a time back with a smaller Universe the speed of light had to be say 250 000 km / sec, then before that 100 000 km / sec and before that it had to have been a 100 km / sec.

It then is madness to place a galactica at 12 000 years or 12 million years because where the galactica is it developed as the speed of light developed but to put reality in place there is no saying how far the galactica is because no one can determine how the speed of light developed since the light we see left the galactica we think we see at present. It is like saying the Universe was 10^{-33} cm at the beginning but the Universe was filled with atomic matter.

Then with the Universe at 10^{-33} cm how big was a neutron at the time when the Universe was 10^{-33} cm? Everything has to be in relevance or not be at all. One can't say the Universe was because what was a centimetre back then? The material forming atoms was already in place and the entire Universe was the displacement intensity of one electron and therefore what was to become each and every neutron had to be fabric and what was then proton material was produced. What was the size of the material grouping to become an atom? It is very fuzzy to

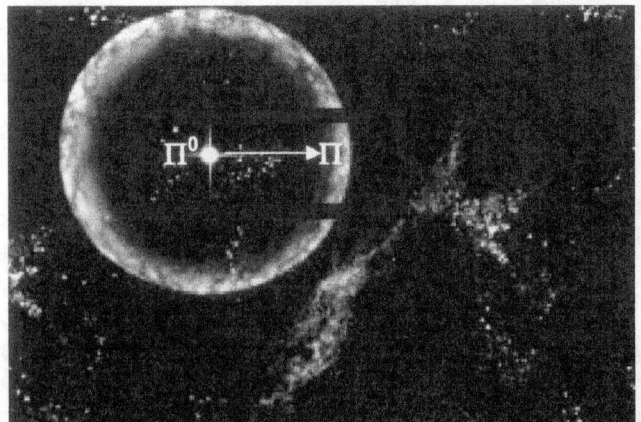

play the role of God and have an overview of the entirety because the scientist then runs out of logical relevancy. The entire Universe then was as big as it is now and was as hot as it now is but applying relevancies changed from then to now as densities and gravity alternated.

If we have an atom then we have the same atom now but the relevancies applying changed in time growing space. What did change was the time in duration as time form space and time then in comparison to now stood still. Time then moved at a rate we now would never understand. We now attach time to movement and movement comes at a rate of the speed of light. How fast back then did light move? If we put the speed of light now in ratio to then we can never understand time back then.

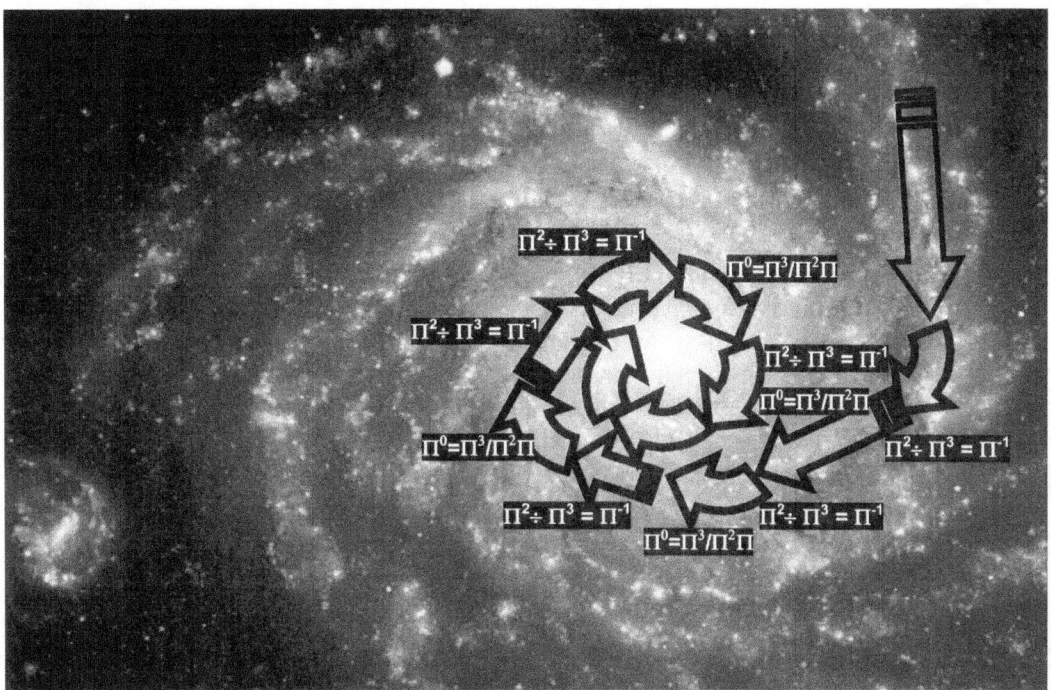

Time places space in relation to every spot that forms space and by becoming a dot becomes space. The first point took an eternity to move from Π^0 to become Π as time Π^2 or gravity forms space Π^3. The Titius Bode alone shows how much time in space or space-time fluctuates as the relevancy of singularity adapt to the location.

By looking at a galactica it is clear that in the centre of the galactica there is a bright bubble holding a sphere of glowing heat. Normally surrounding the centre heat cocoon a ring forms and the ring shows how the galactica went flat as the stars that developed forms rings in which it turns around the centre as the newly developing star spins with gravity around its axis. The star begins to spin around the axis and this releases the star to turn around the galactica centre. As space grows away from the centre the rings grows larger and this space-time development will allow growth that will draw the ring bigger but flatter.

By implementing the Roche limit stars release from the heat cocoon and then the stars develop much the same way as an aircraft starts to fly on earth. The process of release from space heat uses the very same formula I use to indicate how the aircraft flies aster by implementing less dense space. The process by which aircraft fly is the very same process as I explain in the sound barrier process which in fact has nothing to do with the sound barrier and everything to do with the movement of gravity. This movement is the relation between what is horizontal and what is vertical and the newly developing star indicates independence by asserting horizontal movement in defiance of the galactica asserting vertical movement. The process of flying on earth is the same as a new star developing.

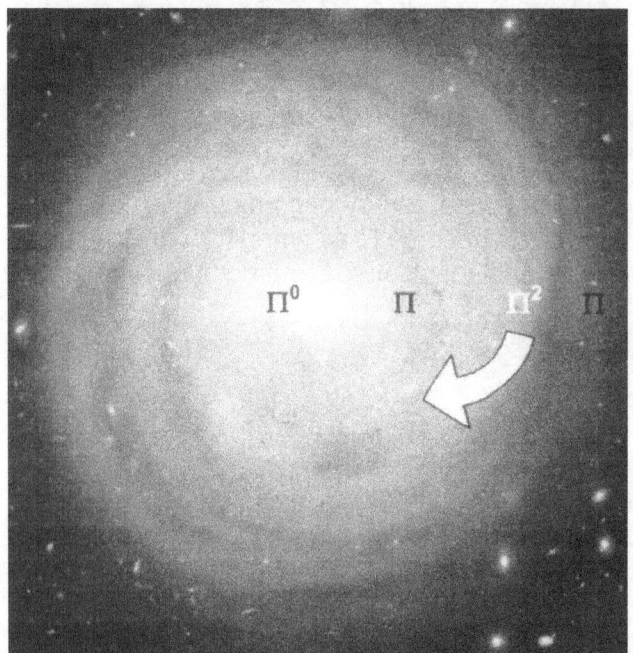

The centre of the Universe is time and all space holds time as the centre reference point. That is why every person that is alive is in time and finds that all the light throughout the Universe uses space to travel in tie to reach the Universe. In the centre of every particle notwithstanding how minute it ever can be forms a spot in the centre that is void of space. That point is time or is singularity and has the value of Π^0 while it forms the value of Π.

So notwithstanding the space it holds the centre value of singularity Π^0 and in being 1 it holds singularity or the single dimension.

Therefore the centre of the Universe is inside the centre that spins or turns around and by doing that creates a centre. The movement is time forming space as a result of time shifting space. Everything there is also is also there because it moves in relevance t every other detail that moves. Movement is the translation of time forming space through turning space back to time and then again back to space. Sped is the value of this interchanging of time to space. Time is what is in the Universe and space is the movement of time in ratio of what was to what is to what will be or going to the future just is the other way around.

There are no planets because there are only

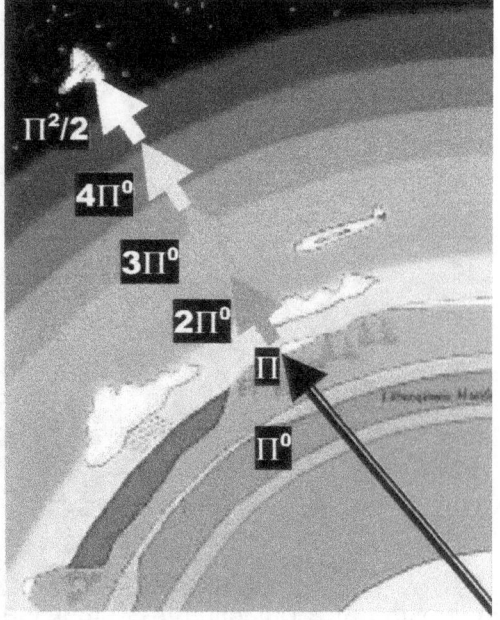

developing stars. The sun is a future Black Hole and the earth is a future black Hole and as density relevancies progress the density of the sun and much later of the earth will transform the structure into an eventual Black Hole. AS the density of unoccupied space evolved the ratio between what is solid and non-solid-material will place the sun into all the stages that a star goes through. The density of materials increase and that shits time in eternity and time in infinity closer until the two forms in time connect leaving all material as just more liquid that has to transform to singularity.

A time will come in the future where one atom will be a star and a star will be a few atoms performs gravity by developing space back to time. However this work is extremely complex to understand. By the end of the Universe time will go eternal because it goes infinitely quick. A star will then be hydrogen, deuterium, lithium beryllium boron and the final atom is carbon. Carbon will serve that same as what copper now does by ending the chain of depletion of space-time from a gas through a liquid to a solid and then to singularity.

If anything did not move it would form a Black Hole since Only a Black hole is within one point that does not mot move and that point that does not move is within not the Universe any longer. It must be understood that whatever is moves that does so in relevance and synchronisation to everything else that does also move and the difference between that movement and space forming forms time. This proves that the relevancy of occupied space increase as the density decrease. The main issue is that it is not the going faster that counts but the slowing down that courts, When travelling at a height of $4\Pi^0$ while going at a relevancy of $7(3\Pi^2)\Pi^0$ this will not happen. The relevancy of the unoccupied space decreases in density the further the movement goes from the centre sauce singularity holds.

Travelling in space would mean to hold any speed the amount of energy will have to increase dimensionally and no earth made object can apply such energy. Therefore the further the object is from the sun the more space would on kilometre hold and eventually say near or just past Pluto one of our centimetre of our earth space will have an equal density of one thousand kilometres. Therefore to go at 31 thousand kilometres earth speed the vehicle must then go say at 31 million kilometres otherwise it will start to stand still. At that point then travelling foreword and standing still will come to the same and the vehicle will then go into a circle orbit where going forewords and going backwards mean standing still and that translates into going in a circle.

This is the table Kepler showed the speed at which planets EQUALLY orbit.

Planet	Mass per Earth unit A	$k^{-1} = T^2 \div a^3$ Movement B	a^3 of space volume	T^2 During time units
Mercury	0.06	$T^2 \div a^3 =$ 0.983	$(a^3)=$ 0.059	$(T^2)=$ 0.058
Venus	0.82	$T^2 \div a^3 =$ 0.992	$(a^3)=$ 0.381	$(T^2)=$ 0.378
Earth	1.000	$T^2 \div a^3 =$ 1.000	$(a^3)=$ 1.000	$(T^2)=$ 1.000
Mars	0.11	$T^2 \div a^3 =$ 1.000	$(a^3)=$ 3.54	$(T^2)=$ 3.54
Jupiter	317.89	$T^2 \div a^3 =$ 1.000	$(a^3)=$ 140.6	$(T^2)=$ 140.66
Saturn	95.17	$T^2 \div a^3 =$ 0.999	$(a^3)=$ 868.25	$(T^2)=$ 867.9
Uranus	14.53	$T^2 \div a^3 =$ 1.000	$(a^3)=$ 7067	$(T^2)=$ 7069
Neptune	17.14	$T^2 \div a^3 =$ 0.999	$(a^3)=$ 27189	$(T^2)=$ 27159
Pluto	0.0025	$T^2 \div a^3 =$ 1.004	$(a^3)=$ 61443	$(T^2)=$ 61703

The space moving which is the time component T^2 in relevance is the same but the actual space that forms outside singularity grows as it locates a position further away from the sun.

This is most significant when understanding the Titius Bode law or the way the cosmos developed. The space it moves through remains the same while the space in volume grows substantially as the orbit is further from the centre of the sun. It is the density that changes and the density requires that time altars the movement so that time in movement remains the same.

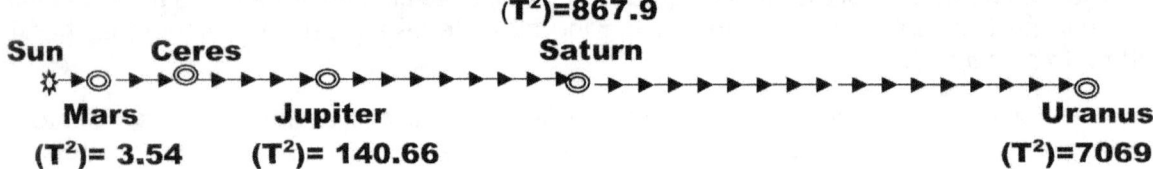

$(T^2)=867.9$

Sun   Ceres   Saturn

Mars   Jupiter   Uranus

$(T^2)= 3.54$ $(T^2)= 140.66$ $(T^2)=7069$

The distance increase therefore the travelling time reduces in time and the travelling time increases in duration. It is the relevancy that changes but the dimensions remain equal because space brings relevancy while time brings equality.

This is the formula suggesting the values Kepler received from the Universe $k^{-1} = T^2 \div a^3$
We must first take note of the fact that k^{-1} indicates a negative or show a direct reducing direction which as a result is showing that space a^3 and not the part that circle T^2 goes into singularity and therefore represents space in which we find material is what goes negative to move to singularity. It is space that will become singular or become one in the direction of travel. That which moves in the circle that forms as a result of the movement coming about can't go singular because going singular which means to become one $a^3 \div T^2 = k$ will always be positive. Mathematically it is impossible for material to "pull" to the sun by "mass" doing pulling.

Mercury	0.06	$T^2 \div a^3$	=0.983
Venus	0.82	$T^2 \div a^3$	=0.992
Earth	1.000	$T^2 \div a^3$	=1.000
Mars	0.11	$T^2 \div a^3$	=1.000
Jupiter	317.89	$T^2 \div a^3$	=1.000
Saturn	95.17	$T^2 \div a^3$	=0.999
Uranus	14.53	$T^2 \div a^3$	=1.000
Neptune	17.14	$T^2 \div a^3$	=0.999
Pluto	0.0025	$T^2 \div a^3$	=1.004

Sun
☼◎
Mercury $T^2 \div a^3$ =0.983

This is the distance from the sun to Mercury and this distance does not require even one arrow if I wish to show any ratio applying in the solar system.
Sun
☼▶

   Venus $T^2 \div a^3$ =0.992

By now at least we can introduce one arrow to show an indication of distance applying as a result of space forming ratio in the solar system.
   Sun
☼▶▶

       Earth $T^2 \div a^3$ =1.000

The earth do not truly double in distance however, the demonstration is not about displaying any form of total accuracy but more about leaving the impression of space becoming more in a defined ratio while time remains in the instant.

Sun

☼ ➔ ➔ ➔
Mars $T^2 \div a^3$ =1.000

Again this ratio of growth too is not precisely representative of what is truthful

Sun
☼ ➔ ➔ ➔ ➔ ➔
Ceres

At this point for the first time we see a more or less accurate representation of cosmic growth.

Sun
☼ ➔ ➔ ➔ ➔ ➔ ➔ ➔ ➔ ➔ ➔ ➔
Jupiter $T^2 \div a^3$ =1.00

Sun
☼ ➔
Saturn $T^2 \div a^3$ =0.999

That's it. That is as far as the ratio goes that this page can present as truthful. More doubling up means going so small this computer can't present that.

This is the ratio applying in the Titius Bode law and that is what I will show is why this proves that gravity forms as a circle to the measured value of Π that moves as Π2'

All gravity rests on movement where movement or gravity or time measured by a circle going straight leaving space as a triangle. This proof we find when studying the Lagrangian points positioning where there is a straight line two equal half circles and three equal triangles all forming the positioning of five points by five positions. In Kepler's formula $k^{-1} = T^2 \div a^3$ where the circle T^2 holds direct relations to a straight-line movement k and this results in space forming as a^3. Movement is a two-dimensional circle going straight in a line and when studying the Titius Bode law this dominates and overshadows every other component in the study.

$k^{-1} = T^2 \div a^3$	Size of the planet	The time Component
Mercury	0.06	$T^2 \div a^3$ =0.983
Venus	0.82	$T^2 \div a^3$ =0.992
Earth	1.000	$T^2 \div a^3$ =1.000
Mars	0.11	$T^2 \div a^3$ =1.000
Jupiter	317.89	$T^2 \div a^3$ =1.000
Saturn	95.17	$T^2 \div a^3$ =0.999
Uranus	14.53	$T^2 \div a^3$ =1.000
Neptune	17.14	$T^2 \div a^3$ =0.999
Pluto	0.002	$T^2 \div a^3$ =1.004

In the Universe and according to the Universe there is a time component and a space component.

Life is in the time component while every atom of the human body is in the space component. Between the moon and me there is about 380 000km space which is worthless history but then there is 1.5 seconds of travel time should I be able to travel at the sped of light. This is not possible because I do not have the relevant size of a photon and therefore this time factor will take on a dimension quality in accordance with my density level. That puts me in relation to my density, which is accordance to the gravitational density the earth allows. This means I will use as much time in space as my density allow travelling to the moon and there the distance of space comes in as a factor of measure. In accordance with the time factor $k^{-1} = T^2 \div a^3$ all space used in the solar system is on an equal time factor.

The reason why Mercury and Venus seem to be slightly smaller in factor (**Mercury** $T^2 \div a^3$ =**0.983**) and in the case of Venus (**Venus** $T^2 \div a^3$ =**0.992**) is that Kepler used the earth as a starting point that produces Π whereas from the earth onwards every other planet position is on par with the earth putting the turning circle equal to the earth. When gauging time not space the planets are in the same circle but space being the history of time uses different means to measure.

Earth	$T^2 \div a^3$ =	1.000
Mars	$T^2 \div a^3$ =	1.000
Jupiter	$T^2 \div a^3$ =	1.000
Saturn	$T^2 \div a^3$ =	0.999
Uranus	$T^2 \div a^3$ =	1.000
Neptune	$T^2 \div a^3$ =	0.999
Pluto	$T^2 \div a^3$ =	1.004

There is Time and there is space between the moon and the earth but the space grow or become more because time produces space as the history of time left to serve as space or as expanding light. The space is the decreasing of density of the liquid or non-material factor and by reducing density the heat increases and the space levels rises accordingly. The Universe do not expand because the Universe can't expand since it has nowhere to go. The Universe

Mercury	0.06	(a^3)=	0.059	(T^2)=	0.058
Venus	0.82	(a^3)=	0.381	(T^2)=	0.378
Earth	1.000	(a^3)=	1.000	(T^2)=	1.000
Mars	0.11	(a^3)=	3.54	(T^2)=	3.54
Jupiter	317.89	(a^3)=	140.6	(T^2)=	140.66
Saturn	95.17	(a^3)=	868.25	(T^2)=	867.9
Uranus	14.53	(a^3)=	7067	(T^2)=	7069
Neptune	17.14	(a^3)=	27189	(T^2)=27159	
Pluto	0.0025	(a^3)=	61443	(T^2)=61703	

However being space we work with time and in time space becomes dimensionally flat or 1 as the ratio of time proves $T^2 \div a^3$ =1.000. We might see as humans space increase it is time that remains the same.

The straight line representing the Titus Bode law.

$$T^2 \div a^3 = 0.999$$

Sun   Ceres   Saturn

☼ ▶◎ ▶ ▶◎ ▶ ▶ ▶ ◎ ▶ ▶ ▶ ▶ ▶ ◎ ▶ ▶ ▶ ▶ ▶ ▶ ▶ ▶ ▶ ▶ ▶ ◎

Mars   Jupiter   Uranus

$T^2 \div a^3$ =**1.000** $T^2 \div a^3$ =**1.000** $T^2 \div a^3$ =**1.000**

Just as a matter of interest and going off the point slightly this destroys any hope of future space travel. From our perspective we go faster the higher we are in the air flying above the earth. That is not the case because we need more drive to fly at any specific altitude and remain at that density level of say $\Pi^2 \div 2$. What this actually proves is that movement or time because that is the same decreases as the relevancy of space increases and in fact we go slower the higher we go. The following figures that I mention was not accurately determined but only serves as a point I wish to make. If an object would travel at 31 000 km / h this vehicle will travel 310 km / hour when it reaches Pluto and by the time it reaches Oord's cloud it will travel at 31 mm per hour.

This is the reason why the circle comes in place of the straight line. According to cosmic time principle all things travel in a straight line but at any given point the travelling away from the centre of gravity will begin to equal any space moving towards the centre point of gravity. By space increasing through loss of density the relevancy between space in place and time forming

movement the directional gravity of $k = \dfrac{a^3}{T^2}$ and $k^{-1} = \dfrac{T^2}{a^3}$ to form $T^2 = \dfrac{a^3}{k}$ so that

equilibrium in relevancy $k^0 = \dfrac{a^3}{kT^2}$ will maintain and this I base on the figures that Kepler

provided when Kepler proved that $a^3 = T^2 k$. Everything in the cosmos goes in circles which is the result of cosmic equilibrium reached when travelling away is as fast as travelling towards and this is the next part of the Titius Bode law I wish to explain. By going forward things are going backwards equally and by rising things fall equal and in that we locate the eternal aspect and that forms a circle but a circle only means equilibrium. In singularity the line maintains the movement applying. That means the line is equal to a half circle is equal to a triangle and on that basis rests everything that brings about any aspect in the Titius Bode law.

The half circle representing the Titus Bode law.

The half circle representing the Titus Bode law.

$(T^2)=867.9$

Sun   Ceres   Saturn

Mars   Jupiter   Uranus
$(T^2)= 3.54$ $(T^2)= 140.66$ $(T^2)=7069$

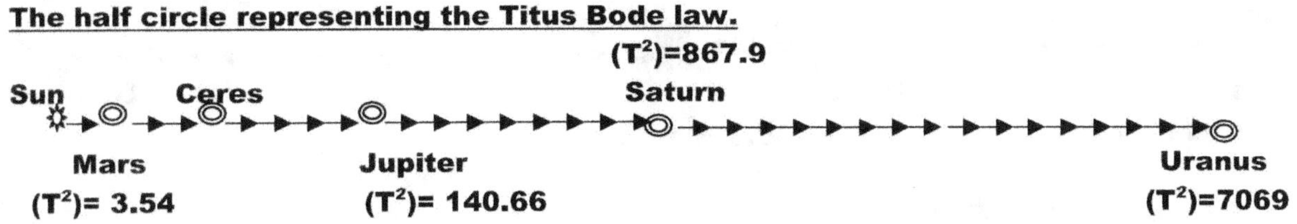

The straight line representing the Titus Bode law.

$T^2 \div a^3 = 0.999$

Sun   Ceres   Saturn

Mars   Jupiter   Uranus
$T^2 \div a^3 = 1.000$ $T^2 \div a^3 = 1.000$ $T^2 \div a^3 = 1.000$

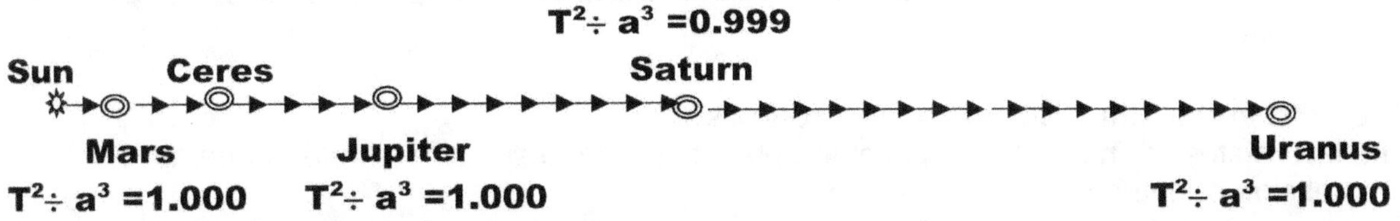

Then the triangle comes in to affect to produce the value of space as a triangle in terms of singularity and by that we find how the law of Pythagoras played an absolute domineering role in the forming of the Universe. Singularity dimensional adapt using the principle of Pythagoras to convert a line to the value of Π and space starts at the value of Π. Mercury and Venus only plays a dimensional part in the earth achieving the value of Π seen from the perspective we have on earth.

What the cosmos apply in singularity is the line moving directionally

To bring about equilibrium it translates a directional equal flow of space moving away and towards as half circle.

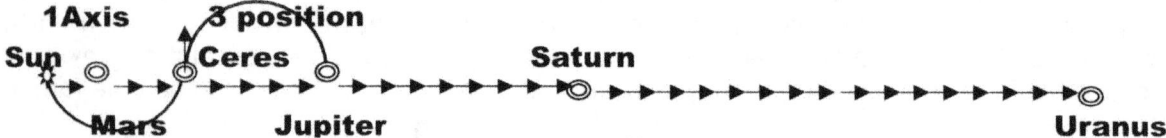

And to translate movement in ratio with space forming the triangle brings about the law of Pythagoras.

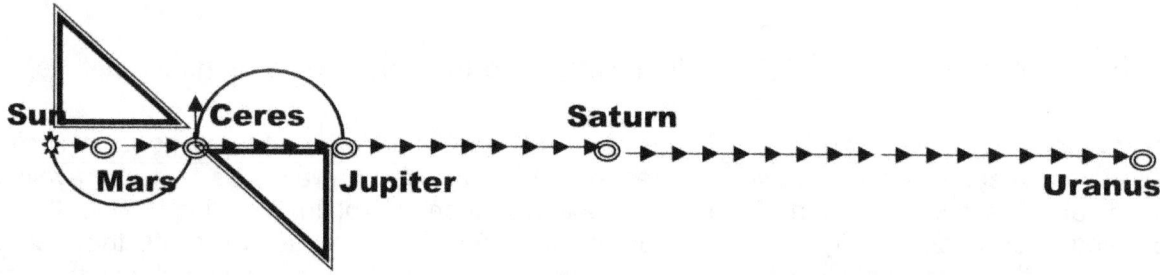

This is all because the triangle is the same (identical) as the half circle and the straight line but the effect on forming space produces results only we living in the dimension of space-time can appreciate.

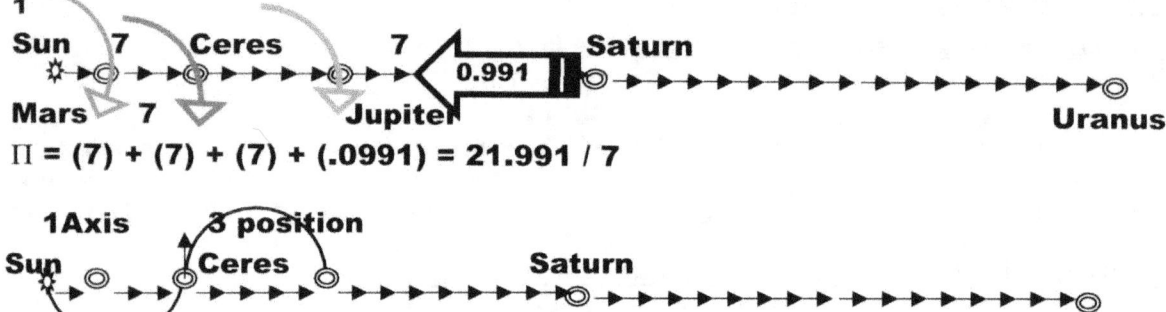

$$\Pi = (7) + (7) + (7) + (.0991) = 21.991 / 7$$

The ratio in the Titius Bode law is as follows:
The law relates the mean distances of the planets from the sun to a simple mathematic progression of numbers.

To find the mean distances of the planets, beginning with the following simple sequence of numbers:

0 3 6 12 24 48 96 192 384

With the exception of the first two, the others are simple twice the value of the preceding number.

Add 4 to each number:
4 7 10 16 28 52 100 196 388

Then divide by 10:
0.4 0.7 1.0 1.6 2.8 5.2 10.0 19.6 38.8

The first number is
0 3 6 12 24 48 96 192 384

Add 4 to each number:
4 7 10 16 28 52 100 196 388

In order to understand how time converts movement into space we have to reverse the process we see applying in the Titius Bode Law.

We start with 3 to find the mean distances of the planets. Then the next procedure is add 4 to each number and that gives a number of 7. If the first distance associates with the value of seven and the next and following planetary ratio holds an equal distance it also must associate with the number of seven.

Although the space is distances apart we see the singularity to space ratio is singular

$$(T^2)=867.9$$

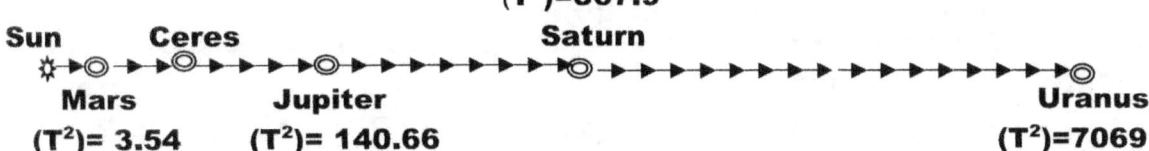

$(T^2)= 3.54$ $(T^2)= 140.66$ $(T^2)=7069$

...And...

<u>The straight line representing the Titus Bode law.</u>

$$T^2 \div a^3 = 0.999$$

$T^2 \div a^3 =1.000$ $T^2 \div a^3 =1.000$ $T^2 \div a^3 =1.000$

Therefore the dimensional equilibrium then must be where the axis forming is 3.

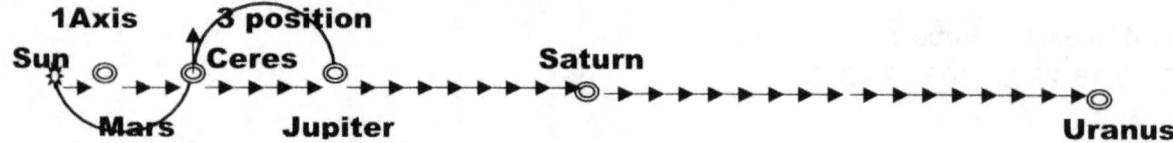

And then the circler movement will form as four.

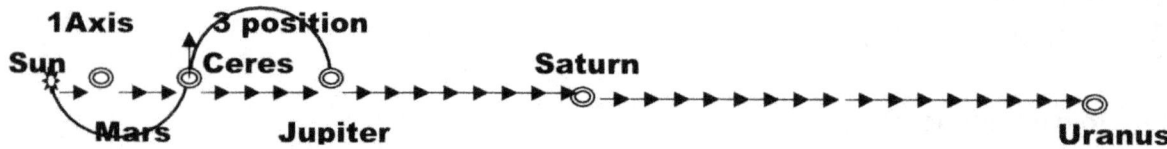

Because the planet forms a location in relation to the sun where the sun is allowing a flow of space towards the sun the planet must maintain a drive away from the sun equal to the flow of space towards the sun to maintain density equilibrium of $k = \dfrac{a^3}{T^2}$ and $k^{-1} = \dfrac{T^2}{a^3}$ to form

$T^2 = \dfrac{a^3}{k}$ so that equilibrium in relevancy $k^0 = \dfrac{a^3}{kT^2}$ to maintain density equilibrium.

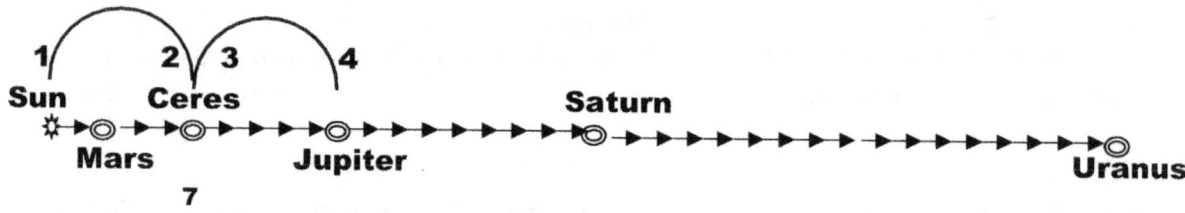

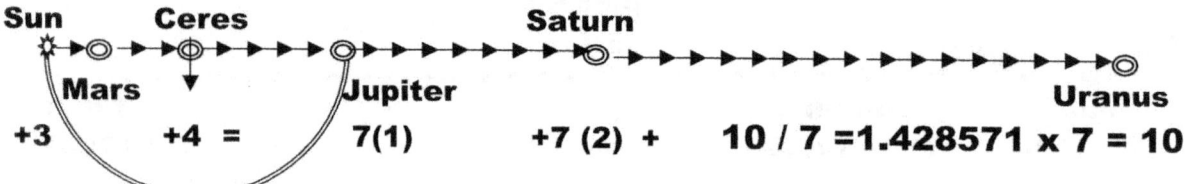

$+3 \qquad +4 = \qquad 7(1) \qquad +7\,(2)\,+ \qquad 10\,/\,7 = 1.428571 \times 7 = 10$

The first 7 pushed the second 7 into a position where it will become 1.428571

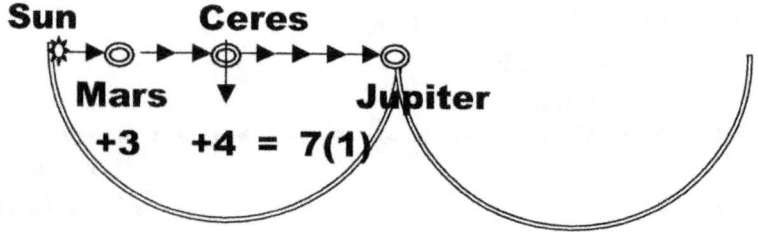

This positions the outside planet in terms of the position of the inside planet because the inside planet spins in relation to 7. The inside planet enforces a position onto the outer plane in relation to forming the value of Π but since this forms

gravity as Π.

The following table shows the time in duration that planets take to orbit the sun. Please AGAIN note that mass has F@%$#Kall to do with anything because mass is as random as a Newtonians' mind and thinking. Everything depends on the control there is between the flow of

space in relation to time in accordance with Kepler's tables putting a ratio $a^3 = T^2 k$ in place. Then Newtonians say that the Titius Bode law has no theoretical basis and therefore they cling to Newton's idiotic rubbish. That holds no proof as basis.

Planet	Distance from Sun (million km)	Mass of Planet (x10 ^22 kg)	Time for 1 Orbit of Sun (days)
Mercury	58	33.0	88.0
Venus	108	487	224.7
Earth	150	598	365.2
Mars	228	64.2	687.0
Jupiter	778	190,000	4332
Saturn	1,429	56,900	10760
Uranus	2,871	8,690	30700
Neptune	4,504	10,280	60200
Pluto	5,913	1.49	90600

From this we can see the compensation is in duration of turn and in singularity duration has no

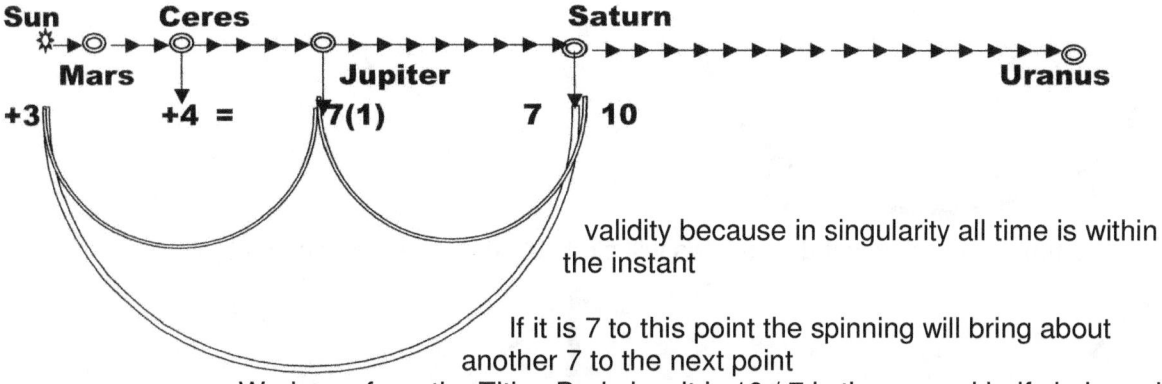

validity because in singularity all time is within the instant

If it is 7 to this point the spinning will bring about another 7 to the next point
We know from the Titius Bode law it is 10 / 7 in the second half circle and that stands related to the first half of the circle developing by 7.

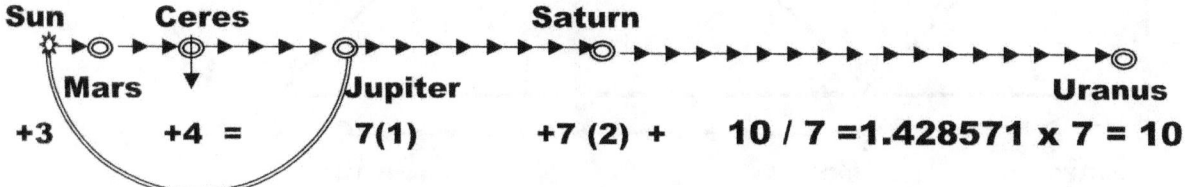

According to singularity the five first planets spinning around the sun according to what would be the Lagrangian points. This we see from the value in singularity of density equilibrium and because of this the fifth planet now a fragment called Ceres and other rock were destroyed as the inner planets pushed the fifth planet into Jupiter. In 2002 I tried to publish a book I titled the **Seven days of Creation** in which I explained how the solar system formed with information I gathered from applying these four laws. Jupiter destroyed the fifth structure as the inner four pushed the fifth into the domain Jupiter already claimed by Jupiter that was already in place and Jupiter produced the Roche limit to fragment the fifth planet. Guess what out of eighty commercial publishers not even one read the work because it violates Newton's integrity.

If I award the positions of the solar inner planets it will not be accurately established but say L1 was awarded to Mercury and so on the fifth planet strolled into the spin range of Jupiter that was already in place and the larger Jupiter destroyed the smaller fifth planet by applying the Roche limit. I explain the Roche limit and how it works in **Exposing Corrupt Science** and this book forms part of the Corrupt Science collection. Every diversion of the Titus Bode law tells the history of events that brought about the solar system. That and other events that took place made Jupiter so extensively bigger than any of the other planets. The first four planets were the core that remained of the dual star that the sun once was and the rest of the material the Sun obtained to give it extensive and unnatural growth in relation to the rest of the development of the solar system.

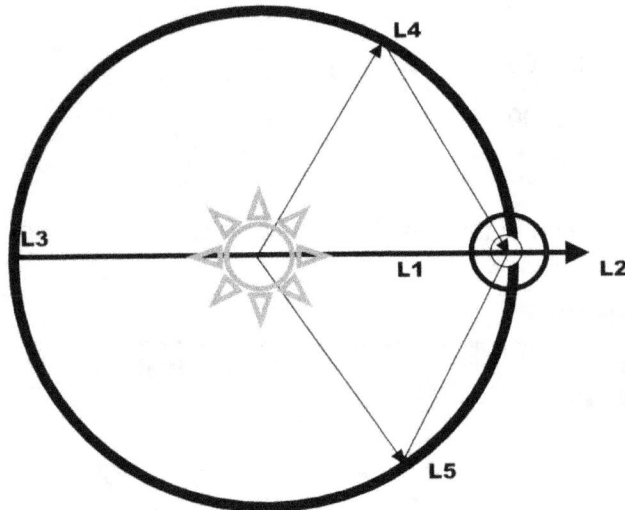

This is the more natural way that satellites will circle around a centre without space

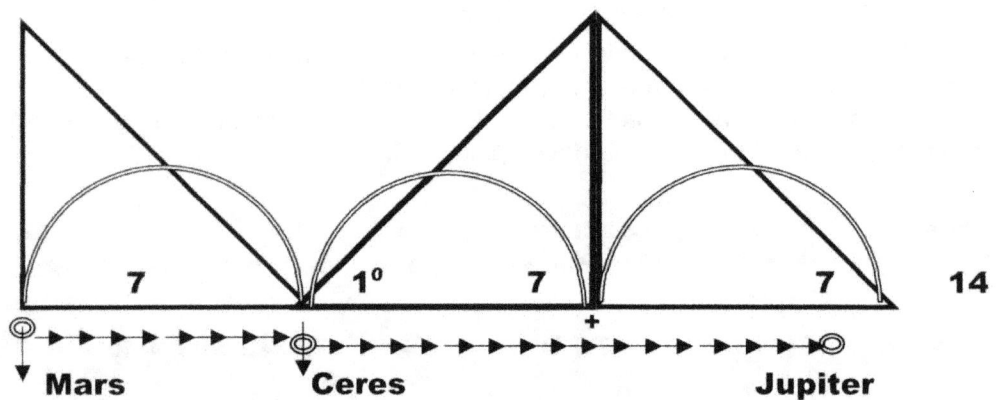

The one is pushing the other into a speed of displacement according to the Roche factor. From this onwards Mars determines the position Ceres has and Ceres puts Jupiter in place while Jupiter holds Ceres as an indicator to the space relevancy Jupiter must have. Jupiter takes a queue from Ceres while cares uses Mars as a reference in relation to Jupiter. Although they use separate tings in singularity the lot are forming a single ring but since all also charge gravity each forms an axis of 3 but the axis of three holds a position only because of the circle of four evoking the spin and therefore the circle of three. By spinning the inner planet holds a circle and by the circle evoking a spin at same movement puts circle in place that puts the spin on the outer planet where the spin of the outer planet enforces a circle to respond to the spin of the outer planet. From the sun there is a seven placing the planet to position a seven spin and that seven places another seven in position to form three sevens rotating where the three sevens together puts a value factor of Π which then completes the ring of Π. However as we can see with the top spinning it is the movement that puts the circle in place that puts the axis in place. Now we have to get to the movement to sort out the equation.

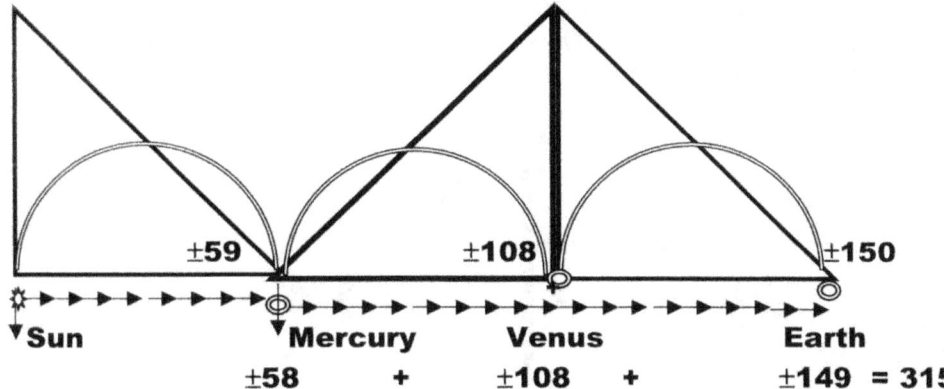

$315 / 100 = 3.15$ and Π is 3.14 which proves that the first three work in tandem to form gravity at Π.

This proves why the accumulative value of space by three plants together form Π and in singularity this results in $7 + 7 + 7 + (1-(1/10 \times 3/10 \times 3/10) = .991) = 21.991/ 7 = Π$

N earth we measure in meters because we use dimensions and distances in time reminiscent of values on earth. If we were on Venus and we applied measurements suited to what applies as space-time on Venus or on Mercury the figure we obtain would be much different and Venus would be $Π^2$ with Mercury being Π. It is not possible from where I stand on earth adapting time in $360°$ and having a meter of water weighting 1000 kg because these factors amount to the density I understand. When being somewhere else things about space-time forming will look much different.

Sun

$3^2 + 4^2 = 5^2$ **This always validates the Lagrangian concept wherever**

$9 + 16 = 25$ **This proves the law of Pythagoras forms space-time always**

That is also the reason why the orbits of Mercury is so different from anything lese in the Solar system. It is because Mercury has to use its own year and day to validate its position because there is no other reference.

Observing the inner planets from our perspective things look skew but that it should do because the mathematics we apply and use do not fit where these structures are located at and values we attach disassociates when we are on these other structures.

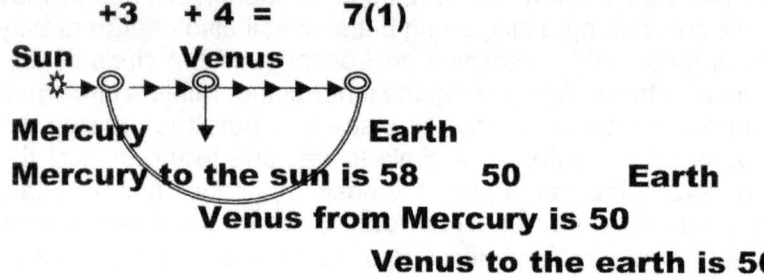

+3   + 4 =   7(1)

Sun   Venus

Mercury   Earth

Mercury to the sun is 58   50   Earth

Venus from Mercury is 50

Venus to the earth is 50

Using our norms we see some formula applying although to us it does not match as it should but if we change whatever we are suppose to change Venus would be Π.

Then being there will change the centre of the Universe to a brand new point just because we were there instead of placing the centre of the Universe here.

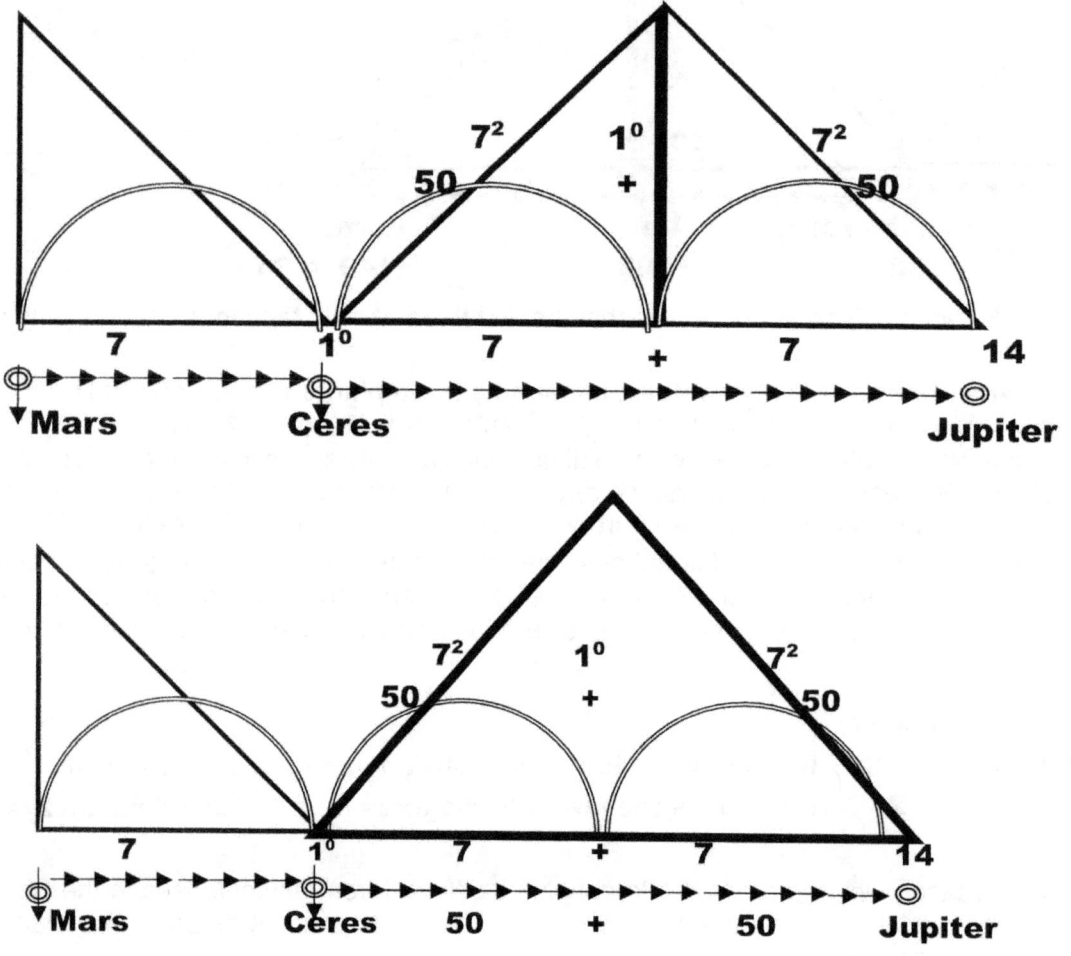

50 + 50 = 100 and √100 = 10

Every seven in material spinning holds 10 in relevance to space because of the spinning. It is the seven that establishes the ten but the seven that establishes the ten on one side is the same seven that establishes another ten on the other side.

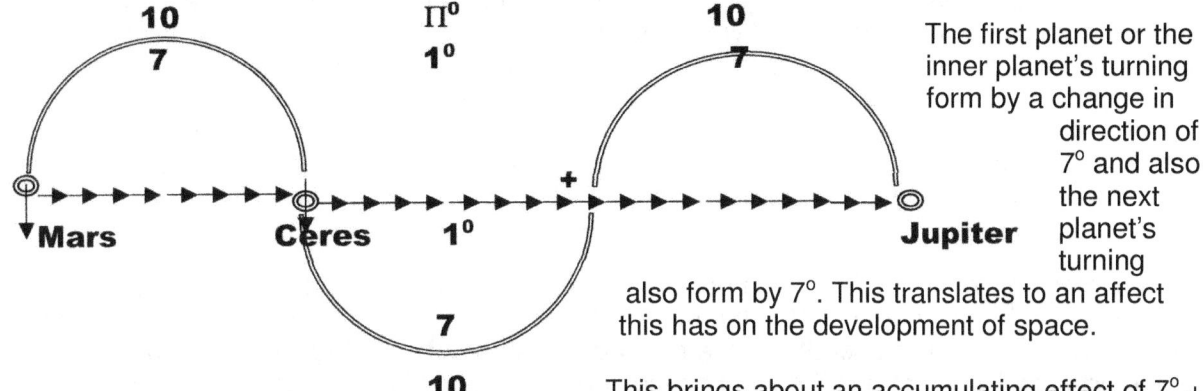

The first planet or the inner planet's turning form by a change in direction of 7° and also the next planet's turning also form by 7°. This translates to an affect this has on the development of space.

This brings about an accumulating effect of 7° + 7° but this is in terms of singularity seven dots and seven dots that add to fourteen dots that are involved in the process of forming unoccupied space. This by the way is exactly the way the first instant occurred when the first moment of cosmic development arrived. Since this was a part of the Universe at the start and only nothing can leave the Universe this process has no where to go but remain part of the cosmos and that is how the cosmos still develop. This process is the process that part one instant

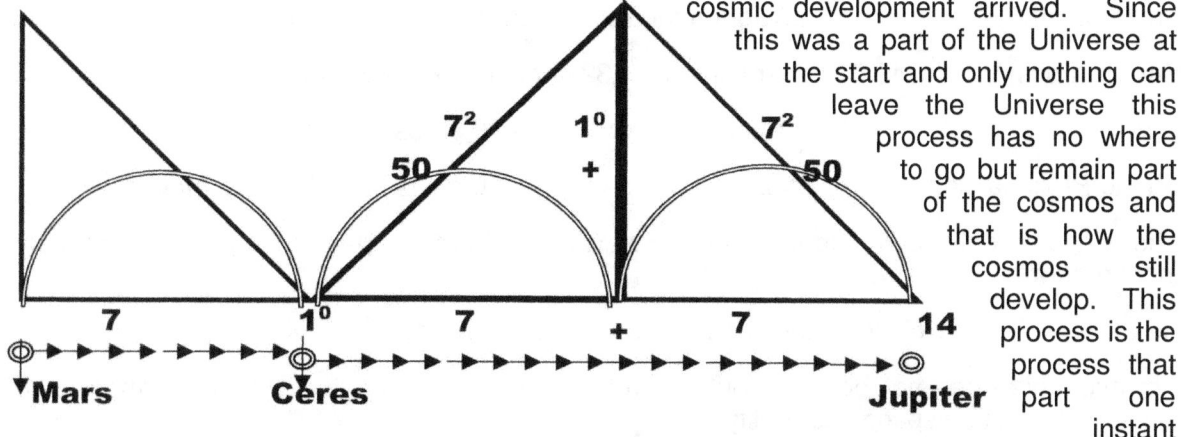

from the next instant by ending at 10 and starting at one again to form the next instant of cosmic birth.

We must realise that this "growing" of space is the depleting of density of the space. As the density capitulates the coldness moves to material and heat moves to non-materials. The "expanding" is a result of overheating and the contracting of materials

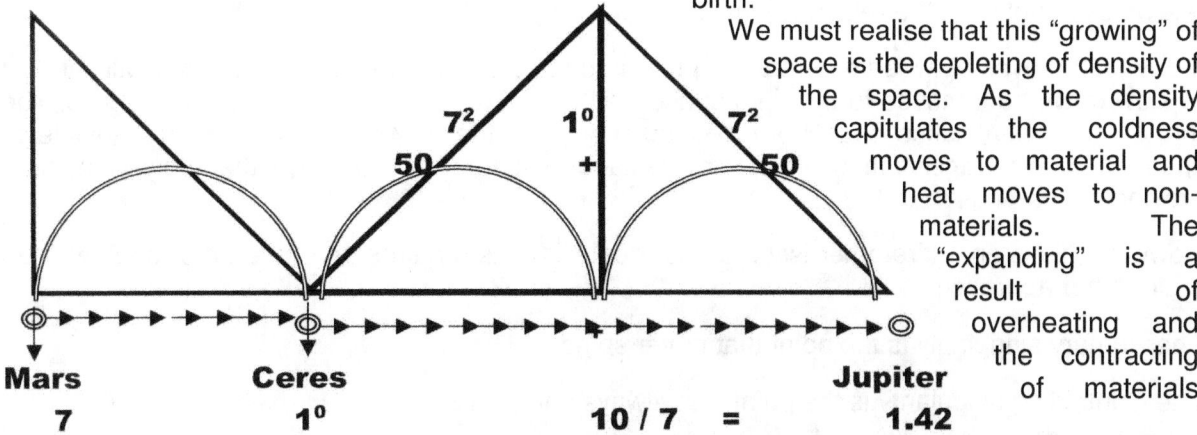

14 / 1.42 = 9.86, which is the numeric square of Π, which proves that gravity is Π going square.

are the result of freezing non-materials onto and than into materials to obtain the cooling required for materials to grow. The growth in becoming larger results not from materials shrinking because materials grow but it is the result of space becoming less dense and with lesser denseness it seem to hold volumetric increase. This is not true either because volumetric space is a hallucination. The ratio of what becomes denser by growth and what becomes much less dense also by growth forms incompatible where the one seems to remain and shrink and the other expands because it overheats. The one freezes into the oblivious and the other expand into obscurity and the poles forming infinity and eternity grows further apart while everything stays exactly the same. It is merely the set limits to existing ratios and relevancies that bring an ongoing change in opposing imitational ends but the limit is there because the next minute it is not.

In the cosmos only "nothing" is fixed because it is not. The speed of light is $3\Pi^2$ because where the speed of light might be 299 792 458 m / s but that is in terms of the earth only. Everywhere else is very different. The speed of movement depends on density of space that translated back to time with displacement and then into space again per relocation position. The relevancy of movement holds light at volumetric occupation 3^3 as well movement $3\Pi^2$.

Moon to Earth	1.28 light-seconds	2.4 inches 0.006096 meters
Sun to Earth	8.3 light-minutes	26 yards or 24 meters
Sun to Jupiter	43.27 light-minutes	132 yards or 121.845 meters
Sun to Pluto	5 1/2 light-hours	1019 yards or 940.6 meters
Sun to Proxima Centauri	4.22 light-years	4000 miles or 6498461.5 meters

These figures only contribute to a time delay issue or forming space as a past in the presence that time forms.

Space is the history of what time left behind in forming relevance.

However this does not begin to put the relevancy of movement of space in space and of movement through space into reality.

This takes a stance that all space is equal as it is on Earth and so shall it be in heaven. The relevancy doubles as the relevancy grows.

If any craft moves between the earth and the moon at a certain speed the relevancy will cause it to slow down as it moves further into space and that is what the Titius Bode law predicts. The further movement takes pace from a singularity point the slower will movement translate to space and this translation of space is displacement in space through the movement time provides.

However this is an entire other issue and to deal with this requires an entire other book with an entire other approach.

The Primary singularity is the point that never spins as Π^0

The Controlling singularity is the point that always spins as 3

The Governing singularity is the point that always turns as 4

This brings about that movement is always in conjunction with 7.

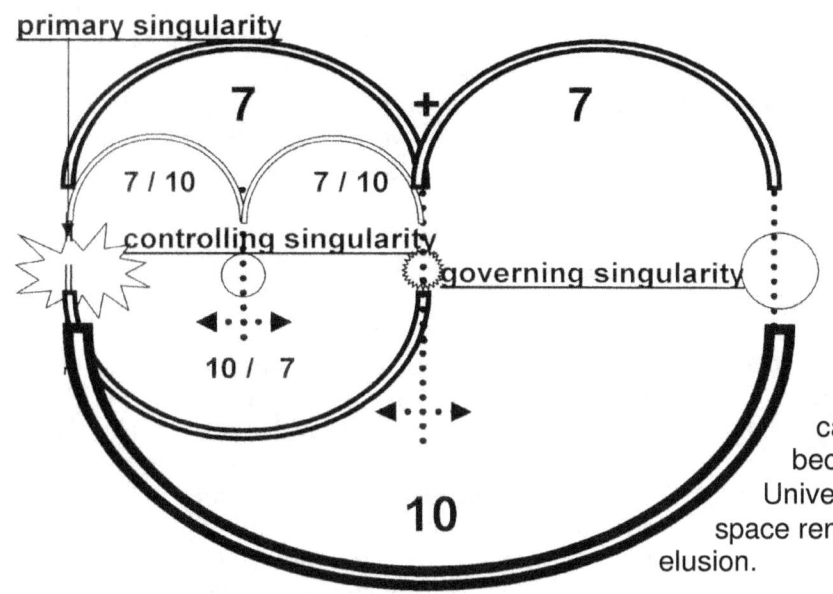

Gravity is $\Pi^0\Pi$ going Π^2
The value of 7 doubles and by applying Pythagoras to the inside of material the ratio that applies is 7 while to the outside where space forms the value the ratio is 10.
This ratio is found in every aspect of the cosmic code and this forms as much the cosmic code as the value and the ratio that $\Pi\Pi^2$ does. One can never refer to space as Π^3 because the entirety of the Universe lies in movement $\Pi\Pi^2$ of space rendering space Π^3 to be an elusion.

Gravity is the relation of objects in movement providing gravity as 7+7 = 14

Gravity is the relation of objects in movement in space gravity as $\dfrac{10}{7} == 1,42$

Gravity is $= \dfrac{14}{1.42}$ = **9.85915492957746478873239433661972 or 9.86 or Π^2.**

This is the reality of physics and when I explain this to those persons in charge of physics principles my calculations are too unimpressive to demand any attention from their side. Not living up to Newtonian expectations and simple nonetheless, as it is it is gravity as nature applies gravity as the Titius Bode law. The square root of **9.86** is Π. **Gravity is $\Pi^0\Pi$ going Π^2**

This is gravity as simple as my simple mind can calculate! There is no grandeur and there is no splendour and there is no impressive greatness. There is a sum total of 1 that forms space as $\Pi^0\Pi$ going by movement duplicating $\Pi^0\Pi$ to form Π^2.

Planet mass comes as random as Newtonian intellect boiling down to insignificant "nothing" precisely on par in value and influence as Newton's corrupt guessing. There is no evidence of Newton even being vaguely correct or having a slight tendency to accuracy and is completely baseless and without a valid theory.

On the other hand the Titius Bode law used in nature has a precise dynamic that only those that don't **_understand Newton_** can **_understand_**. It shows that movement as Π^2gravity formed by Π. This should eliminate Newton forever and another eternity to come and place him in history as the biggest fraud mankind ever encountered.

To understand Newton is a rouge and has been a diversion from the truth for the past three hundred years. There is **nothing** to **understand** or **not understand** about Newton. The "mass" of a cosmic body does not pull and in three hundred years no one in science could come up with one single bit of evidence it does pull. In fact the earth and the moon is growing apart and so will the entire solar system be. When checking on Kepler's finding they will find just how much space came about in the solar system since the time of Kepler. You just have to test the laws and follow the evidence nature leaves. It is much more tricky to understand nature but once you understand nature it is as simple as adding 3 + 4 = 7 and that is how space forms. Those Masters in science won't read my work or maybe can't read my work but they refuse to deal with the information I bring to the table. It is this code of 3 and 4 that I decipher and the reasons why it structurally works by 3 and 4 and it is this growth by 3 and 4 that I unravel. Let's start explaining it this way. To be within the Universe is to move as the Universe. This is the most important aspect we learn from Kepler's finding as $a^3 = T^2k$. All movement is by 7° going straight to form a circle and a circle forming 7°.

The seven degrees of rotation is composed as three points forming the axis and four points forming the circle. Anything that rotates does so around an axis holding three point one on top one in the middle and one below. Then the circle is the rotation of two points changing direction as it turns around another three point. Anything in the cosmic forming gravity holds three plus four points with the circle being 1+1+1+1+4 as well as 2 x 2 =4. There are two points going forward as much as two points reversing on the other side of the Universe. Where is the Universe? The Universe is Time in singularity (1) moving from the abyss (.991). This then forms Π. The Titus Bode is about gravity forming Π. Remember no circle can start at zero because of it started with zero it must continue with zero or then start the line where the line starts with 1. Therefore a circle starts at 1• +1• +1• +1• +1• and that is a line. If it started with 0 then it would be as follows 0 + 1• +1• +1• +1• +1• and that proves that zero has no place to start in any line up.

If Physics does not even know where a line starts where will they locate where the Universe started because the Universe is comprised of numerable lines criss-crossing and lines connecting numerable point. Therefore in singularity (1) everything is a line 1• +1• +1• +1• +1• or then it is forms as singularity Π^0• +Π^0• +Π^0• +Π^0• +Π^0•. The circle is a line and the line is a line and there then is seven points in the line-up forming rotation. It is one line in singularity but in the space in which we are it is a circle and an axis in connected by rotation. In singularity it is Π^0• +Π^0• = 2 +Π^0• = 3 +Π^0• = 4 +Π^0 = 5+Π^0• = 6+Π^0• = 7Π^0•. This is the one part of singularity where 7 connect to 1. The issue here is 3 + 4 = 7.

The three stands in for the measured value of a triangle because it forms one dot in reference to the future and that places the same dot in a ratio of going. This leads to four dots.

past and the coming and

The solution to is because of our sky while represents cosmology as we see it. We see the sun cross our sky because the sun rises in the east and sets west.

encrypt the cosmic code is in the value of 3 and 4. It this value that we see the sun as the sun crosses the earth spins around its axis It is a ratio that

One dot becomes two by leaving one point and going to another point which is in reference two positions for the same dot. But since one dot is coming while the other is going we have dots in two positions representing the coming part and two dots in the going part. This means we have 2 + 2 = 4 as well as 2 x 2 = 4 and the beginning of the second dimension. In this we have the three forming a line motion as Kepler said $a^3 = kT^2$ where 3 represents the **k** part. Then we have four forming the square part or T^2. This is how movement starts space $a^3 = kT^2$ or movement forms space $a^3 = kT^2$.

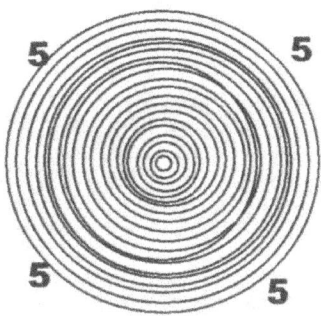

The three of the axis and the four of the circle combine to give a rotation movement of 7°. As the rotation is of a solid object the rotation is 7° connecting by a fixed line and this means that although it is on either side of singularity it still remains the same 7° since it is the rotation of the same object.

In this is where the 7 forms 10. The seven moves from one point to the next point and in mathematics when something moves it is going square. It is not going to become a four sided flat object as Newtonians whish to portray singularity but it is going square such as Π moving and then duplicating $\Pi x \Pi = \Pi^2$. In this case it is 7 going square to be $7^2 = 49$.

With three next point just Therefore with forming a fifth then the value circle is five. Then is 4 x 5 = 20. coming from the When we look at connect time and forming the axis and four forming the circle the outside the circle of four must be five. four points forming the circle and each point point the total value inside the circle is four outside the circle and on the outside of the the full compliment of the circle on the outside Space holds 20, which include singularity future (which I will explain later) is 1.991. space we think we see a circle but when we movement into the equation we see half circles forming as time moves. Moreover we see circles forming that cross singularity going on both sides of the divide forming opposing rhythms or we call it seasons. We do more about this in another chapter.

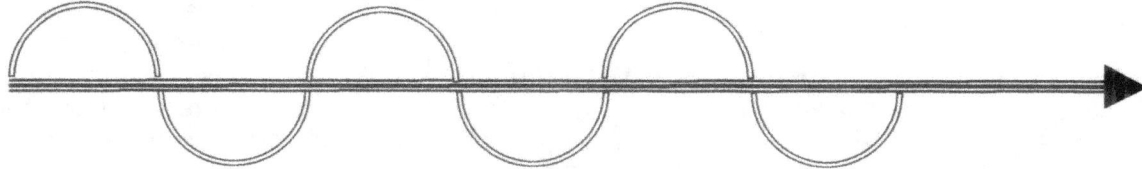

Therefore gravity is half circle flowing with time

The circle turns by 7°.

The circle moves, which also make the turning, go by 7 square, which is 7^2.

This means that the circle goes $7^2 = 49$ on both sides of the centre and that means it goes square on the open divide.

The circle connects by crossing singularity, which is 1.

The one his time remains the very same 1 and it is the line of singularity crossing by movement applying.

On the one side we have 7^2 in relevance to 1^2 and on the other side we have a duplication of the same process forming as 7^2 goes in relevance to 1^2. That means on the one side there is the value of 7^2 in relevance to 1^2 and on the other side it is 7^2 in relevance to 1^2.

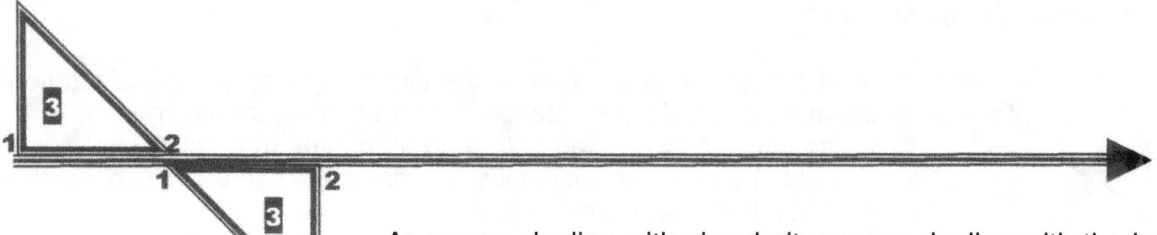

As we are dealing with singularity we are dealing with the issues forming singularity. In singularity the half circle and the triangle has the same value as a straight line forming 180°. Therefore as time moves in a straight lien time therefore moves in half circles crossing the straight line which is the straight line turning as relevancies changes from one side to the next. According to mathematics and mathematical principles forming it is $1^2 + 2^2 = 1 + 4 = 5$. Therefore by movement the value that follows four must be five.

That placed a value of five after four and this brought on the Lagrangian point laws.

The Lagrangian points unite a line (1) a circle (2) triangle (3). That indicates all four cosmic laws act in singularity and

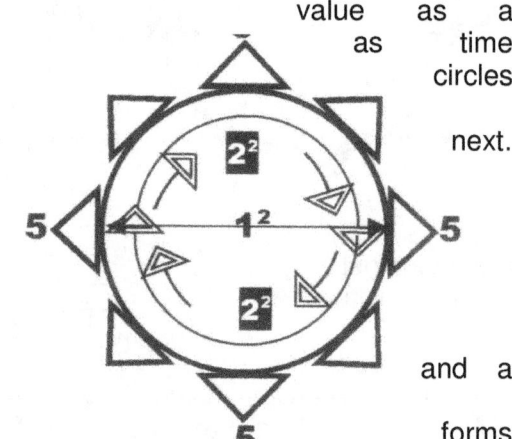

and a forms singularity. This serves to show where moment starts and why movement starts because the Universe is the movement thereof. This indicates where to look for

The two half circles

The Straight line

The two half circles

singularity as much as it shows where we have to learn about singularity.

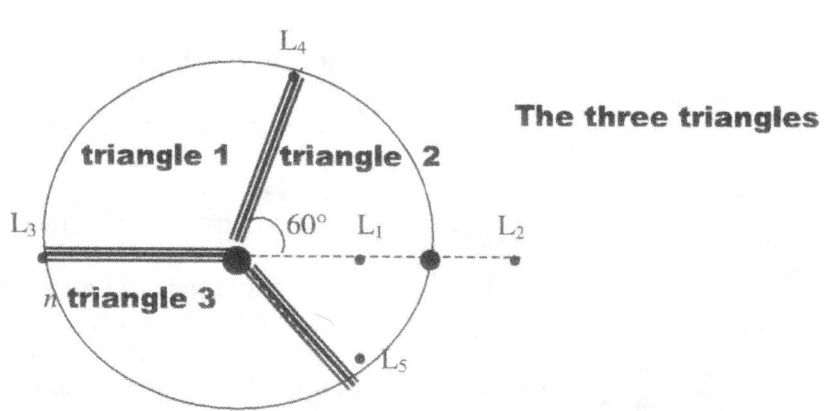

LAGRANGIAN POINT:

The Lagrangian points are five equilibrium points in the orbit of one body around another, such as a planet around the Sun

The three triangles

triangle 1   triangle 2

triangle 3

With the line of three in conjunction with the circle of four the total movement is seven moving. This is where mathematics starts forming physics.

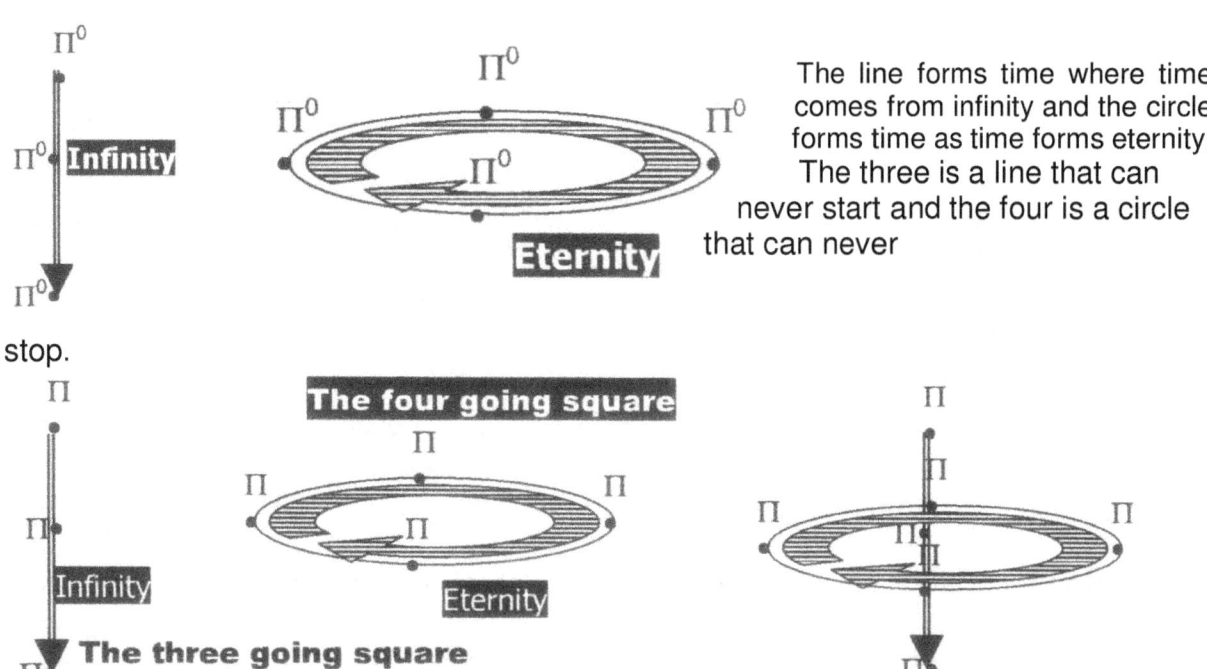

The line forms time where time comes from infinity and the circle forms time as time forms eternity. The three is a line that can never start and the four is a circle that can never

stop.

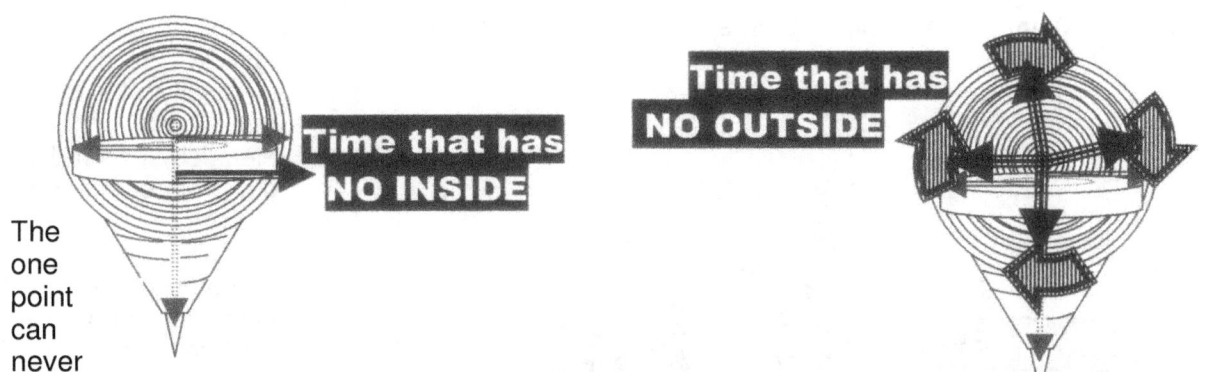

Between these two the line and the circle is a universe filled with movement.

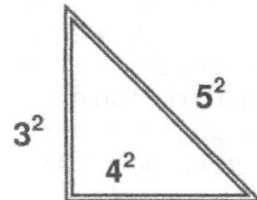

The one point can never move. The other point will forever move. The one point holds the inside of the Universe while the other point captured Universe as the Universe forms the inside of the other point and that way the Universe has no outside ever. In the way that mathematics form we find the basics of mathematics. We find how mathematics employs the law of Pythagoras. First we had $1^2 + 2^2 = 5$, but the next movement is also the result of the Pythagoras triangle where $3^2 + 4^2 = 25$ and 25 is 5^5. This proves that Pythagoras is the way the Universe forms.

Then we have to look for Pythagoras in forming movement.

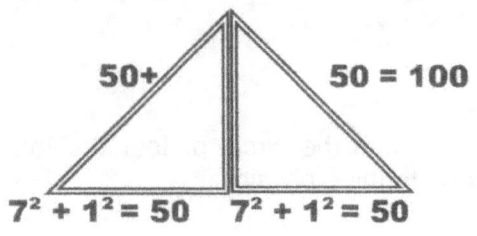

50+ 50 = 100

$7^2 + 1^2 = 50$ $7^2 + 1^2 = 50$

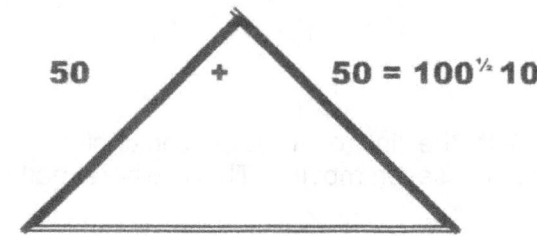

50 + 50 = 100½ 10

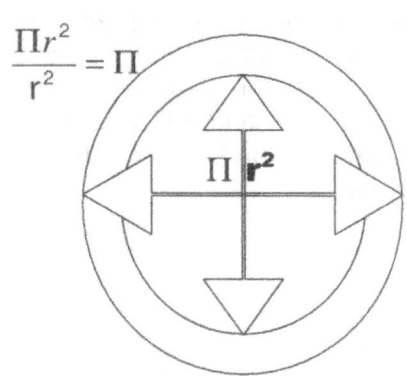

$$\frac{\Pi r^2}{r^2} = \Pi$$

Πr^2

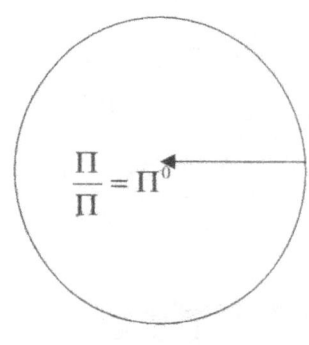

$$\frac{\Pi}{\Pi} = \Pi^0$$

In order to find singularity one must reduce a circle. Everything in the Universe forms by the form of a circle. Anything that is anything holds six sides Π^6 that spin $\dfrac{7}{10}$ in a six sides cube $\dfrac{\Pi^6}{6}$ forms a sphere. Movement in cosmic terms is the Titius Bode law or then $\dfrac{7}{10}$ and space is a sphere $\dfrac{\Pi^6}{6}$, which is a multitude of circles Πr^2. To get to singularity one removes the circle $\dfrac{\Pi r^2}{\Pi r^2} = \Pi^0$ and then you have singularity. That is where everything that is material starts to form space. Anything from Π and larger is space and anything smaller than Π is time within singularity. What can be smaller than space, for one thing the thoughts you use to remember this. We do not use electricity to think but we charge electricity to send to our brain commands to think.

In the centre we have singularity contracting or moving towards to infinity, the void that can't get smaller and it holds the line from which everything progresses. Then outside there is Π which is that which accelerates as time forms space.

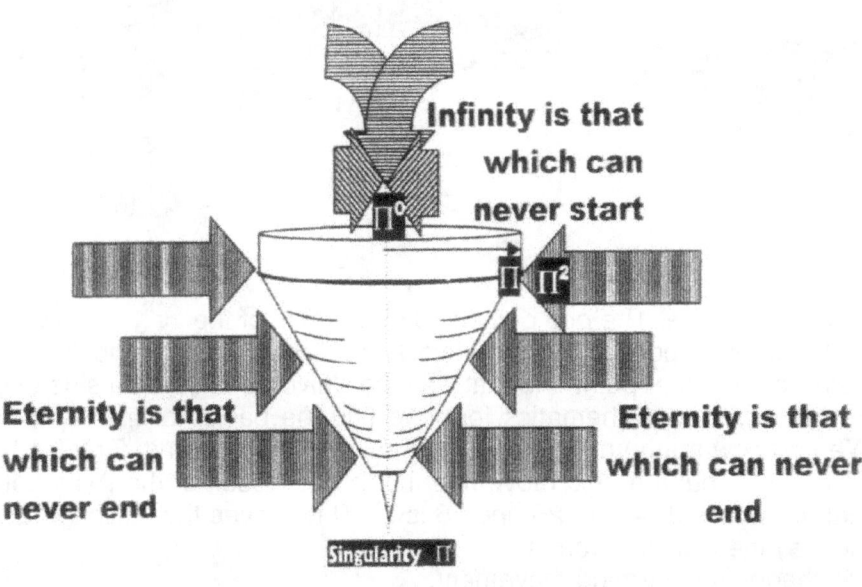

Infinity is that which can never start

Eternity is that which can never end

Eternity is that which can never end

Singularity Π

Kepler with his tables summed up gravity and all

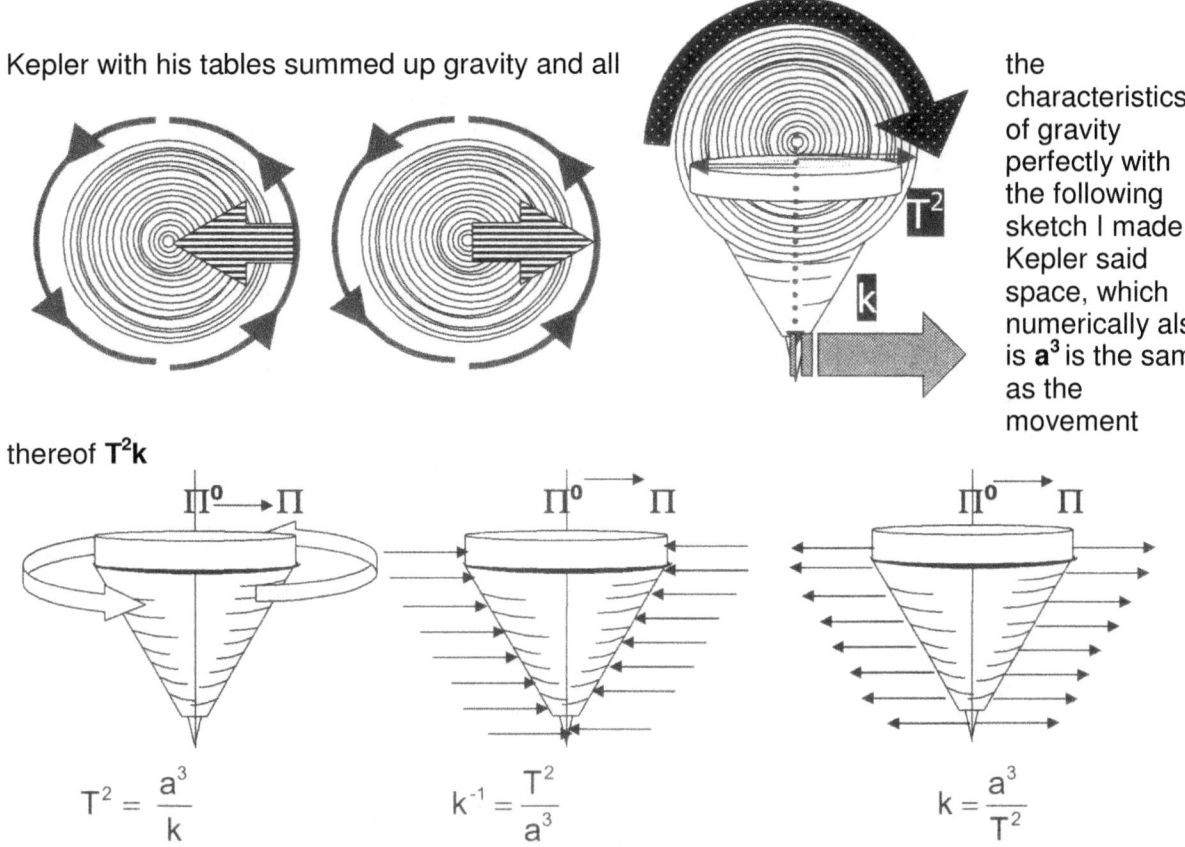

the characteristics of gravity perfectly with the following sketch I made. Kepler said space, which numerically also is a^3 is the same as the movement

thereof T^2k

$$T^2 = \frac{a^3}{k}$$

$$k^{-1} = \frac{T^2}{a^3}$$

$$k = \frac{a^3}{T^2}$$

In order to find independence from the capture and establish individual existence within the influence range of a dominant structure movement is essential. There has to be an axis established to reserve independence but the circle establishes the axis and therefore it is the circle that plays the major role in the establishing of 7° turning.

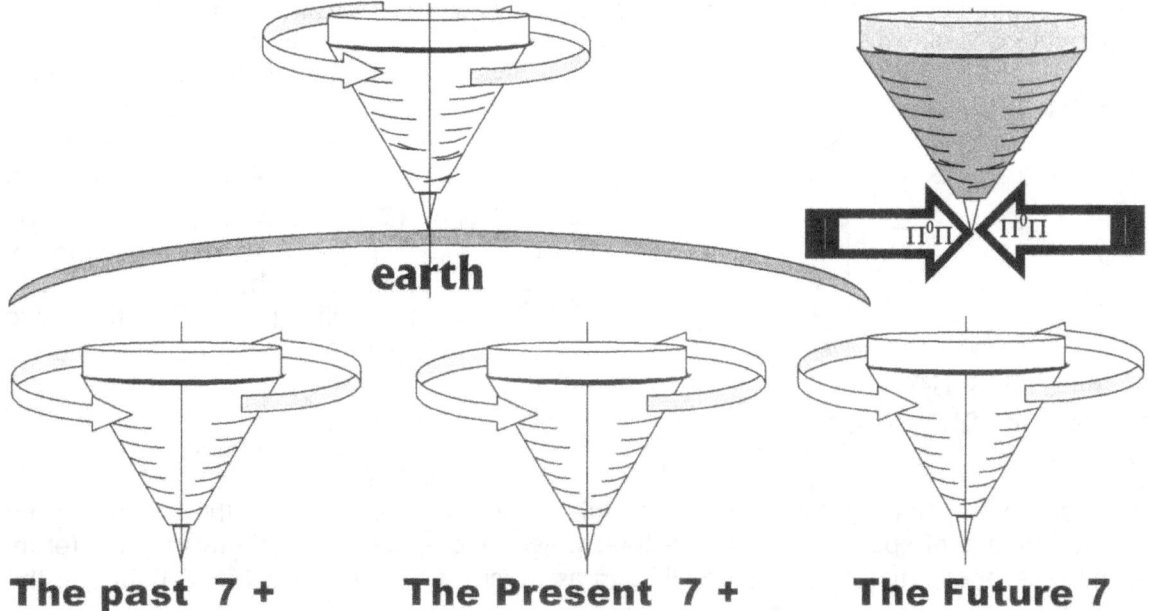

earth

The past 7 +   **The Present 7 +**   **The Future 7**

This is where to locate singularity: In the centre of all moving objects a line forms that we named an axis. If you reduce the circle that turns the circle will get to a point where space is absent. This puts a point inside every spinning axis where space disappears. All turning moves all space towards and in the direction of this point where space disappears.

By within the removing all represents a establishes the the measure of Π no space. The very other side also much completely in was going left the inspecting and investigating this point brings about some understanding of how gravity and therefore how gravity cosmos functions. If one reduces the circle Πr^2 by material only form in the shape of Π remains. This value Π circle with a specific value of 3.1416/1. This value value of time forming space. When entering the circle Π has to even further reduce to become Π^0. Un this there is instant the entering starts that same instant exiting on the starts. However by entering the one side the motion is very the opposing direction on the other side. If the one direction other direction is going right and visa versa. The one point

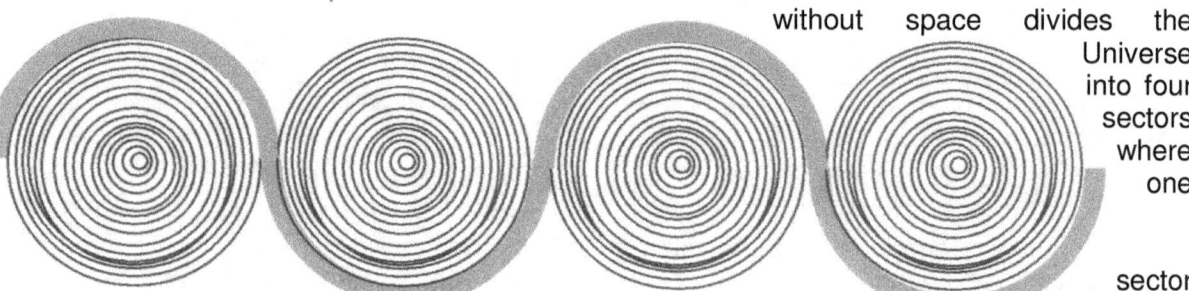

without space divides the Universe into four sectors where one sector is reversing the direction of the opposing side. This is where physics starts. This proves that the entirety we think of as a Universe is controlled through movement applying as time from a point that completely and unquestionable falls outside the realms of what we think of as space within the Universe.

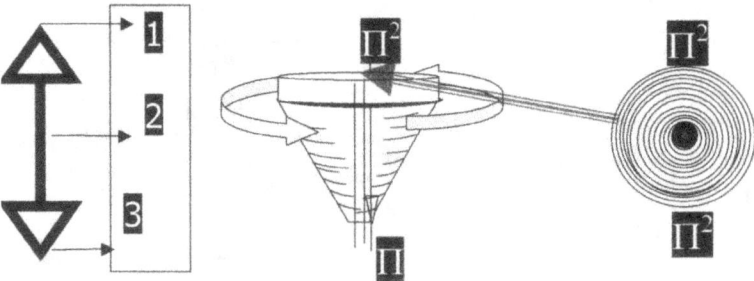

In the atom the electron has a value of 3 which the formula or if you wish to call it an equation this is it $(\Pi^2 + \Pi^2)\ (\Pi^2\Pi)\ \mathbf{3} =$ **1836** proves. This is the displacement or reduction of heat from the gas (electron) to the solid (proton) via the liquid (neutron).

The proton is $(\Pi^2 + \Pi^2)$
The Neutron is $(\Pi^2\Pi)$
The electron is 3.

The atom and every atom individually form a Universe that starts and ends the Universe. The movement in space of space holds two sectors, which places the identity in two total different categories. The one is material driven by time in as much as time forming Π = 3.1412 and the

other is time forming space by the movement Π brings about and that value is time in the future going to time in the present moving to time in the past and that is 3.
The atom forms space within the Universe as Π (Π² + Π²) + (Π² + Π) + **3 = 112.**
This is where space starts a sphere and where the atom forms as a sphere.

Anything smaller than 3 falls outside the realm of space because time forms space by going from the future through the present to the past.

That is how easy it is to prove the existence of God Almighty. There where only a thought can be time forms space as time in the form of gravity shapes space and by a thought in the void of where space can't be a thought drives the Universe as 1 or as Π^0. Everything in the Universe unites in one point that from where we see can never by and since $\Pi^0 = 1^0 = 1^1 = 1^{230759312756299} = 1$ the Universe is within one point forming time. Only by movement can time begin space as 3 but essentially it is one point moving from the future through the present to the past that form time that forms the Universe.

Now we have to investigate how the Titius Bode law forms gravity as **Π².**
This far I have established why and how gravity is the measure and the value of **Π².**

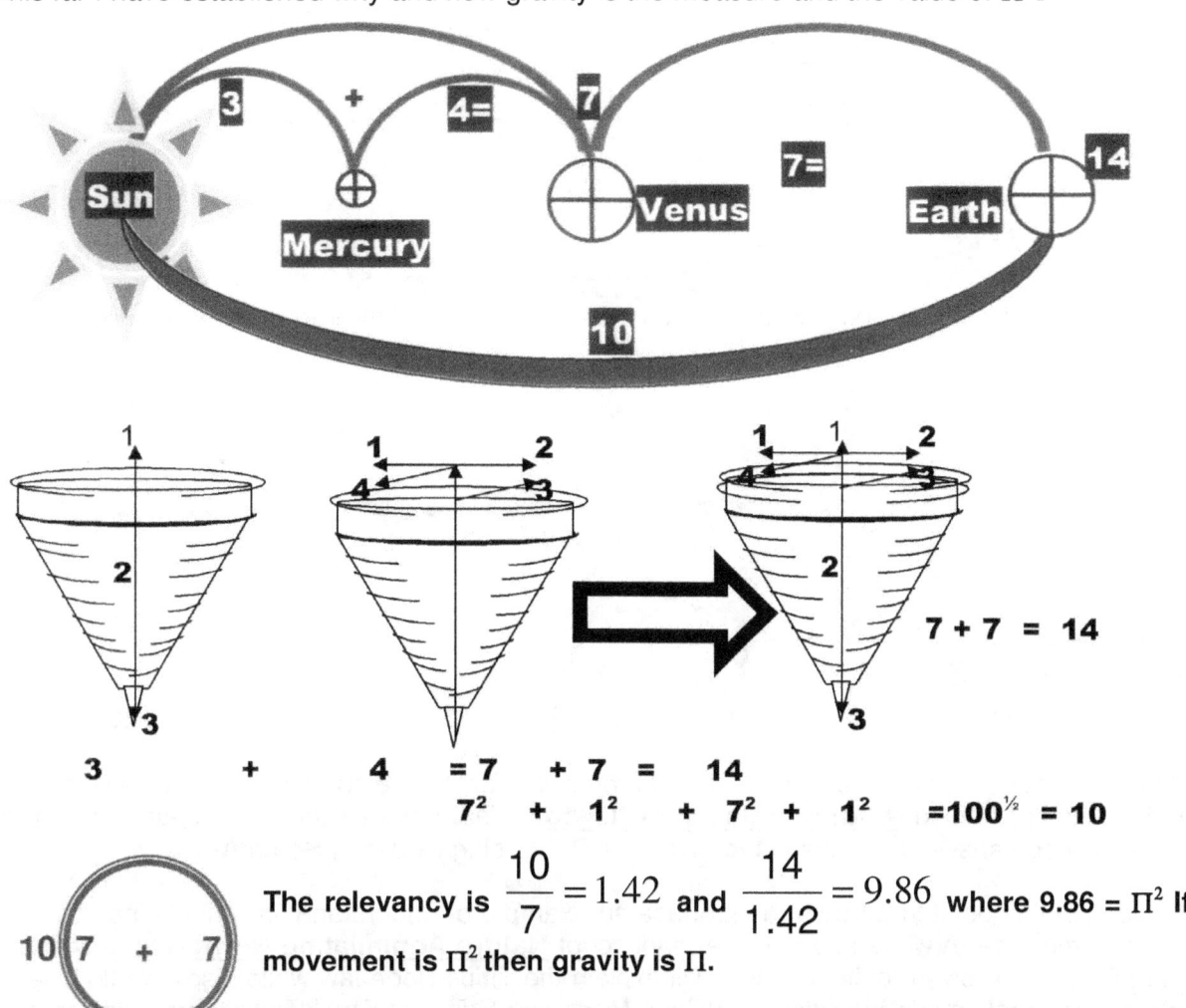

$$\frac{10}{7} = 1.42 \text{ and } \frac{14}{1.42} = 9.86$$

The relevancy is $\frac{10}{7} = 1.42$ and $\frac{14}{1.42} = 9.86$ where 9.86 = Π^2 If movement is Π^2 then gravity is Π.

This proves that gravity is not produced by mass but it is the result of movement forming a circle forming Π that then goes square Π^2.

This proves the value of Π forms gravity where space is a double (not square) of 10 with 7 being in relation to 10. Therefore gravity holds two values in space, which are $\Pi = \dfrac{10+10+1.991}{7}$ $\Pi = \dfrac{7+7+7+0.991}{7}$ and it is in singularity $\Pi = \dfrac{21.991}{7}$

$\Pi = \dfrac{3.1416}{\Pi^0}$ ... **This where the universe and therefore physics start.**

The Universe is the value of Π rotating as Π^2.

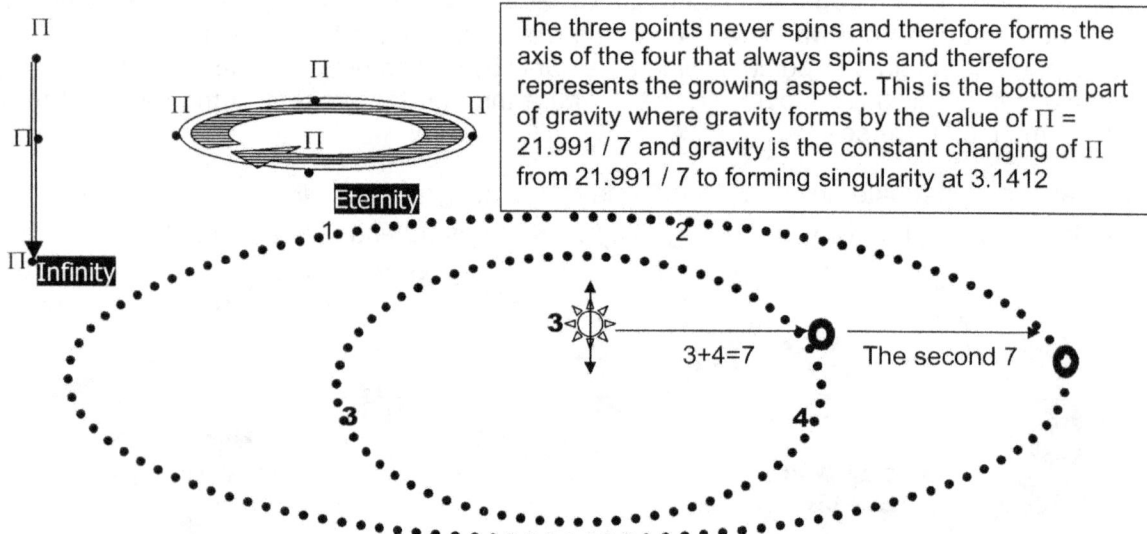

The three points never spins and therefore forms the axis of the four that always spins and therefore represents the growing aspect. This is the bottom part of gravity where gravity forms by the value of Π = 21.991 / 7 and gravity is the constant changing of Π from 21.991 / 7 to forming singularity at 3.1412

The movement of time goes by the rotation of seven in revelation to singularity forming the value of 1

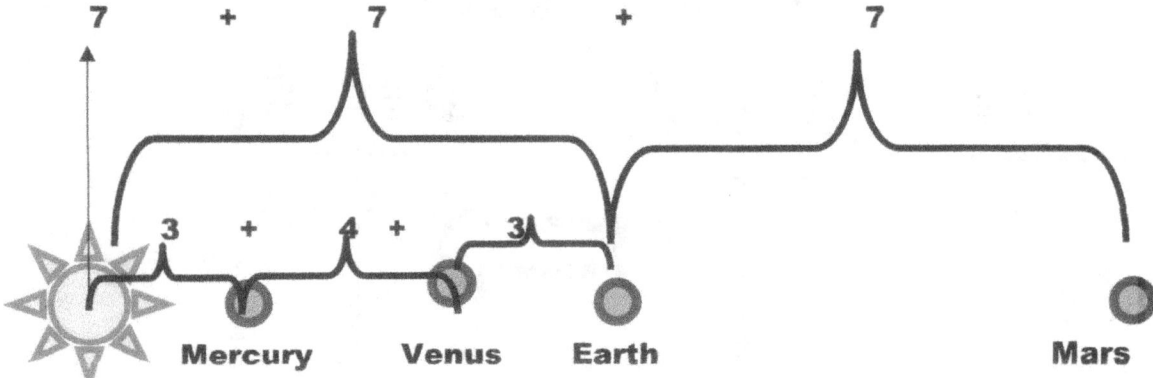

This that I show is a very, very small part of a book written on the subject of the Titius Bode law and the book I named **Nature Annihilating Newton.** That book I sent to eighty-six publishers and this is one answer I received in reply from a Publicising House in South Africa.

It now is at this point that I wish to introduce an example of understanding while being clueless at the same time. A while ago I sent a manuscript **Nature Annihilating Newton** to publishers and this is one reply I received back. I will explain the Titius Bode law while also I would love to address the reply made by Helen de Villiers Managing Editor of Struik Publishers / Nature and clarify what was suggested.

Dear Mr Schutte

We have received your book proposal, *Nature Annihilating Newton*, for which we thank you.

Struik Nature publishes books relating to the world of natural science, as opposed to chemical or physical science – the apparent field of your book. Moreover, our list is ill-suited to contentious argument; a science journal would be a more appropriate forum for such a debate. We have concluded that your submission is not suitable for our publishing list.

Best wishes
Helen de Villiers

Helen de Villiers
Managing Editor | Struik Nature

TEL +27 (021) 460 5400 ○ **FAX** +27 (021) 461 7122 ○ **EMAIL** helend@randomstruik.co.za

Wembley Square, First Floor, Solan Road, Gardens, Cape Town, 8001
PO Box 1144, Cape Town, SOUTH AFRICA, 8000 ○ **www.randomstruik.co.za**

First I did for thirteen years contact journals amongst almost two thousand five hundred others on this matter explaining the Titius Bode law, the Lagrangian point system, The Roche limit as well as the Coanda effect. It is clear that rising beyond and above the fact that Helen de Villiers Managing Editor of Struik Publishers / Nature is uninformed and clueless on the rule of the Titius Bode law or what attaches to this law, she still is very much deeply and sincerely opinionated…opinionated on a topic she has no idea what it contains…still she is fit to conclude an opinion! As she well declares that **Struik Nature publishes books relating to the world of natural science, as opposed to chemical or physical science** and if she was in the least informed about these principles or if she ever read my introduction thereof in any and all my books she would have seen that there can be nothing more natural in the Universe that these four cosmic principles. In fact in the next two chapters I prove that these natural principles form the key to how NATURE no less formed the Universe in the very beginning.

More natural and more in nature as these four principles are, there just cannot be…and she would have been informed about this if she only read the book before getting so opinionated on issues she clearly did not study or read the book.

If she did read the book before drawing conclusions she would have read on many occasions throughout the book that I did send so many articles to so many journals and I showed replies I received amongst which this was one reply I received from Annalen Der Physics. I sent to Annalen Der Physics twelve articles on the four cosmic principles stating how to go about to conclude how the principles form, how the principles form nature, how the principles form gravity as Π and the response I received from Annalen Der Physics in return.
Again I repeat one of the responses.

This is it. This is as I received the e-mail and this is what I received.

Dear Dr. Schutte,

You submitted an article of 15 pages to the Annalen. The content of this paper doesn't constitute a theory in physics. With a lot of words and some simple algebraic relations, there is no way to "explain" the world of physics. You seem to be out of touch with modern developments. This is also shown by the fact that you don't quote any relevant literature.

I am sorry to say, but the Annalen is not able to publish your work.

I am sorry for having no better news for your.

Best regards,

Friedrich Hehl

Co-Editor Annalen der Physik (Berlin)

Friedrich W. Hehl, Inst. Theor. Physics

* University of Cologne, 50923 Koeln _____/\/_____ Germany

fon +49-221-470-4200 or -4306, fax -5159

hehl@thp.uni-koeln.de, http://www.thp.uni-koeln.de/gravitation

* Univ. of Missouri, Dept. Phys. & Astr., Columbia, MO, USA

Now I wish to address those in such remarkable high places where they are totally untouchable and where they can come to any conclusion and form any opinion and we, the lesser of God's Creation can never reach them and therefore they can do what God Almighty allows them to do as it pleases the untouchable, the beyond reach or approach known to us lesser mortals as the High and the Mighty. This is what the journal replied. They said the mathematics I use is too simple to repeat and words in as much as "a lot of words and some simple algebraic relations" was not enough to constitute a theory in physics. For God's sake I am explaining the four cosmic principles that form the building blocks of the Universe and in the next two chapters in this book I prove just that. However, I was informed by no less a person than Helen de Villiers Managing Editor of Struik Publishers / Nature that Nature Annihilates Newton is not worthy of a book and should be downgraded to be used as debating material in a journal of sorts.

Now to turn our attention to what Helen de Villiers Managing Editor of Struik Publishers / Nature suggested when she suggested I should only use such degenerated material and such an unimportant topic in a journal.

Where shall we start with the debate? Shall we start at the importance and the truth about the Titius Bode law? The debate has been going on for many a century as to why the Titius Bode law is in place and that the numerical order goes beyond any human understanding. To degenerate the Titius Bode, science at present questions if the law could ever be explained and if mathematics could ever be applied to prove the law.

To see these questions she should just have clicked onto Google, the search engine and use five minutes of her time before coming to a conclusion about a topic she knows nothing about but still see fit to form an opinion on the matter. This debate has been going on since the law was discovered in 1766 and since then every person in science put the blame for not understanding the law on nature as to cover-up the hoax Newton started and science still maintains.

Shall we in the debate again for the millionth time tell everybody this law goes beyond proof where I in fact prove this law and I am the first person in three hundred years to achieve this breakthrough? Must the debate as inconsequential and indemnified as it is rage on just because our most superior intellectually blessed Helen de Villiers Managing Editor of Struik Publishers / Nature do not understand about my explaining of the Titius Bode law?

I should think not because this debate has been on the table in journals of every nature but this is the first time there is a conclusion brought to the table. Then lets get down to debating the Titius Bode law as it stands in nature and as nature applies the law.

The law relates the mean distances of the planets from the sun to a simple mathematic progression of numbers. Now find the three and the four that produces the seven.
To find the mean distances of the planets, beginning with the following simple sequence of numbers:

0 3 6 12 24 48 96 192 384

With the exception of the first two, the others are simple twice the value of the preceding number.
Add 4 to each number:

4 7 10 16 28 52 100 196 388

Then divide by 10:

0.4 0.7 1.0 1.6 2.8 5.2 10.0 19.6 38.8

Lets get to proving the Titius Bode law by having a debate without a journal but as if we do use a journal. The law says in the beginning and at the start the line-up starts with 3. That is a given. That is beyond any form of discussion in a journal or otherwise and this part is a conclusion made by nature no less.

The sun holds the first inside planet at a certain position

Sun   Mercury **57.9 x 10^6 km**

The law states that in the line up with Mercury holding a distance of 57.9 x 10^6 km then this value constitutes to a relevancy of 3. This is most important to understand and what there is to discuss in any journal about this I would not know for I am not as smart as our super-informed Helen de Villiers Managing Editor of Struik Publishers / Nature. I have no idea what it is that I am supposed to discuss about this given fact. Then we draw our attention to the next given fact.

Let us again look at what the numerical sequence of the Titius Bode law stipulates. With the exception of the first two, the others are simply twice the value of the preceding number.
Add 4 to each number:

4 7 10 16 28 52 100 196 388

We have three or the number of the planet in example (6/2= 3),(12/4 =3),(24/8=3) etc. and this indicates the axis is 3 while the circle then goes 4 to imply the value of the circle in relation to the axis. Can we use this in a debate as Helen de Villiers Managing Editor of Struik Publishers / Nature suggested. I don't truly think such a discussion merits the involvement of a science journal. Placing this in a science journal is a bit like going to war under the pretext that there might be nuclear arms while all evidence is one lorry wreck somewhere in the Arabian Desert. Maybe visiting a primary School close to her will do just fine and be more appropriate and this is the reason why. Someone is this always pretending to be better than the rest while hiding and compromising for a complete lack of understanding what the issue warrants. That plagues and haunts science and forms a veneer of covering deception as the norm is in science at present.

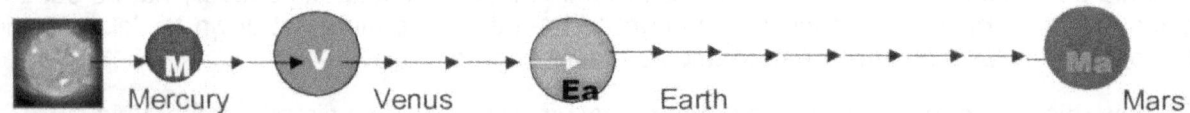

Mercury   Venus   Earth   Mars

If Mercury is holding a position of 3 then Venus will hold a position of 4 and this too goes without discussion because this is a given fact. This is the way the ratio forms. The next part might be in dispute but that can only be in relation to a person's mentality. If there are 3 marbles on the table and you wish to pace another 4 on the table, the total must be 7. I have no idea where to take this to generate a discussion and I do agree that should I wish to put this in any journal it would be fit to send me a reply like "With a lot of words and some simple algebraic relations." Therefore I must suggest that Helen de Villiers Managing Editor of Struik Publishers / Nature takes this argument to her nearest primary school. There she will have to go to the six-year old Grade 1 class to find little people that will discuss this matter with the proper enthusiasm where they would feel challenged and exited to argue about this matter. Don't go to the ten year old because those in Grade 5 will feel bored about the challenge to argue this fact. For that we do not need a science journal but just a Primary school and one as close as she can find.

Then the next given fact is that in the event of Mercury being 4 and Venus 7 in the line up, earth will slot in at a relevancy of 10. That is a given but she might consider to argue that fact, that 7 plus 3 are ten. If that is the case just take this argument to the same primary school and the same grade 1 class that helped her with the 3+4=7 conundrum and the same class will help. They will understand her position because they being the six year olds and in Grade 1 will still remember that not long ago they too did not know about 3+4=7 and 7+3=10 and they will sympathise with her predicament. However, to rush off to a science journal will seem rather very excessive and extreme I would think.

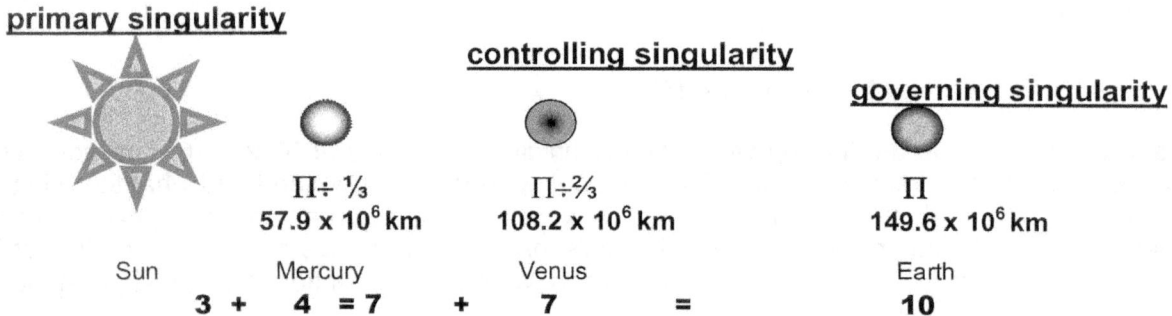

primary singularity

controlling singularity

governing singularity

	$\Pi \div \frac{1}{3}$	$\Pi \div \frac{2}{3}$	Π
	57.9×10^6 km	108.2×10^6 km	149.6×10^6 km
Sun	Mercury	Venus	Earth
3 +	4 = 7 +	7 =	10

This shows that in the manner that the solar system forms and in that the way the entire Universe forms there are relevancies or ratios formed by 3, 4, 7 and 10 and that places all numbers in this code. That is how time builds space and with Venus forming space we have the sun holding 3 Mercury holds 4 on the outside of Venus or the space that Venus forms we have 10. What is there to discuss?

Then if the distance between say the inner planet and the sun forms a value of 7 in relevancy then it would be appropriate I guess to surmise that the distance between the outer planet and the inner planet to form the same value on the principle it holds the same distance that just doubles from the sun to the inner planet and then again to the second or the outer planet. Therefore the circle that the inside planet forms is by rotating at 7° and the planet to the outside thereof also rotates by 7° which then puts both at a ratio or a relevance of 7°.

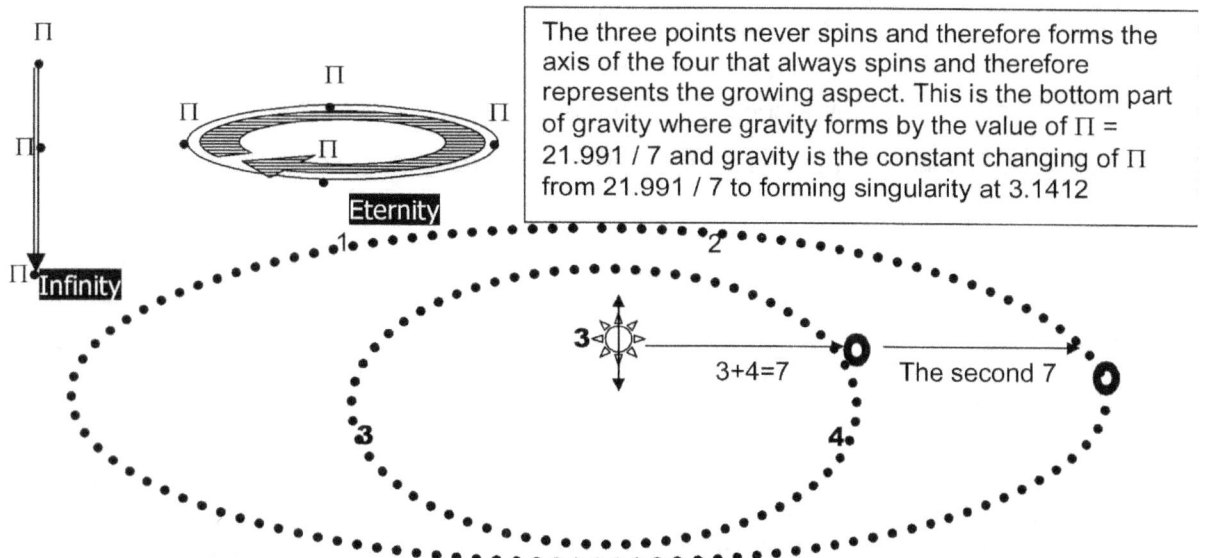

This might become complicated. The next visit to the Primary School will have to involve the 10 to 11 year olds because now the discussion clearly surpasses the six year olds. There is only one issue that bothers me and that is where did our distinguished lady and the most important person Helen de Villiers Managing Editor of Struik Publishers / Nature fall out of the bus. Was it at our visit to the six year olds or could she cope with 3+4 and now she got stuck at discussing 7+7 = 14. I am not sure because she never advised me on this issue.

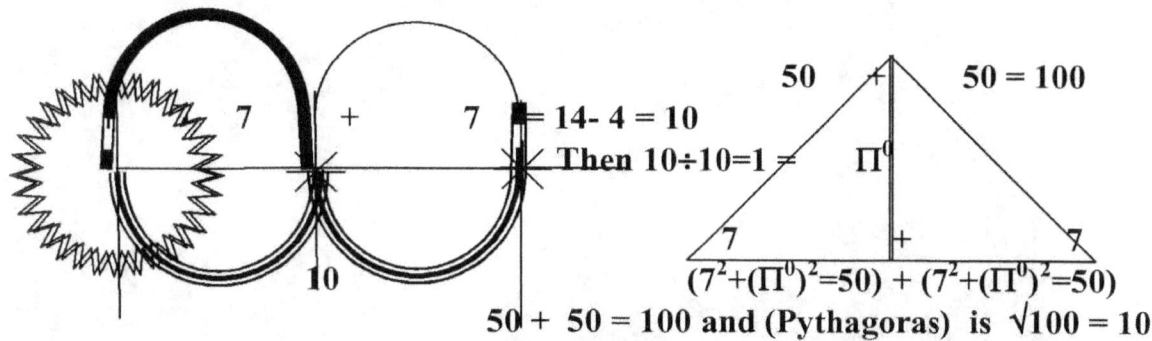

Let's leave the horrible mathematics that the twelve and thirteen year old boys and girls do where we bring in the law of Pythagoras and triangles and complex mathematics. Let's follow a simpler route.

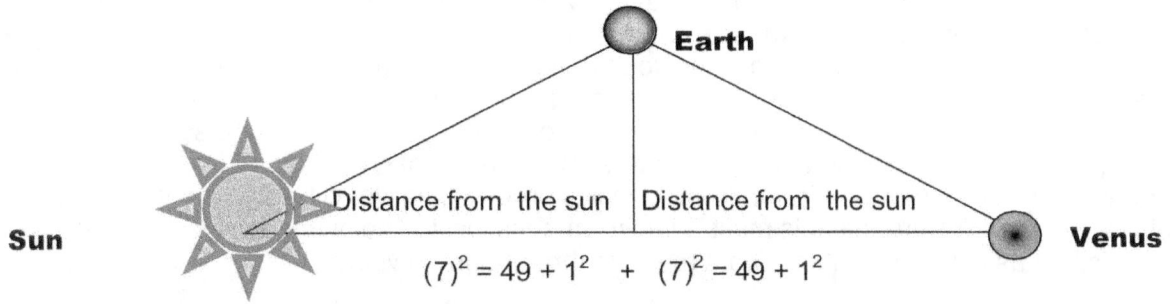

We have established 7+7=14 without involving a science journal. We have seen that the ratio of the Titius Bode law provides a ratio between 10 and 7 and it is 7, the inner circle, that divides into 10 the earth, which forms the outer circle. This now gets complicated and it demands that we visit the 11 to 12 year olds in the primary School.

The one ratio comes about as 7 duplicates a value which is 7+7=14. Then we have the second ratio of $\dfrac{10}{7} = 1.4285$ but let's not complicate the equation result and keep it at 1.42. Then to find

the ratio between the inner and immediate outer planet we divide the $\dfrac{14}{1.42} = 9.86$ and 9.86 is

the value of Π^2. Therefore the ratio in movement is (a value going square) Π and from that we can conclude that the space forming between planets is the value of Π going square and

numerically this is $\Pi = \dfrac{21.991}{7}$ x $\Pi = \dfrac{3.1416}{\Pi^0} = \Pi^2$. The result of this is that the discovering of

the Titius Bode law puts the development of space-time precisely in terms of $\Pi^0 \Pi \Pi^2$. All of this we achieved without involving a journal and this is so simple that primary scholars would be able to handle the complexity of the mathematics. If the debate is about whether Newton's idea of mass is correct and how the Universe works by forming the Titius Bode law we should again not involve a Journal on science but just look at the mass distribution of planets.

The one tells me we have to discuss what I prove by means of a journal and the other editor of a journal tells me the mathematics I use is too poor to serve any purpose notwithstanding that this is what nature uses and this is how nature forms space-time. ...And all around, the lot try to be clever and pose with a lot of dignity while they have no clue what the subject presents.

The Sun and Nine Planets   Copyright © Calvin J. Hamilton

See how totally random the distribution of planetary mass is and how not equitable this line up is mathematically. There is no ratio that can apply. There is no sequence that can predict the one mass value form the rest and there is no formulation that can prove the mass distribution in sequence and yet after three hundred years of discussing gravity and the way Newtonian science detect gravity in journals of all sorts, not once did anyone in any discussion in any journal question the viability of mass playing any role in the forming process of the solar system whereby mass forms gravity. Yet, it is the conclusion of Helen de Villiers Managing Editor of Struik Publishers / Nature that this must again form a discussion point in some journal just to be again falsified as it was done throughout the past three hundred years.

Dear Helen de Villiers Managing Editor of Struik Publishers / Nature keep it up and then with people like you in charge of science books especially on nature we will have another three hundred years where scientists have to bullshit everybody while holding up a veneer of self-importance and absolute enlightenment about their having an informed opinion on matters.

Now I wish to present another reply from a person that is truly informed and while this person is vastly educated on the matter, this person leaves an opinion out of the reply. This is a truly educated person who forms a well-schooled opinion about not having an opinion before studying the book. Please look and see the difference in approach and the self-confidence this one has.

Scott Hoffman scott@foliolit.com

Nature Annihilating Newton

First of all, I'd like to start with an apology for not being able to offer individual feedback as a response to your query. I really wish I could offer each person who takes the time and the emotional energy to send me a query a personalized response, but if I did, I wouldn't have the time to properly represent my current clients. As you've probably guessed by this point, we've determined that, unfortunately, we don't believe Folio is the right agency to represent your work right now.

If you queried about a nonfiction project, the most likely reason we decided not to take it on was because as an author, your platform isn't big enough yet. (If you'd like to learn more about platform, why it's so important, and what you can do to build yours, the best place to start is to watch the video on Folio's website here: http://foliolit.com/nytimesbestseller/) Once you've watched that, click on the link below the video for more free training on platform building from Folio client Brendon Burchard.

As either a novelist or a nonfiction author, though, if you're getting form letter rejects from most or all of the agents you're querying (or if they're just not responding at all) there's actually some good news: chances are the problem is with your query letter itself, and not the underlying book. For some tips on how to improve your query letter, click here:
http://foliolit.com/submissions/basic-information-on-query-letters/

Again, though, I'd like to thank you so much for the opportunity to consider representing your book. Stories can change the world, and every one of us has an important one to tell. I wish you all the best in finding the right home for your story.
See you on the bestseller list,
Scott
P.S. If you've received a duplicate of this email, I apologize. Our query system is still relatively new, and we're still working out all of the kinks.

Such is a dignified response that proves integrity without trying to hide a shortfall in background knowledge resulting in a lack of information to further competence. The one is driven by arrogance and the other is motivated by knowledge portraying self worth. It is clear that the one tries to be clever and come across as superior but fails and the other by modesty shows what superior cleverness is and without pretence succeeds because of a well balanced self opinion. In the world of science bullshitting became such a crime that committing the crime of deceiving and being superior by pretence about posture that it became an art.

Explaining the Titius Bode law is very simple to understand and I just can't see why any discussion in a science journal would be required. However, simple as it may be if you wish to understand the next chapter and see the way the Universe formed from a point holding not even as much as nothing then understanding the Titius Bode law is an absolute requirement. The Titius Bode law started the Universe by starting mathematics in a numerical order.

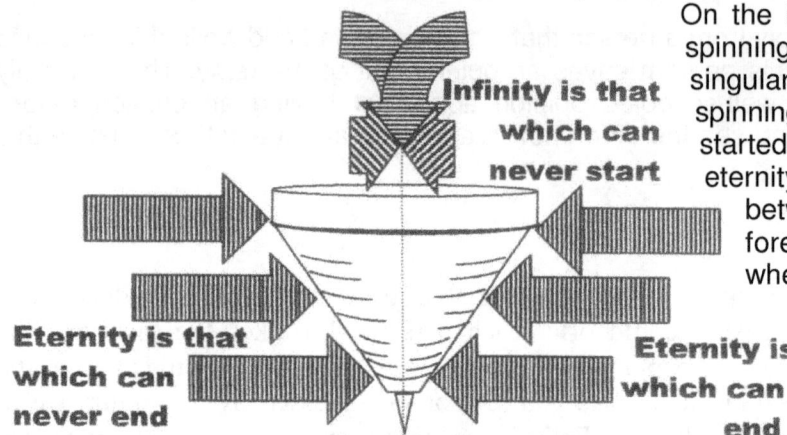

Infinity is that which can never start

Eternity is that which can never end

Eternity is that which can never end

Singularity Π⁰

On the inside in the centre of anything spinning we have infinity holding singularity. On the outside of anything spinning we have eternity and Creation started when infinity parted from eternity to place a Universe in-between. The Universe goes on forever because it can only end when infinity absorbs eternity once again as it happens in a Black Hole. When infinity and eternity meet, time that can never start unite with time that can never end and the Universe in between ends by time uniting. Everything that is can only be because it moves and everything that moves goes in circle following a straight line that ends in a circle once more. In between infinity and eternity there is the axis holding three positions and there is the circle holding four positions and the entire circle expanding or collapsing does so by seven. By seven going square (movement on both sides of the circle of material) this establishes a value of ten. Once any person can understand this then also such a person can understand how Creation came about as a result of the interaction resulting from time in infinity relating to time in eternity.

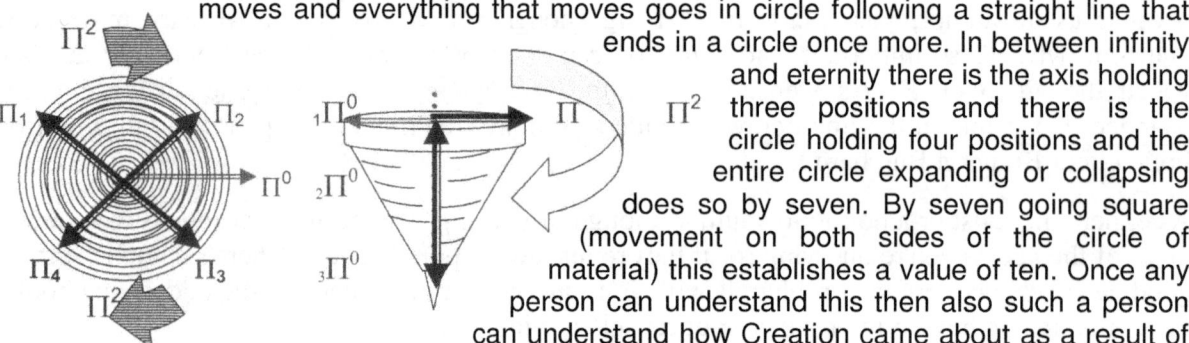

Infinity never moving forms three and eternity circling forever

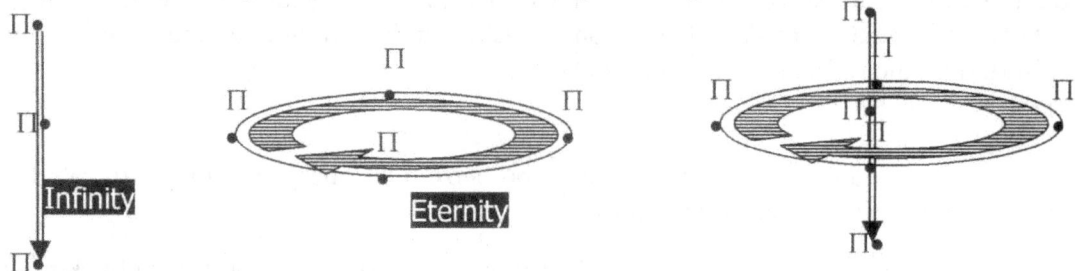

Infinity

Eternity

forms four and together movement is seven and this then puts space by the law of Pythagoras at $(7^2 + 1^2) + (7^2 + 1^2) = 100^{1/2} = 10$ This is the Titius Bode law.

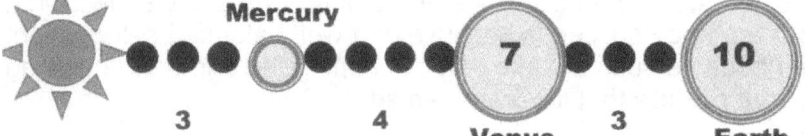

Mercury   3   4   7   Venus   3   10   Earth

This realisation makes the understanding of the Titius Bode law completely pertinent to the understanding of science because this is where science begins by forming mathematics, which form space as time moves on in a circle going straight. The line is always 3 and the circle is always 3 and the line shifting is another three to be added to the 7. This is how Creation started when it started nothing at al but I am getting to that concept in due time.

Part 3:

Setting guidelines as a Concept of Π Forming Gravity

You are about to encounter future physics that was never yet presented. Therefore because it is new please see this chapter as an introduction where in this book I am going to elaborate on everything I say and some paragraphs will be repeated as they are because they have been taken from the text directly as to abbreviate the concept I introduce.

The idea serving the terminology of mass is something humans improvised to serve a purpose when trying to do calculations of whatever nature in physics. The reality is that mass only apply as a creation of man in a quest to find how to calculate and understand how physics work on earth but in the heavens it has no place. In physics applying on earth it is a handy tool man created but in terms of nature used in the cosmos it has no function or place. In short, in the cosmos it has no purpose whatsoever and nature never intended to implement even an idea such as mass. You may use mass as a stick to beat out an answer to suit the purpose of physics in terms of a need on earth but to take mass into outer space is to cheat and present physics as a hoax. Now you think that I have lost whatever sense I still had and you must hide your children because you have come across someone that is dangerously insane and has a murderous tendency.

In the event of any readers who may have questions concerning more facts as it is presented in this book, please feel free to contact me, PEET SCHUTTE. All information divulged came about through independent self-study during the past thirty-two years or so. I have to warn the readers that the topics are showing a very new approach with no quick answers abstaining from proof or holding just a few lines and the information is new in nature but not hard to grasp.

To all those practising physicists claiming to have the mathematical abilities to recalculate the entirety of the Universe and the physical insight to reinvent the cosmos by manufacturing space whirls or other cosmic gismos and explain away occurrences within the Universe such as supernovas' blowing to bits as gravity going "mad", things they clearly have no insight of for them I have bad news. The cosmos does not work with elaborate breathtaking mathematics. It works with singularity, which is one.

Then it expands to Π. It work with numbers not exceeding the value of Π and the quantity of how Π forms by Pythagoras. According to Einstein the Universe is in singularity and in singularity every number of values cannot exceed one. Therefore when you apply your calculations of breathtaking standards you are playing a mind game of self-important soothsaying that is worthless but for your image you create in to your fantasy. There is no mass and therefore there is no force.

There is no energy you can calculate and then fool the world with your marvellous life you lead in your fool's paradise. Prove there is mass by placing the solar system in accordance to mass. Show that the biggest planets are next to the sun and the smallest are at the very far ends and the debris spinning in a ring of their own is only imaginary figments of imagination because those fragments have not the mass to spin in those positions. Show that the cosmos does indeed position planets according to their mass they have and spin accordingly.

If I show that in mathematics according to science the planets' positions are in accordance with Symbolically being: $P^2 \propto a^3$ and therefore $a^3 = P^2$ in position of P and therefore a^3 / (P^2P) I am so far off the truth when I have to accept nature holds the truth ultimately and not Newton. This is taken from the idea that "**Kepler said**", which is totally fabricated by Newton when he said that $\mathbf{a^3 = T^2}$ where Kepler in fact said $\mathbf{a^3 = T^2 k}$ and this is a big difference because $\mathbf{a^3 = T^2 k}$ is the same as $E = mC^2$.

Then, of course, no one could ever imagine Newton being wrong, so to rationalize Newton's baseless unfounded failure of a theory was created where this is the reason behind my new cosmic concept. $\mathbf{a^3 = T^2}$ is not a scientific value forming science as a base. To think of $\mathbf{a^3 = T^2}$ is to give equal meaning to the person looking into a mirror and the image. Look at the charts Kepler produced and numbers form $\mathbf{a^3 = T^2 k.}$

In this equation **k** has a defined value where **k** holds a value. The Newton ratio put forward of $a^3 = T^2$ can't have any substance. Kepler said $a^3 = T^2k$ and with figures and that nullifies Newton's presumed ratio of $a^3=T^2$.

PLANET	PERIOD (Years) (T)	MOVEMENT (T^2)	DISTANCE	SPACE (a^3)	RATIO k
Mercury	0.241	0.058	0.39	0.059	0.983
Venus	0.615	0.378	0.728	0.381	0.992
Earth	1.000	1.000	1.000	1.000	1.000
Mars	1.881	3.54	1.524	3.54	1.000
Jupiter	11.86	140.66	5.20	140.6	1.000
Saturn	29.46	867.9	9.54	868.25	0.999
Uranus	84.008	7069	19.19	7067	1.000
Neptune	164.8	27159	30.07	27189	0.999
Pluto	248.4	61703	39.46	61443	1.004

KEPLER'S LAW OF PERIODS FOR THE SOLAR SYSTEM		
PLANET	SEMIMAJOR AXIS $a\,(10^{10}m)$	PERIOD T (y)	T^2/a^3 $(10^{-34}\,y^2/m^3)$
Mercury	5.79	0.241	$k^{-1} = 2.99$
Venus	10.8	0.615	$k^{-1} = 3.00$
Earth	15.0	1.00	$k^{-1} = 2.96$
Mars	22.8	1.88	$k^{-1} = 2.98$
Jupiter	77.8	11.9	$k^{-1} = 3.01$
Saturn	143	29.5	$k^{-1} = 2.98$
Uranus	287	84.0	$k^{-1} = 2.98$
Neptune	450	165	$k^{-1} = 2.99$
Pluto	590	248	$k^{-1} = 2.99$

With a ratio that Kepler calculated using the formula T^2/a^3 it is very much clear that the formula must read as follows $k^{-1} = T^2/a^3$ and the negative value of **k** proves that space moves in recline towards the sun. The material maintains orbit. The sun contracts space from the outposts of the sun's influence field. These are Kepler's actual numbers before Newton got hold of those numbers and altered it into something completely fraudulent.
Science at present accepts that $a^3=T^2$.

Then Kepler had a specific column for **k** where if you multiply T^2 with **k** it then is equal to a^3. When a person would refer to the columns Kepler said $a^3=T^2k$ but then if you divide T^2 with a^3 it has a value of say 2.96 as it is in the case of the earth **k** has a valid numerical answer that is not 0. Mathematics teaches us that if you have numbers telling you that $a^3=T^2k$ then if division applies those numbers will give you a mathematical result of $T^2 \div a^3 = k^{-1}$. It should then dawn on anyone with a mathematical background that $a^3=T^2$ must clearly be wrong as he suggests in his statement of $a^3=T^2$. Newton couldn't imagine he might have been wrong, so he rationalized his failure to be accepted by creating a theory kept alive by a conspiracy, which is kept alive by those Newtonians practising science to save science.

This that I have written in this what you still must read puts all that was written before, and all of that which you had read, all of that into the past. This that is written in this what you still have to read is the future in science and whether you, as the reader where by that you also become the witness, decide to put this in your present, is up to you to choose. The reader must choose where to stand on these issues that is written in this what you now have to read, because this that is written in these books is the future. When the future will become the present is solely the choice of the reader. Where the reader chooses to not choose what this represents then the choice will condemn the person choosing, as being one of those that will remain with the past.

The choice will place the person choosing not to choose this as the future, to become part of those who will be forgotten as the past gone to darkness and forming part of those that was not worth remembering. Such a choice on the future will put the person choosing not to choose this as the present, part of the past that eventually will become the part that was not worth recollecting. The future forming the present only remembers the past when the past is worth to be used in the present as the present with a value to be remembered by. The future never carries failures in the past through to the present by remembering it in the future there must be a worth to form a value worthwhile as a memory within the past.

Recognising this what is written in this what you still have to read, such insight to choose well forms the future and it then will place the person as a part of the future by removing other failures in the past from the then present and placing the future in the person's future. Realising what this represents and holds replaces the worthless of the present within the present and then replacing this with what was, is the method of making what is written in these books becomes part of the reader's present. These books are the future and no suppressing of the powerful and the mighty or thought of as wise can prevent these books from filling the future. The mighty wise can push their present into forming the past by clinging onto the worthless they represent as worthwhile that will become the past when these books becomes the present. However, that choice will doom them into the past along with all other things and thoughts not worth the burden to take into the present and onto the future from the past. Their adopting the worthless and not adapting to the truth moves them to the past only worth to be forgotten as the worthless part of the past. The choice is theirs to make.

The Universe consists of gravity that forms by the working of the four phenomena never mentioned.

The Titius Bode law has been around for centuries and with all the mathematical splendour available there for all to use, all the brilliant mathematicians could never come close to show any ability to any understanding of this very important phenomena. They could mathematically equate the formula the sequence applying as the formula, but then after that their superior human intellect dries up.

The Roche limit has been around for centuries and with all the mathematical splendour available to apply in order to fathom concepts behind this phenomenon, still with all the computing ability of a machine all those physicists with all the mathematical superiority could not touch any understanding about the concept forming the background. Yet hen using the truth in physics the answer is simple.

The Lagrangian points have been known to science for centuries and with all the mathematical splendour available not one calculation could ever explain why this event is taking place.

The Coanda effect has powered turbine engines and aeroplanes in flight for almost a century and with all the mathematical splendour available to design the most terrific aircraft, not one engineer could mathematically compute one fact to show understanding why this takes place. How sad it is that those claiming of much superior intellect in physics remain just no more than having computing power.

That does not say much for the bountiful prestige that mathematician's claim as their lawful bragging rights in areas where true human intellect is called on. Is it not high time to begin to admit you are playing the game of fools with you arrogance about your achievements using mathematics when designing space whirls and travelling to galactica while not even understanding what movement asks for. You do not even understand the neutron and the neutron is compressing density increasing, which is what gravity is, which is what time is, which is what all movement is…that is why the neutron has no mass because mass is the principle coming about where independent movement ends.

You're mathematics could not get you any closer than playing games in a fairy tale Universe using misguided presumptions about mass forming gravity and living the Universal farce which Newton created because that fairy land is what all the Kings clever heroes and all the King's splendid wise could never prove in hundreds of years. If you feel superior as a scientist practising physics on the highest level having a gloating hail of superior mental capability covering you like an aura, then I have very saddening news for you.

If you have the ability to compute and calculate at the highest level, then look at your computer and see one that machine has abilities as a machine which is equal to you, but it's a manmade machine. Stop playing games by creating fairy worlds making up fairy tales about fairies and little people, mass that can create forces, four of them no less, and come and join the rest of us living in reality that does not need to compute forces to be able to not understand what it is that you compute, but to use human intelligence and in that way to understand what only human intellect could ever understand. Then what in the present is not worth carrying into the future as the past being worthwhile?

If according to the numbers $a^3 = T^2 k$ then a^3 cannot be equal to T^2 because Newton calculated that $k = 0$.

If $a^3 = T^2 k$ then to suggest than $0 \neq \dfrac{a^3}{T^2}$ as Newton did is obviously impossible.

Then $P^2 \propto a^3$ is non-existing and then there is no $a^3 = P^2$.

Also then according to Kepler's calculations $T^2 \div a^3 = k^{-1}$ and that shows that the space, not the planets or material but the space is contracting towards the sun.

Please show how mass by $4\pi^2 a^3 = P^2 G(M + m)$ can produce Planet positions. It is hogwash.

In accordance with reality as reality applies in the solar system there is no big or small because big or small solid or gas massive or fragmented, all the planets are the very same, just as everything falling is the very same irrespective of structure or size. All the planets float in a bowl of liquid that renders the entire lot big or small mass notwithstanding, everyone equal.

The resulting sequence is very close to the distribution of mean distances of the planets from the Sun:

PLANET	Distance from Sun (million km)	Mass of Planet (x10 ^22 kg)
Mercury	57,910	33.0
Venus	108,200	487
Earth	149,600	598
Mars	227,940	64.2
Jupiter	778,330	190,000
Saturn	1,426,940	56,900
Uranus	2,870,990	8,690
Neptune	4,497,070	10,280

This shows more than absolutely that mainstream science are fabricating facts in order to create a make-believe Universe where they can play with their mathematics and act very clever while they are baseless even in the part where physics starts. If mass did form $4\pi^2 a^3 = P^2 G(M + m)$ we must have the biggest planets such as Jupiter closest to the sun and the smallest such as Mercury and Mars way to the outside and very far away from the sun. The tendency of mainstream science has always been to tell nature what nature should be as it then fits the likeness of Newton and Newtonian science. This must end and it could not be soon enough.

This shows that the basis of physics is flawed beyond recovery and a new line of thought is needed most urgently. I have the remedy ready to apply but with that much money going lost and that many accepted theses becoming science fiction the resisting of this new concept is a matter of fact. The truth is that Kepler showed that all movement in the Universe consists of two equal important but not equal valued forms of movement and the one movement cannot be without the other also taking place. The movement everything adheres two and

that forms space $k^0 = \dfrac{a^3}{kT^2}$ is $k = \dfrac{a^3}{T^2}$ and $T^2 = \dfrac{a^3}{k}$.

Another point of an immeasurable many that I can mention is the way that comets and indeed all planets perform. Let's confine the argument to the behaviour of comets. It is said

that mass pulls mass as Newton suggested in the accepted formula $F = G\,\dfrac{M_1 M_2}{r^2}$. This

would find truth when the comet is coming towards the sun. It would seem to carry truth as the comet heads towards the sun. The comet is coming towards the sun but as it comes

closer it seems to divert from the head on collision as the formula $F = G\,\dfrac{M_1 M_2}{r^2}$ suggests.

As it comes closer towards the sun it heads for a designated point way off the line that should head the comet straight towards the centre of the sun and then it follows a route that curve around the sun. That avoids the collision with the centre of the sun, as the formula

$F = G\,\dfrac{M_1 M_2}{r^2}$ would claim by a division of r^2 becoming 1 eventually. This is very much not

what happens when we put reality in comparison with the accepted norms mainstream science upholds.

In this the question arises and which mainstream science fails to answer if it is mass that pulls the comet towards the sun what is it that pushes the comet away from the sun where the comet after orbiting around the sun at close quarters fail to collide but then to disappear into the dark yonder of outer space. If mass "pulls" as Newtonian myth would suggest what then pushes the comet away from the sun into the darkness where the sun almost never shines? It is a definite ignoring of Newton and his laws such as on which physics is founded

by the formula $F = G \dfrac{M_1 M_2}{r^2}$ in the most as well as strongest possible form we can imagine.

The comet acts in defiance to Newton. Once more this shows how mainstream science dictates to nature what is true and what applies in physics instead of getting the information applying within nature from nature and then tries to figure out what is going on in nature. That is precisely what I did and I can show how the Titius Bode forms the solar system. I make this claim and I am the first person in the entirety of the history of the earth that can make such a claim.

For those that does not know it is the Titius Bode law that forms the solar system and not Newton's views on mass pulling mass, which I just showed never was true even for one second.

Bode's Law:

Planet	Mercury	Venus	Earth	Mars	Ceres	Jupiter	Saturn	Uranus	Neptune	Pluto
Bode's Law distance	4	7	10	16	28	52	100	196	-	388
Actual distance	3.9	7.2	10	15.2	28	52	95.4	191.8	300.7	394.6

The Titius Bode's law is a numerical sequence announced by J.E. Bode in 1772, which matches the distances from the Sun of the six planets then known. It is also known as the Titus-Bode law, as it was first pointed out by the German mathematician Johann Daniel Titius (1729-96) in 1766. It is formed from the sequence 0,3,6,12,24,48,96, and 192 by adding 4 to each number. The planets were seen to fit this sequence quite well – as did Uranus, discovered in 1781. However, Neptune and Pluto do not conform to the 'law'. Bode's Law stimulated the search for a planet orbiting between Mars and Jupiter that led to the discovery of the first asteroids. It is often said that the law has no theoretical basis, but it does show how orbital resonance can lead to commensurability.

The importance that becomes known is the sequence the Titius – Bode law saw in the number arrangement of 3; 6; 12; 24; 48; 96 etc. The incorrect application of the Titus Bode law lies in subtracting the figure of 3 from 10 leaving 7. The other way of reasoning is to add four each time to the firs value of three starting with 3 and so on. The true significance of the Titus-Bode law is that it points directly to a circular growth of 7 stages.

The 7 relating to 10 is a precise derogative of the Roche limit or the Roche limit is a precise derogative of the Titius Bode principle because he two systems interlink. This is how I mange to explain the Titius Bode law that is in the solar system by the ratio applying that really form the solar system in the way nature shows space growing by time. What you see on the next page was never been shown but on the other hand Physicist say this mathematics are too simple to apply as physics!

To find the mean distances of the planets, beginning with the following simple sequence of numbers:

0 3 6 12 24 48 96 192 384

With the exception of the first two, the others are simple twice the value of the preceding number.
Add 4 to each number:
4 7 10 16 28 52 100 196 388
Then divide by 10:
0.4 0.7 1.0 1.6 2.8 5.2 10.0 19.6 38.8

The resulting sequence is very close to the distribution of mean distances of the planets from the Sun:

Body	Actual distance (A.U.)	Bode's Law <A.U.)< td>
Mercury	0.39	0.4
Venus	0.72	0.7
Earth	1.00	1.0
Mars	1.52	1.6
Ceres	Many different places	2.8
Jupiter	5.20	5.2
Saturn	9.54	10.0
Uranus	19.19	19.6

The Titius Bode law proves that in the Universe laws apply that positions objects in terms of other rules that mass. That means the Newtonians hides their lack of understanding behind mass that they invent.

The Newtonians gave the Titius Bode law a formula and that explains the lot. To they're under achieving standards that is very satisfactory. Now it is written in mathematics then what more do we need to know. The fact that the distance that Mercury has from the sun doubles by that which Venus has from the sun is completely ignored. In cosmic reality mass plays no part. Then again the distance that Venus has from the sun is doubled by that which the earth has. This clearly has nothing to do with the size or mass of the planets. Explaining that part is completely ignored. Then again the distance that the earth has from the sun is doubled by that which Venus has and inexplicably this forms the layout of all planets in the solar system. Where does Newton's idea of mass fit into what truly applies in outer space. Moreover, why does science never mention this? This is my formulated explanation about how the Titius Bode law forms. The numbers we need to find the key to the mystery of the Titius Bode law is 3, 4, 7, and 10. The 3 is the sun stationary, the four is the inner planet orbit holding a value of 10. I explain this in detail.

However there are three more phenomena that form gravity beside the Titius Bode law.
These are the Roche limit or the Roche Lobe,

The Lagrangian points which secures the rings around the planets in a secure and defined manner,

The Coanda effect that combines the interacting workings of the three above and is known as the Coanda effect all of the above-unmentioned phenomena that form the Universe in every sense is the result of the **Absolute Relevancy of Singularity.**

My very first aim was to find the centre of the Universe where I then could locate singularity.

I had to find the meaning of gravity. Everything performing within gravity spins around while moving straight ahead. That Kepler said almost five hundred years ago but mainstream science has yet to catch up with what Kepler proved. This means everything acting on gravity spins. Spinning is going around according to a circle, which is the form of all objects in the great entirety. By turning as a result of circling the main commitment behind gravity then and therefore must be connected to Π.

I realised I had to find the manner in which Π forms and then I would locate the process whereby gravity forms space. This was one part. If movement was an interaction between a circle and a straight line then a triangle was also a realistic part of singularity as part gravity. Therefore I had to find the manner in which Pythagoras played a union between the half circle and the straight line acting as a unit. Therefore in the entire process I had to look for Pythagoras and I am wistfully aware that such minor mathematics was far below the wise physicist that is so unbelievably clever with mathematics. That is why they never got the answers while I did. I didn't try to prove my cleverness but I tried to prove how simple nature was.

<u>**Albert Einstein**</u> formulated a concept in 1905 he called **The Special Theory of Relativity** and in 1915 he introduced his assessment on the principle of <u>**The General Theory on Relativity**</u>. I do not quite agree with his findings. What I discovered goes far beyond the discovery that Albert Einstein formulated. I have discovered that the Universe is not employing a general relevance of singularity, but throughout the Universe there is a fixed overall state of ***The Absolute Relevancy of Singularity*** that is not only **controlling the Universe**, but is what the Universe **constitutes of**...**it forms the Universe**...**it is the Universe**. However, notwithstanding the magnitude in significance ***The Absolute Relevancy of Singularity*** presents as a breakthrough in science, the influential members of the scientific establishment will not recognise my theory on **The** <u>**Absolute Relevancy of Singularity**</u>. Past encounters taught me that mainstream science in physics will again ignore my ideas that I formulated as ***The Absolute Relevancy of Singularity*** and I don't believe it would even be read, will be seriously considered and much less be accepted by those with the authority to change physics principles. I think the theory I introduce would never be accepted during my lifetime because science is fixated on Newtonian ideas, which makes them bent on believing in the outrageously marvellous, and the unexplainable magical powers with gravity working by mass supplying a pulling power, which is a fact never proven and accepted only on Newton's word and Newtonian cultural bias, although they claim to only use proven facts.

What I ask of readers is to beforehand forfeit the culture of Newtonian bias when reading this by paying attention to what I say and not about the degree in which I stray from mainstream science's thinking.

This way the exercise will present many new ideas and explaining my new concept will become clear. There is so much to benefit from. Science has no idea what a Black Hole is while I can prove what a Black Hole is. I formulate mathematically what **"the sound barrier"** is. I prove what gravity is. By using the four cosmic phenomena, which is what the cosmos uses to form gravity, I show what **"the sound barrier"** is and I go much further than that. I show that gravity forms using the **Roche limit**, the **Lagrangian system**, the **Titius Bode law** and the **Coanda effect**. I uncover these principles by placing Π within the formulating of gravity and when using Π I bring clarity to the misunderstood cosmic principles. The list of the unknowns I can then explain is almost endless.

Gravity forms by movement that establishes singularity initiating a circle in using Π.

I show why gravity is there, how gravity forms and what role stars play in forming gravity. There is no difference between how gravity and electricity forms and that I prove mathematically by decoding the cosmos. I prove mathematically when atoms spin they establish Π that forms the Universe. Whatever forms gravity has to link closely to Π since everything that has anything to do with gravity forms a circle that is Π by the value of the square radius. If mass has anything to do with generating gravity, then mass has to apply Π or otherwise mass has nothing to do with the forming of gravity. Everything using gravity forms a circle of sorts, which forms the curvature of space-time, which is Π and which curves light. The way the planets orbit the Sun and how stars spin has all to do with Π. In spinning in a circle, Π forms gravity as a centrifugal force that condenses space.

I researched the work of Kepler and found science doesn't even recognise his work while it is his formula that forms the basis of all physics. Everyone thinks that Kepler found planets rotating, with Newton being able to explain Kepler, which makes everyone more concerned about how Newton saw Kepler's work. The formula used in physics as a principle is $F=mV^2$ which should be $F^3=mV^2$. $F^3=mV^2$ is replicating Kepler's formula in detail as $a^3=T^2k$. By using Kepler's formula we have $F^3=mV^2$ that is a precise repeat of $a^3=T^2k$.

The duplication is so obvious that we have (F^3 becoming a^3) while (m is k) and (V^2 is T^2). Einstein also only duplicated Kepler's formula by putting $E=mC^2$, which also should read $E^3=mC^2$. Again that is precisely Kepler's formula $a^3=T^2k$. (E^3 is a^3), (m is k) and (C^2 is T^2). In $E^3=mC^2$ Einstein mimicked $a^3=T^2k$, Kepler's formula. (E^3 is F^3 is a^3), (m is k) and (C^2 is V^2 is T^2). So what is so brilliant about Einstein's formula if Kepler had it centuries before? $E^3=mC^2$ is $F^3=mV^2$ which is $a^3=T^2k$.

Newton corrupted the formula when he added $4\Pi^2$ to the formula and removed k that Kepler introduced while $a^3=T^2k$ Newton ignored. Newton changed $a^3=T^2k$ by using the symbols G $(m + m_p)$ to replace k and then declared $a^3 = T^2$. I still wish to see the proof confirming Newton's changes as being correct notwithstanding that everyone thinks physics is entirely based on this conception.

Whether the formula used is $F^3=mV^2$ or is $E^3=mC^2$, it still remains duplicating what Kepler introduced as $a^3=T^2k$. So I changed it back to Kepler's version of $a^3=T^2k$ as to better the understanding of the foundation of astrophysics and mainstream physics. The entirety of physics is not based on Newton. It uses Kepler's findings to a precise duplication while science does not even recognise Kepler.

Giving Kepler the credit due, the entire Universe becomes completely understandable...but then for my audacity to show mistakes in physics I am ignored flat! All I ever ask is prove the truthfulness of $G(Mxm)\div r^2$ because it is $F^3=mV^2$ that forms the basis of physics and that accuracy comes from Kepler's view of $a^3=T^2k$ that became Einstein's $E^3=mC^2$.

By re-implementing Kepler's full formula $a^3=T^2k$ and using Π I was able to prove what I discovered as follows:
  1) The **location, the position** and **the value** of **singularity** as a factor forming space-time
  2) Finding **space-time** by dissecting Kepler's formula in relation to **valuing singularity**
  3) Finding space-time, **proving space-time** and **aligning space-time** with **gravity**
  4) The **working principals** behind and **manifesting of gravity** as a cosmic occurrence.
  5) The **Roche limit** and explaining the resulting of a law coming about from singularity.
  6) The **Lagrangian system**, how and why that becomes the building form of the Universe.
  7) The **Titius Bode law** and I show mathematically how gravity comes about from that
  8) The **Coanda effect** and the producing of gravity through reproducing space-time
  9) The **sound barrier** by proving it **is gravity** generated **by motion** in space becoming independent motion. This I conclude because Kepler said $a^3=T^2k$ but that could also be

$k=a^3/T^2$ and could be $k^{-1} = T^2/a^3$ and that is the Coanda effect. Mathematics says a sphere is $a^3 = 4/3 \ \Pi \ r^3$, **which is mathematically correct, however Kepler said the cosmos told him a cosmic sphere is $a^3 = k \ T^2$** where that puts the cosmos in completely different mathematical dynamics altogether. There are the two distinct possibilities of a^3, which Newton saw and which Kepler saw and both are most valid, but are altogether unequal. Between Newton's $a^3 = 4/3 \ \Pi \ r^3$ and Kepler's $a^3 = k \ T^2$ concepts there are one Universal difference. To calculate the dimensions of the sphere within a three-dimensional Universe the formula is $a^3 = 4/3 \ \Pi \ r^3$ while when working with singularity the measurement is $a^3 = k \ T^2$.

It is true that when measuring the sphere, Newton's method or formula $a^3 = 4/3 \ \Pi \ r^3$ is used in calculating, but **Kepler received his code of calculation $a^3 = T^2 \ k$ from a very high authority,** which **is none other than the Universe** and therefore Newton can't discarded **k**. Kepler saw singularity forming relevancies and Newton knew nothing about that. It is the duty of the cosmologist not to reject Kepler's findings, or as Newton did, try to transform it into something that Newton could understand, because it then strays from the original meaning…but science should dutifully search for the meaning as Kepler received the formula $a^3 = T^2 k$ from the cosmos. We can test any of the following symbolic values in the mathematical expression and also test the principal behind the expression in which Kepler stated them.

By such testing $a^3 = T^2 k$ repeatedly we find that the translations of Kepler's formula into English never required any corrections in translation because Kepler never presented it incorrectly. By taking the formula on face value it can change as follows: $a^3 = T^2 \ k$ can become $k = a^3 / T^2$ or become $k^{-1} = T^2/a^3$. When translating Kepler's mathematical expression into English we can see what Kepler said also could read as $k = a^3 / T^2$ where **k** is indicating one point from a centre point that is space a^3 relating to time T^2. From a centre comes space-time. The centre **k** brings space a^3 in ratio to time T^2, which is space a^3 / time $T^2 k$.

Reading this correctly can't bring any dispute…yet it does…and it's been doing it for centuries! Kepler said $a^3 = T^2 k$ and that correctly translates to a mathematical expression $k^0 = a^3 / T^2 k$ which in the English verbal statement translates that Kepler said that there is a **space a^3** which is **equal =** to the motion in **the time duration T^2** thereof between two specific points which holds a relation onto a centre k^0 where from there forms **a straight line k** that is centred on the spot where space begins from k^0 **that produces k** as well as producing the circle therefore that spot $k^0 = a^3 / T^2 k$ has hold k^0 at a value of having the least space. The line **k** is centred onto a spot where space begins specifically at k^0.

This point not only produces the line **k** coming from a point k^0 but represents also the space a^3 that forms the eventual circle by the rotation of T^2. Therefore from the centre holding k^0, k^0 leads to **k** that forms the revolving space a^3, which is rotating T^2 at a distance **k** where T^2 forms the outer limit of k^0. Mathematically $a^3 = T^2 \ k$ will also be $k^0 = a^3 / (T^2 k)$ because $k^0 = 1$. But $k^0 = 1$ also presents the single dimension where all factors are a product of one. If anyone can locate k^0 then also that person will find singularity. That is where gravity is because gravity is strongest where space is least. Then that suggests that gravity is strongest at k^0 because there space is least.

That is gravity because that is what keeps the orbiting objects in orbit but also that is what Newton completely missed when he changed Kepler's work. Newton failed to recognise gravity as the only ingredient in Kepler's formula. He admitted he missed this because he admitted he did not know what gravity is while Kepler explicitly showed what gravity is. Gravity is what keeps the orbiting objects in rotation while orbiting. $k = a^3 / T^2$ is **distance^1 = space ^{3}/ time^2** forming from a pivoting centre k^0. That is a cycle and moreover it is a cycle formed **by space/time**. What Kepler said is that space is a^3 **being in motion T^2 k.**

As Kepler said $a^3 = k\,T^2$ and therefore $k^0 = a^3 / k\,T^2$ and therefore we have to find k^0. As a result of examining this proposition, I located two principle positions both holding singularity. The cosmos is made up of one type (1^0) that is in two categories where one type moves and the other type does not move. The one is a liquid and the other is a solid.

The condition for the presence of this singularity that forms everything, controls everything and is everything is the centralised to a centre singularity $k^0 = a^3 / (T^2\,k)$ that forms by movement $T^2 = a^3 / k$ of space $a^3 = k\,T^2$ placed in relevancy $k = a^3 / T^2$ that is centrifugally going both ways $k^{-1} = T^2 / a^3$ thereof (Newton's 3rd law). This explains the Coanda effect and the Coanda effect is gravity and gravity "glues" the water to the glass by implementing Π to form singularity! *What is in the Universe is spinning.* **The entirety of everything forming the Universe is spinning inside the Universe** and such spinning are always in the centre of one specific point, wherever such a point might be. In the **precise middle** of all **objects in rotation** is a precise centre where this pre-designated centre is dividing the object in rotation into sectors that will **start the spinning initiation** from that centre point. This is what Kepler's formula confirms in $a^3 = T^2k$. By spinning, the

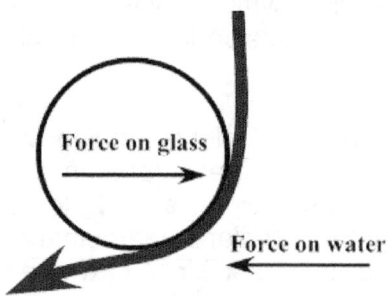

Force on glass

Force on water

one side is coming towards while the opposing side at that time is going away. Thus, the spinning object **will have a middle point**, a very specific **centre point that does not spin** and only holds Π as a specific value because within that centre being that small, no radius can apply. We have named this position or line the axis, but the true meaning of this line has eluded us since the concept was realised. This line that forms holds no space although it directs all the space that it controls by spin. When going toward the centre where the axis form at the very centre of rotation, the space on the one side has to end and the space at the other side has to begin with the line unable to hold space.

On the one side space turns in a completely oppositional direction from the direction in which the space spins on the other side and in between the opposing movement a line forms without the ability to contribute space. But also within the one value forming, such a line **cannot have a value of zero** because the line **is there and holds contact** to the rest of the material bringing about that **zero does not start any** line and therefore the **value of the line must be infinite**, just as described in **accordance** and by **the definition of singularity.** While spinning the points will change direction every $90°$ of spinning and will oppose what it was every $180°$. Although the points had the same characteristics only one instant before, they oppose the characteristics it had just before and just after the very instant in which they are and to which they relate by similar points also in rotation. The fact of the graph proves my point in quarterly opposing dimensions and values. As every point relocates, therefore every point completely changes its attitude from what it was to what it is in terms of what it will be when it is going there. Going down to the centre, as the rotating direction moves inwards, the rings will become smaller and smaller. In dimensional terms, which I explain later on, the value of **2k** relates to T^2. That relation extends to the next value where T^2 relates to **k**, which positions T^2. The first space in the circle T^2 will then be located at point **k**. From the centre being in infinity one can realise by thought that the single dimension factor is not visible, but is present all the same. Extending that into the 3D comes six **k** and any one of the six will further extend to form a seventh point as T^2. All this forms a point that finally refers to the location of one spot holding singularity attached to space by the measure of $k^0=a^3/(T^2\,k)=7$

Let's find $k^0 = a^3 / (T^2\,k)$ and see where it is hidden. The sphere is a circle in many facets and therefore we will approach the sphere as one multi dimensional circle. However, the sphere as such remains one circle to the power of many. When investigating a circle one would draw a line from one edge running through a centre all the way to the other edge. In doing that we would find the measure of the diameter, which is most important when trying to establish the volumetric worth of the sphere. The circle has Π to indicate form and uses r^2 to establish the worth of such a circle by using the radius symbolised as r in drawing a straight line. In any circle or sphere the size only depends on the fluctuation of r in the square as a component to the circle or sphere but that does not affect the form, which comes by indication of Π in any way there may be. The conclusion from this is that no line can start at zero because that will be a mathematical impossibility.

Lines mathematically cannot start at zero because there is no evidence of zero as a factor in mathematics. Should you disagree with my statement the question in need of answering is this: What will the length of the shortest hypothetical line imaginable be and moreover, what would the total overall length be in that case? A line or spot starting at zero would therefore be shorter than the shortest line possible. For obvious reasons can no line, or any line grow or extend from zero because such a line must then quit zero and become something, thus abandon its original value by the adding of the first value. Mathematically said it would be as follows $0+0=0$ whereas if it started with something infinitively small it would be $1^0 + 1^0 = 2$ and then from using something infinitively small it will grow into something immense such as the Universe. In any circle or sphere the size only depend on the fluctuation of r in the square as a component to the circle or sphere but that does not affect the form by indication of Π in any way there may be. The conclusion from this is that no line can start at zero because that will be a mathematical impossibility. If a line started with zero, that would nullify Π ($0^2 \times \Pi = 0$) and that would leave the form without having any form because $\Pi \times 0 = 0$. This statement by itself excludes zero and with zero excluded one then begins to appreciate all the rest of the concepts governing corrected cosmology. If there is a distance, it holds a measured one of whatever norm or value, which is a specific length that apples and that zero or nothing then could never fill.

By saying the distance constitutes of nothing we have to substitute the one factor with a factor of zero to find what mainstream says fills the Universe. Including nothing as to state the presence of that part contained by the calculation delivers the total of zero. It seems as if science has ignored this mathematical principle that $1 \times 0 = 0$ as an issue by simply not thinking about the fact of the matter and therefore simply ignoring that which is measured forming the sole value of space. It is somehow more convenient to put the value of nothing as part of the distance in calculation because that is what is understood. Measure zero and then see how one can multiply when using zero in mathematics to reach a distance holding a value other than zero when multiplying by zero.

I agree that what is filling outer space is invisible, but also it is there, it is present and being present and there while being invisible disqualifies whatever is there from being zero because being zero will mean it is not there and we cannot deny whatever is there of being there. Then what is there will be there, while being invisibly small, but it will still be possible to form a line because every aspect of the Universe forms lines while also it will have the potential to fill space and can still form a measurable unit. That then must be 1 because while $1 \times 1 = 1$, $1 + 1 = 2$ and that qualifies that invisible thing to be present ($1 + 1 = 2$) but at the same time be completely invisible ($1^3 = 1$). When realising this I knew what has to be true about that which I was looking for and that it had to be singularity because singularity can only have one value and that is 1.

To find the invisible I had to locate singularity. I realised that my effort to locate the point holding singularity enabled me to backtrack the exploding Universe to its origins. The Universe is a sphere because it is filled with spheres filling the void spaces (not the nothings) and in that I first had to investigate the visible. Newton's mathematics says a sphere is $\underline{a^3 = 4/3\Pi r^3}$ while Kepler said a sphere is $a^3 = T^2 k$, and both are equally correct because the cosmos gave numbers to support its statement. Where Kepler says $a^3 = T^2 k$ and with mathematics saying that $\underline{a^3=4/3\Pi r^3}$, we think of volumetric size of space in terms of using normal mathematic formulations. We think if it is volume then it has three sides and in the case of a sphere the measure is $\underline{a^3=4/3\Pi r^3}$. Comparing $\mathbf{a^3=T^2k}$ to $\underline{a^3=4/3\Pi r^3}$ is like comparing the equal ness of a triangle and half circle and line to numerical values. $\mathbf{a^3=T^2k}$ predates mathematics, where $\mathbf{a^3=T^2k}$ determines positions at a period in cosmic development when only form was used going before when numbers as value was in place. It shows the half circle $=180°$ is equal to the triangle $=180°$ and both are equal to the straight line $=180°$ notwithstanding the obvious differences used in form. However, the starting point of these forms has to be equal and also has to be not zero to have the end be equal and result in all being equal in value in the end.

Kepler said a sphere is $\mathbf{a^3 = T^2k}$, which also mathematically is $\mathbf{a^3 \div (T^2k) = 1 = k^0}$. In honesty we have to realise that we cannot dismiss the whole formula that Kepler produced just because it doesn't match the scenario set to determine volumetric size as the Newtonian version does. Kepler's version holds a foundation based on movement and it is in the movement we find the measure and not in the size as Newton's mathematical formula does. In Kepler's formula the entire formula is formulating a circle being motion. However, with the correct interpretation we find so much more than just motion. The correct formula is $\mathbf{a^3 = k / T^2}$: That is what Kepler brought into civilization for all time to come. He saw space $\mathbf{a^3}$ being in isolation due to the time it uses to move $\mathbf{T^2}$ claiming such space forming independence according to what the line $\mathbf{k}$ indicates. Let us look at the factors in more detail before we proceed with the rest.

Space $\mathbf{a^3}$ will always be circling around as $\mathbf{T^2}$ is in a position referring $\mathbf{k}$ to the centre $\mathbf{k^0}$. That is what Kepler said when he said $\mathbf{a^3 = T^2 k}$. Kepler indicated space $\mathbf{a^3}$ will forever fight for independence and show separate individuality in remaining apart as identifiable cosmic components by means of motion. Every space will cling to independence indicated by $\mathbf{k}$ through fighting off the integrating of another overall unifying unit by applying the motion of $\mathbf{T^2}$! The problem we have to solve is what will the cosmos use to secure such independence between all particles? What sets space apart from the rest of space? First we have to admit that Kepler was the one that introduced the following: Kepler gave us the answer to the following but no one ever took notice!

Kepler was the one who discovered **space / time** as **space $\mathbf{a^3}$ = time $\mathbf{T^2}$ k**
Kepler was the one who discovered **singularity** as $\mathbf{k^0 = a^3/T^2k}$
Kepler was the one who discovered **gravity** is holding **space-time** relative by the measure of distancing $\mathbf{k}$ as $\mathbf{k = a^3/T^2}$ and $\mathbf{k^{-1} = T^2 /a^3}$

Kepler said gravity in space is about the area $\mathbf{a^3}$ that would always keep equilibrium with the time $\mathbf{T^2}$ it takes to travel the distance of the full circle position placed by the indicator $\mathbf{k}$, therefore adjusting $\mathbf{k}$ as the need arrives. With $\mathbf{k}$ shifting in length $\mathbf{a^3}$ will have to readjust and therefore $\mathbf{T^2}$ will find a new relating value each time. This was the finding of Kepler and came after his intense study of orbiting planets. Translating Kepler's mathematical expression $\mathbf{a^3 = T^2k}$ correctly to the verbal statement in English Kepler said that there is a **space $\mathbf{a^3}$** which is **equal =** to the motion in the **time duration $\mathbf{T^2}$** thereof between two specific points which is a straight line $\mathbf{k}$ that holds a relation from a centre $\mathbf{k^0}$ to an end $\mathbf{k}$ where the two ends run from the beginning of $\mathbf{k^0}$ to connect at the end of $\mathbf{k}$. I might not be the smartest boy on the block but I'm not that stupid either. I know how to translate mathematics into English... and I translate as follows:

$\mathbf{a^3}$ must have a volumetric interpretation because the third dimension is sure evidence of multiple conjunctions of dimensions put together in three sides opposing three sides having the third dimension in place. The fact that any symbol uses a value to the **third power $\mathbf{a^3}$** indicates **space** or a volumetric established and separate unit. Using a cube by three dimensions symbolises a cube, a room, a space to be filled, a unit able to hold other ingredients on the inside when empty or partly filled. It is space because it is volume using the third dimension.

T^2 is an indication of something having a cubic nature other than the square forming motion that is provided by the motion the square indicates, which is where the moving object is representing a third dimensional object that is moving from point to point and it is this point to point that multiplies into the square. The space is moving as a unit from one point to another point and the moving between the points are represented by a flat square or following a flat distance between two points. The cubic space was in one instant in one place and then the second instant in the other and because time can never stand still or become single dimensional (this I am about to prove) insisting that time must always support the motion it consist of or space as well as time in time cannot be. It is motion that is taking time, which is motion in the second dimension moving the space in the cube.

k is the symbol used to indicate a straight line between two points with a definite beginning and a specific end position. It is the location where the form in question is holding space running from where the space was to where the space will be the very next split instant that follows while time by movement repositions the allocations. This indicates points of representing k in different time positions to which the points will then be multiplying to form the square that forms between k_1 and k_2. The movement indicates not a square surface but it indicates movement by the square. This indicates the time the journey took to move the space from one point where k is to where k will be. It indicates the location of the space where from to the point where the next indication of k runs. T^2 will shift k where k

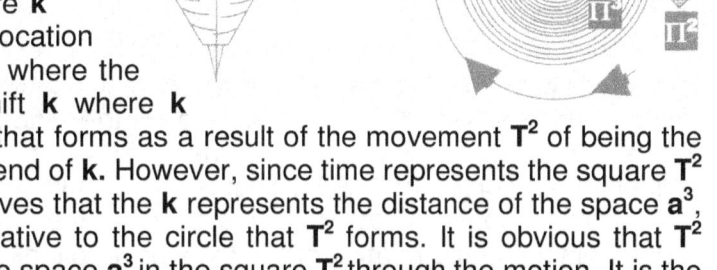

indicates the position of the space a^3 that forms as a result of the movement T^2 of being the space a^3 indicated by the point at the end of k. However, since time represents the square T^2 and with k being the distance that proves that the k represents the distance of the space a^3, and is also representing the form relative to the circle that T^2 forms. It is obvious that T^2 represents the time that represents the space a^3 in the square T^2 through the motion. It is the distance moving space a^3 in the cube to complete time in duration in the square of motion T^2; therefore k is permitted to be in the single dimension.

<u>Let us find the smallest possible line first</u>. We have already reached the conclusion that by reducing the line, the reduced line will eventually leave all sides on the same spot on the condition that the circle spins. Such a spot must be round in form since it still holds Π as a factor next to r^0. We now are entering the domain of singularity where the visible is no longer traceable and only intellect can bring understanding of the scenario. With the line being the smallest line, such a line will start off as a dot Π that moved away from a spot Π^0. With all possible sides being in precisely the same spot we have all possible sides onto one spot. I chose to differentiate the dot and the spot by giving the spot a value of Π^0 while the dot holds Π next to r^0. Mathematically the spot is placing form evenly spread being Π coming from the single dimension Π^0 where the space is one (1) and holding exponentially zero (1^0). There the space moved over to form the spot Π^0 and by introducing form the movement changed Π^0 to the dot Πr^0 forming a circle as a dot. Again I must draw the attention to the fact that we now are reaching into areas only the human mind can venture by understanding and seeing nothing more than with the eye of intelligence. The understanding of this concept demands our reaching the point where the mind of the animal cannot reach. If it starts with a line it then is there where that line only represents two sides being one and as such that is representing rather a flat Universe. At the dot Π we have roundness because we have Πr^0 while at the spot there is not yet any round form because of Π^0 and only when Π being round forms it then is requiring a shape or form and this lies beyond or before space at a point where any form of shape comes into the cosmos scenario. This part of the Universe comes in a place at a point in a location where shape and form is a part of the distant space hidden in and beyond where eternity develops. The spot is located at a point where entering the domain of the spot also at the same time is crossing the spot and landing on the other side of the spot where entering the spot is crossing the spot. Nothing can enter the allocated position the spot holds because entering the spot is crossing over to the other side of the spot. It serves us well to realise that the entire Universe was that small at a point where everything started forming because the spot that developed into the dot is still with every spinning circle...and the Universe is a multitude of spinning circles. It is also very wise to remember that once anything becomes a part of the Universe, it can never leave the Universe since it then has no place to go or no gate to pass through in order to leave the Universe. With the spot becoming a dot, there must have been a time when everything in the entire Universe was that big as the spot is, and that then moved on to form the dot and in that it went on growing in relevance. The point around whichever spins becomes the centre of the Universe by singularity.

We know that dividing the radius into the circumference produces Π. I have the answer and the answer is also the answer to gravity and explains what gravity is. Forming Πr^0 leaves the ratio where every 21.991 lines or dots, the line will bend by 7. The Universe introduced space by introducing the four cosmic laws named **The Lagrangian system,** 2) **The Roche limit** 3) **The Titius Bode law** 4) **The Coanda affect.** Then one can clearly see how gravity forms Π to put an entire Universe in place.

We know that dividing the radius into the circumference produces Π. We know that Πx the radius is the circumference, so $\Pi x r^0 =$ the circumference and when the circle is at its smallest the circumference is Π. But never has anyone gone further and asked why is the smallest circumference $\Pi x r^0$. I have the answer and the answer is also the answer to gravity and explains what gravity is. Forming Πr^0 leaves the ratio where every 21.991 lines or dots, the line will bend by 7. Every 21.991 the ratio will reduce by 7. There will forever be an inclination of 7 for every 21.991. That is the formula that produces the circle, but what brings this formula about? What would bring about the 7 as a factor because the 7 are very pertinent in the entire relation of Πr^0? We have to look at the top spinning.

This relevancy is so pertinent that from this we can surmise how the Universe came about and what happened the very first instant the Universe came in place. The Universe introduced space by introducing the four cosmic laws named **The Lagrangian system,** 2) **The Roche limit** 3) **The Titius Bode law** 4) **The Coanda affect.** The phenomena never made any sense in the past, bust once one attaches gravity to their meaning in the correct manner by implementing gravity as Π, the true function of the phenomena as far as implementing gravity comes in place. Then one can clearly see how gravity forms Π to put an entire Universe in place.

The circle forming Π uses 7 to indicate the roundness of the circle but the 7 holds its roots deep within creation. It indicates how the Universe started because this is the way a star will start moving and it shows how as the infant star starts generating gravity just as the top starts to spin when it is thrown by life. Life can create nothing and that is true but life can mimic all laws in the Universe. Time is eternal movement and will be with us always. The line in infinity is still present while not being a part of the Universe. This line is always ready to be in place when the slightest movement orders it in place. Before the Universe was in place eternity and infinity was in perfect harmony and the line forming singularity validates this fact. Before infinity parted from eternity, eternity met infinity on one spot as eternity came from the past (1) forming the present (2) to go onto the future (3) but also returned to come from the past which was the spot held by the future and this we find in the fact that the line forms 1 when not spinning but as soon as it evokes by spin, 3 points form even now. Then heat and cold differentiated values and space landed in between eternity and infinity. As eternity moved in relation to infinity but not forming a part of infinity any longer, eternity had to follow a path by never going away from infinity (3) and always returning to the point infinity holds but never lash onto the point again. With space parting the points, eternity had two points (the past and the future) before the partition came about and infinity held both the past and the future while infinity had the present as it still gas presently. By eternity also moving, the two points it held opposed each other (the past and the future) and since it moves, by the movement it became the square of the two because movement is the square and not a flat blanket-like surface with squares embroidered on it as Newtonian science depicts it by using grand mathematics to understand singularity.

Then we had two point holding eternity in place going square by movement to form 4 points serving eternity and infinity captured the first three points held by both and since eternity could not release from the two it had but had to duplicate what it had, eternity by movement became a circle captured by the line. With four points captured by the line of three points the circle coming about is eternally returning to infinity but never complying with infinity because if mismatching temperature or movement (3 against four). Material will always be colder than outer space. It is because material spin and outer space moves by expanding due to overheating. This is where I start when I start to explain the first moment but I use a shipload more information to do explaining when I explain the star in the book I do so. I involve the four cosmic pillars to substantiate the claims I make because all four still work the very same way as it did at the beginning of the Universe. The three points serving one part of singularity combined with the four points serving singularity unites as seven to form a circle of either 3.1416 or 21.991÷7. The seven going to one is eternity matching infinity by movement. But since seven moves it is seven that has to produce gravity. How do I know all these facts, because we can see from the top it is still doing what it did the very first second.

 When time started infinity as well as eternity had altogether 3 positions, the top of the line, the centre of the line and the bottom of the centre line called the axis. It is still forming the very line in the centre of the top as it forms all lines in the centre of all things spinning. When the top started spinning eternity parted from infinity when movement separated what is stagnated from that which became moving and eternity formed four more points than before when the axis only had the three points which was part of the earth and not of the top. With infinity and eternity then jointly having 7 points within singularity the top became representative of its entire cosmos because the cosmos came into rotation. In the aftermath of the now rotating top we now see the phase of cosmic development where the two sectors bring about contraction through spinning and the rotation brings along the contraction. This rotation on the outside of the top that keeps the top erect is the same rotation forming and formed by gravity we see that produces the atmosphere. When Π forms it does so on the grounds that 7 rotates in relation with 21.991 points. The circle forms by a change in direction by 7°. Every circle has opposing sides forming in relation to the axis line. If the one side goes to the rite then the other side has to the left. If the rite side goes down then the left side goes up. There is this double presence of a change in direction forming on both sides of the circle. The 7° move and by moving 7° goes square 7^2 and that is the triangle doubling which then forms double seven (7) by the square (7^2) in relation to the line holding singularity or 1 that in turn goes double (2 x 1) and goes square (1^2) and that forms Pythagoras. This results in 50 + 50 = 100.

Gravity is about the reducing of space to singularity. In spinning the sphere the movement contracts by measure of 21.991 in division of 7 going singular reducing to 7 to 1 and 21.991 to 3.1416 while the rotation is produced to reduce 7 to form singularity, but also gravity forms when the 7 comes from the future 7 to the present 7 and then onto to the past 7 and this becomes 21. The future feeding the Universe is the part .991 which is smaller than singularity and that shows it is not yet within or part of our Universe because it is smaller than what the smallest connection can be connecting our Universe to the future of our

Universe. Not only that but with singularity advancing from infinity to become one it proves that even as we see singularity as one, singularity also is multi dimensional but that ability is beyond our scope we have being in the Universe. The dimensional change that Π undergoes shows that singularity repeats into a new location by the value of 0.1416 that lines up with the 3 and the 3 is part of a singular one line forming an axis without space and is therefore 1 Then as there always is a bottom the new 7° as a redirection forms as at first becoming 0.991 that then progresses to 1. That is how the cosmos started. Infinity holding eternity on one spot coming from the past to the present being one spot and onto the future being one spot the cosmos was singular monotonously eternally by repeat before the event of space forming in light that science calls the Big Bang. That is how I wrote the book on the subject as to how the cosmos became created whereas the Big Bang is the cosmic birth.

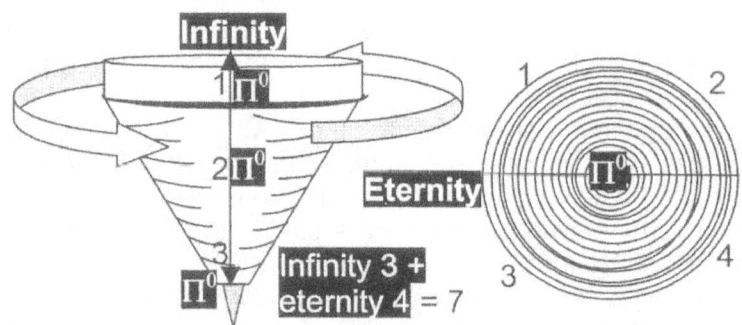

The circle forming Π uses 7 to indicate the roundness of the circle but the 7 holds its roots deep within creation. The line in infinity is still present while not being a part of the Universe. Before the Universe was in place eternity and infinity was in perfect harmony and the line forming singularity validates this fact. Before infinity parted from eternity, eternity met infinity on one spot as eternity came from the past (1) forming the present (2) to go onto the future (3) but also returned to come from the past which was the spot held by the future and this we find in the fact that the line forms 1 when not spinning but as soon as it evokes by spin, 3 points form even in the present day. Movement froze material away from not moving space and space not moving overheated in relation to material freezing. Material also employing space also overheated and grew in stature but the movement captured some of the space, which is the process we think of as gravity or as time. This is your Big Bang. Then heat and cold differentiated values and space landed in between eternity and infinity. As eternity moved in relation to infinity but not forming a part of infinity any longer, eternity had to follow a path by never going away from infinity (3) and always returning to the point infinity holds but never lash onto the point again. Then we had two point holding eternity in place going square by movement to form 4 points serving eternity and infinity captured the first three points held by both and since eternity could not release from the two it had but had to duplicate what it had, eternity by movement became a circle of two duplicating the two to form four and the rotating circle of eternity captured by the line that held 3. With four points captured by the line of three points the circle coming about is eternally returning to infinity but never complying with infinity because if mismatching temperature or movement (3 against four). Material will always be colder than outer space. The three points serving one part of singularity combined with the four points serving singularity unites as seven to form a circle of either 3.1416 or 21.991÷7. The seven going to one is eternity matching infinity by movement. When time started infinity as well as eternity had altogether 3 positions, the past, the present and the future. It is still forming the very line in the centre of the top as it forms all lines in the centre of all things spinning. Then eternity parted from infinity when heat separated what is cold from what is hot and eternity formed one more point than before when it had the three points. With infinity and eternity then jointly having 7 the cosmos came into rotation. The circle forms by a change in direction by 7°. Every circle has opposing sides forming in relation to the axis line. If the rite side goes down then the left side goes up.

Gravity is about the reducing of space to singularity in relevance to the expanding of space in overheating. In spinning the sphere contracts by measure of 21.991 reducing what was7 in relation to 21.991 to 3.1416 in relation to singularity while 7 is the factor that by spin reduces to form singularity, but also gravity forms when the 7 comes from the past to the present 7 and onto the future 7 and this became 21. That is how Π will eternally return to form the value of 21.991/7 that in the very instant of infinity will form 3.1412 /1. It is this process in which Infinity holds eternity on one spot never releasing the position of capture as movement is always coming from the future to form the present being one spot and onto the past being one spot but then forms space in the cosmos. Before the Big Bang when light created space the cosmos was singular monotonously eternally by repeat. Everything relates to a centre. The next big thing I had to look for was the centre of the Universe that connected everything.

In establishing such a centre containing singularity we find the reason why bullets travel more straight when they are fired circling and circling is what gives the bullet the accuracy in its trajectory that then established a cartelise singularity that establishes a value forming Π in relation to the centre singularity being 1 or as I named it as singularity Π°. When a rocket is fired and the spin is not present there will be no stable trajectory. The only way to secure the stability of the trajectory is to allow spin (Π^2) that enables a point holding (Π) as this will locate and establish singularity (Π°). Establishing singularity is the most fundamental principle about gravity we can ever find. This is the one part that is most important when we go in search of gravity secured by singularity that forms the absolute relevance of everything filling Universe we have. Everything is a rotating object that holds any point allocated in Universe to form the centre of the Universe because everything in the entire Universe spins around any given point and that then forms the centre of the Universe. Every centre of every atom forms the centre of the Universe by spin! Again I indicate the precise location of such a point. What is in the Universe, is spinning and therefore what I am referring to, applies to everything holding a place in the Universe and therefore this which I mention directly links everything holding any space whatsoever in the entire Universe to one single point around which all spins. In the **precise middle** of all **objects in rotation** is a precise centre dividing the object in sectors that will **start the spinning initiation** from that centre point. Thus, the spinning object **will have a middle point**, a very specific **centre point that does not spin** and only holds Π as a specific value because no radius can apply. But also the one value such a line **cannot have is zero** because the line **is there and holds contact** with the rest of the material bringing about that **zero does not start any** line and therefore the **value of the line must be infinite**, just as described in **accordance** and by **the definition of singularity.** As I am introducing a very new idea, I wish to explain in better detail what I try to convey. While the toy top is spinning one will find singularity by moving the rotating line or radius progressively to the middle by reducing the length the line has from the edge to the middle. At one point all further reducing must end but the ending cannot include zero or nothing because the rest of the line still attach the rest of the top. As the rotating direction moves inwards, the rings will become smaller and smaller. Then we reach a point everyone thinks of as being the axis around which everything rotates. The line only forms when everything around the line spins by establishing a circle to the value of Π.

Everyone calls this line that forms the axis. Everyone knows about the axis and yet through so many thousands of years of using an axis, no person ever thought to scrutinise the principle behind the axis. Yet in all the millennia everyone was aware of the line that forms called the axis, no one took time to see it holds singularity at Π° presenting Π. The only conclusive value singularity can have is 1 or Π°. The axis controls all particles spinning around the line being the axis while the axis in itself forming the line represents no particles because the axis represents no space. If there was space within the axis, the space had to spin in some or other direction. Having no space would mean occupying no space which means forming no part of the Universe filled with space and yet it controls all the space as wide as the mind can imagine. Without space it does not form a part of the cosmos, but forms the cosmos as wide and as deep as the cosmos goes. The axis could not be seen but with applying intelligence the axis could be witnessed. Having no part in the cosmos in space, the axis could only be understood and never be seen. The axis could be proven but never be shown. The axis is what controls the Universe from end to end because when there is no end there the axis provides one end to what never can have another end and the axis governs whatever spins in relation to such a line. Again I wish to press this issue to form clarity. The line forming the axis is without space and only holds form, and therefore the line represents a point not having any dimensions while it still is there without ever being there. If ever there is a concept I have to introduce, then it is the concept of how important the axis is and how science up to now missed the biggest issue that is responsible for all movement within the cosmos. The line forming the axis is there but only intelligence will ever form the concept whereby one can realise where the line is without ever seeing the line. Anyone unable to understand this concept can never see the validity of space-time. In the axis line there is a something that is there but only intelligence can bring understanding to the understanding thereof. Only motion of space can resurrect the line coming from the point it holds as a dot. Everything in the cosmos spins and everything that spins has to form a line that doesn't exist but yet the line controls everything that spins around this line that never can hold any space or be part of the Universe. Without having space to fill, the line can never form any viable part of what forms the cosmos, which is space.

The point in reference is the line forming the axis and the axis must be a line that never forms in space because if it did, it would have to rotate in either one of the directions space spins in and by not spinning, it has no space. **That point** albeit hypothetical, is also as much a reality none the less and is placed where that point **must be standing still** because every line **running from that point** in **opposing directions** is also **in opposing directional spin the other or opposing side.** In considering the spinning motion in the fraction of time in the detailed instant every aspect of rotation will turn in every instant of change in time. Although the points had the same characteristics only one instant before, they oppose the characteristics it had just before and just after the very instant in which they are and to which they relate by similar points also in rotation. Looking at the graph unfold will explain my point about quarterly opposing dimensions and values unfolding.

The circle can reduce one step more when the circle eliminates r completely by returning r to a point of singularity r^0, but the elimination of r as the factor reduces the major factor to the single dimension in Π^0. That will not reduce the cosmos to zero, but it will only eliminate all potential lines r^0 to potential circles $\Pi^0 r^0$ and from there the circle Πr^0 will come about by manifesting as a line but that manifesting can firstly only establish a circle Πr^2. The only value that singularity can have although the single dimension may host the entire Universe is Π^0. Pick a number and elevate it to the power of zero and in the process one may have established another point holding all points in singularity because that is the value of singularity. Only Π^0 or any other value holding one accompanied by zero as an exponential value can ever be the accurate value of singularity while singularity will then host the rest of all the possibilities in the Universe. This means that the entire Universe composes of and is made up of singularity... this much I am going to prove. Every point occupied or otherwise constitutes of singularity either under control by movement in a form we call atoms or being passive in a location we call outer space. This position one can derive from Kepler's formula $a^3 = T^2 k$. It is just a question of how to fit this sensibly into Kepler's formula $a^3 = T^2 k$ and find a way that will bring much understanding to cosmology and the way that singularity connects one Universe to form cosmology.

Reading this mathematically encrypted coded formula of the cosmos given to Kepler and keeping it removed from Newton it reads as being that the space a^3 is equal to = the motion T^2 of the space a^3 in ratio k to a centre k^0, which is relevant to the positioning of k. If we bring in the full equation it will be $k^0 = a^3 \div (T^2 k)$ which means half of space is solid $k = a^3 \div T^2$ and half of space is liquid $k^{-1} = T^2 \div a^3$ where liquid is moving. However, it is also true that everything through movement defines a value in relation to one point holding singularity k^0 and that is what the formula $k^0 = a^3 \div (T^2k)$ underwrites.

What this proves is that gravity is the motion of space provided by time being the liquid. Please allow me to explain. In the formula $a^3 = T^2 k$ the space forms as the space is in motion. Newton suggested that $\frac{dJ}{dt} = 0$ where he stopped time to have

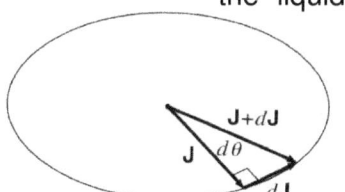

the motion of the circle demolish the work that the circle does. That means he got time standing still or being T^1 and the motion T= 0. Let us ponder on that thought for a while, while we remain with the formula Kepler suggested $a^3 = T^2 k$ and then it will seem that according to Newton $a^3 = T^2$ and in that **k** then becomes **0**. Should that be the case then we have space going flat because $a^3 = T^2k$ where $a^3 = T^2 \times k = 0$ forming a square instead of a cube, and the Universe we have is a three dimensional system in every aspect there is. The concept Newton brought about that $a^3 = T^2$ is putting a person that looks at a mirror equal to the possibility of walking in and out of the mirror by becoming the reflection in the mirror T^2 and then himself a^3 again. It is rediculous to say the very least.

It is quite apparent that Newton saw no difference between the top spinning while the top was standing in an upright position and the top lying down on the Earth. This is a crucial mistake that has such a wide implication that on the one hand it either values the Universe to the value of singularity or on the other side dismisses everything about the Universe to the value of zero.

I am of a very different opinion about Newton's point of view where he declared that forming a circle moving $\frac{dJ}{dt} = 0$, and by doing such the movement then removes Kepler's relevancy factor. This places a value of empty space in which a top would spin and Newton missed the difference there is between a top spinning and a top laying on its side on the Earth. There can be no such a thing as empty space. The fact that space is valid removes an empty connection because space can be anything there is in space except empty space that is filled with nothing. The Universe is time contained in space, which makes it space-time. Space has only one value, and this is to contain time and time provides space with a definite value. **I do not disagree** for one instant **with Newton**'s calculations whereby he came to the conclusion that $\frac{dJ}{dt} = 0$ and therefore I am not going into repeating the entire calculating process. All of the calculations Newton made are very correct except the eventual and final conclusion Newton came to. Newton never understood the mathematical concept of time playing a part in physics. In the time of Newton singularity and the relevance thereof had no feasibility in any concept regarding physics. Newton had the concept that time could stand still and that is impossible in physics or any other place. Time can never stand still because time is forever moving by establishing space in a three dimensional environment.

Being the mathematical genius as Newton is as often portrayed as Newton had very little insight into mathematical possibilities, because when he suggested that $\frac{dJ}{dt} = 0$ he made one huge mathematical blunder. Newton or no other person may place any two objects in a direct relation where the two factors divide and have an outcome that forms zero. Much surprising is that not one mathematical genius that came after Newton drew the correct

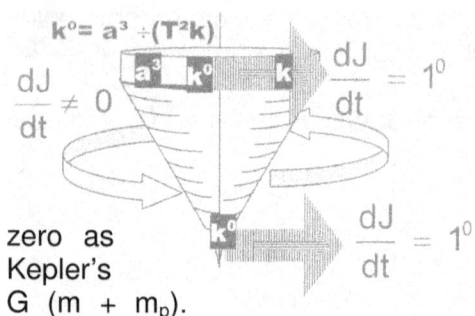

$$k^0 = a^3 \div (T^2 k)$$

$$\frac{dJ}{dt} \neq 0$$

$$\frac{dJ}{dt} = 1^0$$

$$\frac{dJ}{dt} = 1^0$$

conclusion that forming $\frac{dJ}{dt} = 0$ is mathematically not acceptable. Newton saw that dividing something into something else could bring about zero and that is impossible. In concluding that $\frac{dJ}{dt} = 0$ bringing in zero as Kepler's G $(m + m_p)$. a legitimate value Newton found a way to replace symbolic relevancy value of **k** with using the symbols In doing that Newton painted a picture that has no real meaning except where Newton tried and succeeded to put mass into an argument that has no true validity in cosmic principles. This is just a longer and probably a more detailed manner of indicating **k** and better defining of **k** but it symbolises precisely to the point what **k** stands for nonetheless. I wish to draw your attention to the matter of Johannes Kepler's findings that Mainstream science considers as resolved and closed for many a century while it is not. My investigating Kepler helped me to resolve other unresolved matters but it was only possible by using Kepler's work.

Newton never considered why the spinning top stood erect and the top not spinning lay flat and still. Newton did not think that as soon as the gyroscope started spinning, the balance shifted in favour of a position wherein the gyroscope stands upright. What then comes about has the ability in keeping the gyroscope upright. This is rotational movement and in my other books on the **_Absolute Relevancy of Singularity_** I explain how rotational by the square of the double seven forms Π and Π is forming the curvature of space-time and in that bending of space-time is what we call the atmosphere that keeps the gyroscope square with the Earth and through that the gyroscope stays upright. The gyroscope is acting in accordance with the Coanda effect where the Coanda effect is gravity. By spinning it establishes a solid forming as $k = a^3 \div T^2$ and a liquid forming as $k^{-1} = T^2 \div a^3$. By spinning $T^2 = a^3 \div k$. That is evoking singularity $k^0 = a^3 \div T^2 k$ establishing gravity $a^3 = T^2 k$ in relation to the Earth evoking gravity through also spinning.

Newton found mathematically that the movement of the top by spin removed the value of the radius $\frac{dJ}{dt} = 0$ where quite the opposite applies. The spin of the top $T^2 = a^3 \div k$ positions the relevancy that **k** as a factor produces by initiating singularity k^0 on both sides of the relevancy $k^0 = a^3 \div T^2 k$ as well as placing singularity in relation to the spinning top $\frac{dJ}{dt} = 1^0$ because that is the correct mathematical principle coming from the equation. The smallest any dividing can be is one and one is the producing of singularity. The spin of the circle does not eliminate the relevance of **k** but institutionalise the measure of **k** by confirming the space a^3 in terms of singularity k^0. However **k** has no confirmed and specifically applying value but puts a relevancy of space a^3 forming in relation **k** to movement T^2 applying. By trying to find a measured value applying to **k** is showing no understanding about what **k** is. The value of **k** is finding the space that **k** indicates in terms of what moves. The indicator **k** identifies the space a^3 that the circle claims in terms of singularity k^0 that the movement T^2 isolates from the rest of singularity $\frac{dJ}{dt} = 1^0$. The value of **k** is dictated by T^2 as the movement that isolate the space a^3 but also **k** dictates the value of T^2 to form space a^3. The measure of **k** is the relevance **k** is claiming on behalf of the space a^3, which uses the relevance of **k** to put a limit on the space a^3 by spinning in accordance with T^2.

What Newton suggested while never realising he did suggest it is the following, and that is that the rotary movement of objects puts singularity $\frac{dJ}{dt} = 1^0$ in position on the outside of the moving circle. However, by using $\frac{dJ}{dt} = 1^0$ Newton placed emphasis on the turning movement of the circle and saw this as a destroying of the circle while in fact the turning is putting the space that identifies the circle on the cosmic map. That Kepler also found without ever realising what he found. Kepler said $a^3 = T^2k$ which is $k^0 = a^3 \div T^2k$ which is the spin $T^2 = a^3 \div k$ which is the circular movement T^2 that validates the space a^3 in relation k to a centre k^0 which is exactly and precisely what Newton said when Newton said $\frac{dJ}{dt} = 0$ that actually should read $\frac{dJ}{dt} = 1^0$. The location where Newton placed singularity as being singularity established by the movement of space $\frac{dJ}{dt} = 1^0$ If we look at the result of $\frac{dJ}{dt} = 1^0$ in terms of such a conclusion it then would match the findings of Kepler where Kepler said that space moves in relation to what forms singularity. It would mathematically be equated as such

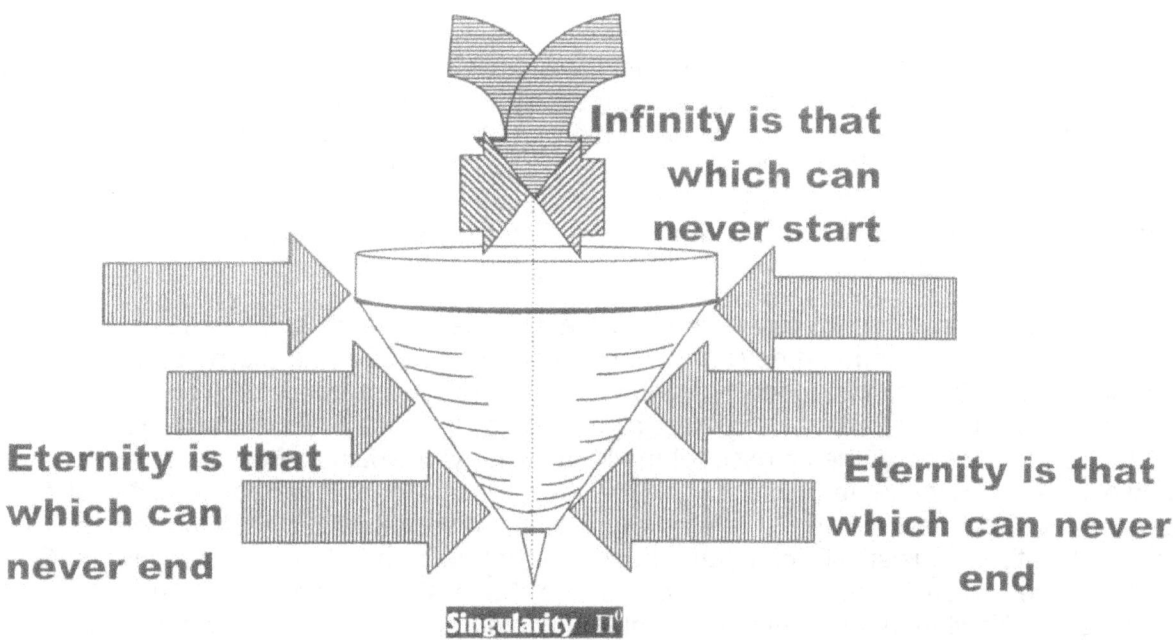

Infinity is that which can never start

Eternity is that which can never end

Eternity is that which can never end

Singularity Π^0

$k^0 = \frac{a^3}{kT^2}$ where it says that when space a^3 moves T^2k it does so in terms of singularity and that is albeit k^0 or 1^0 because $k^0=1^0$. Now we have to find the procedure by which singularity forms space in terms of forming Π, which is the circle. Now comes the explaining how Π forms as a result of movement.

In the picture I give you two time components in the Universe. In the centre there is infinity and on the outside we find eternity. Infinity is space that can't go smaller charged by the spin of material that evokes eternity where eternity is not space but is the spin within space. This is cosmic physics put in place as a reality that controls the Universe. This is the mathematical truth...but now compare this reality to Newton's fiction of mass pulling mass and see the fiction in the latter. In the Newton formula as a mathematical entity the where influence that the mass holds in the formula $F = G \dfrac{M_1 M_2}{r^2}$ as the influence of mass diminishes in respect to the growth of the distance. In that sense the gravity force between the earth and the moon must reduce its ferocity therefore weaken. Therefore the control of force holding gravity is not carried by the value that the mass factor holds but it is within the radius square that divides the value of mass that establishes the force. However, Newtonians only apply their ability to calculate and knowledge for the purpose of upholding Newton and never to provoke Newtonian liability by telling the truth. Never is there any mention of mathematical reality when mathematics is used. However this no one ever knew to be true outside of the intimate upper circles of physics, and this unmasking was seemingly to be prevented at all cost that would allow the public to discover that not only Newton but Newtonian physics was based on fraud. A plan was to be devised because if the public found out Newton was a fraud all along and being Newtonian was the personification of stupidity the entire science world would come tumbling down on the heads of those most important Brainy Bunch. Hubble showed that Newton's formula is bogus and fraud and to cover Hubble a plan was devised.

There is always a point everywhere and anywhere that contains no space but only time. Only movement excites the point that is not to become part of all that which is. In singularity we find a point that I called a spot and the spot develops by movement into that which I called a dot. Space is the leftover of what time left behind after it moved into and onto the future. The Universe is what is between infinity and eternity two valid points forming the markers of the Universe. Looking at something as simple as the children's top we can find infinity and eternity. Infinity is a point that forms only when the top moves because when a top is not spinning it is on its side and only forms part of the spin of the earth. When spinning it forms an entity that holds singularity and forms a spot that became many dots. To become from a spot to a dot it has to form ten seven points. In the centre of the top a line forms that is part of the top but it is not part of the Universe. In the centre of anything spinning a line forms that has no space.

No, better still, to save science a conspiracy was devised. The most intellectuals on earth had to cook up something and devise a plan to save their image and the name of Newton. The most intellectual minds concluded and fabricated to form a conspiracy to withhold the truth and forge a fraud that lasted almost ninety years to the day. Should anyone disagree with the term conspiracy, then please let me know what you would call what happened after Hubble's discovery became prominent on the news. In the past it was accepted that only Newton and God never made a mistake and since the Brainy Bunch Newtonians were mostly atheistic or atheistically orientated they were not that sure about the credibility of God but about the unquestionable accuracy of Newton those atheists were pretty sure of...and now it seems Newton made a mistake. That just could not be. They would rather have God make the mistake than let it seem as if Newton was up to no good while Newton made the incredible mistake.

Then they allowed God to make the mistake. If the Universe was expanding it was God's fault. It sounds much better than have anyone think it is Newton's mistake. If the cosmos expanded while Newton said it must contract then this rebellious behaviour of the cosmos must end. They had to put the blame on the Universe and then ultimately on God for making such a mistake. It must be God that made the error to allow the Universe to expand while Newton claimed the Universe was contracting. The easiest is to put the blame onto God by finding the fault at the door of an uncompromising Universe. To make the conspiracy believable they had to conjoin a concocted story that would have every halfwit on earth believe it. What will make the Universe expand? It must be a lack of the something that makes the Universe contract. Not enough contracting solution will be the cause of the expanding. If it was mass that had the Universe contract in Newton's terms, then a lack of mass will lead to a shortfall in gravity and then the elasticity would not be enough so the elasticity would be tested and before the elasticity failed completely they had to get the Universe back on track. They saw a good measure of darkness splitting small bits of light. This must be it then. Who put that much darkness amongst that little Light?

If it was mass that should contract the Universe then it had to be not enough mass that would to be blamed for the expanding of the Universe. Now to get someone credible enough in the eyes of the public yet foolish enough in ego to provide the cover for the conspiracy to work was another matter. The only candidate must be Albert Einstein. They had to get Albert Einstein to measure all the mass in the entire Universe. Now this is where the joke no longer seems funny. Only fools and idiots are going to fall for that. Look at nighttime at the picture heavens provide and the read picture. There are unrecognisable small portions of a large and overwhelming large Universe. Now go on and count the entirety of all the mass that your nighttime viewfinder will show in your picture. Who is keeping whom for a bloody fool?

Then they resorted to what came in handy throughout the past, they formulated to cheat the public by deception. They proceeded to cover Newton's deception with more elaborated Newtonian deception. Newton started the trend and Newtonians are still doing at present, and when facing all other concluding evidence showing that the Universe was expanding since time began they come up with the utmost unrealistic garbage only an idiot can devise. The Universe started expanding and never stopped since and yet they wish to promote the critical density theory just to cover up Newtonian misconception about cosmic reality. No one ever told them to bring proof with evidence that the cosmos is contracting as Newton said. In the critical density conspiracy all they say is that they are waiting to see when the cosmos would stop its criminally insane behaviour and start to listen to the laws of Sir Isaac Newton. Newton said the Universe contracts and so it must. They shove all the blame of wrongdoing onto the cosmos and take away all error from Newton. If the cosmos does not contract as Newton said then when will the cosmos mend its ways and follow what Newton said and to start contracting! It is a conspiracy to cheat and lie and crook the human race in order to keep Newton untouched. With The Critical Density shambles the modern Newtonian set out to defraud the world in the same manner as Isaac Newton has did centuries ago. Newton said the cosmos is contracting. When Hubble proved the cosmos is not contracting, Newtonians looked where the cosmos went wrong by not following Newton guidelines he so clearly set the cosmos to follow. It has to contract and not expand. Those in academic positions fabricate non-existing material no one can detect to cover the real conspiracy they try to hide. When the argument arrives of contraction versus expanding they wall this down by referring to the search for a substance that can't exist and therefore could never be detected. It is not the dark matter issue that is the real conspiracy but the dark matter forms a conspiracy to hide the facts that the true conspiracy covers up. It is this mother conspiracy I am gong to uncover and present.

This is not rocket science; this is mathematics at its most basic. If the Universe was expanding from the beginning then the measured value of the mass was declining ever since, that is if mass was responsible for producing contracting gravity. They are the ones that are the masters in mathematics. They are the ones that know mathematics better than anyone else on earth...and they missed this truth. This missing the basics was as deliberate as it was swindle the hide Newton's incompetence and with it their failure to understand physics. This is where the second conspiracy started. The first conspiracy was the idea of mass that produces gravity and Newton's fraud to convince the world to believe him. Then the Mathematical Physicist devised a plan to protect Newton's image and therefore their academic standings. They got Albert Einstein to hunt for the presumed missing mass and thereby distract attention away from Newton's oversight and their failure. Albert Einstein carried the heavy burden of being acclaimed the title of the best mathematical mind of all times. If you bend mathematical laws you will get a distorted Universe, as distorted as the Newtonian Universe is. If the Universe doesn't agree with Newton's principles and Newtonian science the Newtonians have to bend all aspects to get it to fit.

They invented dark matter. In astronomy and cosmology, **dark matter** is a hypothetical form of matter that is undetectable by its emitted electromagnetic radiation, but whose presence can be inferred from gravitational effects on visible matter. This is totally fiction and is as fabricated as modern science could be. According to present observations of structures larger than galaxies, as well as Big Bang cosmology, dark matter and dark energy could account for the vast majority of the mass in the observable Universe. This means if they can't see it they can't show it and that is brilliant to fool all the sceptics. The vast majority of the dark matter in the Universe is believed to be nonbaryonic, which means that it contains no atoms and that it does not interact with ordinary matter via electromagnetic forces. The nonbaryonic dark matter includes neutrinos, and possibly hypothetical entities such as axions, or supersymmetric particles. This is the perfect tool to fool all those begging to be conned.

Dark matter was postulated by Fritz Zwicky in 1934, to partially account for evidence of "missing mass" in the universe, including the rotational speeds of galaxies, gravitational lensing of background objects by galaxy clusters, and the temperature distribution of hot gas in galaxies and clusters of galaxies. Fritz Zwicky is the "Father of Dark Matter," coining the term itself, as well as gravitational lensing and the sky survey technique. He devised it but I can't say if he was part of the conspiracy or if his ideas were hijacked and misused by the conspirers. But in the end these ideas came in pretty handy to use in the rest of the conspiracy. Dark matter is believed to play a central role in structure formation and galaxy evolution, and has measurable effects on the anisotropy of the cosmic microwave background. All these lines of evidence suggest that galaxies, clusters of galaxies, and the Universe as a whole contain far more matter than that which interacts with electromagnetic radiation: the remainder is frequently called the "dark matter component," even though there is a small amount of baryonic dark matter. The largest part of dark matter, which does not interact with electromagnetic radiation, is not only "dark" but also, by definition, utterly transparent. Most impressive but here is the catch... If they are asked to show it they already admit they can't... because it is utterly transparent. If they are asked to prove it they already admit it is illusive and therefore they can't utterly transparent. They can make up the story on the trod as we run along because no one can prove them wrong because they can't prove they are correct.

Unlike baryonic dark matter, nonbaryonic dark matter does not contribute to the formation of the elements in the early Universe ("Big Bang nucleosynthesis") and so its presence is revealed only via its gravitational attraction. In addition, if the particles of which it is composed are supersymmetric, they can undergo annihilation interactions with themselves resulting in observable by-products such as photons and neutrinos ("indirect detection"). This is the same as saying there are "anti matter" eating up matter. If they are asked to say what is "anti matter" or what is "anti matter" made of they can't say. It is a name and naming nonsense is a great Newtonian pastime. They do it to relax and to become social with other Newtonians, which may or may not be part of some mating ritual. If you have no idea what you are talking about but named something you then it is as if you explained something because naming it and creating mathematical formula goes hand in hand. There is no need to be realistic because the end of any Newtonian intellectual capacity is to give a mind blowing mathematical formula of which the practicality remains a mystery and then give it a name. The name must be so impressive that just to remember it would take up all the effort any Newtonian has in reserve so getting to the point of proving it taxes the Newtonian's stamina beyond breaking limits and then there is no need for it. You only name it.

The light back then could not have travelled 299 792 458 m / s 12 billion years ago because 12 billion years ago the entire Universe was a couple of centimetres across, well that is according to Newtonian wizardry. How fast was the speed of light when the Universe was the size of a pea. Then light travelled 10 million years to cross one yocto (y)(10^{-24}) of a mille meter. In terms of us today, light back then stood still. If it took the light 12 billion light years to cross, then the first part of the journey was pretty slow which makes the measure of time travelled versus space crossed rather ridiculous in every aspect it is portrayed. This means a lot of the light years it took to cross had to be billions of years just to gain one millimetre of space. Then the Universe must be trillions upon trillions of years older that they reckon it now is. It took a lot more darkness to form all the space is present in the entire Universal than the 12 billion years they say it did That brings us to the dark matter bit and the conspiracy that carries on undeterred. At present the conspiracy went as far as forming dark matter with (I suppose) dark energy.

One thing it does not answer is if the dark matter does have mass it must have pulling power as gravity. Then what is the dark matter waiting for to unleash the gravity by mass to pull the Universe back into forming contraction. According to Newtonian religiosity the dark matter is there although not visible. The dark matter has the mass although the dark matter is very spiteful to hide it from Newtonian view. With presence visible or not and having mass why is it not pulling now, if it is going to pull at all? Either it within the Universe being active or not active and pulls by mass or it does not pull by mass but it can't have some retarding switch preventing the mass to pull now but will kick in at a time when it pleases the Newton's. What makes the dark matter slumber mysteriously while waiting to jump on the poor defenceless little Universe and force it to comply with Newton once more. No one can prove the dark matter is not there since no one can prove it is there. This is how one go about to devise a conspiracy. You keep it quiet and while everyone smells a dead rat but no one even think of looking in the right direction. The conspiracy is a success if everyone accuses anything accept detect the true conspiracy.

They think they know gravity to such split detail and to where the detail goes to such precise extend that they are able to calculate how much gravity the missing dark matter will provide to change what is expanding into what then will contract into all the Black Holes. Black matter must be to provide a force allowing the cosmos in experiencing the next big implosion that is coming somewhere in the future while it turns the exploding direction of the cosmos into an imploding direction. That the implosion must come even in the face of insufficient gravity is a certainty otherwise Newtonian physics is completely inadequate in their cosmic vision about the Universal future! With their having this qualified virtue of intellectual splendour spawning such phenomenal abilities they then would have to know what gravity is!

This dark matter hides in places we can't see. This dark matter is what now forms the lost matter that protected Newton's image of correctness. Still Newton is untouchable because now in the present time the undiscovered dark matter is waiting to contract the Universe and this dark matter hanging suspended is what protects Newton. Is that not that sweet? Is that not the bedtime story every five year old would wish to hear every night? Every child will go to sleep feeling secured and in comfort. We can't attack Newton because the unseen matter that is dark is protecting Newton. Newton becomes untouchable by undiscoverable, unseen, untraceable material that lives like a fairy tail in fantasy. No one asks that if there is dark matter, what is the dark matter waiting for before it unleashes its incredible mass deployed force of gravity on this little unsuspecting Universe tiny as it is and pull it into redemption. What is preventing the dark matter from forming gravity that will do the job at this point in time? Why is it that the matter must be dark and must be seen in order not to form gravity. If the matter is present and forming a part of the Universe, albeit dark or not, seen or unseen, detected or not, if it has mass and if mass does bring about a force and the force is contracting gravity, then it must employ gravity. What is suspending this dark matter from kicking in and clocking in for duty? What prevents the dark matter from starting to get pulling? This is as big a scam as all the rest of the fraud they use to cover up Newton's fraud. I am showing all of this to prove how much deception there is in cosmology. Everyone in astrophysics is living a fantasy and everyone can make as they please, as long as the mathematical calculations seems to be in order. The reality and the viability or the lack thereof is no one's concern. As long as they can come up with stupefying formulated mathematics any dream will do. And the conspiracy carries on as long as it avoids reality and is void of constructive argumentative facts.

One night so many moons ago I have I don't wish to remember the number, I was sitting outside staring at the night sky while anticipating about the riddles of the Universe while I was attempting to solve the part that riddled me and with my meagre abilities since I don't even understand Newton it did not take that much to riddle me. Sitting outside and staring at the night sky gave me a break from all the confusion that faced me as I was again rejected by one of the so many academics rejecting my work and at the time I was still taking their rejection seriously and took their replies to heart. Then I saw the darkness of the night sky and compared that darkness with the brightness we find the stars portrait in order to inform us of its location. That idea made me wonder about light and the manner in which light travels. The question on which my entire work is founded is if the Universe is that immensely big and my eye can manage to accommodate about one electron then how is it possible to see the entire Universe through that small space. In this answer is hidden all the secrets ruling the Universe. When we go in search of what principles applies to form the building material in the Universe we better look and see what is it that the Universe shows us most graphic and we better stop telling the Universe what it is that we want to see and what the Universe should offer us that we wish to see. We better stop telling the Universe it must get mass and start to see what the Universe tells us what it has to offer us to see. If stars burst by releasing heat then stars are constructions that confine heat or cram heat into a small space. If this is true then gravity must be the process of freezing heat by turning movement and displacing space into compacted heat making gravity a process whereby space freezes as it condenses

Is there anybody that will seriously try to convince me or any other sober-minded person that any human being can measure what is to be considered as mass even in the picture your viewfinder shows? Go to the any picture showing the cosmos and look at what is presented as material in that picture. Is any sanity left in the suggestion that any body may even think of attempting to calculate what is in such a small portion of a fragment of a sideshow of the Universe? Can there be any person in his right mind that will think he could have the ability to measure only what is in a picture, let alone what might be available in the entire Universe? If anything can ever bear testimony of how mad those Newtonians got, then this must represent their total loss of mental coordination about what is reality and what is hallucination of a mind gone missing of reason. Their arrogance at that instant of deciding to say they measured the Universe or in following such a direction of thought grew into mindless stupidity. Put the earth in any of these pictures hat show a part of the cosmos and the task of picturing the earth within the cosmos true to size are senseless to perform. Put the sun in as a visible star within the cosmos and the task is ludicrous. Put the solar system in and it is not a freckle. The entire Milky Way might come about as a speck somewhere, but not big enough to be noticed. Then tell me please how many Milky Ways might fit into this small part of a huge Universe…and Albert Einstein was prepared to measure the entire Universe. I know the formula he used but using that formula indicated just how lubricous the attempt was. Still, they say Einstein determined the average mass of the entire Universe. This is a small part of what he said he could achieve. This is a conspiracy as blatant as ever there was one. Believing this first require the drinking an bottle of rum and then getting high on a barrel of cannabis. Either they were fools or they thought the entire human race was brainless fools incapable of thoughts.

But this blunt arrogance has its foundation in the way that they cheat when they teach how mass pulls by gravity. If that hoax can work any bloody attempt on fraud must also work. Please do try and fit any of Newton's gravitational equations into this or the other pictures

$$\left(\frac{P}{2\pi}\right)^2 = \frac{a^3}{G(M+m)},$$

ands then remain sensible and sane. Please put in place, using the mass in the picture above and come up with a realistically believable and sensible answer as to explain the positioning of planets in the solar system! Fit any of Newton's formulas to find either the mass in the entire picture or the force of gravity holding this lot together. For the mathematical genius Einstein said he was this shows he is pretty stupid in understanding the basics of physics. He said he tried to calculate the mass of every object. Even in taking a representative part or fraction of what is the cosmos and declaring that as what would be equal to measuring the Universe still declares his uncompromising ability to measure the Universe. To use the formula he did use would bring no results so I am ignoring that wasteful attempt he tried in using the formula he did. If I explain that one in this article the exercise would become a comic or a funny and that I don't want. I try to give his attempt some realistic credibility for the sake of Einstein trying to achieve an effort worth some dignity. However, all the blatant ignoring of human respect is not showed nor epitomised by this ludicrous ignoring of fellow human intellect because with what they tried to hide by showing what they did with what they tried to hide, with that they got away. Everyone this far except me as far as I know never kicked against the critical density theory and therefore they were successful in fooling yet again all intellectual person within the context of the human race. At the present everyone that launched the critical density conspiracy died with not one on earth being any the wiser about the role anyone had in this conspiracy! That is success storey in every sense as far as a conspiracy goes.

The crudeness that they got away with was to give the Universe the blame for Newton's misjudgement. It was that they succeeded to put the blame of the error on the part of the Universe. They turned the facts around that the expanding of the Universe was blamed and the cosmos. Instead of contracting the Universe is at fault to expand and it is time to find out how the Universe will find a remedy for its lost way. The Universe expands contradicting Newton's contraction and it therefore is to be understood that it is the Universe that has to mend its ways and correct the misconduct to befit what Newton held as correct. Newton can't take the blame for the Universe expanding as this goes against goes nature since it has to contract in precisely the manner that Newton said it should. If it is not following Newton's orders then there has to be lost mass. Where the lost mass will go is beyond me but again the fact that the Universe could end also exceeds my understanding. But with the Universe being the wrong party there has to be mass that is unaccounted for. If they find the mass they can replace it (why else would they bother to locate the missing mass) and get the Universe back on tract and contracting once more.

Think how scandalously farfetched the presumption goes. Where in any of Newton's formula or cosmic principles is their any indication that there is a limit to the expanding in accordance to what Newton declared. The elasticity or stretch ability and the limit thereof they made up as they went along while formulating this conspiracy. The deduction they make is on the grounds of having a "flat" Universe that is "slowing" and losing momentum. This too is reckless deduction about singularity they have no perception of. They wanted the Universe to slow down so that the Universe would stand still and form a zero expanding change direction and then start to contract. Why would I have this conclusion? Because if $\Omega = 2q_0 = (2/3\Lambda)(c^2/H^2)$ was correct the Universe had to have a limit. This idea is preposterous because with every end there has to be some beginning of something else. The cosmos cannot end because Kepler showed that every ending of a straight line is a circle and every circle follows as a straight line. This means that every circle forms a straight line when an object orbits another object and every object in the Universe orbits some other object while being the centre of something else around which that objects orbits. This is why the Universe can never end because going in a straight line puts such a straight line in an immediate circle. However reading this mathematical equation one see that movement T^2k of space a^3 centres around singularity k^0 because if space $a^3 =$ is equal to movement then movement T^2k forms space in singularity k^0 since $a^3 / T^2k = k^0$.

No object within the solar system that orbits the sun can ever leave the gravitation of the sun to go to another system of orbit. To be stronger than the centre of the sun k^0 the object has to create space a^3 by movement T^2k to overcome the gravity $a^3 / T^2k = k^0$ with which the sun holds the object in gravity. In this principle we find the eternal time position that captures space forever. The planets in the solar system go in a straight line and while they are going straight they divert from direction and circle around the sun at a specific location. This is energy distribution all work has to comply with. The wheel of the car turns as it goes straight down the road as the car turns around the earth by going straight down the road as it goes straight with the earth as the earth turns around the sun as the sun goes straight to turn around the Milky Way and there will be more going straight while spinning as the picture grows. That is the way that all power drives because this movement connects time to space and that is one thing Newtonians have no idea about...they have no idea what time is in relation to what space is because in all of this we find singularity which is above anything any Newtonian this far could manage. Energy distribution is $a^3 = T^2k$ where linear movement $a^3 / T^2 = k$ depends entirely on circular movement $a^3 / k = T^2$. By the way the most successful formula ever devised by any man is the formula Einstein came up with as $E = mC^2$. This should read $E^3 = mC^2$ and that is a translation of Kepler's formula of $a^3 = T^2k$ **in that $E^3 = a^3$, m = k and $C^2 = T^2$** This information they will never release because then they agree that the Newton idea of changing $a^3 = T^2k$ to $a^3 = T^2$ is very ridiculous and if Einstein redeployed it then it is Kepler's formula that proves to be the most successful formula ever devised. ...And this formula is the basic formula that Kepler received from the cosmos when he translated what the cosmos told him in mathematics to a verbal language used by man. Kepler had to translate cosmic mathematics into verbal language and Newton disagreed then about the finer detail the cosmos confirmed to Kepler. The cosmos said $a^3 = T^2k$ by using mathematical numbers to solidify the proof and remove any doubt where Newton then changed it to $a^3 = T^2$.

When Newton changes the tables and the findings of Kepler's work, how bright did Newton think he was to change what the cosmos gave Kepler to something that befitted Newton's ideas about the cosmos? Newton's idea of $a^3 = T^2$ is as false as a three dollar bill because this means that all third dimensional values are equal to square values making $2^3 = 2^2$ which means that 8 = 4 or that a flat surface is equal to the same surface being three sided. It also means that you being in three dimensions in front of your mirror looking at the reflection of the square image can have a debate with yourself while your image in the mirror replies on your argument. When this happens, don't call the police it is too late. Just jump from a building and make the end easy on everyone including you. Newton's idea of $a^3 = T^2$ is clear that the man had no understanding of any mathematical principles.

I am not going into this further because I have books dealing with the issue but the so called Hubble constant has so many variations and changing values that they had to call a halt to the number of people investigating this field and all of those reached another totally different value. That throws the values connected to the q_0 = Deceleration Parameter, Λ = Cosmological Constant and the H = Hubble Constant out the window. You can cheat to reach a number and to promote your point but when honesty prevails the outcome is always just a joke. Even the speed of light is no constant because the very same Einstein proved that "gravity can bend the flow of light" and "slow light down" and therefore light holds speeds according to the gravity applying in that specific part of the cosmos.

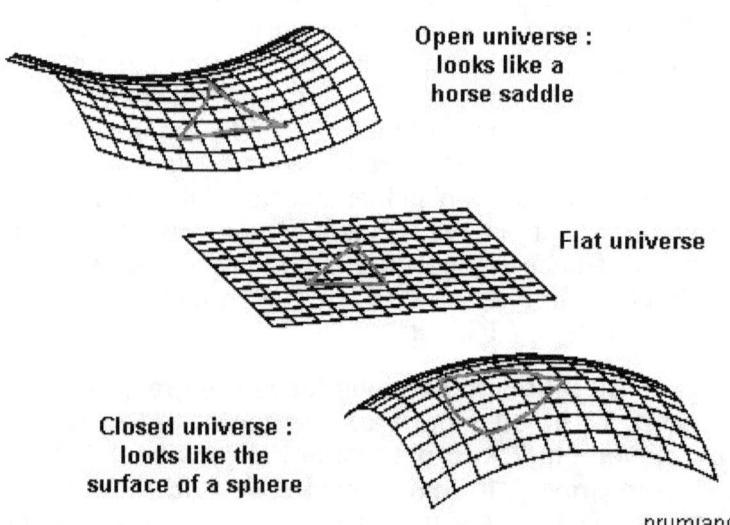

Open universe : looks like a horse saddle

Flat universe

Closed universe : looks like the surface of a sphere

nrumiano

I do not wish to go past the understanding level that easy reading is required to appreciate this article because then it would make this article only enjoyed by experts. That is the last thing I want. I want to unmask these crooks called the cosmic experts and the way they hide behind senseless mathematics. Look how far I have been thrown off the track and in that we find the purpose and the success of any conspiracy. Get the argument to divert into a million other non-consequential arguments and in the end the conspiracy had success because nobody got anywhere. ...And in that I am a sucker every time but I learned to keep the argument heading on track again after such a diversion.

Now nobody would be surprised to learn that somehow Einstein found not enough material to bring about a contraction that would save Newton and get Newtonians to save face as well as get Newtonian fraud believable again. Einstein did what he was told but clearly he was just a puppet in the conspiracy because he came back with the numbers that clearly showed the Universe was heading for eternal expanding in view of what they thought applied. He called his model the Open Universe as if there was a Closed Universe also as an option. I am not getting into that because I have devoted many pages in more serious books on this matter where I show how futile this view of the Universe is.

There can be no open or closed Universe because open and closed brings about limits that has to stretch and anyone in view of a Universe with growing limits or confined limits are not looking at the Universe in which I live. We can't look at the Universe in an entire picture because we are inside serving a place as part of the Universe. We can't view space as if we are looking at space because the space we think we see is time in eternity developing as space but we have to look at space being time within space forming. We can't look at a

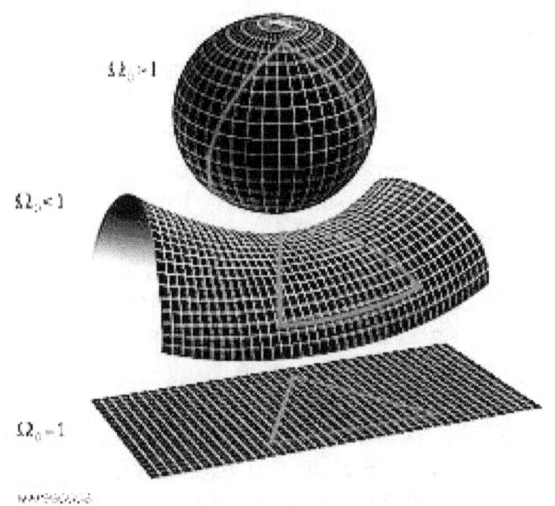

quasar and say that is what it is. We have to look at a quasar and realise that at the time we see the quasar that was what was as we were too small to be. We were at that time so small we were not even part of space or time. We also are at this instant in a developed quasar of which the ends we will never see so forget about finding the ends to what hold the quasar. We are not God looking from above but we are humans looking from within at what we are. We are the part of the Universe, a very small part of the Universe. We look not at space but at time because space is what history time left behind in light. We look at space and look at our history at a time we, or something as small as the sun we now regard as huge but then had no room to be within that what we now look at because at that instant the sun did not exist. That is how small the sun was back then. The Universe is not getting bigger but is getting smaller by allowing small things within to get bigger in relevancy of growth.

$$ds^2 = -R^2(t)\left[\frac{dr^2}{1-kr^2} + r^2 d\theta^2 + r^2 \sin^2\theta\, d\varphi^2\right] + c^2 dt^2$$

Applying a formula such as these confines the Universe to the limitations carried by the formula and does not put claim to borders within the Universe. The use of this carries the limits of the human mind onto the Universe that has no limits.

Notwithstanding how impressive the mathematical reasoning is, there is one fundamental flaw in all the suggested possibilities. The Universe according to Newtonians can develop in three possible categories with two more being a cone and a sphere. I am not making it my task to promote or to argue the pros and the cons of each system because all these suggestions are ultimately completely flawed. There is a Universe in singularity but it is a not even close to what they wish to see. I explain that Universe in many of my books.

The Universe only has an inside and can never have an outside.

Every one of the pictures they present and the others I do not show have the Universe holding an outside. The Universe only has an inside in which I am. Time in singularity forms the outside and because space is the departure of infinity from eternity the space that we think we see is ultimately time and time never ends. Because what we see is time where space is the presentation of the history of time we see what we see without having an outside or a stop to what we can see. Time moves to the past by forming a hologram we see as space and that I prove. We are inside as small as we are and we are looking at the smaller end we are able to look towards a direction that has no end.

Looking at the Universe we look at the past. We can't see the future; we can only see where what we are in also were so many trillion years ago. But what we now see looking at galactica back in time was so small that what we now think of as big had no relevance to exist in terms of what then filled space. We can't look at a galactica and presume we hold equality in time. Trillions of years filled space from what we see to where we are. We are in time in the present but seeing the galactica does not put the galactica in the present. It puts us a trillion years back before there was any possibility for a sun and an earth to be at all with no chance back then of even finding a thought of life as a factor. We are in time in the present while what we see is the past we look at being in the present billion years ago. What we see grew as much as where we are and time moved on just as time moved on where we now are. We can't project anything into that space or situation and we can't project us into what we see because what we see had no "us" and what we have has evolved so much we can't even realise how it evolved. It's like looking at Julius Caesar and asking why did he not make radio contact with Rome just as he crossed the Rubicon. It could have settled the aftermath in so much better fashion. We can't ask why did Rome not put Attila the Hun under tank fire and prevent him from taking Rome because we know that in the future the tank will become the most decisive weapon man has invented. It is different eras and what we see is what we were and not where we are. If we were there we will still would be going trillions of years to get to where we are and it takes time to be going further from there to where we now are. We are in time and space forms the history of time as every instant go by.

Don't confuse time and space or space with time. When we look at the Universe we look at our past. That which I see is a hologram written in light of something that was that never again will be. It is not something as substantial in substance as the moon or Mars but is a thought God provides of an era long gone into the past. I can't look at the Universe in which I am because I am the Universe in which I am. Every atom is the end of a Universe and the start of the next Universe as Kepler said $a^3 = T^2k$ and those Universes made entirely of atoms are divided by time formed as space by the history of time. When we wish to look at the future we have to look into the darkness because we are heading in that direction. It is the darkness where we are going so we have to look at the darkness around us to try to see where we are going and also look at the darkness to see where we came from.

We are a speck coming from the past being in darkness heading to the future also in darkness.

Either we look at where we are and only find darkness going smaller and going bigger because that is where we are in the Universe or we look at a quasar and find a non-existing spot too small to be within the quasar and in that point where there is still no point in the quasar try to find our position while knowing we are too small too have any position in what the quasar was at that time a couple of trillion years ago. We did not have a position in the quasar back then and we now are in our own quasar but we are so small in the dark vastness we are undetectable within where we are. This issue is much too complex to debate in a book as light hearted as this because this issue borders on testing sanity and renders not the cosmic limit but our mental limit. To find where we are going we can't look to the past because we are the future of the past we are looking at when we gauge at space filled with material trillions of years ago. Where we now are we find our position is so small that wherever we look we find eternity covered in darkness surrounding us from everywhere.

But beware of a conspiracy because a good conspiracy has in it a program to allow any person to divert from the route of tracing the essence by starting to argue the content and then land in a completely new debate that has no essence of the conspiracy one wish to detect. So therefore let's get back to where we were before we were here. Let's catch the conspiracy as it ran pout of breath and into a new idea. Einstein came up with some bad news. He did not only saw a Universe expanding endlessly, he draws a catalogue of how the Universe will work. The man again did not soothe Newton's shortcoming but built an argument with models included to exacerbate the issue. Now they had to come up with a plan that will kill all further questions. They now longer could relay on Einstein because Einstein was cashing in on Newtonian despair. Remember, the Universe is supposedly going to crash into itself according to Newton and now these models all show a Universe doing exactly what the Universe should not do.

In the past it was accepted that only Newton and God never made a mistake and since the Brainy Bunch Newtonians were mostly atheistic or atheistically orientated they were not that sure about the credibility of God but the unquestionable accuracy of Newton those atheists were pretty sure…and now it seems Newton made a mistake. That just could not be. They would rather have God make the mistake than let it seem as if Newton was up to no good while Newton made the incredible mistake.

Then they allowed God to make the mistake. If the Universe was expanding it was God's fault. It sounds much better than have anyone think it is Newton's mistake. If the cosmos expanded while Newton said it must contract then this rebellious behaviour of the cosmos must end. They had to put the blame on the Universe and then ultimately on God for making such a mistake. It must be God that made the error.

The easiest is to put the blame onto God by finding the fault at the door of an uncompromising Universe. To make the conspiracy believable they had to conjoin a concocted story that would have every halfwit on earth believe it. What will make the Universe expand? It must be a lack of the something that makes the Universe contract. Not enough contracting solution will be the cause of the expanding. If it was mass that had the Universe contract in Newton's terms, then a lack of mass will lead to a shortfall in gravity and then the elasticity would not be enough so the elasticity would be tested and before the elasticity failed completely they had to get the Universe back on track. They saw a good measure of darkness splitting small bits of light. This must be it then. Who put that much darkness amongst that little Light?

If it was mss that should contract the Universe then it had to be not enough mass that would be to blame for the expanding of the Universe. Now to get someone credible enough in the eyes of the public yet foolish enough in ego to provide the cover for the conspiracy to work was another matter. The only candidate must be Albert Einstein. They had to get Albert Einstein to measure all the mass in the entire Universe. Now this is where the joke no longer seems funny. Only fools and idiots are going to fall for that. Look at the picture on the previous page and the picture above. This is unrecognisable small portions of a large and overwhelming large Universe. Who is keeping whom for a bloody fool?

This is the mathematical truth…as the dividing factor increases; the influence that the mass will project in the formula $F = G\ \dfrac{M_1 M_2}{r^2}$ will diminish in respect to the growth of the distance.

In that sense the gravity force between the earth and the moon must reduce its ferocity therefore weaken. However, Newtonians only apply their ability to calculate and knowledge for the purpose of upholding Newton and never to provoke Newtonian liability by telling the truth. Never is there any mention of mathematical reality when mathematics is used.

However this no one ever knew to be true outside of the intimate upper circles of physics, and this unmasking was to be prevented at all cost. A plan was to be devised because if the public found out Newton was a fraud all along and being Newtonian was the personification of stupidity the entire science world would come tumbling down on the heads of those most important Brainy Bunch.

No, better still, to save science a conspiracy was devised. The most intellectuals on earth had to cook up something and devise a plan to save their image and the name of Newton. The most intellectual minds concluded and fabricated to form a conspiracy to withhold the truth and forge a fraud that lasted almost ninety years to the day. Should anyone disagree with the term conspiracy, then please let me know what you would call what happened after Hubble's discovery became prominent on the news.

Is there anybody that will seriously try to convince me or any other sober-minded person that any human being can measure what is to be considered as mass even in the picture above? Go back to the previous page and look at is presented as material in that picture. Is any sanity left in the suggestion that any body may even think of attempting to calculate what is in such a small portion of a fragment of a sideshow of the Universe?

Can there be any person in his right mind that will think he could have the ability to measure only what is in these pictures, let alone what might be available in the entire Universe? If anything can ever bear testimony of how mad those Newtonians got, then this must represent their total loss of mental coordination about what is reality and what is hallucination of a mind gone missing of reason. Their arrogance at that instant of deciding in following a direction grew into mindless stupidity.

Put the earth in any of these pictures and the task is senseless to perform. Put the sun in as a visible star and the task is ludicrous. Put the solar system in and still it will not show as a freckle. The entire Milky Way might come about as a speck somewhere, but not big enough to be noticed. Then tell me please how many Milky Ways might fit into this small part of a huge Universe...and Albert Einstein was prepared to measure the entire Universe.

I know the formula he used but using that formula indicated just how lubricous the attempt was. Still, they say Einstein determined the average mass of the entire Universe.

This is a small part of what he said he could achieve. This is a conspiracy as blatant as ever there was one. Believing this first require the drinking an bottle of rum and then getting high on a barrel of cannabis. Either they were fools or they thought the entire human race was brainless fools incapable of thoughts.

Please do try and fit any of Newton's gravitational equations into this or the other pictures ands then remain sensible and sane.

Please put
$$\left(\frac{P}{2\pi}\right)^2 = \left(\frac{a^2\sqrt{1-\varepsilon^2}}{\ell}\right)^2 = \frac{a^4(1-\varepsilon^2)}{\ell^2} = \frac{a^4(1-\varepsilon^2)}{a(1-\varepsilon^2)GM} = \frac{a^3}{GM}$$
or
$$\left(\frac{P}{2\pi}\right)^2 = \frac{a^3}{G(M+m)},$$
in place, using the mass in the picture above and come up with a realistically believable and sensible answer!

Fit any of Newton's formulas to find either the mass in the entire picture or the force of gravity holding this lot together. For the mathematical genius Einstein said he was this shows he is pretty stupid in understanding the basics of physics. He tried to calculate the mass of every object. To use the formula he did use would bring no results so I am ignoring that wasteful attempt he tried in using the formula he did. If I explain that one in this book the exercise would become a comic or a funny and that I don't want. I try to give his attempt some realistic credibility for the sake of Einstein trying to achieve an effort worth some dignity.

However, all the blatant ignoring of human respect is not showed nor epitomised by this ludicrous ignoring of fellow human intellect because with what they tried to hide by showing what they did with what they tried to hide with that they got away. Everyone this far except me as far as I know never kicked against the critical density theory and therefore they were successful. Everyone that launched the conspiracy died with not one on earth being any the wiser about the role anyone had in this conspiracy! That is success storey in every sense as far as a conspiracy goes.

The crudeness that they got away with was to give the Universe the blame for Newton's misjudgement. It was that they succeeded to put the blame of the error on the part of the Universe. They turned the facts around that the expanding of the Universe was blamed and the cosmos. It therefore is to be understood that it is the Universe that has to mend its ways and correct the misconduct to befit what Newton held as correct. Newton can't take and therefore the blame of the Universe expanding goes to nature and all the while it has to contract in precisely the manner that Newton said it should.

If it is not following Newton's orders then there has to be lost mass. Where the lost mass will go is beyond me but again the fact that the Universe could end also exceeds my understanding. But with the Universe being the wrong party there has to be mass that is unaccounted for. If they find the mass they can replace it (why else would they bother to locate the missing mass) and get the Universe back on tract and contracting once more. Think how scandalously farfetched the presumption goes. Where in any of Newton's formula or cosmic principles is their any indication that there is a limit to the expanding in accordance to what Newton declared. The elasticity or stretch ability and the limit thereof they made up as they went along while formulating this conspiracy. The deduction they make is on the grounds of having a "flat" Universe that is "slowing" and losing momentum. This too is reckless deduction about singularity they have no perception of. They wanted the Universe to slow down so that the Universe would stand still and form a zero expanding change direction and then start to contract. Why would I have this conclusion? Because if $\Omega = 2q_0 = (2/3\Lambda)(c^2/H^2)$ was correct the Universe had to have a limit. This idea is preposterous because with every end there has to be some beginning of something else. The cosmos cannot end because Kepler showed that every ending of a straight line is a circle and every circle follows as a straight line. In mathematical equations it reads as $a^3 = T^2k$ and then $a^3 / T^2 = k$ and $a^3 / k = T^2$. The formula reads as $a^3 = T^2k$ where gravity is space a^3 that moves in a circle t T^2 hat goes straight k.

This means that every circle forms a straight line when an object orbits another object and every object in the Universe orbits some other object while being the centre of something else around which that objects orbits. This is why the Universe can never end because going in a straight line puts such a straight line in an immediate circle. All driving in the cosmos be it steam, combustion engine or electric or even cosmic driving of orbiting objects has to be in a circle followed by a straight line followed by a circle and this continues indefinably.

The driving value of $a^3 / T^2 = k$ becomes $a^3 / k = T^2$ that becomes $a^3 / T^2 = k$ that becomes $a^3 / k = T^2$ repeating the process indefinably. However reading this mathematical equation one see that movement T^2k of space a^3 centres around singularity k^0 because if space $a^3 =$ is equal to movement then movement T^2k forms space in singularity k^0 since $a^3 / T^2k = k^0$. No object within the solar system that orbits the sun can ever leave the sun to go to another system of orbit. To be stronger than the centre of the sun k^0 the object has to create space a^3 by movement T^2k to overcome the gravity $a^3 / T^2k = k^0$ with which the sun holds the object in gravity. In this principle we find the eternal time position that captures space forever.

The planets in the solar system go in a straight line and while they are going straight they divert from direction and circle around the sun at a specific location. This is energy distribution all work has to comply with. The wheel of the car turns as it goes straight down the road as the car turns around the earth by going straight down the road as it goes straight with the earth as the earth turns around the sun as the sun goes straight to turn around the Milky Way and there will be more going straight while spinning as the picture grows.

That is the way that all power drives because this movement connects time to space and that is one thing Newtonians have no idea about...they have no idea what time is in relation to what space is because in all of this we find singularity which is above anything any Newtonian this far could manage. Energy distribution is $a^3 = T^2k$ where linear movement $a^3 / T^2 = k$ depends entirely on circular movement $a^3 / k = T^2$. By the way the most successful formula ever devised by any man is the formula Einstein came up with as $E = mC^2$. This should read $E^3 = mC^2$ and that is a translation of Kepler's formula of $a^3 = T^2k$ in that $E^3 = a^3$, $m = k$ and $C^2 = T^2$. This information they will never release because then they agree that the Newton idea of changing $a^3 = T^2k$ to $a^3 = T^2$ is very ridiculous and if Einstein redeployed it then it is Kepler's formula that proves to be the most successful formula ever devised.

…And this formula is the basic formula that Kepler received from the cosmos when he translated what the cosmos told him in mathematics to a verbal language used by man. Kepler had to translate cosmic mathematics into verbal language and Newton disagreed then about the finer detail the cosmos confirmed to Kepler. The cosmos said $a^3 = T^2k$ by using mathematical numbers to solidify the proof and remove any doubt where Newton then changed it to $a^3 = T^2$.

From this load of rubbish they try to play god by envisaging how the Universe will grow in the future.

Looking at the Universe from this angle is most beautiful and by applying the magic of mathematics it seems to be so real except for one problem we encounter and that comes when we use a dash of logic. When using equated mathematical formulas one include a certain part of the Universe by excluding the rest of the Universe. There is no rmula that has the ability to contain the Universe in its inclusive entirety because the Universe is eternal no matter how one would look to appreciate what one sees. Looking at any part of the Universe the distance we don't see removes the quantification of the picture that we see. What ever forms a picture of the Universe excludes untold many, many times more than what the picture we see reveals because it is shear stupidity to think what we see in any picture represents everything there can be.

Beyond what we see is eternity looming outside of the view we have. If we could get as far back as this picture would allow we would encounter as much space in all directions and that space will be filled with as many stars as this entire picture reveals and then even much, much more. If we shift to the back of this picture we take the centre of the Universe with us because the centre of the Universe according to our position would be where we are. Where we go we move the centre of the Universe with us everywhere we go.

What I tell must be told by looking at pictures in order to understand the failures of Newtonian thinking and the flawed concepts they arrive at by using mad mathematics they don't even understand. Let's say we can look back in this picture to the point that is 12.5×10^9 years from us and is where the Universe started according to Newtonian wisdom. We then look at 12.5×10^9 years of space forming. When we get to the point that is 12.5×10^9 years from us we will be in the centre of the Universe and standing at that point looking to where we now are we will be saying oh, that is where the Universe are because that is the point where the Universe started, which in fact then will be the point where we now stand and look at the limit we can see and that makes what we see not the end of the Universe that we see.

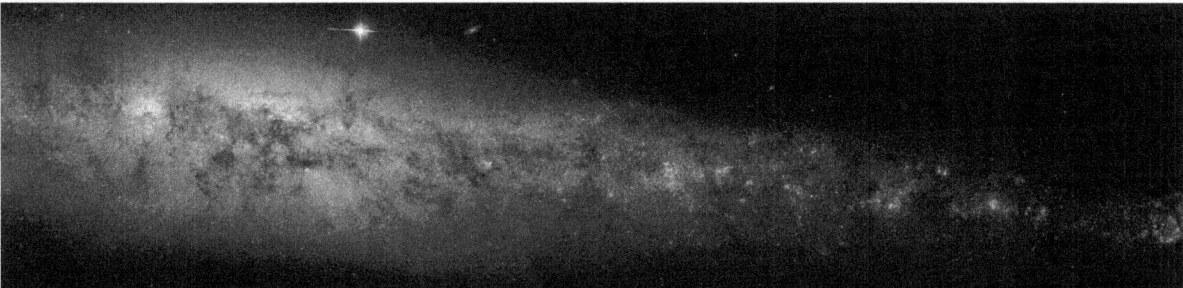

When we see this picture this is not all we see. This is what the electronic and mechanical driven machine and telescopes bring us and those limitations allow us to see what we see in this picture. Behind what this picture shows are an eternity of more stars, which we can't see. For the Newtonian to admit his devises constrains his views will be too much to contemplate because they know everything.

But it is so Newtonian to reduce the Universe to fit the simple-minded approach that the Newtonian attach to the Universe and by shrinking the Universe to their mathematical abilities as being 13×10^9 years old it shows not how young the Universe is but how limited is the physics of the Newtonian simpleton. It is the Newtonian's mathematics and their ability to apply mathematics and their ability to understand what no one can understand that holds the limitations and the ends they portray. They can "understand" Newton and dare I say more than that. Newtonian wisdom let us to believe there are three ways of cosmic development. I have to explain that Newtonian wisdom fills all the space in the distance covered by outer space with loads of "nothing". Newtonians say what forms outer space is "nothing" and outer space comprises in total of "nothing". "Nothing" in outer space is what the space in outer space is according to Newtonian intellect. Newtonians think that between the moon and the earth there is "nothing".

The way Newtonians also think is that between the sun and Mercury there is lots of "nothing". Would it then surprise you to know that Newtonians think that between the sun and Pluto there is lots and lots and lots of "nothing". From the earth to the moon the Apollo space mission travelled through lots and lots of "nothing" and the "nothing" stretched all the way up to and even further than the back of the moon because they found "nothing" even there. It is important to know that on the moon surface there is "nothing" and on the back of the moon there still is "nothing" but that "nothing" that is on the surface of the moon is very much different from the "nothing" that the Apollo missions went through on their way to the moon. Between every planet and the sun the "nothing" is escalating to the point that the "nothing" is doubling between every planet that is separated by "nothing" and this doubling of "nothing" is continuing beyond Pluto where there is even more "nothing"! Can you match that Newtonian argument?

In the Universe three possible forms are under construction. One is the closed Universe, bringing boundaries in the shape of roundness or ultimately a sphere. Then there is the open Universe   idea   allowing the Universe to form in the shape of a saddle.

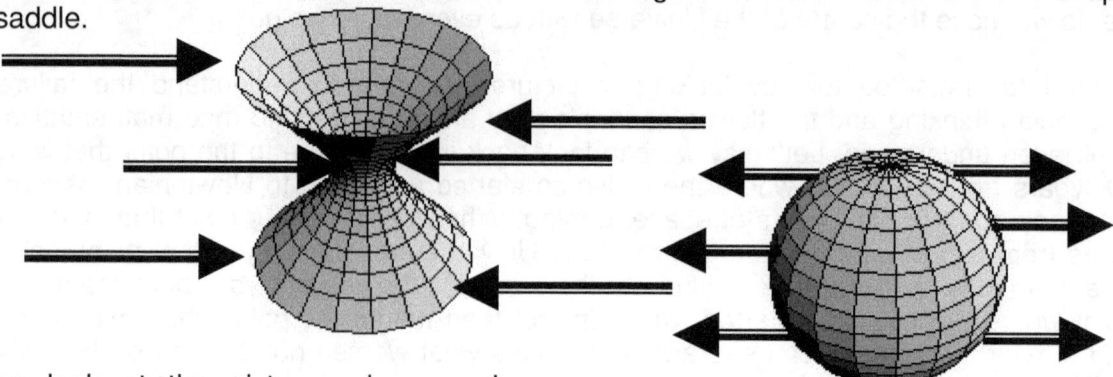

Please look at the pictures above and answer my   question with your clear mind thinking logic. If that is the end the lines indicates then what is beyond the limit where the lines end. Our Newtonian wizards say it is "nothing". That brings one of the biggest problems I ever faced to mind. Looking at the examples they hold of the Universe there is a point beyond which there is no point. At the end of some point is a limit after which there is "nothing".

However, the most brilliant minds man can bring forward already filled an entire Universe with "nothing" and at a point where "nothing" ends we have "nothing" more going on. What is in place of the point where "nothing" ends by a wall (I suppose) of "nothing".

The "nothing" that ends at that point begins with more "nothing" not continuing but ending in "nothing".
In order to "be", there has to be "something" but if they fill that which should be filled with something with "nothing" then what is in the place of something when they say they run out of "nothing". I think there is a lot of "nothing" filling their thoughts and then they translate the "nothing" they think of to outer space and into the Universe and that is how they got all the "nothing" with which they are filling the Universe.

How bright did Newton think he was to change what the cosmos gave Kepler to something that befitted Newton's ideas about the cosmos? Newton's idea of $a^3 = T^2$ is as false as a three dollar bill because this means that all third dimensional values are equal to square values making $2^3 = 2^2$ which means that $8 = 4$ or that a flat surface is equal to the same surface being square. It also means that you being in three dimensions in front of your mirror looking at the reflection of the square image can have a debate with yourself while your image in the mirror replies on your argument.

When this happens, don't call the police it is too late. Just jump from a building and make it easy on everyone including you. Newton's idea of $a^3 = T^2$ is clear that the man had no understanding of any mathematical principles.

Using $(2/3\Lambda)(c^2/H^2)$ puts the entire Universe equal taking away development of space in time. It says that the quasar they see at this moment never changed in say the three billion years it took the light to get to us and the space holding material that they see is what there is at the very moment they see it.

They never gave tolerance for Universal development. This is the crap you get when following Newton $a^3 = T^2$ without thinking. Then $(2/3\Lambda)(c^2/H^2)$ is ridiculous because every second space changes at any place in the Universe and that is just because the moon will never crash into the earth or the earth crash into the sun.

Another formula I wish to explain is

$\Omega = 2q_0 = (2/3\Lambda)(c^2/H^2)$ where ...

Ω = density

q_0 = Deceleration Parameter (*where they get that from only they would believe the fantasy*)

Λ = Cosmological Constant (*every spot in space has a different gravity value so how a Newtonian would get a constant applying is a fairy storey come true but that is Newtonian motto*)

c = speed of light (*can never have a fixed value because gravity slows down the speed of light even to become a minus as it is in a Black Hole*)

H = Hubble Constant (*this too is a dream Newtonians cook up to give Newton some form of legitimacy*)

I am not going into this further because I have other books dealing with the issue but the so called Hubble constant has so many variations and changing values that they had to call a halt to the number of people investigating this field and all of those reached another totally different value. That throws the values connected to the q_0 = Deceleration Parameter, Λ = Cosmological Constant and the H = Hubble Constant out the window. You can cheat to reach a number but when honesty prevails the outcome is a joke. Even the speed of light is no constant because the very same Einstein proved that "gravity can bend the flow of light" and therefore light holds speeds according to the gravity applying in that specific part of the cosmos.

Should anyone wish to read about the working of singularity then download the book **The Absolute Relevancy of Singularity The Website** it will serve as a starter. I do not wish to go past the understanding level that easy reading is required because then it would make this book only enjoyed by experts and that is the last thing I want. I want to unmask these crooks called the cosmic experts and the way they hide behind senseless mathematics.

Look how far I have been thrown off the track and in that we find the purpose and the success of any conspiracy. Get the argument to divert into a million other non-consequential arguments and in the end the conspiracy had success because nobody got anywhere. ...And in that I am a sucker every time but I learned to keep the argument heading on track again after such a diversion. Now nobody would be surprised to learn that somehow Einstein found not enough material to bring about a contraction that would save Newton and get Newtonians to save face. Einstein did what he was told but clearly he was just a puppet in the conspiracy because he came back with the numbers that clearly showed the Universe was heading for eternal expanding in view of what they thought applied. He called his model the Open Universe as if there was a Closed Universe also as an option. I am not getting into that because I have devoted many pages in more serious books on this matter where I show how futile this view of the Universe is. There can be no open or closed Universe because open and closed brings about limits that has to stretch and anyone in view of a Universe with growing limits or confined limits are not looking at the Universe in which I live. We can't look at the Universe because we are inside as part of the Universe.

We can't view space as if we are looking at space because the space we think we see is time in eternity developing as space but we have to look at space being time within space forming. We can't look at a quasar and say that is what it is. We have to look at a quasar and realise at the time we see the quasar that was what we were while we were at the time so small we were not even part of space or time. We also are at this instant in a developed quasar of which the ends we will never see forget about finding the ends to what holds the quasar. We are not God looking from above but we are humans looking from within what we are. We look not at space but at time because space is what history time left behind in light. We look at space and look at our history at a time we, or something as small as the sun we regard as huge had no room to be within that what we look at because at that instant the sun did not exist that is how small the sun back then was. The Universe is not getting bigger but is getting smaller allowing small things within to get bigger in relevancy of growth.

$$ds^2 = -R^2(t)\left[\frac{dr^2}{1-kr^2} + r^2 d\theta^2 + r^2 \sin^2\theta\, d\varphi^2\right] + c^2 dt^2$$

Notwithstanding how impressive the mathematical reasoning is, there is one fundamental flaw in all the suggested possibilities. The Universe according to Newtonians can develop in three possible categories with two more being a cone and a sphere. I am not making it my task to promote or to argue the pros and the cons of each system because all these suggestions are ultimately completely flawed. **The Universe only has an inside and can never have an outside.**

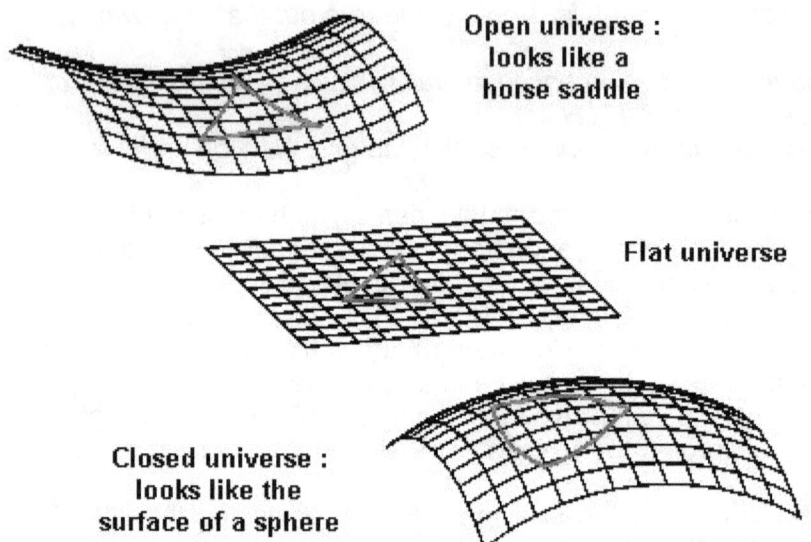

Open universe : looks like a horse saddle

Flat universe

Closed universe : looks like the surface of a sphere

nrumiano

Every one of the pictures I present and the others I do not show have the Universe holding an outside. The Universe only has an inside in which I am. Time in singularity forms the outside and because space is the departure of infinity from eternity the space that we think we see is ultimately time and time never ends. Because what we see is time where space is the presentation of the history of time we see what we see without having an outside or a stop to what we can see. We are inside and looking at the smaller end we is looking towards a direction that has no end.

Looking at the Universe we look at the past. We can't see the future; we can only see where what we are in also were so many trillion years ago. But what we now see looking at galactica back in time was so small that what we now think of as big had no relevance to exist in terms of what then filled space. We can't look at a galactica and presume we hold equality in time. Trillions of years filled space from what we see to where we are. We are in time in the present but seeing the galactica does not put the galactica in the present. It puts us a trillion years back before there was any possibility for a sun and an earth to be at all with no chance back then of even finding a thought of life as a factor. We are in time in the present while what we see is the past we look at being in the present billion years ago. What we see grew as much as where we are and time moved on just as time moved on where we now are. We can't project anything into that space or situation and we can't project us into what we see because what we see had no "us" and what we have has evolved so much we can't even realise how it evolved. It's like looking at Julius Caesar and asking why did he not make radio contact with Rome just as he crossed the Rubicon. It could have settled the aftermath in so much better fashion. We can't ask why did Rome not put Attila the Hun under tank fire and prevent him from taking Rome because we know that in the future the tank will become the most decisive weapon man has invented. It is different eras and what we see is what we were and not where we are. If we were there we will still would be going trillions of years to get to where we are and it takes time to be going further from there to where we now are. We are in time and space forms the history of time as every instant go by.

Don't confuse time and space or space with time. When we look at the Universe we look at our past. That which I see is a hologram written in light of something that was that never again will be. It is not something as substantial in substance as the moon or Mars but is a thought God provides of an era long gone into the past. I can't look at the Universe in which I am because I am the Universe in which I am. Every atom is the end of a Universe and the start of the next Universe as Kepler said $a^3 = T^2k$ and those Universes made entirely of atoms are divided by time formed as space by the history of time. When we wish to look at the future we have to look into the darkness because we are heading in that direction. It is the darkness where we are going so we have to look at the darkness around us to try to see where we are going and also look at the darkness to see where we came from. We are a speck coming from the past being in darkness heading to the future also in darkness. Either we look at where we are and only find darkness going smaller and going bigger because that is where we are in the Universe or we look at a quasar and find a non-existing spot too small to be within the quasar and in that point where there is still no point in the quasar try to find our position while knowing we are too small too have any position in what the quasar was at that time a couple of trillion years ago. We did not have a position in the quasar back then and we now are in our own quasar but we are so small in the dark vastness we are undetectable within where we are. This issue is much too complex to debate in a book as light hearted as this because this issue borders on testing sanity and renders not the cosmic limit but our mental limit. To find where we are going we can't look to the past because we are the future of the past we are looking at when we gauge at space filled with material trillions of years ago. Where we now are we find our position is so small that wherever we look we find eternity covered in darkness surrounding us from everywhere.

But beware of a conspiracy because a good conspiracy has in it a program to allow any person to divert from the route of tracing the essence by starting to argue the content and then land in a completely new debate that has no essence of the conspiracy one wish to detect. So therefore let's get back to where we were before we were here. Let's catch the conspiracy as it ran pout of breath and into a new idea. Einstein came up with some bad news. He did not only saw a Universe expanding endlessly, he draws a catalogue of how the Universe will work. The man again did not soothe Newton's shortcoming but built an argument with models included to exacerbate the issue. Now they had to come up with a plan that will kill all further questions. They now longer could relay on Einstein because Einstein was cashing in on Newtonian despair. Remember, the Universe is supposedly going to crash into itself according to Newton and now these models all show a Universe doing exactly what the Universe should not do.

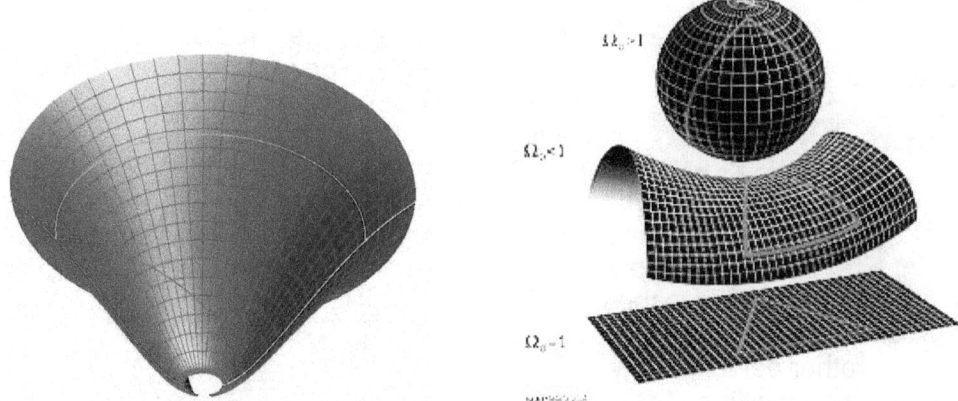

Again the incorrectness of every model is showing it has an outside. Outer space has not and can never have an outside. To correct the model we have to put where we are in the middle with as much going smaller that what is going larger. We will never see the smallest and we are unable to see the biggest. The sphere we search for forming is not formed as a result of outer space forming a spherical limit to space, as outer space is not representative in the largest structure because there is no largest structure. This is because the Universe has no outside. The sphere that forms represents the smallest part, which is within singularity Π^0 becoming space Π and forming space Π^3 by the movement $\Pi^2\Pi$ where that forms the sphere that forms singularity as singularity return the favour to form a sphere. In that we find the eternal value of space $\Pi^3 = \Pi^2\Pi$ by the relevancy of singularity trapping space in the movement thereof. Everything forming space $\Pi^0 = \Pi^3 / \Pi^2\Pi$ will loose all the space that movement forms $\Pi^0 = \Pi^3 / \Pi^2\Pi$ when movement of space is lost. That is what happens in a Black Hole where movement of space stopped and time reclines directly into singularity producing the opposite of what the Big Bang represented. That is the Big Crunch not that big and not very crunchy where it all ends. The Black Hole is the smallest and the largest at the same time of what the Universe offers as an object filling space (or not filling space). In the Theses the explaining gets a lot more technical but now we have to get back to the conspiracy and show the silliness of those hiding the conspiracy and posing the funny part as reality. The conspiracy and what it represents in information becomes a silly joke…and it works, as a conspiracy because as far as I can trace I am the first not to be fooled by the fantasy of the feeble theory underwriting the serious part! All the models and theories show the Universe expanding while every one needs a Universe to contract. This was not helping Newton while in purpose it should be helping Newton. If everyone saw that Newton was a blubbering fool that was mistaken about the cosmic principles that the Newtonians underwrite as better-than-Evangelic-Gospel then they all were a pack of idiots that new nothing about what they professed that they know everything about. They had to get around this because now all theorists were ganging up to abandon the drowning ship representing Newton and came up with theories how the Universe work by expanding. All the models had one thing in common they didn't represent Newton's contracting principles because the growth supported Hubble by "being bigger", abandoning Newton's "getting smaller" idea.

In astronomy and cosmology, **dark matter** is a hypothetical form of matter that is undetectable by its emitted electromagnetic radiation, but whose presence can be inferred from gravitational effects on visible matter. This is totally fiction and is as fabricated as modern science could be. According to present observations of structures larger than galaxies, as well as Big Bang cosmology, dark matter and dark energy could account for the vast majority of the mass in the observable Universe. This means if they can't see it they can't show it and that is brilliant to fool all the sceptics.

The picture portrays the calculations resulting in a cone formation that the Universe should form. This is nonsense and if the Newtonian cosmologist uses only a child's common sense they would see that this shape could never be used or detected within the Universe. The Universe can't have borders while the cone shape only depict limitation carried from the formula they calculated that should represent the cosmos. Whenever anything ends something must start. What starts where their Universe show the cone shape ends where the Universe then has its end with the form that gives the cone shape reality. This rather depicts a few power-crazed people going mad with self-importance trying to fool the world as they allow the world to see what fools they truly are in their concepts they manufacture from ideas they clearly have not.

Again the incorrectness of every model is showing that the Universe has an outside. Outer space has not and can never have an outside. To correct the modal we have to put where we are in the middle with as much going smaller that what is going larger.

We will never see the smallest and we are unable to see the biggest. The sphere we search for forming is not formed as a result of outer space forming a spherical limit to space, as outer space is not representative in the largest structure because there is no largest structure. This is because the Universe has no outside. The sphere that forms represents the smallest part, which is within singularity Π^0 becoming space Π and forming space Π^3 by the movement $\Pi^2\Pi$ where that forms the sphere that forms singularity as singularity return the favour to form a sphere. In that we find the eternal value of space $\Pi^3 = \Pi^2\Pi$ by the relevancy of singularity trapping space in the movement thereof. Everything forming space $\Pi^0 = \Pi^3 / \Pi^2\Pi$ will loose all the space that movement forms $\Pi^0 = \Pi^3 / \Pi^2\Pi$ when movement of space is lost. That is what happens in a Black Hole where movement of space stopped and time reclines directly into singularity producing the opposite of what the Big Bang represented. That is the Big Crunch not that big and not very crunchy where it all ends. The Black Hole is the smallest and the largest at the same time of what the Universe offers as an object filling space (or not filling space). In the Theses the explaining gets a lot more technical but now we have to get back to the conspiracy and show the silliness of those hiding the conspiracy and posing the funny part as reality. The conspiracy and what it represents in information becomes a silly joke...and it works, as a conspiracy because as far as I can trace I am the first not to be fooled by the fantasy of the feeble theory underwriting the serious part! All the models and theories show the Universe expanding while every one needs a Universe to contract. This was not helping Newton while in purpose it should be helping Newton. If everyone saw that Newton was a blubbering fool that was mistaken about the cosmic principles that the Newtonians underwrite as better-than-Evangelic-Gospel then they all were a pack of idiots that new nothing about what they professed that they know everything about. They had to get around this because now all theorists were ganging up to abandon the drowning ship representing Newton and came up with theories how the Universe work by expanding. All the models had one thing in common they didn't represent Newton's contracting principles because the growth supported Hubble by "being bigger", abandoning Newton's "getting smaller" idea.

In the centre of any spinning object a line forms that can never start because it is the start of everything we call a Universe. That point albeit hypothetical, is also as much a reality none the less and is placed where that point must be standing still because every line running from that centre point towards such a centre will reach the very extreme centre and crossing that centre line that in truth does not exist finds a point where crossing this point the directional travel will be then in opposing directions and

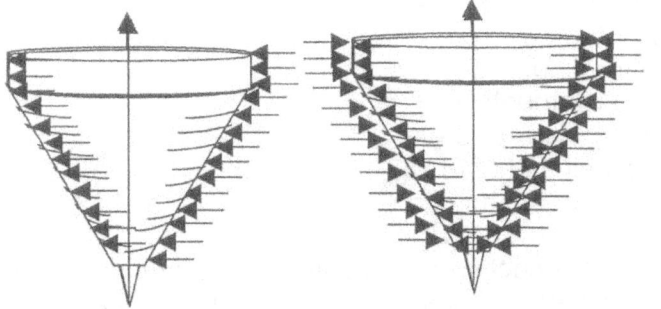

"the other sides" is also in opposing directional spin forming the other or opposing side. Thus, the spinning object will have a middle point, a very specific centre point that does not spin and only holds Π^0 forming Π or the roundness of that which spins as a specific value because no radius can apply. But also the one value such a line cannot have is zero because the line is there and holds contact to the rest of the material bringing about that zero does not start any line and therefore the value of the line must be infinite, just as described in accordance and by the definition of singularity. It is 1^0 or

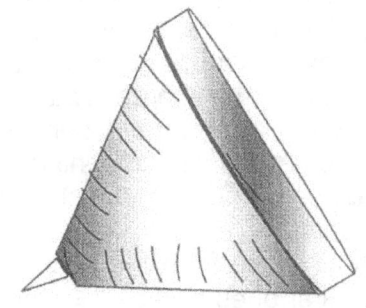

it is 254^0 or it is infinity, which is that which can go no less that what it already is. In the very centre of the sphere the form of the sphere dictates that the shape will relinquish space as the line run from the outside towards the very centre. With this natural state of affairs in the form of the sphere and is naturally inclined to dismiss all space that it can form in the form as the sphere holds space inside and the form will finally be without dimension. However this line without space can only come about where spinning movement takes place. All that I attribute to the radius line shrinking by reducing the radius that actually takes pace in every sphere as the diameter reduces towards the centre. In the centre where the radius line goes single the form relinquish the three dimensional form it has inside. Crossing the line is instant and there is no space to cross at the very centre. Being without dimension in the very centre means that at a point in the extreme centre of all spheres there are a point that holds singularity because this point with no space has a mathematical position of Π^0 although it is invisible since there is no sides to such a point to give that point any dimensions. The shape of all spheres are calculated by using the formula $4\Pi (r^3) / 3$. By reducing r to a point where r is r^0 singularity steps in because only the form remains as Π without measured space. Going even further we find that there then comes a point where Π goes singular Π^0. At that point absolute singularity is present but so is absolute gravity also present at that point. When holding the strength of the shape of the sphere in mind as well as taking into account that all cosmos objects of importance is in the form of planets or stars and they are all in the form of a sphere, we therefore may contemplate that it is where gravity originate. We now only have

to find the reason why gravity will hold a base in a space less ness as Einstein predicted. It is clear to be seen that gravity is in the centre of the sphere controlling from the centre everything that is outside the space less centre. We can reason with confidence that gravity is the strongest where space is the least and that confirms the absolute gravity within a Black Hole. We can further reason that it is gravity that is holding the sphere in true form and since the sphere allows gravity the best working opportunity, gravity can form the sphere in as strong a shape and form as the sphere seems to have. From every point on the surface of the sphere is where that point connects with the other side of the surface of the sphere by a line that runs through the space less ness of such a centre of the sphere. Such a line also connect by an angle of 180^0 as well as 90^0 to six other lines running from top to bottom, right to left, and back to front, where all join and cross in the centre of the sphere. There are therefore six lines crossing and connecting by a centre from any given point on the surface of the sphere. Such points connects in total six surface points on each side of the sphere while they all support one another through the space less centre connected by one point in the centre. In that absolute space less ness in the centre holding singularity we find gravity supporting and controlling all space within the sphere as well as space connected to the sphere. That is where gravity control and guide the space, which falls in the parameters as well as under the influence of the form of the sphere. In the gravity centre space goes singular meaning space becomes space less or flat or without dimensional substance we find in the rest of the Universe.

Inherent to the form the sphere offers, there is a specific location of singularity where the radius first goes single $r^0 = 1$ and then form goes into the realms of singularity $\Pi^0 r^0$. As the rotating direction moves inwards, the rings holding Π will become smaller and smaller. The reducing of the radius r will eventually end where the spin direction ends at Π^0. However that point where the directional spin ends is the point where the actual spin takes place. The spinning is on the precise location the point is not spinning. The cube also may have such a point but having such a point does not connect directly to six points located on the edges of the cube or any other form the is where all points are held in conformity. It is from the layout that the sphere uses as natural form that we are able to locate singularity. Keep in mind where we are at the Universe can go no smaller.

It is where the Universe forms space from points forming singularity and every point I refer to forms a point carried by singularity. It is the smallest there is not the biggest that we see. It is where the expanding of the Universe takes place as the Universe grows out of singularity

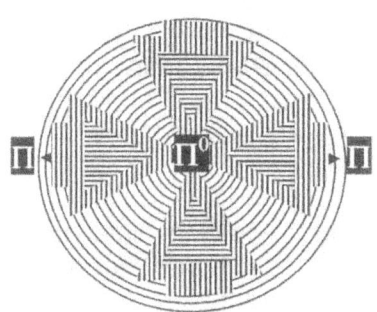

into space by time flowing. When we observe the sphere in the case of the sphere the material naturally reduces by measure of the radius becoming smaller to a point where the radius is r^0.

At that point the line that will form the radius has gone single dimensional r^0 and that is equal to 1^0, which is singularity. Also it is true that the entire form that is the sphere is controlled from a centre within the sphere and that point is r^0 or k^0 or Π^0, which has all identical values. That centre holds the sphere in form and shape that is Π. Therefore the strong form the sphere has is dictated from that space within the centre where there is no space and no form left. The natural inclining is in the form of the sphere. It is part of the roundness that the overall shape of the sphere represents and this structural strength is carrying down to the very centre. Because the circle could forever reduce that reducing which is inherently part of the form of the sphere becomes a tool in distorting of space in the sphere and is eventually removing all forms of space from within the centre of the sphere.

The very centre within the sphere where lines cross over to the other side ends up as having no space because of the reducing that continuous down to become the space less inner centre. The all roundness is the ingredient that forms the backbone of the absolute strength that the sphere has and that is the component that the sphere is so famous for. However at the smallest as space forms from singularity it is the roundness we encounter that forms that gives the Universe an indestructible quality. The form the sphere has allows the sphere to have a control that is coming from the centre deep inside the sphere where the space vanishes and being without space seems to keep the entire structure rigged. From the centre the sphere shape shows strength that the shape as tough as it is. How does it work in its most basic analyses? The moving of Π^0 to Π involved relegation and not motion as we consider what forms motion.

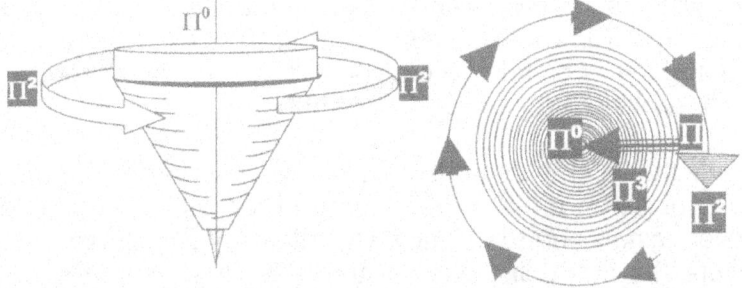

There are no correlation between mass and elements prone to space or prone to be solids. Mass do not create gravity and again on one more point Newton was wrong. Mainstream Science would rather ignore such compelling evidence as well as my writing about the matter than to admit that Newton could ever be mistaken.

Time came from eternity, where it stood still in eternity, as it still does in singularity. Remember that little line I indicated previously that is running through the centre of the spinning top and that holds time apart, while it in itself is motionless, eternal. That being without motion represents eternity, a state of being timeless. As space moves away from that point the duration of time begin to rise and the further the extending the larger the time factor will become.

Any fire represents many stages of time, where one part will burn quickly and the other slowly saying this means that science should recognise any fire on earth represents many conditions of heat where the smoke is a solid –gas, the flame is a fluid going on to be gas, the coal are a solid heat going to a liquid as it simmers producing photons which in itself is the dispensing of liquid heat turned to singularity particle dividing. The range that heat forms are so vast one may never appreciate it. Chernobyl showed the world how many forms of burning and burning injuries can come from radiation. Some heat was in the grass, undetectable until a bicyclist past and gained wounds from it killing the person a few weeks later.

The role that heat plays goes beyond (I suspect) what we may ever come to realise. Heat is an eternal fluid in relevancy to matter being the solid and space being the gas, but what is space to the one is solid to the next or a fluid to the other. For instance the proton is dimensionless to the neutron, yet it is fluid to singularity allowing heat to flow.

The neutron is solid in appearance yet it is fluid allowing dimensions to concentrate. Because heat is a liquid in relevancies, it is the father of specific density, allowing heat to flow differently in different forms of matter. To realise the correctness of that, one may gauge the heat relation there are in the first ten elements, and how each stand so different from the next element. Consider neon and boron, where boron has many times the density of neon, yet only half the mass, or where oxygen has more mass than doe's lithium, yet lithium has a much different relation to heat than does oxygen.

Time is the motion of heat in space, and producing more motion, the duration of time will extend.

Every dot insignificantly small as it may be, is a part of another universe as much as it is part of the accumulative universe and every dot in the infinity holds singularity, which we translate as " nothing" being " darkness". There cannot be "nothing" just as much as there cannot be "darkness". There cannot be something big or small, but it into relevancy of perception, and then the relativity of perception becomes the question. There cannot be hot as much as there cannot be cold. The sun FREEZES hydrogen to a liquid at six and a half thousand degrees Celsius and universe boils over in the form of the Hubble constant at the temperature (we presume from our vantage point) at minus 273 degrees C. If we Humans cannot or will not abandon our human perception and our manly perspective, we may as well return to astrology for all its worth.

Every point in the infinity we may observe at is not merely part of the universe in not being "nothing", but is the point where the universe started representing singularity. It is the very first point where everything began so many eternities ago, because after all, how can we ever determine where the first point was, as they were very much equal and alike at the beginning. Every aspect of the universe started with the fundamental fact that no point in the universe can represent "nothing" as a number, because every aspect in the universe represents singularity in what ever form it may hold in that specific spot forming space-time. If man does not reach a conclusion where that conclusion is matching the universe and stop to match the universe with man (and man's incapability), we may all go back to caves and become starving hunter-gatherers again, because we will never find a way to progress to the ultimate understanding of the universe.

Experience taught us that there is a definite precise and secure corresponding that can only result from a direct connection in as much as the line being the same line. In that case the line must then come down to infinity and release from infinity as the same line still connecting to a point of re-bouncing to either side. The graphic cross is he results of singularity applying opposing sides but still maintaining connection through the application of Pythagoras that will connect and always bring about a direct relevancy. The graph does not hold zero because information derived as result of a relation prove a contact remaining when the line crosses singularity without applying detachment.

The Brainy Bunch holds the view that in a graph the line crossing amounts to breaking the zero mark. That cannot be the case. Now I wish to refer once again to one of the academic letters, which I already used as a referral. This is very shortly my theoretical proposal

When the Universe started there was one spot that released a dot. The spot as large as a thought grew into a dot. This too is exactly as the Bible says it happened and it still happens exactly like this. The Universe starts from infinity and grows into eternity just as it happened at the start and this is what Newtonians confuse with what they know as the Hubble constant. To check if I am correct look at anything spinning and you will see conformation. The dot that released was by such release relevant to other dots released because there was to be motion that measured many dots. The dots in release were relevant since they were the same. Only time being in delay by cycle of infinity interrupting eternity to form space, as space is the history of time gone to the past. But since we are looking at things as it started I put it into the past tense.

That cycle brought about time delay as every cycle drifted further from the original singularity while the original was still responding as well. That was the first space. It was time being one infinity part. Since the dot was also singularity and was the very same as singularity with a small difference that the spot was $\bullet\Pi^0$ and the dot was $\bullet\Pi^1$. At first, at moment-Alfa, there was no space nor time for only relevancies came about. Relevancies acted as motion to bring change to eternity by changing the flow of time in eternity. There was the perfect spot in which time moved while remaining the same. Then heat brought expansion and expansion brought space and space brought movement and movement brought about a Universe. That is how it started and that is how it still is. This was before the atom and the atom was before light and that too is exactly what the Bible says in Geneses 1:1.

Time will forever remain eternity but space or time distortion, which will forever remain infinity, interrupted the flow of eternity. Space breaks the monotony of time in eternity by parting time from in infinity. Yet, the relevancies did imply motion except for the fact that singularity is very much incapable of motion. Every dot had a purpose to fill a position in relation to the other dots that the spot excited. All dots had a line of three where two was one, each on every side of the spot. The spot was one with the two dots forming two, which improvised for motion that would later come to space and the three was what space was going to become. How do we know this: 0.1416 x 7 = .991 and that is by the value that the Universe grows.

Time was four because the four would bring about motion as heat separated infinity from eternity or hot from cold. Five was space because space was one removed from time. Space is the distortion of time and one outside time would bring a time delay or a time distortion of four plus one which is five, hence the principle behind the Lagrangian system. Because material was the square of space material was a crossing of three plus three forming six.

Space-time is the four of time, plus the three in singularity around which the four of time turns, therefore space-time is seven.

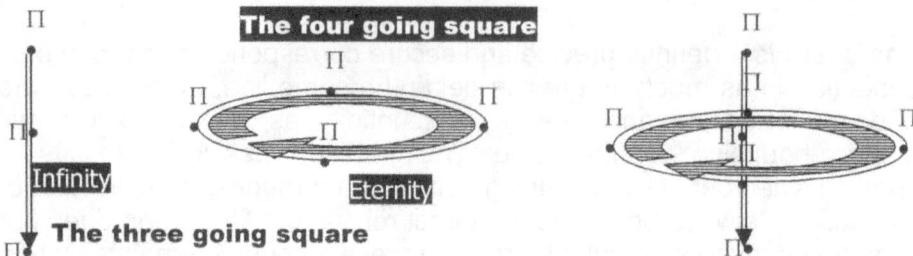

In the circle using $r^2\Pi$ the r has to have distinctive qualities placing it as a factor apart from Π. Where the growth shows no separate distinction but a continuous flow from the precise centre to the precise edge the flow would become in relation with Π depicting the circle and Π replacing r as reference to any point on the circle.

By using r, distinction in the circle is possible but by using, Π there is no distinction possible. Therefore, in the beginning when time formed space there was only relevance coming about from $\bullet\Pi^0$ and the dot was $\bullet\Pi^1$ with no mention of any possible r. The fact of r representing a radius represents space and what we refer to be long before the Big Bang introduced space or mathematics using space.

Before the Big bang the lot was form without dimensions playing any part. Then the atom came. Only after that did the Big Bang come. Even before the atom was the point lining up and forming positions that was spinning faster than the speed of light can ever achieve. However every point today still serve the role it took on at that stage and serves in the position that it had during the time it had no space with eternal time. These relevancies developed as part of a Universe we shall never understand. The Universe had no sides and a line was equal to a triangle, which was equal to a half circle. Singularity holds the double space-time position of five times two (matter and space duplicating singularity) which then is ten.

How did the Universe liberate material and heat forming space from singularity because with singularity comes an unchangeable eternal condition that is non-changing-everlasting in all conditions and aspects that is remaining in absolute equilibrium. This equilibrium maintains because all development extends form precisely in a detailed equal equilibrium throughout. Think about what brought the cosmos out from the eternal rest in which it was.

The eternal rest still maintains and is therefore our detection. What inspired the eternal rest the cosmos was in and inspired change to the state of eternal rest? What evoked change? That is the question the Atheist will never be able to answer but that too is the most basic and ever-lasting fundamentals of the Universe. Singularity Π^0 is not substance but it is a thought establishing substance Π. What changed in this split second start before the official start? I do not wish to ponder on this matter in the letter I am writing at this minute, as there are other books where I delve into this matter. It is called <u>The Absolute relevancy of Singularity</u>

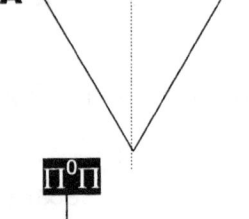

Where there is no space, there must be singularity 1^0 because the space is present although in singularity 1^0. If zero were a factor where all space finally halted in zero as the value, then zero would be able to remove the space from the centre and such removing would continue to remove the space until all space was removed. It will finally abolish all space in the sphere and it would remove the sphere. Zero removes all possibilities of anything coming about. Since the sphere is there, a zero factor in the centre cannot be present. Only infinity can be a factor from where space may grow because infinity can extend and grow into and up to eternity.

The moving of Π^0 to Π involved relegation and not motion as we consider motion. It was Π^0 getting a side and that is all. There was no true side but only a form that came into place. Singularity (A) received singularity (A) and no more of anything but the shift to comply with having a relevancy forming in relation to singularity. The dots had no sides, had no length or diameter. There was not measurable space or measurable time involved.

The time could have been a micro, micro second as much a trillion millennium because time had no relevance. It was eternity interrupted by infinity, as it still is the case, however the line that eternity followed was no line because there was no space to hold the line. The line was momentarily interrupted by infinity, however with no one there, there was no one to notice. The lines were not lines but relations to sides being formed. The relevancy that had the power to set Π apart from Π^0 is the only relevancy that still has the power, to set particles apart or join particles. It is heat in variation from cold. In order to excite singularity, singularity must establish a basis of heat that sets such a heat basis apart from cold. From there the form the atom will take on, however, the atom was still enumerable eternities to the development side.

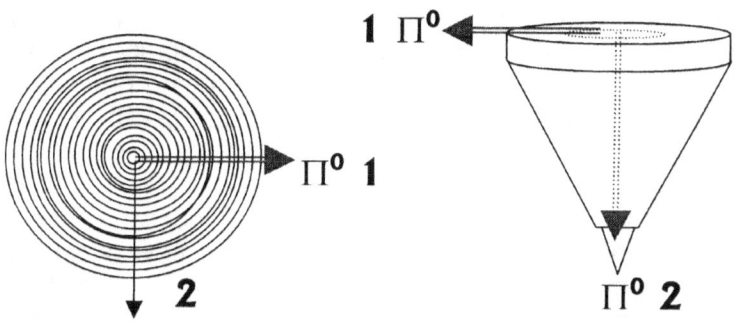

Where they are equal in value we must test the reason why this then is valid.

What is in the Universe is spinning. In the precise middle **of all** objects in rotation **is a precise centre dividing the object in sectors that will** start the spinning initiation **from that centre point.**

$k^0 = a^3 / T^2 k$ states that whatever is, is also spinning in order to be present.

Thus, the spinning object will have a middle point, **a very specific** centre point that does not spin **and only holds Π as a specific value because no radius can apply. But also the one value such a line** cannot have is zero **because the line** is there and holds contact **to the rest of the material bringing about that** zero does not start any **line and therefore the** value of the line must be infinite, **just as described in accordance and by** the definition of singularity. As I am introducing a very new idea, I wish to explain in better detail what I try to convey. While the top is spinning, one will find a line that formed in the centre where no line can form. It comes from spin but can never participate in spin.

That line must be singularity because if one moves any point on that line one position on, such a movement will land the point that then form on the line, on the other side of the line. The line is where the radius ends and starts because the line divides what is spinning in innumerable sectors and when reducing the radius progressively towards the centre of the spinning top at the centre where no line can be there is a line dividing the entire spinning top. At that centre point all further reducing must end because the next movement however slight will fall on the other side that is completely contradicting the one side. One movement further will change whatever is, so completely every aspect of that characteristic will contradict what it was before.

There is one point that is neither left nor is it right but any point next to that point must be either left or right. The only value that point may not have is zero because albeit so small that it is not part of our Universe, still the point is there for all to witness and that point is a reality as much as the entire Universe is a reality. Whatever one attaches to the top either in the line of being material or a concept, such a concept or material has to start at the spot in the centre of the top because every aspect of the top changes in contradicting from that point onwards in all directions. That point albeit hypothetical, is also as much a reality none the less and is placed where that point must be standing still because every line running from that point in opposing directions is also in opposing directional spin the other or opposing side.

When the Universe started there was one spot that released a dot. The spot as large as a thought grew into a dot. This too is exactly as the Bible says it happened and it still happens exactly like this. The Universe starts from infinity and grows into eternity just as it happened at the start and this is what Newtonians confuse with what they know as the Hubble constant. To check if I am correct look at anything spinning and you will see conformation. The dot that released was by such release relevant to other dots released because there was to be motion that measured many dots.

The dots in release were relevant since they were the same. Only time being in delay by cycle of infinity interrupting eternity to form space, as space is the history of time gone to the past. But since we are looking at things as it started I put it into the past tense. That cycle brought about time delay as every cycle drifted further from the original singularity while the original was still responding as well. That was the first space. It was time being one infinity part. Since the dot was also singularity and was the very same as singularity with a small difference that the spot was $\bullet \Pi^0$ and the dot was $\bullet \Pi^1$.

At first, at moment-Alfa, there was no space nor time for only relevancies came about. Relevancies acted as motion to bring change to eternity by changing the flow of time in eternity. There was the perfect spot in which time moved while remaining the same. Then heat brought expansion and expansion brought space and space brought movement and movement brought about a Universe. That is how it started and that is how it still is. This was before the atom and the atom was before light and that too is exactly what the Bible says in Geneses 1:1.

Time will forever remain eternity but space or time distortion, which will forever remain infinity, interrupted the flow of eternity. Space breaks the monotony of time in eternity by parting time from in infinity. Yet, the relevancies did imply motion except for the fact that singularity is very much incapable of motion. Every dot had a purpose to fill a position in relation to the other dots that the spot excited. All dots had a line of three where two was one, each on every side of the spot.

The spot was one with the two dots forming two, which improvised for motion that would later come to space and the three was what space was going to become. How do we know this: 0.1416 x 7 = .991 and that is by the value that the Universe grows.

Time was four because the four would bring about motion as heat separated infinity from eternity or hot from cold. Five was space because space was one removed from time. Space is the distortion of time and one outside time would bring a time delay or a time distortion of four plus one which is five, hence the principle behind the Lagrangian system. Because material was the square of space material was a crossing of three plus three forming six.

Space-time is the four of time, plus the three in singularity around which the four of time turns, therefore space-time is seven.

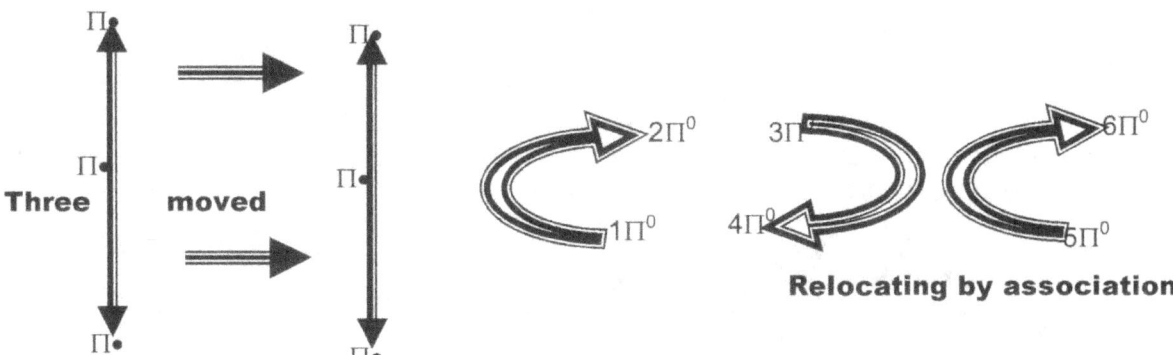

Relocating by association

In the circle using $r^2\Pi$ the r has to have distinctive qualities placing it as a factor apart from Π. Where the growth shows no separate distinction but a continuous flow from the precise centre to the precise edge the flow would become in relation with Π depicting the circle and Π replacing r as reference to any point on the circle.

By using r, distinction in the circle is possible but by using, Π there is no distinction possible. Therefore, in the beginning when time formed space there was only relevance coming about from $\bullet\Pi^0$ and the dot was $\bullet\Pi^1$ with no mention of any possible r. The fact of r representing a radius represents space and what we refer to be long before the Big Bang introduced space or mathematics using space.

Before the Big bang the lot was form without dimensions playing any part. Then the atom came. Only after that did the Big Bang come. Even before the atom was the point lining up and forming positions that was spinning faster than the speed of light can ever achieve. However every point today still serve the role it took on at that stage and serves in the position that it had during the time it had no space with eternal time. These relevancies developed as part of a Universe we shall never understand. The Universe had no sides and a line was equal to a triangle, which was equal to a half circle. Singularity holds the double space-time position of five times two (matter and space duplicating singularity) which then is ten.

How did the Universe liberate material and heat forming space from singularity because with singularity comes an unchangeable eternal condition that is non-changing-everlasting in all conditions and aspects that is remaining in absolute equilibrium. This equilibrium maintains because all development extends form precisely in a detailed equal equilibrium throughout. Think about what brought the cosmos out from the eternal rest in which it was. The eternal rest still maintains and is therefore our detection. What inspired the eternal rest the cosmos was in and inspired change to the state of eternal rest?

What evoked change?
That is the question the Atheist will never be able to answer but that too is the most basic and ever-lasting fundamentals of the Universe. Singularity Π^0 is not substance but it is a thought establishing substance Π. What changed in this split second start before the official start? I do not wish to ponder on this matter in the letter I am writing at this minute, as there are other books where I delve into this matter. It is called <u>The Absolute relevancy of Singularity</u>

As the rotating direction moves inwards, the rings holding Π will become smaller and smaller. The reducing of the radius r will eventually end at r^0 but the top does not end there because the top still then is Πr^0. The form we attach to the spin still applies as Π and the top finds directional contradicting change at a point that never moves because it can never move being in the centre where the spin direction ends at Π^0. It is the only aspect in the entire Universe that can be and still be motionless because it is not within the Universe.

It is the centre of the Black hole because it is the centre of the Universe. It is 1^0 and only the centre of the Universe in singularity can have 1^0 However that point where the directional spin ends is the point where the actual spin does not takes place because if its immovability.

The Spot

becoming

the Dot

When space brought division between eternity and infinity

It is at a point in singularity $k^0 = a^3 / T^2k$. It is where space ends because motionless ness ends space there. The spinning is on the precise location where the point is not spinning because the Universe ends in its not spinning there.

That line running through the centre of the spinning top divides every possible side from the opposing side in innumerable points that are divided by angles and degrees. Moreover, it changes the future position of the point from the present and from the past as it redirects every point every time k moves to reposition in a location where T^2 ends. In the end it proves that both k and T^2 confirms a^3 just like Kepler stated before Newton interrupted with dishonesty.

Another huge factor that favours the use of Π as a measure in singularity progressing is that any expanding by any mathematically sympathising method will have to use Π since Π is the only route that a spot of no significance can develop into a dot that represents a Universe of development in waiting. Only by promoting through the measure of Π can all possible sides progress on equal terms in all directions simultaneously without bias. The development must progress by measure of equality to the smallest indication and that purpose only Π can serve.

The progress must be generated so that it can flow equally to all sides in all directions spontaneously where not one side will favour the growing process as such. Time in it's flow does form a bias but explaining that at this stage will also involve too much other concepts which I would rather leave to the books. There is only one way to permit such a flow and have a mathematically correct outcome and that would be using Π for such expanding. The use of Π would ensure that a dot rises from the developing that comes from the first spot. The dot would have to form a sphere and a sphere is Π in relation to seven. This bring as back to the Titius Bode concept where ten is one half of a sphere relating to seven points forming the sphere where the seven with singularity puts form to the double ten that totals (including singularity) 21.991 and that is the overbearing dominating issue.

When the cosmos came into motion, motion was not yet defined. When the cosmos brought about motion, the first motion was relevancies. Cold parted from hot. Eternity parted from infinity. Motion parted from motion absence. Infinity broke the laboriousness of eternity for the duration of infinity. The spot became and grew into the dot.

From what the spot was to what the dot now is might be just a mathematical implication of going from 1^0 to 1^1 but in reality that first motion was the creating of and establishing of an entire Universe which was with all possibilities that now is it. Never again can that much growth become a reality, although to us the growth is beyond what we ever can notice. But it is because the growth is so massive and we are so small that we are unable to notice such almighty growth. When the spot Π^0 became functional and established all relevancies possible, heat parted from cold as eternity parted from infinity.

The expansion was not clear motion but more a parting of relevancies where a centre formed a relevancy because the centre could not provide motion. Without being capable of motion, the centre established four points, which also served singularity. From the inverse square law we know that the centre doubled by producing the four points holding singularity. We have to presume there is a time line because the Universe has this as evidence. The fact that light travel from there to here and from here to there proves of such a time line because there is no distance in outer space except in the Newtonian's misconception they have about the cosmos. Any line shows direction and the direction implicates positions according to the line having dimensions in the Universe we have in space and time. The line brings in Pythagoras and Pythagoras implicates mathematics.

$$1^0 \text{ going } 1^1 \text{ where } 1^0 \Rightarrow 1^1$$

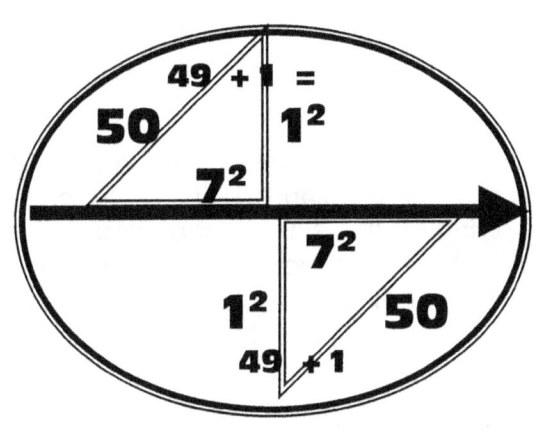

If there were progress that developed from singularity in the form of the first spot and we are the evidence of such progress, then the mathematical conclusion must be that a line formed where the line developed two sides and we have the evidence of that still present in our Universe. That brought about that three markers formed in relation to one and by admitting to the law of Pythagoras we find that what formed was 3^2 in relation to singularity 1^0 that became 10 on the one hypotenuse and 10 on the other hypotenuse which forms the square of space. Therefore mathematically space has ten positions and material has seven.

From the three the four (2 on both sides of singularity allocating the cosmic divide) the two in square developed as a mathematical consequence and that brought about the five.

The five was duplicated as a response on the other side of the divide and having five as a result of $(2^2) + (1^2) = 5 \times 2 + 10$ we find that the square of space holds the value of ten in place. But 5 is in pace because $3^2 + 4^2 = 5^2$. And $2 \times 5 = 10$ is in place because $3^2 + 1^2 = 10$. I just can't go in depth showing how the Universe formed mathematics as the Universe built what is within the Universe because that boo (if my memory serves me correctly) is about 1300 pages in all. There is a reason why some part of mathematics show how a line and a half circle and a triangle is equal and why $1 = 1 = 2$ and at the same time $1 \times 1 = 1$. It all has to do with building the Universe in stages of development.

We read from the way mathematics formed how the Universe formed because the Universe formed mathematics as it formed what is presented as a Universe. Before the Universe there was no mathematics so mathematics did not form the Universe but the Universe formed mathematics and above all this came about exactly as the Bible says it Happens.

If you divert one glitch from how the Bible presents it the entire process goes astray, but mathematics form the key in the presentation of how it all came about. At the very beginning there was a spot. How do we know that? The spot is still with us and holds a value of 1^0. This spot is in the centre of all spinning objects. Then time came into motion and 1^0 moved to 1^1. Since from where we stand we see 1^0 and 1^1 as being the same and therefore with the moving of time in the very beginning such moving must contribute to an increase of space.

Therefore 1^0 and 1^1 has to have a difference where the one is 1 and the other is one point in singularity smaller making the 1^1 coming from infinity and rising into eternity 1^0, making infinity forever one point smaller than what we find as the value one will associate with the one we find as a measure in infinity.

Therefore we can judge that singularity combines to have a total of 1 + .991, which then becomes 1.991 or whatever because the one going smaller is running into infinity and since infinity is one less than eternity we are in eternity 1^1 looking at infinity 1^0 which is one point reduced in infinity. However this moving from 1^0 to 1^1 involved 1^2 as well as 1^2 on the other side of the divide. As a result of the form the sphere holds, there is a centre connecting the sides and the centre holds singularity.

However, by presenting a centre where all lines cross on a point that cannot distinguish sides since that point has no individual sides, the centre holding singularity is inactive. Motion makes it active and the motion of space in time activates singularity to charge gravity that we find as a factor in the Coanda effect. That motion that establishes the purpose of space a^3 as a result of motion k through time T^2 was what Kepler presented as a formula. Gravity is $k^0 = a^3 / T^2 \, k$ and to install k^0 the motion of space-time a^3 / T^2 is required to complete

Producing singularity sets the divide because singularity splits the Universe apart in separate equal components that in combining form the duplication of singularity, being Π. Since the split brings about equality it, means that what is applying on this side must be applying on that side. When motion changes Π^3 to the proton $\Pi^0 \, \Pi^2$ it will happen on both sides of the divide of singularity Π^0. In effect it means that, that which combines the proton also parts the proton as it combines the proton because the proton becomes $\Pi^2 \, \Pi^0 \, \Pi^0 \, \Pi^2$ where the adding is the divide being Π^0.

The circle motion comes from space being dismissed by ending the motion and such ending of motion compromises the space it forms. By returning to where it is coming from it is ending the motion that began the space and as space is motion that is duplicating space motion returning is also motion that is ending which is destroying of space. THAT IS GRAVITY! Gravity is the balance between motion forming space by duplication space in motion forming time and time ending motion by destroying the space.

Gravity is about space duplicating space in relation to space destroying space and some particles are more prone to duplicate than destroy not withstanding mass or proton numbers. Those we call gasses. Then there are others that are more prone to destroy space that duplicate and those we refer to as metals. Then there are a few that destroy as much as the create space by duplicating and that we call fluids. When heat is added to some elements we consider as solids, the heat helps with the duplicating of surrounding space and brings about a balance restoring the difference there are in the destroying of space and the re-establishing of space. The metals become liquid and the heat forming the liquid brings about an adding to the material where such material diminishes space.

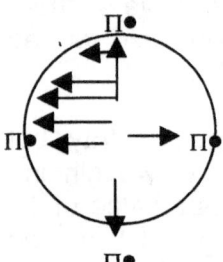

By moving from the spot to the dot is a process in which space les ness forms space and forms a process whereby Π^0 getting a side Π and that is all. There was no true side but only a form that came into place. Singularity holding the axis at 3 points received singularity holding 4 points and no more of anything but the shift to comply with having a relevancy forming in relation to singularity. The dots had no sides, had no length or diameter. There was not measurable space or measurable time involved. The time could have been a micro, micro second as much a trillion millennium because time had no relevance. It was eternity interrupted by infinity, as it still is the case, however the line that

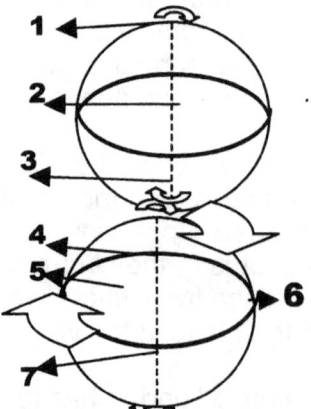

eternity followed was no line because there was no space to hold the line. The line was momentarily interrupted by infinity, however with no one there, there was no one to notice. The lines were not lines but relations to sides being formed. The question now to answer is if gravity is Π forming how does Π form? The forming of Π by perfuming a spin constitutes to gravity applying.

When the rotation takes place a line holding 3 points in Π^0 are excited into forming by a circle holding 4 points in Π^0 and this gives the 7 points in Π^0 that is the one aspect associated with Π. In the centre a line forms with a centre (Π^0) and a top ($2\Pi^0$) and a bottom ($3\Pi^0$) that forms 3 points of Π. This is evoked by a circle that spins around the top's axis of 3 points and the circle consists of 4 pints crossing the centre and this centre makes the six sides that form by spin a unified structure. That gives the 7 points tat in Π goes singular by rotation. The circle that commits the axis into forming also has a specific role to play. The four by crossing the centre unifies as five points and therefore each point is five [points moving as one.

This movement does not involve the centre line as the centre line is charged but not moved with this rotation that is taking place. Every point Π^0 becomes five point $5\Pi^0$ that moves in other words that goes square $5\Pi^0 \times 5\Pi^0 = 25\Pi^0$ and there are four locations meaning then the circle charging the line into forming is $4 \times 25\Pi^0 = 100\Pi^0$ points. After the movement ends the $100\Pi^0$ has to become rooted again and the square root of 100 is 10. In that 10 forms on the one side from where the circle is coming and 10 forms at the other side where the circle is heading. Again I wish to remind the reader that we are at a location in space where one spin results in the circle crossing the Universe at that point. We are where Π^0 becomes Π. By one rotation without having a radius because the only radius at such a point is $\Pi^0\Pi$. That gives Π a value of 10 going to 10 and in total it forms 20 of the 21.991 we have in a

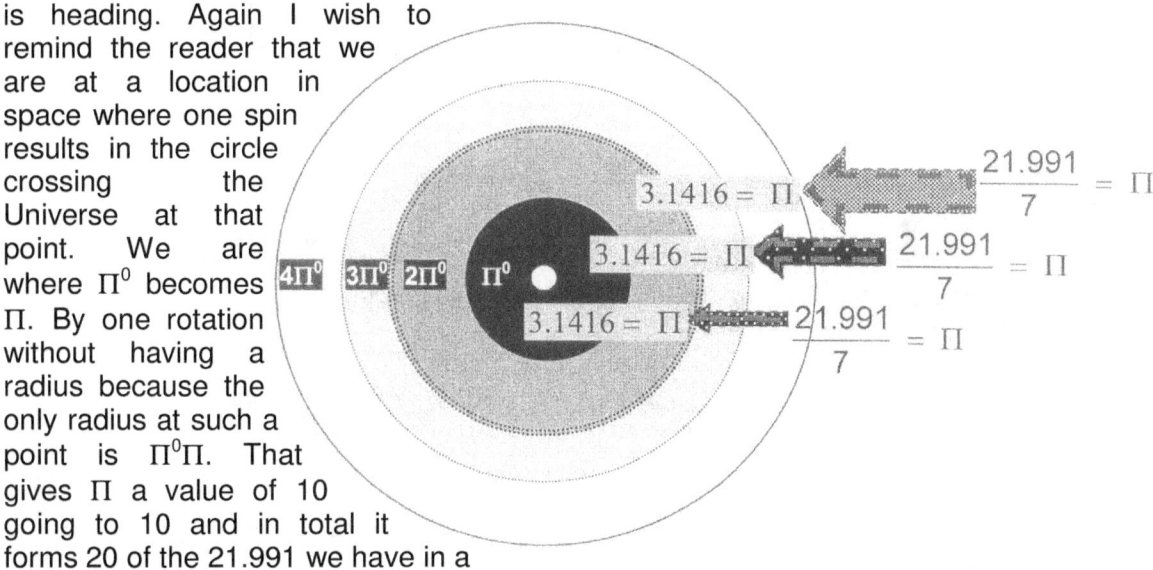

circle that holds Π. When the circle rotates it is 7 rotating to form 1 and thereby reducing Π from $\dfrac{21.991}{7} = \Pi$ to form $\dfrac{3.1412}{1} = \Pi$ and this is gravity as we find gravity to be. By spinning the spin reduces the space making the space more compact and that is the process thought of as gravity. There is no mass pulling anything by magical means. As the object turns is realigns a value of $\dfrac{3.1412}{1} = \Pi$ by changing to $\dfrac{21.991}{7} = \Pi$.

I show why the triangle and the straight line and the half circle are all equal to 180° and in the world using space as form by using dimensions this fact about mathematics is bizarre. The triangle and the straight line and the half circle are all unequal in form while mathematics proves the three equal. It is obvious that the triangle and the straight line and the half circle are as wide apart as the sea and the Sun is, and yet there was a period in cosmic development when the three were mathematically equal as much as they still are. I have mathematics telling me this fact beyond doubt. Please use a formula and your brilliance in mathematics and using no words to prove to me why the triangle and the straight line and the half circle are all equal as they all are 180° while explaining details because on this rests one entire pillar of mathematics.

The answer about this we find in the Lagrangian point system, which is one of the four cosmic phenomena, I explain when using the four cosmic phenomena to explain gravity. This becomes clear when using the law of Pythagoras to prove how this very law became the basis for mathematics and I do use mathematics in the law of Pythagoras to prove how

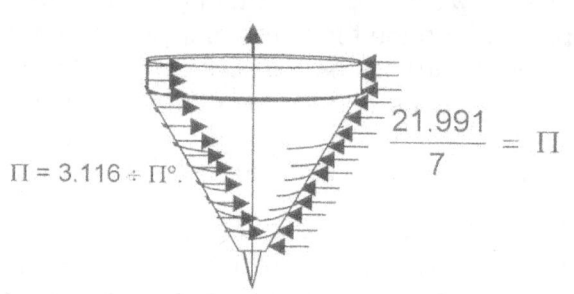

$$\frac{21.991}{7} = \Pi$$

$$\Pi = 3.116 \div \Pi°.$$

mathematics started when the Universe started mathematics. However, I don't prove that in the article because the space allowed in the article is much to little to prove anything.

In the article however, I show why did Π become $21.991 \div 7$ or then $\Pi = 3.1416$ or why is a circle Πr^2 or why is a circle circumference Πr or $\Pi d \div 2$. I show why a circle begins with Π and don't just surmise it. In my books I show why the phenomenon called the Titius Bode law is responsible for Π as a cosmic form and value. In my books I explain just as I claim in the article how the Roche limit come about and how the Roche limit is responsible for the sound barrier and what is the true cosmic value of the Roche limit as it plays a part in gravity on stars...that I show when I enter the era of singularity when calculations were still not yet developed.

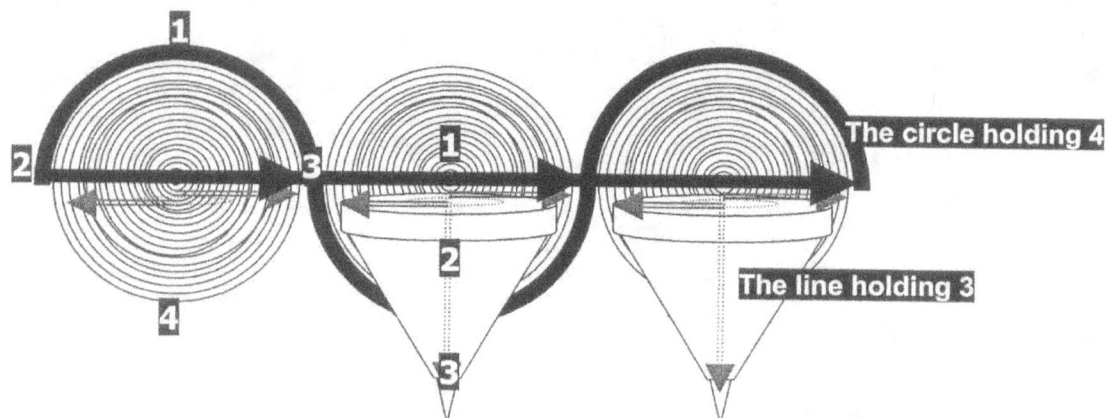

I show why a sphere in calculating the volume of space is represented as the formula $a^3 = 4\Pi r^3/3$ and why it is used to calculate the sphere when using these specific interpretations and how this is different from Kepler's $a^3 = kT^2$, which is the way to calculate volumetric space in applying singularity. The basis of this formula is derived from singularity finding form and that too I prove, but I have to use words because prior to when volumetric space came about, singularity prevailed and singularity is single dimensional. I pertinently state this over and over in the article. In the article alone I have no space to show all these facts and therefore in the article I only show why a circle uses Π to begin with. I show where and why did gravity start and what the true value of gravity is as gravity kick-started the Universe into a beginning because the beginning began with gravity. That I don't show in the article because printing space available will not permit me the opportunity to do so, but I introduce a book where I show exactly why, how and by which factors did the Universe start by using singularity. I show how the Universe evolved by singularity before space developed and at that time it implemented the four cosmic phenomena that later became part of space when space developed.

The gravity forming results in rotating that forms 7° it takes space of 21.991 to be in line with a centre that holds 3 and values a specified centre at .1412. I am not going into detail at this point about that because explaining the .991 or the .1412 will take up far too much space than what is available. In the books however I do just that. By the rotating circle forming 5 points I explain how the Lagrangian points come to form and by not moving the centre axis the five points rotates around the centre axis permanently.

The five points are a result of the four points moving (going square) and the result of that is realigning with the three (going square) and where 4 goes square $4^2 = 16$ in relation with 3 going square $3^2 = 9$ the net value of the points shifting is $9 + 16 = 25$ and 25 finding its root in relation to Pythagoras (triangle with one side 3 and the other side 4 in the square and there fore the third side must be $\sqrt{25} = 5$ we find Pythagoras playing the part as a triangle.

Newtonians uphold their law of physics without showing mercy. The very first things the Newtonians use to beat us into submission are to blast us with incomprehensible mathematical formulas. Incomprehensible they are but it is to scare anyone with the mathematical equations to get everyone hiding. They bewilder you with equations that put the fear of God into you; used simply to make you feel inferior so that they can feel superior and frown down on your inferiority from a dizzy height. They are masters at manipulating anyone into a state of senselessness...but mostly they do it onto themselves. That they do because it forms the backbone of the fraud. They do not wish you to read closer and to find the fraud they hide to protect Newton. Ignore their mathematics because it only shows their incompetence to understand physics or Newton and see the fraud they propagate...They employ mathematics to bewilder and that is all. I am going to show what we can uncover underneath what they cover. Look at what the mathematics supposedly says and then wake up, they are using maths as a scare tactic for three centuries to scare the daylights out of you and all this while its been working! Looking at the formula shows just how little Newton understood physics. Please allow me to show you how they scare you to become fooled and suckered. Don't run and hide when you see the mathematics; it is meaningless although it was used as a scarecrow for more than three hundred years forming the backbone of the conspiracy. In this case I am referring to the so-called Kepler's Laws that has nothing to do with Kepler.

Use this picture below to show me where the planets are positioned according to mass or where the orbit going around the sun goes according to mass. The entire Newtonian idea of mass creating gravity by pulling is the complete misrepresentation of the truth. I show what principles are in place do give the reason why. It is easy to talk about "mass" and never get "mass" part of reality when hiding the truth.

Newtonians make a statement about "mass" holding the solar system in place. No matter how much this is corrupt, nevertheless they put it down as a given fact so much so that they will show doubt in a living God being present but that mass pulls planets goes beyond doubt. The proof of mass pulling to form gravity can never be tested because it is beyond doubt. If you doubt in it they throw a Newtonian made formula they call Kepler's laws at you. "Kepler's third law" supposedly is "the square of the orbital period of a planet is directly proportional to the cube of the semi-major axis of its orbit." In mathematics it is Symbolically: $P^2 \propto a^3$ and therefore $a^3 = P^2$ in position of P and therefore $a^3 / (P^2P)$. This is taken from the idea that "Kepler said", which is totally fabricated that $a^3 = T^2$ where Kepler said in fact $a^3 = T^2 k$ and this is a big difference because $a^3 = T^2 k$ is the same as $E = mC^2$. Look at reality. $a^3 = P^2$ is total garbage and as big a hoax as is the idea of mass being any form of factor in

gravitational physics or that gravity applies in accordance with $F = G \dfrac{M_1 M_2}{r^2}$. Look at the

picture below. Look at how the planets are sorted and that is not by size. There is a ratio applying called the Titius Bode law and this law puts planets in terms of size or mass at a precise equal base notwithstanding that Jupiter is many time bigger than Mercury. Everyone is so taken by the accuracy of Einstein's formula that $E = mC^2$ but this is exactly Kepler's formula where $E^3 = mC^2$ is taken from Kepler's formula when accurately used as $a^3 = T^2 k$. There is no $a^3 = P^2$.

Do not get scared as everyone usually does when see and get frightened then consequently as a reaction to find survival you turn on your heels and run...

Don't run, just read on and see how simple it is to prove Newton was a backward dark aged sod!
You don't have to be a mathematical mastermind to see that it is not mass that applies to allocate planet or star positions. Here is an example that any person can understand when they don't succeed in bewildering people with frightful mathematics and comprehensive formulas.

I know and realise that you are disgusted by my attitude when I degrade the name on which physics are founded. In this introduction part I am going to show you just some minor deceptions all students are forced to believe since all physics students are forced to believe in Newton, Sir Isaac Newton that is. I am giving you a choice. You can say I am going to commit fraud or Newton has committed fraud.
If I am judged to be the culprit that is guilty of deception then it is because Newton misled me. You can choose.

You are expected to believe the following:
Newton stated under the nametag of Kepler that there are so called Conversions for "Unknown" factors.

$4\pi^2 a^3 = P^2 G(M + m)$ Newton introduced this concept because he said mass brings about gravity.

From the top formula Newton devised the next formula $P = \left(\dfrac{4\pi^2 a^3}{G(M+m)}\right)^{0.5}$, which he named after Kepler. Kepler had nothing to do with the entire idea and every incorrect aspect is a Newton contribution. Students don't shy away from the mathematical aspect because the formula is complete bogus fraud.

Ask your physics professor to put in the mass of the sun and any or every planet and from that determine the allocated position in accordance to the calculations derived from the method that the formula dictates. The formula is fraud and keeping the formula in place and used by students all form part of the conspiracy to hide the incompetence.

Any person that upholds Newton's ideas and principles abut mass forming gravity use the next formula Newton introduced $P = \left(\dfrac{4\pi^2 a^3}{G(M+m)}\right)^{0.5}$ and then go on and put in the allocated position of the planet in the pace of **P.**

Then put in this formula $P = \dfrac{1}{G(M_1 + M_2)}$ the mass of the sun plus the mass of each planet to show how the position is in place in accordance. This is no less than fraud and yet, on this the entire conception of Newtonian science rests. Let any physicist I challenge take me on by proving this or any other Newtonian formula correct.

Let any physicist I challenge **$4\pi^2 a^3 = P^2 G(M + m)$** prove that it is mass $\dfrac{1}{G(M_1 + M_2)}$ that **allocates the poison reserve by the planet $P^2 = 4\pi^2 a^3$ which is what this part of the formula.**

If "mass" did form gravity by a value that commits a force then the large planets must be on the inside next to the sun having the small planets way on the outside. Instead because we have a random allocation, that destroys the idea of mass forming gravity to pull.

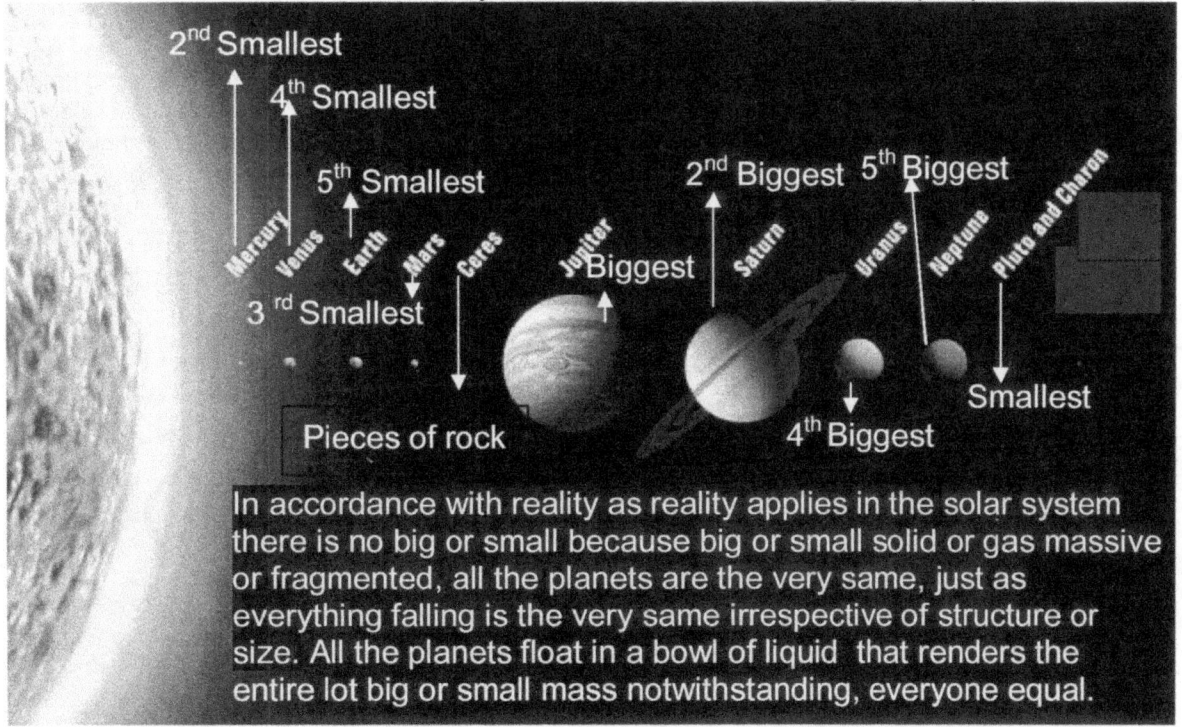

2ⁿᵈ Smallest

4ᵗʰ Smallest

5ᵗʰ Smallest

Mercury Venus Earth Mars Ceres Jupiter Saturn Uranus Neptune Pluto and Charon

2ⁿᵈ Biggest   5ᵗʰ Biggest

Biggest

3ʳᵈ Smallest

Pieces of rock

4ᵗʰ Biggest

Smallest

In accordance with reality as reality applies in the solar system there is no big or small because big or small solid or gas massive or fragmented, all the planets are the very same, just as everything falling is the very same irrespective of structure or size. All the planets float in a bowl of liquid that renders the entire lot big or small mass notwithstanding, everyone equal.

The smallest planets holding the least "mass" are at either end and the largest planets holding supposedly the most "mass" are in the centre. This disproves both arguments that the pulling force forming gravity by the value of "mass" to establish the orbit goes according to "mass" or that the locations of planets are adhering to Newton's ideas of "mass forming gravity by pulling".

The way that the present cosmology shows gravity forms is by telling everyone about mass and then this is how gravity forms the Universe. That is the way they put the Newtonian model forward.

Everyone is as gullible is the preconditioning would allow the people to hold the mindset in which the people are in. All knowledgeable persons know the sizes of the planets and yet no one thinks about the bigger planets being in the centre and the smaller planets circling near the sun or far from the sun. No one ever took it to task to confront those cheats in physics about the claims that it is mass that holds the planets in place. The world of people wants to be cheated as long as no one is asked to think and apply personal wit.

In the Universe all thing are equal in size because Neptune spins around the sun equal to mercury's time and Jupiter floats around the sun equal to mars or Neptune. Notwithstanding what "size" or "mass" they grant the planet to have the rotation happens equal and without mass bringing any favouring in positioning or in speed of movement. So where the hell is mass a factor in gravity forming?

PLANET	Mean Distance from the Sun (AU)	Equatorial Radius (km)	Mass of planet (Earth=1)	Mean density (grams/centimeter^3)	
Mercury	0.3871	2439	0.06	5.43	2nd smallest but closet to the sun
Venus	0.7233	6052	0.82	5.25	4th smallest and 2nd closest to the sun
Earth	1.000	6378	1.000	5.52	4th smallest and 3rd to the sun
Mars	1.524	3397	0.11	3.95	3rd smallest 4th closest to the sun
Jupiter	5.203	71490	317.89	1.33	The biggest with "most mass" and bang in the centre and position 5
Saturn	9.539	60268	95.18	0.69	The second biggest
Uranus	19.19	25559	14.53	1.29	The 4th biggest
Neptune	30.06	25269	17.14	1.64	The 3rd biggest
Pluto	39.48	1160	0.002	2.03	The smallest of the lot.

From the Titius Bode that forms the solar system I have compiled the following formula by which gravity forms to the value of Π

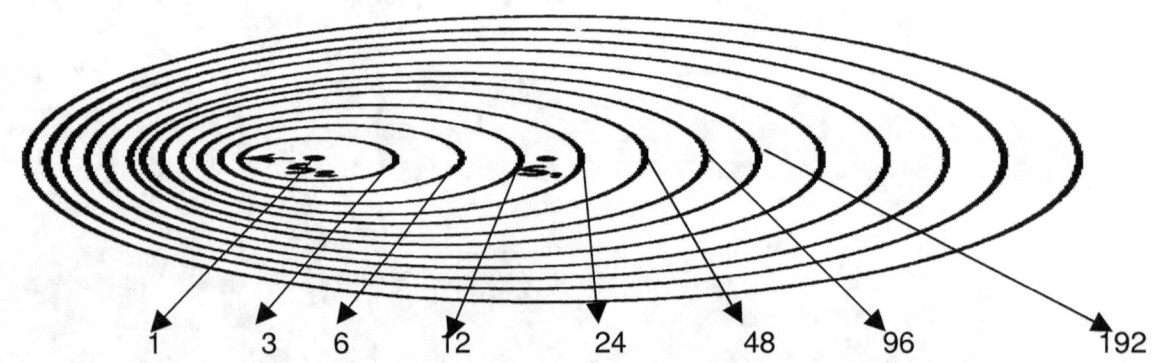

1   3   6   12   24   48   96   192

Planet	Mercury	Venus	Earth	Mars	Ceres	Jupiter	Saturn	Uranus	Neptune	Pluto
Bode's Law distance	4	7	10	16	28	52	100	196	-	388
Actual distance	3.9	7.2	10	15.2	28	52	95.4	191.8	300.7	394.6

Bode's Law: A numerical sequence announced by J.E. Bode in 1772, which matches the distances from the Sun of the six planets then known. It is also known as the Titus-Bode law, as it was first pointed out by the German mathematician Johann Daniel Titius (1729-96) in 1766. It is formed from the sequence 0,3,6,12,24,48,96, and 192 by adding 4 to each number. The planets were seen to fit this sequence quite well – as did Uranus, discovered in 1781. However, Neptune and Pluto do not conform to the 'law'. Bode's Law stimulated the search for a planet orbiting between Mars and Jupiter that led to the discovery of the first asteroids. It is often said that the law has no theoretical basis, but it does show how orbital resonance can lead to commensurability.

The importance that becomes known is the sequence the Titius – Bode law saw in the number arrangement of 3; 6; 12; 24; 48; 96 etc. The incorrect application of the Titius Bode law lies in subtracting the figure of 3 from 10 leaving 7. The other way of reasoning is to add four each time to the firs value of three starting with 3 and so on.

The true significance of the Titus-Bode law is that it points directly to a circular growth of 7 stages. The 7 relating to 10 is a precise derogative of the Roche limit or the Roche limit is a precise derogative of the Titius Bode principle because the two systems interlink.

This is how I mange to explain the Titius Bode law that is in the solar system by the ratio applying that really form the solar system in the way nature shows space growing by time. What you see on the next page was never been shown but on the other hand Physicist say this mathematics are too simple to apply as physics!

Science would rather deny there is cosmic principles that is in place in the solar system, which are
The Roche limit,
The Lagrangian points,
The Titius Bode law and
The Coanda effect
They would not admit to Newton's failings because then the entire world will see they know less about science than does a pig know about history. They would rather put the error on the solar system than they would commit to the blatant mathematical cheating that Newton committed. It is the Universe that is always at fault when Newton becomes incorrect because without Newton's fabrication of science they have nothing to show for all the wisdom they try to pretend they have. Newton fabricated "Kepler's laws" has some correctness mixed largely with a farce and blending the truth with total fabrication of reality hides the lie behind something presenting the truth.

This sounds simple but when delving deep into this we find a lot of meaning.

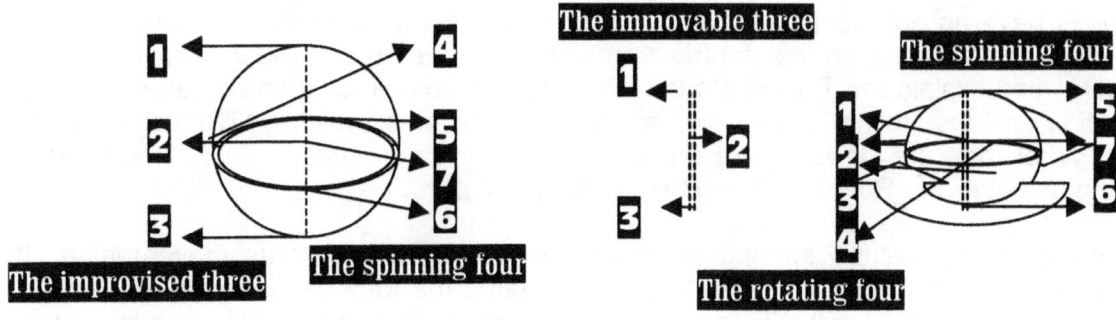

This confirms the value of 5 filling the next position of five (5+5=10) on both sides of the circle 20 adding the centre in singularity Π^0 and the net result is $\dfrac{21.991}{7} = \Pi$ which is gravity contracting. The .991 or then .1412 in the case of adapting to singularity is time growing or what Newtonians refer to as the Hubble constant where "space expands".

It is time .1412 coming from the future x 7 = .991 and then by moving into the Universe to find or link with space it then in the next instant becomes Π^0 =1.

The next step is involving not only the circle forming but also connecting the linear movement to the circular movement which is the essence of all movement or gravity forming in the Universe and time forming the Universe. This means to show that the axis that can't move can reposition according to time. In the circle that forms there is 7 points taking singularity 1 across the divide time forms from coming from the future to the present and the movement is the future 7 + the present 7 + the 7 going to the past which forms 21 plus time moving from beyond the Universe .991 and the concludes the Π as 21.991 and this is space forming as relevancy to 7 dots going singular as Π^0.

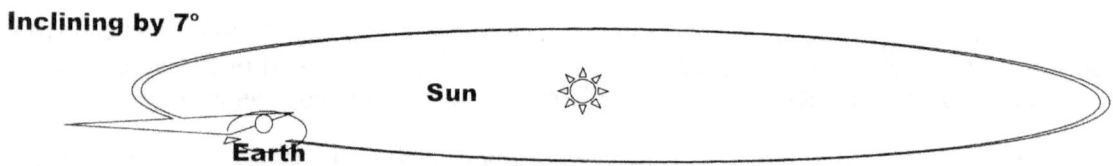

Inclining by 7°

Sun

Earth

Inclining by 7° as the Earth goes around its axis.

By moving the line forming the axis or singularity 7 forms that involves 10 and by these two applying we find gravity forming as Π where material (7) interacts with space (10) resulting in gravity or time applying space. This relation between 7 and 10 forms the Universe as cosmic physics and gives explanations to what was never answered before.

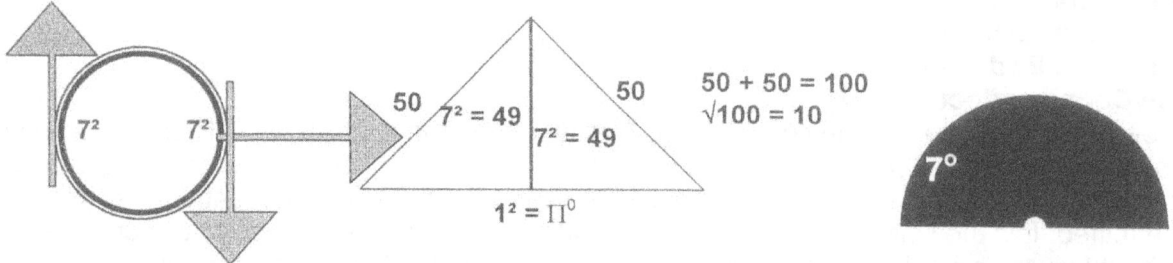

Also there is a movement of 7 which is 7^2 forming one side of the Pythagoras triangle and Π^0 or 1 forming the other side giving 50 the value of the third side. Since movement is the duplication of what is the triangle doubles to give a value of 50 + 50 and that forms 100 where when rooted it is 10 in space. In this we find the 7 and the 10 where the seven hold a spin value of 4 that is the movable component in the Titius Bode law and the 10 is space forming. I have no agreeing on Newtonian definitions and I worked out that gravity is the difference between material that contracts and space that expands and in that you find how to resolve and explain the Titus Bode law, the Lagrangian points of which I already laid the foundation and the Roche limit as well as the Coanda effect which I will touch on just as briefly. What form the Universe are the four pillars, which   are the Titus Bode law, the Lagrangian points the Roche limit and the Coanda effect.

This indicates four factors forming singularity that absolutely dictates movement as the cosmos. Holding that in mind, I therefore had to name the four positions that equally form singularity by dictating gravity. To argue this concept of singularity guiding movement let's take the Sun that provides a centre k^0 for the Earth a^3 forming a centre where k points a line that forms the orbital circle T^2 where from the edge of the line k is pointing at the position of whichever planet a^3 forms a circle T^2 in relation to a line coming from a centre k^0 of the Sun. The line k indicates the distance from the Sun's centre to the planet that orbits and this forms the circle as the planet a^3 orbits T^2 around the Sun.

The line k will provide a line from the Sun's centre k^0 and the line k will provide a spot where T^2 produces a circle holding space a^3 in a located position by running around the centre of the Sun k^0. In this view the space a^3 of the Earth rotates and in that forms the **controlling singularity** indicated by k forming between k and k^0 being singularity. The Sun holds singularity in the centre, which is forming the **governing singularity** that forms the circle T^2 that forms the orbit. That means every single point that k indicates there are positions forming space a^3 implicating sides of a double dimension. In the same manner is k not limited to distance or is T^2 lesser by size.

If Kepler said $a^3 = T^2k$ then also $k = a^3 / T^2$ is what Kepler said. There are three dimensions a^3 between any two points T^2 flowing as time from the centre of the sun, which is indicated by the line k. However in the next scenario the Earth holds the **governing singularity** running from the centre k^0 to k forming the edge while the circling rotation T^2 then forms the **controlling singularity**. There are also two other points holding **mutual singularity** and the **primary singularity**, which both I do not explain in this presentation but without which the four phenomena would not be understood.

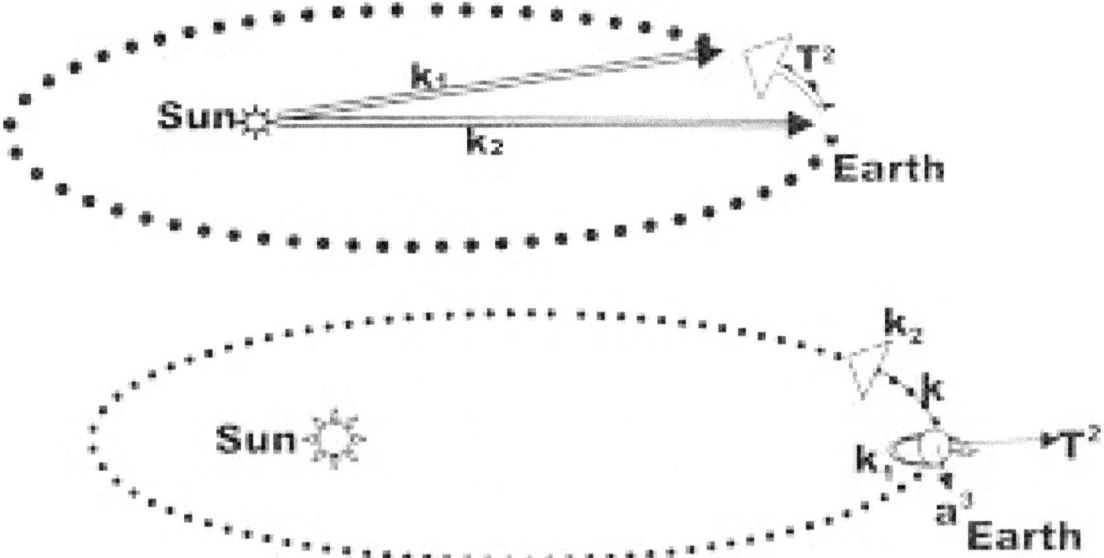

The value of k is not to be put in place as a measured value, but to bring a reference to the location of singularity $k^0 = a^3 / (T^2k)$ applying as to place a specific singularity in as the **governing singularity** and acknowledge the position of another singularity in place as the **controlling singularity** because there always has to be a **controlling singularity** determining the orbit while there has to be a **governing singularity** determining the spin of the body in relevance performing as the space a^3 in question in the formula $a^3 = T^2k$ where in that formula k determines the relevance of k^0 as in $k^0 = a^3 / (T^2k)$. In that we find eternity within the Universe. By going straight the object going straight forms the next circle by taking the smaller circle that formed, straight into the next larger circle that forms and this endless rotation is the eternal circle that keeps whatever moves in a circle going straight bound to singularity.

However, this burdens k forever with the responsibility of forming a line and a line is what places the Universe in place while the circle T^2 is forming the Universe a^3 at the same time. Every space a^3 in question puts singularity k^0 in position by the motion T^2 in relation k to the position allocated **k** in the Universe **a³**. Nothing in the Universe can move without moving straight **k** that is also going in a circle **T²** to form space **a³** in relation to a centre **k⁰** while in orbit around another centre **k⁰**. In this point **k⁰** time forms space and space develops as the history of time running from **k⁰**.

a³ symbolises in a mathematical interpretation of implicating the three-dimensional space holding a specific centre in relation to another specific centre indicated by **k** that could apply to either centre points in question. This is always a straight-line **k** representing the position of the **controlling singularity** moving in a circle **T²**. The space forming **a³** is a **positional validity** of the space indicated by **k⁰ = a³ / (T²k)**.

Our Newtonian-Wise Physicists use the elaborate formulas such as $\frac{d}{dt}\left(\frac{1}{2}r^2\dot{\theta}\right) = 0,$ and $P = \left(\frac{4\pi^2 a^3}{G(M+m)}\right)^{0.5}$ as well as $T^2 = \frac{4\Pi^2}{G(M+mp)}a^3$ to explain and prove what? If this is true then Jupiter must spin 317 times faster around the sun than the Earth does and be almost next to the sun while Mercury and Pluto in comparison must hardly move being cast into the darkness of the oblivious. The formula $T^2 = \frac{4\Pi^2}{G(M+mp)}a^3$ says that the spin or time in which the circle comes about **T²** holds a space allocation relation **a³** directly proportionate to the mass (G(M+m) of both the sun and the applying planet multiplied by placing the gravitational constant also in relevancy.

It says planets spin in accordance to mass...lets see. The truth about physics and the correctness within physics is so far from what this statement this statement $T^2 = \frac{4\Pi^2}{G(M+mp)}a^3$ upholds as Newton is from being correct and yet it is applauded and upheld as correct as serve as the truth for centuries while it is a conspiracy to hide the truth.

I say prove it! Then $\frac{d}{dt}\left(\frac{1}{2}r^2\dot{\theta}\right) = 0,$ says that the Planets don't move at all because with $\frac{d}{dt}\left(\frac{1}{2}r^2\dot{\theta}\right) = 0,$ the movement acceleration is zero. Zero indicates no movement and that is as corrupt as the rest of Newton's ideas. Every person associated with physics to whatever extent has been and had been conspiring to hide the truth and the truth is that no one knows (not even Newton and even less Einstein) what physics is.

What they say is precisely what the solar system proves as completely incorrect and void from all truth and either the solar system has no idea about what is driving gravity or they have no idea of what drives gravity, but far from their wisdom it is definitely not mass. If they don't know what forms gravity and it is clearly not mass, then they know nothing about the way that gravity forms.

They hide their incompetence under a blanket of ignorance and all commit to a conspiracy to hide the truth from the public. Let any one of those So –Wise-In-Mathematics show just how does Saturn arrive at the position it holds in relation to the mass it has by applying the Newtonian formula $P = \left(\frac{4\pi^2 a^3}{G(M+m)}\right)^{0.5}$, which they claim is data confirming Kepler's Law of Periods of the motion of the planets in relation to mass. Newton could not understand Kepler and so he corrupted Kepler's work.

If only Newton understood Kepler and not tied to reinvent to cosmos to fit into Newton's vision the cosmos would be so much better understood. The Solar system and therefore the Universe uses the Titius Bode law to form space.

Again just because Newtonians have no idea about how the Titius bode law apply and the fact that the Titius Bode law does not confirm Newton's standing on how the Universe forms the Titius Bode law is shifted into darkness never mentioned in well-to-do company and is blocked out of being a critical part of the solar system. Now for the first time ever you will se how and why the Titius Bode law forms and that it proves Π forms gravity. It proves that gravity is vested in Π forming as Π^2, which is by spin.

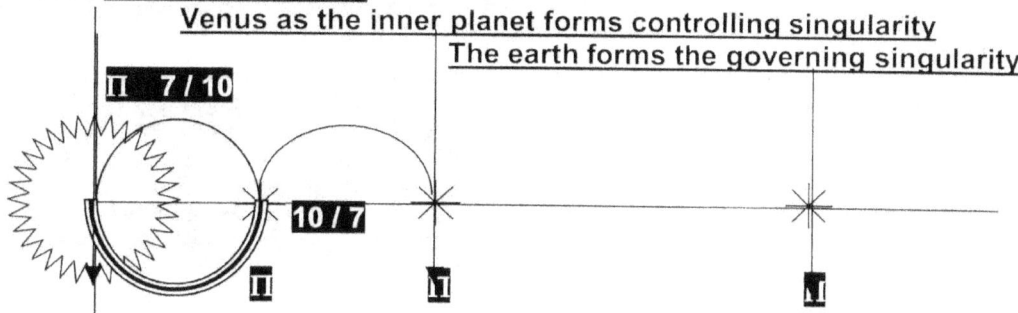

T^2 is representing the circle that goes around the **governing singularity** k^0 that forms in relation to the line **k** in reference to the centre k^0 The space that forms holds the orbiting planet a^3 in direct circular contact with the space in relation to a very specific centre k^0 moving from point T_1 to T_2 in relation to a precisely placed centre k^0.

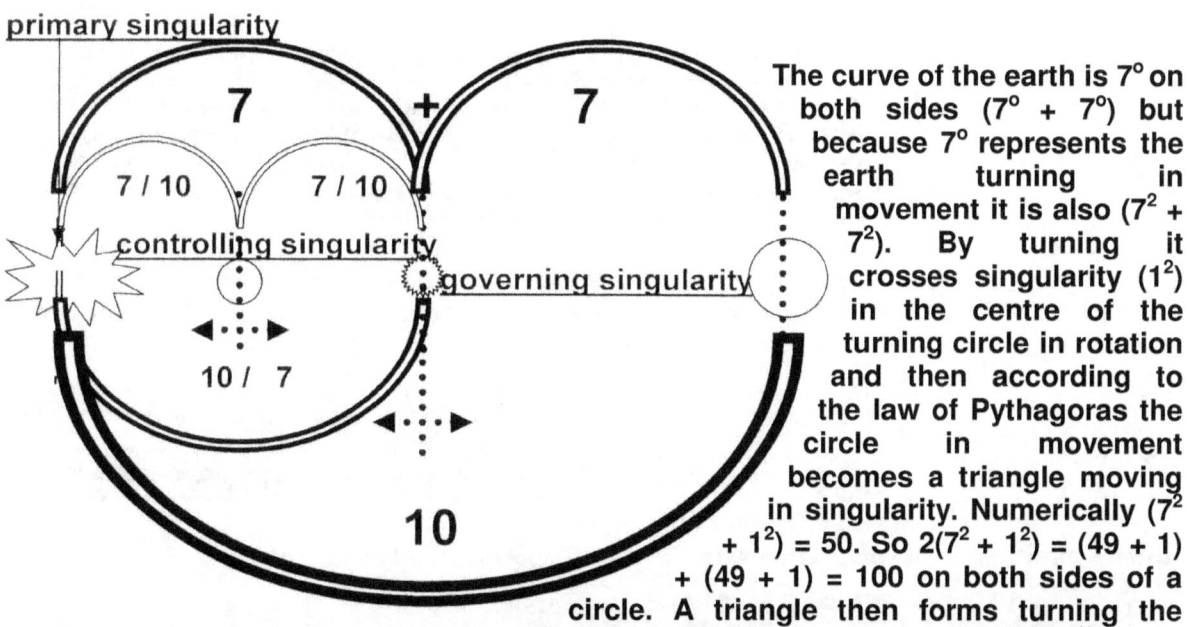

The curve of the earth is 7° on both sides (7° + 7°) but because 7° represents the earth turning in movement it is also (7^2 + 7^2). By turning it crosses singularity (1^2) in the centre of the turning circle in rotation and then according to the law of Pythagoras the circle in movement becomes a triangle moving in singularity. Numerically (7^2 + 1^2) = 50. So 2(7^2 + 1^2) = (49 + 1) + (49 + 1) = 100 on both sides of a circle. A triangle then forms turning the direction = 50 + 50 = 100. Therefore the space in which the circle turns is $100^{1/2}$ to the root thereof = 10 and therefore the Titius Bode law shows the inside of the circle factors forming Π as gravity where 7 goes related to twice times 10. That is Π and that is why 7 goes double squared in a circle adding the centre part of the circle, which is 1.991 divided by the space in which the planet orbits around the sun that forms singularity at 1.991. This adds to the 10 +10 placing each planet in the allocated singularity position, which is then derived. It is implementing Π as gravity. The Titius Bode law shows that gravity is when (10 + 10 + 1.991) / 7 = Π.

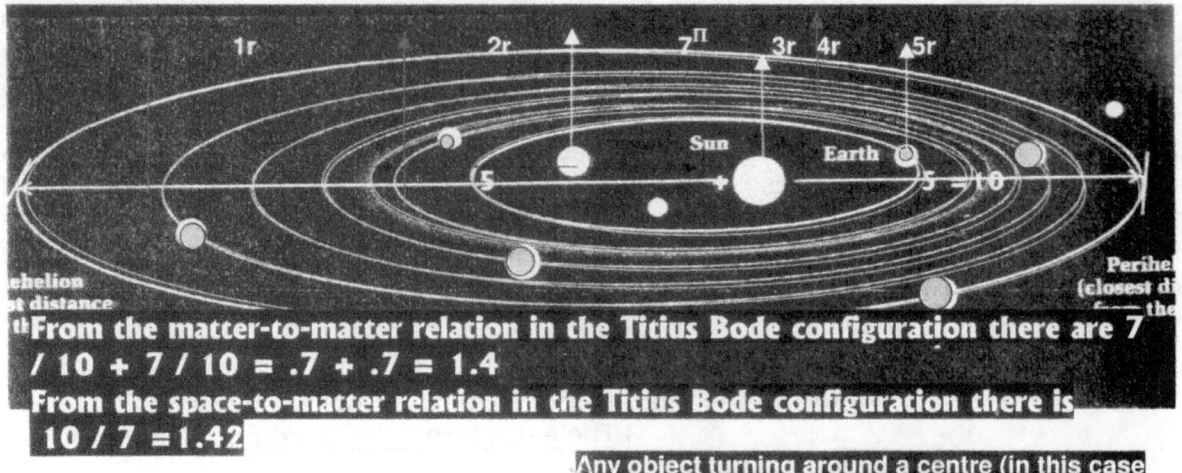

From the matter-to-matter relation in the Titius Bode configuration there are 7 / 10 + 7 / 10 = .7 + .7 = 1.4
From the space-to-matter relation in the Titius Bode configuration there is 10 / 7 =1.42

$$(7 + 7) / 10 = 1.4$$

$$(7 + 7) / 10 = 1.4$$

Any object turning around a centre (in this case planets turning around the sun) goes in a straight line by diverting from a straight line by 7°. On a later occasion in this book I show how the 7° forms a direct link in becoming 10. This is the ratio that the space between the planets grows by taking 7° and forming 10 from doubling that

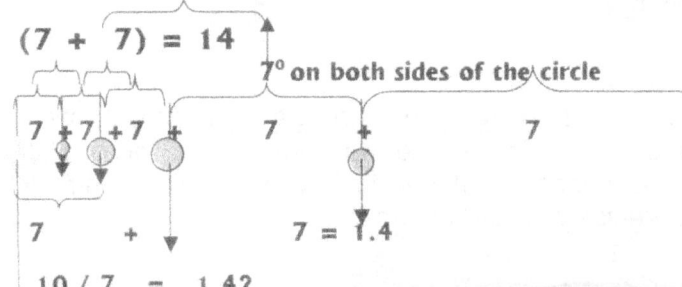

$(7 + 7) = 14$

7° on both sides of the circle

In using very simple mathematics and I also dare also say too simple for the extremely intellectual Newtonians Physics Academics I prove how the Titius Bode law works as the Titius Bode law forms space through applying gravity.

7 + 7 + 7 +   7   +   7

7   +   7 = 1.4

10 / 7 = 1.42

10 / 7 = 1.42

While there is no hint of mass as a factor in the solar system, this Titius Bode law is what is present and is what is applying. The only thing new is that I am the first that prove how this law works and no physicist is interested this far in what I have to say because I belittle Newton and his corrupt principles. Every physicist in office fights this because with this I prove Newtonian views are no more than science fiction

10 / 7 = 1.42

= .7 /⁄\ 1.42
= 1.4 /⁄\ 1.42 **Because the space-to-matter is in the square at 10 placing the matter-to-matter at a square of .7 + .7 = 1.4 the space-to-matter forces the matter-to-matter to double the distance b number as structures are place father from the main$\Pi°$ maintaining singularity.**

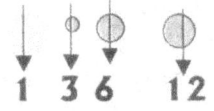 

1 3 6   12   24   48

I'm correct. If they admit that I'm correct the entire world of physics becomes recognised as the joke it is in reality and they are recognised by all as being the

Later I show that 7^2 in conjunction with singularity applying the law of Pythagoras forms 10.

$14 \div 1.42 = \mathbf{9.859}$ **or** $\mathbf{9.86}$ **or** Π^2

I use primary school level mathematics and for that being so simple them say it's too simple too apply as physics. I will show you one the many letters of rejection I received later on in the book. This is what there is and that is all there is in the solar system. Look at any picture and try to finds mass. The measure of mass forming gravity clearly plays no role in allocating the positions of planets as Newton declared it must do. The entire idea that gravity is a magical force created by the value of mass is as unbelievable as the dogma is of those presenting this idea. Please use what the solar system provides to confirm what Newton says is in place when he says mass forms gravity. Science would rather accept Newton where there is no proof of Newton ever being correct than to admit Newton's incorrectness. I prove mathematically the reality of all four cosmic principles that are in place in forming gravity but because I do that and because I then make a mockery of Newton and their Newtonian principles they ignore me.

The circle coming about from T^2 is the **controlling singularity** which is always a circle at the centre that is poisoned by the line **k** in relation to the centre k^0 and by forming a circle it holds reference to the **governing singularity.**

Where **the governing singularity** is the centre of a spinning object such as the Earth, the centre of every atom holds **mutual singularity** that collectively puts a mutual value of all the atoms' singularity as a combined equal to the **governing singularity** and then the solar system will provides a **primary singularity**.

The one would represent T^2 the other forms **k** that then produces the third singularity forming space a^3. The sun spins through 7° and the marker planet spins through 7° while the positioning planter spins through 7° where then the combination of such spinning form (7+7+7 = 21) and sins it builds space coming from the future there is an additional .991 of space which forms the future that is not part of the Universe yet. This proves that the Titus Bode law produces gravity by measure of Π.

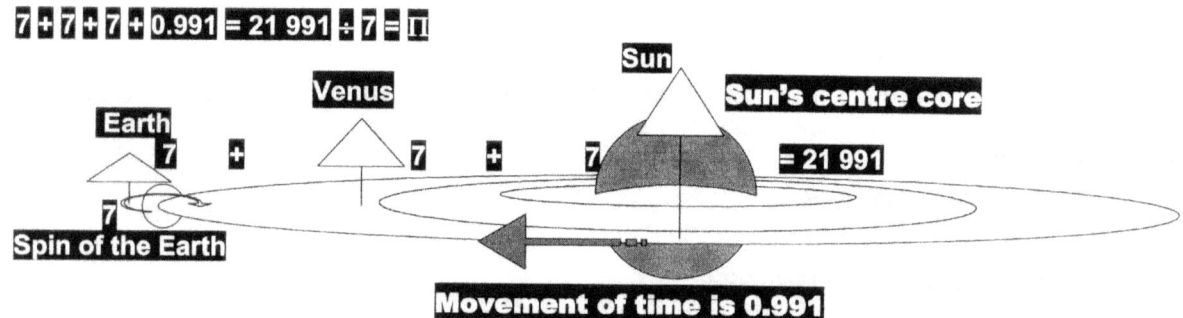

$$7 + 7 + 7 + 0.991 = 21\,991 \div 7 = \Pi$$

k is the space taken from the centre k^0 to the end of the line **k.** This line shows where the location is around which planet circles. The specific value about the centre is most important because from the specific centre gravity indicates a positional worth. The line forming **k** is pointing the circle or the **governing singularity** formed as a line that eventually forms a circle running from the centre k^0 to where the space a^3 is indicated.

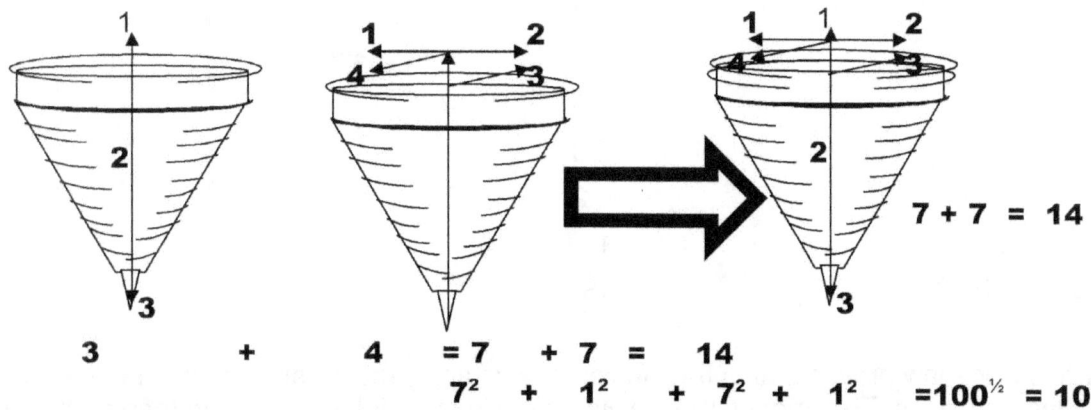

primary singularity	= 7
the governing singularity	= 7
controlling singularity	= 7
mutual singularity	= .991
total development	21.991 or Π

Total producing Π to the value of = 21.991 by the measure of spin = 7
In that we find the ratio that forms the Titus Bode law.

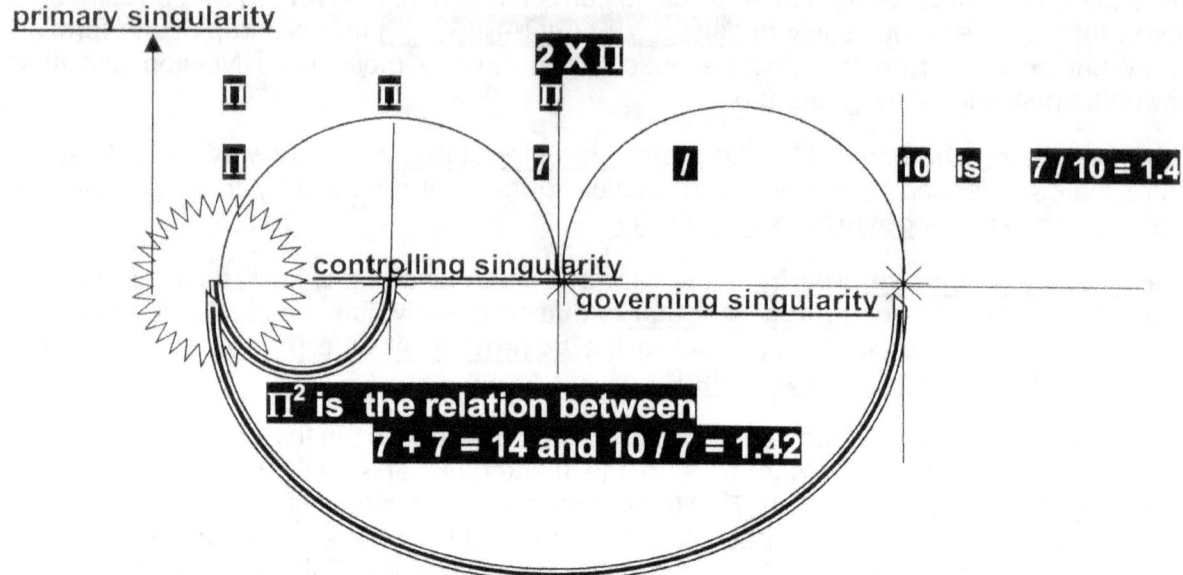

To find the mean distances of the planets, beginning with the following simple sequence of numbers:

0 3 6 12 24 48 96 192 384

With the exception of the first two, the others are simple twice the value of the preceding number.
Add 4 to each number:

4 7 10 16 28 52 100 196 388

Then divide by 10:

0.4 0.7 1.0 1.6 2.8 5.2 10.0 19.6 38.8

The resulting sequence is very close to the distribution of mean distances of the planets from the Sun:

Body	Actual distance (A.U.)	Bode's Law <A.U.)< td>
Mercury	0.39	0.4
Venus	0.72	0.7
Earth	1.00	1.0
Mars	1.52	1.6
Ceres		2.8
Jupiter	5.20	5.2
Saturn	9.54	10.0
Uranus	19.19	19.6

The Titius Bode law proves that in the Universe laws apply that positions objects in terms of other rules that mass. That means the Newtonians hides their lack of understanding behind mass that they invent.

The Newtonians gave the Titius Bode law a formula and that explains the lot. To they're under achieving standards that is very satisfactory. Now it is written in mathematics then what more do we need to know. The fact that the distance that Mercury has from the sun is doubled by that which Venus has from the sun is completely ignored. In cosmic reality mass plays no part. Then again the distance that Venus has from the sun is doubled by that which the earth has. This clearly has nothing to do with the size or mass of the planets. Explaining that part is completely ignored. Then again the distance that the earth has from the sun is doubled by that which Venus has and inexplicably this forms the layout of all planets in the solar system. Where does Newton's idea of mass fit into what truly applies in outer space. Moreover, why does science never mention this? This is my formulated explanation about how the Titius Bode law forms.

The numbers we need to find the key to the mystery of the Titius Bode law is 3, 4, 7, and 10. Every time the allocation is pointed the sun forms the 3 of the axis and the planet forms the four of the circle locating the first 7. Then to form the next 7 the distance of the first 7 has to double to allocated the poison at 7 + 7 which in terms of crossing an axis twice (the sun's and the first inner planet) there forms a crossing of singularity 1^2 twice and the compliment is $(7^2 + 1^2 = 50) + (7^2 + 1^2 = 50) = 100^{1/2} = 10$. That explains the sequence involving 4 (the first circle forming singularity) and 10, which is the total space in ratio of the allocated point the outer planet, locates planets according to the space that build according to Π at a point we find double seven producing 10.

The Titius Bode is in position

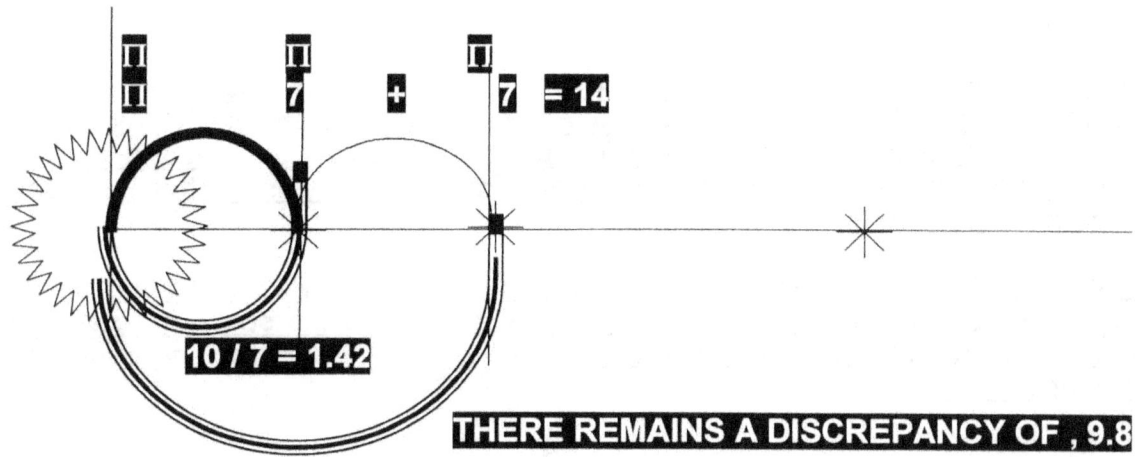

The turning T^2 of any circle holding space a^3 is valid only if forming a reference **k** to a centre k^0. $k^0 = a^3/(T^2 k)$. This depicts a position a domineering singularity k^0 fills in relation to another point serving subordinate singularity **k**. There are always a dominant and a serving singularity interacting.

If **k** indicates the centre of the Earth then T^2 rotates to form the **governing singularity k^0** where then the centre of the Sun **k** will form the **controlling singularity.** When the Sun rotates, the Sun's centre k^0 forms the **governing singularity** giving the Earth in orbit **k** holds the **controlling singularity**.

The measure of **k** is not a specific value but serves only as an indicator to which space rotates or applies by the space rotating in a circle.

This role of singularity being **controlling** or **governing** is playing part in movement of gravity forming and is very important when trying to understand the role that the four phenomena play in the forming of gravity.

It is most important to understand what happens in the event of an object going through the "sound barrier" or when escaping from the Earth's atmosphere. Where the object is standing still holding a position that allows the object to have mass, the object is part of the Earth while the Earth has the **governing singularity** and the Sun has the **controlling singularity**.

As soon as any object moves on Earth, the movement switches singularity by allowing the object to obtain the **governing singularity** while the Earth then fore fills the directional circular control in forming the **controlling singularity.**

All four phenomena interacts in a manner forming this role where for instance in the solar system the Sun holds the **controlling singularity** and Milky Way forms the **governing singularity. In this is the connection between singularities that we find all four Phenomena holding relevance.**

It is secured in every area of cosmic creation. It is what binds the cosmos and it is what rules the cosmos and that is what links the cosmos. No object that is within the solar system forming a part of the solar system can ever leave the solar system without destroying everything in the solar system.

The cosmos is a unit secured and united by movement going around the sun.
Because **the sun forms primary singularity** everything forming a relevancy as **the inner rotating object around the sun that forms controlling singularity** in terms of an outer rotating object that in relation **forms the governing singularity** the solar system will remain a unit intact. Only by breaking the sun will this bond break.

Primary singularity holds the initial axis with 3 within the Titius Bode law that forms by adding the four of the controlling singularity circle and the object pivoting around the object that represents the controlling singularity holds the second three in relation to the four that is the circle of the governing singularity and in this the means to build the solar system.

This in total shows 7 moving by singularity within relevancy of 7 moving also in context of singularity. The 10 forming from this represents the space in total of all movement up to that point. That is the ratio of the Titius Bode law and shows the order that persist.

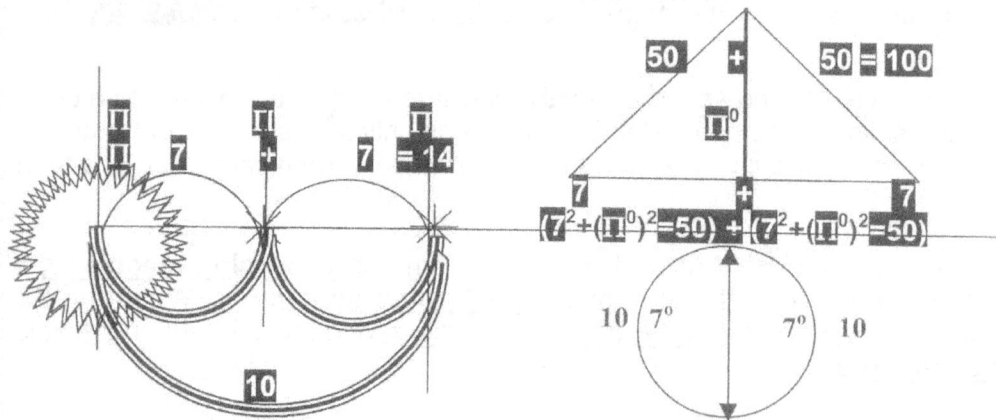

To find validity in my argument one must draw this statement of motion back to the point where singularity is getting sides or said mathematically Π^0 going Π. This is the **controlling singularity** and Π forming Π^2 is the **governing singularity.** When there is singularity there can be no sides. The one forming singularity Π^0 by measure fills no space while form Π develops Π^2 into space. The space that even the dot fills being Πr^0 does not really exist in the manner we humans see space to exist. It is a spot that is there without being there. It does not visually exist because it is not filling any substance and it cannot be recognised since it is not three-dimensional. The spot and the dot have no dimensional worth of any measure.

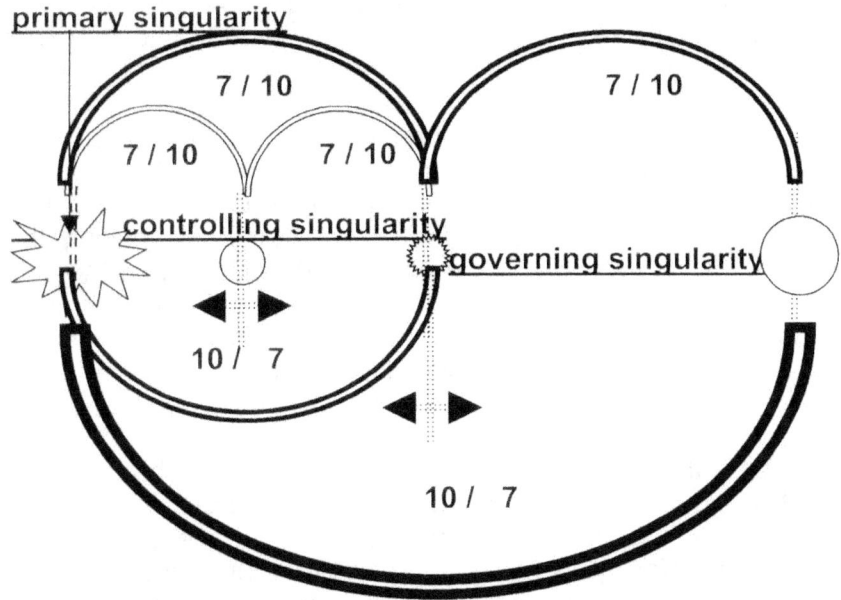

In the Titius Bode law movement in rotation of the bodies in question confirms the value Π. The bodies holding material contracts or destroys space by compressing space into singularity within ever atom rendering material the ability to remain cool and not to excessively expand as it happens in a Super Nova. By reducing space holding a value of Π material expands up to a point where the gravity is less or equal than the speed of light. After the gravity within a star exceeds the speed of light the material or the atoms reduce space until only singularity remains and such a star is named a Black Hole. That is where infinity takes charge of eternity and combines time into the point as it started but in reverse this time. As material removes space by compressing Π space outside the rotating influence of material expands Π to maintain equilibrium throughout. If that were not the case the Universe would only hold material that would form a lump of atoms that dissolves all forming creation. As material compresses space by reducing Π in space then again Π forms by allowing space to become greater. In that process we have the **primary singularity** the **controlling singularity** and the **governing singularity.** The process of the Titius Bode law works on singularity maintaining equilibrium I can explain how space forms as much as I can explain how compresses where the one is the reverse of the other process. The sun being stable in relation to the other holds 3. The immediate inner planet maintains the space distance from the sun and holds 4 in terms of the sun and the location of the outer planet in terms of that relation forming. The sun having 3 and the inner planet holding 4 forms the first 7 in the value of Π being formed as a location to the outer planet. Then from the inner planet to the outer planet forms at the same distance the other 7 where then in accordance with Pythagoras $7^2 + 1^2 = 50$ locating the inner planet and $7^2 + 1^2 = 50$ locating the outer planet and the two distances holding the value of 7 as a constant maintains a collective 100. The square of 100 in accordance to Pythagoras is 10 and that is why in the ratio the Titius Bode law forms we have 4 confirming the inner planet circle and we have 10 confirming the outer planet circle forming the end result of the location of the planets in accordance to the Titius Bode law. But this is because gravity works on Π that forms and not non-existing mass that pulls more mass. As much as material contracts space it is within that much material expands in terms of space occupied but it reduces the density of space overall and that allows space not occupied to expand in ratio of space contracting. That is the Hubble constant that could never be a constant because that constant depends on stars and the gravity stars form. Therefore the planets moving while contracting secures the stability of the solar system while the contracting space

moving towards the sun $k^0 = \dfrac{a^3}{kT^2}$ and $k^{-1} = \dfrac{a^3}{T^2}$ as Kepler's figures prove forms the ratio by which space in the solar system grows in countering to the sun that depletes the moving space. As much as the Universe contracts that much the Universe expands and the solar system is proof of this statement.

I admit that at this point you may even find a remark as innocent as the one I just made very offensive and rowdy, but as you progress through the volumes you will find I am not in the least exaggerating. I ask you not to dismiss the book after the first few pages just because you have the opinion that I do not understand Academic's laws. Again by making the following remark, it will come across as very presumptuous on my part, but I may be the only person in the world that understand Academics and not accept Academics. There is a huge difference between the two notions. The formulas you find is the diversion that an object in movement holds as the movement relates to the line singularity holds in relation to the moving object.

Have you as you sit reading this part at this minute sat back and gave a thought about the light enabling you to read? Such a thought brings to mind the most simplistic answer one can imagine. The light hits the page bounces from the page and contact the lens of my eye where the lens conveys the photons becoming electricity to a part of the brain that translate the electricity to an understandable message and that makes one read. It is as simple as that! Ever gave a broader thought about light streaming across the night sky, coming from ends of the Universe we do not even realise it is there?

How does the photons manage to convey one complete picture coming from as far apart and as wide an area as it does? With a few photons connecting the eye or lens no one ever noticed the wonder of light. The photons reflect a view that seems as if coming from all the billions upon billions of stars. But most is coming from darkness covering an area no man can measure. Yet how many photons can actually connect to the lens of the camera or to the eye?

Still a few photons coming from a single direction directly ahead eventually tell the entire storey. It is very simple to take the process of seeing by means of photon conducting very lightly and I have never heard one of the Brainy Bunch really in sincerity dissect the process to its potential. It is impossible that light from such an array of assorted sources can simply come together at the eye lens and show a picture of objects spanning across a Universe as wide as our mind can receive where the objects they reflect is beyond human measurement and the quantity is inconceivable many. Light is much more than the medium science takes it to be. Light connects the Universe in a way we cannot contemplate. Light being far apart originating from regions not in the same time or universal space connects in a way that present us with a picture holding the Universe in an understandable content. From the point we stand and we watch the Universe the significance of what we see surpasses the sense of understanding of what we are experiencing.

How can the few photons that our lenses catch coming from such an area as the night sky cover transmit the complete picture of what we see. Take a few seconds and gleans at the picture of the night sky then rethink the picture applying the full content in the picture to what the size of you eyes is. Think how big the picture is that your eyes take in and translate that area to the size of your eyeball in an effort to determine a ratio. One will be forgiven if one thinks of the ratio as eternal to nothing. Yet a few pages back I showed that according to mathematics there couldn't be anything as nothing. Consider the path the light followed from the source connecting to light from all other sources where all particles of the other light may come from and bringing a full picture to the lens one use to look through. In your mind connect a line from every atom producing light and connect the lines to your eyeball and see how you can manage to fit all the lines, as small as the lines may be.

Scientists think of outer space as geodesic zero, with nothing in outer space but space. Geodesic zero means the light travels in a straight line from where it originates unhindered all across space to where the light connects the eye. Such an idea by itself is outrages because the stream of photons reduce in space to such a minute quantity that taken the area the photons travel and the space in vastness it covers, the chances of one photon coming across many hundreds of light years through billions upon trillions of cubic kilometres of space and selecting my eye to convey the electricity is less than infinite. Yet such conveying takes place every second of every minute.

The position of the location of the second singularity, which is the precise duplication of the first singularity but in a diminished capacity, is obvious to miss when one is not applying a detective mentality, as one should in scrutinizing the cosmos. Culture will have us believe that when one sees a colour shining from an object the colour is associated with the object. Logic tells a different storey. A yellow dot is all the colours in the spectrum but yellow because it is disassociating with the yellow. That goes for red blue and all other colours we may visualise. I think the norm accepts this as scientific fact with very little argument or substantiating proof about that required.

If light came as individual streams of photon flurries our visage would translate that as such shown in the fragmented picture above. It would be a picture unconnected bringing across some photons in the manner where every object stands apart not being related in any way and that will be what we see, if it is anything that we see. That we know is not the case but that means geodesic zero is as much rubbish as anything Scientists regard with simplicity and with careless thought. Geodesic zero means nothing and how can I see nothing as darkness because "nothing" is not darkness, nothing is "nothing" and the darkness I see is darkness showing the darkness as something.

What then about colours that are technically not colours as is the case with black and white? White is simple. By spinning all the colours in the spectrum the colour white shines through. Black is quite another matter. A friend of mine whom is one of the best painters I have ever come across told me that one couldn't paint black but have to make black a dark blue to show shade on the canvass. That apparently is his success in achieving the realism.

He also went on to explain how many variations of dark blue form the shadows in one simple tree. This remark set my mind in motion. One cannot see black because black has no colour to show, but black is the colour most prevalent in the universe. One can see only by colour and since black is not a colour we should not see black, but we do.

If the darkness was the representation of "nothing", then that should be exactly what we must see, nothing but the stars. Taken from the top picture some stars and leaving the rest to nothing is what we see in the picture below. A blind person sees nothing but when we look at space, we see something that we think nothing of as we see as space. One cannot have the ability of sight and see nothing except by closing your eyelids and then you see nothing. But in that case you do not see "nothing" in contrast of "something" you see "nothing" without it contrasting to "something".

 **Nothing is all about not being and not "not seeing".**

By the ability to see the darkness renders the darkness something other than nothing and that changes the acquired value of the darkness from nothing to something. There is an eternal difference between something in infinity and nothing.

The arguments introduced up to this part of the introduction prologue only touches the most basic aspects of my work and by no means can such an introduction secure an opinion. Yet, not once through all my long investigation in the past thirty or more years have I found any other person claiming such views that I have brought about even in this skimpy way as I do in the prologue. The arguments introduced up to this part of the introduction prologue only touches the most basic aspects of my work and by no means can such an introduction secure an opinion. Yet, not once through all my long investigation in the past thirty or more years have I found any other person claiming such views that I have brought about even in this skimpy way as I do in the prologue.

As it applies with all things, so it does in this case as well that when delving deeper into any issue, the complexity of the issues truly come to the fore ground when analysed in more detail. I wish to advise the reader to treat the seven books as seven different works and in that light I have separated each work in volumes of seven separate books with individual I.S.B.N. numbers with adding one part, the one you are reading, with one sole purpose and that is to bring about an academic introduction to clarify a quick perspective.

Then the next three parts being of a general introductory nature there are overlapping in some sense but each highlighting issues in different manner as to clarify facts used in the last three parts bringing conclusion to different cosmic perspectives. Yet the work is seven parts of one thesis and as such it serves.

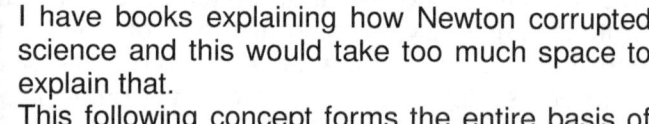

I have books explaining how Newton corrupted science and this would take too much space to explain that.

This following concept forms the entire basis of everything forming anything in physics that is part of science. If ever any thought represented physics then this is the most fundamental start of physics.

This is so impotent I wish to run through this again because this forms the basis of all physics. Have you, the person reading this, ever thought how it is possible to see that much information that you see at night when looking at the sky. Ever thought about how you are able to see when you see everything in the night sky and how that much light information can fit into such a small space as your eye? Have you ever sat back and think what the amount of information it is that you see when you see the entirety of the Universe when looking at the Universe at night and what the size is of everything of that which you are able to see?

The one star you see seems to be a near visible dot in the picture while the dot might be hundreds or might even be many thousands of times the size of the sun...and we think of the sun as big. The dot is then that much bigger than the sun because the star we think we see could be a galactica hundreds of times the size of our Milky Way galactica but that shows as in the sky as one little dot and yet that entire structure as big as it is, does also fit into our eye socket. But that is not all...there are trillions of such light images and they all fit into one eye socket. What we see is immeasurable and yet we see it effortlessly in the space our eye holds...how can that be?

How is it possible to fit what we see into the space of our eyes we have? Think how much is the entire information that is visible at night and think about how all of that fit into the space your eye holds?

Consider how big is what is visible and put that space into the size is of what your eye can hold and ask your mathematically educated Professor in physics to find some ratio between what you observe and the size of your eye. The ratio is astonishing, but more-over what is truly astonishing is the arrogance of man to think of his position, as being important while the space man holds is beyond any comparison in ratio to everything we see in the Universe we see. Think how small we are when we are able to see the entirety out there! Even if there was other life out there, what is the worth of it in comparison to what there is that we see?

In this idea about how you are able to see the entire Universe you will find all the answers to the questions about how physics use time to employ gravity. Mass and anything Newton ever said has no implications on the explaining. It is about all the information of the entire Universe presented in one electron contacting a nerve in your eye. The question about the Universe is how can whatever is in view, come stored as a parcel in an electron, and tell the entire story about the entirety out there locking all that data into the space of an electron. That is physics and tries as you may, not one person Newtonian or otherwise can have mustered the ability to calculate that part. Newtonians can pretend to play God and live in their fool's paradise as long as they are King of the Universe of fools while keeping the

conspiracy alive to hide the truth about Newton's corrupt formula of $F = G\dfrac{M_1 M_2}{r^2}$. There

is a mad conspiracy in physics to prevent anyone not in physics to learn about the truth hidden to all about Newton's Gravitational principle fairy tail. The fact that you can see the entire Universe and everything in it through one optical nerve tells everything that physics and cosmology up to now missed completely.

Forget about the fanciful corrupt mathematics that proves nothing when the cosmos does not confirm Newton's crooked mathematical arguments. Newton's religiosity might corrupt science but who in science would cares about correctness when it simplifies the ongoing brainwashing of students studying science.

They confuse everyone about what *weight* is, what *mass* is and what *gravity* is because they wish to have everyone think of "mass" in terms of weight while they then deny weight and "mass" is the very same thing and then they confuse "mass" and gravity because they never distinguish between what "mass" does and what gravity does. This is only the tip of the iceberg and you will see when reading this book.

We humans are cursed by all the conspiracies that we have to endure. To break this we better start with science. We are all so entangled in a society filling our senses with one conspiracy upon the other that we can never find freedom in thought without disentangling our mental state from the brainwashing that furthers conspiracies by which we are taught to behave like those in power wish us to behave. This manipulating they subdue us too by rendering our thinking power to become computerised slaves is part of the general public's mental state and proves how much John and Jane Dow and Mr and Mrs Nobody care to be subjected to a comatose condition subdued to having no brainwaves functioning. We are told how to think and what to buy and how to vote and in which to believe by conspiracies whitewashing us so that those in power and those that are rich and those that are intellectual can control us on a daily basis.

It is the point singularity which is better smaller and point is where forming the very centre that plays the part as the **controlling** within the Universe I have **named** as **Infinity,** known as the axis. It is where nothing can go anything within that point can never reduce. That the entirety called the Universe begins and where everything holding substance begins. Once one accepts the fact of singularity being present in that location, that accepting of singularity then is contradicting all the things we know and we can measure and we recognise that point being present by merit of the fact that the point referred to is not being formed by any of the things we can recognise. It is made up of everything we don't know and constitutes of everything we are unable to recognise or visualise. In that spot there is no space.

Infinity is that which can never start

Eternity is that which can never end

Eternity is that which can never end

Singularity Π³

That spot holds **Infinity.** In that space there can be no motion because there can be no space to have the motion within. It is formed as a line that is so small that our human reality by perception declare that point as not being there and the only reason why we know it is there is because of the results it left as an imprint of its not being there. We cannot detect it but notwithstanding our failure to note it we can recognise the dot on the merits of its absence and while in our Universe it is always absent, reality disallows the dot ever to be absent, because it is never absent. It cannot be absent. It cannot go absent but it can never be there where it should be in a place from where the third dimension forms and it is always present if I wish to locate it. It is infinity that can never go away.

I named that part forming space **eternity** because in that area there is nothing that ever can go bigger or become more or find an end to the outside. Whatever was and is and will ever be is locked in that space I named **eternity** and it is eternity that never ends because eternity can never end moving. What we think of, as expanding is never ending movement giving eternity the eternal motion that will go on forever. The "so called expanding" of the Universe $T^2 = a^3 \div k$ is where singularity is shifting relevance k from liquid $k^{-1} = T^2 \div a^3$ to solid formulated as $k = a^3 \div T^2$ and the process whereby this happens is precisely the same as the Coanda effect. Getting back to my first argument about a line and that no line can start at zero but has to use singularity as a starting point, this is all the proof I require. The line k coming from the centre (singularity k^0) forms by forming an initial dot Π^0. However, I went on to say that whatever the line used to start with has to continue in order to repeat the same that began the line.

Therefore the line started with Π^0 and it has to continue with Π^0 until such a point, as it must end with Π. Whether the line is Π^0 or is r^0, or uses 1^0 the outcome all refers to singularity being used. By reducing the line we come to the end of the mathematical equation of the circle but the circle does not end there. That is what Newton did not recognise from the figures the cosmos represented to Kepler. The circle only secures the final cosmic figure and the value to singularity where all things have equal value. The movement of the circle splits singularity in two sectors. By forming Π the circle has to form Π^2 due to the movement coming about in securing the space Π^3. Kepler chose to use different symbols too those being valid, but the concept remains the same. Kepler said that $a^3 = T^2k$ while I show that Π^3

$= \Pi^2\Pi$. It still confirms that movement Π^2 = is the forming space by three dimensions Π^3 in relation with the movement Π^2 being relevantΠ to singularity Π^0. I shall try and explain what this concept holds in terms of a piston moving while working inside an internal combustion engine.

The piston goes up to a point we call top dead centre where the piston stops and according to the crank the piston halts in directional movement. Then the piston starts to accelerate to a point we call bottom dead centre where, again it comes to a dead halt. The piston stops directional movement at T.D.C. and at B.D.C. or that is what we see without seeing anything. This is not the case because if this was the case the engine must vibrate at those two points of stopping. We reason that the piston stops twice and starts moving on the two occasions (at the very top and bottom) but if that was the case of stopping at two points without stopping anywhere else, the vibration that the stopping will cause will have the engine disintegrating completely. To us favouring positions the piston stops at two locations but the fact of the matter is that the crankshaft stops every $7°$ of rotation and if the crankshaft stops, then so does the piston stop.

The stopping is a continuous and is an ongoing process that happen every $7°$ of rotating. The crankshaft moves in a straight-ahead position going straight and then it stops and redirects by $7°$ and then it turn by going straight again. It is $a^3 = T^2k$ and then it stops (a^3), it turns (T^2) and then again goes straight again (k) while holding reference with singularity $k^0 = a^3 \div T^2k$ all the time. One cannot part the redirecting and the going straight T^2k because it is the same movement since the space forming a^3 is equal = to the turning T^2 and the going straight k. This is evident when dissecting Kepler's formula $a^3 = T^2k$ that $T^2 = a^3 \div k$ and $k = a^3 \div T^2$ while honouring Newton's 3rd law $k^{-1} = T^2 \div a^3$. Please believe me that this puts movement as a much complicated dimension because this has the material $a^3 = T^2k$ moving $T^2 = a^3 \div k$ in terms of ($k = a^3 \div T^2$ as well as forming $k^{-1} = T^2 \div a^3$) while always referring to singularity $k^0 = a^3 \div T^2k$.

Kepler gave his formula symbols $a^3 = T^2k$ that do not quite represent gravity in its true symbolic nature and that then was the reason why I came on the idea that gravity has to link to Π more than any other value or symbol. It is because everything holding gravity or representing gravity (not mass) is round. Gravity connects by the use ofΠ. We have to part what mass does and what gravity does. Mass is where the object connects to one point on Earth and being at that point with mass the Earth does the moving by spinning. The spinning of the Earth then represents the movement or the intention to move because the Earth spins byΠ. This movement gives mass its qualities because mass does not possess the influential value of Π since mass is a quantity representative of the amount of atoms and not the spin of the atoms within the mass quantity.

If we look at the way the Moon connects to the Earth, committing movement in a circle does it. That representsΠ. When we look at the way the solar system connects to the Sun in circles every planet holds an individual symbolic value to Π that circles in relation to the Sun. If we look at the roundness of galactica, the formation represents Π. Every solar star holds roundness and roundness only represents one value, which isΠ. The connection gravity has is not by mass but it is byΠ. When we go in search of a cosmic resolve to find gravity, we better start looking for the influence Π has on the subject or leave the entire subject alone because the gateway in understanding gravity goes by the meaning of Π relating to Π^0.

The condition for the presence of this singularity that forms everything, controls everything and is everything is the centralised $\Pi^0 = \Pi^3 / (\Pi^2 \ \Pi)$ singularity that forms by movement $\Pi^2 = \Pi^3 / \Pi$ of space $\Pi^3 = \Pi\Pi^2$ in relevancy $\Pi = \Pi^3 / \Pi^2$ going both ways $\Pi^1 = \Pi^2 / \Pi^3$ thereof (Newton's 3rd law).

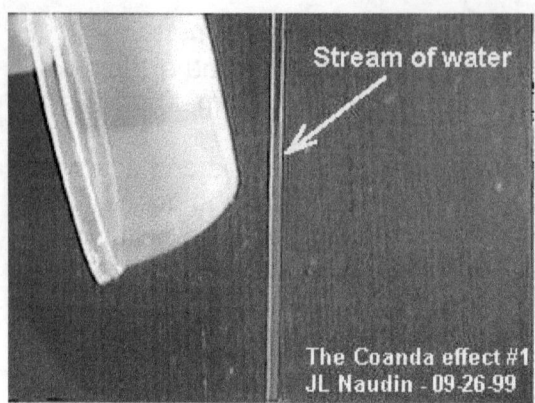

The Coanda effect #1
JL Naudin - 09-26-99

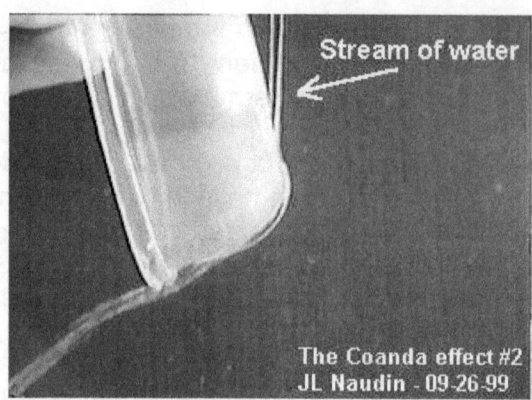

The Coanda effect #2
JL Naudin - 09-26-99

The condition for the presence of this singularity that forms everything, controls everything and is everything is the centralised $k^0 = a^3 / (T^2 k)$ singularity that forms by movement $T^2 = a^3 / k$ of space $a^3 = kT^2$ in relevancy $k = a^3 / T^2$ going both ways $k^{-1} = T^2 / a^3$ thereof (Newton's 3rd law). Now put this formula in terms of gravity and we can see the gravitational picture of the Coanda effect come to life.

This explains the Coanda effect and **the Coanda effect is gravity** and gravity "glues" the water to the glass! The water forms a value of $\Pi^1 = \Pi^2 / \Pi^3$ while the glass forms a value of $\Pi = \Pi^3 / \Pi^2$ This process happens to all spinning things and as much as it happens to a piston connected to a crankshaft, just as much this will happen to an atom spinning an electron in a similar manner as the crankshaft is spinning holding a piston connected. This proves that gravity is the Coanda effect and in another

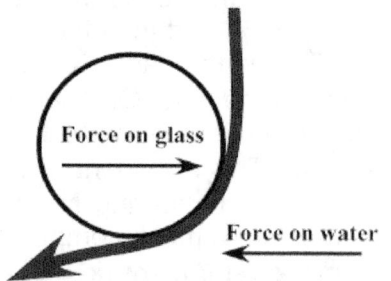

Force on glass

Force on water

book I prove that the Coanda effect has its origins in Π forming a value and that value forms gravity.

In order to understand physics applying in cosmology I had to start by dissecting the set-up forming pi.

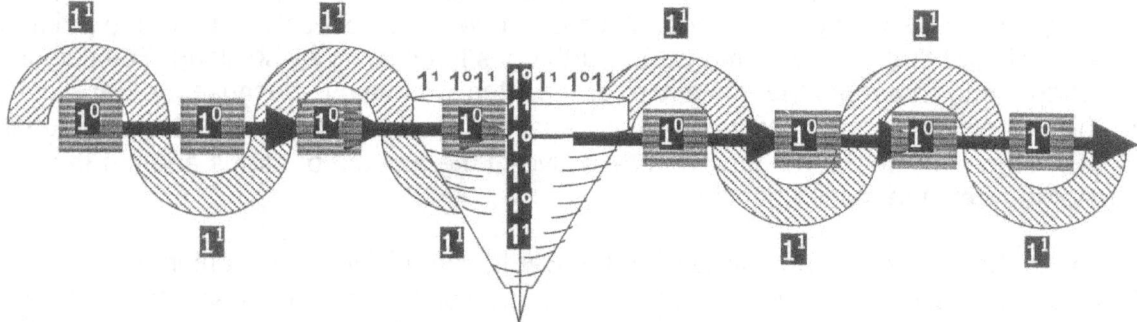

At this point I can introduce my theory on the ***Absolute Relevancy of Singularity*** At the point in the centre of the circle a line must start. In the beginning when I explained the way I figured how the line starts I said a lot of dots has to continue in order to form a line. It would be 1 + 1 + 1 etc. because the line must form by holding singularity. After that point does mathematics begin but in the line that forms representing space as other all factors, then time holds 1. The line can only form when all the points forming the line have the value of 1 being 1^0. In that conclusion one realises something must separate singularity from all other factors because singularity hosts all other factors but is by own initiativeΠ^0. Only when singularity meets the end value can the end value have Π where the final ring of the spinning circle forms Π. That will be the spot of origin forming the relevance in Π. That will hold the eternal spot...the smallest spot ever because all spots that ever can be were secured in a position in the centre of that spot that must continue as a line that forms.

Because of the progress singularity follows from the single dimension singularity only allows mathematics a start at Π^0 progressing further onto Π^0 and from there the line is born as $\Pi^0\Pi^0\Pi^0$ and to $\Pi^0\Pi^0\Pi^0$ Π^0 etc. where Π^0 then may form the concept and value of r. But the line starts at $\Pi^0 = r^0$. This forms because cosmology is singularity based and the value is $\Pi\Pi^0$. This line $\Pi^0\Pi^0\Pi^0$ of singularity can only continue because every spinning atom preserves Π^0 in the very centre and since $\Pi^0 = \Pi^0 = \Pi^0$ the line is the same without finding conclusion except at the end where it forms mass at Π. At the point where Π forms, the movement Π^2 of the circle defines the space Π^3 of the circle and it confirms the centre Π^0 of the circle through the rotation. Let's call this the solid forming or if you wish, let's call it Kepler's singularity.

After that singularity forms a line $\Pi^0 = \Pi^0 = \Pi^0$ where this forms another line again as Newton stipulated it by $\dfrac{dJ}{dt} = 1^0$. Let's call that the liquid singularity or Newton's singularity and the relevance of singularity having a solid base compared to the singularity holding a liquid base comes about by the movement of gravity.

There is no substance difference between 1^0 and 1^1 and it is a relation where one moves as the liquid partner and the other stands still as the solid factor. Both are not as much equal as they are precisely the same. Infinity cannot move and eternity cannot stop moving. By parting, infinity had to remain motionless and eternity had to remain moving as it introduced a part of the cycle to one line (point) where it stops moving in relation to the other factor that cannot move but does start moving.

Time is a graph that never begins since it never ends and while everything never repeat everything always never remains the same. The factor that shows motion forms the liquid while at that moment the factor that does not show motion forms the solid. The measure of 1^0 is transformed to 1^0 and which ever are 1^1 is passing the extending of space on to 1^0.

Time spins around a centre because everything spins around some centre somewhere in order to secure the centre singularity. But also time moves and in that there is the linear that always are part of cosmic motion. The centre is referred to by heat but heat also secures the centre by reconfirming the centre in the lateral. But in both cases singularity is reinstating singularity by confirming it as it is referring one another. In the manner that 1^1 confirm a position in singularity 1^0 it is supporting 1^1 by generating 1^0. By generating 1^0 it is repositioning and reallocating 1^1 as a position by confirming 1^1.

From all of the above one can deduct that outer space is something viable through which objects travel. It is clear that something filling the space between Jupiter and its first moon because of lightning interaction between the two structures. There is a reference between a structure holding material and the space above as well as the space between two objects. There is electric lightning travelling between Jupiter and its closest planet. If there is lightning there is electricity and electricity means a very distinct interaction that connect the space inside the material structures through the space parting the material structures putting the space in between as a conducting medium which nothing can never do. Understanding this nothing not existing is fundamental in understanding the relevancies applying.

From these conclusions I prove that gravity is the result of four cosmic phenomena interacting to form the value of Π which by movement becomes the value of gravity Π^2 and gravity is equal to cosmic time applying. In order to understand the development of the cosmos and moreover the start of the cosmos and the progress in the cosmos as the cosmos formed, one has to understand the measure of Π. One has to see that Π is not merely 22 over 7 or that Π is a ratio that no one ever bothered to clarify, but Π is the key that unlocks every lock that hides a secret in the Universe. One has to microscopically dissect the measure of Π to find the cosmos in measure. One has to understand where 7 fit in Π. The fact that Π is 7 at the bottom and that 7 relates to a double value of 10 is a key issue. Furthermore, it is very important to see why Π is 10 times two by adding 1.991 on the top part of the equation. In this measured value is what holds the building blocks of the entirety we call the Universe. It is behind Π that we will find the four phenomena, which I named the four pillars performing as gravity as they form gravity. It is by the actions of Π that the Universe develops. The Hubble expanding goes by implementing gravity as Π in the square through the four pillars on which gravity and time rests. It is behind Π we discover the meaning of singularity and how singularity forms the absolute and only building block as a form that forms the Universe. It is in Π we find the Cosmic Code unlocking the meaning of the Universe. Time is centralised in Π^0 that forms Π as space's limit that becomes space by gravity being Π^2.

Space is time gone to the past in which time confirms its presence it had in the cosmos by moving from the present time into space and then onto the future leaving space behind as the past. By forming a present, time is in infinity forming singularity that then has to move on and in doing so it leaves a legacy behind being space. Time is the movement of everything forming the Universe where in time the movement of time relocates everything in space by moving from the present onto the past leaving behind space. As time becomes the past by going to the future it forms space as it confirms the past, and in that space is what time forms by going to the past leaving space behind. Space becomes what time was at the point where time formed the particular space in relation to Π. As time becomes the present coming from the past, time has to move on to the future at the same time and as time moved on it left space that represents that instant in time in relation to other space that was in some position at a specific location at such a point in time wherever that point in relevancy might be. The fact of Π not only refers to form but also validates the Universe by splitting infinity from eternity.

By forming space when creating Π, time is using Π^0 in establishing movement Π^2. It is in the process of relocating Π to new positions by establishing Π^2 and connecting this as it forms a network consisting of Π^0 by forming space Π^3 in relation Π that establishes infinity Π^0 that always stays motionless. If not for movement, the Universe would be one line holding time by repeating singularity Π^0 uninterrupted and it is in the diverting of eternity to a position away from infinity that the Universe comes about. This is what happens in a Black Hole where no movement within the Black places eternity that always in a standing position to never moves. Without

Singularity also continuing to the value of Π^0

Singularity continuing to the value of Π^0

It is the movement of Π^0 that alters the line of singularity parting material from time Π^0

Hole moves infinity that movement the entire Universe will fall back into and onto one point and everything we thought is real and solid will disappear into that one point holding infinity onto eternity where infinity and eternity then reunites. The Universe is an unreal concept with nothing being a reality but for the movement whereby Π confirms everything in a location in relevancy to all other things in a specific time slot or space.

The Roche limit works on a principle coming from singularity and in that we find the foundation that carries the sound barrier as a cosmic principle. The Roche limit work on the method as explained above where in the circle Π^0 forms Π and in between $\Pi^0\Pi$ the is a continues line of $\Pi^0\Pi^0\Pi^0\Pi^0\Pi^0\Pi^0\Pi^0$ until and up to a pint is reached where Π^0 ends by forming Π.

This line holds singularity throughout to the end of the circle $\Pi^0\Pi$. At the end of the circle where an object connects to the circle as to form the en $\Pi^0\Pi$ and then becomes $\Pi^0\Pi$ mass is awarded to the object holding a solid position in terms of forming $\Pi^0\Pi$. If the object does not form a valid connected line coming from the centre as a continuing $\Pi^0\Pi^0\Pi^0$ the mass cannot be awarded and the object will maintain a status of being part of the liquid surrounding the circle.

At the point where singularity ends by forming a circle, it does so by circling $\Pi^0\Pi$ $\Pi\Pi^2$ to form Π^3 and that end the circle. The circle can only form by both movements coming about one forming a singular circle and the other forming as singularity repositions the circle. The value of Π^2 is in contexts of Π and the point $\Pi^0\Pi$ holds value is the ferocity of the spin formed by $\Pi^2\Pi$. This is the Coanda effect and the faster that the circle spins Π^2 the further would the solid claim space into the liquid forming the gravitational borders of Π^3 or the limited end of space forming the solid.

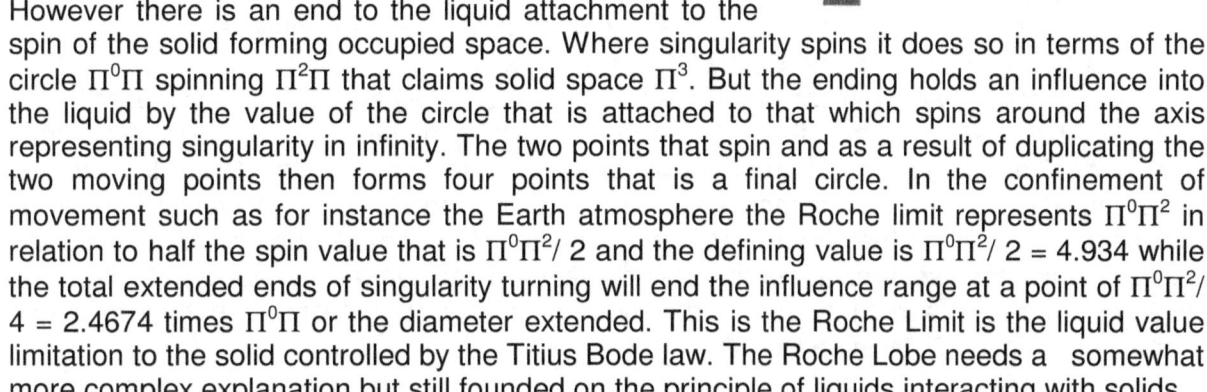

However there is an end to the liquid attachment to the spin of the solid forming occupied space. Where singularity spins it does so in terms of the circle $\Pi^0\Pi$ spinning $\Pi^2\Pi$ that claims solid space Π^3. But the ending holds an influence into the liquid by the value of the circle that is attached to that which spins around the axis representing singularity in infinity. The two points that spin and as a result of duplicating the two moving points then forms four points that is a final circle. In the confinement of movement such as for instance the Earth atmosphere the Roche limit represents $\Pi^0\Pi^2$ in relation to half the spin value that is $\Pi^0\Pi^2/2$ and the defining value is $\Pi^0\Pi^2/2 = 4.934$ while the total extended ends of singularity turning will end the influence range at a point of $\Pi^0\Pi^2/4 = 2.4674$ times $\Pi^0\Pi$ or the diameter extended. This is the Roche Limit is the liquid value limitation to the solid controlled by the Titius Bode law. The Roche Lobe needs a somewhat more complex explanation but still founded on the principle of liquids interacting with solids.

When I, as a person forms a part of the Earth by the virtue of having mass that connects me Π to the Earth Π^2, stands on the Earth Π^3, my position in relation to the Earth gives me a specific positional relation to time Π^0 and the Earth. That gives the Moon a future of say one point five seconds being the past in relation to the Earth and that gives the Earth a past in reference to the Moon's future of one point five seconds. Where I am at any specific point in the present, that point I am holding is that which secures my present point in time. The Sun is eight and a half minutes into my past with all the space being in between the Earth and the Sun and by my view of the Sun I have a present time slot, as it also gives me a past of eight and a half minutes in relation to the Sun since the light travelled eight and a half minutes through space to confirm my past during that present instant.

That secures my past by eight and a half minutes at the point of giving me a present location in time. However, that also secures my future I have from the point I now have in the present by the margin of eight and a half minutes because that establishes a flow of light that would last another eight and a half minutes of filling a presence worth eight and a half minutes while travelling through space by moving with time and every spot filled on the way would secure a position that I will have in a future presence for the next eight and a half minutes, which then becomes my future as it fills my past. Looking at this scenario in a view from Alfa Centauri the allocated position Alfa Centauri holds in space relating to the Earth, gives the Earth a past of say four point six years while this secures the present and having that present secure the Earth to a future of say four point six years by forming time as space between Alfa Centauri and the Earth and this is confirming time to the tune of four point six years. By securing movement it forms time in having a past in relation to the present that by the same margin also secures a future in relation to a definite past. This is how the Universe builds space in establishing time. This applies to all allocated positions of rotating objects throughout the Universe. This means that every point away from Π° serving as Π, wherever that might be, secures my past I have by giving me a future in terms of the present Π°.

Take this in relation to Kepler's formula we then find the Earth (a^3), which is in relation as viewed from Alfa Centauri (**k**) four point six years (**T²**). That secures the three dimensional status the Earth has ($a^3=T^2\,k$) within the space from the Earth to Alfa Centauri (a^3) forming the Universe in terms of a present (k^0) being in the Earth centre which then depends on a location (**k**) secured by a future (**T²**) that will come by movement where the future also doubles as a past ($k = a^3 \div T^2$ and $k^{-1} = T^2 \div a^3$). That is time and that is how time forms space and that is how space-time forms the Universe and that is the ***Absolute Relevancy of Singularity***. That then forms time in the centre in infinity in relation to space in eternity in singularity where time that moves forms space by holding time that does not move secured in positions in relevance to where every point was in time gone by. Π **Divides** **infinity** from **eternity** where **infinity** can't **move** and **eternity** eternally moves as time.

If we put this in terms of singularity (Π^0) we find the Earth (Π^3) is in relation as viewed from Alfa Centauri (Π) four point six years (Π^2). That secures the three dimensional status the Earth has (Π^3) in terms of a present (Π^0) that depends on a location (Π) secured by a future (Π^2) that will come by movement where the future ($\Pi = \Pi^3 \div \Pi^2$) also doubles as a past ($\Pi^{-1} = \Pi^2 \div \Pi^3$). That is space formed three dimensionally by keeping time in infinity apart from time in eternity. The relevance (Π) that forms in relation to the present (Π^0) will relate to movement (Π^2) and the movement is circular which ensures that the relevancy forming is circular (Π) by securing that the movement is circular (Π^2) in terms of one specific point (Π^0) in infinity which then secures a roundness (Π^3) that forms an everlasting eternity ($\Pi\Pi^2$) which validates an never ending circle. In this time in infinity (Π^0) secures that there is an everlasting eternity ($\Pi\Pi^2$) in space (Π^3).

The **governing singularity** (Π^0) holds a **positional validity** (Π^3) of three dimensions in terms of any **relevance** (Π) formed by the **controlling singularity** (Π^2) thus mathematically it equates to $\Pi^0 = \Pi^3 \div (\Pi\Pi^2)$. If a **relevance** ($\Pi$) did not validate a **positional validity** (Π^3) securing a **primal singularity** (Π^0) in terms of movement formed by **the gravity** (Π^2) that produces the **controlling singularity** (Π^2), a three dimensional status, then space (Π^3) would not be obtained and thereby the Universe would not be secured.

Time is the movement of space in relation to any one centralised point not spinning securing such movement. Everything in the Universe moves in relation to any one single point that forms in any location that then has to stand still to form the centre of the Universe wherefrom that point must be motionlessness to allow everything else movement. In that manner the Universe is constructed and there is no valid solid Universe because the Universe is constructed from singularity (Π^0) that holds no valid space (Π^3) other than being in position (Π) while having gravity (Π^2) that forms the time (Π^2), which is also the movement (Π^2) of space (Π^3).

The flow of time being the present in singularity forms space by moving time in relation to space as much as relocating the present in terms of a past that is determined by the movement of time whereby that action of time moving by the same token is establishing space that confirms the past as it secures the future as time moves on to leave a positional legacy, a footprint of time gone by in the presentation of space. From this we can deduct that the Universe in a three-dimensional form starts at $7/10(\Pi^6) \div 6 = 112$, which is a value forming the start of the element table and that I explain in the Cosmic Code. In the **Cosmic Code** there are numerous values consisting of Π forming relevancy by which certain rules comply throughout the cosmos. One is 7/10 is the interaction of gravity spinning a sphere (Π^6) within a cube ($\div 6$) and that is how the cosmos forms using Π. In this I prove that for instance amongst so many other things that electricity and gravity is the same thing.

Outer space forms as $7/10(\Pi^6) (\div 6) = 112$ **and as cosmic gas it is** $\{10 /7 [4(\Pi^2+\Pi^2)]\}$ **=112.795**

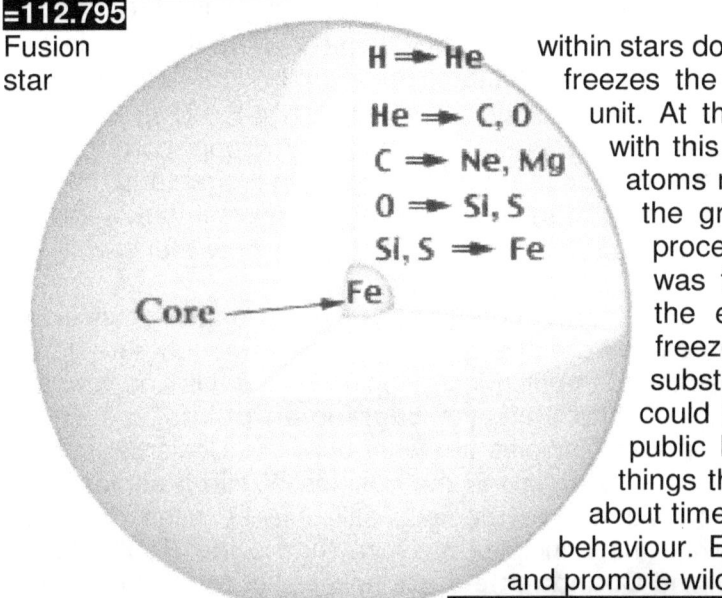

Fusion star

H ⇒ He
He ⇒ C, O
C ⇒ Ne, Mg
O ⇒ Si, S
Si, S ⇒ Fe
Core ⟶ Fe

within stars does not come about "pressure" but the freezes the atoms into forming a new element unit. At this point I must also say I disagree with this idea that atoms freeze together but atoms rather grow into... The earth has not the gravity to "freeze" any type of fusion process and therefore all attempt this far was futile and will remain futile because the earth must first find the gravity to freeze hydrogen into a water-like substance as the sun does before fusion could be a possibility. Newtonians rob the public blind with their falsified ideas about things they clearly don't understand and it is about time the public put a stop to this criminal behaviour. Either they stop to bullshit the public and promote wild schemes or they go to jail.

Iron core is $\{7/ 10 [4(\Pi^2+\Pi^2)]\} =55.27$

Copper core is $\{\Pi [4(\Pi^2+\Pi^2)]\} = 62$ **and this ends space-time flow into unification with singularity.**

This proves that movement in the atom freezes cosmic liquid into cosmic solids and each arrive at a specific but different density to produce cosmic development in the Universe.

Gravity is the cooling of space by material; duplicating the positions material holds in a specific time duration thus distributing space over a wider area into smaller segments. The faster the movement is of material spinning the smaller the units of space will be that occupies space as a init in a given period and this means the wider space material holds that are distributed over a wider area. This brings the heat within down while it intensifies the density of the material. As space is being cooled it is placed in transformation, while the time component, which then can increase the spin value, or time by the value thereof. Therefore, it will transform more unoccupied space-time to material, when spinning more rapidly and increase the density of the material within the star making the star go smaller in space as the gravity within the star increases. Flowing the cosmic code one can see from where gravity comes to where gravity ends. This is a very short example of the root gravity follows.

Gravity forms when a solid holding an Iron core in relation to copper$_{62}$ turns in a cosmic gas, which is the name that I gave to the singularity substance forming outer space. The turning of what is cosmic gas which is outer space $\{10/7[4(\Pi^2+\Pi^2)]\} = 112.795$ to a cosmic liquid $\{7/10[4(\Pi^2+\Pi^2)]\} =55.27$ that as the same as electricity the solid $\{\Pi[4(\Pi^2+\Pi^2)]\} = 62$ inside the cosmic singularity and in this process the gas re-forms to become cosmic liquid which is also electricity when charged on a minute scale or light when charged with gravity or what holds the name as the sun's prominence the cosmic gas into a cosmic liquid to cool the star as it then adds as lubricant to serve within cosmic solids or also called material.

Gravity is about outer space $\{10/7[4(\Pi^2+\Pi^2)]\}$ = 112.795 conformed by iron $\{7/10[4(\Pi^2+\Pi^2)]\}$ =55.27 the displacement value of $iron_{55}$ in relation with singularity presented by $\{\Pi[4(\Pi^2+\Pi^2)]\}$ = 62 and that is the displacement value of $copper_{62}$ which is the element that has this proton value according to the periodical table. This is exactly how electricity is generated and that proves that gravity and electricity is the same thing charged on different levels

All stars have an $iron_{56}$ core spinning in relation to a copper core inside the star centre. Without an $iron_{56}$ core it cannot charge gravity because gravity and electricity is the same but produced by spin yet on different dimensions. Both electricity and gravity using the process of $iron_{56}$ core $\{7/10[4(\Pi^2+\Pi^2)]\}$ =55.27 in relation to $copper_{62}$ $\{\Pi[4(\Pi^2+\Pi^2)]\}$=62 that in the process freezes cosmic gas $\{10/7[4(\Pi^2+\Pi^2)]\}$= 112.7 into cosmic liquid $\{7/10[4(\Pi^2+\Pi^2)]\}$=55 that then translates to become cosmic solid's coolant or lubricant $\{\Pi[4(\Pi^2+\Pi^2)]\}$ = 62. At present, the Universe is in the iron peak era. Every star has an iron / copper core to form gravity and the relevancy between outer space on the outside holding hydrogen and the iron forming gravity or electricity in relation to $iron_{56}$ by spinning around copper as it forms gravity. In this Π going Π^2 proves all of this. I called this process explaining how stars work by formula the cosmic code.

I show how Newtonians fabricate Newton's ideas about gravity. This is ongoing since the end of the dark ages and since Newton came up with a vision by which he sought after fame for the name Newton. There is no mass that can pull. Most people reading this and who are schooled in physics never heard of the **Roche limit**, the **Lagrangian points**, the **Titius Bode law** and the **Coanda effect** and these principles are what build the Universe. I show that there is no factor such as mass used by nature in the cosmos. While it serves their purpose notwithstanding never finding evidence to the fact, still science uses *only* and *exclusively* Newton's idea of mass while the principles in place that is the **Roche limit**, **Lagrangian points**, **Titius Bode law** and **Coanda effect** are never ever mentioned. They sometimes put referring to these principles as a law in brackets to deny the status that any of the above law have. I show you within the next reading of my work the silliness Newtonian principles hold. While I discuss the principles please see where I am incorrect or going wrong and convince yourself whom is wrong.

This is because by using the disciplines underlining the **Roche limit**, the Lagrangian points, the **Titius Bode law** and the **Coanda effect** such use of these phenomena disputes Newton and science would rather discard what the Universe uses than to put a question mark behind the fabrication Newton put in place. Where everyone knows the fabricated information and hiding the reality, which is in place within the cosmos, and doing that is the conspiracy I show to all. Science stupidity ensures they don't understand the working principles that are in place and that was known for centuries in some cases as the **Roche limit**, the **Lagrangian points**, the **Titius Bode law** and the **Coanda effect** and therefore not knowing how the principles should be interpreted they hide the concepts due to not wanting to be seen as the ignorant fools not able to know how the cosmos implements the principles as reality. Science hides their limitations and incompetence behind providing the public selective of information. Take for instance the edge of the Universe they talk so much about. There is no edge of the Universe because there is only an unlimited everlasting Universe out there. What the limits are that they see as the edge of the Universe is the limitation of their equipment that can't trace time back beyond what they see and that serves as their limit in understanding what the Universe offers and how the Universe unfolds.

I am the first to admit that there is no substantiating proof presented in this article and I don't even begin to claim that I deliver any proof in this article. There is no room to present even the least bit of proof in any form possible in the space given to this article. With the limited space available to publish information in a presentation by way of a small article such as this and having so much information at a premium I decided to release some vital information and the required proof about my claims in other small but comprehensive works that can be studied before tackling the more comprehensive work.

I made it my quest to correct cosmic physics. At first I tipsy-toed about how mainstream science would react on my work. The more I tried to be humble the more they disregarded my work and everything I said. Now I call then by the name I use and by which I refer when I address their work and the double standards they apply when conducting physics. There is no mass and this is the way that physics apply:

The Lagrangian points hold a centre point attached to four points circling the centre. The four points spinning around the centre point confirms the status of singularity and moves in relation to a static fifth point at the centre but never moves in terms of the axis. It is the centre point and the four points turning around the centre point without recognising the axis.

The Titius Bode law is four points confirming an axis holding three points. In this the seven points regard the shift of one point and from this one point representing singularity that shifts another seven points grow. That movement takes the seven to a square value because of movement as well as the one to a square because the point in singularity shifts by time. Doubling seven to the square relating to one to the square confirms the movement of the entire sphere as a unit and this builds space within the Universe.

The Roche limit extends the solid spin into what ends the liquid influence of the solid that spins.

The Coanda effect captures space forming liquid and by increase of spin or increase of gravity it extends the limit on the border of the solid further into the realms of the liquid, making liquid by gravity turn to solid.

In this book I show briefly that what you think of as Science, is only ideology based on rumoured disinformation made up by those making up facts, as they hide what they don't know, hide what they don't want to know and conceal what they don't want others to know. By proclaiming they know everything while pretending to know all there is to know, they preach what they falsify to sustain their religiosity in the form of unproven dogma. Science tells what they wish and reserve what they wish not to tell and hide what they don't wish others to know that they don't know. My biggest undertaking is to prove that what science presents as credible facts is unfounded and incorrect. Science is believable because culture sustains Science. My problem is busting a culture people trust as Science and be a whistleblower about accepted religiosity people unconditionally believe in as being trustworthy. The public never once thought to question what they were taught as science. This forms a problem since what is presented as science is falsified information and that which we need instead is a new approach that is truthful about science. Therefore no one realises what big a problem there is in the level of correctness there is in the approach to reality science now presents. Science never shows the support or proof in what they present as ultimately credible facts. Bluntly put; Newton was never proven to be correct but only accepted. I firstly need to promote the need we have for the truth to replace what we have as science before I can promote what the truth is that we need to form science. Science hides their flaws under a profile that they present as the uncompromising truth while in fact what they present are lies covered by a veneer of truthful facts and this then neglects the urgent need for the truth by denying there is a problem. If you don't believe this statement I make on the fact of mass forming gravity then I challenge you to read on and challenge me on any fact I present about the untruthfulness you think of as science and if you disagree with me send me a e-mail, but if you believe me then help me to challenge the establishment about their scandalous disinformation that they hide to commit the conspiracy they call science. Everyone that has contact with physics is unbelievably naïve while everyone is getting hoaxed and your children are being brainwashed to unconditionally believe in science, which is totally without credit. Everyone accepts without questioning the science because they believe in the persons teaching them and not necessarily in the credit of the information they are taught. Read what it is that science hides to put in place this notion that they know everything and what they know are without blemish.

There is an obvious conspiracy in science that everyone misses and about this fact there is no doubt about it. It runs as deep as it can go and involves every person that studied science worked with science teaches science and thought about science. In short it touches very person that lives notwithstanding culture, ethnicity, race or religion. Everyone on earth believes the science taught at schools and everyone is working with fake science and conspires to promote science fiction passing it off as Newton's science.

Moreover what will you do when you know the truth or are you satisfied to be told what to believe is the truth without ever investigating by personal standards as to finding the truth about science. Then go on reading and confront the truth, as you never had ever before. The truth about a conspiracy is that everyone involved with the conspiracy will fight tooth and nail to stop the conspiracy to get leaked and leave those behind the conspiracy and all those feeding from the corruption of the conspiracy exposed.

Those that have the power to maintain the conspiracy would keep the waters as still as possible as to draw no attention to such conspiracy.

A true conspiracy has to be as quiet and as unseen as it could be.

A true conspiracy must involve everyone without anyone detecting even a hint of what the conspirers hide.

There are many conspiracies going on such as the banks involvement with crime and the bankers profiting from gamble rackets and drug selling.

The same goes for the Insurance business profiting from lenient sentencing of courts holding very merciful judges in office so that the crime cases and the burglaries, car theft, hi-jacking and all other forms of crime will shoot through the roof every year and grow by thousand percentage points from which insurance will sell ever-more cover-policies.

The more crime is about and committed daily, the more people need insurance covering and thereby the more money insurers bank giving bankers profit to spend on the stock exchange by controlling the economy.

Bankers buy democracies with money they give politicians to write laws that protect the rich against the poor. In time I might write about this but now I cover another conspiracy, a bigger case file about a conspiracy everyone on earth participates in...it is about…

Should anyone desire to contact me with an opinion about my views on science then please do by using mailto:info@questionablescience.net or mailto:info@singularityrelavancy.com but please do so when coming to the end of this book at least and then see if you understand the entire concept that I introduced. In most cases all your questions are answered further down the line and I am not fond of repeating what is already said in the book. The book forms a line in explaining and the concepts are best understood if the reader follows the designated line.

This book started off as a website to inform about a science conspiracy but although reduced still it grew into a book that serves much more information than what I first intended to supply. You will see many new aspects about gravity please make sure you understand what you read. It grew into a comprehensive study on cosmology. At times you may observe while reading this book that it seems as if my frustration will ring through like the chiming of the Big Ben Bell.

For that there is a reason.

At times my frustration and anger will boil over drowning my politeness and that is true, which I admit. For twelve years I have had the answer that would correct the philosophy that has a stranglehold on cosmological science.

I discovered the building blocks of nature where my discovery puts all other cosmic aspects of science into science fiction.

Those who force-feed non-existing dogma do so to brainwash students to hide the incompetence of "modern science" so they can rule supreme while ignoring the truth that they deliberately hide by concocting a conspiracy.

To keep everyone unguarded they practise a conspiracy by which they perform an accepted practise of thought control on students to further the false dogma presently in place.
I try to blow the whistle on such a practise but accepting my resolution makes every thesis ever written science fiction.

Therefore no one in science dare to read my work leave alone appreciate the revolutionary nature thereof. Whatever now is deemed to be accepted science would then become what is the past tense in science because the flaws that those in power of science principles kept coated for centuries on end as untouchable truth will then be rust that breaks the surface to show the holes!

They try to silence me but surely somehow somewhere I have to break through with my message! I bring you a true form of science as never seen before in all of history and I do that when I dispose of the conspiracy that hides all the incorrectness and the failures that haunts science today. Science is accepted as the most righteous information available to man and that is a scam.

You are going to read about a conspiracy but people think of a conspiracy in many terms. Let us define not by definition but by interpretation to what a conspiracy constitutes. What do you think is a conspiracy?
All the conspiracies you know about is known about because someone somewhere makes money by allowing the revealing of the conspiracy. Silencing the conspiracy does not make money but informing a suspicious public loosens the flow of money. If it were a true conspiracy no one would know about the conspiracy because the powerful would make money from not revealing the conspiracy. The revealing of the facts about any conspiracy would be stopped before it leaked because it would kill the flow of money.

A conspiracy is thought to be a gossip story that makes money and by not revealing it or revealing it goes in line with making money or not making money. You can download this book free of charge because I don't make money by revealing the conspiracy. I truly want to find an audience to divulge the truth. I want to make money but it is by showing how I can correct the flaws in science, not by hiding it in a conspiracy.

People put a conspiracy in the same realms as a gossip story, an old wives tail, which is going about but does not intend to harm and mostly serves as amusement to many. Hearing about a conspiracy tests your intellectual comprehension. It is some quiz that you match your truth against the truth that the conspiracy reveals. It is a funny, but it is not funny until you catch the funny part hiding behind the conspiracy and only when you measure the catch behind the conspiracy are you treated to be amused.

If the conspiracy does not touch the person directly then no harm is felt and no harm is intended. Every one holds this view that a conspiracy is on a slightly higher level than gossip. It is a gossip story about someone living in the neighbouring village known only to some people next door but has no direct linking to me or has no threat to the safety of others directly associated to me. Everyone treats a conspiracy as if it is something amusing that holds no threat at all. It is something that goes around as a joke of sorts.

We all live in a bubble we call civilisation and we all run after a dream called peace and we all preach a fantasy we named democracy and every one excluding no one lives a fantasy we love to believe but never can. Those that say they lead us are programming us in an assortment of ways to have us believe that we are happy content with our fate and we are lucky to be as prosperous as we are and this they do by a process of programming us mentally. They call it advertising. They call it politically socialising. They even call it following teaching procedure. The teachers tell us what we must believe and then force us to write examinations about what we must believe and according to how we believe what they say we have to believe in, they set our future.

The only form of fuel there is in the cosmos is something that science at present does not even recognise. Science placed "nothing" where the drive for cosmic gravity initiates and this shows in what despair science is when it comes to understanding the cosmos. In the cosmos there is liquid and there is solid and there is no more. Compressing liquid into solid or releasing liquid to form gas or to destroy solids to form liquid is the only fuel that drives the cosmos. The cosmos is a variation of density formed my relative movement and this movement results in turning density a notch up or down. That is what the Titius Bode law proves and that is what Kepler's figures are all about. There is not "nothing" between the earth and the moon or the earth and other planets but there is liquid that formed space. Time leaves space and what is between the solid objects and other solid objects is heat forming space as time leaves space as a reminisce of times gone by. This is the fuel that drives the cosmos and this becomes most important when proving in accordance to the Bible how the cosmos started.

These are the facts I found applying in the cosmos and with these phenomena working in nature and Newton telling the lot out there are forces that works on "mass" pulling "mass" it makes science fiction. If something is not applying and I tell you it is applying I am not conveying science but fiction. If I tell you cats fly away from birds trying to run the flying cats in it is fiction. Although Newton has never been supported by cosmic evidence, still everyone is sharing the Newtonian vision of a contracting Universe where the lot would one day come together and Creation will end where Creation started some time ago.

The supposition is that Universe has ends and the ends are drawing closer by the mass that is pulling mass towards one another and we are in the centre of an ever-shrinking Universe where this process is named the Big Crunch. The earth pulls us closer while the earth pulls the moon closer while the sun pulls the moon, the earth and us closer. That is what the lot of us can see... we are forming the centre of the ever contracting cosmos where every Newtonian can vividly see with his or her eyes through any telescope that all Newtonians minded scientists are sharing the centre stage of the ever collapsing Universe. That is the conspiracy holding science at ransom and caged in, locked in a cocoon of ignorance for almost going on to be three centuries.

Try as I may while no one I approached can prove a force called mass has pulling power, I could convince no one there is no mass anywhere because there is no such evidence, but in that is the devastation of the conspiracy. I am fighting a religiosity called science based not on fact but on accepting culture with a head priest called Newton and a cult or sect that has a demonic hold on the minds of the masses. If you believe in mass (not how much you weigh but mass that pulls) then prove it.

They confuse everyone about what *weight* is, what *mass* is and what *gravity* is because they wish to have everyone think of "mass" in terms of weight while they then deny weight and "mass" is the very same thing and then they confuse "mass" and gravity because they never distinguish between what "mass" does and what gravity does. This is only the tip of the iceberg and you will see when reading this book.

This is what there is and that is all there is. The measure of mass forming gravity clearly plays no role in allocating the positions of planets as Newton declared it must do. The entire idea that gravity is a magical force created by the value of mass is as unbelievable as the dogma is of those presenting this idea. Please use what the solar system provides to confirm what Newton says is in place when he says mass forms gravity. Science would rather accept Newton where there is no proof of Newton ever being correct than to admit Newton's incorrectness.

Science would rather deny there is cosmic principles that is in place in the solar system, which are **the Roche limit, the Lagrangian points, the Titius Bode law and the Coanda effect** than to admit to Newton's failings. They would not admit to Newton's failings because then the entire world will see they know less about science than does a pig know about history. They would rather put the error on the solar system than they would commit to the blatant mathematical cheating that Newton committed. It is the Universe that is always at fault when Newton becomes incorrect because without Newton's fabrication of science they have nothing to show for all the wisdom they try to pretend they have. Newton fabricated "Kepler's laws" has some correctness mixed largely with a farce and blending the truth with total fabrication of reality hides the lie behind something presenting the truth.

This is the solar system. Can you the reader see that the smallest planet is on the very outside and the second smallest planet is on the very inside? Can you see the two largest structures called planets are in the very middle? Notwithstanding planets or dwarf planets that is making dust to hide the truth! I am not only fighting a myth but I am fighting the brainwashing we all had to endure to force us to believe the myth they call Newtonian science. Whatever I say can't penetrate the layers of psychological abuse students went through and the mind control they were submitted to in order to believe it is mass that pulls any person and it is mass that creates gravity. The person that believes mass is the factor that positions planets never saw the solar system as it is and any person with a clear mind that sees the solar system would not be able to fit mass as a factor of allocating structures into it. Between Mars and Jupiter there is a band of debris we named the asteroid belt. It is a lot of rocks that remained in a group after Jupiter destroyed the planet by implementing the Roche limit because this planet at one time was exceeding the Lagrangian points.

Please look at the picture. While looking at the picture tell me how is it possible that the "mass" of Mercury could put it in the place it is and the "mass: of Jupiter could put in the centre where both hold positions according to "mass". I have been saying this so many times I am getting sick of repeating it but this formula they say positions planets. They say or then Newton said and they confirm that planets are in space in accordance with the formula $4\pi^2a^3 = P^2G(M + m)$. Now it happens just what they want to happen because now you are scared witless. When looking at this you are intimidated beyond your senses and you feel so stupid your inferiority takes control of your thinking and the feeling of utter stupidity wants to make you run and hide. This was the scare tactic Newton put on everyone he came across. No one dared ask him a question because the man was bewildering intellectual. That is bullshit because I as stupid as I am can look at this and see right through his menacing bluff.

This says $4\pi^2a^3 = P^2G(M + m)$ that the circle ($4\pi^2$) formed in the three dimensional space of the orbit of the planets (a^3) is equal or is the same as the location in which the orbit is () and this is determined by the gravitational constant (G) directly in measure wit the mass of the sun and the mass of the planet (($M + m$)). Go on and put in the mass of the sun and the mass of Jupiter and show how this can locate the position the planet is as they try to prove the formula $\left(\dfrac{P}{2\pi}\right)^2 = \dfrac{a^3}{G(M + m)},$ proves. You can read these formula as if it says @#%$&^* and it will have the very same meaning because it is completely invalid. ...And if that won't scare you socks off they put you in a mental torture chamber by throwing in the formula $$\left(\frac{P}{2\pi}\right)^2 = \left(\frac{a^2\sqrt{1-\varepsilon^2}}{\ell}\right)^2 = \frac{a^4(1-\varepsilon^2)}{\ell^2} = \frac{a^4(1-\varepsilon^2)}{a(1-\varepsilon^2)GM} = \frac{a^3}{GM}$$ 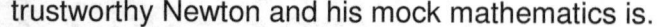 and one look at that will make you run and hide behind your mother hoping she will get these clever physicists away from you! Think how this formula will have any first year student scared witless and he or she would be great full NOT to apply it but just to learn this off by heart and repeat it one million times to secure the validity thereof as the truth. This is fabricated lies big enough to be worthy of any politicians commitment to honesty and if any physicist or Astrophysicist want to disprove what I say use it and show everyone how big a liar am I and how trustworthy Newton and his mock mathematics is.

Let's look at what the cosmos uses instead of the mass as Newton said is there.

The presenting of the cosmos in this picture is used to falsify the truth. They get away with it by saying the solar system can't be picture perfect presented in the ratio that it truly is but they never say this when showing a picture. The truth is that a planet doubles the distance every time a new position is allocated.

Mars is as far from the earth as the earth is from the sun as is from the sun.

The asteroid belt is a far from Mars as Mars is from the sun.

Jupiter is as far from the asteroid belt as the asteroid belt is from the sun.
Saturn is as far from Jupiter as Jupiter is from the sun, not from Mars but from the sun.
Uranus is as far from Saturn as Saturn is from the sun.
Neptune is as far from Uranus as Uranus is from the sun.
How can I put it more vivid than what I just did and yet I can't break through the brainwashing science put on people? Newton is not there...it is the Titius Bode law that is there in place in the solar system. If you still wish to accept Newton then you are too damaged by brainwashing to repair your mental status. You can accept Newton and remain part of the hoax or you can accept not me but nature. In the cosmos the reference point in doubling the distance is the first inner planet and because Newtonian science is incapable of producing any type of explanation they cheat like the mafia would by telling everyone

Newton's formulas
$$\left(\frac{P}{2\pi}\right)^2 = \left(\frac{a^2\sqrt{1-\varepsilon^2}}{\ell}\right)^2 = \frac{a^4(1-\varepsilon^2)}{\ell^2} = \frac{a^4(1-\varepsilon^2)}{a(1-\varepsilon^2)GM} = \frac{a^3}{GM}.$$

This phenomenon I just described is named as **the Titius Bode law**

Most persons reading this has never heard of this but it is what truly is in place in the solar system when one study the layout of the solar system. There is no place for mass and you can see how I present the mass of every planet as you read later on. Pick an argument about this and I must refer you to the cosmos because I only show what is in the cosmos. Now we get to the point where I reveal what it that I discovered is. I discovered how **the Titius Bode law** in principle works. The Titius Bode law is the way that the Universe grows and it is how plants position in accordance with a precise ratio that is so accurate one can put values in a cyclic formula as follows.

$P_n = P_o A^N$

P_n = **period of orbit of the nth planet**
P_o = **period of the sun's rotation**
A = semi major axis of the orbit.
This is what the cosmos uses so if you have a problem with me condemning the Newtonian version take it up with the cosmos because I only show what is used by the cosmos and that which the cosmos uses is not what Newton said. So you have the dubious honour to stick with Newton and discount what the cosmos uses or you can side with me when I explain how this layout forms.
I am not going to reveal why it uses this system because that information the reader must purchase when they purchase the books I offer in the partnership. That information has never been known for as long as man is on earth and knowing that took me a life-time of study to accomplish. This next part is most vital to find truth about cosmic science.

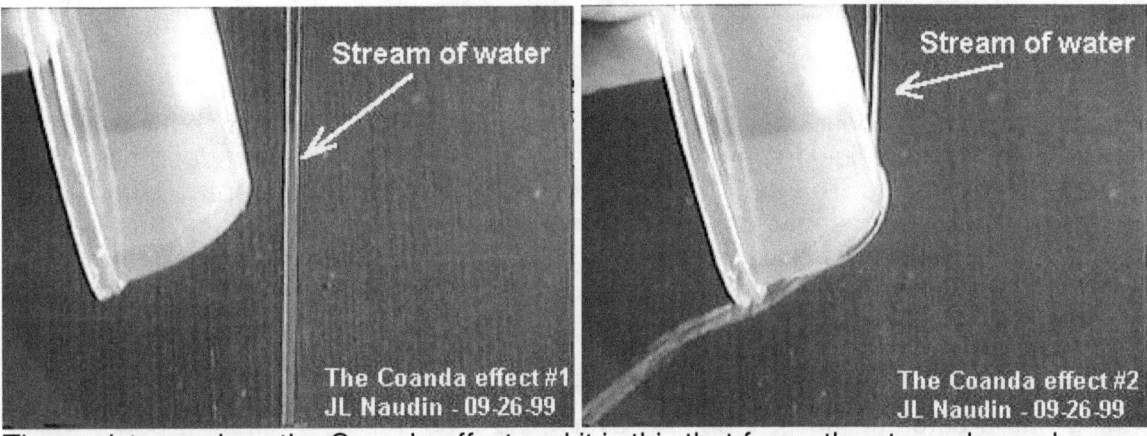

These pictures show the Coanda effect and it is this that forms the atmosphere when gravity solidifies outer space from a gas to a liquid in cosmic terms. This is what drives the Universe and not recognising this or not knowing this nullifies ever concept that science in its present form holds. Not recognising this is denouncing reality and accepting mythology as science.

The spin of the earth forms the gas of outer space to compress into a liquid and that compressing of the space forms the increase in density in atmospheric layers.

What this is goes by the name of the Coanda effect. I prove that this represents gravity as a principle more than any other form or factor could. Yet, with this so prominent in physics, you will never see any explanation about the Coanda effect in any physics handbook because the Coanda effect puts a serious question mark behind Newton's idea about physics allowing this.

The Coanda effect is the very reason why the Earth has an atmosphere, but you will not learn anything about the Coanda effect… because with the limited view that science at present portray they have no explanation for the Coanda effect or the atmosphere being there other than the mass of the atmosphere pulls the atmosphere down.

What a lot of unproven Newtonian gargle that is; what mass could the atmosphere have? By going to Lulu.com and then downloading my books you will see how the spin of the earth compresses the space and by compressing with movement, not with mass pulling, the turning produces the Coanda effect and the Coanda effect by gravitational motion condenses space to become the atmosphere.

The Coanda effect is the principle that also proves the sun is not a gas giant but it is a liquid as can be seen from the liquid spewing from the sun's surface. That is a dead give away about what gravity really is. The earth spinning contracts the atmospheric space surrounding the earth and that process cause gravity to attract and not the mass of objects as Newton insisted. It is the space holding the object falling that moves downwards and not the object that falls. That is why Galileo was correct when he said all things fall equal under the same conditions notwithstanding size differences.

What Newton says is that things fall by the value of mass bringing on gravity. This means if everything has a different mass, therefore everything must fall at a different pace, which doesn't happen…and in that is where science is making the biggest mistake. I ask any one in science to prove the fact of mass applying, not as weight but as a force. Read on and I will show you how those in physical science are brainwashing students into believe that mass is responsible for gravity as a force.

I discovered how **the Lagrangian points** in principle works. The Titius Bode law is the way that the Universe grows and it is how plants position in accordance with a precise ratio that is so accurate one can put values in a cyclic formula. These are the points used by moons and safelights in orbit as they circle around planets. Again I repeat, those with a problem about Newton not working in science you better go to the cosmos and find out why the cosmos rejects Newton as much as I do because I follow the cosmos.

I presume Newton was completely unaware of the rings around Saturn and the other planets but these rings alone put Newton's claim on mass pulling mass in serious doubt.

This one of the four of the pillars I introduce by which the Universe works. You can say I introduced it because the Universe introduced these four comic principles when the first spot exploded into a dot. The four works together and never can they be separate. However in certain conditions in movement one might stand out more prominent in the relevancy in which the phenomenon is at that very place in that space. The phenomena are there used by the cosmos but science never understood why the phenomena are there and so science not wishing to show how stupid they are acted even more stupid and ignored the phenomena altogether.

I wish to quickly show one example of the mathematics by which these phenomena is calculated. A sphere has a line forming the axis. That is three points. The circle by which the spin runs holds four relative positions and that brings four. When moving this lot goes square because the points displace space and locate new positions. Therefore there are 3 going square = 9 and four going square = 16. Gravity is Pythagoras so according to the law of Pythagoras we have 9 + 16 = 25 and the root of 25 is 5. Therefore the next point to become part of the Universe when the Universe started in singularity is five and that position is 5. Everything is so incredibly simple if the correct lines in thoughts are used.

The stars show **the Roche factor** as the result when the Roche limit is bridged. The Roche limit is one of the four pillars I explain. Newton could never come close to even guessing why the rings are forming and why the star acts in the manner it does and as we see it does. You either side with Newton and know nothing that is reality in the Universe or understand the Universe and know why the Universe uses what is there that applies as physics. The gravity left behind as circles is Π or space. Gravity isn't mass. Gravity is the forming of space as the factor Π. Therefore gravity is not mass related but it is Π and the circle that's prominent is gravity that remains behind as Π in space. If you disagree with me take it up with God.

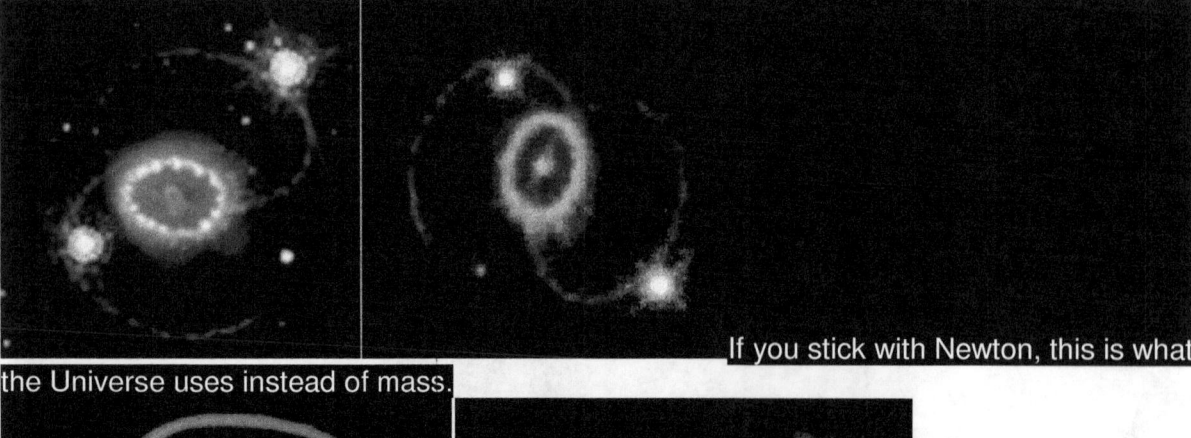

If you stick with Newton, this is what the Universe uses instead of mass.

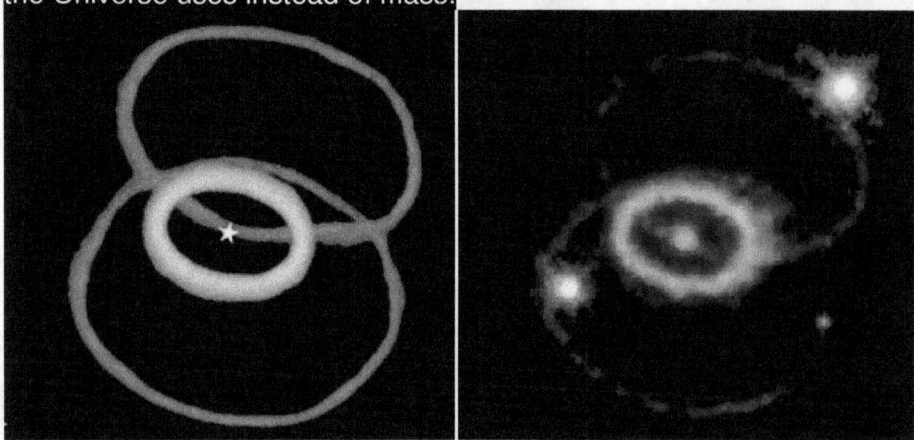

The Roche limit is the region surrounding each star in a binary system, within which any material is gravitationally bound to that particular star. The boundary of the Roche lobes is an equipotent surface, and the lobes touch at the inner Lagrangian point, L_1, through which mass transfer may occur if one of the components expands to fill its lobe. This puts the Lagrangian points system in the very heart of the Roche limit and made me realise that all these four phenomena are interlinked and moreover interwoven. It names after the French mathematician Edouard Albert Roche (1820-83). This means that since the 19th century Newtonian science was aware of the Roche phenomena but never took any effort to try to explain this because this kills Newton's mass idea off more than most other cosmic phenomena. This phenomenon prevents stars from colliding and that totally jeopardises the entire idea of Newton and his mass. I remember thinking that the behaviour of the stars are connected by some cosmic connection that requires urgent research. Then I started dipping

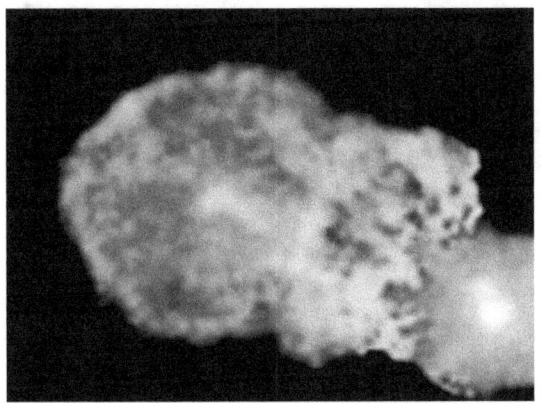

into cosmology with unrest and this is how it started if my research started at any given point. This made me realise gravity is something to do with Π.

Everything in these pictures is heat and it clearly shows that heat is space gravity compressed by applying Π. Everything there is in these pictures and even the blowout happening is a circle applying gravity as Π. How can any one that pretends to be serious or professional in cosmology or even interested about science miss it? This started me off looking for something better than was in use.

There are two concepts involving the Roche phenomenon. The one is the Roche limit and the other is the Roche lobe, which is what happens when material move faster through space

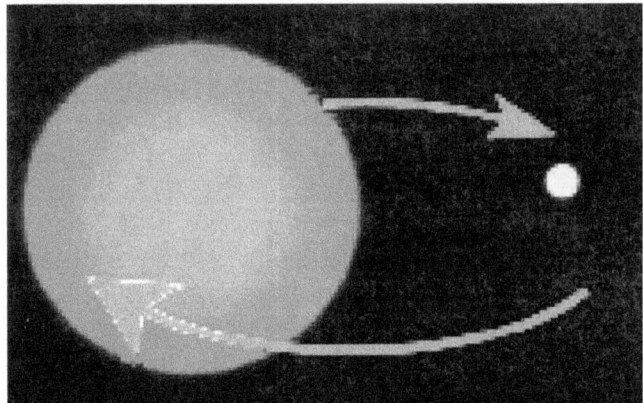

than the applying relevancy would allow. When I studied this I came to the conclusion that there is no mass pulling mass. The entree Universe must work another basis that what science propagates.

The system work as follow: There are two or more stars involved but lets keep it simple and only explain a "big" star and there are a "small" star and the "small" star is circling the "big" star just like the moon circle the earth.

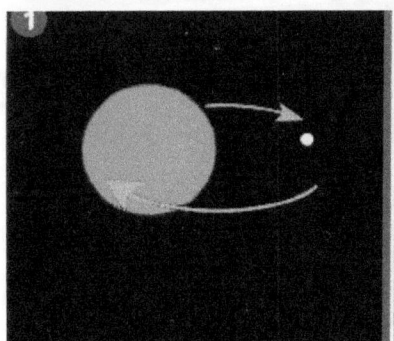
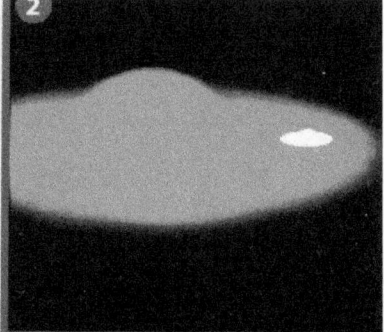

For the Roche limit to become effective the "small" star must be with in the radius of 2.4674 the diameter of the "big" star. This puts the "small" star within the atmosphere of the "big" star.

Then the "big" star starts to resolve the component forming the "small" star. The "big" star starts to spin the "small' star and this helps to convert the "small: star to become the density of the space surrounding the "big" star.

Because of the difference in movement the density of the "small" star is much less than the density of the "big" star.

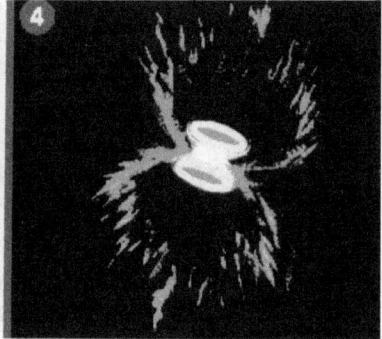

Therefore the "big" star vaporises the material forming the "small" star to the same status as the atmosphere of the "big" star is.

Since gravity is the contraction of space and the "small" star becomes the liquid atmosphere of the "big" star this then forms the space within the "big" or forms what we see as the atmosphere.
The "sound barrier" is the Roche limit applying.

Coming to the conclusion about gravity being motion and mass being the restriction of motion was the easy part. The facts that presents the understanding of what produces the motion and what prevents the restriction from overcoming the motion was the part that required thinking. Figuring out why was everything on the move and where did the motion stop, well that was the part that took some figuring and some explaining.

What makes gravity move and why does gravity move...the answers are in the four phenomena never yet explained to satisfaction but now turns out to be the cradle of gravity. The answer can only come when the full content of gravity is fully understood as being the unexplained phenomena that produce in conjunction with one another the totality of gravity as we experience it.

The TITIUS BODE Principle

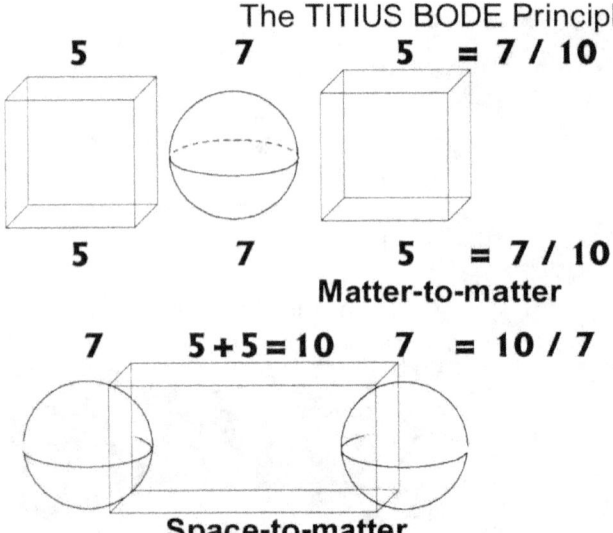

5   7   5 = 7 / 10

5   7   5 = 7 / 10
Matter-to-matter

7   5 + 5 = 10   7 = 10 / 7

Space-to-matter

They are the following:
Gravity is The **Roche limit**,

Gravity is The **Lagrangian system**

Gravity is The **Titius Bode law**

Gravity is The Coanda affect

In the Roche limit the space factor provides occupied space-time and therefore the value of r is replaced by the value of Π bringing about a square in half of Π.

Gravity is the dimensional changing of heat holding r as reference to the sphere holding Π as the reference. Heat occupying space has the cube that can apply r, as a straight line bringing about the cube with all its other names than may find attachment to specific form but nevertheless still remains only a six-sided cube with angle changing in some cases.

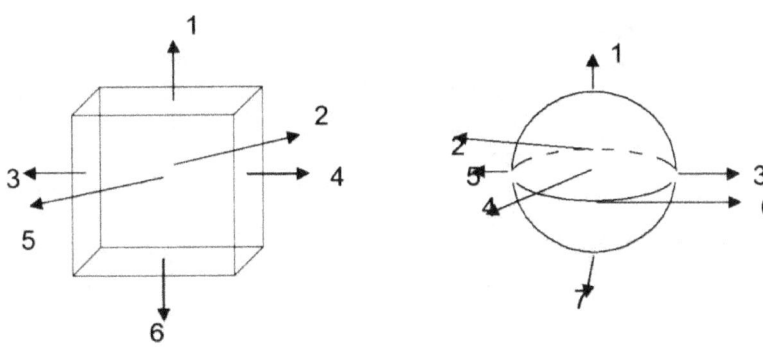

In the sphere there is no radius but only the extending of **k** from the centre **k** in six opposing directions relating to one another by the square but remaining Π because of the unity the matter holds in relating to space. In every sphere there then are the seven Π relating in precise dimensional and positional equality to the centre Π as well as to one another by 90^0 and 180^0 implicating the dimensional positioning.

Therefore the sphere holds 7 points and the cube holds 6 sides.
This puts material as the eternal sphere in a relation with space that without bonding forms the everlasting cube. The material as a sphere performs movement and the space as a cube allows the movement to take place.

By coming into contact with the sphere the cube loses on dimension to the seven dimensions dominating six bringing about that the cube then has 5 sides to the seven of the cube. That is the Lagrangian system with five cosmic atoms holding relevancy to the centre

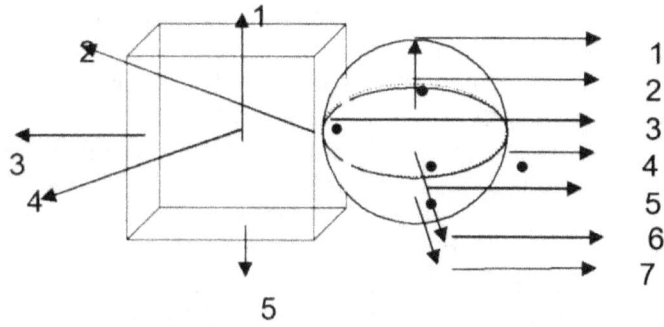

cosmic atom where the centre cosmic atom stands in for seven and the orbiting cosmic atoms standing in for five positions in space. There is a more explicate explanation about this somewhere else in this book.
The Titius Bode law is an extending dynamic deriving the law from the gravity dimensional factor where the space factor in a square of ten relates to a matter factor in the square by half (half since nothing can be in two places in the universe simultaneously) of the matter factor of Π or the square of space (10) relate to the matter factor of 7

The Roche limit is:
The region surrounding each star in a binary system, within which any material is gravitationally bound to that particular star. The boundary of the Roche lobes is an equipotential surface, and the lobes touch at the inner Lagrangian point, L_1, through which mass transfer may occur if one of the components expands to fill its lobe. It names after the French mathematician Edouard Albert Roche (1820-83).

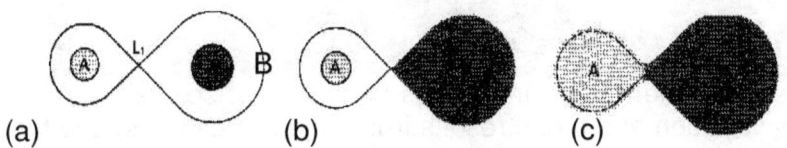

(a)   (b)   (c)

THE ROCHE LOBE: In a binary system, the Roche lobes of components A and B meet at the L_1 Lagrangian point. (a) In a detached system, neither star fills its Roche lobe. (b) In a semidetached system, one massive component, B, fills its Roche lobe. (c) In a contact binary, both components overfill their Roche lobes and share a common envelope.

The ROCHE LIMIT

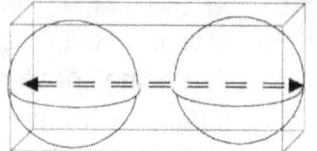

$$5/2 = (\Pi / 2 \times \Pi / 2) = 2.4674$$

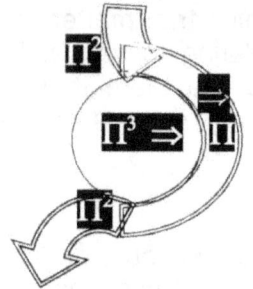

The COANDA AFFECT

$$\Pi^3 \times \Pi = \Pi^2 \times \Pi^2$$
$$= (\Pi^2 + \Pi^2) \text{ FIRST FORMING}$$
$$(\Pi^3 \times \Pi) \text{ SECOND FORMING}$$

Without resolving the concept that these four comic pillars present I would never have been able to figure out how the cosmos works, why the cosmos works and how everything within the cosmos fits together.

I compiled **a new cosmic concept** by which I eliminated all the incorrectness that Newton has burdened science with but with this being my opinion I did not find a garage full of academics supporters waiting to applaud me and to uphold my views on the matter.

Gravity rests on movement of material in relation to other material also moving. The movement of the sun provides the earth with movement but not only that al movement going straight becomes circular movement and circular movement takes place within the circle in which it moves going forward as a straight line and in that idea of a circle becoming a straight line and a straight line becoming a circle the entire concept of cosmic gravity is vested. According to the Big Bang theory the Universe expands and there is no evidence of pulling bringing about a Universe contracting or becoming smaller. The Hubble constant is sole evidence of this proof of expanding. Therefore I challenge the concept they build on the fact that mass attracts mass and everything is pulling everything else. Yet still I was not going to be ambushed by their relentless stonewalling my efforts and blocking my efforts in introducing both the incorrectness and the new cosmic theorem I concluded. My cosmic concept is that the Universe is about heat forming densities. It is the density of hydrogen making it a gas as much as it is the density of the massive Krypton, Xenon and Radon that makes these elements gasses although the gasses are the heaviest inert gasses and Lithium as a solid being the lightest solid. The gas I mention is many times over as massive as the lightest solid is and yet with all that massiveness, it is gas and gas floats in the air. The gasses form what they are because they are a mixture of heat and material putting a factor such as mass completely out of the picture. All materials are solids as they are liquids as they are a gas because they can be frozen into solids and melted into liquids and vaporised into a gas. In each case the density of the material changes from forming a solid to going into a liquid or becoming a gas. It is about density putting a relevancy between the statuses   of   materials.

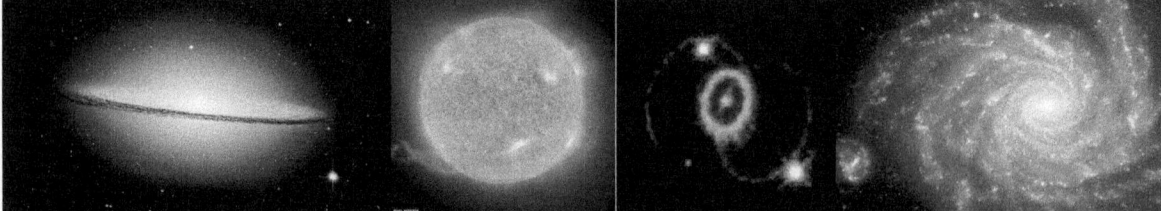

Everything that these pictures tell has one theme and that is heat being displayed in many forms and colours. It is a density variation and those results from objects holding space that moves at different rates in relation to space surrounding objects. Every picture of a Supernova tells a story of the loss of density in the liquid part of the star because materials can only lose density when it bursts by a nuclear explosion. In every cosmic picture we see heat flowing from a less dense and bigger space to a denser and reduced space and that is what gravity is. It is cooling the Universe by applying movement. It is all about moving space to preventing material from overheating where movement brings about density relevancies

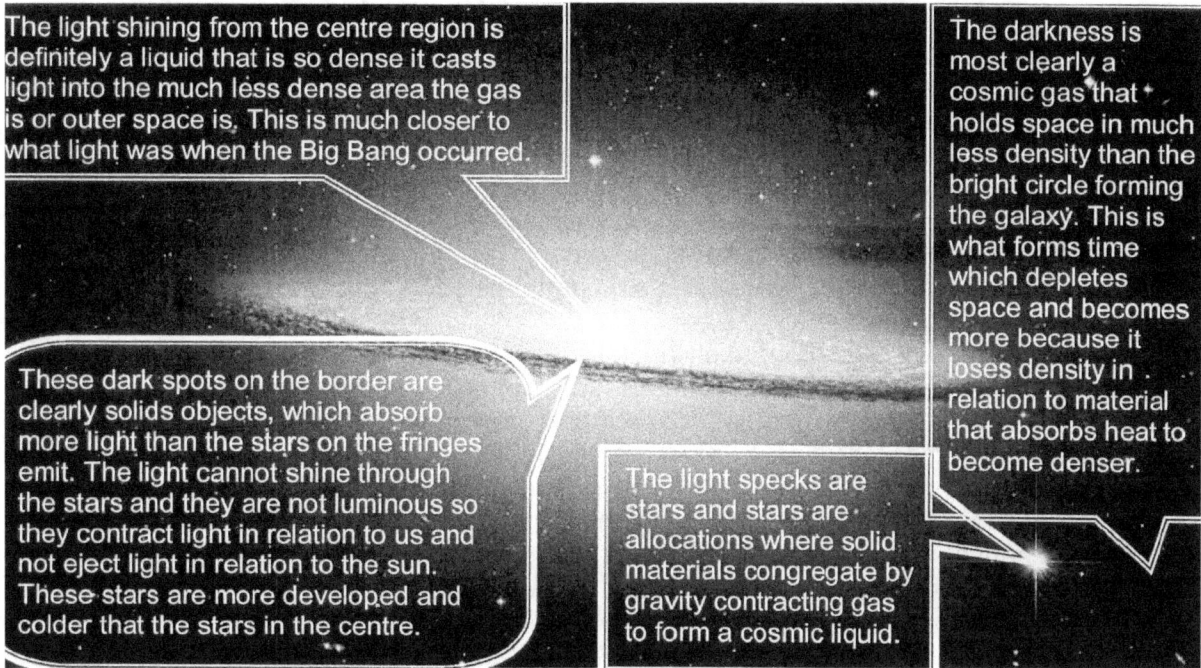

The light shining from the centre region is definitely a liquid that is so dense it casts light into the much less dense area the gas is or outer space is. This is much closer to what light was when the Big Bang occurred.

The darkness is most clearly a cosmic gas that holds space in much less density than the bright circle forming the galaxy. This is what forms time which depletes space and becomes more because it loses density in relation to material that absorbs heat to become denser.

These dark spots on the border are clearly solids objects, which absorb more light than the stars on the fringes emit. The light cannot shine through the stars and they are not luminous so they contract light in relation to us and not eject light in relation to the sun. These stars are more developed and colder that the stars in the centre.

The light specks are stars and stars are allocations where solid materials congregate by gravity contracting gas to form a cosmic liquid.

The Universe is about heat finding concentration through movement applying by varying density levels. Whatever is in the Universe is in motion too. If it does not move by moving towards then it moves by moving away but seen from any point every point forming the Universe is on the move. That movement is gravity and albeit moving towards or moving away, it still is gravity. The movement can come from material moving faster than the speed of light or outer space expanding as time moves on and time forms space and therefore space expands in accordance with time growing which is what expanding in the Universe is and this expanding is movement. There was a time before the Big Bang, there was the Big Bang and there was a time after the Big Bang. Before the Big Bang everything there was, was in singularity as it is at present also. Then the Big Bang came and what was then overheated and exploded. But then everything that is was already inside the Universe and nothing but time came afterwards.

There is a problem I see with the Big Bang concept! At instant 10^{-43} second the temperature was 10^{32}K.
At instant 10^{-35} second the temperature was 10^{27}K and then at 10^{-6} seconds the temperature was $\mathbf{10^{13}}$K. With all this being true where was it zero degrees K. If there was 10^{32} K and 10^{27}K and $\mathbf{10^{13}}$K where was it 0K? To have these phenomenal temperatures it had to be 0K somewhere because if not then 0K was equal to any of these temperatures at the time of that specific event. That concept we then carry to the time factor. If it was 10^{-43} and it was 10^{-35} and it was 10^{-6} then there had to be a place where it was 1 hour and 1 day and one year. You can't have 10^{-6} without at the very same instant had another place where the time would be 60 seconds. When aircraft refuel in flight it happens at high speed but in relation to each other the aircraft stands still. To travel at four 400 k / h somewhere then something else had to stand still in order for the aircraft to fly at four hu400 k / h or else everything stood still according to the aircraft. For the pilots flying the refuelling aircraft the one was flying the speed of the other without thinking of the earth passing by at 400 k/h. everything is in relevancy. To fly at similar speeds the speed difference, which is the same, had to be the relative speed is 1 or equal.

That means there is no hot or cold in the Universe just because if it is 42° C on my farm it must be −20° C in New York. I can't have 42°C on the farm and that is it. On the farm it might be 42° C in the shade but believe me in the sun it is 65°C because there are differences. If it was 10^{-43} according to the time applying to the earth then where was the earth? Did the earth rotate 1/24th time around its axis by the time the first hour come to be? Did the earth rotate once when the first twenty-four hours came in place? By the time the first year came about I guess this was when the earth rotated around the sun for the very first time otherwise it could not have been one year or minute or second or a part of a second. This is the incoherent rubbish one gets when using mathematics with a scale applying to science putting the earth in the centre of the Universe. One can't be on Jupiter and apply the spin of the earth to form a concept of time because the time on earth is in relation to the degrees the earth turns in a specific period of time. This humanising the cosmos shows the narcissistic backwardness in the argument formed by using mathematics to draw conclusions.

When temperature was 100 billion Kelvin at Time ~ 1/100 second what then at that time period was thought to be cold or hot. If this is stated in terms of today's zero Kelvin which is the scale they use and that was outer space then this statement is on the verge of insanity because then the temperature of 0 Kelvin was 100 billion Kelvin at Time ~ 1/100 seconds. That means using that scale there had to be regions hotter and colder than other regions were. It is as if they put the Universe functioning in regard of life and life into the thick of things and because we are alive at 37 °C and that is when outer space is 0 K. That was the Universe then and this is the Universe now and by putting it to an inclusive scale is indicating the universe is comparable. That is insane and can only come as a result of mathematical formulating that does not reason or put values mindlessly together. The cosmos was 100 billion Kelvin and that means there was no 1° K anywhere, which means there never was 100 billion Kelvin. What then was the freezing point of water at that time? Or then answer this: If space was 100 billion degrees at what temperature did water freeze and if there was no water to freeze that Universe couldn't be 100 billion degrees K because then water did not freeze at 0° C. That shows time stood still. Then at Time ~ 1/10 second the temperature was 10 billion Kelvin which is a little over 10 billion times more than now as the Universe continues to expand. If this means water was 10 billion Kelvin then there was no water and mentioning this proves madness because there is a relevancy that could not have existed at that point. This goes on and on but this says little to nothing because at that point when it was 10 billion degrees in the shade at noon in the summer without a cloud or wind blowing at what temperature did water turn into vapour to bring about rain as to relieve the Universe from this overbearing drought. This is what you get when you design your personal Universe in accordance with the needs you see befitting a Universe. The same applies for time. Time is the rotation of the earth in relation to the space the earth moves through. How big was the earth at that time to give the influence in seconds that they measure by? With formulas they personalise the Universe by putting them on earth as the applying measure of everything

The Universe works by relevance and to fathom what is big we have to refer to what is small in relation to what is big. In the Universe there is no big as there is no small. The biggest star we know of is a Black Hole and with the density in a Black Hole all the material within the normal Galactica will fit into the size of one atom. In that case an atom is bigger than an entire galactica. Whatever size the Universe was at a time it had a temperature of 10 billion Kelvin, it still is that very same temperature and size that is present today because whatever was that was filling the Universe at that point is still in the Universe and all of that will be in the Universe for as long as we can think. This shows in cosmology there is no hot and there is no cold. There is no far and there is no near. There is no young and there is no old. The youngest star is just as long been in the Universe as the oldest star is because "mass" does not pull "mass" and no star formed a unit by "mass". All material formed when dots formed and space was not yet even a thought. These are man's misconceptions because man makes science revolve around what conditions man would prefer. Man is no factor in cosmology and deserves no place in thinking in terms of the Universe.

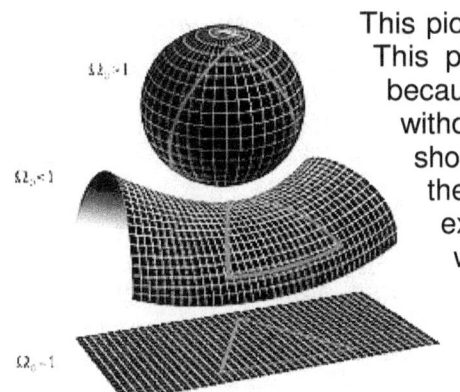

This picture does not show singularity or a Universe gone "flat". This picture does not show singularity in any state or form because it makes space inside a picture and singularity is without space because the Universe goes "flat". However, it shows the shortcomings of the formulation of the equating of the Universe by applying breathtaking mathematical expressions. Where is the border that Universe has that we see in the picture? Where is the flat side where the flat square the Universe forms end. Having a top there has to be a bottom and that is space. There is no limit to the Universe but the incorrect mathematics puts a limit on the viability of the formula and it is this limit they transfer onto the Universe. If this is absolute brainpower I thank my Creator for making me as incomprehensibly stupid as I am!

I show do where to find space less singularity The Universe started when singularity heated from the eternal zero it was to the infinite heat it developed. Before heat came there was no cold as there was no hot. That means in the cosmos there still is no heat in hot or cold, only margins forming borders. In nature space expands only when heat increase. Even the Universe at first was not strong enough to contain the heat that rose. Why did the heat expand? This process has a name. It is called an explosion.

It is when heat becomes contracted in the process of developing new space where such containing will light up to produce photons and in the process turn cosmic liquid to cosmic gas. Think of what thermal nuclear reaction came about when far more than half the Universe formed light in heat. It cracked and split the cosmos in two. The other part was contained by gravity to counter act the expanding but much more development came to be light with the heat forming the first photons. Only by heat creation can space grow. There is no other process that can form space other than heat overheating into space that expands into more space to cool down. The initial heat must have been infinite to bring a parting between what became cold and what became hot placing material in between what is hot and what is cold or that which can never start and that which cannot end. To humanise this affair Mainstream science accepts the concept only when it comes to applying their human values of a very manageable and understandable 10^{34} K. Science has "mass" to be clever as to play God to charge the admiration of all other little people. They calculate so that they can admire the mathematical ability of one another as to miss their stupidity of not understanding all the reality that is out there and which they have to miss by looking at mathematics. Their mathematics doesn't promote the Universe. It reduces the Universe to the size of their mathematics.

By moving material removes heat from liquid to solids by leaving behind a cosmic gas. Look at what the inside of stars held before the blowout. Stars are filled with heat on the inside and what the star released with the blowout is heat. It is heat that comes out and that is what a star is. It is an unsealed container of heat and gravity is freezing space into a condensed heat. The reason why we think space is hotter within the earth is because we experience the heat the space releases form the confined space within the earth. But by releasing the heat we experience the space is colder because it is finding relief from the heat it cannot carry. It cannot carry the heat because it is cold. Outer space can absorb as much heat what is cast into it because it is so hot it can absorb whatever overflow comes its way. The cosmos is a variety of density caused by differences in movement within any specific confinement of space in relation to any and all other space formed by heat. I show what forms the Universe is small spots forming dots and these dots when compressed as it was during the Big Bang form heat in abundance but when it expands it relieves heat by expanding into abundant space. That is what the Big Bang was about. Space overheated and formed a means too cool and that became space.

The Universe is formed by heat that we call light and light is dark when it moves away from us (expanding or becoming bigger thus moving further apart and by that it is drawing visible light inwards) and light is bright when it moves towards us (concentrating by gravitational contraction because contraction makes the light denser) but everything in the Universe is a form of heat.

I am about to show that in the Universe there are two substances filling the Universe. One is material that forms clusters in atoms or stars and the other is non-material and there is no such a thing as vacuum or nothing. The Entirety of the Universe is about the relevance of density variation. The faster any object moves the slower will the solid move becoming intensely more dense until it stands still in relevancy to the liquid space that increases in movement where the density will rise to appoint the liquid is as dense as any solid is and it then reaches the speed of light. However at that point the solid will go into singularity and singularity will go into time that is immovable. That then becomes a Black Hole where the solid is immovable and the density removes all forms of what space is in the cosmos. It is the ratio between occupied space and unoccupied space that increases.

There is no mass in the Universe but it all comes down to specific density. Because the

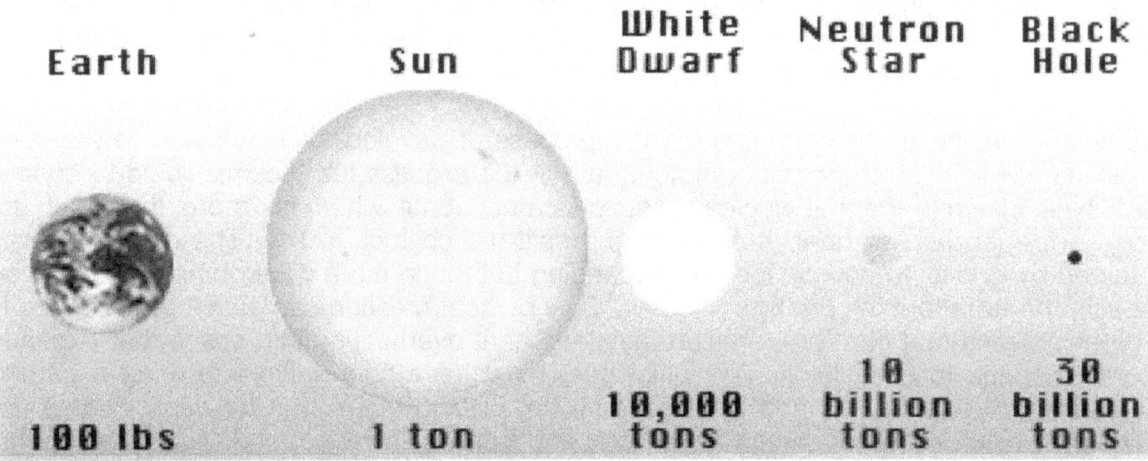

faster moving stars associate with more space per time unit the overall size of the star decreases because the space the star runs through increases. However, this allows the density of the star to increase because the material then becomes more solid as it moves through more space. It becomes more massive because it becomes denser but because it becomes denser it becomes smaller in relevance. As the object moves faster it also moves through more space and moving through more space makes the space that material hold become less in comparison to the space that the object goes through while moving that fast. As the space the object goes through increases the temperature of the object decreases and the object cools down because of the movement that increases the cooling of material. It is the same as putting a fan on that blows over the radiator of a car whereby the blowing of air increases the space the water circulates through and the water cools more because the increase in the volume of space moving in relation to the water cools the water more. When any object moves it duplicate the allocated position more as the movement increases. It fills more space in duration of time but during the instant in time it holds less space per single time unit. In this we have space-time because time going singular produces space. The shorter the period of time is that an object fills space the less space it would fill in the instant but the more space it will fill in the total duration. In other words it shrinks the space it is in by extending the space it holds relevance too. The space material moves through increases as the space material holds shrinks. This will go on until only singularity is left. This means the size of the atom inside a star reduces as the star movement in spin increases and this is why stars that move faster or is gravitationally stronger will reduce in size but increase in density.

As any one can see a star is a container of pure heat but it is not a container that holds pressure. Pressure is associated with space and not with heat. When the star or heat container gets pressure it explodes as we all see in Supernova outbursts. The moment the star gets pressure the space it holds increases and that indicates a sudden increase in temperature. If the star explodes by increasing the temperature then gravity is all about reducing temperature because as temperature reduces the space will reduce accordingly. Temperature is not some scale on some meter some scientist devised in human terms but heat expands when it gets hotter and heat contracts when it gets colder. The space the earth holds might seem hotter to us but that is because it is more contracted. To be contracted (not pressured because it is not a metal container such as a boiler is) the heat has to reduce and increase shows heat levels rising. If space expands, the heat increases and if space reduces the heat decreases and that is cosmic science. The levels of heat drops because the heat transmits away from the space in which it are. We can't transform heat values we have to the cosmos. We must adapt to science laws and not the other way around. True science rules say when things expand it is hot and when things contract it is cold.

A star is a cosmic atom and a star is filled with little pumps called atoms. Every atom in the star pumps heat by spinning faster than the speed of light pumping heat from the outside we call outer space to the inside of stars. This process of contracting and condensing outer space, which is expanded heat, we named gravity. The star spins slower than the speed of light in order to harbour atoms that spin faster than light. Every atom is a pump and the protons condenses space or heat or singularity in conjunction with the neutrons and how this works is it compresses expanded heat to smaller space that condenses heat by confining the heat into a smaller area which we see as a star. The star has the condensing ability or the pumping capacity equal to the combined effort of all the protons and all the neutrons and all the electrons within the container we see as a star and in that capacity we see how gravity works.

To unlock scientific truth we first have to dispose of scientific misconception

In the two pictures we are seeing disposing or releasing heat creates space. We may call it plasma or shock waves or what ever, but in the final analyses it is heat turning to space. Whatever you wish to call that which lies between the particles comes from being a solid, then with adding heat, the solid *"whatever"* becomes liquid and that is the white and orange plasma that we find. That white and orange is heat in a liquid form, just as all flames and smoke is heat in a liquid form. But that liquid does not remain liquid because the governing singularity cannot enforce a commitment ensuring the liquid heat remains liquid. The liquid *"whatever"* you call the heat in fluid form then further overheats turning the heat to space. The space created must be equal to the heat reformed. That is a law of energy where energy equals equality everywhere it is. The only "energy" is to transform heat by expanding or by contraction to movement.

If there is movement it can only come from an exchange of cosmic liquid turning to gas or taking cosmic gas and converting that to liquid. But driving anything or moving whatever comes about as a result of cosmic fluid we call plasma and a many other names but it is condensed heat. Heat differentiation translates to movement and gravity is movement. Gravity applies when an object drops by movement or objects keep on tending to move from its position on the surface of the earth to the centre of the earth. The frustrating of movement by stopping further descending brings about mass and that is measured by weight notwithstanding any other definition to try to correct or hide the misconception science attaché to mass. Mass is a measure and not a factor while gravity is a movement and forms a factor that results in mass.

Galileo proved that all things fall equal and by all things falling equal this idea eliminates mass completely notwithstanding the corruption of a feather falling with a hammer in a vacuum. Yes the feather will fall equal in vacuum because the feather holds a different density to air because it is larger and that changes the relevant density it has to the density the hammer has. But a car falls at the same rate as a person and just as fast as the person carry bag. To stop the falling one has to open a parachute and the parachute alters the density of material in relation to the space the parachute confines and it is the density that changes because whatever is tied to the parachute did not lose mass in the process.

By going into the air as a hot air balloon is heated is pure evidence that gravity is because of density and density is the changing of mixture between solids and non-solids. By increasing the air ratio in a form of heat which is what air is and confining this increase within the parameters of a balloon acting as a container the density of the solid material changes in favour of the air and the increase in air brings about a change in density and not in mass and with an increase brought about in the density the solids and the bag becomes a gas where the lot rises into the air as if it is a gas. That shows clearly that gravity is the cooling of space because when lifting the balloon becomes ant-gravity and lifts up. If gravity is "pulling down" then anti-gravity must be "lifting up". Let us humans first detach culture from facts. Take the argument to iron, which we know well. Iron cannot boil, iron cannot flow or bend and iron cannot brake. Iron is an element like all the other elements we know, not one element can do any of the above, in sharp contrast to human belief. As indicated in this book the limits we should find to guide us we ignore for the reason that we cannot see it. We may not be able to ever see singularity, but with intelligence guiding mankind, we do not have to see everything to believe everything. It is because we could not see religion, but still practised religion that set us apart from the other animals. At the start one would find iron and iron in a "natural state" as we find iron on earth being a human produce on the surface of the earth it will be a solid, suitable for man to handle with bare hands. When such a piece of iron is left in a desert in the midday heat, the human hand cannot handle the iron any longer without aid of covering the skin of the hand. Our perception is that the iron became hot, but that is not the case and our view is a culture contribution and not scientific fact. The sun mixed cosmic liquid known to us as sunlight, which is heat into the iron that upgrades the heat part in the iron. By heating the iron artificially with combined gasses (acetylene and oxygen or what ever) we now can over heat the iron to a state of flowing like a fluid. Increasing the heat in the iron increases the non-material section in relation to the material section and by heating it to a point of vaporising the heat part lowers the density of iron so much it forms a gas.

Our human culture tells us the iron now is melting. That is a misconception! Lines forms singularity and it is these lines that holds singularity that forms concentrated heat or expanded space as condense or expand heat and singularity expands heat or concentrates heat. After introducing artificially even more heat with more heat releasing gasses we may artificially form a condition where the iron would become a gas. Again it is not the iron that becomes a gas; it is the space the iron finds itself in that became hot enough to become a gas.

The iron particles remain the same; it is the condition surrounding the particles that changes form with overheating. Important to note is the fact that iron in a solid state will surround itself with solid matter in space applying a solid space. By introducing conditions producing _more overheating_ the space or connecting between the particles become concentrated heat forming a liquid substance! It is not the iron that turned liquid but the wrapper containing the iron that concentrated so much it formed liquid fluid by the introducing of more heat to a point where the overheating created a fluid. It is considered that the oxygen burn and by that the iron heats up. NOT TRUE! If oxygen burns no oxygen would be left on earth by the time man arrived on earth to use it to the benefit of intelligent life. The oxygen remains oxygen while the oxygen merely does a task in nature where oxygen carries heat to a specific space. On the other hand it is the task of nitrogen removing heat from the point of overheating by means of flames whereby it creates space. One can feel the "wind blowing" as the flames generate created space. In the extreme the creation of such space we call an explosion. In the process where the space between the iron particles still further overheats, it becomes a gas. It cannot be iron that becomes gas, because depending on mixing heat with iron it will be as much a gas as iron will be a liquid or a solid.

It is the space covering the iron particle separating the different iron particles, which will convert and sustain form. The gas is as invisible as space because the gas is the form space holds. It is the relation that materials form not by heat or cold but by linear or circular motion that forms density. There are two forms holding substance earth (solids) created and heaven (heat or gaseous/liquids) created. These are the only two forms of substance that is the Universe solids and non-solids, where non-solids are what increases to form liquids and gas. It is not the solids going liquid but it is more of the liquid in ratio with the solids in between the solids that make a structure go solid or gas. There are cosmic solids and comic non-solids. It is movement going at or slower that the speed of light or solid atoms going faster than light.

Iron is a solid. Introducing more heat the iron becomes more a mixture between liquid heat that reduces the density to the point of concentration where it became a fluid. The iron remained what it is, neither a solid, nor a fluid nor a gas. By introducing more heat it becomes a gas. The gas we cannot see because the gas is space. But so was the fluid space.

The introducing of heat brought about the turning of a solid to a liquid to space and every time more space becomes part of the picture. Iron is in its normal form a solid. That means the space, which the iron particles are in is solid and that disallow the iron to alter the form in which it is. By introducing considerable heat the iron melts changing the form of the iron from solid to liquid. One will find that whatever group one chooses there are gasses and there are solids. If mass was attracting mass then the strongest mass must be attracted to the strongest mass and the least mass must float in the air. $F = G (M \times m) \, r^2$ hardly can even begin to explain the fact that there is a gas that is more massive than iron but floats in the breeze just as hydrogen which is the least massive element.

Nitrogen 7	melts at -210°C	boils at −195.8°C
Oxygen 8	melts at −218.8°C	boils at -183°C
Fluorine 9	melts at −219.6°C	boils at −188.2°C
Neon 10	melts at −248.59°C	boils at −246°C
Sodium 11	melts at 97.85°C	boils at 892°C
Magnesium12	melts at 650°C	boils at 1107°
Aluminum13	melts at 660°C	boils at 2450°
Silicon14	melts at 1412°C	boils at 2680°C
Phosphorus 15	melts at 44.25°C	boils at 280°C
Sulphur 16	melts 119°C	boils at 444.6C
Chlorine17	melts at −101	boils at −34.7 C
Argon 18	melts at −189.4°C	boils at −185.8°C
Potassium 19	melts at 63.2°C	boils at 760°C
Calcium 20	melts at 838°C	boils at 1440°C

Every element clearly has a different density point It is not mass that plays any role in the measured formation of elements. Elements can be on earth prone to form a solid better than it holds a liquid form but that is related to gravitational conditions we have on earth. If man would land on Jupiter there would be no man landing on Jupiter because the conditions on Jupiter could never sustain any form of life. All the above elements would have total different for limitation on Jupiter than they would have on Earth. If one looks at mass there is no connecting point putting a realistic ratio between what is heavy and what is light? If gas floats in the air we have to presume it is lighter than air because if something floats on water the density the object has to float makes it less dense than water is and therefore it is lighter than water is. In this same manner we have to look at gas and irrespective of the "mass" value Newtonians grant the element to have, if it floats and it is airborne like a gas is then it is less dense than air which makes the element lighter than air. The hot air balloon fills with more hot air and that reduces the density that allows the balloon to float in the air. The balloon is lighter than air notwithstanding it never lost mass.

Looking at stars the way Newtonians do they relive the coal burning boilers. They see coal furnaces being stoked to burn and heat boilers. In the days of Newton coal stoves were the nuclear science of the day and while all other departments in science moved on and away by developing away from 17th century values and from coal stove principles Astrophysics and cosmology remained true to Newton by reinventing the coal stove in so many ways not even the coal stove could think of the facets it can go through. Newtonians see stars being fuelled like coal boilers and such steam boilers can run out of fuel. This is so much Newtonian backwardness as mass forming gravity and the moon coming closer and the cosmos shrinking and we falling into the sun because of non-existing dark matter making up what is required to make Newton not to seem the idiot that Newtonians are because they make him and his contraction theory to be less foolish that what it apparently is and they overbearingly are. What is of vital scientific importance is that there are three fundamental dimensions controlling the universe. The three are beyond intermingling and one confirms a status in relation to the others but not intermingling in status. From singularity comes matter and forming space-time in own accord. By matter not controlling time, space grew uncontrolled and the third dimension came about. That dimension birth we now recognize as the Big Bang, but the Big Bang is the last of a three prong cosmic growth. Science has to recognise the dimensions of densified (singularity), occupied (matter behind the electron) and unoccupied (space-time outside the orbiting electron boundaries) forming three points of cosmic recognising space-time.

Every atom holds (I am guessing), as many dots as the sun has subatomic particles per atoms and that would still be a very conservative guess. Every dot is a controlling centre selecting a regional centre where every regional centre selects a centre. This goes on as long as there are spots forming groups as individuals unable to survive independent. The others that was unable to group formed heat that became space, which became the broken dots. The dots form groups to survive and as a group, the survival depends on doing what the group has to do to remain cool. In another book, I reserve one chapter to explain the phenomenon what I called the Lagrangian atom. These dots arrange in a manner that they could favour either the space duplicating aspect or the space dismissing aspect.

This can only be the result of the fact that even in the case of the sun, the inner space is almost entirely liquid heat and the liquid heat produces sufficient space to dismiss as the centre that holds the heavy metal particles, where all the dismissing is done. The liquidity provides motion while the solidity removes motion in the centre of the star. The dismissing going on is in the space factor where the space leads to a denser heat within that space because there are insufficient material to accommodate all the heat by the dismissing factor T^2. In that case motion far outweighs dismissing $k>T^2$ but a time comes in every star that the dismissing takes absolute charge. $k<T^2$ That is when the star goes dark. The Earth is mainly

about duplication of space much more than dismissing of space and so is every structure in the solar system.

I would suggest we think of stars in the following terms. A star that generates and transmits a lot of light is weak on gravity because their progress started recently. They command a lot of space-time but the demand they have to keep their cooling acceptable is very low. In that they can generate a lot of light but with the demand on cooling low and the gravity in the centre not very developed, those stars cast a lot of light back into outer space. It is just because of the size the stars holds that tell that the stars are still young and have a weak developed governing singularity. The stars will have very prominent hydrogen and helium layers, with the inner core not very prominent. The control of the star is still very much in the individual atoms and in that the motion the atoms have to produce in order to maintain their individual singularity will only come about through motion. The atom has to make contact with as much space-time through motion as possible since it has a very poor ability in contracting space –time in support of the cooling system. There cannot be something big or small except in the relevancies of perceptions and then the relativity of such perceptions becomes questionable. There cannot be hot as much as there cannot be cold The sun freezes hydrogen to a liquid at 6500 ^{0}C and outer space boils over at 0 K. If we humans cannot or will not abandon our human culture driven perceptions and our mankind's pre-programmed perspective we may as well return to astrology for what the future hols. There are so many boundaries out there ready to destroy us because of our lack of insight, as did the challenger disaster. Creation birth started off with one dot so small eternity met infinity within. Then came one more, and another and they continued coming until there were a countless number of dots. The accumulative size of the dots were the same size as one dot because in the true Universe big and small plays no part. The dots were infinitely small and eternally big at the same time because size is a relevancy and without one the other has no size. So in the true perception, there is no difference in size.

The idea than humans make aircraft that can fly has no meaning in the cosmos. Life as an entity is totally alien to the cosmos and life can only be on earth and nowhere else unless the atheists can prove otherwise. Everything that moves in the cosmos moves because of gravity and gravity is a difference between densities in space. A gas is a lesser density because it spins at a higher velocity and not because it holds more or less mass. A gas floats in space because like the hot air balloon it has a higher relevancy to heat or it captured more heat in-between the atoms forming the compound of the element. When anything moves fast it is colder and because it is colder it captures more heat to maintain the speed of spinning it has. On the other hand when objects spin slower the elements tend to be more solid and the density is much higher. Every element is made up in form by innumerable many dots that form singularity. Every dot was by itself as well as the accumulation as it currently is the present universe. The earth in itself is a Universe standing apart from other universes such as the moon as well as the space between the moon and the earth. The moon is a universe. Rules applying on earth do not apply on the moon and visa versa. When considering conditions with in the oceans and applying space-time another set of rules apply therefore the sea places a body in another universe. It takes the same engendering technology going underwater in deep sea diving that going into outer space. Every dot was a Universe in its own and the accumulation was a Universe. The earth in itself is a Universe as the moon is a universe, because rules applying on earth do not apply on the moon and visa versa. When considering the conditions with in the ocean and applying space-time another set of rules apply, therefore being in the sea places a body in another universe. The number of universal entities is still countless, as much as it was in the beginning, before dots formed atoms. Every dot insignificantly small as it may be, is a part of another Universe as much as it is part of the accumulative Universe and every dot in the infinity holds singularity, which we translate as " nothing" being " darkness".

The very first instant when the cosmos started the perfect became imperfect. When what was perfect became imperfect the Universe moved as time in eternity split from time in infinity and I show where time in infinity is and where time in eternity is. When the spot differentiated and became differently allocated from the dot the Universe started. When infinity moved away from eternity the Universe started. When the perfect overheated, hot and cold formed relevancies that put space in between time in infinity and eternity. Even today this is the fuel that drives the Universe as liquid that parted from solid reunites with solid to form a density difference. Every dot insignificantly as it may be is a part of another Universe as much as it is part of the accumulative Universe and every dot in infinity holds singularity, which we translate as "nothing" but it cannot be nothing. Singularity is what forms the Universe and is the smallest that can be.

The light specks we see scattered throughout space at night are stars and stars are allocations where solid materials spinning and with that are by gravity or movement contracts gas to form a cosmic liquid.

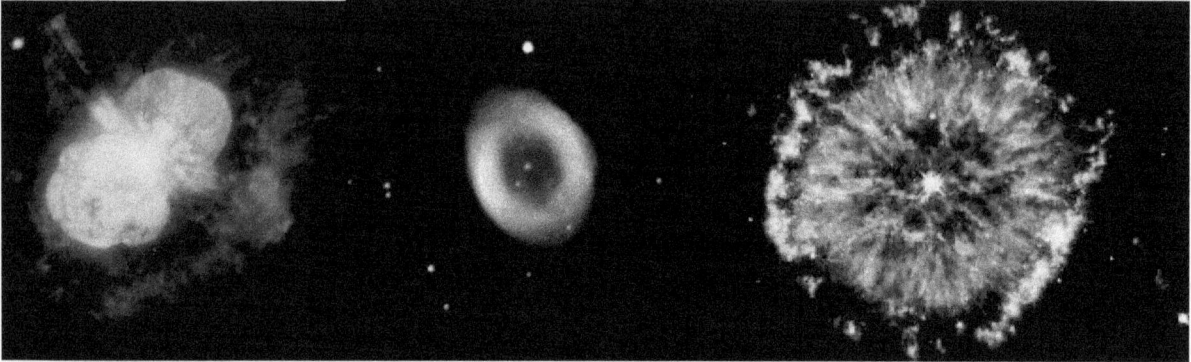

It is clear from all the images that the liquid that was inside the star before the explosion froze in liquid form when the explosion thrust out what was inside the star and clearly it was heat forming space in the form of liquid that came about because what was remained on the rim of the circles. The liquid that was inside clearly remained liquids that froze in space resulting in liquid forming the outer edges of the layers and as it turned to gas the gas formed more outer space. Clearly the density variation is visible as

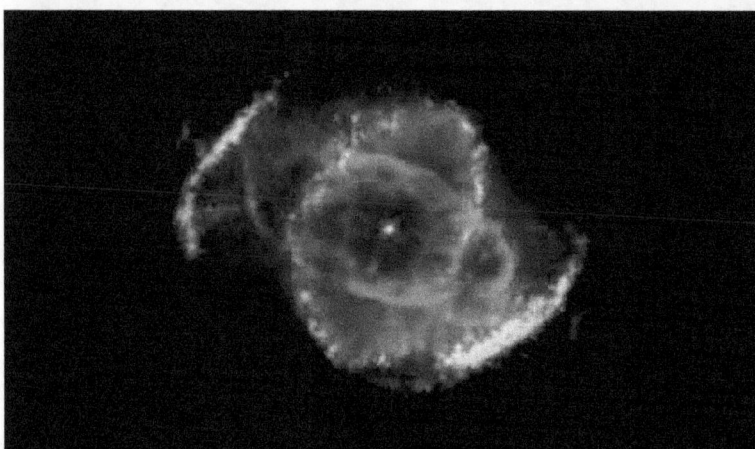

what was inside formed either a denser liquid or a darker gas. It is also clear that one layer after another layer overheated and expanded by causing an explosion as we can see from what remained. This explosions and what comes about from it as "shock waves" carry many other Newtonian names but in the end it is heat going into gas as the cosmic liquid forms cosmic gas and cosmic gas becomes more cosmic space as it expands. In the picture of the sun the sun shows liquid and so does the galaxies and it is all too obvious that space in the form of gas freezes into liquid as gravity reduces the space into frozen liquid. Moreover still, we can see the point of the remaining core holding the incredibly dense material that forms the singularity within the star. Every star is a Black Hole in development. As the Universe grows the density of material increases by the same margin as outer space loses density and in the end the core of the star unites all atoms within the star where the atoms by movement becomes singular in one unit. Inside every star spinning there has to be an iron core to produce spin because I prove mathematically that electricity is the very same as gravity but it is dimensionally charged on different levels.

In every pigment of every picture we see heat being in contrast to heat setting differentiating levels in any volume of space one compare to another. The entirety is heat being space and heat being condensed but in the end it is heat forming a different contrast to other heat that forms more contrasts. Every atom forms a Universe. The sun concentrates heat towards its centre and the galactica concentrates heat within its centre and then everyone believes mass pulls mass without ever realising what drives the Universe is heat contracting and heat expanding. The driving force if I may use such a primitive word is relative density. The star collects heat by turning around within heat and that turning makes the star denser than outer space because by being denser the star is also colder. The star grows by collecting heat because the star is a unit of dense material and outer space is a nonbonding substance of non-materials. Material grows while space expands and it is about density relevancies and the density applying varies between elements, atoms and gravity. As time moves on it forms space and that process is the fuel the Universe applies to return the Universe back to singularity as it was.

Look at the image of this Nebula and see how it glows with heat. It is clearly show the star was a heat container that opened up and had the liquid inside burst into gas. My eyes tell me this picture is a showpiece of heat contained and the container overheated where the liquids turned to gas. In order to cool gas the gas has to flow and our refrigeration and air-conditioning work on this principle. Even the engines of the space shuttle apply this very principle to cool the heat. The star is hydrogen gas frozen to liquid by the spin of the star and in that the star is filled with liquid because the pumping of the star froze the gas into liquid. By contracting the heat surrounding the star the heat condenses and it releases heat because it became cold but due to the spinning slowing down it couldn't maintain the cold and the supernova overheated.

At first when I started a study to find out more about "gravity" and the ever, elusive graviton, I came across lazy gravitons, and eager gravitons. It seemed that the lazy gravitons produced little gravity and the heavy elements such as Xenon and Radon are the heaviest inert gasses and therefore must be the laziest gravitons around while Lithium as a solid being the lightest solid must have gravitons so potent you can use them as Superheroes that replaces Tarzan on weekends and Public Holidays. This remark is pure trash only because the graviton is pure trash.

Gravity is a reality because atoms pump heat into material as to avoid overheating that result in an explosion as we can see when looking at Nebula. There are so many facts that accepted science know about, but do not fit into the perspective of accepted science, after which science blatantly ignored these facts. I have indicated but a few examples and the examples are the most basic science offers. All information I have disproved thus far is science taught to children at school. Therefore, it is not hard to imagine how much nonsense they propagated in the more complex issues. Can you imagine for one second, a star that "collapse under its own "gravity!" ...And this rubbish resulted in that two Nobel Prizes were awarded to two Nobel winners because they proved it!

Only in Newtonians' Science as primitive and backwards the outlook is will a coal-stove fire need a thunder- storm to start the fire. Please allow me to explain, and believe me, I am not making this up as I go along, just to ad humour to this otherwise very dry letter. I did not invent this fiction in order to create a bit of comedy; this is truly Xepted Science.

There apparently is a point at which point the star performing like a coal stove starts because gravity lit a fire and the fire is burning the coal (core) to get the star going. Take into consideration that every idea science clings onto today was an idea put forward when ships were driven by winds blowing and in Africa there were believed to be dragons.

This is pure hot science as was practised in the 17th century but very current. At the time everyone saw steam changing as advanced as nuclear science is viewed today. They new the boiler and steam was going to change life and the essence of it but that was then projected this massive new knowledge of the day into astronomy because only the brightest minds could understand the working of steam. If it formed the edge of science astronomy had to use it. So steam was the fuel of the Universe because Britain was the centre of the Universe. As in the case of all fires, the star (or stove) ignites with a small spark. It had to be that the fire grows as the hydrogen heats up, just as it would do with the coal that has to heat before it can burn.

Depending on the amount of fuel available, and the heat to ravage the fire, this star can burn from anything between a few million years and eternity. What a lot of Neanderthal bollocks fit for fairy tales and other fantasies. The fuel the Universe holds can never run out because it is time forming space. They thought it is the mass that determines the life cycle of the stars and the bigger the star the more mass it had so the shorter would the lifespan of the star be. Today we know big stars are very soft in terms of mass while the smaller the star is the more potent it would be in terms of mass. This is because the relative density of the star increases as the overall size of the star reduces. This is all determined by the "weight" of the matter in the star. In the case of the sun, the star will "die" leaving behind the ashes of a helium core, with the density of 10^{17} g/ cc. How helium can remain as helium with such density, is yet another unsolved riddle. Every idea Newtonians believe in is centuries old. Whenever they discover something new such as the Titus Bode law, the Roche limit, the Lagrangian points, the Coanda effect, the Hubble expanding or nuclear physics it is burdened by either covering the importance, hiding it by preventing it from becoming common knowledge or turn it into a prank. Hot air lifts balloons and when this clashed with Newtonian mass pulling things down it became a joke by referring anything that overshadow its purpose to be hot air. Can you believe that they can believe a star can die? Only life and what comes from life can die because life holds time at a limit but not the Universe. To their view the Universe is limited but the cosmos shows it is eternal. Black Holes prove that stars can't run out of fuel!

So ignoring hot air they stuck to the story of mass pulling things down instead of thinking that if hot air makes things rise it must be cold air that makes things drop and that would have brought my theory to the forefront two centuries ago! To put this into context, they had the Newtonian stove burning coal and the stove starts with a small match, then it simmers, while the heat produces more weight and pressure and the coal burns until much coal burns bright in all splendour and glory.

With the fire raging, the flames will light up the night sky, and the light is visible for miles around. The more the coal the stove has the more the glow is of this red, hot fire. After a while the stove runs out of coal, and with no one to stoke it, it starts to simmer, after which the glow disappears altogether and the stove dies out. Believe it or not but this in all the out datedness it represents, this is still modern cosmology, preached by people that has so many degrease they can put it to use as wallpaper to cover walls! It is clear that the twenty first century has not reached cosmology, because if ever I heard a Medieval, old wives tale, then this must be it.

However, this is not where the stove comparison ends. I have books in my possession, where one of the worlds most accomplished and renowned cosmologists, is of the opinion that the hydrogen falls on the red, hot core of the star and then ignites. This will happen to cooking oil that drops on the red, hot stove, but not on the inside of a star. Science only chases money! Somehow, somewhere someone found traces of minerals on the red planet, and this unleashed the feverish interest that is taking place currently. For many decades and years, there was a lack of interest in manned space flight albeit to the moon or Mars. The only interest was getting into space and manufacture semi conductors in "micro gravity" again to profit the Hoggenheimers and the John Dows of the world fit the bill with Tax money.

I might sound cynical, which I am not. On the contrary, I applaud every move NASA makes all the way. All I am asking is for honesty. A small amount of earnest will go a long way. In fact, to my thinking, they can take 100 percent of all the money spent on arms each year and dedicate it to NASA. That should not be that hard to do. Tax all sales of arms by one hundred percent, no matter who is baying be it the American, Russian or Chinese government, or the African governments living on charity, yet having enough money to conduct the most horrifying wars. Less people will die and mankind will profit much greater from such an action. Whatever we see in pictures about the Universe we see density differences everywhere.

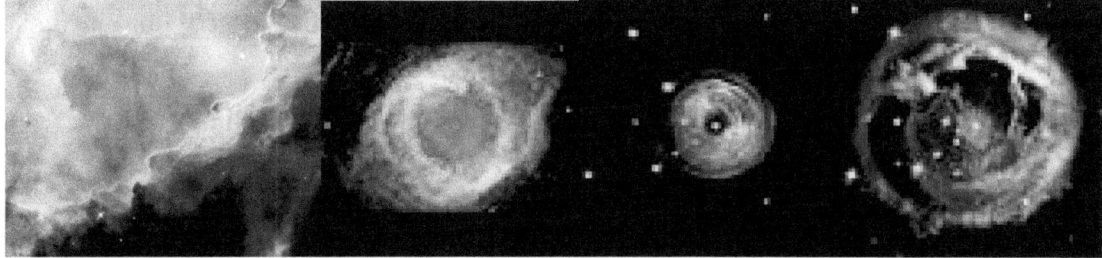

The density by gravity shifts from where the substance is nonbonding to where the heat as a substance is bonded by movement forming materials. The density shifts from non-materials to materials and this flow of heat forms one part of gravity. As the liquid or cosmic gas loses substance to materials contraction the density in the non-material division loses compactness by losing compound and therefore it expands into space while materials collect heat and therefore grows in space. That is what the Hubble shift is about. As materials form density the non-materials loses density and also therefore value. A star does not grow into a Black Hole but outer space envelopes the star into a Black Hole because as the star seemingly shrinks so it is outer space than reduces the effective space that it holds. The more outer space loses compound and gain space by losing density the denser does material get but as it gains in density it loses in relative space. The cosmos is a shift in density running from non-materials to materials and this evidence is proven by the fact that the earth grows in size while the moon drifts further away and this is correlated.

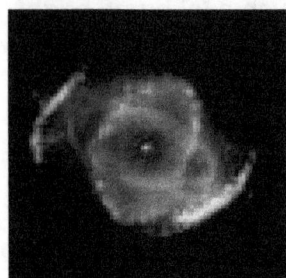

In the star heat serves to keep the atoms cool. Without sufficient heat flow the atoms will get hot and expand. When the atom expands it not only holds more space depreciating the relevancy of space to material within the star but it spins less and the heat within the tar rises even further. This forms more space as heat that expands by overheating produces space and when the space reaches an ultimate critical level the Roche limit is reached and following the Roche limit come the Roche lobe. I explain this process in much better detail elsewhere in this book. Therefore the fuel that a star runs on is not coal producing fire that can finish when the stoker retires from his job with old age and it leaves the star to "die". No star can ever "die" because a star is not a coal stove burning as it cooks oil or burning hydrogen. The atoms are pumps pumping liquid into the star and the liquid allow the material to grow while the liquid produces the rise of relevant density where the relevant density matches the overall density applying in the Universe at that moment. A current Black Hole was a star a very long time ago and our sun is a future Black Hole a long while from now but it is in the process of developing.

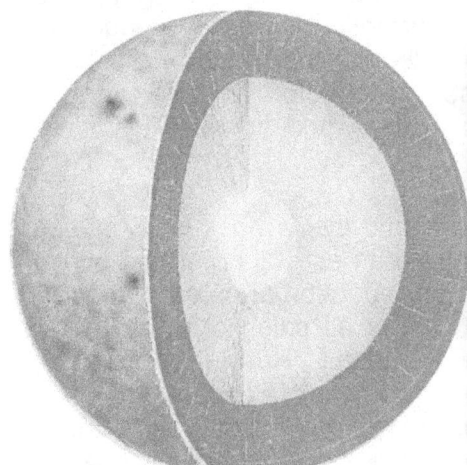

Looking at the stars and galactica through a Newtonian perspective Creation rumbles on without perspective, purpose or destination. It is a tragedy that people will be so obstinate in their programmed mentality to click on one thing and miss the entire picture. Only a definite relation between two balancing values forms the complete Universal relevancy of SPACE-TIME $a^3 / T^2 = k$ as Kepler got it from the Cosmos. It's clear that the centre concentrates heat.

I compiled **a new cosmic concept** by which I eliminated all the incorrectness that Newton has burdened science with but with this being my opinion I did not find a garage full of academics supporters waiting to applaud me and to uphold my views on the matter. Gravity rests on movement of material in relation to other material also moving. The movement of the sun provides the earth with movement but not only that al movement going straight becomes circular movement and circular movement takes place within the circle in which it moves going forward as a straight line and in that idea of a circle becoming a straight line and a straight line becoming a circle the entire concept of cosmic gravity is vested. According to the Big Bang theory the Universe expands and there is no evidence of pulling bringing about a Universe contracting or becoming smaller. The Hubble constant is sole evidence of this proof of expanding Therefore I challenge the concept they build on the fact that mass attracts mass and everything is pulling everything else. Yet still I was not going to be ambushed by their relentless stonewalling my efforts and blocking my efforts in introducing both the incorrectness and the new cosmic theorem I concluded. My views are founded on what is there and what I can calculate as it is applying.

Their mannerism in blocking and frustrating my opinion when showing the mistakes in science convinced me about a **Conspiracy in Science in Progress** and this spurred me on to tell the entire world about their brainwashing students minds. By the manner they selectively withhold information when teaching science, amounts to deliberate brainwashing of students in physics by "normal" education practises. The new concept I wish to introduce puts all emphasis on space ands material is only space filled with material substance while other space is filed with non-material. In the end all space are equal but the movement it has makes the difference it presents in relevancy.

All space structures hold in the centre singularity concentrating heat and from that centre that all material holds comes all the drive. I can go on and on but heat in the centre couples gravity to space-time, just like Kepler said before he was spoken for on his behalf and without his permission or his agreeing to it. It would have been much more palatable if the Newtonian views were based on some form of a possibility but it works on total fabrication of facts that has one purpose and that is to mislead and to mesmerise by concocting untruths in the use of untrustworthy mathematics and meaningless formulas that no one can legitimise or prove! Newtonian physicist know all their mathematics are senseless because it proves only that they understand nothing except it makes them feel equal to God Almighty because of a feeling of total superiority the mathematics allow them too experience. This we can see from how they tell the Universe what they want it to be. The flowing is the result of using much superior mathematics to prove they know nothing about what they know.

If I made a statement that Newton is wrong about gravity, which person would believe me? If I said all those in science know very well that Newton is wrong about gravity but is hiding this fact for personal benefits in order to ensure their work remains to be accepted, who would believe me. If I said that everyone in science are aware that the formula on which all science are based $F = G\ \dfrac{M_1 M_2}{r^2}$ is as false as a politicians' honour, this fact will then come as an astonishing surprise to everyone and I get blamed for smearing the characters of the most honourable group of persons God ever thought to put on Earth. .

This $F = G\ \dfrac{M_1 M_2}{r^2}$ is the formula judged to form the basis on which the entirety of physics rests. Yet, nothing can be more inaccurate than this formula that science forms its entire basis on. As students learn this formula off by heart it starts off a brainwashing process and this procedure becomes everything that science represents. Should students not accept this formula as the gospel truth and as if it forms the only concept that could represent accuracy found in the entire Universe, that student would be sent off branded as not to be capable of understanding the fundamental basis of science. That student will go home labelled as stupid without any further possibility of studying physics in the future. I prove not only that this formula is rubbish but also that there is no mass at all.

Every one sharing the Newtonian vision of a contracting Universe is dreaming of a Universe where the lot would one day again come together and Creation will end where they say Creation started some time ago. The presumption is that the Universe has mass that is pulling mass towards one another and we are in the centre of an ever shrinking Universe. That is what this formula $F = G \dfrac{M_1 M_2}{r^2}$ represents. That is what the lot of us must think we know... we think we are forming the centre of the ever contracting cosmos where every Newtonian can vividly see with his or her eyes through any telescope that all Newtonian minded scientists are sharing the centre stage of the ever collapsing Universe. Newtonian science holds the view that the Universe is about to end where all mass contracts into one huge lump of material or that is the basis for Newtonian science.

If you want to go anywhere you better use heat to give you "energy" which "energy'" is another word for heat that will allow you to move. You move with using heat. You need hydrogen but you don't

burn hydrogen. You need propane but you don't burn propane or the oxygen that you also need. Hydrogen, oxygen, propane are gasses and because they are gasses they hold much more heat in association than does a solid such as tungsten. You can't burn tungsten as a liquid because tungsten as a sold holds very little association with heat in comparison the mentioned gasses.

It is the heat that the gasses hold captive that releases and that turn to fuel but it is cosmic heat. The gasses as a liquid in association with the solids atoms hold cosmic heat in relevance to a specific density and by releasing the heat the gasses hold in liquid and allowing the liquid to expand to gas it forms space and this release of space is the cosmic driving force used to bring about movement. If you want to produce movement you have to produce heat by converting cosmic liquid into movement that changes the cosmic space in the process. You have to convert cosmic liquid into cosmic gas but first cosmic gas has to convert to cosmic liquid and that is where the sun supplies us on earth with the only fuel we can use. We can use elements that naturally capture the comic liquid and store it or we can use fossil carbon that as life converted the cosmic liquid to reusable fuel back before stupidity ruled the earth but when it releases from the carbon it converts flames (cosmic liquid) into heat (cosmic gas) and it is this converting that is the process we harvest but it still remains cosmic liquid turning to cosmic gas or the other way around. Electricity is converting cosmic gas to cosmic liquid. It is the same as gravity.

In cosmology there is only one method of creating energy and that is by a displacement of heat in terms of cosmic liquid spinning and allowing material to spin (7/10) within a liquid. If the liquid moves the relevancy changes to (10 / 7). It carries four different titles but the application is the same as the four are the same transferring of heat and when Einstein searched for the common denomination between gravity and electricity he was on the right track. He was correct but never found out why because (I guess) he was looking for mass and mass related ideas while that is because every atom is a generator of electricity. The atom has an electron and an electron is what provides the flow of electricity. There is and there can never be any free electrons because there is no such a thing as an electron. An electron is a gaping hole into an atom and it is the spin of the atomic liquid neutron that pumps the cosmic liquid into the atom by reducing the space as it increase the speed of flow from what was below the speed of light to a point where it was an electron flowing at the speed of light and from the increase the velocity of the flow of displaced heat to become higher than the speed of light. Where an electron is we find space-time accelerating from as fast as light to faster as light and yes Einstein is wrong because there is a transfer of heat exceeding the speed of light and going faster than light. The inner atom spins faster than the speed of light and the release of this heat we call nuclear energy. That is the first form of gravity or electricity or electromagnetism, which is nuclear energy. All these mentioned above are the same thing.

When we get to the Bible part remembering these pictures are important

I show stars are only transferring heat and driving whatever transfer heat.
If the atoms individually generate a flow of heat we call electricity because when charging electricity with any dynamo it is iron spinning in copper ending the Ferrous process or the conducting of electricity from iron being magnetised to copper and in copper the flow ends where the line that flows is cosmic heat. To explain this in detail by mathematics is too cumbersome and time consuming. When all the atoms in a star spin altogether with unification acting as one cosmic atom (a star) and therefore acting as one unit we call it gravity but it is the concentration displacement uniting in centre- singularity that forms the difference to the measure of the very same applying principle. If it is a flow of electro magnetism it is one electron. If it is the flow of a generating of electron by spinning iron in relation to copper we call it electricity. If it is the flow of all the atoms forming the planet or star joins we call it gravity. What is needed to get the flow of heat conducted is the spin (7/10) of material in space. That is gravity, electricity, magnetism or electromagnetism and by name it sound different but it is the same thing but operates under other dynamics. This flow of heat is a liquid cosmic substance that we call air if it is condensed and outer space if it is cosmic gas and if it is pure liquid we call it lightning, light photons or electricity or just simple flames but it is a flow of heat. When it is condensed by electricity of gravity it condenses from a space valued at outer space that has a 112 proton-displacement to the

inner space, which is a 56 proton-displacement, which is the proton value of iron. This limit is set by light that holds a space value of 3^3 in combination to a displacement value of $3\Pi^2$ and these are the two forms of light that forms the Universe. I can't press this issue strong enough because all these pictures prove Newtonian science to be science fiction and proves the Bible as totally scientific facts. Look at these pictures and remember this.

The light value of $3^3 = 27$ is space that builds the Universe which Newtonians can't see and therefore named it as "nothing". It being "nothing" or not it still is light moving away form infinity and towards eternity and therefore moving towards the future as it has been doing since time split infinity from eternity fills a Universe to an overflow. The building of the Universe since the Big Bang where the heat release came about as three dimensions formed $112.79 = \dfrac{7\Pi^6}{10(2 \times 3)}$ as space. This is outer space that formed at a value of 112 and inner space displaces into singularity at half that which concludes when light $((3^3 = 27) +(3\Pi^2 = 29.6))$ combines to form 56.6 and because of that iron $_{55}$ and copper$_{62}$ combines as a displacement of 118 and as the average value $(\Pi \div 2)$ which in this case is 118/ 2 and then forms a displacement limit at Cobalt $_{59}$ and the light forming a total value of 56 halves the space-time value of the displacement of 112. I can give you the other formulas but it would be meaningless if I don't explain the entire process mathematically and that is tedious and involves a lot of mathematical principles applying. When a planet or a star does this it is gravity because then the planet or star becomes the value of all the turning of atoms in the star and the flow of heat from cosmic space or outer apace to the centre iron core forms what we see as gravity.

Einstein knew there was a connection but was never able to figure this connection out and he died a frustrated man feeling he was a failure in his own eyes (not in mine but according to his interpretation) and photos taken of him with his hair standing and all messy was from this era. He knew there was a correlation and it formed a direct link but he could never solve the problem and consequently died as a frustrated individual. I was fortunate to formulate the mathematical equation that showed why and how electricity and gravity is the very same thing and to put electricity on par with gravity is the only step in the correct direction. The main issue is material mist spin in relation to each other and this displacement albeit concentrating or expanding is the fuel that drives everything we think of as or we think form the Universe. The process where comic fuel is being heat applied to create movement and that it is the only fuel in the Universe is a fact of life because it even sustains life. Life drives along by using cosmic fluids. Cosmic liquid is what life uses as fuel to function for a short while. Yes, you eat food and you breathe oxygen and that is so Newtonian I choke when hearing this simplistic childish reasoning. It is because of such simple-mindedness that atheist thrive and prosper and feel kings with no one taking on their stupidity. Even the essence of life depends on the capture of sunlight and removing the sunlight as cosmic heat from our surroundings and supplies that to life where life as carbon with life's ability to manipulate what the cosmos applies, life stores this heat in its carbon fibre to be reused by other life forms so that life can sustain life. In short animals eat grass and we eat animals

In order for human life to live, life has to devour what other life forms left as their heritage and the result of what they did with their time they left behind. Other life leaves their time they lived behind as material with carbon they structured according to what they are while being alive either as plants or as animal meat. The grass grows by accumulating and processing heat from sunlight taking from the air carbon and mixing this recopy to feed animals grazing on the grass. We more intellectual life forms then consume animals that live from grass because with our intellect to sustain we can't consume grass alone. We have not got the time. Time is life...there is just so many heart beats per minute and we have just that many heartbeats that forms a life-time and that translates to time equalling life. Grass won't feed intellectual life and in essence the heat that I am referring to is actually time which is the substance or the result of what time deposited as time moved on. That is why the moon and earth is moving apart. It is not moving apart it is time leaving space behind as a result of time being in the universe at that point. The space that time left behind as space we incorporate as a product when plants assembled structures to form fibre and the material we eat is the time of the plant we eat and the time the animals spent on feeding and to use the time that animals spend on feeding is what we eat to stay alive ourselves. We eat lesser-developed forms of life because we with more developed life can't find the time to accumulate enough heat and carbon and still fill our intellectual life. When you eat sunflower you don't eat sunflower. You use the sunflower plant as a dispensary in which cosmic fuel is stored within the carbon of the plant or meat. By having life the body takes in what other had time to accumulate and then to leave behind. This concept fits everything that forms the concept we have of life notwithstanding whatever rule any one wishes to fit to life.

Life which is what you are, because you are not a decomposable body that the Newtonian intellect describes you as, started accumulating material by thought when you were sperm and egg, and after the Unification of sperm and egg you as a form of life and not a sperm floating around started accumulating useful building material. Take away the thought or life that holds material we see as life and whatever remain destructs automatically because the thought sustaining life and the body life occupies will dysfunction and self-destruct. This is how life as the energy captured in a thought functions...it accumulates carbon to use as building material notwithstanding the form or dimension we think should befit that life form. The instant life leaves the sperm or the body the cosmic material start a decomposing or destructive process that instant. Having life present brings about building material but losing life destroys the fabric that contains or holds life. This is proof that material forms a vehicle by which life moves in this Universe but the body is just a commodity and when life discards the cosmic material it decompose to mere atoms. Life accumulating of material is done by mind controlling the body and I see plants living with a massive intellectual drive.

The plant has the purpose to take carbon from the air then process it by mixing it with heat coming from time or space by the way of frozen heat we call sunlight and build a body we call plants. Then we have life that has the purpose to accumulate carbon from the plants that took carbon from air and use that carbon in plants to fabricate carbon that forms animals. This ability to form the living structure albeit plants or animal is vested in a thought within singularity and singularity is only big enough to hold one thought. Then we humans take animals that accumulated carbon because that is all the animals do all day long and we remove life and use the structure that the previous occupying life-form built in order to build our own body. We use what the animal or plant left behind as time left to form and fill space and apply that then to sustain our life by giving life building material to replace and replenish carbon used in our own body. By devouring what life left behind can we live on what other life left as a result of time spent to form space. Don't allow the atheistic senseless stupidity tell you different. Life is the only energy that can leave the Universe because it is only what life left behind as space filled with carbon that remains as atoms. Life that was never part of this Universe can leave the Universe and leave behind what belongs to the Universe. If your body was what is in charge which is you, then when you are dead someone with life can pump some oxygen into you and shock you with electricity until you bounce around like a ping-pong ball and you will begin life again. Then giving you oxygen and electricity for the brain to function has to replenish the life you had in your body except if life is no more and has gone somewhere beyond this Universe into singularity. That is total rubbish. If life leaves the body there is no structural formation left to control and maintain the structural I integrity of the body. When they shock your heart with electricity it is to get blood flowing and not to put electricity back into your brain. Only blood can do that and to have blood flowing you need life to perform the vital flowing. If you blood stops flowing then you are dead.

Take a close look at a cadaver. It is not something to be scared of because it is a body NOT containing life, as anyone of us will be someday. So it is the same as you being scared of you as you are going to be somewhere in the future and that is pretty silly. On the condition that you were born the only thing you will be someday is dead. If you are alive then you will face death and your only human rite you will ever have is to die and not to vote as politicians shout during an election. What we have to answer is what is the difference between this cadaver in the mortuary and me. One is that the cadaver has no life and I show vital signs filling me with life. The cadaver can't move by it self and I can move by myself. Even if I only have the ability to have my blood to flow it is movement, which is what the cadaver lost. The biggest factor is movement and that movement is linked to thought. Considering the implication of this is vital if you wish to enhance your physical strength and build your body. There are persons in hospital in a coma for years and they apparently show no thought because their muscles don't move and therefore they wither away. The thought gives control over the body and the thought form the muscle and the thought form the size of the muscle. The cadaver or dead person can't get up and walk as I can.

Why can't this dead body get up and walk, it is because the dead has no thoughts. If you think the Newtonian idea is correct that life is part of the body then rethink. I dare you to conduct some tests. If life is electricity as they say it is, then why can you shock that cadaver until it hums like an electric transformer and life will not return? If life is as they say it is electric convulsions then try and shock the brain with electricity and you will find no response. The fact that you can manipulate muscle spasm with electric convulsion shows that life controls the brain by charging electricity and that process is done by thought in life. Life generates electricity that life then implements to control the body life extends for the purpose of serving life. Life is in charge of the body and of thought and not the body being in charge of life. By electrocuting a body with life you merely short circuit life's actions with a stronger jolt of electricity but the electricity is just a modem through which life controls muscles and growth in the body. Then you burn the electricity conducting connection that life has as life controls the functions of the body and do that long enough and life may not find a manner to form conduction of electricity whereby the organ control will become suspended Let us look at the definition of energy. Energy is, as I understand it, indestructible, which means it cannot be destroyed. Energy can only be transferred from one form to another form. At present when energy is measured in work science apparently does not take into account energy losses brought about by anger; fighting and frustration brought about by tribulation but these are as much energy draining as everything else. These are also energy losses and not only work done in the sense of the body being a machine fuelled like all machines. I will have to declare that Newton's statement of energy and work being the same thing is utter nonsense. This is the simplest example we teach children in school and in that they with their utter simplistic views get away with intellectual murder.

Newtonian gossip they call science puts all of life into the realms of the body. Ask what is life and the Newtonian has a lot to offer but put what they offer from an atheist stance and it all falls apart in a sorry way. When does a person die? A person dies when his blood parts the red particles from the white particles where the red particle becomes solid and the white particles become fluid. When being alive the blood flows and that makes the solids interact with the liquids and doing that complies with the Coanda effect on gravity. This meant nothing in thousands of years to the medical paternity but then the Coanda effect still doesn't mean anything to science because science understands nothing about the Coanda effect. Go and try to find an explanation about the Coanda effect (except in my books) and see how far you may get. A person dies when the blood flow stops and the blood stalls because then the Coanda effect using gravitational movement no longer applies as the blood as a liquid does not interact by depositing cosmic heat into the arteries carrying the blood. You don't use oxygen being alive you use the heat oxygen associates with being very volatile and the 6 of carbon with the 8 of oxygen releases the heat to the 7 of nitrogen and then the oxygen going without heat unites with the 6 of carbon to form a unit compound that breathing dispose of. However the collective sense in the maintaining of life sin the supply of cosmic heat to the fibre structure forming the body that life formed.

This process form aging and God knows how Newtonian atheists try to convince the public with a less than honest suggestion that science in the Newtonian manner could in the future keep a person alive for a possible millennium. Eating puts not only cosmic heat into the blood but it supplies carbon fabric to life but also it supplies life-carrying-carbon by which life can replace cosmic tissue. This is one part of life' aging process. Then the next part is the disposing of the replaced carbon 6. Life sustains the body by removing carbon and disposing of it by mixing it with oxygen. The removed carbon that previously formed your body structure as tissue is disposed of through breathing and uniting the removed carbon with two oxygen particles that lost its heat supplement the oxygen as a gas was associated with is the method of doing. In God's creation God never intended man to live forever but to forever become wiser by accumulating knowledge. That is the purpose of life. You learn to live to live to learn and dying is the result of living. By eating you supply carbon with heat as previous life and by breathing you supply heat that burns away old carbon that is removed by paring it off with two part of oxygen for every part in carbon. This is why life ages and this breathing destroy life. I wish to see how they are going to end this process because it is breathing oxygen that supplies heat that allow movement to form life but also aging.

The carbon that replaces the carbon by which the body is renewed and sustains the body in healing and growth is done in a process whereby breathing is inhaling oxygen and exhaling is carbon removed from tissue and then united with two oxygen particles. The oxygen that delivered cosmic heat also removes the used carbon life deposit into the flowing blood. The blood flow is the process of bringing cosmic heat to the body to enable life to apply movement. Life is capable of functioning by movement of a body and movement within a body and any loss of a delivery of heat to the body brings instantaneous death. When oxygen deposited the heat life has to use to move it thereby also remove the used carbon life replaces. Every seven years every atom is replaced in the body. By removing the carbon life rebuilds the body but as gravity changes with the expanding Universe so does the size of the fibre or atoms with which life has the ability of replacing that life has changes to maintain the body and also the fibre size of the atoms that life uses. The body never stops growing and that is why wrinkles form but the body degenerates according to atoms becoming bigger and therefore fewer atoms fit into the original structure. The body outgrows its initial size and thereby collapses under its own growth. It has to perform this way while being part of a body within the growing Universe. This is aging and no medicine can ever stop this process.

This is extremely important to realise that from the first second where life accumulates cosmic material to form life this process of forming life collects tissue that will become a human body. It is not like your halfwit Newtonian professor believes that the human body represents life. From the first moment it is life that forms the body by using cosmic fluid to form movement and it is not the body that forms life. Therefore you with your life forms your human body and it is not your human body that takes the responsibility for life. You are going to age and you are going to die and the only reality in life is death that follows. What is important is not that you have life but it is what you do with life that gives life a purpose.

It is by the thought process that life collects material to form the human body and the human body does not collect life as it goes along. Doing so the ability to do so is the supply of cosmic fluid that is the remaining form that time left behind as space hence space-time or as Kepler put is $a^3 = T^2k$. Everyone in modern science think it is the brain that controls the human body but they are so completely wrong. You use your thought process to control your body and in this thought process you form your body to be as strong as you wish it to be. Even in the very beginning life formed the sperm and the sperm did not represent life because it was life that made the sperm wiggle and if the sperm did not wiggle the sperm was dead or then life-less. You can't have a tube filled with sperm and when you find the lot are dead you then are able to revive the sperm with an electric jolt. You can't have a jar filled with D.N.A and by applying electricity this will regroup the composition and then you build the body of the person once more. You build your body through thinking with your mind. You construct your body cell by cell by using your mind to do so. In the process you take in air that gives you cosmic heat allows you the ability to perform this task. The main issue of life is life needs cosmic fluid to perform as life or welter into death.

How can I prove this? The instant your life vacates the body the body's ability to restructure is gone and the ability to again perform the structuring leaves that very second. The moment life vacates the body, the body degenerates by fragmenting the structure until the entire construction disassembles into forming atoms again. It is life that keeps the body into form and without life the body de-fragments into atoms once more. Your body does not hold life but your mind by thought controls your body and that allows your body life. Your body and life does not run on food and oxygen but on cosmic fluid carried by carbon and oxygen. Life applies gravity to control the body by implementing the Coanda effect, which allows thinking. In the Universe there is only one fuel and that is cosmic liquid. When anything moves it is because heat is traded for movement. When coal and fossil fuel is burnt the process depends on the release of solar energy stored in the fossil fuel for millions of years. The sun compressed light and the compressing was more than what the sun in its atoms could absorb. By not using all the heat condensed some of the heat escapes into outer space because the space in which the heat is can't manage the heat since the space got cold. A star is not a coal stove but by pumping hydrogen and other gasses on the outside is much more an air-conditioned than a boiler burning coal. When fire burns oxygen it is not the element of oxygen that burns but it is the association it has with heat making the density into a gas that burns when oxygen releases the heat it carries as a gas. If fires did burn oxygen there would be no more oxygen left to burn. The sun does not "burn" hydrogen but pumps hydrogen and by pumping a gas such as hydrogen this action removes heat from the interior to the outside. However it is sunlight that forms the fuel part in fossil fuel as the heat of the sun was trapped in the fossilised remains of carbon life that goes back billions of years.

There are three forms of adaptation forming a relation between cosmic substances where there are two substances, one being material and the other being non-materials. When it has no space in-between atoms we think of it as a solid and to be a solid the material must be cold, which means it, lacks heat. When the mixture of heat or non-materials and materials rise we think of the substance forming a liquid. The non-material heat is then more represented in the mixture than was the heat present when it was a solid. When it forms a gas there are much more heat present in the space than there are solid atoms and the density is very low because the non-material substance is overwhelming more. .

Everything in the Universe moves and to be within the Universe forming part of any idea within the Universe there has to be movement going straight and movement going in a circle. That is where we find the essence of Creation as the cosmos informed Kepler mathematically. To understand the cosmos we have to understand why 1 + 1 is 2 and the cosmos started with 1 growing to 2 and therefore becoming more than what was before. We therefore must know that the following value must be 1 + 1 + 1 is 3, but why would it be 3. We have to know why 2 + 2 is 4 and moreover why is 2 x 2 = 4. We have to know why 2 + 2 = 4 + 1 = 5. Understanding this that 2 x 2 = 4 and 2 + 2 = 4 and 4 + 1 is 5 is the most basic but also the most important aspect of creation. Not knowing why 2 x 2 x 2 = 8 while 2 + 2 + 2 = 6 indicates a total lack of understanding the dynamics why the cosmos is what it is in all the dimensions it holds.

One must realise that time forms space as "space" forms the history of time left as light in "space" and that the "space" between the earth and the moon is not "space" but it is what time left behind as space. To know the age of the earth and the moon one must take the expanding that happens every year putting "space" between the earth and the moon and put that in relevance to by what time leaves as "space" and from that find the true age of the earth. As the earth "grows" by becoming "bigger" so the moon and earth forms distance by "space" according with time moving. This movement "away from" and "going bigger" is the true and only measure of how much time developed the Universe and that we see as "outer space"

In perspective to singularity relating to the centre of the earth it is the earth that stands still because the connection from the centre to any point on the surface never moves as the line that forms a connection between singularity and the surface remains still. That means all movement is in the liquid aspect. This puts everything that changes as a part of liquid movement although it forms a "mass" connection. Although the line ends at the surface the line running from the centre of the sun is connecting the relevance, which extends to more than a third of the distance going all the way to the next stars, which are Alfa and Beta Century. This line connects every planet to the sun and that is why the Titius Bode law positions all planets. However from the perspective of the sun it is Alfa and Beta Century that moves at a rate the sun can't actually cope with while the sun is standing dead still. The moon by moving in a twenty-four cycle (according to the earth centre the earth is holding still and the moon is moving in a double shift) as well as forming a cyclic connection of $7° \times 4\Pi^0$ it also rotates once every (about) twenty eight days but it is the moon that moves on both accounts and the earth is dead still.

The sun reduces space to a liquid and that the pictures show with liquid and not gas squirting from the star. Instead of looking at what they see they look at the 6500° C and according to that scale declare the sun as hot. Hydrogen is a liquid on the surface of the sun and as it comes into contact with outer space the friction caused by the movement makes the hydrogen or some of it overheats again where the hydrogen turns from liquid to gas. The rest of the material that squirts into space does not overheat and returns to the sun as a liquid and as cold as a liquid. The sun moves extremely fast in comparison to all other planets and by moving so fast it freezes the hydrogen in outer space from forming a gas in outer space to becoming liquid within the sun. Gravity is about movement freezing space and we better forget we feel heat and start to think as the cosmos operates. The cosmos is not human and holds no human concepts. Kepler shows in the tables that space a^3 reduces k^{-1} as the sun spins T^2. When space being three-dimensional divides into movement, which is square, space, declines or reduces indicating the relevance (k) goes negative or the distance becomes smaller k^{-1}. This is a

Mercury	$T^2 \div a^3 =$	0.983
Venus	$T^2 \div a^3 =$	0.992
Earth	$T^2 \div a^3 =$	1.000
Mars	$T^2 \div a^3 =$	1.000
Jupiter	$T^2 \div a^3 =$	1.000
Saturn	$T^2 \div a^3 =$	0.999
Uranus	$T^2 \div a^3 =$	1.000
Neptune	$T^2 \div a^3 =$	0.999
Pluto	$T^2 \div a^3 =$	1.004

mathematical statement showing physics reality that not even Newton can break because this is physics and not that three dimensions a^3 is equal to two dimensions T^2 as Newton stated by declaring $a^3 = T^2$. Anyone stating this as accurate has no mathematical sense or has no inclination about sensible physics even if the idiots name is Isaac Newton! The Table I show is Kepler's finding and that shows who is correct, I or science that upholds Newton's views that $a^3 = T^2$.

Individual movement of material occupying specific space is forming density in relevance to all other material moving at various but specific speeds and the faster any atom or material moves, the denser form the movement will make the material to be. Seeing relevancies apply in the picture above it is not the mass that increases but it is the density of the material within the star that increase by claiming less space to hold more material in an denser environment. As cosmic gas or also known as outer space expands it moves slower while the density decreases. The increase of the density of stars reducing space while becoming denser with more material in less space comes about by more material within less space spinning faster because of reduced space bringing about faster circling of material within a smaller confined space. In contras outer space again is moving slower because the space increases through expanding and more space moves slower. This puts the applying relevance on material to move faster in relation to outer space moving slower and thus

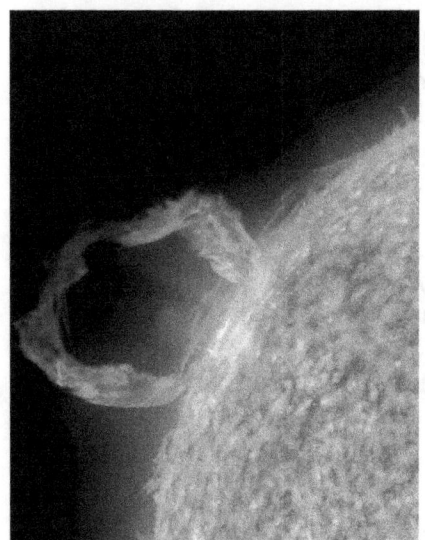

material becomes denser as it moves faster while it is in ratio with outer space expanding and thus moving slower. This ratio allows material to move faster and then contract more space in the form of heat from outer space, which is filled with non-material heat. As material compact it absorbs heat from outer space that loses density. That secures material growth and by reducing density secures outer space expanding. The star stays the same as outer space expands and that makes that the space the star claims to occupy remained as it was when the Universe in outer space began to expand in terms of the star contracting.

The prominence squirting from the sun can only be liquid heating up as it touches the much more hot cosmic gas. The fact that it rises can only be because it is heating up or getting hotter. If the sun had "pressure" it will release that "pressure" in a cloud of gaseous steam and the sun would go supernova in an instant. The sun spins and this movement contracts the space that by spin or gravity becomes a liquid air that came in as a gas then turns to heat being in a liquid state because of the density increase changing the sun's inner space or atmosphere on the inside from a gas to a fluid that surrounds all the atomic solid particles. Our Earth's atmosphere has all the characteristics of a liquid and is then as such also a liquid and that makes the sun's atmosphere so much denser and therefore so much a denser liquid. At one very specific point gravity compresses outer space from a gas into a liquid and heat in that space then becomes liquid. In the photo's we see the heat returns back to space as it cuts through the sun's curving surface wall bringing about "an explosion".

The main issue to realise is that the pumping produces a density increase and the density increase turns the inside from gas to liquid. It is not the oxygen or the hydrogen or whatever fills the container that is a gas or a liquid but it is the amount of space that turns to liquid heat that turns the container from a "planet" into a "star". Even the earth has already some flimsy liquid atmospheres in comparing to outer space. This is the only difference between planets and stars if you insist on having planets and having stars. The sun has no pressure but the excessive movement freezes the gas in the sun into a liquid because the idea of expressing values in terms of temperature is a Newtonian made custom.

When we go in search of what principles applies to form the building material in the Universe we better look and see what is it that the Universe shows us most graphic and we better stop telling the Universe what it is that we want to see and what the Universe should offer us that we wish to see. We better stop telling the Universe it must get mass and start to see what the Universe tells us what it has to offer us to see. If stars burst by releasing heat then stars are constructions that confine heat or cram heat into a small space. If this is true then gravity must be the process of freezing heat by turning movement and displacing space into compacted heat making gravity a process whereby space freezes as it condenses.

Whatever the Universe is it is made of it is made of heat forming light that causes heat density. When I see a star burst open I should take note of what the star releases and look for the principles applying that should form such a release of heat. Newtonians are forever copying Newton's style by telling the Universe it holds the planets in formational alignment because of their mass while not size nor invented "mass" plays any part in the process. We must stop playing God and create a non-existing Universe and begin to confirm what there is. The biggest concept of being a Newtonian is to tell the Universe what it is instead of looking what the cosmos says told what the cosmos is.

When a star burst open it releases massive amounts of heat into outer space. If it exposes heat bursting out then the reason that would apply is it must be because it froze heat into a state of solidity. If the star bursts as it explodes by releasing heat it then clearly overheats. The question never asked is why would stars overheat? We can blame pressure, but pressure would not bring about a star disintegrating from the centre, as the star depicted here clearly does. A burst from pressure should blow the sides out.

When a Super nova goes bang Newtonians say it is because *"**gravity has gone mad**"* and

then they still see their position as being intellectual. Since when has gravity got emotions that can go array or "mad". Still more off the point is how can that be to their ability the best answer they can get up to while remaining satisfied with the effort! Stars we call Super Nova has blowouts. It can't be a pressure burst because there is no material wall enveloping the heat and thereby Stars we call Super Nova has blowouts. Pressure release comes from when material containing compressed space bursts its limit. That we humans know since before writing began, but since of late this phenomenon becomes more and more seemingly misunderstood. If stars blow as stars should and as we can clearly see from the picture just above, then the explosion happening to the star we know as a Supernova comes about from other principles, surely.

It is very obvious the two occurrences are not a result from the same basic method the Universe uses in destroying stars. When heat surges and becomes too high, it turns into space. That process we call an explosion. It is frequently seen, yet never acknowledged by science. When heat reduces, it relinquishes space in the producing of more concentrated heat, this process we see as cooling.

.

What ever the terms used there must be a recognising of the inter relation between heat and space where the reducing of the one will lead to the increase of the other. The star does not apply pressure to bring about fusion it freezes the elements into fusion by applying millions and even billions of degrees Celsius. It is our conception of hot and cold bringing total confusion about the principles of cosmology.

I do not think that I or any other person is at liberty to try to calculate any on goings with in the star but from what is clear from the outside one may come to some measured idea of the stars position in space –time. Gravity is the cooling of space by duplicating or moving space, albeit filled or not filled. When the star spins too slowly it does not cool sufficiently and then it becomes warmer inside. As it reaches a point where it overheats because it moves to slow it burst and by expanding the space it regulates the temperature. At a point when it can no longer contain the confined heat it expands and such expanding we call a Supernova occurrence. The contraction of space must be equal to what amount of heat the total number of atoms spinning within the star can retain and gravity is that balance. The fact that it can freeze heat to liquid surrounding hydrogen while holding a temperature of 6500^0 C should be an indication it is not what we seem to acknowledge as normal. The sun is freezing hydrogen to a dense liquid at 6500^0 while space is boiling (expanding through overheating according to the Hubble Constant) at -273^0. Science academics have to review there thoughts on relevancies because what seems to be hot is cold under certain circumstances and what seems to be cold to a point of freezing is boiling hot. There are no standard issue and fit all through out the universe. Every singularity attaché different criteria to borders controlling the space-time with in it rule. What fits humans on earth does not even suit conditions on the moon, yet science cannot appreciate that the moon applies very different standards to that of every structure and every structure is a cosmos on its own turf, supplying its own turf.

We must look at nature to find what is hot and what is cold. Something hot is that which can expand no more because that which is hot expands. When something is cold it can contract no more because it reduced space to the utmost limit.

Outer space is the hottest because it eternally expands while the Black hole is infinitely cold as it contracted what it could contract and keeps on contracting. It is not the specifics that are of importance because the specifics change considerably when taking into account that hydrogen remains in a frozen state at 6500^0 C therefore it is obvious we have to look at other clues to give some indication of what is in process. On earth in the time we have as a duration we find hydrogen freezing at 269 ^{0}C as where it freezes on the sun at 6500^0 C, which implicate the reduction of space to an enormous increase in time duration.

In conditions on earth the rotating velocity of the electron is 3×10^5 km / sec. With conditions being that different it can not nearly be the same in the sun. As space reduces time increases. By having the space reduced to such an extent that it matches near Big Bang relevancies (a period where heat flowed like water and which is the very same conditions we find within the sun) the space would apply accordingly. We also know that relevancies is all about conditions showing similarities under variables and therefore the space and heat component may seem altogether incompatible but is almost the same given the singularity presence within the sun and comparing that to the earth.

What is applying to stars inside the galactica centre is applying to particles inside the sun. Science sees the nuclear reaction but do not recognise and therefore do not admit that the nuclear reaction is three different phases. At the beginning of the process all the heat is solid, placed in a container by nature and the container has a human name called the atom nucleus. In the atomic explosion there are three ingredients that are distinctly apart. When the solid melts down, it becomes a fluid. The fluid we gave the name of light. There is not enough space to explain the detail of the argument, but light is not a gas, it is a fluid. The first step of the nuclear explosion is converting the solid to liquid. In the liquid state the star does not overheat. The overheating becomes part of the second phase. That phase involves the turning of the heat-fluid to a heat-gas we call space. Space is heat overheated creating space, as heat is space concentrated creating a fluid or liquid not yet correctly named.

Every one knows that a gas is one dimension HOTTER than Liquid as liquid again is one dimension HOTTER than being solid. If the star is liquid on the inside, and the liquid evaporates when coming into contact with outer space, then outer space is the hottest, notwithstanding what ever boundaries and values we humans' attaché to the dimension. Our human standards have to change to accommodate the rules layer down by the cosmos and not apply the cosmos to suit our rules of hot and cold, big and small, near and far. In the case of the Super Nova, smoothing prevented the liquid turning into gas, therefore overheating. He liquid froze as a liquid becoming a cosmic lollypop. That which prevents the overheating turned the layers into frozen identities not overheating therefore it became a liquid outside the star. This star was turned into a miniature galactica, sustaining billions of individual singularity, because the governing singularity did not destroy, but the singularity of every nature is still in support of one another. From this picture (and others of Super Nova) one can learn a lot, if one is truly interested in applying cosmic law to the picture and not some human response to what we think would apply to an earth-like star that holds gas as an ingredient. Again I have to press the thought that it is singularity determining space-time form through conditions that bring about the state between matter. Matter can be solid liquid or gas, but it is the condition of the space-time derived from singularity that places the form and conditions valuing the form of the elements. Hydrogen can be as much a solid as gold can be a gas. In the next scenario the overheating core is hotter than outer space and that brings about that the heat will flow to a colder region. In this case the star is overheating and with that can no longer protect its individual singularity. The part the official verdict and mine is in agreement on of creation is that it all started small, but I go one step further by saying the Universe was at the same time eternally large.

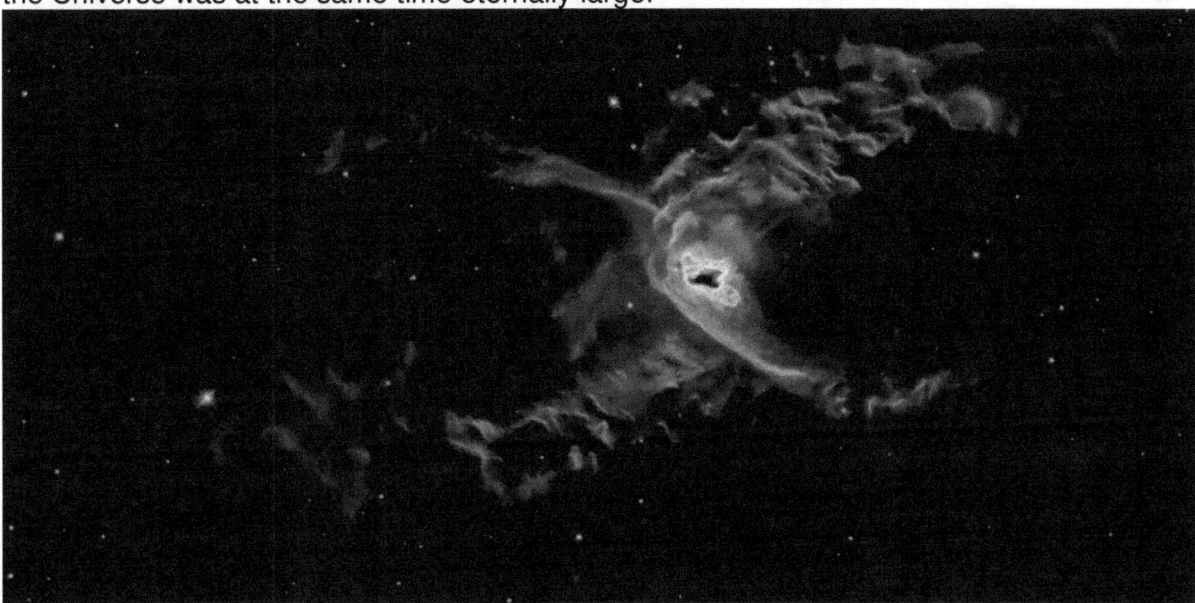

The pictures show clearly the difference of a star NOT overheating being "normal with liquid pouring from it and then becomes a gas as it evaporates. We see the sun exerts heat but still we think of the sun as hot. The sun gets rid of the heat, which means the sun is cold and that is why it removes heat from its surroundings. But because we feel that which the sun rejects we then contribute this heat to the sun by attaching that which the sun removes. It is all about relevancies forming as singularity applying matter in relation to the overheating it started to combat.

It started with singularity producing matter and the matter changed in relevancy to one another by becoming solid or liquid in relation to each other. Space still was at a premium because the space we know and we see as gas, was not yet part of creation. Since the Big Bang the fluid heat is in a process of converting to space enlarging the role of time as the Universe still systematically overheats. The entire purpose of gravity is to combat overheating to allow time growth. From the offset of the first dot dividing as it became the first two dots, it was bringing about the second dot and the eternal number of dots growing from that means that the splitting of the dots assumed as the dots were growing from infinity in size, which is in fact only part of the relevancy because at the same time the infinity presented eternity, where both locked the same value to the dot. This long sentence structure is an effort to explain that everything is linking to another either directly or through other particles and everything came about precisely simultaneously being eternities apart.

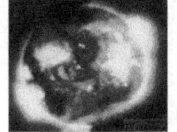

The stars are in relevancy part of the growing cosmos, where the growing cosmos presents a liquid covering all solid strictures. The structures are no more or less particles irrelevant of size, since time places the value and space is dependent on time. But by the continuing process of the eternal overheating, the geodesic cosmos overheat gradually which presents as the Hubble Constant and this process changes time in space. Since every star holds an individual singularity, separating its singularity from the galactica singularity it is within, it remains as a relative liquid while the cosmos changes its side of the relevancy becoming more a gas.

The difference between the star being the dot we can see in the picture on the outer edge and the star being the dot we cannot see on the inside is the time in promoting the individual singularity. First the star in the centre core changes, starting to collect liquid heat, while the outer part remains part of the cosmic structure. As the Hubble Constant grows the star distinguishes its singularity as it protects the singularity from overheating with the cosmic geodesic space-time. The geodesic space-time is also the outer space, but I prefer not to use outer space. At a point the star becomes a separate structure from the liquid cosmos that turned to gas, and then starts using the liquid to promote the generation of matter in a solid state whereby that matter then later turns to space less singularity as space-time completely breaks down forming neutron stars, pulsars and eventually a completely space less point of singularity in the cosmos being an ancient dot once more we think of by using

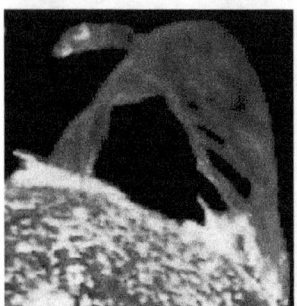

the name a Black Hole. One has to differentiate between heat and overheating because a star represents the coldest space in the Universe and not the hottest space. Heat and cold are relevant dynamics forming in appreciation of singularity. The sun is the coldest place in the solar system and that is fact. Looking at evidence the sun provides contradict everything science wishes to believe about cold and hot. Science wish to see the cosmos through the eyes of what fits the needs sustaining life on earth and what benefits maintaining surroundings in support of life as one find on earth whereas life has no part in the cosmos except for the speck of dust we call earth. Looking at the cosmos impartial to life the evidence supports another view. Every aspect in the cosmos is the very opposite of what science believe it is. The sun is not a ball of gas but a giant sea of liquid, frozen without any form of gas or air in the interior. Having a liquid interior the sun has no pressure but has the very opposite of pressure to which there is yet no name given. The liquid comes from singularity freezing space-time within the atmosphere of the sun, and such is the case with all stars still in the shining phase. Stars more developed than the sun is frozen solid causing fusion. Isaac Newton was an alchemist and was not a puritan as Newtonians make him out to be. Isaac Newton did believe in magic but this now is conveyed into the 21st century.

The Newtonians measure the surface temperature; test the material on the surface of the sun exactly like the Druids did that came just before Newton did during the Dark Ages they decide that by the magic force of gravity this gas "pulls" into a ball. **Space and heat directly relates being the one form of the other**. As material contracts the space it spins in this absorbs heat by gravitational condensing which cools the material and the size of the material increase. However as material absorbs cosmic liquid the density of the non-material decreases because of a loss of non-material substance going into material substance and materials grow in size while space expands because of loss in density. That is why material grows (the earth seems to grow bigger) and outer space expands (outer space expands) as Hubble indicated. It is all about relevancy changing cosmic dynamics every instant in time. This is why the earth becomes larger and the moon goes "further away from the earth".

The density of material increases as the volume size of material grows bigger and as the liquid is concentrated into the spinning object in order to keep it cool by controlling the heat. By removing the liquid from outer space the density of outer space reduces as the heat concentration decreases the relevant space increases by expanding. Material grows by removing heat from outer space to the atom and into singularity as outer space expands by losing the quantity of singularity that concentrates heat. In short it is heat that moves by material concentrating the heat from where the density is less to form more density in materials.

In the picture to the left we find not withstanding whatever name we attach to the red liquid substance flowing from the sun into space and back to the sun, that liquid is heat in a very direct form. If outer space was the coldest place in the solar system the heat should immediately escape to outer space and not return to the sun as it clearly does. If outer space were colder the heat would not return to the sun. All elements forming matter in as much as the heat forming an atom is as much a liquid as it is a gas and a solid. There is no hot as there is no cold. It's about storing energy in space or in heat, which is another cosmic equal being opposing similarities.

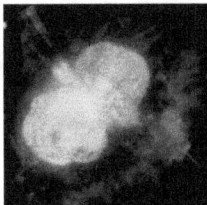

Hot and cold are **relevancies brought about by singularity valuating space-time** and during **the Big Bang** the Universe was **freezing cold** at **three billion degrees C**. It is the relation matter has with heat that provides the form the particle has at that moment. The increasing or decreasing the heat will alter the form of the element. Therefore all elements forming **matter is as much a liquid or not than it is a solid or a gas**. **It is the space surrounding the atom which provides the form the atom find its relativity to the rest of the atoms it share space with**. **Hydrogen is as much a solid as tungsten is a gas depending on the heat in relation to the space matter is within.** Should **you reply** that it is **the gravity pulling the heat back to the sun**, then that **confirms** my theory that **gravity is all about collecting heat onto matter** with outer space being the hottest place. **It is the concentration of heat in space being relevant to form. When overheating a star turns its liquid to gas whereby it merely transforms its interior to a relevancy it has**

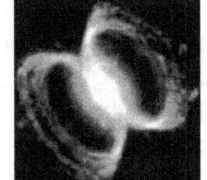

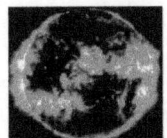

from pre- to post- Big Bang. We humans on earth think that hydrogen is a liquid at -259^0 C but that only apply to the earth. The picture clearly shows the **heat in a liquid** flowing **from the sun** and **back to the sun**. In the **sun the hydrogen holds enormous quantities of heat in a liquid at a temperature of 6500^0 C.** When a star has its singularity secured the star is bitterly cold because it has heat in a liquid form flowing back to the point of singularity although we may regard the star to be rather on the hot side. The sun (fore instance) freeze hydrogen in a liquid form at 6500^0 C.

If hydrogen remains a liquid at 6500^0 C, just think how cold it must be as the star's interior approaches the point of singularity. Therefore fusing protons comes from cold and not from heat or pressure.   By allowing the singularity to overheat the star overheats and heat within the star flows from singularity to outer space freely. In such an event outer space is then colder than the star because the heat releases to outer space with no intention of returning whereas in the sun it returns as soon as it leaves. There are two ways to reduce heat; one is to bring about expanding space, as the photographs clearly show. The second one is where heat will reduce when in motion by spin. When withholding or retarding motion matter will overheat. Gravity is the motion of unoccupied space through the dimensional transformation to occupied space. Motion and space therefore is the anti-, the opposite the negative to heat being the positive. With singularity overheating the expansion of the singularity drives heat into space, creating space to compensate overheating

That is a natural phenomenon. The only reason why **heat will** rather **flow back** to the star than **escape to outer** space once the star released it into outer space is **if outer space**

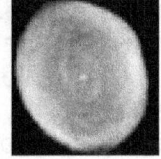

presents more heat than does the star, because **heat always flows from hot to cold** no matter what influences may arise. Outer space must hold more heat than does the star but the accumulation of space in relation to heat makes it seem colder bringing expanding of heat to become space. The cosmos is all about **converting space to heat** which we see **as gravity** and **returning heat to space** as a **control mechanism** always **keeping** a very delicate **balance** which we see as **a star shining or being normal.** The purpose of the converting of space to heat is to supply the core where singularity is with heat. **It turns space to heat** sustaining matter but sometimes singularity overheats and then matter converts to heat allowing heat to convert to space. That we call many names amongst others exploding into super nova. Whatever the names used is less important because the process rests on space and heat interacting to form energy. That was what the Big Bang was and the Hubble Constant is all about where matter converts heat to space. I show that space and heat is the very same thing and there is no such a thing as pressure but releasing heat produces space and concentrating heat reduces space with the two interacting on singularity demand setting time to space with time being the spin or motion of heat in space. Heat and space form the second singularity caused by the fragmenting of singularity to compensate overheating during the pre- Big Bang matter forming era. That is what we see as light and space, which again is the same thing and is fragmented singularity forming radiation and heat, where the star re-transfers heat back to space due to an overload.

Since the first instant that time began to convert movement into space the cosmos grew away by forming space and not towards points to destroy space. Material grows and non-material expands. Planets never moved closer, are not moving closer and will never move closer to each other and this is backed by all information collected this past century. **This is how gravity contracts.** In spinning the sphere contracts by measure of 21.991 reducing to 3.1416 while 7 is reducing to form singularity, but also gravity forms when the 7 comes from the past to the present 7 and onto the future 7 and this became 21. Not only that but with singularity advancing from infinity to become one it proves that even as we see singularity as one, singularity also is multi dimensional but that ability is beyond our scope we have being in the Universe. The dimensional change that Π undergoes shows that singularity repeat into a new location by the value of to conform to the roundness of Π changing in value as the circle goes singular. By the revolving of a sphere the space surrounding the sphere compresses as seven changes to one every time twenty-one point nine, nine one becomes three point one four one six. This contraction reduces space in outer space and in that the cosmos grows under a process known as the Titius Bode law. The Universe grows according to the Titius Bode law and not mass and stars condense according to the Coanda effect and not by mass. In this we find the Lagrangian points and the Roche limit forming l limits in this relation between that which expands and that which contracts and in this we have the keys by which the Universe formed. Gravity is the contraction of space density taking Π from a value of $\dfrac{21.991}{7} = \Pi$, which is what is in space to the rim of the earth, which is $\Pi = 3.116 \div \Pi°$. This indicates contraction by the earth's change in direction by 7° to alter the relevancy applying from $\dfrac{21.991}{7} = \Pi$ to form $\Pi = 3.116 \div \Pi°$.

The reason why there is something such as gravity is there is a transfer of heat by material movement.

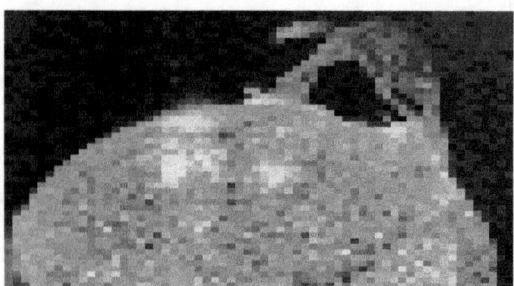

Studying the tables Kepler left us it is very clear that space and not material is moving towards the sun and therefore the sun is contracting space where space is in a process of reducing in volume while the sun is compressing space by a similar margin. Material can't move towards the sun because the Titius Bode law prevents material to come closer to the sun. It is this and three other laws I explain for the first time since the science started. Look very good at every picture of the prominence flowing from inside the sun. What you see is not vapour forming a cloud of mist but a liquid squirting from an even cooler inside. On the inside the movement of the sun freezes the gas forming cosmic gas, which is what is between the earth and the moon to a liquid filling the sun. Again it comes down to enforcing standards our Newtonians apply to life onto the cosmos and the result is stupidity only a Newtonian wise man could be capable of. See the fluid push out of a bowl of liquid, spilling both sides as it falls into liquid. The inside of the sun is not gas but it is fluid. In all of nature there is no NATURAL GAS as much as there is no NATURAL SOLID. Hydrogen is as much a liquid as iron is a gas and neon is a solid. It depends on the element relating to the space/heat in the circumstances surrounding the substance at that very precise instant in time. We have to stop telling the cosmos to show us what we wish to find and start accepting what the cosmos is telling us to find. This shows science must place much less emphasis on life and much more on reality. Life as a comic reality is non-existing and life not withstanding corruption plays no part in the Universe.

Then one final point that would explain the niftiness they craft to protect Newton's fallacies and fables: the formula Einstein introduced being $E = mC^2$ is that that original and did not that great and is even less admirable seeing it is the formula Kepler introduced and which Newton raped because Newton never had the brains to evaluate and understand the formula Kepler gave to the world. It is just the first truthful formula that applied meaningfully and science could work with seeing nobody ever made the effort to investigate Kepler's formula and I would think Einstein did it for the first time. But if he did he never gave credit to where credit is due. The formula $E = mC^2$ should read $E^3 = mC^2$ and because if E is equal to m^1 and C^2 then E must exponentially be $^{1+2=3}$ or then E^3 which in full is $E^3 = mC^2$ and in that it is that is a copy of Kepler's formula $a^3 = kT^2$. Now today it is claimed that this Einstein formula $E = mC^2$ opened the Universe to the understanding of man and all that shouting is a deliberate sidelining of the truth. The truth is that Newton's claims on Kepler's work is a hoax and to divert the attention away from this fact everyone gave

Mercury	$T^2 \div a^3 =$	0.983
Venus	$T^2 \div a^3 =$	0.992
Earth	$T^2 \div a^3 =$	1.000
Mars	$T^2 \div a^3 =$	1.000
Jupiter	$T^2 \div a^3 =$	1.000
Saturn	$T^2 \div a^3 =$	0.999
Uranus	$T^2 \div a^3 =$	1.000
Neptune	$T^2 \div a^3 =$	0.999
Pluto	$T^2 \div a^3 =$	1.004

great credence to Einstein formulating nuclear power as $E = mC^2$ and all the while this formula dates back directly to Kepler and Kepler's formula the cosmos gave him as $a^3 = kT^2$.

Then the only great thing about Einstein's formula is that science managed to cover the truth bout $a^3 = kT^2$ being the same as the Einstein breakthrough of $E = mC^2$ that must be $E^3 = mC^2$ and is just Kepler's $a^3 = kT^2$. It is in small detail that the lie covers the truth by blind sighting people as some manor detail misleads the correctness covering a lie to protect the cover-up they have to protect to keep them in office and get the billions of funding rolling in.

By positioning the point where I prove singularity is I managed to prove many aspects in cosmology that is still unclear or not understood. I prove and formulate the Roche limit as having two factors, both of which play a most dynamic role in the cosmos. I have also managed to prove, formulate and define the Titius Bode principle and in that principle also comes with two factors. The principle is a derogative of the Roche limit and in amongst that there is another principle I have discovered concerning the dynamics as well as the role of light in the universe. The Lagrangian principle also flows directly from the Roche limit as the Roche limit is a ratio in conjunction to the point which I claim singularity is. This is the normal manner in which science presents the solar system and this is as false as the value of money. Look at the way the planets are arranged and then see how it is possible that "mass" can put the position of planets in accordance to the way the planets are arranged. How can science in the light of this evidence still maintain planets are positioned according to "mass"?

To intimidate and discourage people science places formulas that have no purpose at all but to put the fear of God into in place to prove…nothing because they and nobody else can use these meaningless formulas. Now the question is that if the "mass" was pulling the comet closer, what then is pushing the comet further away. Never do I get any answers about these and many more matters science can't explain and yet they teach the children that "mass" pulls "mass" and to that effect there is no evidence proving this statement. There is no way in hell that any person can use the mass of Jupiter to pinpoint its position in the Universe and all these formulas this use is total criminality because the formulas are nothing more than the cover –up to conceal the fact they have no idea why the solar system and indeed the entire Universe is what it is. This book is the first book in the history of man that will explain why the solar system and in fact the Universe form in the manner of the Titius Bode law. Getting away with all the deception they scare people witless with the most impressive mathematics they can think of so please when you see the mathematical formulas I introduce don't let it scare you because they're using of it is only to scare you and that is all.

Gravity and electricity is the precise same principle applying on different dimensions.

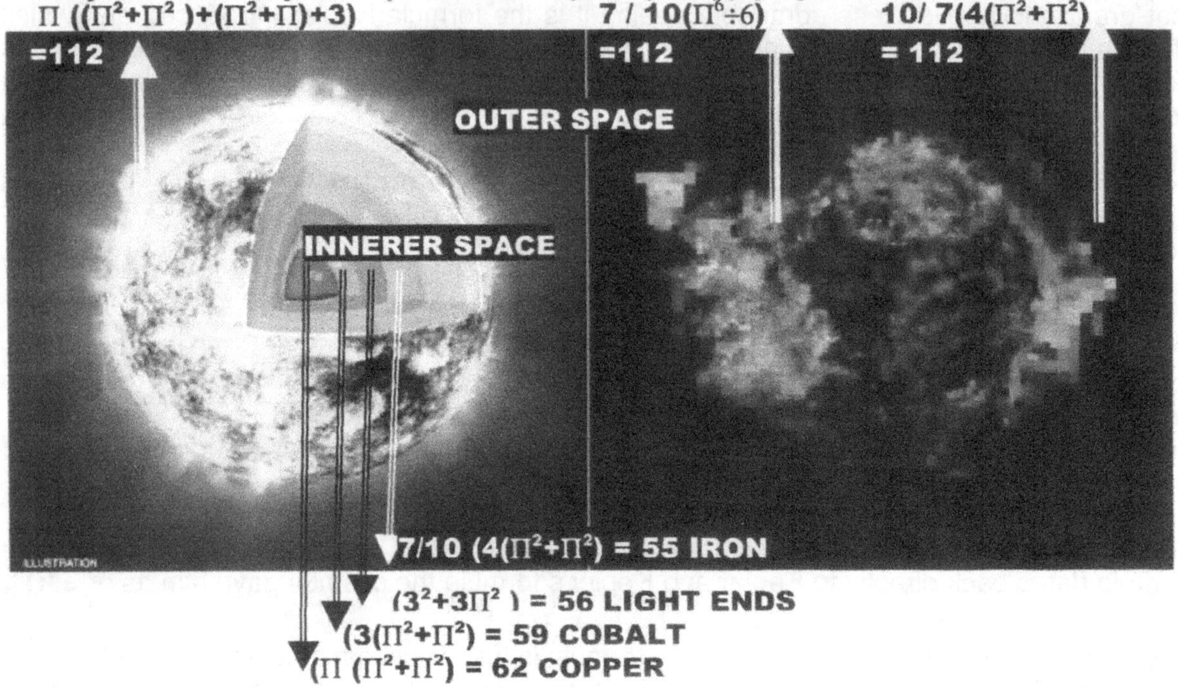

$$\Pi\,((\Pi^2+\Pi^2\,)+(\Pi^2+\Pi)+3)$$

$$7\,/\,10(\Pi^6\div6)$$

$$10/\,7(4(\Pi^2+\Pi^2)$$

=112

=112

= 112

OUTER SPACE

INNERER SPACE

$$7/10\ (4(\Pi^2+\Pi^2) = 55\ \textbf{IRON}$$

$$(3^2+3\Pi^2\,) = 56\ \textbf{LIGHT ENDS}$$

$$(3(\Pi^2+\Pi^2) = 59\ \textbf{COBALT}$$

$$(\Pi\ (\Pi^2+\Pi^2) = 62\ \textbf{COPPER}$$

Outer space forming a sphere is **7 / 10($\Pi^6\div6$)**

Outer space forming cube **10/ 7($4(\Pi^2+\Pi^2)$)**

Space allowing movement limitation is $\Pi((\Pi^2+\Pi^2\,)+(\Pi^2+\Pi)+3)$ = 112

Inner space gravity moves to Iron at 7/**10** ($4(\Pi^2+\Pi^2)$) = 55

Inner space light ends dimensional value at ($3^2+3\Pi^2\,$) = 56 **LIGHT**

Inner space dimensional collapse at **COBALT**

Complete space collapse at inner space dimensional at Copper ($\Pi\ (\Pi^2+\Pi^2)$) = 62

These formulae prove that gravity and electricity is the one and the same thing. It proves that movement condenses space and that there is interaction between gas, liquid and solids.

The formula $(\Pi^2 + \Pi^2)(\Pi^2\Pi)3$ = **1836** proves how much does liquid in terms of density **condenses liquid to solids.** This proves the atom condenses liquid into solids is explained by the formula but moreover this more than anything proves the Bible in Genesis 1 verse I as absolutely correct leaving no room for any doubt. These formulae serve one purpose and that is to prove how gravity apply through the atom and element' structure to ring about condensing of space. It proves beyond anything else that gravity is the condensing of heat into smaller or reduced space. This proves gravity is the displacement of heat throughout the Universe.

Moreover, this proves that what the Universe is made of and whatever forms the Universe is heat going through movement to form density levels. The cosmos is heat and that leaves no doubt. The Universe overheats. To curb the overheating material forming the Universe moves. The material contracts the non-material to regulate the overheating and by the contraction the density of the non- material reduces in favour of the material side gaining.

Part 4

The Where and The Why and What

In opening the last part of the book I wish to address the entire worlds scientist: You are the brainpower of the world. In this letter I make fun of you and with good reason. You may impress one another with mathematics that can burn off the hair on a dog's back. You may investigate the smallest of quantum physics that will stun an elephant to a stand still, but what is the use if children at school find the most basic of science a laughing matter. What is the purpose if it is isolated from nature? Why use it if nature contradicts it. Science can only use nature because it is nature that creates this Universe and not you using Newton. It is the choice scientists have to make. The choice is to go on with senseless science because changing to a new view will offend too many and demolish an institution of collected knowledge, or change and become sensible by using nature.

I do not use the term sensible to self praise because every reader will see the science I apply comes from one that is academically very poorly developed. I do not use great quantum references or have the academic power to involve myself with the enormity of developed research on "the cutting edge" of the twenty first century and I am under no elusion that I impress any one with a vast field if accumulated knowledge. However being totally ignored by you because of poor academic reference does not make you lot big but it uncovers you poor equipped thoughts. The examples I use are basic. The method of my thinking is crude. The science I refer to is every day and almost child-like and I am the first to admit that. I might not agree with your work but you impress me to a point of questioning my personal self worth but that is in vain if nature kick you lot off the table. I do not wish to be anybody than who I am and least of all be foolish enough to think I can impress the greatest minds on earth with my work but then again you cover Newton and Newton is pure mythology. My work throws Newton's mythology out the room and that I have not because I have but because I have not the training you have and my lack of basic training is the reflection of my thinking and that is using the basic that nature uses; I do not wish to present myself in any other way because then I shall not only be ill littered but also on top of that be a fool. I do not have the illusion that I have the ability to even try too fool the BRAINPOWER of our time, which is you but I have nature and you lot tried to fool nature for three hundred years. To you with high academic achievement my work may seem trivial, as I was unceremoniously and in an extremely crude manner told by the editing team of Annalen Der Physics but to me it is a mammoth task. Then again think of it in this way…I was the first one in three hundred years that could add 3 plus 4 and get seven and on that I could solve the Titius Bode law.

Still this did not impress any academic in fifteen years and therefore I accept that my work will not be accepted as science during my lifetime although it is the only science there is and that has a small significance to bear. I do admit that I would enjoy the acceptance because the work I have done might not be of tremendous academic force and vigour, but that is not who I am and is not my capabilities.

What I do feel strongly about is the small-minded view science hold about religion and a person as small as I can see their greatness they think they have comes from their smallness that their views have. See the greatness out there and think for one second that with all the mighty brains there is, no Newtonian can explain what came before $t = 10^{-43}$ seconds the most crucial and no one knows what came before that which we know. Before the Big Bang started is what is the big issue. We may see the event in progress, but what actually started the event? Answer that one first before you spit on the Bible. If you cannot, then at least admit you are less than the Bible because the Bible gives answers man is incapable of understanding.

To the atheist I give the challenge use your vast intellect and calculate the space the earth holds only in the circle it relates to as the earth orbits the sun. Calculate the cubic meters of space between the earth and Venus and between the earth and Mars. Then calculate the ratio the earth occupies holding matter to that space not holding matter. Then calculate the space you hold in cubic meters in relation to the space the earth occupies. Use your great mantel power to see your place, not in the Universe, not even in the Milky Way, not even in the bigger solar system, but only in the area you cover each year while on the joy ride the earth provide, Position your greatness in relation to your immediate surrounding and find your space you have. If that cannot bring you in relation to You creator, the one that created the immeasurable, then see yourself in the space you have and then admit to your greatness, Then see who you scowl.

The way man is taking space in time in killing and destroying other species also in the space of the earth we share time, will have dire consequences. I maybe small comparing to others, the role I play may have little to offer in glamour but I also may (just maybe) have some way enabling me to see what others miss. If that is the case, then I can assure you that WE LEARN TO LIVE TO LIVE TO LEARN, IN MORE WAYS THAN EVER WE IMAGINE AND WE SHALL PAY THE PRICE, THE PRICE TO PAY.

That puts the cosmos on one side of the argument but the only thing that can have movement in gravity and not necessarily as gravity is life. All movement in the cosmos and on earth beside life is gravity. Wind is a product of gravity and earthquakes are a product of gravity and the growth of the earth and therefore the entire Universe is gravity. First of all it is true that gravity or lightning produces life because everyone with a swimming pool will know about the green slime forming of moss and algae in the swimming pool and this proves that gravity and light and electricity closely associates with life. As a farmer I knew that after the first lightning storm weed and unwanted plants would start growing.

The different species on earth all holding life is different because of gravity and the form life takes on is dictated by gravity. Life shows how the Universe formed and to look at life one can see precisely how the earth went through stages forcing life to adapt according to gravity. But as I showed gravity or time or 1 is formed by God's thought and will and that makes God and nature in charge of how the earth and the cosmos developed.

Man tries to change nature because man thinks he can beat nature but man can't. Fighting fungi and fighting germs will only create a super bug the man can't destroy but it will destroy man. Man now develops poisons to destroy life that opposes the creation of wealth but it is putting in place a danger that can wipe out man. It will not wipe out life but if there is something like twenty billion people on earth it becomes a rat-infested pit where these micro creatures can grow and develop. Once the tied turns ands one of these creatures form a chain that can't be defeated by man it will come and defeat man and that day will bring an end to human life.

One question never asked is why so many lines of human development never survived to continue on earth. The get a piece of scull here and declare that a new species and they get a bone there and declare they found a new specie and the some tooth by which another new finding come to light. No one ever bothers to ask why did they not continue to live and what happed to them by which they died out completely. Why would one entire species die out and not leave one specimen just to be followed by the next development. If this demise came from some micro form of life that then went into retreat once it killed off an entire species that micro animal is still around but is dormant until it find the host it has been waiting for all its existence and if it could wipe out the entire specie back then it will be able to do it now. There is a disaster looming that man has not got the brains to realise the hardship coming. Every doctor in research thinks he or she produce evidence that changes the course of man and every paper written wants to bring human life on a new health path and the more the shouting the more the money. It is al about money and no truth and the more glamour the more money there is to be made. Life started with lightning and lightning still introduce micro life. In South Africa after the first lightning storm swimming pools go green with algae growing and as farmer I knew but insects come to life only after the first lightning storm brought rain. There is so much more too life than what science wishes to admit and horse playing lose and fast with is going to bring penalties man can't endure.

The gorilla  the orang-utan and the girl is what they are because of life forming a body and not due to the closeness of the body they form. Let me explain but I realise that this would be unbelievably hard for the Newtonian atheist to understand because the Newtonian atheist understand Newton that makes no sense while the rest that can think and that do not depend on understanding Newton will follow but it is beyond the Newtonian atheist just plainly because the Newtonian atheist is able to understand Newton. If it is not a hoax Newtonians don't get it. As life formed it produced a different body befitting the requirement to suit that specie holding life and life is about the intellect in reasoning that forms a body by inhaling oxygen and exhaling carbon dioxide and using the heat the elements carry or transport. The ability to produce a body is life and not the body holding life that supports life to use space within the Universe. Life is about intellect and not DNA that forms a body.

This picture shows man carrying a stick with which to hunt whereby suggesting this gave man his first hunting edge. According to my view, which as usual is different from everyone else's opinion, man developed intellect and the use of the brain by two things that placed man above and beyond other beings. This first development of man's ability was to raise his arms above his head and that is the very foremost. They can come with spears and rock-splinter-cutting flint-knives and hunting techniques but the positioning of the arms holding the hands outstretched above the head was the first development that put man in a category above and beyond the rest of life. This placed man in the top as the most proficient hunter. That made man the top species. One tiny hand movement allowing better hunting made man cleverer.

All apes can hold an arm square to the shoulder and all apes can take something and hit it on the ground to break it and get to the content within whatever container or shell. All apes can crack the shell of a fruit or nut by slamming a rock on it and get to the inside of the edible nut but only man can raise his arm above his head and throw a stone accurately at an object. The ability to aim with a stone in hand above your head and throw instead of it being square with the shoulder to crack a nut brought about humanity. Ape became man with the ability to throw the stone at something in order to kill it. That action made man a hunter like no other before. This meant man could throw a stone and injure an animal at a distance, which then provided an advantage above running fast or being strong enough the break its neck or to kill while the animal ran. Man could stand a few meters adrift and throw a stone to injure and then even effortlessly walk to the prey and kill. That advantage gave man a jump on every other hunter. Today every human man can throw something and throw rather accurate to become (with some practise) deadly accurate. No other species can achieve this and through that ability man got wise. Lifting his arm from square to his shoulders to above his head in a throwing action as in raising the arm another 90° is what brought intellect and civilisation to humanity. It was only moving the arm a little further that enabled man to hunt perfectly and more deadly.

The difference that made man an intellectual hunter and not merely a strong-armed killer was lifting the same stone he used for cracking nuts a little higher and throw the stone while directing the flight thereof to a target that was a distance away from the hunter. A big issue in this development is that all men can throw and I have never seen woman throw naturally. Women shout and squeal naturally to call men but men always pick something up to throw. The ability to throw kept man opportunistic vigilant in hunting and man had to walk upright as to allow his hands to be free for any action at any second. To kill, man had become upright to be most opportunistic in order to have the advantage to survive and feed the best. So those that survived the best were those that walked upright the longest to have arms freed.

This where the arm comes from a chopping position to a throwing position is where the road to civilisation begins but this is still far from presenting a human from being ape. It began by lifting the arm. A lot of other inputs contributed. In that idea man formed intellect and civil order and left the others as ape. Throwing gave man more energy and time to think about ideas and less time spent running and hunting. By incapacitating animals before killing reduced man's dependence on body strength and encourage cunning in order to kill more swiftly with better weaponry. But hunting in a tribe will produce better results and also give much more leisure time, which brought along a closer social order. There was more time to interact that produced intellect. Nights still had to be profusely dangerous but with more idle time man then had the ability and energy using leisure time to research many things such as other tools with which to improve life and living conditions. Just think how much boredom it took and what idle time was available to allow a person to rub one stick against another stick until it smoked and burned to start fire for the first time. Reposition one arm brought a new stance that found intellectualism.

In many thousands of years man then formed better development of improving a throwing arm with better throwing techniques that produced better tools with which to throw and to kill quicker and devour more food. The next step was fire. Those that had the ability to roast food had the advantage of quicker and more sufficient digesting that also allowed those to have more energy and more strength to hunt even better. This is common knowledge but sitting with full bellies around a fire and many persons all trying to convince others who are top dog made the stronger seem weak in stupidity which then had to allow the more intellectuals to entertain the tribe with more ideas and thereby gain most respect within the tribe. The ones that developed verbally better earned more attention and demanded more respect. The respect pushed aside the fear for stronger and gave way for the clever to step to the forefront. This was all the result of sitting at a fire and planning hunting strategy and the intellectual had the advantage to convince others of planning better hunting and applying more cunning.

This I take from facts that still have a place of prominence even in modern society. The silver tongue devils can cheat more but hold better respect because they can talk faster and so they are more convincing although they are better at cheating. It is no longer the ape with muscles that leads society but it's using intelligence and convincing others to follow his lead. Those with better planning get more followers and the one with the plan gets people to listen. The leaders are the intellectuals and this is part of all cultures in humans but only in humans. The safety having nighttime fires gave people chance to think and to exchange ideas without being scared and chased into hiding in darkness at night. Sitting beside a fire in conversation brought about much more emotions of understanding and enduring and companionship and from this I think family-bonding ties became more appropriate than merely ritual mating. It gave special bonding and connecting in better ways than other primates do even to this day.

Then the second function that developed man was the ability to recognise stars and to philosophise what stars were and to give stars a meaning and a purpose exceeding the ordinary use of either eating it or using things as tools. It made man think of what could be outside the need of man to just survive but to name stars. It gave man reason to think about the cosmos and about what is greater than man and what could control the destiny of man. This made man human and set man intellectually on a road where man developed spiritually as well. All this came from having, more time in hand which came from throwing with the arm.

They sat at the fire at night and saw objects they could not reach or touch and yet it clearly drew their attention. Maybe in the very beginning they thought of stars as eyes looking down on them and this made them identify with what they saw up in the sky. The wise found new avenues to speculate and could convince others about powers outside man's mode of reality, much as the press does in the modern age with UFO which they connect to aliens visiting. The fact that man gave prominence to something that was outside his reach and he had no use for made this aspect about man just as unique as the ability to hunt better. This might be speculating but it would explain many things that man's behaviour is different than we find in any other animal on earth, even those we crudely associate with. This is where my speculating ends and this is where pure factual evidence and proven arguments starts again.

We investigate the role of physics in life and how life by exerting thought controls the body. I prove gravity forms thought. Thought controls the brain but I also prove that thought, as a factor is not in space. The body is in space according to life while life is in thought, which is in time. This is an enormously complex issue but is so simple to understand the idea is child-like simple. The entirety of the Universe is in two aspects where one is time and the other space.

Like most other things Newtonians got this wrong. Let's get rid of some misconceptions. The Universe does not expand because whereto can it expand if it is everything that can ever be. The Universe can't grow to become more because it is everything that ever was, that can be and that ever will be and therefore this mythical misconception that the Universe expands is more Newtonian mythology wrapped in a coating of unbelievable stupidity. I am going to show what expands and that which expands is growth outside the Universe that we know. We investigate the role of physics in life and how life by exerting thought controls the body.

In time lapse we are able to see many moons although we know there is only one but the significance behind this serves as the only indicator we may use to understand physics. This picture answers all our questions about the Universe and life within the Universe. There is a make believe to pretend it is the Universe that we see and that Universe we see and know but that is a Universe that doesn't exist but for our visible light focus. The Big Bang brought light and that was when God Almighty rewrote the cosmos in dimensions using light as ink and space as paper. However, don't kid yourself; that is not the real Universe. By giving us

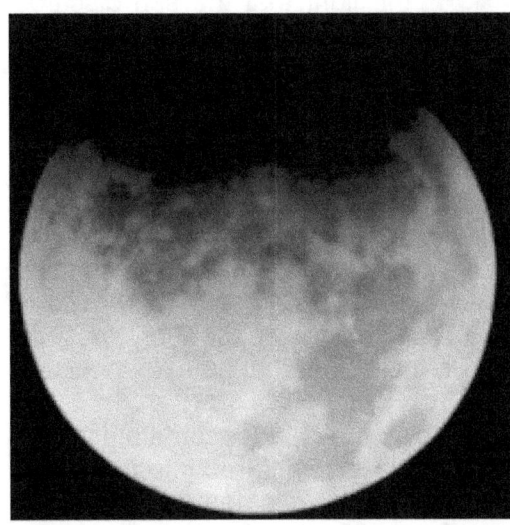

less perspective on reality it is what we wish to imagine. There is one Universe that really exists and that Universe is in singularity. The rest is just space that does not exist but in our vision. We think we see many things forming the most beautiful cosmos but all we see is light painting a false picture of a hologram that doesn't exist. What is out there is dark and invisible.

Allow me to explain that because this is most vital and any concept we must have must use this as a basis reference or we will never get to grips with understanding God's Creation.

We all look at night sky and see the moon. We see the moon rise and we see the moon set. However is it the moon that we see or are we allowing our imagination to run wild? We can't see the moon because there can't be a moon that we can see. The moon is pitch black void of all colours because it is void of all light. The moon has no colour by which it can present an image to leave us with a view. We see light bouncing from

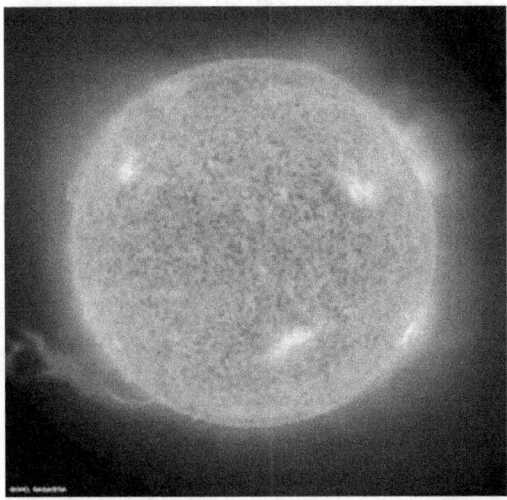

what we believe is the moon and we use that light to form an image that by culture alone we recognise as a moon. It is light we see rushing from a part of space bringing as an image we then interpret as something our teachers and parents told us we all call the moon. The moon is totally invisible and therefore the moon we will never see.

...But there is plenty more... It takes light about one and half seconds in time to cross the space between the moon and me to reach me on earth. The moon is in space one and half seconds into my past and the moon I see was the moon one and half seconds previously but by the time I see it, the moon is in another space. I don't see the moon.

With that in mind we can only see the sun. The rest we see is a picture painted in light on a canvass of space. That is what I said the one part of the Universe is. So all we can see is the sun...and that too is a myth. We can't see the sun. We see light in photon rays forming heat that the sun discards and we see the photon rays carrying an imager we also by culture associate with what we believe is the sun. It takes light about eight and half seconds in time to cross the space and reach me on earth. Again I do not see the sun but the way the sun was in the past using light that is not the sun. But even more is that I can't even see the sun as it was in the past because I see light.

The sun by rotation freezes heat we call outer space by gravity onto its surface, which by

concentrating so much heat in so little space it forms light. That frozen heat the sun then reflects and it discards what it does not use in the form of a concentrated white light. The sun froze dark expanded light into a concentration of white visible light and then discards this visible light back into the dark light, which absorbs the white light gradually. We see the light that the sun discards which is not the sun but which is an optic illusion with which we see. There are two forms by which the Universe forms and we use the one that is light-illusion to avoid reality. We don't see the sun but we see light it discards and again our culture steps in to make us believe we see what we think we see. We think of what we see in terms of the immediate while in cosmology space is time's history. Where space is light forming space and if time crosses space then space in light is part of the history of time, which is what space is. Space is light that time left behind.

The only thing forming the Universe is time moving along and space is the records of time as time went on and left space to keep record of where time went as it ventured into the future while leaving the past behind as space. However make no errors there is no such a thing as space except the light that forms a hologram of what time formed as time kept moving into a future placing the present in terms with the past. Space is the past that is written in light on space in three dimensions but has no other value.

Can I see another person? When I look at someone a few meters from me I can't see the person although all my senses tell me I see the person. I see the person with light again that bounced off the body I think of in terms of the person. The image the person conveys is in time phase behind what I experience because it took time to pass from the point the light bounced from the person to where my eye is. That puts the place the person holds both into my future and part of my past. The person that I am looking at is at a point that is a part of the greater Universe of which I am part of because the entirety forms one big future that will arrive the very next instant. The person forming the past is the light that connected on the person and from there took time placing that to the past because that happened in the past of the point in time I am now experiencing. Therefore whatever I think I see is not what I see but it is the past I see in the present in which I am. This might sound like splitting hair but we have to realise that there is time being in the here and the now and then there is space that forms a past and the past is not part of the present which is in the reality. Time is now and space is then. There is a now and then there is back when. This is physics forming a reality splitting space and time. That is why there is space-time.

Can I touch another person?
The person that sits there might imagine he touches something but he does not because he is not able to touch. He uses his arms to touch, which is an extension of his body. To touch anything he must use an arm extended by a hand extended by a finger. It takes time for the

nerve impulse to move as electricity from the finger through the hand and the arm then through the body to the brain to get recognition about the validity of the touch. Travelling through any space at any speed uses time and that means time moved on through space becoming the history of time. There is the now that is in time and there is the when back then that is in the past and in that history of time we get space. As time moves on it leaves behind light that forms space.

In order to think cosmic we have to think in terms of space-time and not the idiotic nonsense Newtonians connect to this concept. The moon is in time and I am in time and space is between us therefore space puts the moon in my past time frame as well as my future time frame because light will come to me providing a future. However the space disconnects the time the moon holds from the time I have and the moon is in a different location when the light that tells me there is a moon reaches me. By associating and disassociating time and space we find the reality about life. Since this is a website I can't get too involved in this argument but in the book I can. Kepler formulated this entire concept he got from the Universe perfectly before Newton raped Kepler's work and destroyed it with his (Newton's) senseless stupidity.

Everything in this picture is beautiful, marvellous, beyond what words can describe and most of all it does not exist. What you see is not what you see because what you see is time that formed space as time moved on to the next instant while leaving space behind that forms an image in many colours. It is an array of light formed in colours mixed with what are not technically colours. It is an image drawn in light on a medium canvass called space and is only there as a supplement of my imagination. It is a lot of light that is dark as it moves away or expands and light streaming towards me in a concentrated form colouring the canvass many colours but in all it is light and it is light alone. But there is one more significance in this picture and that significance underwrites the cosmos as it defines cosmic physics and in that also every aspect of physics.

Light moving towards me places me in the centre of the Universe in time but I can't be in the centre of the Universe in space because being in space and forming space presents a mythical historical light-illusion and being in space is fiction there can't be a centre in fiction.

Space is the history written in light and a history of events gone by can't have a centre because of no conclusion seen from where and when it started.

Kepler gave a formula, which reads $a^3 = T^2 k$. Do not think of this formula lightly because this is the correct formula that carries all physics. From this formula Einstein got his so famous $E=MC^2$ that should read $E^3 = mC^2$ which is a carbon copy of Kepler's

$$a^3 = kT^2.$$

This is a far cry from Newton's idiotic $a^3 = T^2$ where Newton says a thee dimensional view equals a two dimensional view. Kepler said $a^3 = T^2 k$ which places the one sector which is space $a^3 = space$ in equal terms with movement or time, which is, $T^2 k = time$ and in this we have $a^3 = T^2 k$ that is

$$(space)\, a^3 = (time)\, T^2 k.$$

There is a location in time and there is a location in space and time as it moves on converts the past time into space and in doing that it uses light. Life is in time while my body is in space and my body can never be in time because there are always a few billion atoms between my thoughts (time) and my body (space)

In this we investigate the role of physics in life and how life controls the body. It is very easy to say the brain controls the body, but what controls the brain? When life ends the stomach immediately stops to process food. The stomach stops to use the acid to formulate what we had for a meal into what we will use as stuff feeding our body. When life seizes the body loses smell and that means life gives off a distinct smell. In death the smell changes and the acid no longer digest food.

If it was acid that digested food it would carry on processing the stomach content but it stops immediately and therefore we know it is life that digests the food using acid as a tool. The moment life leaves the body it stops discarding body material. This body material floating in the air holding the smell of the person's identity is what life uses to rebuild the body and we call that process aging. Life builds the body atom-by-atom in healing using carbon and oxygen. When life exists the body all vital aspects change. Then the cosmos atomises the then cadaver back to atoms.

Then a cadaver dog can trace the smell as death because the body as a cadaver is without life and became just more cosmic rubbish that breaks down into atoms. The body disintegrates the second life no longer secures the structural integrity of the composition of the body and therefore life in time removes from the body holding space.

Newtonians has all this down to just showing no electrical activity in the brain but that is a stupid argument. Only a senseless atheist can have such a meagre thinking ability. Electricity charges the mind or so science says but that is a very outdated and backwards small-minded look on the entire matter. If that is the case I can plug the brain into a supply of volts and the brain will put life back into the body and the brain will continue to control the body. Life is a separate issue that builds the body and the body is not what host's life.

Can I touch my body? Science holds the opinion that the brain controls the body. It's easy to  say the brain controls the body but then what controls the brain? It is Newtonian to even think the brain controls the body because Newtonians put everything in simplicity because they can understand nothing so good they are able to fill a Universe with nothing overflowing making the Universe expand because of nothing growing and becoming even more nothing than the nothing it was. Thank God they are atheists and that shows their stupidity is endlessly big while increasing every second. It is thought that governs the brain but what then produces those thoughts that control the brain that moves the body?

There is partition between singularity and space. Singularity can hold a thought whereas space is controlled by thought. Thought via electricity or gravity or electromagnetism or electric field all being the same and that takes charge of body material and controls the behaviour and abilities of material. That is life. It is establishing a channel of electric flow from what life starts the fields being where cosmic gas at $10 / 7(4(\Pi^2 + \Pi^2)$ is going spherical $7/10 (\Pi^6)6$ and forming an electric field around the brain whereby it generates electric thought into the brain.

This is a controlled operation.

If it were the mind controlling the electricity what would direct the specific flow and what will control such flow. Something more must take charge of the release of electricity to control what is in the mind. When they tested the flow of electricity in the brain by cutting the human scull of a person alive it was proven that this control of electricity forms thought in the mind but shocking the entire brain with a jolt of electricity did not do it.

The medical doctor that took charge of the process took the place of the person controlling her or his body. Then by directing the electricity the doctor in charge released electricity just like life would do of the person in control of the human body. If that happens the electricity will unleash involuntary impulses much similar to the shock with a stinger or porter that forms spasm.

The muscles will go into spasm, as all the muscles will get electricity at the same time but without control. Charging the muscles is done with the flow of blood by establishing the Coanda effect. The second, the instant the blood flow stop the generating of electric field stop and the person is dead. It is the blood holding plasma or liquid in relation to flowing solids (red blood) that charges electricity via the fields surrounding the brain.

Practising an athlete or anything with life is just establishing fitness better electric flow to the body that results in having better body control. The better flow of electricity forms more muscle control and then the higher standard of achievement wills the athlete accomplish.

This is my two
sons   battling

My sons are internationally acclaimed wrestlers and they have to practise eight hours at least per day for four days a week to be on the top level of the required performance. If Newtonian stupidity is correct then when they require more fitness they don't have to practise more; that is if the Newtonians are correct because all they need is a jolt of extra electrons running through the brain matter and they get the strength Superman has. To get them to wrestle we have to condition the mind to accept orders from life. Life does this by practising to induce current flow to the brain that conducts the flow through the brain to the muscles in the body. If the current is not strong enough the muscles cant do the job and if the current is strong enough any muscle of any size can do the job. If the electricity impulse is not strong enough a person can't lift a heavy weight but in circumstances of life and death I saw a man lift a car to get another person out that was trapped underneath the car. The emotion of fear and desperation was so massive it charged an electrical current that allowed the brain to order the body to pick up the car and free the person stuck underneath. Both my sons can lift a person of 120kg from the floor up above and over their heads one with movement. I have many such moves documented on film taken during their fights. As strong as they are and they are strong, they will not be able to stand next to a car and flip it over on its side with one movement. That's why we shout and encourage sportsman to get maximum electricity charges and to get the most voltage emotion can muster to the athlete's body.

We have to associate thoughts with life while disassociate thoughts from being a product of the brain. When being dead the brain are present but empty with thought. When life or the thought process ends in death the stomach acids immediately stops processing food. The stomach stops formulating our last meal into something that produces life's energy. Newtonians and other simpletons say it is the electric charge that makes the brain think but put any voltage through the brain and the stomach won't start digesting food even with the brain charged with electricity. The absence of life stops stomach acids from digesting food and how does that change the process of acids digesting food?

Typical Newtonian atheistic small-minded research tries to show that the brain applies electricity to convey messages to the human body. Certain parts of the brain sends messages to perform certain acts in movement and some part of the brain controls some characteristic aspects to allow certain behaviour to take place. The brain through electricity controls the body by performing manipulation of muscles that create movement ability. Well this is like saying to breathe is to take in air. That is hardly science but comes down to observation of elementary facts. However, this is where Newtonian atheists stop the argument because as usual they run out of brainpower. At this point of researching life we find that the argument hardly started.... And now it becomes complicated to the extreme.

Newtonians are convinced that thoughts are electrons within the brain. By emitting electrons the brain performs the manipulation of the body through electrical charges and that when released controls the muscles in the body and that is life performing. Well yeah and also Newtonians say that mass pulls mass to supply gravity to a small edge-filled Universe and in this simplicity it is only the Newtonians' small brain that can understand so much complexity with such minuscule simple approach. It's easy to say the brain controls the body but then what controls the brain? If those Newtonians were correct sportsmen would have so little to do and reap so much awards.

To get them to better perform during a tournament I can put a belt sporting a stun gun around my sons' head and every time any one of them needs more strength of essence during a wrestling match I put a surge of electricity through the brain while that will make them many times stronger. I stand with a release mechanism and when they require a jolt of electricity to beat the opponent badly I give him a jolt and he breaks the other person's arms with the strength he then has. When more strength is required we up the voltage and increase the amps. No more practising is necessary because we just up the supply of electricity and the mind will supply so much muscle power to the body my son will have the power to lift not only his opponent but the entire tournament above his head and break the venue. Then everyone will need to have stronger battery packs and we can do without the hours and hours of tedious practising sessions. You just plug any person into a strong battery and let the current flow.

Can you see how utterly childish such an argument is but then again what can you expect from a simpleton that admits to being animal-like in mentality as an atheist. By the way my dog is the biggest atheist I have ever met and my dog is not an atheist because he is a clever scientist but because he is too simple to understand physics and science. My dog is an atheist because of stupidity and that goes for every other mindless atheist but because they elevated Newton's corruption, the lot sit with a hoax they try to give reality and because they can create their own truth they make it up as they go along while pretending to be the clever atheists. They always were too stupid to see that Newton's concepts are a hoax, so how can they be bright enough to see a God that created everything we call a Universe only by God's thought? Life is in thought and thought is in singularity.

Again they say it is the acids in the stomach that digests the food but death does not remove acids absorbing stomach content. It has to be life that digests the food because when life removes after death the stomach content and acids are still present but the processing part stopped and by life vacating the body the digesting process stopped. With life gone notwithstanding whatever outside intervention, life's functions will remain absent. Electricity in the brain doesn't reinstate life again by controlling the body.

With the human body in its entirety forming space that the mind controls by forming gravity it puts the entire human body in space and brings the mind in time from where it conducts electricity. It is the mind or life or call it what you wish that generates electricity outside the body and then by forming a circle or band of electricity around the brain it charges electrons within the brain and life directs the electrons to manipulate certain parts of the body in the manner life wants the body to perform. These are all part of finding the process we call life. Life isn't as simple as putting a few electrons that runs through brain matter and therefore instigates life. Life is a lot more complex than such simple ideas about the most complicated process. I prove that gravity and electricity starts at $\dfrac{7\Pi^6}{10\Pi^0 x 6\Pi^0} = 112$ and 7/10 is movement of gravity of the sphere Π^6 in terms of a six-sided cube (6) where outer space or the element table starts at 112. This I explain much more extensively but this is where life that develops electricity forms an electric sphere around the brain. This requires much better explaining than what the limits of the length of this article allows for. The 7/10 is the Titius Bode law forming gravity and this is a sphere turning in a square and time forming space.

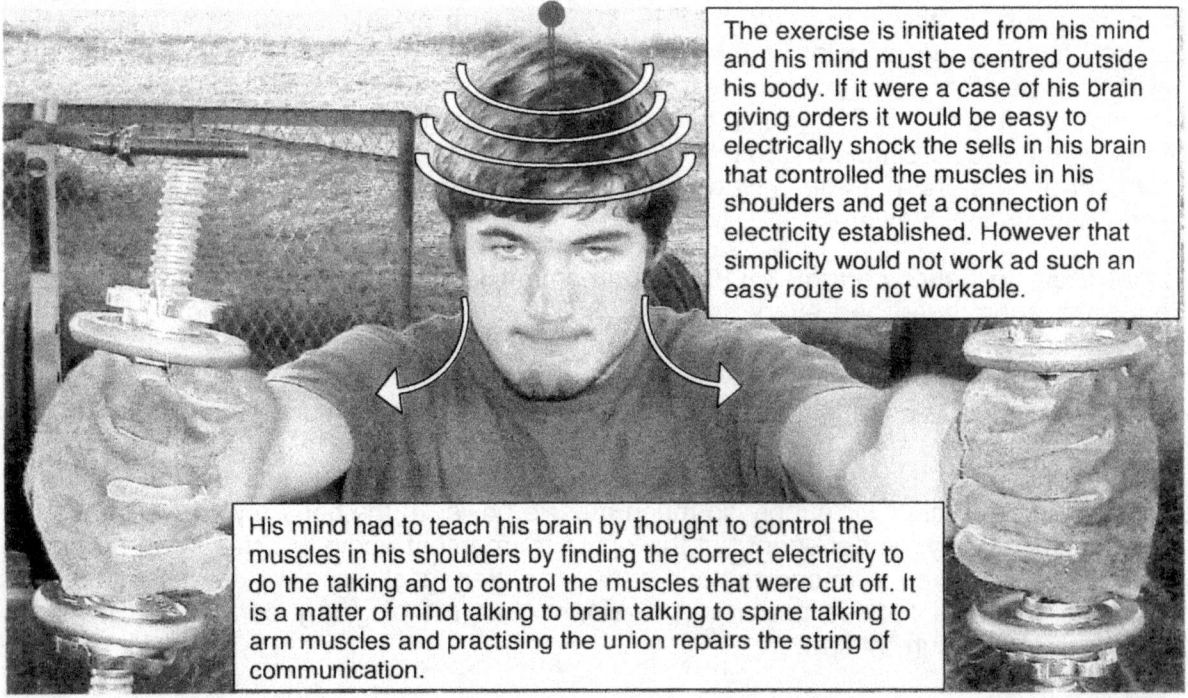

The exercise is initiated from his mind and his mind must be centred outside his body. If it were a case of his brain giving orders it would be easy to electrically shock the sells in his brain that controlled the muscles in his shoulders and get a connection of electricity established. However that simplicity would not work ad such an easy route is not workable.

His mind had to teach his brain by thought to control the muscles in his shoulders by finding the correct electricity to do the talking and to control the muscles that were cut off. It is a matter of mind talking to brain talking to spine talking to arm muscles and practising the union repairs the string of communication.

When Willem recouped from an operation we had to get his mind to again talk to his muscles correctly to give him the required control over the manipulation of the movement within his muscles. Life reinstates the ability to control the brain to manipulate the body movement. These muscles do not represent physical strength as much as it indicates mind control and the ability to charge by the strength of thoughts that generates muscle power.

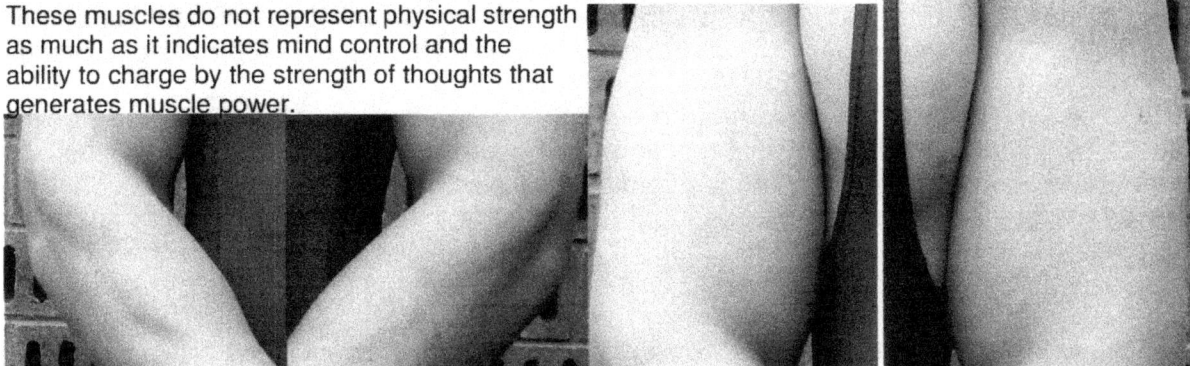

That concludes that with electricity the brain doesn't instate any functions we attribute to life. Therefore Life by thought generates electricity that forms around the brain as the Coanda effect. I prove it in another book. The electricity controls the brain but what charges electricity? In order to improve his muscles I purposely increased the electricity flow to his muscles from his mind. People call this concentration but a rose by any name still remains a rose. Without practise the current weakens and the muscle power fades. You have to charge the battery.

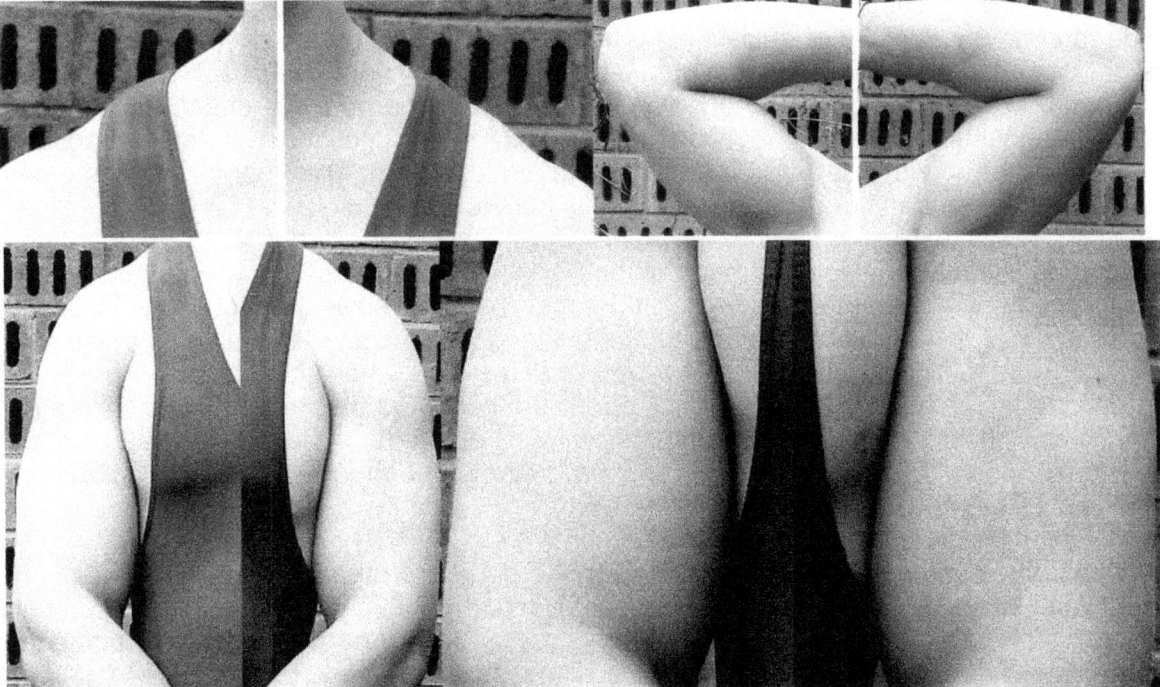

These are the muscles we built when Andre was 19 without lifting one weight in the process. Through practising the mind and the body I allowed the electricity to develop that controls the mind and the muscles and then by practising the brain to endure lots of pain and suppress the pain this method of stretching the limits of endurance developed the muscles as well as the control of pain in the brain. By developing the mind the body forms muscles and these are the arms of my son when he was 19. It is not the muscle that does the job but the mind controlling the muscle that produces a muscle strong enough to do a job. This is no cheap oriental philosophy but true raw cosmic physics placing gravity and electricity in charge of the brain to charge the muscles in the body.
When the body and life parts ways we call it death. When life seizes the thought process the body loses smell and that means a by-product of life is to leave an identifiable unique smell. After death it is gone and that is why we use cadaver dogs. The cadaver dogs tell the story of the body being lifeless throwaway reusable recycled material ready for the next person.

Electrocuting the cadaver won't bring back odour the body had before death. That means life leaves small particles of the body behind as it rebuilds and replenishes the body while being alive. The body also leaves particles behind but clearly it is of a different nature because the cadaver dogs recognise that sent as a cadaver, which is a cosmic body without life and is breaking up. Science says electricity charges the mind but this perception proves outdated.

If it was electricity alone the brain had no control over the flow, much like a person being stunned by a stun gun. The body just reacts without control because behind the flow there is no message about what to do or whereto to direct the electricity conducting. It is thought that governs the brain but what produces thought that controls the brain that moves the body? The control is in the thought and the thought is what produces life and in the book I show where exactly life is when it directs control over the brain that manipulates the muscles.

We have to place thoughts in another position because it is clearly not in the brain. The brain acts on electricity impulses it receives. When jolting the brain with electricity we can produce movement in the muscle but that shows the brain receives and relays electricity and the brain does not generate electricity The brain is merely a conducting relay of an impulse that direct the flow to the station that should respond to the wishes of life's commands. In that we have hypnotic commands because with hypnosis we remove the character that life has and replace that with the hypnotisers orders.

There is a string of commands coming from a source that is not part of the body but is outside the body. Electricity as such will not become a hypnotic replacement. The hypnotist does not use electricity to convey commands but also put any voltage through the brain and the stomach won't start digesting food even with the brain charged with electricity. With life gone notwithstanding whatever outside intervention, life's functions will remain absent. Electricity in the brain doesn't reinstate life again controlling the body. Life enforces electricity that life as a factor generates and life is not electricity.

That concludes that with electricity the brain doesn't instate any functions we attribute to life. When the body and life parts ways we call it death. When life seizes the thought process the body loses smell and that means a by-product of life is to leave an identifiable unique smell. After death it is gone and that is why we use cadaver dogs when searching for the dead. Electrocuting the cadaver won't bring back odour the body had before death. Science says electricity charges the mind but this perception proves outdated. If that's true I can plug the brain into electricity and the brain will continue controlling body movement by putting life functions back into the body.

Life or thought is a separate issue that rebuilds the body and the body isn't what host's life but functions according to life being in control of the body. By thought life controls the brain that manipulates the body. Life is an entity that's in thought and thought is in singularity, which is time. That I prove. I show Life is in a space that isn't space because life is outside where we think space is. If you wish to find life's place follow my argument and find life's place in the Universe. Life or thought is a separate issue that rebuilds the body and the body isn't what host's life but functions according to life being in control of the body.
By thought life controls the brain that manipulates the body.

Life is an entity that's in thought and thought is in singularity, which is time. That I prove.

I show Life is in a space that isn't space because life is outside where we think space is. If you wish to find life's place follow my argument and find life's place in the Universe.
All light throughout the Universe meets at the point you are. All light comes from everywhere to meet you wherever you are. If you weren't the centre light won't flow directly to you.

Life by thought assembles the body from even before birth and when being without thought the cosmos dismantles the body again after death. If a lifeless body is a total different thing from a body filled with life then after death and with the body still usable but not functional it is totally Newtonian or atheistic or insanely stupid (which is all the same) to say that the body is what confirms life meaning the body is life. The body is still present after death but the decaying process starts immediately to dismantle the tissue structure that life maintained and that life preserved while the body was in use of life.

It is madness to discount life, as a presence in body while without life the body is cosmic material going into a process of dismantling. If the body held life then we can re-install life by electricity and the body will return to all its normal functions. However that is not possible and so it is clear that life constructed the body from before birth and maintained the body by an aging process until the final departure of life where the body is no longer maintained and has no further purpose and so the cosmos in space goes into destruction of a useless accessory but no longer necessary of life. What walks with time and presides in time as part of time went with time to where time forms the future while the human remains remain cosmic atoms and remains in space as a part of the history of what time once was when life occupied and built a human body.

If life is electricity or if is as simple as any of the functions Newtonians atheist stupidity try to contribute to life then it can revitalise a cadaver by shocking it until it hops like a ping – pong ball. I then can replace the brain's electrons and produce thinking within the cadaver mind. The cadaver will again be able to lift its arms and start to walk around.

If I am not able to replace life within the human body that puts life in thought and I prove singularity is what holds thought but that issue is far too complex to touch in this website. I aim to keep the website very simple and in the understanding of any person except atheists who will be too simpleminded to make sense of anything that is a reality and not part of the Newtonian hoax. If this sounds complicated then by me using 480 pages the explaining gets much better while the topic becomes very logic. If life removes from the body life was never part of the body.

The entire Universe is light and that is why we can focus on one image with lenses and with another lens view a picture as wide as we wish to see. However, what is out there is light either expanding or contracted but it is light we see that fills the cosmos. We see darkness as light expanding thus moving away and the white light we see is light coming towards us. Have you sat back and think about how light enables anyone to read?

Newtonian science says the light bounce of the page and I use my eye to see the words. What can be simpler than this explaining…only the mind of a simple atheist and has no idea about what reality is because they believe in Newtonian corruption. How does the light take the words away from the page and put the image of the words printed in ink on paper into my eye so that I could see the image of the words on the paper.

How does light take what is in the image and produce a picture of that image and transport that image to my eye so that I don' see the paper because I can't see the paper but I see the light that presents me with an image takes of the paper as a photo of what the paper was when the light bounced on the paper. How does the process work whereby light can carry a photographic image of what the paper is and convey that to the ball of my eye and where that specific photon that hits my eye is the very one that contains the image of what I wish to see. If all the photons carry the image then how does that work and how do I explain that every photon contains the entire image?

Before science got so wise science back then had the earth in the centre of the Universe. Everything in the Universe was spinning around the earth and hell was where it was very hot in the earth centre. If you went to heaven you rose into the sky and when you went to hell you descended to the centre of the earth. The most valid part of this was that the clergy got everyone so scared they paid huge sums of money to the Church so that the Church would tell God that the person in question had no sins because the Church had his money...and every one knew exactly where the centre of the Universe was. Now we only know where the centre of the Universe is not, because according to science it is not in the centre of the earth.

Without knowing where the Universe centre is no one will understand how the Universe layout works. Where the centre is was always a question that was nagging humans since humans saw a Universe in the sky. In the Ptolemaic Universe the earth was the centre but then "modern science" rubbished this idea. If you read the book you will see that the centre of the Universe is even much closer to home than what that idea allows. Ptolemy said the sun and everything within the Universe spins around the earth, which then places the earth in the centre of the Universe, and I can see the sun and the stars rotate around the earth and me but me mostly.

FIX STARS
SATURN
JUPITER
MARS
SUN

VENUS
MERCURY
MOON

EARTH

Everything about physics and about mathematics and the cosmos cradles behind this question where to find the centre of the Universe and "modern science" has no clue where to find it. If there was one point where the Universe started then the entire Universe will put all focus there is on that point with all light streaming to that spot. I discovered that point because I went about researching nature and using how nature applies the Universal laws I discovered where to find the centre of the Universe. The principle is simple. The centre point shows us where the Universe started. The Universe started where singularity is and that I prove in **Uncovering Corrupt Science**. Therefore to find the centre of the Universe we have to locate a point where the Universe started when it started with one single point. If I embroil on this idea in this small space it will seem to be exceptionally complicated but when explained correctly with a few pages to spare the answer is laughably simple. To find out where the centre of the Universe is follow my argument. It is simple to do because I can just as well say follow the trajectory of the light.

Such a thought brings to mind the most simplistic answer. Science says the light hits the page bounces from the page and contacts the lens of my eye where the lens conveys the photons becoming electricity to a part of the brain that translate the electricity to an understandable message and that makes one read. Sure sounds simple to me. Is it as simple as that? Ever gave a broader thought about light streaming across the night sky, coming from ends of the Universe we don't even realise it is there? Have you ever given it a thought of how big it is out there and how small we are down here and how far did the light we see travel to meet us where we stand. I have to be in the centre of the Universe if all light comes from all over and find me forming the point to which light travels and where all travelling light eventually ends.

All light throughout the Universe meets at the point you are. All light comes from everywhere to meet you wherever you are. If you weren't the centre light won't flow directly to you.

First I had to put life where life belongs and life isn't in the body but in the Universe. Then I had to connect life to the body by life conducting electricity. I prove how life exerts control over the brain to manipulate the body in movement but it starts where one put life in relation to the Universe. Life is alien to the

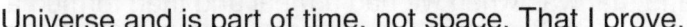

Universe and is part of time, not space. That I prove.

Step outside into the night sky see where you are. You see an entire Universe. Every sparkle of light coming from where ever is directed at you. All the light released from any point in the Universe comes to where you stand. This puts you in a location where whatever light there is will come to you. You can see all the light coming to you and it's beaming towards you at the speed of light. This positions you in the centre of the Universe.

The Universe is what we see as the entirety when we see everything in the Universe. How does the photons manage to convey one complete picture coming from as far apart and as wide an area as it does? With a few photons connecting the eye or lens no one ever noticed the wonder of light. The photons reflect a view that seems as if coming from all the billions upon billions of stars with all the information present in one photon. Most information comes from darkness covering an immeasurable area. Yet how many photons can actually connect to the lens of the camera or to the eye to transport such vast information? All the light released from any point in the Universe comes to where you stand. This puts you in a location where whatever light there is will come to you. You can see all the light coming to you and it's beaming towards you at the speed of light. This positions you in the centre of the Universe. But being in the centre of the Universe you then can't be in space therefore you have to be in time because time is part of any and every point that forms space according to time placing space in whatever location. Light formed our culture. It is said that because we ate meat or because we threw a spear or because we ran upright we got clever.

That is rubbish.

We got civilised when we sat next to a glowing fire and started discussing the stars and what influence stars have on our culture and that became our culture we developed. If we didn't see the sky we wouldn't have formed intellect or culture. It is good and well to couple everything to fire and cooking food but that didn't bring us to embrace intellect. What formed our intellect is when we formed speech and that was when we had a meal and then formed a society sitting around glowing coals. That is when speech developed but looking at the basis of all cultures in antiquity the wisdom was locked in the manner that the culture developed along the lines of the studies of star formations.

First I had to put life where life belongs and life isn't in the body but in the Universe. Then I had to connect life to the body by life conducting electricity. I prove how life exerts control over the brain to manipulate the body in movement but it starts where one puts life in relation to the Universe. Life is alien to the Universe and is part of time, not space.

That I prove. Step outside into the night sky see where you are. You see an entire Universe. Every sparkle of light coming from where ever is directed at you.

Am I my body? I know for a fact being the father of Andre that he is not inside his head. Inside his head are girls in bikinis and girls not in bikinis and girls wearing much less than tiny bikinis and girls chasing him while he chases girls wearing very small garments. With that many girls filling his head I am surprised that there is room for brain matter let alone leaving space for his self assured and arrogant self taking up all of the 120kg his body fills and that self worth are so prominent that can fill all the vacant cracks and spots between the grey-matter. In his body he is not and that I showed and so between what we see and where he is in space and where space is we can't have time while in truth he and you and I am in time.

Why is he in time? Why is he not in his body because that is where everyone think they are.

If I am in time where the F@%$^K am I then? This is how I figured out where a human is…

This is my nearest and dearest with my two sons. Both were awarded prizes for achievements in wrestling and this is a high but the wrestling is so small part of the practise and the practise was to control the mind to control the brain to control the body.

This is the upside of sportsmen but the following is the downside of sportsmen. Both came back from series injuries and were out of practise for months when this photo was taken. I include this to show muscles that might look big but the control of the mind is clearly not present and the muscles are completely out of practise and therefore out of control.

This was taken just before (a day or two) they both started training again after many months

of injury lay off. The muscles are there but the mind control of which controls the muscles are very clearly absent. Now we start to start to get the mind to deliver messages to the body via the brain and that is what practising is all about. It is about getting electricity charged around the brain that charges a current within the brain and this current produces electricity supplying the muscles with not only control but also most of all the strength.

I got the idea of mind-muscle-control when witnessing a person that was in shock lift a car from another person that was trapped underneath the car. In shock this person took hold of a car and rolled it over from where it was on its roof right to where the car landed on its wheels. Seeing this that day so many decades ago made me realise that muscle strength is not in the muscles and is not in the brain but is pure emotion and to unlock that pure power I have to excite the emotions that control the muscles should I wish to tap into human power. The human is in emotion and emotion is electricity charging the mind to produce power to the muscles. So where is that then? From that I formulated the practising method I applied on my children that made them sportsmen.

Remember I said you are in the centre of the Universe but it is the Universe time fills in singularity and not in space because time fills the centre of the space in the Universe.

Try to mathematically fit all of this space into this human eye.

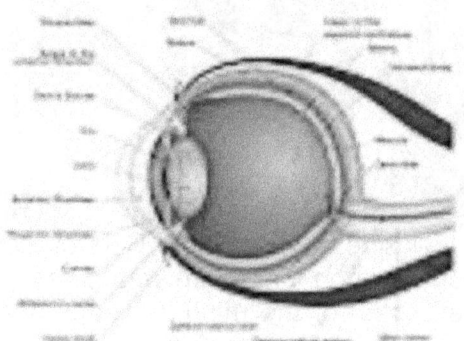

We live in an imaginary Universe built in light in three dimensions while time moving as gravity that moves in darkness. Not darkness I am able to see as light but darkness I am unable to see. Singularity is one spot equal to 1 or 2468123489^0, which comes down to one and in one the entirety that is reality forms gravity. This concept is so much more intricate than what Newtonian stupidity could ever realise and that says why they are all stupid atheists without having the ability to understand reality and how the cosmos works outside their lies.

We are able to see the entire Universe because life that does the viewing by only using the eye to see everything is in the singular dimension of singularity which is one spot Π^0 that is equal to 1 and it is also equal to as many as 2468123489^0 because being part of singularity brings equality $1^0 = 1^1 = \Pi^0 = 2468123489^0$ and that is why I can see al space by applying one single electron that enters my eye and dhows me the entirety of universal cosmic dimensional equality where everything is in singularity. Mathematics show that $1^0 = 1^1 = \Pi^0 = 2468123489^0$. Singularity is time and time is in a dimension we can't understand because although we are part of time in singularity we can't witness it as we reside in the centre of the entire Universe.

But being in the centre of the Universe you then can't be in space therefore you have to be in time because time is part of any and every point that forms space according to time placing space in whatever location. There is singularity controlling time in which life is and then there is the human body material in which space is and the body is to life just more space. By dying life removes from space and the body goes cosmic by getting destroyed back to atoms.

It is the way we appreciated stars that made us develop beyond animals and brought us to have an intellectual group of people. Yet as far as I know I am the only one who asked AND answered the question as to how can we see what is out there in relation to what we have. Man has been living with light from before man became intelligent if man ever became intelligent and yet we have the most small-minded approach to the most complex issue man could ever devise: how can we see all the space through something as small as the space of an eye? Space is never-ending and so is our view thereof looking through an eye so small. Newtonian wisdom currently dictates the idea that it is the brain that controls the body and as usual just before the reasoning can become complicated they stop to avoid issues they can't deal with. If it is the brain that controls the body there has to be a controlling aspect that decides which part of the brain should be active in any particular instant. If there weren't such a mechanism controlling the brain and dictating which part of the brain should send messages in that instant wherever in the body then brain would release all the electrons instantaneously and then cause an uncontrolled electricity shortcut.

This will cause instant death because every motorised action within the body will try to do what the rest does because there is no control to direct which part of the brain must be active when. There is control over what the brain does and that is in the mind. I show that life is part of time and the body is part of space but singularity is time and space is the Universe we see. This is all new and this line of thought was never before pursued because I am the first to show singularity forms gravity by implementing four cosmic laws. Taking control from singularity to space follows paths we would consider as dimensions and dimensions are a product of mathematics.

There is singularity controlling time in which life is and then there is the human body material in which space is and the body is to life just more space. By dying life removes from space and the body goes cosmic by breaking down to atoms.

Everything about physics and about mathematics and the cosmos cradles behind this question where to find the centre of the Universe and "modern science' has no clue where to find it. If there was one point where the Universe started then the entire Universe will put all focus there is on that point with all light streaming to that spot. I discovered that point because I went about researching nature and using how nature applies the Universal laws I discovered the centre of the Universe. The principle is simple. It shows where the Universe started. The Universe started where singularity is and that I prove in **Uncovering Corrupt Science**. Therefore we have to locate a point where the Universe started. If I embroil on this idea in this small space it will seem to be exceptionally complicated but when explained correctly with a few pages to spare the answer is laughably simple. To find out where the centre of the Universe is follow my argument.

The entire Universe is light and that is why we can focus on one image with lenses and with another lens view a picture as wide as we wish to see. However what is out there is light either expanding or contracted but it is light we see that fills the cosmos. We see darkness as light expanding thus moving away and the white light we see is light coming towards us. Have you sat back and think about how light enables anyone to read? Such a thought brings to mind the most simplistic answer. Science says the light hits the page, bounces from the page and contacts the lens of my eye where the lens conveys the photons becoming electricity to a part of the brain that translate the electricity to an understandable message and that makes one read. Sure sounds simple to me. Is it as simple as that? Ever gave a broader thought about light streaming across the night sky, coming from ends of the Universe we don't even realise it is there? Have you ever given it a thought of how big it is out there and how small we are down here and how far did the light we see travel to meet us where we stand.

The Universe is what we see as the entirety when we see everything in the Universe. How does the photons manage to convey one complete picture coming from as far apart and as wide an area as it does? With a few photons connecting the eye or lens no one ever noticed the wonder of light. The photons reflect a view that seems as if coming from all the billions upon billions of stars with all the information present in one photon. Most information comes from darkness covering an immeasurable area. Yet how many photons can actually connect to the lens of the camera or to the eye to transport such vast information?

Light formed our culture. If we didn't see the sky we wouldn't have formed intellect or culture. It is good and well to couple everything to fire and cooking food but that didn't bring us to embrace intellect. What formed our intellect is when we formed speech and that was when we had a meal and then formed a society sitting around glowing coals. That is when speech developed but looking at the basis of all cultures in antiquity the wisdom was locked in the manner that the culture developed along the lines of the studies of star formations. All great societies embraced stargazing as a mythological culture and formed the basis of antique religion development. The basis of nation greatness was star gazing intellectual development.

It is the way we appreciated stars that made us develop beyond animals and brought us to have an intellectual group of people. Yet as far as I know I am the only one who asked AND answered the question as to how can we see what is out there in relation to what we have. Man has been living with light from before man became intelligent if man ever became intelligent and yet we have the most small-minded approach to the most complex issue man could ever devise: how can we see all the space through something as small as the space of an eye? Space is never-ending and so is our view thereof looking through an eye so small?

Look at the earth versus the sun on the next page and picture yourself on earth compared to the sun. Then place your self in the picture as you are on earth. That dot that you don't see is the entire earth that is so big to me its size goes beyond my grasp. Now picture me trying to grasp the size of the sun.

The following concept is in many ways dedicated to atheists and moreover directed to those physicists with larger-than-life complex taxed with the more-than-equal-and-even-better-than-God-Almighty syndrome which all small-minded Newtonian astrophysicists have. Then see your eye in the picture. Then see the nerve ending of your eye that you use to see in the picture in space with the sun. See the photon entering the eye and remember the photon is so small no one could ever measure it and yet being so small it still has the able ness to convey all the space that you see throughout the vastness of the Universe. Do you get the picture?
See the reduction there is and the information converted into vision?
Now that we know our human position in the solar system let us go cosmic.

The Milky Way is as big as an atom when compared to space holding another hundred billion Milky Ways in space. If you think you are an accomplished astrophysicist working with real numbers and with a better-than God complex and you want to have reality in your work then put the eye in space in ratio with one hundred billion Milky Ways in space. The tiny light specs you see as far-off galactica drown in the overwhelming darkness of space that fills the real picture. Remember with such vastness the galactica, as material in ratio in space not being shiny material when compared the material doesn't even show up as dots in the space viewed as darkness in the space holding the hundred billion Milky Ways. So as an accomplished Newtonian astrophysicist don't find the edge of the Universe but calculate the space reduction in terms of how the cosmos fits into your eye and get you place and position you have in our perfect God-Created cosmos where the cosmos is a mere thought God Almighty has. That the cosmos is a mere thought of God Almighty I also prove mathematically but that work is complicated beyond mere explaining and is advanced to the extreme.

So how did the Universe start mathematically from one point as Einstein said it did? Where you might think this is old news I have news for you because Einstein never got further and I did. This is where the entire concept changes because I prove that gravity and electricity is the very same thing and so is electromagnetism and nuclear energy but each is on a different level of intensity.
What would I mean by that. Let's take mathematics.
There is the dimension of adding 1+1 = 2.
That is one dimension because at that dimension one will have 1x1 = 1 and $1 \times 8756543^0 = 1$.

The there is multiplication where this is 2 X 2 = 4 while this also is where 2 + 2 = 4 and this alternates at 3 x 3 = 9 and 3 + 3 = 6. This is a dimensional change. Then a dimension comes about when 2 x 2 x 2 = 8 and 2 + 2 + 2 = 6. Therefore at 2^3 another dimension intervenes. This is exactly how the Universe progresses and new dimension changed the formation of the Universe. Life applies electricity to control the brain that manipulates the body into movement. In that the gravity plays a huge role in our ability to think.

I shall quickly and briefly explain but only in the books does these arguments find proof.
7×2^0 is a gravitational border and therefore round objects spin at $7°$ directional change.
7×2^{1} is a gravitational border where gravity forms space by also moving linear (7 going circular and 7 going straight and every straight is a circle forming).
7×2^2 is where light controls space by forming a circle in pi.
7×2^3 is where light ends as space and iron can charge electricity as well as gravity.
7×2^4 is where the element table starts by putting gravity and electricity into movement.
At 7×2^4 is why time becomes the fourth dimension that provides movement to a three dimensional Universe.

So electricity is on another dimension (2 x 2) than gravity (2 + 2) but is the same thing. I am not going into the proof of this in this article but that I do in the book. However one cannot discard the influence of gravity in favour of electricity or visa versa and that is why we have to test how electricity is conducted in terms of gravity.

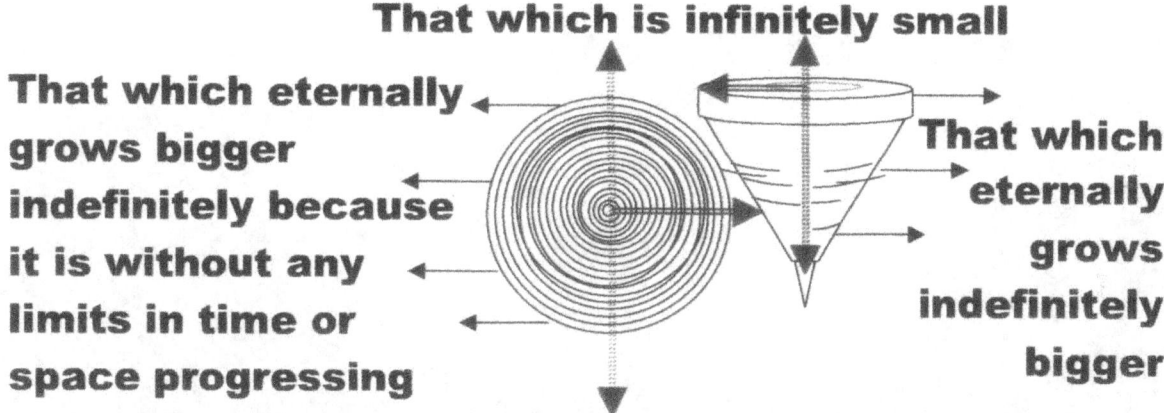

Whatever is in the universe in space forms between two points, one, which forms time that can never start and time that, will never end. That which is the Universe is in between these two pints holding time.

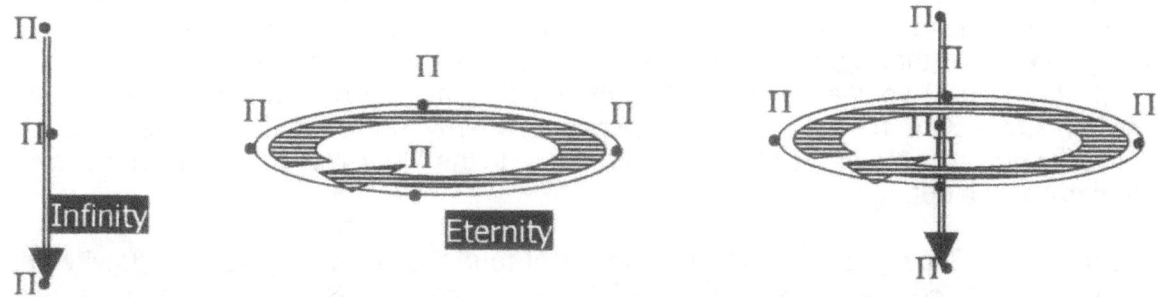

Whatever is in the Universe spins while it moves by displacing its previous position from the past through the present to the future. This form a line (3) that shifts (3) while the line circles (4) and this forms space by the value of 10.

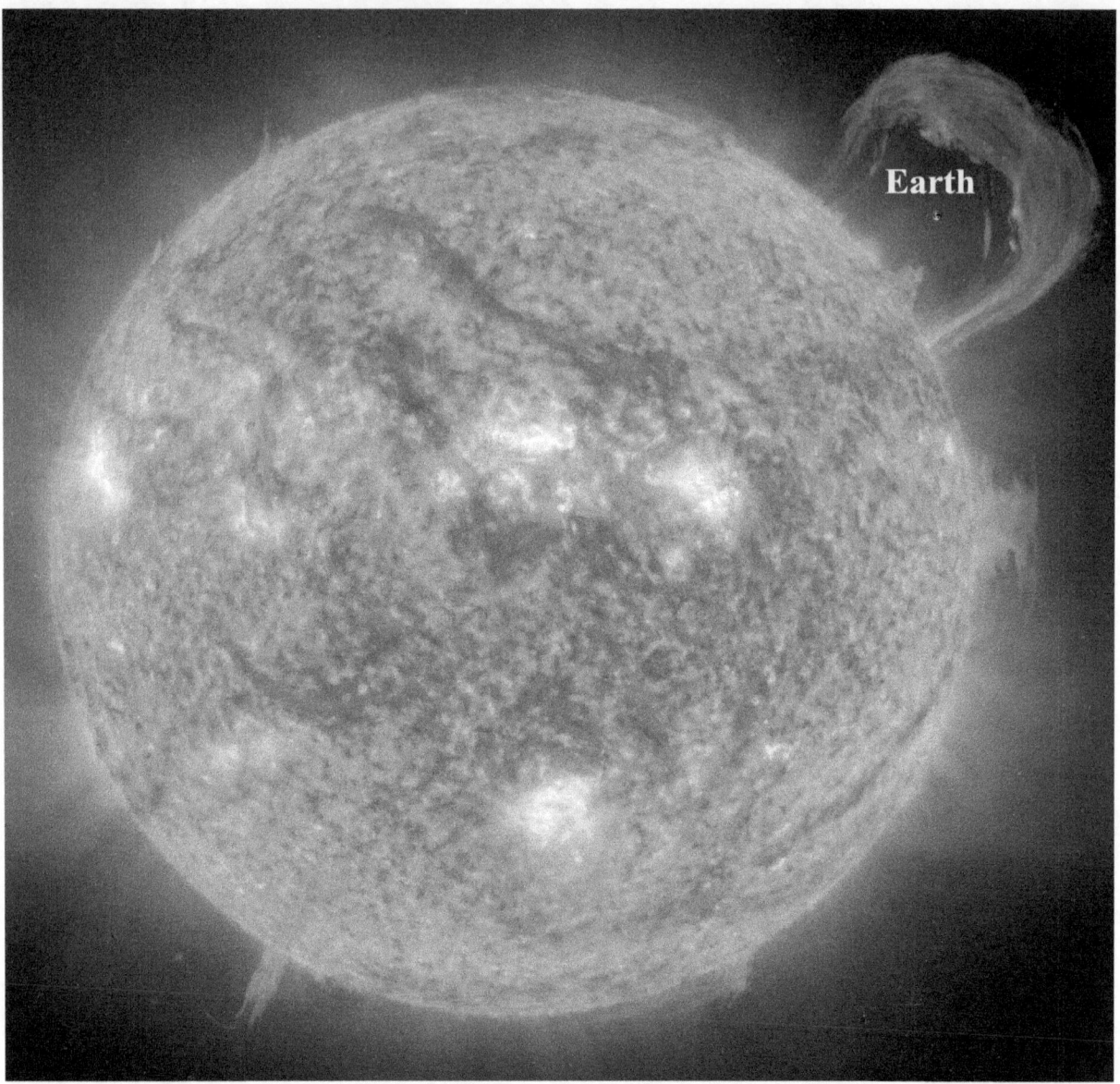

Look at what is the smallest cosmic object in the sky, which is the moon. Try to vision that you are able to see an entire country such as Britain on the moon and you might be able to entertain such a thought. Now put England within Britain. The view that you can see becomes very narrow. Put London in England in Britain on the moon and there is no chance that you will se London. Then put yourself in London and put your eye sock in your head in your body in England in Britain on the moon at the distance the moon is from the earth and calculate the scale that space has to reduce just to accommodate the reduction from the moon to the size of your eye. Interpreting this mathematically no number will ever make sense. This reduction in size is beyond comparing when you put the moon in terms of the sun and the sun in terms of the Milky Way. We know in the Milky Way the sun does not even come across as a dot.

To do this I show how gravity conducts in terms of four cosmic laws governing gravity and this same process conducts electricity. The Titius Bode law, the Roche limit, the Lagrangian points and the Coanda effect form the cosmic structural code. I show how these four form the cosmic code and the cosmic code is in the way that Π forms gravity and how gravity is Π.   Where these laws form gravity by forming Π and measuring Π as a value of

$$\Pi = \frac{10 + 10 + 1.991}{7}$$ as space and $$\Pi = \frac{7 + 7 + 7 + 0.991}{7}$$ as time which puts

time which is movement in time and space in movement and this is crucial when going into something as massively complicated as how life controls the body.

I manage to answer these never asked before questions because I resolved the mystery behind the four Cosmic Pillars. I wish to make one fact very clear. I base my work on formulating the working process of four cosmic principles in Nature. These are:

1) The Coanda effect
2) The Titius Bode law
3) The Roche limit
4) The Lagrangian points.

I did not discover these phenomena because science knows about these phenomena for a very long time and in some cases even for hundreds of years. Science knows they apply and where they apply. When science discovered or allocated missing planets they used the law applying such as the Titius Bode law from which they deducted positions that they knew in that circle according to the planetary layout that the law predicts there had to be a planet according to the law. Science did not apply Newton's formula to discover and locate planets but they applied these phenomena and especially applied the law of planetary allocation to discover the precise location the planets discovered after Galileo.

Everyone in science knows these phenomena are there and are in place and they rule the orbit set-up of the planets. The solar system functions according to them. These four laws on planetary motion that are used by nature at this moment and have been in place since time began, are what apply and they dismiss Newton. If you argue with me about Newton being correct you better take your case to God or the solar system because the four cosmic phenomena are working in nature and nothing Newton said is applying in nature. This is a truth and a fact and a foregone conclusion and can never to be in doubt.

Brainwashed as you may be in believing Newton you can't either side with my view or decide on Newton because it is not a case of choosing between Newton and me. I'm out of the picture! It is either telling nature to listen to Newton and change what is in place or read and see what is in place and what is applying in nature all along. The phenomena are what we find to be used in the cosmos while Newton is in the imagination part of the minds of scientists and nowhere else. If you don't believe me and if you wish to discredit me first find out a little more about science. Then deflate your ego as to what you think you know.

Science never mentions these phenomena because science can't use Newton and explain these phenomena or use these phenomena to prove Newton. These four phenomena that the cosmos uses as we speak have been in place ever since the Universe formed. Since science can't explain the phenomena and the phenomena destroy the credibility of Newton science avoids these phenomena as if it brought the plague. You can't choose between Newton and me because I did not put the cosmic phenomena in place. All I did was doing a study since 1977 to formulate how and why these phenomena work and how these phenomena keep the cosmos and the solar system working. I am the first in history to show why they work. We are all been brainwashed for centuries to believe Newton. Should your brainwashing kick in and you have an axe to grind with me about what I say, then first prove these phenomena are not in the cosmos and are not applying to form the laws that the cosmos put in place as gravity. I only found out how they work and why they work and I did not make the phenomena work. All those clever stooges that have so much to say even before you read first learn what is in place before getting so opinionated.

This is book # 2 of
A COSMIC BIRTH...
STARTING BEFORE
ZERO

...and that is precisely what it does.

ISBN-13: 978-1499786132
ISBN-10: 1499786131

This forms part of Naturescosmicconcept

WRITTEN BY Peet (P S. J.) SCHUTTE

©KOSMOLOGIESE EN ASTRONOMIESE TEGNIKA

All rights are reserved.
No part, parts or the entirety of this book may be reproduced by publishing, electronically copied, duplicated by whatever means that form reproduction or duplication, without the prior written consent of the copy rite owner.

This publication aims to put truth into science!

This reveals how the Universe formed and how nature works in accordance with a vision about Creation and in all of that it excludes Newton.

Representing the view presented in the following websites Naturescosmicconcept or Titius-Bode-Law-Explained and www.sirnewtonsfraud.com Also ANaturesCosmiConcept

PLEASE TAKE NOTE OF THE FOLLOWING SINCERE WARNING

Before even attempting to start with A COSMIC BIRTH... STARTING BEFORE ZERO # 2 the reader is strongly advised to read A COSMIC BIRTH... STARTING BEFORE ZERO # 1. The science you are going to encounter in A COSMIC BIRTH... STARTING BEFORE ZERO # 2 is new and was never yet introduced. To reach a level of understanding the arguments set in this book as conclusions one has to read the founding principles dealt with in A COSMIC BIRTH... STARTING BEFORE ZERO # 1. If not, then the information in A COSMIC BIRTH... STARTING BEFORE ZERO # 2 will not be understood at all notwithstanding the level of your education in science. Again I repeat: this approach to science has never previously been done.

There are 4 laws applying in nature and that is the part that science ignores because with these laws Nature is Annihilating Newton

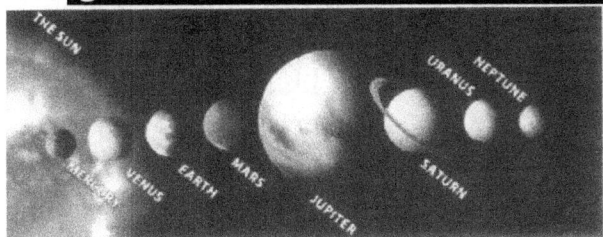

This is the Titius Bode law

The Titius Bode law proves that mass has no place in science. See in the picture how random mass is and with such randomness how can mass place planets in the positions they hold. By my effort to solve the mystery of the Titius Bode law I prove that gravity forms not by mass but gravity forms by Π forming in movement Π^2. Solving the Titius Bode law and proving from that how gravity works opens up a new view on the cosmos.

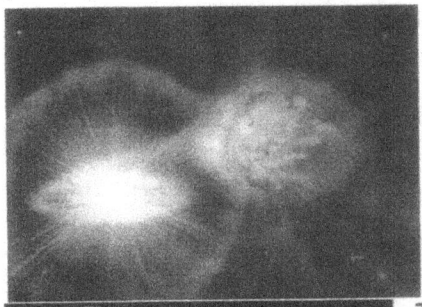

credit: NASA/JPL

This is the Roche limit. The Roche limit proves amongst others how the sound barrier applies and works. It also proves that cosmic structures with an atmosphere can never collide because the Roche limit that produces the atmosphere prevent foreign object from moving faster that $\Pi^2 / 4$ within the boundary limitation of that atmosphere. The Roche limit brings further proof that using the truth about gravity in physics the answer is simple; it is that gravity is Π.

This is the Lagrangian points

The Lagrangian points have been known to science for centuries and with all the mathematical splendour available not one calculation could ever explain why this event is taking place. The satellites form precise locations positioned around the major planet and never comes closer while remaining the their positions.

This is the Coanda effect

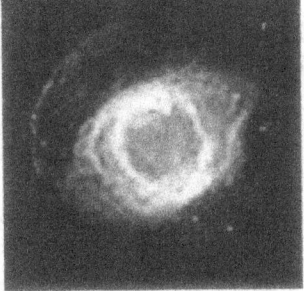

The Coanda effect has powered aeroplanes in flight for almost a mathematical splendour available to aircraft, not one engineer could fact to show understanding why this

turbine engines and century and with all the design the most terrific mathematically compute one takes place.

How sad it is that those physics remain just no more understanding is not

claiming of much superior intellect in than having computing power. The complex.

I have to warn the readers that the topics are showing a very new approach with no quick answers. Understanding is in the proof and that does not come by reading just a few lines and then forming conclusions. The information is new but not hard to grasp. I did not put these phenomena in place and these phenomena nullifies Newton's correctness and that proof I bring goes beyond any doubt. I prove the Titius Bode law. Go to the Internet and see how science doubt the Titius Bode law and the correctness thereof while to solve the problem you add **3** plus **4** to get **7** that is if you want to find a solution. I have published the Titius Bode law in four already published books but in this one I go deeper than the four already published. In each of the books I present I disclose how the Titius Bode law forms gravity

These 4 laws are part of nature. Go look it up before you go on...and why don't science recognize these laws...because these laws brings the entire industry of science to an abrupt end and will stop everything science put in place as science. Recognizing the importance of these laws will kill an industry worth trillions... I now introduce you to the Titius Bode law or how the solar system forms.

There are these four laws or phenomenon or principles in the Universe. Newtonian science can't explain it but I can. In doing so I shatter the myth called Newtonian science and I reveal the hoax science portray for three hundred years or more as the truth. I tried to get published but I was unable because then I break the strangle hold physics has.

Those physicists formulating science are a mafia-gangster club controlling the dishonesty that formulates science... I am one person trying to correct what is a joke but what is also sold as the truth.

Science force humanity to accept the hoax Newton founded and science brainwash everybody to disbelieve nature while clinging on Newtonian hogwash. That is mind-control and that is brainwashing!

I have sent this book with six other books to eighty-five publishers and e-mailed this book to thirty something more. I had no response...not from one. This book opens a new era in understanding how the cosmos works...and not one publisher found it interesting enough to publish or to reply reasons why they found no interest in this as a publishing project. No one is prepared to break the hoax and publish the truth...

For the first time in all of human history there is a method deciphered to show how NATURE no less forms the solar system...and in eighty five DVD's sent plus another (about) thirty six or seven e-mails going via sendspace and not one was interested to publish. Go to the Internet and see it is said this code can't be deciphered but I did find a way to decipher.

That science says is impossible and yet you will read how I did it. Science plainly ignores nature while nature is the reality.

Nature is the only reality but science brushes nature off the table, as if nature is madness. To so many publishers I sent the entire book...I sent it as a unit with two chapters more than the book you read and found no publisher prepared to take on science and correct the hoax Newtonian science is.

Now you can find out how to crack the code by which nature (not Newton's fiction) forms the Universe in the manner that it forms the solar system. It is simple; it is adding 3 plus 4 to get 7! Finding the Titius Bode is 7 also turns what you thought was cosmology into an explanation of the truth while the truth turns cosmology into truthful science. This Titius Bode law, its 7 / 10 or 10/7.

In this book I explain in detail the layout working process of Titius Bode law.

I explain this in detail because what there is in the solar system is this; the Titius Bode law forming in conjunction with the Coanda effect, the Lagrangian points law as well as the Roche lobe / Roche limit that is what forms gravity. This is it!

Newton and his ideology is as absent as the correctness Newtonian science hides and if you do not believe me go and research this by yourself. Giving this truth I contacted (about or more or less) 150 publishers among which there are about fifty or more Universities and not one had any interest in publishing this book.

This is the first time in human history that any person had the inclination to explain the forming of the solar system and nobody is interested in this venture?

This is how space forms and out there amongst all not one shows interest in publishing?

Some just believe because they accept and others have to understand to accept truth and you can choose where you are. Please be warned about the following:

Reading this book will **intellectually** find the reading to be **very challenging** to any person since what I say was never yet published. Everything you are about to read is new! That I in principle disagree with science's accepted principles on very basic issues is a fact that is undeniably true. What you read about the principles I propose is new to everyone alike. However, I found that the ordinary persons with a scholastic physics background cope with the difficult explaining much better than does Super-Educated-Masters. The Super-Educated-Masters have information stored by culture and if they can't bring the information to mind by recognition of it they fail to understand new science concepts. You are going to read this in a letter that was sent to me.

The purpose with which I wrote this book is to get around the network of Super-Educated-Masters who strangle any information that forms of science in the form I propose and therefore that does not fit their views or match their liking. If what anyone says does not stroke with what the Brainy Bunch says who controls physics and agree with "Mainstream Science" or echo their thinking, they just smother all intellectual publication on the grounds that it is not fitting their profile on science. If you ignore culture and read with using your logic you will find many accepted norms as ridiculous. With most concepts I disagree most strongly and I disagree because those concepts lack proof but I do also supply detailed proof of my views and that is where Mainstream Science blocks the publishing of my views on science that does not compliment their views. Read this and wake from the culture you believe in; that which science has lulled you into and made you accept science as the absolute undeniable rock fast truth by instating it as a religiosity then stop reading or get your tranquillising anti depressants next to you with a large bowl of water and a big glass. You will find some mathematical equations, if you are not familiar with it ignore it because it shows the silliness of "Mainstream Science" but if you don't read it you will still understand the explaining by reading the language where I explain it. "Mainstream Science" hides behind maths. I need help to fight their fraud and I need you to help me fight them. What you read I prove to even every last detail and even in this book and therefore I dare anyone to prove otherwise or reprimand me.

You are not going to read a book but you are going to travel a journey. Everything written in this assembly is collected from numerous articles and papers I wrote to individuals, journals, science magazines, Universities, academics in administrative-teaching positions as well as many on line physicists which I tried to interest in my view and my findings on science. I did not remove parts or sections of the articles to disguise this as a book that is ready-written but formed it to be a road on which I travelled and the way I found never-ending rejections. I never tried to make a name in physics but tried to get some money to make a living from and take care of my children. I always knew there were so many smarter brains out there than what I have and who could see what I saw and take the challenge of correcting science from what I brought to the table. I was stonewalled by a bunch of corrupt conspirators trying to lay claim to rubbish Newton presented as truth and in some cases their efforts to justify Newton's lies became pathetically poor. I realised there are those that has brains and then you have those that understand Newton and that is why astrophysics remained backwards as it got stuck promoting forces flying all-over pulling to form gravity. I say this straight: no amount of words can describe my utter disgust I have for those in science thought of as flawless performing beyond blame because they are criminals hiding their criminality.

Why would knowing the Titius Bode improve our understanding of physics?

An in-depth overview of my what my work involves you may visit naturescosmicconcept to go there click on
naturescosmicconcept **or**
ANaturesCosmiConcept : **For a broader basis of the content concerning the concept and what it involves go to** ANaturesCosmiConcept

NaturesCosmiConcept-E-Z: To be informed about what information involves the new antural cosmic concept go to NaturesCosmiConcept-E-Z

NaturesCosmiConcept-E-Z-R: Should you not have time and only need being informed about the basis got to NaturesCosmiConcept-E-Z-R

Part 5 Introducing the Titius Bode Law.

This is true astrophysics. It is not mathematical misrepresentation of what those that has no idea about astrophysics try to sell as astrophysics. This is what the Universe presents as astrophysics. When a study is done on cosmology one would find the Titius Bode law in place ands although it never was acknowledged as such this is what is astrophysics.

It might not be what the Newtonian astrophysicists wanted to find because it does not apply such illustrious and breathtaking physics as they wish to place in the Universe but it is what is in the Universe. This is how the Titius Bode law applies as it is in the solar system.

It is not even for the first time since Newton or for the first time since Galileo or for the first time since Archimedes or who ever is your favourite all-time famous scientist that changed everything about science but this is the first time ever since there was life on earth that the Titius Bode law is explained and is defined by an **explanation.**

If you are smart can you see there is no way in heaven or hell that mass will fit into the factor that is

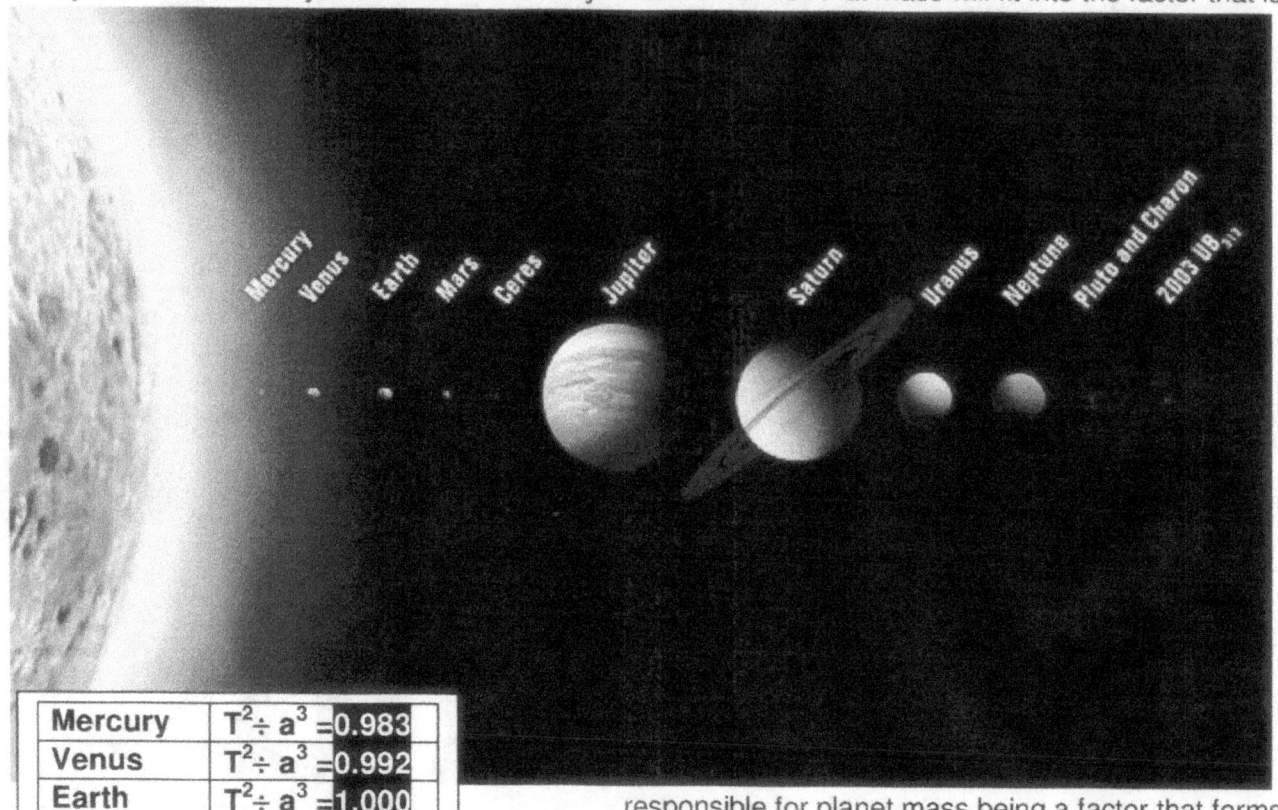

Mercury	$T^2 \div a^3$ =0.983
Venus	$T^2 \div a^3$ =0.992
Earth	$T^2 \div a^3$ =1.000
Mars	$T^2 \div a^3$ =1.000
Jupiter	$T^2 \div a^3$ =1.000
Saturn	$T^2 \div a^3$ =0.999
Uranus	$T^2 \div a^3$ =1.000
Neptune	$T^2 \div a^3$ =0.999
Pluto	$T^2 \div a^3$ =1.004

responsible for planet mass being a factor that forms gravitational layout positioning. This presentation of how the solar system lay out is and then connecting mass as the building block forming the solar system is as false as what money as a currency ever can be. By gauging the tables Kepler left us it is clear that the space **k** is moving in decline or reducing by going smaller in value. This means the sun is drawing all the space not material towards the centre.

The table that Kepler provided shows the time that the orbit of every planet takes according to the distance the planet travels in the same time lapse and considering all the planets it is very much the same thing and in that there is no provision in the table for any idea that might form mass. This ratio is the indication of speed travelled. The idea that mass exists is a Newtonian invention made up by Newton and is completely groundless except for the value that Newtonian science gives it in order to maintain the Newtonian principles. The entire idea of mass is a myth. The idea that mass pulls mass is complete mythology and is as baseless as any fairy story. But even more deceiving is that notwithstanding that every planet has a value when $T^2 \div a^3$ =0.983 which is $a^3 \div T^2 = k$ Newtonian science completely ignores the values and declares that $T^2 = a^3$ whereby they ignore the values in the column. That is cheating the truth into submission to corruption to say the least. Newtonians fabricate their truth.

Using the actual table the mathematical formula says $T^2 \div a^3$ =0.983 which is $T^2 \div a^3 = k^{-1}$. That puts the flow of space including all space directed towards the sun.

$k^{-1} = T^2 \div a^3$ says mathematically in a formula that the circle moves towards the sun because the direction being negative is negatively directed. What the table also say is that the Newtonian presumption that a square can be equal to a cube or $T^2 = a^3$ is colossal madness because the table says clearly in mathematics that $T^2 \div a^3 = k^{-1}$. **Therefore what Newton suggested does not apply?**

What we do find in the solar system and what does apply is the Titius Bode law.

Planet	Mercury	Venus	Earth	Mars	Ceres	Jupiter	Saturn	Uranus	Neptune	Pluto
Bode's Law distance	4	7	10	16	28	52	100	196	-	388
Actual distance	3.9	7.2	10	15.2	28	52	95.4	191.8	300.7	394.6

A numerical sequence announced by J.E. Bode in 1772, which matches the distances from the Sun of the six planets then known. It is also known as the Titus-Bode law, as it was first pointed out by the German mathematician Johann Daniel Titius (1729-96) in 1766. It is formed from the sequence 0,3,6,12,24,48,96, and 192 by adding 4 to each number. The planets were seen to fit this sequence quite well – as did Uranus, discovered in 1781. However, Neptune and Pluto do not conform to the 'law'. Bode's Law stimulated the search for a planet orbiting between Mars and Jupiter that led to the discovery of the first asteroids. It is often said that the law has no theoretical basis, but it does show how orbital resonance can lead to commensurability.

The importance that becomes known is the sequence the Titius – Bode law saw in the number arrangement of 3; 6; 12; 24; 48; 96 etc. The incorrect application of the Titus Bode law lies in subtracting the figure of 3 from 10 leaving 7. The other way of reasoning is to add four each time to the firs value of three starting with 3 and so on. The true significance of the Titus-Bode law is that it points directly to a circular growth of 7 stages. The 7 relating to 10 is a precise derogative of the Roche limit or the Roche limit is a precise derogative of the Titius Bode principle because he two systems interlink. This is how I mange to explain the Titius Bode law that is in the solar system by the ratio applying that really form the solar system in the way nature shows space growing by time. What you see on the next page was never been shown but on the other hand Physicist say this mathematics are too simple to apply as physics! This is what there is. This, the Titius Bode law is a given as it is fact.

With the exception of the first two, the others are simple twice the value of the preceding number.

In singularity the axis is always 3 and the circle is always four making the total of the circular space forming always 7

primary singularity   controlling singularity

The first part of the planet radius distance the inner planet = 7

governing singularity   + 7

The second part of the planet radius distance = 7

The inner is 7 and outer planet is 7

To find the mean distances of the planets, beginning with the following simple sequence of numbers:
0 3 6 12 24 48 96 192 384

To explain this is as follows: the axes line holding 3 doubles because in singularity all axis lines are 3 (1 bottom 2 centre 3 top)

Add 4 to each number:
4 7 10 16 28 52 100 196 388
Then the circle value that always has four point.

Then divide by 10:
0.4 0.7 1.0 1.6 2.8 5.2 10.0 19.6 38.8

Travelling in a straight line or a half circle or a triangle in terms of singularity is equal because it is all 180°. By taking 7 (the first or inner planet) in terms of Pythagoras 7^2 breaking the centre line 1^2 the result is 49 + 1 = 50. The second circle also values 50 and since singularity unites the movement it totals to form 100. The square root of 100 is 10 and dividing the travel by 10 the allocated position becomes valid. Is this not far better and truer than the following Newtonian accepted rubbish?

The Titius Bode law is a part of the Coanda effect and the Coanda effect is the interaction there is between what is material and that which holds material. That which is 7 is part of material and the 10 part is liquid / gas. In the picture the outside planet holds relevance with the line that the direct inner planet has and that in turn holds relevance to the axis the sun holds. The outer planet has the **governing singularity** , the inner planet holds the **controlling singularity** and the Sun has the **primary singularity**. As soon as any object moves on Earth, the movement switches singularity by allowing the object to obtain the **governing singularity** while the Earth then fore fills the directional circular control in forming the **controlling singularity.** All four phenomena interacts in a manner forming this role where for instance in the solar system the Sun holds the **controlling singularity** and Milky Way forms the **governing singularity**

The outer planet is the **governing singularity** that forms a governing position in terms of the location and that has a place in terms of the second 7 that will eventually form the 10 of space. This is the outer 7. The planet on the inside of the planet holding the governing position has a **controlling singularity** since it delivers the four points in rotation that positions the last 7 with which the governing singularity finds position. This hold the value of the inner 7 of which the **controlling singularity** forms a value of 4 as that forms the **controlling singularity** circle in the Titius Bode law. The sun axis holds the **primary singularity.**

The axis around which the Sun turns forms the **primary singularity.** This **primary singularity** is the axis that draws all the space ($T^2 \div a^3$) towards the sun and in that it has the absolute **primary singularity** role of forming the value of 3. The first 3 (the sun) and the second 4 (the **controlling singularity**) forms the first 7 while the **governing singularity** is the positioning point to form the two 7 point that forms the 10.

I am able to explain what is there by using "With a lot of words and some simple algebraic relations, there is no way to "explain" the world of physics." Well where I am out of touch with the world of physics it seems that the world of physics is completely out of touch with reality. Furthermore because "Your seem to be out of touch with modern developments." How improper it might seem when compared to the grandeur of mathematical splendour, mine works because it is there and Mainstream science lives in fantasy because what they see is not there.

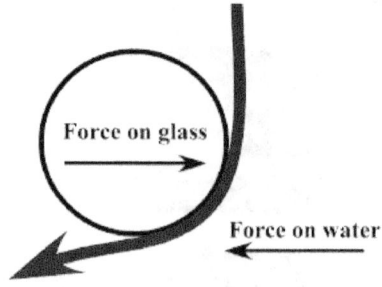

Force on glass

Force on water

What it is that these two phenomena has in common except sharing a Universe because he should know, after all he is the Newton-physics expert ... but I will bet you although being a physics expert, your professor will have no idea...mainly because he is the Newton-physics expert. This prevails because of singularity, which clashes with Newton head on As Kepler said $a^3 = k\,T^2$ and therefore $k^0 = a^3 / k\,T^2$ and therefore we have to find k^0. As a result of examining this proposition, I located two principle positions both holding singularity. The cosmos is made up of one type (1^0) that is in two categories where one type moves and the other type does not move. The one is a liquid and the other is a solid.

The condition for the presence of this singularity that forms everything, controls everything and is everything is the centralised to a centre singularity $k^0 = a^3 / (T^2\,k)$ that forms by movement $T^2 = a^3 / k$ of space $a^3 = k\,T^2$ placed in relevancy $k = a^3 / T^2$ that is centrifugally going both ways $k^{-1} = T^2 / a^3$ thereof (Newton's 3rd law). This explains the Coanda effect and the Coanda effect is gravity and gravity "glues" the water to the glass by implementing Π to form singularity! What is in the Universe is spinning. The entirety of everything forming the Universe is spinning inside the Universe and such spinning are always in the centre of one specific point, wherever such a point might be. In the **precise middle** of all **objects in rotation** is a precise centre where this pre-designated centre is dividing the object in rotation into sectors that will **start the spinning initiation** from that centre point. This is what Kepler's formula confirms in $a^3 = T^2 k$. By spinning, the one side is coming towards while the opposing side at that time is going away. Thus, the spinning object **will have a middle point**, a very specific **centre point that does not spin** and only holds Π as a specific value because within that centre being that small, no radius can apply. We have named this position or line the axis, but the true meaning of this line has eluded us since the concept was realised. This line that forms holds no space although it directs all the space that it controls by spin. When going toward the centre where the axis form at the very centre of rotation, the space on the one side has to end and the space at the other side has to begin with the line unable to hold space.

It is not you being glued or not being glued to the Earth that I discard when I question the mass idea. It is the definition holding this whole idea that I do not share in the least. What the definition describes is magnets pulling and it is the total opposite of what I experience. Breaking the first millimetre of gravity clampdown is the easiest and not the most difficult. The difficulty increases as the radius grows and not as the radius decreases. When I say there is no gravity everyone thinks I say we all are going to fall off the Earth at random and with me thinking that way then it is obvious that I must be a nut. Everyone thinks of me as the clown acting mad when I say gravity is not to be found in nature. But I do not say we are not standing on the Earth. I do not say there is nothing that is keeping me glued to the earth. I say there is no attraction between two bodies by the force of the mass that is such doing is diminishing the radius parting the bodies by the inverse square law. I say there are a connection by motion between the centre of he body and the material surrounding the centre. This is what I say when I say there is no gravity.

The condition for the presence of this singularity that forms everything, controls everything and

is everything is the centralised $\Pi^0 = \Pi^3 / (\Pi^2 \Pi)$ singularity that forms by movement $\Pi^2 = \Pi^3 / \Pi$ of space $\Pi^3 = \Pi\Pi^2$ in relevancy $\Pi = \Pi^3 / \Pi^2$ going both ways $\Pi^{-1} = \Pi^2 / \Pi^3$ thereof (Newton's 3rd law).
This explains the Coanda effect and the Coanda effect is gravity and gravity "glues" the water to the glass! The water forms a value of $\Pi^{-1} = \Pi^2 / \Pi^3$ while the glass forms a value of $\Pi = \Pi^3 / \Pi^2$. This process happens to all spinning things and as much as it happens to a piston connected to a crankshaft, just as much this will happen to an atom spinning an electron in a similar manner as the crankshaft is spinning holding a piston connected. This proves that gravity is the Coanda effect and in another book I prove that the Coanda effect has its origins in Π forming a value and that value forms gravity.

In order to understand physics applying in cosmology I had to start by dissecting the set-up forming pi. At this point I can introduce my theory on the **_Absolute Relevancy of Singularity_** At the point in the centre of the circle a line must start. In the beginning when I explained the way I figured how the line starts I said a lot of dots has to continue in order to form a line. It would be 1 + 1 + 1 etc. because the line must form by holding singularity. After that point does mathematics begin but in the line that forms representing space as other all factors, then time holds 1. The line can only form when all the points forming the line have the value of 1 being 1^0. In that conclusion one realises something must separate singularity from all other factors because singularity hosts all other factors but is by own initiative Π^0. There are always a line of atoms made up of spinning subatomic particles also spinning holding dots Π^0 without space and the line runs through the dots not having space but still connecting by Π^0 or 1^1 forming a line $1^0 \times 1^1$ up to Π.

Only when singularity meets the end value can the end value have Π where the final ring of the spinning circle forms Π. That will be the spot of origin forming the relevance inΠ. That will hold the eternal spot...the smallest spot ever because all spots that ever can be were secured in a position in the centre of that spot that must continue as a line that forms. Because of the progress singularity follows from the single dimension singularity only allows mathematics a start at Π^0 progressing further onto Π^0 and from there the line is born as $\Pi^0\Pi^0\Pi^0$ and to $\Pi^0\Pi^0\Pi^0 \Pi^0$ etc. where Π^0 then may form the concept and value of r. But the line starts at $\Pi^0 = r^0$.

This forms because cosmology is singularity based and the value is $\Pi\Pi^0$. This line $\Pi^0\Pi^0\Pi^0$ of singularity can only continue because every spinning atom preserves Π^0 in the very centre and since $\Pi^0 = \Pi^0 = \Pi^0$ the line is the same without finding conclusion except at the end where it forms mass at Π. At the point whereΠ forms, the movement Π^2 of the circle defines the space Π^3 of the circle and it confirms the centre Π^0 of the circle through the rotation. Let's call this the solid forming or if you wish, let's call it Kepler's singularity. After that singularity forms a line $\Pi^0 = \Pi^0 = \Pi^0$ where this forms another line again as Newton stipulated it by $\dfrac{dJ}{dt} = 1^0$. Let's call that the liquid singularity or Newton's singularity and the relevance of singularity having a solid base compared to the singularity holding a liquid base comes about by the movement of gravity.

a^3 symbolises in a mathematical interpretation of implicating the three-dimensional space holding a specific centre in relation to another specific centre indicated by **k** that could apply to either centre points in question. This is always a straight-line **k** representing the position of the **controlling** singularity moving in a circle T^2. The space forming a^3 is a **positional validity** of the space indicated by $k^0 = a^3 / (T^2 k)$.

T^2 is representing the circle that goes around the **governing singularity** k^0 that forms in relation to the line **k** in reference to the centre k^0 The space that forms holds the orbiting planet a^3 in direct circular contact with the space in relation to a very specific centre k^0 moving from point T_1 to T_2 in relation to a precisely placed centre k^0. The circle coming about from T^2 is the **controlling singularity** which is always a circle at the centre that is poisoned by the line **k** in relation to the centre k^0 and by forming a circle it holds reference to the **governing singularity.** Where **the governing singularity** is the centre of a spinning object such as the Earth, the centre of every atom holds **mutual singularity** that collectively puts a mutual value of all the atoms' singularity as a combined equal to the **governing singularity** and then the solar system will provides a **primary singularity**. The one would represent T^2 the other forms **k** that then produces the third singularity forming space a^3.

k is the space taken from the centre k^0 to the end of the line **k**. This line shows where the location is around which planet circles. The specific value about the centre is most important because from the specific centre gravity indicates a positional worth. The line forming **k** is pointing the circle or the **governing singularity** formed as a line that eventually forms a circle running from the centre k^0 to where the space a^3 is indicated.

The turning T^2 of any circle holding space a^3 is valid only if forming a reference **k** to a centre k^0. $k^0=a^3/(T^2 k)$. This depicts a position a domineering singularity k^0 fills in relation to another point serving subordinate singularity **k**. There are always a dominant and a serving singularity interacting. If **k** indicates the centre of the Earth then T^2 rotates to form the **governing singularity** k^0 where then the centre of the Sun **k** will form the **controlling singularity.** When the Sun rotates, the Sun's centre k^0 forms the **governing singularity** giving the Earth in orbit **k** holds the **controlling singularity**.

The measure of **k** is not a specific value but serves only as an indicator to which space rotates or applies by the space rotating in a circle. This role of singularity being **controlling** or **governing** is playing part in movement of gravity forming and is very important when trying to understand the role that the four phenomena play in the forming of gravity. It is most important to understand what happens in the event of an object going through the "sound barrier" or when escaping from the Earth's atmosphere.

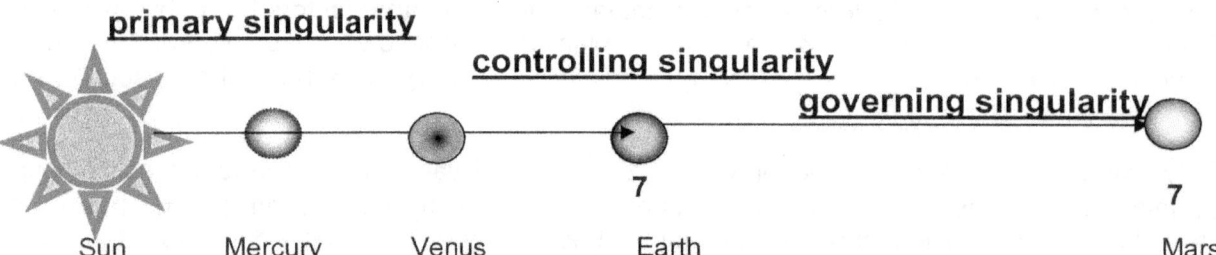

Where the object is standing still holding a position that allows the object to have mass, the object is part of the Earth while the Earth has the **governing singularity** and the Sun has the **controlling singularity**.

As soon as any object moves on Earth, the movement switches singularity by allowing the object to obtain the **governing singularity** while the Earth then fore fills the directional circular control in forming the **controlling singularity.** All four phenomena interacts in a manner forming this role where for instance in the solar system the Sun holds the **controlling singularity** and Milky Way forms the **governing singularity** .

To Venus forming the **governing singularity** Mercury is the **controlling singularity** and the sun is the **primary singularity**.

To the earth forming the **governing singularity** Venus is the **controlling singularity** and the sun is the **primary singularity.**

To Mars forming the **governing singularity** the earth is the **controlling singularity** and the sun is the **primary singularity.** That is why this table forms.

Planet	Mercury	Venus	Earth	Mars	Ceres	Jupiter	Saturn	Uranus	Neptune	Pluto
Bode's Law distance	4	7	10	16	28	52	100	196	-	388
Actual distance	3.9	7.2	10	15.2	28	52	95.4	191.8	300.7	394.6

Where there is anomalies we can read into it events that happened in the past we were unaware of.

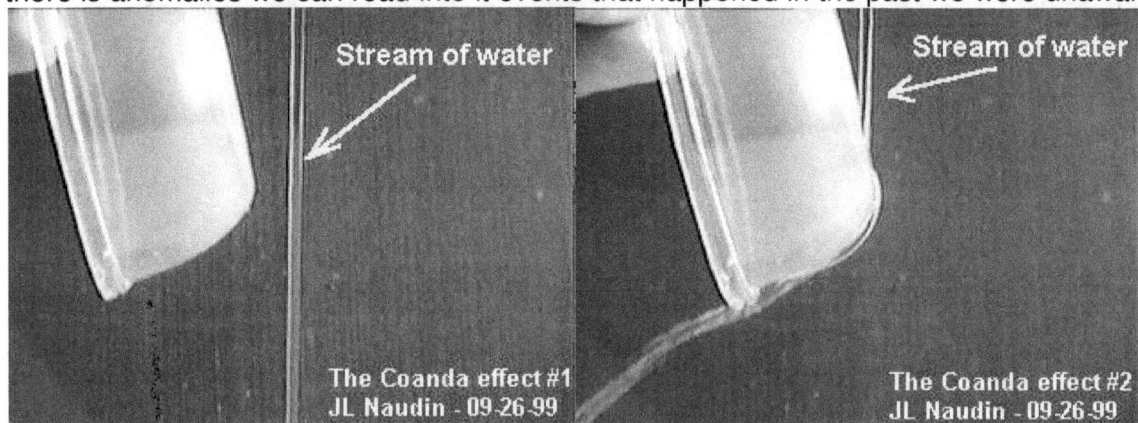

I have **written twelve articles** in which I explain the Titius Bode law, why it is in place, how does the Titius Bode law apply gravity, what keeps the Titius Bode law structurally in place and why is it in place as it is.

The gravity it should hold in distributing movement will bring along a certain amount of linear gravity to maintain the circular gravity position it wants to hold in the cosmic balance. The duel movement of gravity forming the allocated relevant position provides ratio of singularity Π accompanied by the same value but in movement by the square thereof Π^2 and repositioning the structure as a star. As the influence of the axis (3) extends it increases the circle and its speed of circling and that is the Coanda effect and the Coanda effect is gravity.

$$\Pi^0 = \Pi^3 \div \Pi\Pi^2$$

$$7\Pi\Pi^2 \times \Pi^0 - 4\Pi^0$$

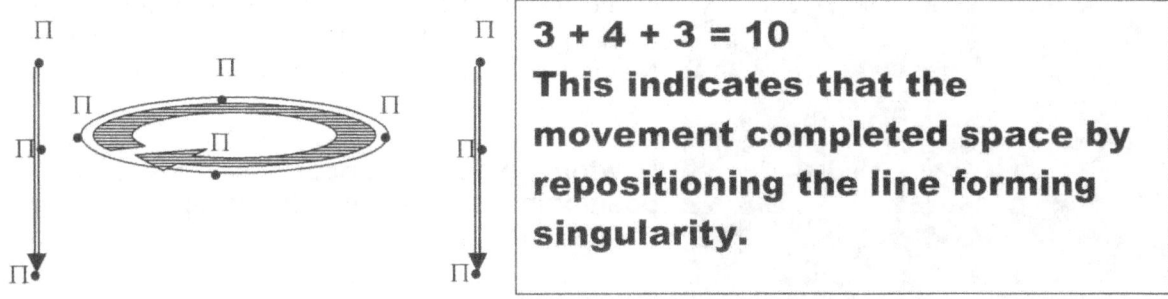

3 + 4 + 3 = 10
This indicates that the movement completed space by repositioning the line forming singularity.

This is gravity in so far as the spin (4) dictates the axis (3) and that is why we can see the sun form an arch as the sun passes over our heads. This is how the Universe formed according to mathematics many stages before the Big Bang even became a concept. Every dot in the Universe either form as part of 3 or 4 in relation to every other dot and every dot is either the axis with a circle turning around it or is part of the circle turning around the axis. This is how creation started when it started long before it started with light becoming a dimension and a factor. In that we have the primary singularity, the controlling singularity and the governing singularity that places everything there is relevant to everything there is.

In short this is my explanation. Now please compare this to the accepted mainstream version and see why all my work up to now is ignored as if it holds the plague. See now what is science and what are fiction and why my work is ignored.

That is the "With a lot of words and some simple algebraic relations, there is no way to "explain" the world of physics." "You seem to be out of touch with modern developments." **of which I am accused of.**

Please allow me to show you how they scare you to become fooled and suckered. Don't run and hide when you see the mathematics; it is meaningless although it was used as a scarecrow for more than three hundred years forming the backbone of the conspiracy.

This picture is as big a hoax as Newtonian science is when Newtonian science presents mass to be a factor that produces a pulling force called gravity. The question shouting for an answer in the picture is if mass is a factor that produces gravity as Newtonians claim it is then why are the planets not positioned according to mass as Newtonians declare.

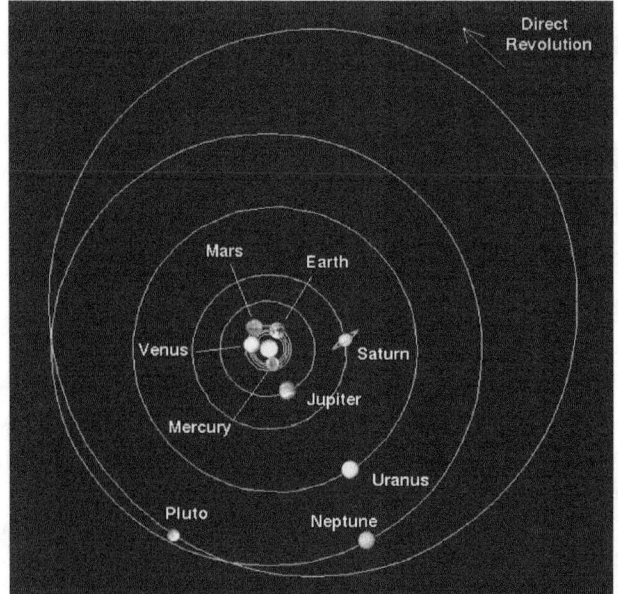

The claim is as bogus as the entire philosophy. They present the proof that planets orbit according to mass in the following "Kepler law" which in its entirety had nothing to do with Kepler at all. It is all devised by Newton because Newton had no inclination of what Kepler's work was about.

Newton brought about the idea of mass positioning the planets in the formula $4\pi^2 a^3 = P^2 G(M + m)$

Do not get scared as everyone usually does when seeing the mathematics and then as a result get frightened. Those physicists expect you to turn on your heels and run as fast as your legs can carry you. Then consequently as a reaction to find survival, you turn on your heels and run… but this time don't. Don't run, just read on and see how simple it is to prove Newton was a backward dark aged sod!

This time, don't run because I am about to show how meaningless this entire mathematical statement in reality is! This formula is total garbage and there is no sign of evidence that this formula forms any part of the solar system, even in the least.

 Lets test this formula and see how truthful it is. $4\pi^2 a^3 = P^2 G (M + m)$ indicates that the circle in which the planet orbits ($4\pi^2 a^3$) is the result of (=) the position of the body (P^2) positioned by the mass of both bodies ($M + m$) in terms of the gravitational constant (G).

The best way to find clarity is to test this statement with what is happening in the solar system just as it is, wouldn't you think.

A picture such as this provides much credence to the idea of gravity by mass since the lines drawn does not even begin to represent what is truly out there ands what is used by the cosmos in place of Newton's mass concept.

This is a table indicating the **mass** that **every planet** has in relation to the **distance every planet holds** in terns of the sun. If mass positioned planets then why did no one bother to inform the Universe about this because it is clear the Universe did not receive the memo from Newton's office to act accordingly.

Body	Mass (10^24) kg	÷	Orbital Distance(10^6 km)	=	ratio
Mercury	.3302	÷	57.9	=	0.0057
Venus	4.869	÷	108.2	=	0.045
Earth	5.975	÷	149.6	=	0.039939
Mars	0.6419	÷	227.9	=	0.00281
Jupiter	1898.6	÷	778.3	=	2.439419
Saturn	86.83	÷	1427	=	0.060847
Uranus	102.43	÷	2869.6	=	0.03569

This picture shows the hoax the Newtonian conspiracy pampers to keep the rest of Newtonian physics believable. They never mention the Titius Bode law and try to explain the Titius Bode law while it is the Titius Bode law that is really in place in the solar system. If *you wish to learn the truth then think again.*

Mass as a factor does not present or apply in one instance anywhere in the entire Universe and yet that is all theta physics says applies...but why would they cheat? It is because what is there applying between the planets is called the Titius Bode law and although this law is in place you have almost a hundred percent chance that you have never heard of it.

This is how it would apply if Newton was correct and mass did position planets. There is not even a remote chance that the positioning of the planets go in accordance with mass or **4π²a³ = P²G (M + m)**. Do you realise there is much more "gravity produced by mass" in the space your feet has contact with the earth than there could ever be between Jupiter and the sun?

You that can calculate it
$$\left(\frac{P}{2\pi}\right)^2 = \left(\frac{a^2\sqrt{1-\varepsilon^2}}{\ell}\right)^2 = \frac{a^4(1-\varepsilon^2)}{\ell^2} = \frac{a^4(1-\varepsilon^2)}{a(1-\varepsilon^2)GM} = \frac{a^3}{GM}$$ so then do it.

Show that $\dfrac{a^3}{GM^2}$.

Put the orbit of Jupiter in relation to the mass of Jupiter and in relation to the position Jupiter holds. Forget getting swept away by the fancy Mathematics; just get to the task of putting the mass in relation or ratio with the position that any of the planets hold.

Take the mass of the earth and your mass you have and then divide that with the square of the distance there is between your feet and the earth by the square thereof then divide that square with the product of the mass of the earth and your mass you have. Keep in mind your distance between your feet and the earth is about 10^{-11} meters going square!

You that can't calculate it, the value would be meaningless but it is so much it will crush your atoms into a pulp, leaving you not even in a blood blob. It will leave a force to the value of about one Zetta 1 000 000 000 000 000 000 000 g per square meter. There is no chance in hell that any object of whatever size and formed by whichever method of construction could manage such a force of pressure.

Newtonians uphold their law of physics without showing mercy. The very first things the Newtonians use to beat us into submission are to blast us with incomprehensible mathematical formulas.

Those Physics-cheats always want to have the radius down to 1 meter because it makes their argument look sensible, and then they "forget" to use the correct mass of the earth not to stun any student into realising reality. Using one meter makes good sense when you wish to cheat but no body floats on meter above the earth. When anything stands on the earth, the distance between such a body and the earth is less than what could be sensibly measured.

Newtonians uphold their law of physics without showing mercy. The very first things the Newtonians use to beat us into submission are to blast us with incomprehensible mathematical formulas.

Incomprehensible they are but it is to scare anyone with the mathematical equations to get everyone hiding. They bewilder you with equations that put the fear of God into you; used simply to make you feel inferior so that they can feel superior and frown down on your inferiority from a dizzy height.

What Newton show that should place planets according to mass is not used by nature. I show what nature uses namely the Titius Bode law, The Roche limit, The Lagrangian Points and the Coanda effect and how this forms gravity as well as place the positions of the planets in accordance with singularity. Because I trash Newton's rubbish that does not fit and that can't apply no publisher of science books or science magazines will publish my work I show what goes on in nature while Newton's contribution of mass applying is total rubbish. Because I call it rubbish and I rubbish Newton I am ignored.

Please look at the information next to the planet showing size and then with the use of the mass and the allocated position of the planet use the Newtonian formula that is set in such a high esteem

$$\left(\frac{P}{2\pi}\right)^2 = \left(\frac{a^2\sqrt{1-\varepsilon^2}}{\ell}\right)^2 = \frac{a^4(1-\varepsilon^2)}{\ell^2} = \frac{a^4(1-\varepsilon^2)}{a(1-\varepsilon^2)GM} = \frac{a^3}{GM}$$ to conclusively mathematically

pot the that each planet is according to its mass.

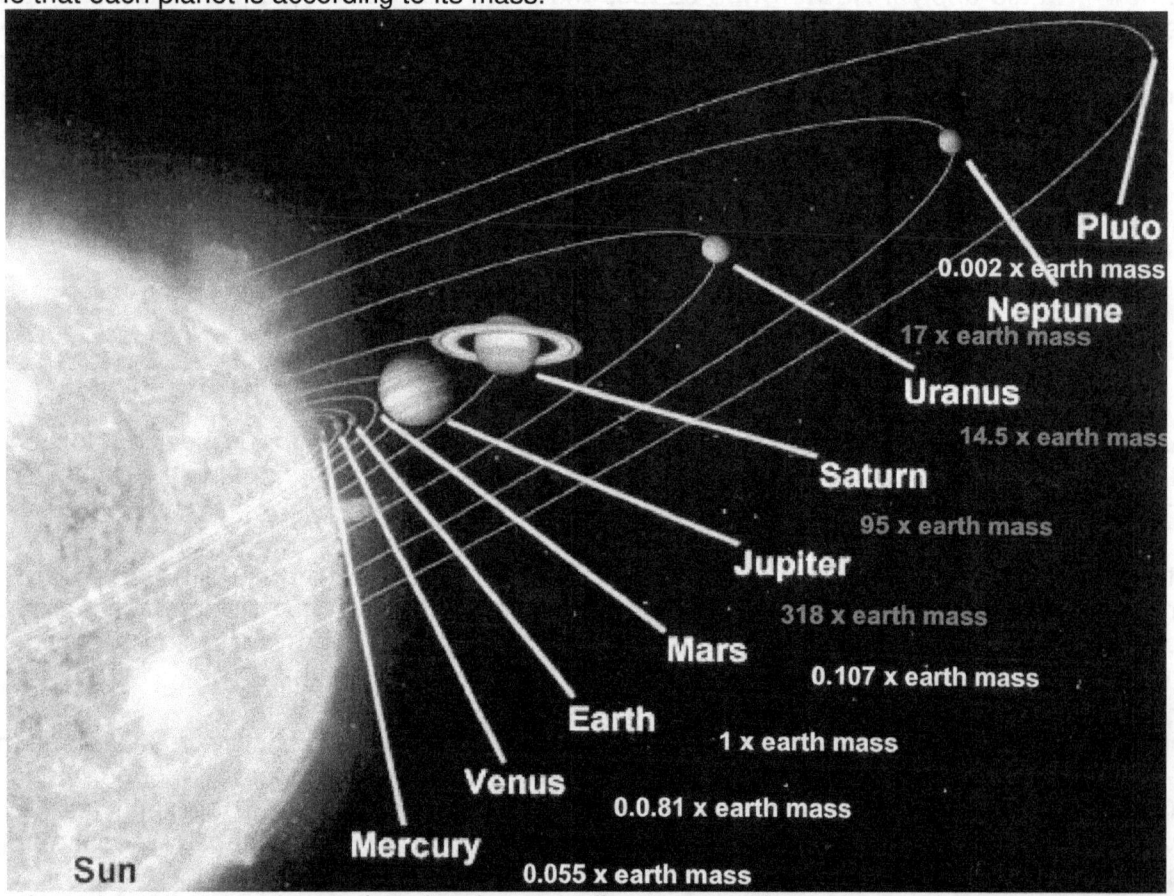

If the planet layout was as I now show it to be according to mass then this was the order that is in place in the solar system and if it is true according to the solar system that mass do produce the position of the planet:

1) Jupiter	318 x earth mass	at a distance of	57.9 million kilometres
2) Saturn	95 x earth mass	at a distance of	108.2 million kilometres
3) Neptune	17 x earth mass	at a distance of	149.6 million kilometres
4) Uranus	14.5 x earth mass	at a distance of	227.9 million kilometres
5) Earth	1 x earth mass	at a distance of	778.3 million kilometres
6) Venus	0.81 x earth mass	at a distance of	1427 million kilometres
7) Mars	0.107 x earth mass	at a distance of	2871 million kilometres
8) Mercury	0.055 x earth mass	at a distance of	4497 million kilometres
9) Pluto	0.002 x earth mass	at a distance of	5913.5 million kilometres

They are masters at manipulating anyone into a state of senselessness...but mostly that they do onto themselves. That they do because it forms the backbone of their fraud. They do not wish you to read closer and to find the fraud they hide to protect Newton. Ignore their mathematics because it only shows their incompetence to understand physics or Newton and see the fraud they propagate...

They employ mathematics to bewilder and that is all. I am going to show what we can uncover underneath what they cover. Look at what the mathematics supposedly says and then wake up, they are using maths as a scare tactic for three centuries to scare the daylights out of you and all this while its been working! Looking at the formula shows just how little Newton understood physics.
 Do not get scared as everyone usually does when see and get frightened then consequently as a reaction to find survival you turn on your heels and run... Don't run, just read on and see how simple it is to prove Newton was a backward dark aged sod!

I am going to show you how miserably incompetently incorrect this formula of Newton is when shown what the cosmos has in place. However, since I don't support Newton's blunders my work goes unpublished by science magazines and science publishers.

Bring me any mathematical formula Newton devised where he used mass as a factor and I show you how far the cosmos discards all of his claims. Newton is hugely wrong.

What the cosmos does use Newtonians reject because they can't explain it, they know too little about physics and secondly it rubbishes whatever fraud Newton thought up.

$$P = \left(\frac{4\pi^2 a^3}{G(M+m)}\right)^{0.5}$$

If the cosmos supported Newton's claims of [above] then the planet arrangement would have been much more likely as I show above, but the picture indicates the mass as well as the planet formation. You must judge; it is either the cosmos that is incompetently wrong or it is Newton that is incompetently wrong because what the cosmos has in place Newton knows nothing about and what Newton claims the Universe uses, the cosmos knows nothing about. Who would you say knows more about the cosmos' method of workings, Newton or the cosmos? If Newton is correct then the planet layout must be as I show with Jupiter very close to the sun. It seem the cosmos is just as unaware of Newton's ideas as Newton is of what is happening in the cosmos. Who would be correct about cosmic principles applying, the cosmos or Newton?

Our Super-Educated-Mathematical-Wise use the elaborate formulas such as $\frac{d}{dt}\left(\frac{1}{2}r^2\dot{\theta}\right) = 0,$ and $P = \left(\frac{4\pi^2 a^3}{G(M+m)}\right)^{0.5}$ as well as $T^2 = \frac{4\Pi^2}{G(M+mp)}a^3$ to explain and prove what? If this formula statement is true then Jupiter must spin 317 times faster around the sun than the Earth does and be almost next to the sun while Mercury and Pluto in comparison must hardly move being cast into the darkness of the oblivious.

The formula $T^2 = \frac{4\Pi^2}{G(M+mp)}a^3$ says that the spin or time in which the circle comes about T^2 holds a space relation a^3 directly proportionate to the mass $(G(M+m))$ of both the sun and the applying planet multiplied by placing the gravitational constant also in relevancy. It says planets spin in accordance to mass...so lets see. I say prove it! Then $\frac{d}{dt}\left(\frac{1}{2}r^2\dot{\theta}\right) = 0,$ says that the Planets don't move at all because with $\dfrac{d}{dt}\left(\dfrac{1}{2}r^2\dot{\theta}\right)=0,$ the movement acceleration is zero. Zero indicates no movement and that is as corrupt as the rest of Newton's ideas. Every person associated with physics to whatever extent has been and had been conspiring to hide the truth and the truth is that no one knows (not even Newton and even less Einstein) what physics is. What they say is precisely what the solar system proves different and either the solar system has no idea about what is driving gravity or they have no idea of what drives gravity, but it is definitely not mass. If they don't know what forms gravity and it is clearly not mass, then they know nothing about the way that gravity forms. They hide their incompetence under a blanket of ignorance and the lot commit to a conspiracy to hide the truth from the public. Let any one of those So –Wise-In-Mathematics show just how does Saturn arrive at the position it holds in

relation to the mass it has by applying the Newtonian formula $P=\left(\dfrac{4\pi^2a^3}{G(M+m)}\right)^{0.5}$, which they claim is data confirming Kepler's Law of Periods of the motion of the planets in relation to mass. Please prove

this $m\ddot{\mathbf{r}}=-\dfrac{GMm}{r^2}\hat{\mathbf{r}}$ assumption Newton made. The go better and prove how Newton could have

formulated $\ddot{r}-r\dot{\theta}^2=-GMr^{-2}$, and be realistic aboput what trully happens.

Now I challenge anyone out there to show Newton is not rubbish and Newton is correct. Please use the

formula $T^2=\dfrac{4\Pi^2}{G(M+mp)}a^3$ to prove Newton and then explain why the Universe does NOT apply

Newton in the event where Newton is so admirably correct.

The formula that they say must allocate planet positions are $4\pi^2a^3 = P^2G(M + m)$ just as Newton introduced this concept because he said mass brings about gravity.

In this he had to force the issue even to the point of committing fraud and here comes the fraudulent part because there is no evidence of mass playing a part or forming an actual presence in the solar system.

$P=\left(\dfrac{4\pi^2a^3}{G(M+m)}\right)^{0.5}$ What hogwash does the factor $\overline{G(M+m)}$ indicate? The same can be

said in the formula $M=\left(\dfrac{4\pi^2a^3}{GP^2}\right)-m$ when $P=\left(\dfrac{4\pi^2a^3}{G(M+m)}\right)^{0.5}$ that the factor $\dfrac{P^2}{}$ is

senseless and $\left(\dfrac{P}{2\pi}\right)^2=\dfrac{a^3}{G(M+m)},$ has no foundation other than fraud. It says the position of

the planet is derived from the mass $M=\left(\dfrac{4\pi^2a^3}{GP^2}\right)-m$ that each planet has and when viewing the reality that is totally and complete fraud. The Cosmos does not support the Newtonian formula even in one place where it could apply.

Position as a function of time $P = \left(\dfrac{4\pi^2 a^3}{G(M+m)}\right)^{0.5}$ This is what Newton said is in place and with no evidence ever founding this ridiculous proposition, all Newtonians that ever come after Newton. This is what Newton and his Newtonian followers tell the solar system it has in place and tell the cosmos it uses to operate.

I have indicated that mass has no place or use in the solar system according to what the solar system puts in place. According to Newton $P = \left(\dfrac{4\pi^2 a^3}{G(M+m)}\right)^{0.5}$ puts the location or position $\dfrac{P^2}{}$ of a

planet in relation to the mass $\dfrac{}{G(M+m)}$ of such an individual planet. I am coming back to this and then you will choose which of us Newton or me, is committing blatant fraud.

I say I can prove Newton correct by showing the formation of the planets in orbit going around the sun.

With implementing Newton's formula $P = \left(\dfrac{4\pi^2 a^3}{G(M+m)}\right)^{0.5}$ the planet distribution are as follows:

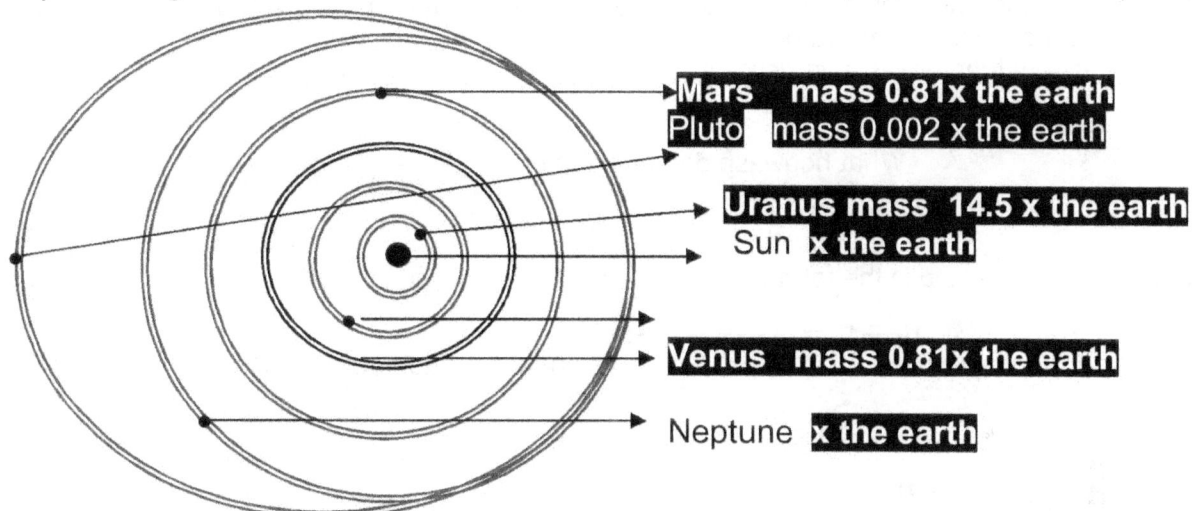

Mars   mass 0.81x the earth
Pluto mass 0.002 x the earth

Uranus mass 14.5 x the earth
Sun x the earth

Venus   mass 0.81x the earth

Neptune x the earth

The inner circles that are very close to the sun we find the big gas planets with so much more mass forming the force of gravity that these planets are almost on top of the sun so close they are to the sun.

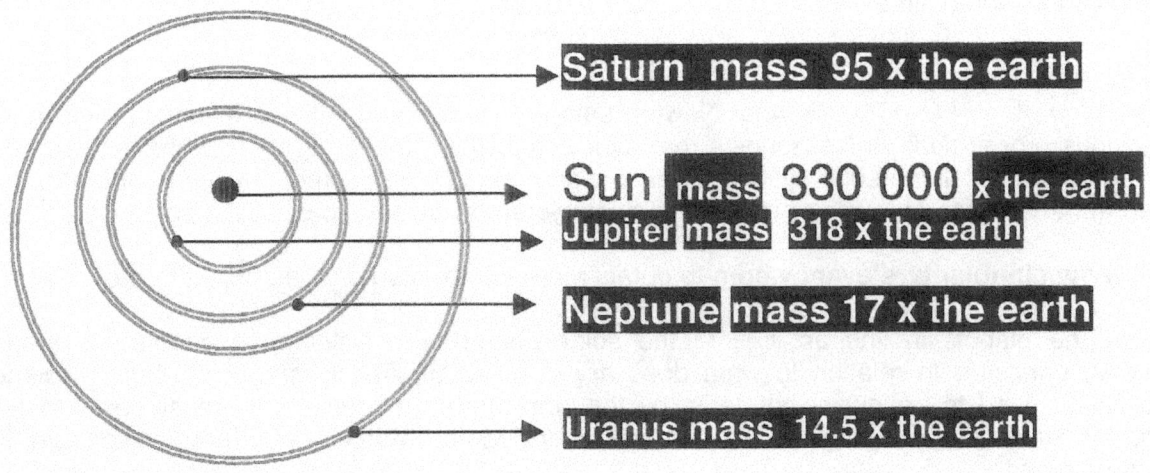

Saturn mass 95 x the earth

Sun mass 330 000 x the earth
Jupiter mass 318 x the earth

Neptune mass 17 x the earth

Uranus mass 14.5 x the earth

Then in the outer circles that are very far from the sun we find the smaller solid planets with so much less mass the force of gravity just can't pull these planets closer to the sun.

Closest 1) Jupiter mass 318 x the earth
    2) **Saturn mass 95 x the earth**
    3) **Neptune mass 17 x the earth**
    4) **Uranus mass 14.5 x the earth**
Then come the smaller planets with less mass and therefore less pulling force called gravity
    5) **Earth 1 x the earth**
    6) **Venus   mass 0.81x the earth**
    7) **Mars   mass 0.81x the earth**
    8) **Mercury mass 0.055 x the earth**
    9) **Pluto mass 0.002 x the earth**

The above can be the only designated outlay of the planet position according to the sun when applying

$$P = \left(\frac{4\pi^2 a^3}{G(M+m)}\right)^{0.5}$$

mass. Remember, if I am wrong then Newton and his formula is wrong but if I am correct and Jupiter is the closest planet then Newton is correct. However facts show my suggestion is incorrect and so is Newton and then the academics in physics are indulging in Newton's fraud by forcefully brainwashing students to believe in Newton notwithstanding the fact that it is the solar system that disputes Newton's ideas altogether. If you disagree with my layout, then you better disagree with Newton and his ideas

Tell me, can you find any credence in the "Conversions for "Unknown""
$4\pi^2 a^3 = P^2 G(M + m)$
In this comes the fraudulent part because there is no evidence of mass playing a part or forming an actual presence in the solar system.
  Lets put Mercury as a yardstick and see how Newton's formulas pan out.

$$P = \left(\frac{4\pi^2 a^3}{G(M+m)}\right)^{0.5}$$

What hogwash does the factor $\dfrac{}{G(M+m)}$ indicate?

The same can be said in the formula $M = \left(\dfrac{4\pi^2 a^3}{GP^2}\right) - m$ when $P = \left(\dfrac{4\pi^2 a^3}{G(M+m)}\right)^{0.5}$ that the factor

$\dfrac{P^2}{}$ is senseless and $\left(\dfrac{P}{2\pi}\right)^2 = \dfrac{a^3}{G(M+m)},$ has no foundation other than fraud.

$$M = \left(\frac{4\pi^2 a^3}{GP^2}\right) - m$$

is complete fraud. The Cosmos does not support the Newtonian formula even in one place where it could apply.

Position as a function of time

$$P = \left(\frac{4\pi^2 a^3}{G(M+m)}\right)^{0.5}$$

This is what Newton said is in place and with no evidence ever founding this ridiculous proposition, all Newtonians that ever come after Newton. This is what Newton and his Newtonian followers tell the solar system it has in place and tell the cosmos it uses to operate. I have indicated that mass has no place or use in the solar system according to what the solar system puts in place.
  Visit **www.singularityrelevancy.com** to obtain more information on the subject free of charge.

There are the planets in line as it is in the solar system. Put anything connected to Newton's gravitational principles in relation to what does apply in reality. At this stage Newtonian science is science fiction used to brainwash students by thought control in pressuring the students to accept Newton. It is more effective mind control than any other dogma enforcement of any religion ever in the history of mankind.

These are the closest because these are the massive giant gas plants and having the most mass must put them the closest to the Sun. However, the location is random and not by mass in any way. Forcing students to accept the truth about mass is creating a staged science platform that annuls all other concepts that might sprout from this dogma. This is the most condemning religiosity that enforces a make believe in fantasy like no other religion can offer.

Now we show what is really in place as the cosmos has the validation. The Titius Bode law is in place holding every available piece of evidence of what the solar system uses. The Newtonian mathematical mongers will put the mathematics in place to scare everyone out of their wits should anyone ask questions.

The mathematics is there to indicate the non – Newtonians inadequacy of understanding the issues. But guess what, it is the mathematics that show the stupidity raging amongst the Newtonians. They approve the use of mass and was too stupid the question or to verify mass forming gravity.

Later on I shall go into what the cosmos uses in forming the solar system. I shall explain how the Titius Bode works which is what is in place. There is not a sign of mass but for my effort to explain how singularity forms gravity in applying singularity as Π therefore I am rejected as mad.

$$P = \left(\frac{4\pi^2 a^3}{G(M+m)} \right)^{0.5}$$

Get your professor to prove Newton correct in the face of and if he can't let him admit he has been conducting in a fraudulent practise all the time he was teaching.

Gravity is the cooling of space. As space is being is placed in transformation, the time component which then can increase the spin value, or time by fore times the value. Therefore, it will transform fore times more unoccupied space-time to densified space-time, which is the name by which I call material.

Gravity forms when a solid holding an Iron core in relation to copper turns in a cosmic gas, which is the name that I gave to the singularity substance forming outer space. The turning of the solid inside the cosmic gas re-forms the cosmic gas into a cosmic liquid.

Science has to realise that the cosmos formed as the bible says and if there is any atheist idiot out there trying to go get smart about religion then read this book and see how wrong you lot are that believe in science and then order **The Veracity of Gravity** from Lulu.com and see how correct the Bible is.

Creation started to a T exactly like the Bible says and precisely according to Genesis 1. Time split into what is infinity forming time and into time we see as space, which is a singularity substance, controlled by movement exceeding the speed of light in relation to cosmic liquid / cosmic gas moving below the speed of light.

Gravity is a relative movement that has a relation between what is solid and what is not solid.

Gravity starts at Outer space as {10 /7 [4(Π²+Π²)]} = 112.795

Gravity moving to
{7/ 10 [4(Π²+Π²)]} =55.27 and ending at {Π [4(Π²+Π²)]} = 62

Gravity is about outer space $\{10 /7 [4(\Pi^2+\Pi^2)]\}$ = 112.795 conformed by iron $\{7/ 10 [4(\Pi^2+\Pi^2)]\}$ = 55.27 the displacement value of iron$_{55}$ in relation with singularity presented by the relevancy of $\{\Pi [4(\Pi^2+\Pi^2)]\}$ = 62 and that is the displacement value of copper which is the element that has this proton value according to the periodical table. This is exactly how electricity is generated and that proves that gravity and electricity is the same thing charged on different levels.

Fusion within stars does not come about "pressure" but the star freezes the atoms into forming a new unit. At this point I must also say I disagree with this idea that atoms freeze together but atoms rather grow into... The earth has not the gravity to "freeze" any type of fusion process and therefore all attempt this far was futile and will remain futile because the earth must first find the gravity to freeze hydrogen into a water-like substance as the sun does before fusion could be a possibility. Newtonians rob the public blind with their falsifies ideas about things they clearly don't understand and it is about time the public put a stop to this criminal behaviour. Either they stop to bullshit the public and promote wild schemes or they go to jail.

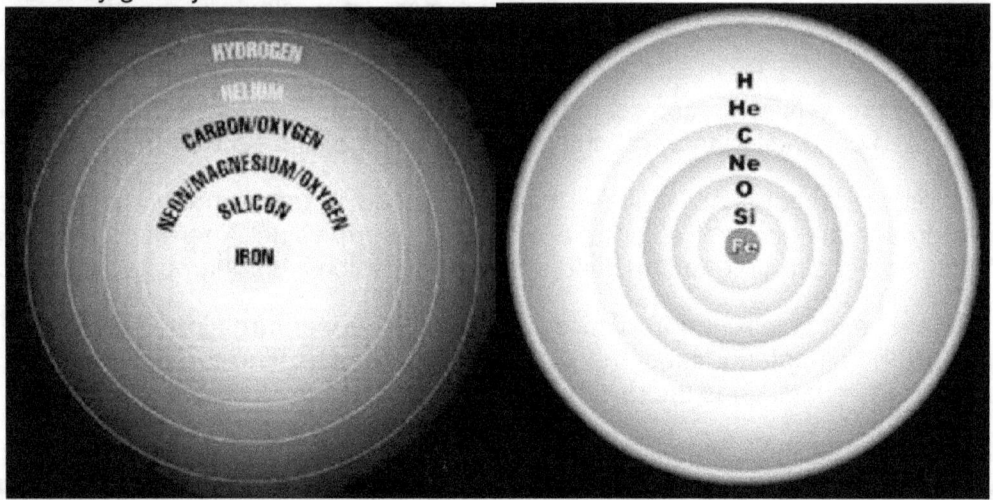

Gravity and electricity is the precise same principle applying on different dimensions.

$\Pi ((\Pi^2+\Pi^2)+(\Pi^2+\Pi)+3)$ $7 / 10(\Pi^6\div6)$ $10/ 7(4(\Pi^2+\Pi^2)$
=112   =112   = 112

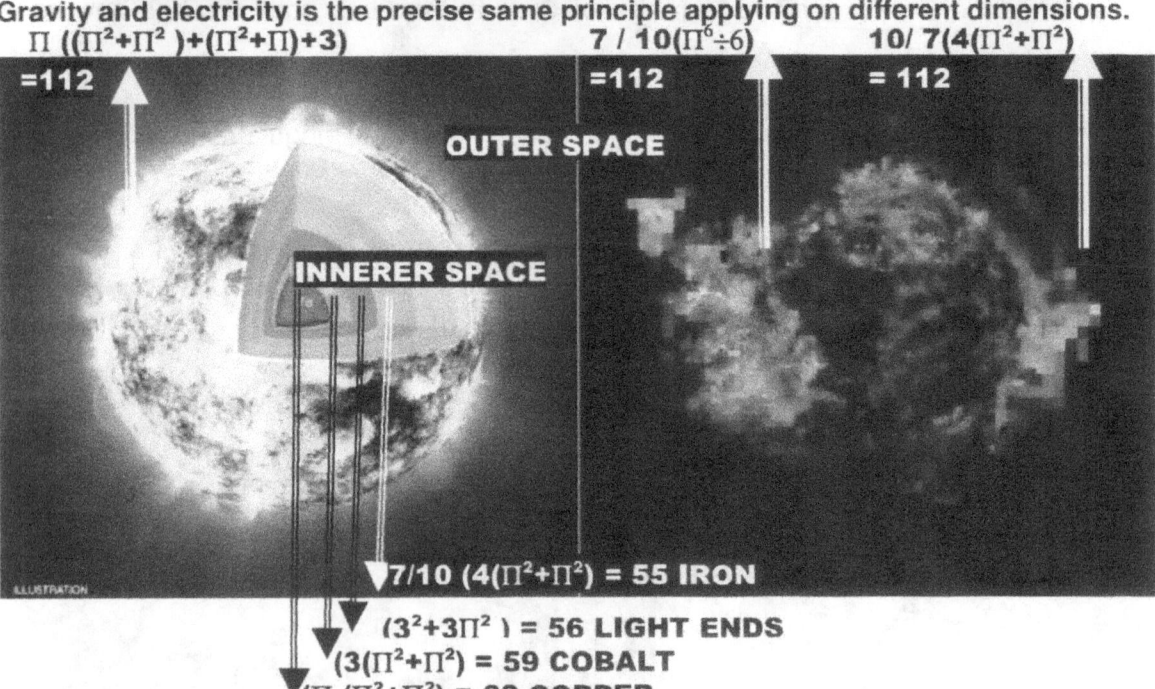

OUTER SPACE

INNERER SPACE

$7/10 (4(\Pi^2+\Pi^2)$ = 55 IRON

$(3^2+3\Pi^2)$ = 56 LIGHT ENDS
$(3(\Pi^2+\Pi^2)$ = 59 COBALT
$(\Pi (\Pi^2+\Pi^2)$ = 62 COPPER

Outer space forming a sphere is **7 / 10($\Pi^6\div6$)**

Outer space forming cube **10/ 7(4($\Pi^2+\Pi^2$)**

Space allowing movement limitation is $\Pi((\Pi^2+\Pi^2)+(\Pi^2+\Pi)+3)$ = **112**

Inner space gravity moves to Iron at 7/10 **(4($\Pi^2+\Pi^2$)** = **55**

Inner space light ends dimensional value at ($3^2+3\Pi^2$) = **56 LIGHT**

Inner space dimensional collapse at **COBALT**

Complete space collapse at inner space dimensional at Copper ($\Pi (\Pi^2+\Pi^2)$ = **62**

All stars have an $iron_{56}$ core. At present, the Universe is in the iron peak era. Every star has to have an iron / copper core and the relevancy between outer space on the outside holding hydrogen and the iron forming gravity or electricity in relation to spinning around copper forms gravity. In this the Titius Bode law proves all of this. No star can function with out having this layer balance. Should this layer balance go array in any one of the layers the layers above it would not be able to withstand the time development, the matter will overheat and the matter will be transform to unoccupied space-time. This disaster is not a natural growth process, but a disaster of catastrophic proportions. The movement of iron $\{7/\ 10\ [4(\Pi^2+\Pi^2)]\} = 55.27$ in terms of the singularity product copper $[\Pi(\Pi^2+\Pi^2)=112.795$ **allows heat that is totally expanded at the value of** $0\{10\ /7\ [4(\Pi^2+\Pi^2)]\} = 112.795$ **to reduce. Gravity is the movement of liquid in relation to a solid.**

This again was proven by the very first ever experiment concluded scientifically. This fact of space descending does not come as a surprise because Empedocles proved this fact back in 450 BC. Empedocles showed that space displaces water from the clepsydra, which was a sphere shape container with a sprout on the top and small holes in the sphere through which water ran in small streams out at the bottom.

 When the flow of air or space was blocked in the spout by a finger covering the hole at the top of the sprout at the entry, the water stopped flowing from the clepsydra. They concluded in 450BC that it is the empty space that pushes the water out of the clepsydra because the moment one restricts the empty space or air to flow into the clepsydra from the top, the water will stop flowing out of the bottom of the clepsydra.

Why would the flow of the water stop if the mass did pull the water down? When the finger blocks the sprout and stop the space entering from the top, the water does not fall to the ground but it is the empty space that pushes the water out at the bottom to fill the clepsydra from the top. When the finger blocks the sprout and stop air to come in through the sprout opening the water should still run out at the bottom by the mass of the water pulling, if mass was doing the pulling. If mass was the force giving factor, then the water must keep on flowing because the mass of the water did not disappear when the sprout was covered and therefore it still has to produce the pulling by forming gravity. All this evidence was known to science about 2500 years ago but since "With a lot of words and some simple algebraic relations, there is no way to "explain" the world of physics" it lacked mathematical communication and it should therefore surprise anyone very little that physics could not fathom this result 2500 years onwards.

Forget the example always used about the hammer and the feather falling equal in a vacuum because the hammer and the nail and the elephant falling together will also fall equally notwithstanding falling in a vacuum or not falling in a vacuum. The vacuum part is conspicuously in place to purposely confuse reality as it is brought in to flagrantly spread misunderstanding of the issues in hand about the falling that takes place. With everything always falling equally when the same the condition applies to all objects falling and therefore with such falling happening under the very same variation of natural conditions applying, this shows it is the space in which the object is that falls and not the object falling while leaving the space it holds behind. The lack of relevant density in relation to air moving down stops the feather from falling equal just as gas does not fall with the space at the rate that space does

descend. All space falls by the compressing of the atmospheric space.

The rotation of the earth moves the space sideways and this brings the space to move downwards by increasing the density of space or air as it comes closer to the earth. This results from the Roche limit applying to fix atmospheric layers varying in density. In my books I explain that principle applying mathematically. Notwithstanding using your mathematical marvels, science has not got any vague idea to explain any of the phenomena mentioned above. To understand these phenomena one has to understand singularity.

I am able to explain the how the Universe started only because I discovered the building blocks used to build the Universe. The building blocks are the Titius Bode Law, The Roche limit, The Lagrangian points and these all culminate into the Coanda effect. Using these I can and I do prove exactly how the cosmos was built dot by dot.

Gravity is TIME forming SPACE = Π^2 = 9.8696 = MATTER HOLDING THE COSMIC LIQUID COUPLING THAT TO THE NEUTRON TO COMPLETE THE COSMIC GAS to the liquid part of the association between solids and liquids. The earth is the solid while the atmosphere is liquid and outer space is gas.

Every person associates gravity applying as "The natural force of attraction exerted by a celestial body, such as Earth, upon objects at or near its surface, tending to draw them toward the centre of the body" with what we think of happens to solar bodies having mass which is "the natural force of attraction between any two massive bodies, which is directly proportional to the product of their masses and inversely proportional to the square of the distance between them" and science cheats everyone into believing the two is the very same. The one I experience every day and with the other there is no evidence of applying anywhere. Yet students are fooled into believing the two are exactly the same issue, which is untrue. If there is any academic feeling insulted by me calling the lot fraudsters, bring evidence of the second form of gravity working on the principle of mass and I will withdraw my statement, otherwise if no proof can be brought, then you lot that ignored me for ten years are the fraudsters I accuse you to be. You are villains brainwashing students to corrupt their thinking.

Einstein's Critical Density lacks the accepted matching facts we need in proving the critical mass factor, which makes the entire idea silly and bogus. You can't force the Universe to conform into something Newton said because Newton said the Universe works on mass contracting... But our inability in securing such required evidence defies the most basic logic. It seems all new evidence we receive from outer space is disputing all Newton laws and new findings disprove **Einstein's Critical Density** as the answer.

The Universe will not reach a point of contracting, not withstanding whatever dark matter astronomers try to locate in the vast space. A rush is on finding the black-matter that will be applied to force the cosmos too retract back to where it came from. But what if our view of the cosmos was as incorrect as our views at present is about the sun? I prove that contraction is at present as much part of the cosmos as is the expanding is that we focus our attention on and it is our culture we carry from generation to the next generation that leaves the human view obscured in admitting the truth. The Sun is not a coal stove burning fuel. The Sun is an air-conditioned pumping gas (hydrogen and helium with electricity. I prove by applying the Titius Bode law that gravity is electricity and the two are the same thing. By using gravity or electricity the sun is a huge air-conditioned pump freezing cosmic gas, which is what the Universe is into a liquid that we see squirting from the sun.

Newtonian science has NOT developed from the idea that the sun and the rest of the Universe are orbiting the earth. Everything applying on earth and befitting human standards they think must be transmitted directly to the Universe. If I stand in the sun and feel hot, then "the earth is hot" and we have "global warming" just because we humans think it is hot and we humans feel slightly bothered. Every point in the Universe applying singularity has a different measure and the entire Universe is NOT made to cope with life but is completely alien and destructive to life. Newton started to tell the Universe what it must be and the Newtonians never stopped. Newtonians better stop with the Ptolemaic concept of putting the earth or life in the centre of the Universe and start to place singularity in the centre of the Universe where the centre of the Universe is..

Why would the expansion turnaround and do a reverse by going back to where it came from. Consider the momentum alternation such a change will bring about. The sun is not a gas-filled sphere holding hydrogen in its "natural gas" form, but it is all fluid and is in a liquid form where singularity is liquid-freezing hydrogen at 6500^0 C while outer space is boiling over at -276^0 C. **The Absolute Relevancy of Singularity** book explains the Roche limit as well as the other four cosmic principles in the practical sense... when applying cosmic laws instead of improvising cosmic laws uncovers that reality then becomes awesome. It becomes clear the Universe is as much expanding as it is contracting and contracting by expanding. As there is no hot or cold, no big or small, no grand opposing but relevancies in ratio to one another. If you do not believe me, then believe your eyes when looking at the picture. What ever the sun is it is fluid falling into fluid.

Because hydrogen is a gas on earth and we think of 6500^0 as hot on earth, therefore the sun must be hot and the sun must be gas at the same time. It is obvious what we see is liquid squirting and when hydrogen is in a liquid form it then must be cold because my eyes tell me that! Newtonians still have the entire Universe apply the standards befitting life on earth and that is why they wish to locate life.

 When something is hot it expands. When something is cold it contracts. Outer space expands to the very limit and keeps expanding therefore outer space must be the hottest there can be, notwithstanding scientific Newtonian stupidity. The sun contracts every bit of space that forms the solar system and therefore the sun must be the coldest place in the solar system notwithstanding whatever Newtonian ego whish to declare. If I feel the sun is hot it is because it is diverting all the heat my way and it diverts the heat to me then where there is no heat it must be cold. It may sound incorrect and unscientific madness but with my applying of Kepler's formula in alignment with the position I located and valuated singularity it clarifies the possibility of the above statement... but please do not take my word for it, use your eyes and make sure you look past the culture bias of past incorrectness. See the fluid push out of a bowl of liquid, spilling both sides as it falls into liquid. The Hydrogen inside of the sun is not gas but it is fluid. In all of nature in all elements found through out science there is no NATURAL GAS as much as there is no NATURAL SOLID.

Hydrogen is as much a liquid as iron is a gas and neon is a solid. It depends on the element relating to the space/heat in the circumstances surrounding the substance at that very precise instant in time. We have to stop telling the cosmos to show us what we wish to find and start accepting what the cosmos is telling us is out there that we should look for and find. In creation there are two substances that formed the Universe. One was earth or solids and the other was heaven or uncontrolled heat. Between all solid we have heat parting the solids and being a solid or a liquid or a gas depends on the ratio between the solids and non-solids. Under conditions suiting life certain elements may be a gas, but in stars conditions don't suit life and in outer space conditions don't suit life so therefore life cannot be a barometer for conditions applying in the Universe. Kepler gave us **solids as $a^3=T^2k$ as liquid or gas.**

The earth, just as all other cosmic objects do, contracts outer space by the movement of the rotation. Material spins in gas to reduce the volumetric size and by that it reduces the concentrated space around the star / planet / earth which then reduces the cosmic gas forming outer space to the cosmic liquid forming the atmosphere. By rotation the earth "pumps" gas from outer space to the core within the centre of the earth by applying centrifugal pump action. As the space becomes denser the heat level rises but this is because the rotation of the earth reduces the heat, not the heat that we as humans feel or experience but the heat level within the space. We have to maintain science laws. When any cosmic substance overheats it becomes larger and when cosmic substance becomes colder it shrinks.

That is cosmic law.

The heat we feel on our skins is heat escaping from where it is cold and contracted to where it is hot and expanded because the levels that reduces has to dispel the heat from the location it is within and move it to where heat is excessive and expanded. It is the heat moving away that we feel and then think of it as hot. As the earth turns it cools the space in which it is and the space reduces in heat and therefore the heat levels can reduce in order to make the space move towards the centre. As a matter of fact the very first experiment in science ever conducted and recorded was the discovery that it is space that reduces and not things (water) that falls but notwithstanding Newtonians have gravity as mass falling.

But as Professor Friedrich W. Hehl, Inst. Theor. Physics With a lot of words and some simple algebraic relations, there is no way to "explain" the world of physics. Your seem to be out of touch with modern developments. I guess that is why Newtonian science can't conclude they are mistaken even after three-hundred years.

Again and as for e so many time in the past I repeat the warning once more of the Newtonian Brilliant-Brainy-Bunch to please take note of a conscientious warning about the gravity of the misgiving there is on the part of the most respected Academics in physics about a much concerning matter. As you can see why I state it emphatically that science accuses me to be not schooled to the point where I am able to have any form of an opinion on any matter concerning Sir Isaac Newton.

Notwithstanding that my research proves I did my private studies and through which I skipped the indoctrination and mind control academics place on students goes unrecognised by their standards and so too my ability to have any insight on matters regarding physics. However my skipping their methodical and systematic brainwashing enabled me to see and allowed me to be able to express the incorrectness in Newton's teachings and allowed me to show in clarity what destructive force Sir Isaac Newton used to corrupt the laws of mathematics, corrupting to science along the way and mostly raping to the work of a great man, Johannes Kepler and what Sir Isaac Newton did can only be expressed as being blatant criminal fraud. What his deeds amount to is to corrupt the laws of mathematics, to render the laws of cosmology useless and to rubbish all of science. Should you find this to be unbelievable, then I am glad to announce that this book is more for you than any other person, so go on and read what academics guarding science never wanted published. I challenge any one that disputes any claim I make to prove me wrong by proving me wrong and not merely suggesting claims in that direction.

We have to realise that it is about heat and cold where moving cools down and stationary space expands. Creation started with differentiating between what is cold and what is hot. That is exactly what the Bible says in Geneses 1:1. The Bible says God made heaven and earth. This means God made what is solid and what is not solid.

It means God made material (the earth) and heavens (the sky or non materials). Back then as is the case now with our most brilliant Newtonians there was no other names for solid as earth and non-solid sky as heavens. At least those back then realised the difference and did not put sky down to "nothing" as our incompetent Newtonians of our modern era does. There is singularity controlled by movement (material) and there is singularity not controlled by movement (non-material).

In terms of mathematical equating the very first recorded instant happened at:

This is exactly what the Bible says, if only the Theologians and the Newtonians were not so preoccupied in fighting for their own small-minded egos instead of looking for the truth.

When the Universe started there was one spot that released a dot. The spot as large as a thought grew into a dot. This too is exactly as the Bible says it happened and it still happens exactly like this. The Universe starts from infinity and grows into eternity just as it happened at the start and this is what Newtonians confuse with what they know as the Hubble constant. To check if I am correct look at anything spinning and you will see conformation. The dot that released was by such release relevant to other dots released because there was to be motion that measured many dots. The dots in release were relevant since they were the same. Only time being in delay by cycle of infinity interrupting eternity to form space, as space is the history of time gone to the past. But since we are looking at things as it started I put it into the past tense. That cycle brought about time delay as every cycle drifted further from the original singularity while the original was still responding as well. That was the first space. It was time being one infinity part.

Since the dot was also singularity and was the very same as singularity with a small difference that the spot was $\bullet \Pi^0$ and the dot was $\bullet \Pi^1$. At first, at moment-Alfa, there was no space nor time for only relevancies came about. Relevancies acted as motion to bring change to eternity by changing the flow of time in eternity. There was the perfect spot in which time moved while remaining the same. Then heat brought expansion and expansion brought space and space brought movement and movement brought about a Universe. That is how it started and that is how it still is. This was before the atom and the atom was before light and that too is exactly what the Bible says in Geneses 1:1.

Time will forever remain eternity but space or time distortion, which will forever remain infinity, interrupted the flow of eternity. Space breaks the monotony of time in eternity by parting time from in infinity. Yet, the relevancies did imply motion except for the fact that singularity is very much incapable of motion. Every dot had a purpose to fill a position in relation to the other dots that the spot excited. All dots had a line of three where two was one, each on every side of the spot. The spot was one with the two dots forming two, which improvised for motion that would later come to space and the three was what space was going to become. How do we know this: 0.1416 x 7 = .991 and that is by the value that the Universe grows.

Time was four because the four would bring about motion as heat separated infinity from eternity or hot from cold. Five was space because space was one removed from time. Space is the distortion of time and one outside time would bring a time delay or a time distortion of four plus one which is five, hence the principle behind the Lagrangian system. Because material was the square of space material was a crossing of three plus three forming six. However to find out what this means you have to read the entire theses called **The Absolute relevancy of Singularity.**

Space-time is the four of time, plus the three in singularity around which the four of time turns, therefore space-time is seven.

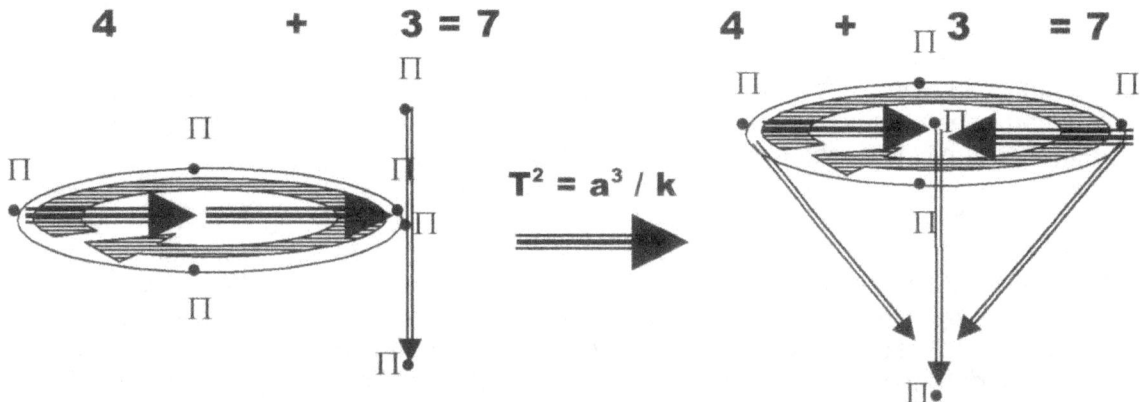

In the circle using $r^2\Pi$ the r has to have distinctive qualities placing it as a factor apart from Π. Where the growth shows no separate distinction but a continuous flow from the precise centre to the precise edge the flow would become in relation with Π depicting the circle and Π replacing r as reference to any point on the circle.

By using r, distinction in the circle is possible but by using, Π there is no distinction possible. Therefore, in the beginning when time formed space there was only relevance coming about from $\bullet\Pi^0$ and the dot was $\bullet\Pi^1$ with no mention of any possible r. The fact of r representing a radius represents space and what we refer to be long before the Big Bang introduced space or mathematics using space.

Before the Big bang the lot was form without dimensions playing any part. Then the atom came. Only after that did the Big Bang come. Even before the atom was the point lining up and forming positions that was spinning faster than the speed of light can ever achieve. However every point today still serve the role it took on at that stage and serves in the position that it had during the time it had no space with eternal time. These relevancies developed as part of a Universe we shall never understand. The Universe had no sides and a line was equal to a triangle, which was equal to a half circle. Singularity holds the double space-time position of five times two (matter and space duplicating singularity) which then is ten.

How did the Universe liberate material and heat forming space from singularity because with singularity comes an unchangeable eternal condition that is non-changing-everlasting in all conditions and aspects that is remaining in absolute equilibrium. This equilibrium maintains because all development extends form precisely in a detailed equal equilibrium throughout. Think about what brought the cosmos out from the eternal rest in which it was. The eternal rest still maintains and is therefore our detection. What inspired the eternal rest the cosmos was in and inspired change to the state of eternal rest? What evoked change? That is the question the Atheist will never be able to answer but that too is the most basic and ever-lasting fundamentals of the Universe. Singularity Π^0 is not substance but it is a thought establishing substance Π. What changed in this split second start before the official start? I do not wish to ponder on this matter in the letter I am writing at this minute, as there are other books where I delve into this matter. It is called **The Absolute relevancy of Singularity**

From the deep freeze of creation came the Hot Big Bang and the 3D Universal displacement came about with the relatives being $10 \div 7(4(\pi^2 + \pi^2)) = 112.795$ and then a second one established 3D by introducing the six to seven sides Universe at a density point of $7 / 10\ \pi^6 / 6 = 112.162$. There is of course a lot more information about this establishing of the Universe than what I mention at this point. The question is what made the Universe freeze, to form the Universe in space and through time. It had to start with a specific reason applying, which brought about space-time. Once the process started there was no stop to it, but there is no chance that the initiation of the start was spontaneous by nature. With everything being in one spot, all within that spot was in a state of eternal rest. While all remains the same and nothing changing, what brought on the sudden change of everything, shocking

It is the time the space takes to bring about new positions in space occupied through motion applied and through motion applied, it takes space-time to duplicate and in the relevancy of duplicating, the duplicating takes certain duration in time to move from point to point. Gravity is motion of space towards and in relation with a centre and the time is the period such motion takes while that centre is attempting to produce space by doubling space through motion leading space away from a specific controlling centre to another specific controlling centre. The time it takes to complete such an attempt provides space the opportunity to double its status. Moreover, the time stands affected by this motion material creates to duplicate space-time by generating singularity and activating different locations holding singularity. Gravity is speed and speed in space in motion through time duration. Gravity is motion combating heat expansion and supplying space with space producing through motion providing the space the opportunity to expand while remaining in relevance.

At the start, the gravity part invested heavy in material. However, in space light came about since there is no gravity in space. Light is the attempt to establish motion not controlled by any centre and the time it takes to establish space between such a centre of space control and the light finding an ability to dispense the space by reactive motion. Gravity produced space by allowing as much as producing overheating with a feeble attempt to combat the overheating. What is the meaning of heat if there is no cold to set the standard for the heat to become a value, which is then related to the other end, where such another end must be the limit in cold? The moment singularity produced space-time heat distanced from the cold factor. Space produced a cold base to have heat within. How did singularity part the shared principle of being the unification of heat and cold as a unit? Singularity froze in applying gravity bringing about particle separation within singularity. Heat and cold parted to produce frozen heat by atoms forming and captured heat by unleashing uncontrolled material that ended as space in time.

We can see from what is available how everything fell in place. There was space filling with heat where the heat was compacted by the time delay. Every point established a relevance of three crossing singularity a motionless forming a line and four points formed the square of time. Crossing the four points that supply the time or the motion aspect is a line with three points supporting singularity by not spinning but they maintain singularity in being as motionless as singularity is. That four plus the three is seven. Then all this, which is the body of material requires somewhere to be within because to be is to move within something. That too is time but we regard that time to be space. That time is the heat that became space during the event of the Big Bang. That gives the atoms seven point as plus the electron or space having three points in relation to the seven points there is in time. If singularity is at the limit the value of ten but is the tenth portion of the value then ten must be divided by the tenth portion and from that the tenth portion must be distracted.

This as a group forms a relation with heat unlike any other. If mass did the trick these must have been the group having the second least density, but they form the group with the least density as a five point group. The next group holding the Pythagoras five or the Lagrangian five plus two or three, four or five enabling their relation with heat to be quite remarkable. This must be some indication of events during the period just preceding the Big Bang at say 10^{-7}, 10^{-6}, 10^{-5}, the time when the fuel that would ignite the Big Bang turning heat into space turned material into heat. The relation of five plus one, plus one and five, plus one plus one plus one is just too uncanny to ignore.

Following the process and seeing the influence of singularity should bring about a pattern that may lead one to a pattern of how the required heat formed and how the intended heat transformed to space.

Density depends more on proton number arrangement producing specific form in relevancy as to merely and only having mass as factor that contributes to the forming and development of stars in the cosmos. The evidence is so clear that mass has nothing to do with gravity but density has everything to do with gravity. Density is the volume of space in numbers used to fill material in ratio with numbers of space per volume not filled with space. It is matter versus space in every sense there are. This came about before the Big Bang took place and before space was formerly space and time was formally motion. It was a time when singularity set relevancies moving from Π^O to Π

The spot
becoming
the Dot

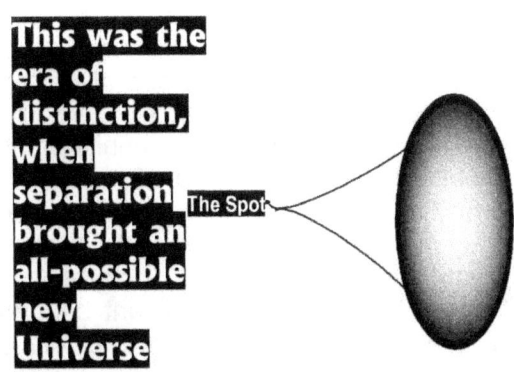

This was the era of distinction, when separation brought an all-possible new Universe

The Spot

In that manner we know that that was the way particles formed combinations just after the arriving of moment-Alfa. Singularity brought the Universe but also singularity brought the divisions between the many Universes that followed the immeasurable many Universes that came after the flooding of Universes to follow the leaders. The term "moment-Alfa" is the way I refer to the moment when singularity changed, not when space formed or time began or space exploded but even before anything including mathematics became definitive. At this point mathematics renders it useless. There was no space or time to calculate because relevancies came in place. Form took shape but space there still was not because Π^O moved to Π. Every slightest point in space became an opportunity of establishing a Universe with most different functions and ingredients there might form. This is apparent from the fact that it still takes place at the present moment by motion attaching new singularity through duplication and through duplication releases previously attached singularity from serving the purpose of duplicating by motion.

When the cosmos came to motion, motion was not yet defined. When the cosmos brought about motion, the first motion was relevancies. Cold parted from hot. Eternity parted from infinity. Motion parted from motion absence. Infinity broke the laboriousness of eternity for the duration of infinity. The spot became and grew into the dot.

From what the spot was to what the dot now is might be just a mathematical implication of going from 1^O to 1^1 but in reality that first motion was the creating of and establishing of an entire Universe with all possibilities now in it. Never again can that much growth become a reality, although to us the growth is beyond what we ever can notice. But it is because the growth is so massive and we are so small that we are unable to notice such almighty growth.

When the spot Π^O became functional and established all relevancies possible, heat parted from cold as eternity parted from infinity. The expansion was not clear motion but more a parting of relevancies where a centre formed a relevancy because the centre could not provide motion. Without being capable of motion, the centre established four points, which also served singularity. From the inverse square law we know that the centre doubled by producing the four points holding singularity.

By exciting the centre spot, the centre spot came to be because of the heat that formed in relevancy as heat parted from the cold bringing about the division that followed and that was the motion that formed. Therefore the heat had to move but being singularity it could not get singularity to move. In an attempt to establish growth, singularity activated six spots of which four was having motion drawn into relevance four spots that was providing what was to be motion and three that was to be securing the position the centre holds. There were four forming a ring around singularity with two forming in locations we will refer to as above and as below or north and south.

The three in line was in singularity not being able to move but the four was also in singularity and just as incapable of moving. All the points came as relevancies applying the forming of more of what was to come but only the four committed to time were expected to move. The four points that came as a result of discrepancies that became time that produced form and that established the relation with the one but had to perform the motion by expanding was as much incapable of motion as the centre was that charged the four with motion in the first place. As they were incapable of motion, it still required a tendency to apply motion that did separate Π^O from Π. This not only involved form but it involved all relevancies that did come or may in the future come about as a result of the attempt to commit motion. If mass was a factor contributing to gravity the cosmos would have frozen back to singularity without ever releasing singularity to relevancy.

Mass does not establish gravity. There is no magical graviton. In the beginning there was no mass but boy was there gravity! The only means that the cosmos could find a way to break from the grip of eternal eternity was to expand into relevancies. Such a feat can only go to task by forming opposing hot and cold. Becoming hot produces more of what is heating. That implies motion or a moving away from where it was by generating more of what is available. Only where hot released from cold could whatever was repeated once again and duplicate what was before into what then is more. Secured by motion T^2 in relation to a specific centre **k** from where singularity holds the Universe true to form. The **k** was an intention to place apart and by today's standards will not even qualify any noticing.

All that are is in singularity. From singularity comes the motion and the space we call space-time. Singularity is dimensionless, time less and space less and because of all this features, it carries the value of Π^0. By expanding, singularity applies a relation coming about that reforms singularity from Π^0 to Π. Only when extending Π^0 to Π, the extending creates motion and the motion creates space that then doubles through motion applying which cuts the space in motion in half by matching the space as a duplicate. Motion creates another dimension or another level reforming singularity from Π^0 to Π or from Π to Π^2 or from Π^2 to Π^3

As said before we now know Π came about since Π is achieving form and not space. Only **r** can establish space as size will accumulate and as it had with everything else singularity had **r** covered by one as in being $r^0 = 1$. By reducing the circle radius **r** by half continuously will lead to an infinite small circle and an infinite number holding r would place **r** to the power of one as a factor. Then as a factor **r** would not contest any change when change is introduced into any future equation but Π will remain because the circle as a form remains even being infinitely small. By reducing r indefinitely to the tune of half each time, r would become infinitely small, beyond human calculating means, however as mentioned in the case of the smallest dot holding one spot, r would become insignificant beyond human comprehension even, but never reaching zero and still Π would remain intact and dictating form. To amplify by dimension a value has to be set to r but if r remained covered by singularity all alterations that could possibly come about was in the form, which was Π.

This expanding can be a problem one can wrestle with for one lifetime and never reach any conclusion. How can something grow without getting more that what was before? Then it hit me like a ton of bricks. The answer is in heat but not heat, as we know heat. It is heat in getting relevancies between outer limits. Only heat could break the monotony of singularity. Heat in the form we now know heat as heat is now. Since the Big Bang heat is material transforming from one state to another state.

The change that took place involved singularity but singularity was 1^0 and being $^0 1$ could not grow. The growth came about. Heat rose from singularity, but if heat rose from singularity. Singularity as a factor changed from 1^0 to 1^1, which means a relevancy came in place that no one could detect. It is true that 1^1 are still one, but one could then escape from singularity by producing factors other than 1. Heat came about but only as a relevancy to utter cold. If there is heat, there is cold or if there is no heat there can be no cold. Space came into forming a relevancy that brought form. Since it is a relevancy and not a generation by accumulation, the form produced was Π.

The spot formed a dot by heat and cold establishing relevancies and from that singularity was broken to allow all other forms of relevancies to come about. The cosmos did not start because of gravity. The cosmos started with heat and cold coming into a relevancy and in the cosmos there is no hot as much as there is no cold. The cosmos broke, put from the confinement of singularity by establishing a singularity in a relation of heat and cold. The heat that came about was beyond measure because the cold that held the heat was also beyond measure. The immeasurable heat was on the outside of the dot that formed and the cold was on the inside of the dot that formed.

The cold contracted because in nature cold contracts. The heat expanded into a dimension of form and heat by expansion is in nature about motion. Motion is duplicating that which is and heat is what is duplicating by motion. But only heat by expansion was possible because in affect singularity cannot move. The motion became contraction, as the motion was the result of heat expanding which was forming four points in the rim of the dot. The expanding of the points created motion in relevance of a centre that formed because of the motion, which established an immovable centre as the Coanda effect, placed more dots in relation to more dots that formed.

Every dot was Π and every dot formed Π^3 because of the expanding heat, which produced Π^2. With that a new relevancy came about forming a centre in between the four points of expansion that was resulting in time. But since the points were in themselves singularity, which is immovable and space-less, they still heated forming a cold centre with the heat bringing about motion. It became a repetition where infinity broke eternity by producing a centre because of space (or rather form) forming the motion to enable the space to form in relation to the heat applying motion. This brought about a Cosmos being conceived.

The spot forms a full circle, but the line running through the circle is forever present because that is the future radius of the circle that will one day develop the circle, which is equal to the present diameter. The fact of the presence of such a possible line in such a possible circle dividing the possible circle into two parts makes the centre line equal to the half circle. The line forms the half circle but not only that the line presents the half circle as much as the line is the half circle. The line then is 180^0 and the half circle is 180^0 because in singularity the two factors are the same.

The same value is of course $\Pi^0 = 1$. The issue of concern is o understand that singularity cannot move. Singularity has no space. Singularity is no only part of the Universe but singularity is the Universe. By establishing motion singularity has to be charged with the time delay we find space to be. The space is time taking a period or a duration while moving from one singularity point to another singularity point while conducting the heat and the accumulation of heat that built up due to the retarding of the time to conduct the heat forms the space that is conductor to bring about the motion of the space.

It takes heat time to entice singularity and singularity can only entice. Singularity cannot move and neither can singularity form space. By enticing from one relevancy to another there is a bridging of heat that has to be crossed in order to send the gravity or the enticing or the relevancy to depart the space and reconnect the space to the next singularity. Bridging all the accumulated various time delays that formed an accumulation of heat through time distorting brings us the space we see and have. However there is no true space or motion but it is eternal motionless space is singularity charging time to provoke heat into forming space.

Three points formed a line covering singularity where the centre singularity recovered heat to grow and two points served as an axis to allow the rotation and to assist the duplication. There is one centre connecting the duplication of three as well as the recovery of one (the fourth one) that is applying the tie aspect. Therefore, motion consists of three positions in relation to a centre, which forms as space in relevancy to the motion and the space receive a controlling centre.

The duplication comes about as singularity is exciting another singularity in precise relevancy of 3 to 3 to 1, but the points charged is as space less and as motionless as only singularity is. The heat it requires to carry the exciting between points forming space and the space excites heat and the time delay it takes to excite singularity between points forms space-time.

That is why the Universe is Π
Where motion conducts electrical charging which is equal to gravity the charging of motion is to entice duplication of singularity. This is the basis, the heart and the sole ingredient of the Coanda principle that includes the Roche limit ($\Pi^2/4$). The charging of gravity $((7/10) +(7/10)) / (10/7) = \Pi^2$ and the charging of space-time $\Pi^3 = \Pi^2\Pi$ is all due to the relevancy brought on by the Coanda principle. The value of motion came from singularity exciting singularity and that is the duplication while the duplication or motion presents the space.

The development came into eras as the relevancies brought about new relevancies that spawned even newer relevancies that all remained in touch with the original singularity centres. Every one focused a new time delay that eventually brought about space and every distortion of time brought more. That concentrated between singularity points that charged the points to form space. When the charging became overdue in some sectors it erupted in forming the Big Bang. By the time the Big Bang erupted there was such a huge backlog in heat and time corrupted and delayed the next result was the employing of space as a commodity in the Universe. The relevancy was C the gravity was C^2 and the space was C^3. That left what was inside atom still spinning faster than the speed of light applying the relevancy of $k = C$ where the electron applied the relevancy of $T^2 = C^2$ and that formed the atom which then became the cube of the speed of light $a^3 = C^3$. That left the atom at the relevant size of what the speed of light permitted at the time but since the Universe from that the relevancy expanded as the Atom grew in space to the extent it has now. The purpose of the star is to recapture the space the atom grew into and from there dismiss the space by spinning faster than whet the speed of light will be on the outside of the star.

· This form came about when only form was present in the cosmos. It was in a time era where form featured in relevancies that would lead to one day becoming the atom. The atom forms a dual purpose of duplicating as well as dismissing and some prefers the one better to the other. This relevancy came in place when time was not time and space was form. Time is forever eternity being interrupted by form in infinity to bring about eternity ticking as infinity ticks. Before that singularity took on stages in forming relevancies between duplicating and dismissing space-time, which incidentally was not yet truly space-time in the sense we think of as space-time. At first a dot moved from the spot leaving the spot but taking with the spot as part of the dot to remain in the dot. The two never separated but the one allowed the other to be.

As the dot confirmed a discrepancy between infinity and eternity by defining infinity as an interruption of eternity cold and hot parted a union. The dot that formed was not space but a relaying of time to form a new point of singularity where eternity was interrupted by infinity. Time took form from 1^0 to 1^1 or from Π^0 to Π. It brought form into differentiating between interrupted eternities with infinity doing the interrupting

Then a true distinct relevance came about that positioned a time differentiation outside the realm of time by four. In this realisation we can assume that space had some meaning at this point and the formula used to investigate suggests just that. Even in grouping, there are characteristics, which make a certain group of atoms more perceptible to duplicating and others more perceptible to dismissing

The lagging of exciting one point in relation to another point takes time. It takes time to send the message across to get singularity at that point excited. It takes effort to bridge from the dominating singularity to the independent singularity and that effort slows time down. The crossing of the divide is space formed by pushing time into duplicating. When time brought in a five points to the four points it took time to be, that fifth point became more than only form, it became space because it was one point outside the Universe of four or of form. One must see the three points established as motion duplicating singularity in relation to one dismissing singularity. This always has to strike a balance in order to establish space-time. It began as a relevancy and developed into space-time flowing or space-time displacement.

What the Coanda effect proves is that the rotating motion is acclimating a centre that exemplifies all phenomena in nature as we use nature to our advantage. All of nature including gravity uses the same method of motion forming around a circle in rotation and in the centre of the circle a point of no motion holding no space comes about. This is what Kepler taught us when he taught us $\mathbf{a^3 = k\ T^2}$. With the Coanda effect forming the basic principle of all natural phenomena we can see from that, that the motion of liquid in the presence of a solid forms a centre that excites as it establishes singularity. From that rotation, space flows to a controlling centre but because of the lack of motion in that centre, there is a lack of space in that centre. Therefore, there is proof of a flow towards such an established centre and there is control from that point of singularity. In every case, the singularity controlling space-time sets standards for space dismissing in relation to space duplicating.

The duplicating stands in regard to the flow that the liquidity of the atom in relation to the solidity of the atom can reproduce. This forms density and mass but mass has little influence on the scenario.

There is a balance between the duplication in relation to the dismissing of space and the relation extends to the number of atomic elements present which then creates the balance applying within the star. As the liquid heat subsides in the centre of the star and the heat density is dissolved by the dismissing-prone elements the motion or moving ability of the star as a unit fades away as the star becomes static and solid with less space providing the star with less motion.

It is the way the atom formed before the atom took on space-time. It is in the formation, that space-time relates to motion. We have some elements being quite massive but also lighter than air and others are quit light but as dense as they come. This can only be a contribution from the way the atom relates to heat, which make the atom volatile (movable) or dense (motionless).

Those elements being volatile are also very movable and in that we find the role that such elements play in the star. Stars that are predominantly made up of hydrogen and helium with very slight support from the metallic inner core are those stars that duplicate by producing motion. However the point I wish to press is that mass and being massive and being heavy do not support the fact that some elements have more gravity they produce because their protons are more numerous than others. The fact that mass generates gravity is a myth.

One will find that whatever group one chooses there are gasses and there are solids. If mass was attracting mass then the strongest mass must be attracted to the strongest mass and the least mass must float in the air. $F = G (M.m) r^2$ hardly can even begin to explain the fact that there is a gas that is more massive than iron but floats in the breeze just as hydrogen which is the least massive element.

Humans in science on earth still put humans in science on earth in the centre of the Universe. It is accepted all around that if an element is a gas in the "natural state" on earth then that is a gas. If the element is a solid on earth also in a "natural state" then that is a solid. In the centre of the earth we find iron in a molten sate floating as a liquid because it is within a state of liquid. In the centre of the earth every element that is there is a liquid and if it becomes gaseous it burst to the surface, through the surface and we then call it magma. All elements are either a solid, a gas or a liquid depending on heat.

Nitrogen 7	melts at -210°C	boils at −195.8° C
Oxygen 8	melts at −218.8 °C	boils at -183° C
Fluorine 9	melts at −219.6° C	boils at −188.2° C
Neon 10	melts at −248.59° C	boils at −246° C
Sodium 11	melts at 97.85° C	boils at 892° C
Magnesium 12	melts at 650° C	boils at 1107°
Aluminum 13	melts at 660° C	boils at 2450°
Silicon 14	melts at 1412° C	boils at 2680° C
Phosphorus 15	melts at 44.25° C	boils at 280° C
Sulphur 16	melts 119° C	boils at 444.6C
Chlorine 17	melts at −101	boils at −34.7 C
Argon 18	melts at −189.4° C	boils at −185.8° C
Potassium 19	melts at 63.2° C	boils at 760° C
Calcium 20	melts at 838° C	boils at 1440° C

Ignoring these facts, Mainstream science will hardly answer the problem we do not understand and such ignoring brings strong doubts about the quality and sincerity of science.

Excluding Argon, which is six (carbon's number) times two and suddenly that is a less dense material. The four times five plus... group are the following:

Scandium 21	melts at - 157° C	boils at -152° C
Titanium 22	melts at 1670° C	boils at 3260° C
Vanadium 23	melts at 1902° C	boils at 3400° C
Chromium 24	melts at 1857° C	boils at 2665° C
Manganese 25	melts at 1244° C	boils at 2150° C

Iron being the five times five plus one is the only generator of electricity and therefore the producer of gravity making five times five plus one the ultimate relevancy to heat in reducing space. Still Krypton is much more massive and turns out to be a gas.

Krypton 36	melts at 1539° C	boils at 2730° C
Iron 26	melts at 1536.5° C	boils at 3000° C
Cobalt 27	melts at 1495° C	boils at 2900° C
Nickel 28	melts at 1453° C	boils at 2730° C
Palladium 46	melts at 1552° C	boils at 3980° C
Silver 47	melts at 1412° C	boils at 2680° C
Cadmium 48	melts at 321.03° C	boils at 765° C
Xenon 54	melts at −111.79° C	boils a-108° C

How can science promote their image of establishing honesty when they are confronted by such truths but choose to ignore the truth so long as a lie will bring them some respectability.

Since the star is the total configuration of the atom's characteristics, the atoms will tell us what we should know about every layer from what is applying in such a layer to what characteristics such a layer would show when it provides the function of what it has to for fill within the star.

The concept still is about singularity linking to time and that is a distortion of time. At point, five of

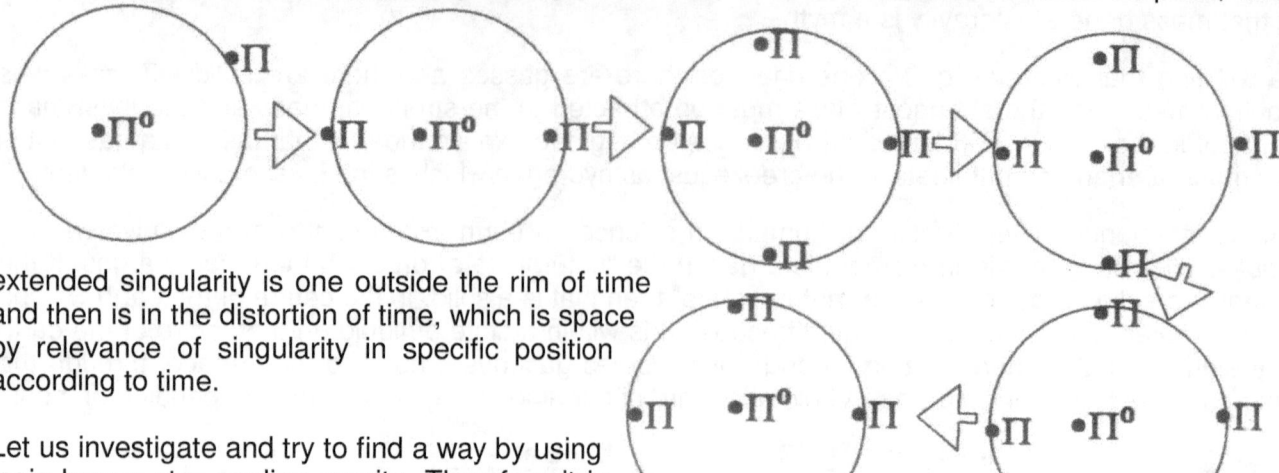

extended singularity is one outside the rim of time and then is in the distortion of time, which is space by relevance of singularity in specific position according to time.

Let us investigate and try to find a way by using logic how a star applies gravity. Therefore it is not the number of dots that is important. It is not the size of the number of dots occupying the position or the size of the space the dots occupy that is prominent. It is the relation in the dismissing of space and the duplicating of space that becomes important. The less space there is the more the favour will be to reduce the space because of the advantage the dots have in securing space-time that will prevent overheating. On the other hand the more space secured will also prevent overheating and therefore those will opt to duplicate space in order to find space to secure and prevent overheating.

Since the Earth has no singularity demand that is much better developed than the Universe sustains, we find on Earth a relevancy of Π to $(\Pi^2+\Pi^2)(\Pi^2\Pi)3$ is adequate. But in bigger units the space-time displacing relating to space duplication presents much more demands on atomic structures occupying space within the star containing through set boundaries. In the presumed to be bigger stars there is much space filled with atoms occupying much space. In the stars more massive but holding lesser space the atoms must also hold lesser space but they also hold more protons by number in the lesser space.

The space the particles hold is directly in relation to the particles the containing structure duplicate. The more space that is relevant to the structure that the star duplicate by motion is then in turn once again relevant to the space the structure destroys by proton action in space less units. The more space the particle claims in relation to the space the container holds that relates to the space the container duplicate is relative to the space the containing structure destroy. From that mass derives value. As individual occupying space the atom is an individual container by own merits and as such duplicate space in this regard within the specific confinements of atoms.

This we will classify as normal applying structure values the atom has in outer space or in structures with very little atmosphere. Please note there is no pressure involved because the motion involved creates conditions naturally instead of unnatural pumping that causes pressure. Pressure is an artificial creation as part of life but has no role in the natural cosmos. Pressure is a condition where the retaining of particles has to be confined in a patrician made of material where the outer wall does the retaining of the substance within. This obviously cannot be in a star because the "pressure" is regulated from a condition applying and space-time controlling inner centre that needs no solid walls to contain whatever is inside. With that one can see there is a Universal difference between the concept of pressure forming due to human action inside a container and what comes about as secluded space-time within a star.

As the demand of singularity in such units grow stronger some relevancies within the atom come into play and I developed a system whereby I can arrange the space-time merits of space-time curtailing within the confinement of the star borders applying in the star to place such a demand in relation to singularity where the ultimate demand sets the standards. In the sun, for instance, which is a minuscule small star a relevancy in the outer region might be 3^3 relating to singularity and with the atom having, a sustaining displacement of $(\Pi^2+\Pi^2)(\Pi^2\Pi)3$ there is no danger of the atom demising. The electron in the sun will have a diminishing factor of 27 whereas the atom can sustain $(\Pi^2+\Pi^2)(\Pi^2\Pi)3 = 1836$. The relation in the atom degenerated by 27 leaving the atom a sustaining value of the electron plus the neutron applying space-time without involving any of the neutron aspects at all. That is the mass of the space-time that the electron will consume in the space reducing flow of space-time.

The flow is the result of heat distributed where the heat is delivers to the dismissing sector and producing of the duplicating of space by mass within the star that then forms a favouring of duplication in comparison to dismissing. The star is a bright little boy shining by dismissing pebbles of light-photons into space. When a demand on space-time displacement reaches an accumulated general displacing or movement to the value of what 56.6 protons can achieve in a general flow of conducting space-time that would be the requirement for such accumulated displacement within that space forming the motion of the space or the time aspect of space.

The star accumulated more heat by consumption applying direct dismissing without accumulating space-time in liquid form beforehand therefore there is no heat remaining to dispose of by producing light. When the general displacing flow of space-time within that sector of the star or the star in total reaches 56.6 displacement the natural state of absolute solidifying becomes the norm within the star. From then on, the star will exclude all electron functions and stop shining as the demand on space-time duplication and diminishing reduced the atom to space without a heat envelope that will be electrons or a liquid/gas jacket. Only the nucleus will be able to sustain the diminishing and the reducing of space by increasing of time. The entire star becomes a solid structure by reducing space-time directly freezing the space-time from a gaseous state to a solid state. By motion, speeding up the tempo of the flow of space-time the liquid state of space-time is by passed going from gas to solidity in one motion. The atom would shrink to such little space it will have space within the star that only the centre nucleus will fit. More reducing by applying motion in creating space differentiation will leave a star with so little space the space will be insufficient to secure a position for the neutrons and the star will then have the name of being a neutron star. Going even further will find the proton rejected from the star.

Every atom holds (I am guessing), as many dots as the sun has subatomic particles per atoms and that would still be a very conservative guess. Every dot is a controlling centre selecting a regional centre where every regional centre selects a centre. This goes on as long as there are spots forming groups as individuals unable to survive independent. The others that was unable to group formed heat that became space, which became the broken dots. The dots form groups to survive and as a group, the survival depends on doing what the group has to do to remain cool. In another book, I reserve one chapter to explain the phenomenon what I called the Lagrangian atom. These dots arrange in a manner that they could favour either the space duplicating aspect or the space dismissing aspect.

This can only be the result of the fact that even in the case of the sun, the inner space is almost entirely liquid heat and the liquid heat produces sufficient space to dismiss as the centre that holds the heavy metal particles, where all the dismissing is done. The liquidity provides motion while the solidity removes motion in the centre of the star. The dismissing going on is in the space factor where the space leads to a denser heat within that space because there are insufficient material to accommodate all the heat by the dismissing factor T^2. In that case motion far outweighs dismissing $k>T^2$ but a time comes in every star that the dismissing takes absolute charge. $k<T^2$ That is when the star goes dark. The Earth is mainly about duplication of space much more than dismissing of space and so is every structure in the solar system.

I would suggest we think of stars in the following terms. A star that generates and transmits a lot of light is weak on gravity because their progress started recently. They command a lot of space-time but the demand they have to keep their cooling acceptable is very low. In that they can generate a lot of light but with the demand on cooling low and the gravity in the centre not very developed, those stars cast a lot of light back into outer space. It is just because of the size the stars hold that tell the that the stars are still young and have a weak developed governing singularity. The stars will have very prominent hydrogen and helium layers, with the inner core not very prominent. The control of the star is still very much in the individual atoms and in that the motion the atoms have to produce in order to maintain their individual singularity will only come about through motion. The atom has to make contact with as much space-time through motion as possible since it has a very poor ability in contracting space –time in support of the cooling system.

When what was perfect became imperfect the Universe started. When the spot differentiated and became differently allocated from the dot the Universe started. When infinity moved away from eternity the Universe started. I show where infinity is as much as I show where eternity is and any person can put his or her finger on the spot. I show where space ends and where time begins and any person can look at the point. I sent an article to Annalen Der Physics and Professor Ibn Christianson explaining this in a 15 page dissertation. They came back to me advising me to While it is possible that a lay person hits on an insight that has been overlooked by academic trained in the field over many years, it is unlikely. We assume that work offering something new would be related to existing theories, either by building on top of them or by showing how and where they fall short (Professor Iben Maj Christiansen) and With a lot of words and some simple algebraic relations, there is no way to "explain" the world of physics (Friedrich W. Hehl, Inst. Theor. Physics). Maybe our Academic elite including the two I quote had no idea what I was talking about, understood not a word of the detail and was fat too brilliant to accept and admit this state of affairs. Maybe if our distinguished Professors including the two I mention read my work with more attention they would have seen what I try to show and what I try to show is where physics starts. Professor Christiansen since when is Π not science?

Creation started off with one dot so small eternity met infinity within. Then came one more, and another and they continued coming until there were a countless number of dots. The accumulative size of the dots were the same size as one dot because in the true Universe big and small plays no part. The dots were infinitely small and eternally big at the same time because size is a relevancy and without one the other has no size. So in the true perception, there is no difference in size.

It started with the fact that there is no place or part in with which one may associate zero or nothing. There are no room for a number such as nothing. Next to the one dot (infinitely close) one will find the next dot, and if nothing was a factor then that is precisely what one will find between the two dots. Nothing of space, a non existing entity, taking up no space, and much more important, no time, therefore the dots are infinitely close to one another, being the same space, eternally big as much as infinitely small. If we as humans cannot find a manner in comprehending this notion, there can be no manner ever understanding the cosmos as much as the start to the cosmos.

Every dot was a Universe in its own and the accumulation was a Universe. The earth in itself is a Universe as the moon is a Universe, because rules applying on earth do not apply on the moon and visa versa. When in the ocean another set of rules apply, therefore being in the sea places a body in another Universe. The number of universal entities is still countless, as much as it was in the beginning. Every dot insignificantly small as it may be, is a part of another Universe as much as it is part of the accumulative Universe and every dot in the infinity holds singularity, which we translate as " nothing" being " darkness". There cannot be "nothing" just as much as there cannot be "darkness".

There cannot be something big or small, but it into relevancy of perception, and then the relativity of perception becomes the question. There cannot be hot as much as there cannot be cold. The sun FREEZES hydrogen to a liquid at six and a half thousand degrees Celsius and Universe boils over in the form of the Hubble constant at the temperature (we presume from our vantage point) at minus 273 degrees C. If we Humans cannot or will not abandon our human perception and our manly perspective, we may as well return to astrology for all its worth.

Every point in the infinity we may observe at is not merely part of the Universe in not being nothing, but is the point where the Universe started representing singularity. It is the very first point where everything began so many eternities ago, because after all, how can we ever determine where the first point was, as they were very much equal and alike at the beginning. Every aspect of the Universe started with the fundamental fact that no point in the Universe can represent "nothing" as a number, because every aspect in the Universe represents 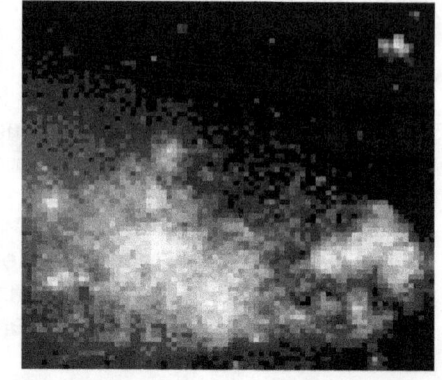 singularity in what ever form it may hold in that specific spot forming space-time. If man does not reach a conclusion where that conclusion is matching the Universe and stop to match the Universe with man (and man's incapability), we may all go back to caves and become starving hunter-gatherers again, because we will never find a way to progress to the ultimate understanding of the Universe.

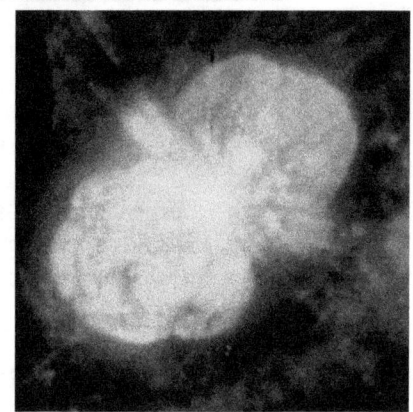

Looking at stars Newtonians see coal stoves being stoked to burn. In the days of Newton coal stoves were the nuclear science of the day ands while all other departments in science moved on and away from coal stove principles Astrophysics and cosmology remained true to Newton by inventing the coal stove in so many ways not even the coal stove could think of the facets it can go through. Newtonians see stars being fuelled like coal stoves and such stoves can run out of fuel. This is so much Newtonian backwardness as mass forming gravity and the moon coming closer and the cosmos shrinking and we falling into the sun because of non-existing dark matter making up what is required to make Newton not to seem the idiot that Newtonians are because they make him and his contraction theory to be less foolish that what it apparently is and they overbearingly are.

What is of vital scientific importance is that there are three fundamental dimensions controlling the Universe. The three are beyond intermingling and one confirms a status in relation to the others but not intermingling in status. From singularity comes matter and forming space-time in own accord. By matter not controlling time, space grew uncontrolled and the third dimension came about. That dimension birth we now recognise as the Big Bang, but the Big Bang is the last of a three prong cosmic growth. Science has to recognise the dimensions of densified (singularity), occupied (matter behind the electron) and unoccupied (space-time outside the orbiting electron boundaries) forming three points of cosmic recognising space-time.

Every dot was by itself as well as the accumulation as it currently is the present Universe. The earth in itself is a Universe standing apart from other Universes such as the moon as well as the space between the moon and the earth. The moon is a Universe. Rules applying on earth do not apply on the moon and visa versa. When considering conditions with in the oceans and applying space-time another set of rules apply therefore the sea places a body in another Universe. It takes the same engendering technology going underwater in deep sea diving that going into outer space.

The number of universal entities are still countless as much as it was in the beginning matter as atoms and even much smaller. Every dot insignificantly as it may be is a part of another Universe as much as it is part of the accumulative Universe and every dot in infinity holds singularity, which we translate as "nothing" but it cannot be nothing. There cannot be nothing as much as there cannot be darkness. There cannot be something big or small except in the relevancies of perceptions and then the relativity of such perceptions becomes questionable. There cannot be hot as much as there cannot be cold The sun freezes hydrogen to a liquid at 6500 ^{0}C and outer space boils over at 0 K. If we humans cannot or will not abandon our human culture driven perceptions and our mankind's pre-programmed perspective we may as well return to astrology for what the future hols. There are so many boundaries out there ready to destroy us because of our lack of insight, as did the challenger disaster.

Creation birth started off with one dot so small eternity met infinity within. Then came one more, and another and they continued coming until there were a countless number of dots. The accumulative size of the dots were the same size as one dot because in the true Universe big and small plays no part. The dots were infinitely small and eternally big at the same time because size is a relevancy and without one the other has no size. So in the true perception, there is no difference in size.

It started with the fact that there is no place or part in with which one may associate zero or nothing. There are no room for a number such as nothing. Next to the one dot (infinitely close) one will find the next dot, and if nothing was a factor then that is precisely what one will find between the two dots. Nothing of space, a non existing entity, taking up no space, and much more important, no time, therefore the dots are infinitely close to one another, being the same space, eternally big as much as infinitely small. If we as humans cannot find a manner in comprehending this notion, there can be no manner ever understanding the cosmos as much as the start to the cosmos.

Every dot was a Universe in its own and the accumulation was a Universe. The earth in itself is a Universe as the moon is a Universe, because rules applying on earth do not apply on the moon and visa versa. When considering the conditions with in the ocean and applying space-time another set of rules apply, therefore being in the sea places a body in another Universe. The number of universal entities is still countless, as much as it was in the beginning, before dots formed atoms.

Every dot insignificantly small as it may be, is a part of another Universe as much as it is part of the accumulative Universe and every dot in the infinity holds singularity, which we translate as " nothing" being " darkness". There cannot be "nothing" just as much as there cannot be "darkness". There cannot be something big or small, but in the relevancy of perception, and then the relativity of perception becomes the question. There cannot be hot as much as there cannot be cold. The sun FREEZES hydrogen to a liquid at six and a half thousand degrees Celsius and Universe boils over in the form of the Hubble constant at the temperature (we presume from our vantage point) at minus 273 degrees C. If we Humans cannot or will not abandon our human perception and our manly perspective, we may as well return to astrology for all its worth, because that is the only boundaries we will find in the cosmos.

To unlock scientific truth we first have to dispose of scientific misconception

In the two pictures we are seeing disposing or releasing heat creates space. We may call it plasma or shock waves or what ever, but in the final analyses it is heat turning to space. Whatever you wish to call that which lies between the particles comes from being a solid, then with adding heat, the solid *"whatever"* becomes liquid and that is the white and orange plasma that we find. That white and orange is heat in a liquid form, just as all flames and smoke is heat in a liquid form. But that liquid does not

remain liquid because the governing singularity cannot enforce a commitment ensuring the liquid heat remains liquid. The liquid *"whatever"* you wish to call the heat in fluid form then further overheats turning the heat to space. The space created must be equal to the heat reformed. That is a law of energy where energy equals equality everywhere it is.

Let us humans first detach culture from facts. Take the argument to iron, which we know well. Iron cannot boil, iron cannot flow or bend and iron cannot brake. Iron is an element like all the other elements we know, not one element can do any of the above, in sharp contrast to human belief. As indicated in this book the limits we should find to guide us we ignore for the reason that we cannot see it. We may not be able to ever see singularity, but with intelligence guiding mankind, we do not have to see everything to believe everything. It is because we could not see religion, but still practised religion that set us apart from the other animals.

At the start one would find iron and iron in a "natural state" as we find iron on earth being a human produce on the surface of the earth it will be a solid, suitable for man to handle with bare hands. When such a piece of iron is left in a desert in the midday heat, the human hand cannot handle the iron any longer without aid of covering the skin of the hand. Our perception is that the iron became hot, but that is not the case and our view is a culture contribution and not scientific fact. By heating the iron artificially with combined gasses (acetylene and oxygen or what ever) we now can over heat the iron to a state of flowing like a fluid. Our human culture tells us the iron now is melting.

That is a misconception!

Like the fact of "nothing" we inherited the idea from our past. After introducing artificially even more heat with more heat releasing gasses we may artificially form a condition where the iron would become a gas. Again it is not the iron that becomes a gas, it is the space the iron finds itself in that became hot enough to become a gas. The iron particles remain the same; it is the condition surrounding the particles that changes form with overheating.

Important to note is the fact that iron in a solid state will surround itself with solid matter in space applying a solid space. By introducing conditions producing ***more overheating*** the space or connecting between the particles become concentrated heat forming a liquid substance! It is not the iron that turned liquid but the wrapper containing the iron that concentrated so much it formed liquid fluid by the introducing of more heat to a point where the overheating created a fluid. It is considered that the oxygen burn and by that the iron heats up. NOT TRUE!

If oxygen burns no oxygen would be left on earth by the time man arrived on earth to use it to the benefit of intelligent life. The oxygen remains oxygen while the oxygen merely does a task in nature where oxygen carries heat to a specific space. On the other hand it is the task of nitrogen removing heat from the point of overheating by means of flames whereby it creates space. One can feel the "wind blowing" as the flames generate created space. In the extreme the creation of such space we call an explosion.

In the process where the space between the iron particles still further overheats, it becomes a gas. It cannot be iron that becomes gas, because iron will be as much a gas as iron will be a liquid or a solid. It is the space covering the iron particle separating the different iron particles, which will convert and sustain form. The gas is as invisible as space because the gas is the form space holds. This confirms the Biblical view of earth (solids) created and heaven (heat or gaseous/liquids) created. There are only two forms of substance that forms the Universe solids and non-solids, which is liquids and gas. It is not the solids going liquid but it is more of the liquid in ratio with the solids in between the solids that make a structure go solid or gas. There are heavens (non solids) and earth (solids) and this has to do with movement applying control or non-movement allowing non-movement control.

Iron is a solid. Introducing more heat the iron becomes wrapped in a cover that concentrates the wrapper to the point of concentration where it became a fluid. The iron remained what it is, neither a solid, nor a fluid nor a gas. By introducing more heat it becomes a gas. The gas we cannot see because the gas is space. But so was the fluid space. The introducing of heat brought about the turning of a solid to a liquid to space and every time more space becomes part of the picture.
Iron is in its normal form a solid. That means the space, which the iron particles are in, is solid and that disallow the iron to alter the form in which it is. By introducing considerable heat the iron melts changing the form of the iron from solid to liquid.

Considering the evidence we find it is not the iron that melted and that became liquid, but it is the space in which the iron is that became liquid. The iron particles are still as solid as they were. By introducing more heat the iron would eventually turn to gas. It is not the iron that turned to gas, but it is the space in which the iron particles are that has increased to the extent that the space now has so much heat, the heat turned to more space. The iron as particles remain the same, they are just elements confined to a nucleus with electrons spinning about. The space between the particles increased to such an extent it first became a liquid or a fluid and with more heat introduced the heat increase brought about that heat turned to space. That means by overheating the particles surround with heat as a fluid the heat increase then add space as a gas. The gas is the ultimate form of overheating but where one is unable seeing the gas.

1 Firstly the iron is cold enough to be a solid. Replace the word iron with cosmos and forget the colour we associate with heat being white and note the solidness of the centre of a galactica. This must have been the state of galactica that contained large parts of the Universe when time rolled away from eternity.

2 By introducing overheating the space between the iron and not the iron as such turns to liquid. The same apply as more matter (iron) produce more space forming as some matter turned to heat by overheating. The matter increased spin and in that way went out of sequence where it then became softer and softer in relation to other particles, where the loss of the matter released more of the third cosmic component we named heat and space.

3 Some of the heat introduced with the overheating by means of congestion then forms space while other remain in the form of heat allowing space to seam liquid. The matter could not breath and overheated by the enormous gravity the overheating created

4 As the area between the particles still further overheat certain parts of the area overheats to the extent that the space becomes an invisible gas allowing the congestion of matter to separate from one another and allow the stars' individual governing singularity growth. 5 From the soup of heat galactica come about allowing stars to rise out of the dense liquid cradle from where they can establish singularity growth. The process continues as more space becomes introduced through space overheating turning heat into space

6 Should star development come about as suggested it is foreseen that the Milky Way once was a liquid from which the sun developed the singularity in which it then form self-sustaining. The only pre-condition was that it captured individual space-time where the captured space-time remained a liquid

frozen (as it was back then at the time of parting) by the governing singularity while outer space further overheated into a thin gas

7 The sun captured so much space by the intervention of singularity when released from the Milky Way that it produced space so concentrated today at present it clearly remained a liquid inside as it froze the interior in time the liquid it now is while outer space is still overheating as a gas with no visibility. From this overview one can judge just how far science is behind the time in their views on creation and the beginning of time including the universal establishing. Cosmology still hides behind medieval ideas that other faculties and scientific departments forgot long ago.

See the fluid push out of a bowl of liquid, spilling both sides as it falls into liquid. The inside of the sun is not gas but it is fluid. In all of nature there is no NATURAL GAS as much as there is no NATURAL SOLID. Look at the liquid squirting from the surface of the sun. if it is liquid the sun is liquid on the inside and if the sun is hydrogen then the sun is so cold through movement that it turned hydrogen into a liquid at 6500°

Science a under the impression that water will boil at 100° C when it is on the sun. Let us take this formula back to the accepting of the Big Bang and find sensibility amongst a lot of confusion that I can see.

As a solid	Forming a liquid	Steaming as a gas
Hydrogen 1	melts at -259^0 C,	boils at -252^0 C,
Helium 2	melts at -269^0 C	boils at $-268,9^0$ C
LITHIUM 3	melts 180^0 C	boils at 1300^0
BERYLLIUM 4	melts at 1287^0 C	boils at 2770^0 C
BORON 5	melts at 2030^0 C	boils 2550^0 C
Carbon 6	melts at 804^0 C	boils at 3470^0 C
Nitrogen 7	melts at -210^0 C	boils at -195.8^0 C
Oxygen 8	melts at -218.8^0 C	boils at -183^0 C
Fluorine 9	melts at -219.6^0 C	boils at -188.2^0 C
Neon 10	melts at -248.59^0 C	boils at -246^0 C
Sodium 11	melts at 97.85^0 C	boils at 892^0 C
Magnesium 12	melts at 650^0 C	boils at 1107^0
Aluminium 13	melts at 660^0 C	boils at 2450^0

Hydrogen is as much a liquid as iron is a gas and neon is a solid. It depends on the element relating to the space/heat in the circumstances surrounding the substance at that very precise instant in time. We have to stop telling the cosmos to show us what we wish to find and start accepting what the cosmos is telling us to find. The culture that I am referring to is all about **nothing.** At present, we find that there is something we think of as nothing in outer space. Because nothing is what we wish to find and nothing is precisely what we are getting because we think of outer space as nothing. If you accept the cosmos to be nothing, then please define nothing to yourself and find the definition in the cosmos.

As a solid	Forming a liquid	Steaming as a gas
Nitrogen 7	melts at -210^0 C	boils at -195.8^0 C
Oxygen 8	melts at -218.8^0 C	boils at -183^0 C
Fluorine 9	melts at -219.6^0 C	boils at -188.2^0 C
Neon 10	melts at -248.59^0 C	boils at -246^0 C
Sodium 11	melts at 97.85^0 C	boils at 892^0 C
Magnesium 12	melts at 650^0 C	boils at 1107^0
Aluminum 13	melts at 660^0 C	boils at 2450^0
Silicon 14	melts at 1412^0 C	boils at 2680^0 C
Phosphorus 15	melts at 44.25^0 C	boils at 280^0 C
Sulphur 16	melts 119^0 C	boils at 444.6C
Chlorine 17	melts at -101	boils at -34.7 C
Argon 18	melts at -189.4^0 C	boils at -185.8^0 C
Potassium 19	melts at 63.2^0 C	boils at 760^0 C
Calcium 20	melts at 838^0 C	boils at 1440^0 C

Ignoring these facts, Mainstream science will hardly answer the problem we do not understand and such ignoring brings strong doubts about the quality and sincerity of science.

Excluding Argon, which is six (carbon's number) times two and suddenly that is a less dense material. The four times five plus... group are the following:

Scandium 21	**melts at - 157^0 C**	**boils at -152^0 C**
Titanium 22	**melts at 1670^0 C**	**boils at 3260^0 C**
Vanadium 23	**melts at 1902^0 C**	**boils at 3400^0 C**
Chromium 24	**melts at 1857^0 C**	**boils at 2665^0 C**
Manganese 25	**melts at 1244^0 C**	**boils at 2150^0 C**

Iron being the five times five plus one is the only generator of electricity and therefore the producer of gravity making five times five plus one the ultimate relevancy to heat in reducing space. Still Krypton is much more massive and turns out to be a gas.

Krypton 36	**melts at 1539^0 C**	**boils at 2730^0 C**
Iron26	**melts at 1536.5^0 C**	**boils at 3000^0 C**
Cobalt 27	**melts at 1495^0 C**	**boils at 2900^0 C**
Nickel 28	**melts at 1453^0 C**	**boils at 2730^0 C**

How can science promote their image of establishing honesty when they are confronted by such truths but choose to ignore the truth so long as a lie will bring them some respectability.

Palladium 46	**melts at 1552^0 C**	**boils at 3980^0 C**
Silver 47	**melts at 1412^0 C**	**boils at 2680^0 C**
Cadmium 48	**melts at 321.03^0 C**	**boils at 765^0 C**
Xenon 54	**melts at –111.79^0 C**	**boils a-108^0 C**

If anyone puts anything as hot or as cold that person puts him or herself in as the centre of the Universe. Then from such a stance as the person has that person finds at the centre things go colder or hotter because that person is stupid enough to think the Universe was created with life in mind and moreover for that person in particular. There is no hot and there is no cold. As it is important to realise the above it is just as important to realise that heat is another form of material and a separate form of material. The two developed on an equal basis and as a result of the other. The one produced to save the other and what the one produced saved the other .The one principle brought the incentive for motion while the other took the incentive by providing the motion. The one produced what the other captured and the one retained what the other delivered. Eventually the motion did not bring the required relief and another form had to be devised. By overheating and increasing space it counteracted overheating and by removing the expanded material and retaining it onto the contracting of the other, did the two form a synopses where by all received benefits in the form of cooling.

Only when further requirements developed, did the need arise for more to be made available. The first demand on motion asked no further changes because one change brought on satisfaction to all that suited all. The second was more general and on an ad hoc basis that was established to fit the need of individual places and not groupings in the broader perspective to fit individuals at large. At first the establishing of motion set a trend that brought on required results but afterwards the space required, in which to move became a demanding issue as the heat levels raging out of control. The heat had to be stored in space by becoming space to retain heat for later consumption. The number in ratio that produced the heat providing particles that offered to release their form in contribution to have those that retained form, do so to save those others retaining form. Those on offer became those ones that became the danger of destroying Creation instead of saving Creation.

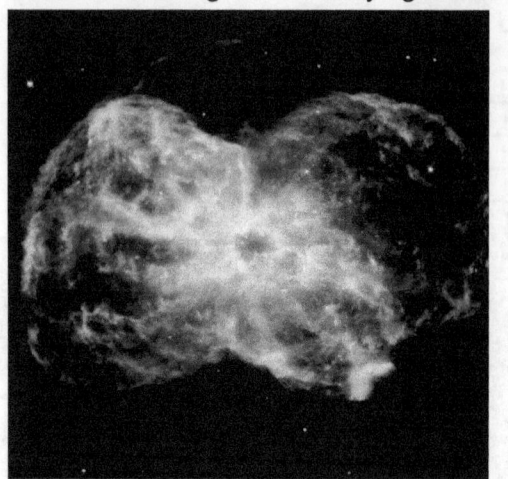

If heat comes out of a star when exploding then the inside is filled with heat and a star is then a heat container. If a star opens up and the inside of the star becomes the outside of the star then we only have to look and see what came out to know what is inside a star. This is so basic it should not be something worth mentioning and yet Newtonian science doesn't think that far. Is heat is contained by movement and then the heat burst out we should know that the movement was too slow to keep what was inside contained on the inside. If the inside comes out and the space it holds surges we should know that the density of the container went down. The star holds heat within and contained not under "pressure" that idea is shit but the density inside the container lapsed because of reduced intensity and then as the density decrease and the heat expands the density lapses and reduces according to the movement slowing down.

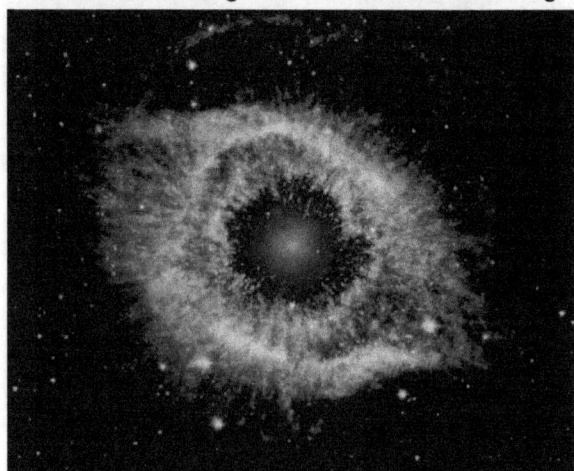

When we see a Super Nova going ballistic we see heat spewing from the star to the outside regions. We see what is inside the star and what is inside the star is heat, not "mass". We see heat bursting out in rage human minds cannot absorb. If heat comes out then a star is filled with heat. That means a star is a cooler of heat because when the star overheats, as it no longer can retain the heat on the inside by gravity it explodes into unleashing the inner heat.

There might even be some areas and regions in far off places in our modern day where an imbalance may evolve and some particles become unsuccessful to save those more successful. By going less successful, the singularity places a demand on another bringing about the command on space-time so that support can be accomplished to save singularity. Therefore by losing density, was gaining security to survive as part of a bigger relative. Density is the distributing of heat in specific relative space and by having less material in more space; the density is the offering for the common survival of the lot. The relevancy brings a contribution in whatever role to secure the survival of the lot in relations. No relevancy therefore can be "nothing" notwithstanding Newton's opinion about the matter as Newton had the opinion a relevancy acquired by rotation brought about an accumulation resulting in nothing.

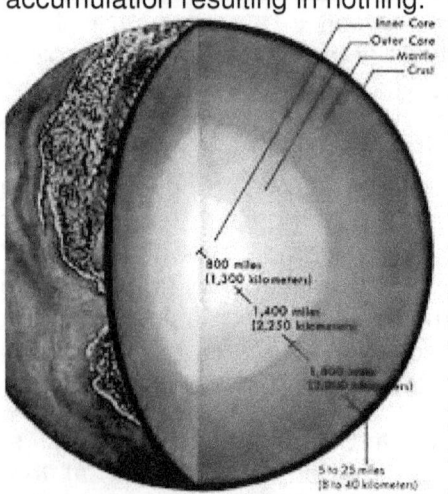

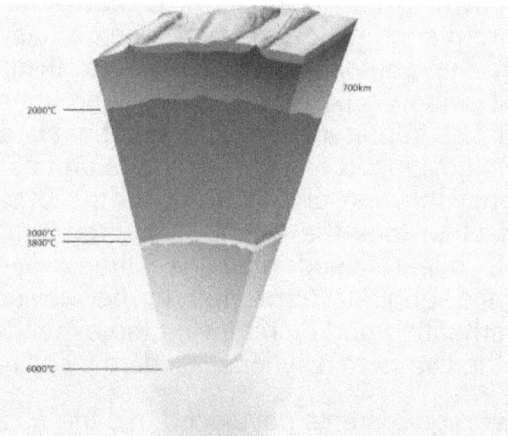

It is the way the atom formed before the atom took on space-time. It is in the formation, that space-time relates to motion. We have some elements being quite massive but also lighter than air and others are quit light but as dense as they come. This can only be a contribution from the way the atom relates to heat, which make the atom volatile (movable) or dense (motionless). Those elements being volatile are also very movable and in that we find the role that such elements play in the star. Stars that are predominantly made up of hydrogen and helium with very slight support from the metallic inner core are those stars that duplicate by producing motion. However the point I wish to press is that mass and being massive and being heavy do not support the fact that some elements have more gravity they produce because their protons are more numerous than others. The fact that mass generates gravity is a myth.

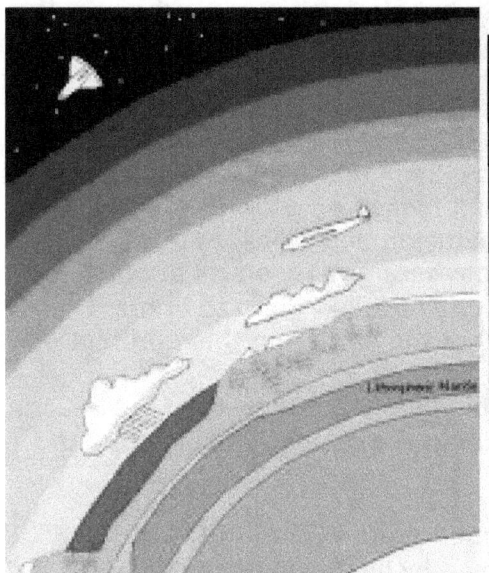

We are not the cosmos and the way we regard heat is not the way the cosmos regard heat to be. "Global Warming" is not about the heat humans feel on their skins or the heat that make

Hottest
2nd Hottest
3rd Hottest
4th Hottest
5th Hottest
Getting colder
Getting more colder
Getting freezing cold as it gets liquefying cold in the centre

It is not the heat that a human feels on his or her skin that forms cosmic heat because life is not part of the cosmos but merely an accessory or a small after thought here on a short limited time span. Cosmic heat is about the laws of nature and not thermometer scale that some person developed in a lab to serve Newtonian intellect. The cosmos applies its rules.

When space holding whatever expands it is hot. When space holding whatever contracts it is cold. Air expanding is hot and air shrinking is cold. It is not cosmic cold because the human can feel heat but on the contrary it is cold when the human feel heat because it is getting rid of the heat since the space got cold as it contracted. That is science law and not the meaning of some stupid scientist. Heat in the centre of the earth is a liquid and heat in outer space is a gas. The liquid is much more contracted than the gas in outer space is and that is why we can fly in space and not swim in a volcano. We must start to realise how the cosmos translate values and stop thinking like humans. When the balloon gets hot it expands and it goes up. When the anything loses heat in shrinks as it becomes smaller. That too must apply to the earth where the inner core space gets colder as it loses heat and the heat rises to the outer space region. It is not what we feel that makes anything hot or cold but it is what the cosmos does for instance increase in size or decrease in size that forms heat.

One will find that whatever group one chooses there are gasses and there are solids. If mass was attracting mass then the strongest mass must be attracted to the strongest mass and the least mass must float in the air. $F=G(M.m)$ r^2 hardly can even begin to explain the fact that there is a gas that is more massive than iron but floats in the breeze just as hydrogen which is the least massive element. Let's look at gravity and anti-gravity and see where the wind blows the balloon.

The saying goes "it fills with hot air" meaning it fills with nothing and yet while the balloon fills with hot air the balloon gets air borne. The balloon gets into the sky by being filled with "hot air" and the idea that this might happen is senseless to science or so they pretend. Still it happens but why does it happen?

It is because science has no idea about heat and cold. When filling the balloon with heat the balloon takes of to where science thinks it is cold. Why would the balloon go to a cold place when it is filled with heat? It is because where science think it is cold it is hot. Forget this idea of pressure because if it was pressure in the air the pressure will escape from the bottom where it is open and being pressure it will escape or release from that opening. By heating the balloon with hot air must increase the heat level and increasing the heat level takes the balloon into the air. This means hotter is higher up by holding more space than down below.

When something fills with "hot air" it goes up and by going up it forms anti – gravity. If gravity is what takes bodies down to the earth then lifting it up into the air must be anti – gravity. Anti means opposing or counter acting and when going down forms gravity an anti of that then must be going up. If it is mass that pulls down then by applying more heat it forms anti-mass but it can't be anti- mass because the persons in the basket seems just as they were when they were on the ground wit all the mass intact.

Gravity is about movement containing heat that is most expanded **where gravity starts at outer space as $\{10/7 [4(\Pi^2+\Pi^2)]\}$ = 112.795. Then by movement gravity reduces the size of the expanded space in accordance with the turning (in this case) of the earth and this contraction can only be as a result of cooling.**

Then by gravity or electricity)it is the same thing with a dimension difference) it moves to **$\{7/10 [4(\Pi^2+\Pi^2)]\}$ =55.27 which is the in core and this is in relegation to the copper jacket where the confinement ends space ending at $\{\Pi [4(\Pi^2+\Pi^2)]\}$ = 62**

Gravity is about outer space $\{10/7 [4(\Pi^2+\Pi^2)]\}$ = 112.795 conformed by iron $\{7/10 [4(\Pi^2+\Pi^2)]\}$ = 55.27 the displacement value of iron$_{55}$ in relation with singularity presented by the relevancy of $\{\Pi [4(\Pi^2+\Pi^2)]\}$ = 62 and that is the displacement value of copper which is the element that has this proton value according to the periodical table. This is exactly how electricity is generated and that proves that gravity and electricity is the same thing charged on different levels

Newtonians are never very clear on this issue but if it is mass that pulls then it must be gravity that moves. Mass supplies the magical force but the actual movement is derived from gravity. So the movement in itself is gravity and that gravity pulls bodies down. To have bodies lift into the air then this action represents what goes anti and anti is filling a basket with hot air. By adding hot air the gravity goes anti and then gravity must be the cooling of space. It is definitely heating space up that makes the balloon go into the air so going into the air forms anti gravity because gravity is going down. This is an argument that science cannot dispute except when they go into another cheating and dismissing mode of the truth. By heating the basket something happens and this science for many years try to avoid by reducing it to a joke. A joke it is not because it is a fact nature substantiates with forming anti- gravity. When heating a basket makes it lift than gravity is making things cool and that science can't deny.

When the earth spins there are always a relevancy applying between that which spins are double 7 and that which moves straight continuing at 10. The part spinning at seven we think of in terms of the diverting of direction it applies at 7. The liquid / gas holds the 10 factor.

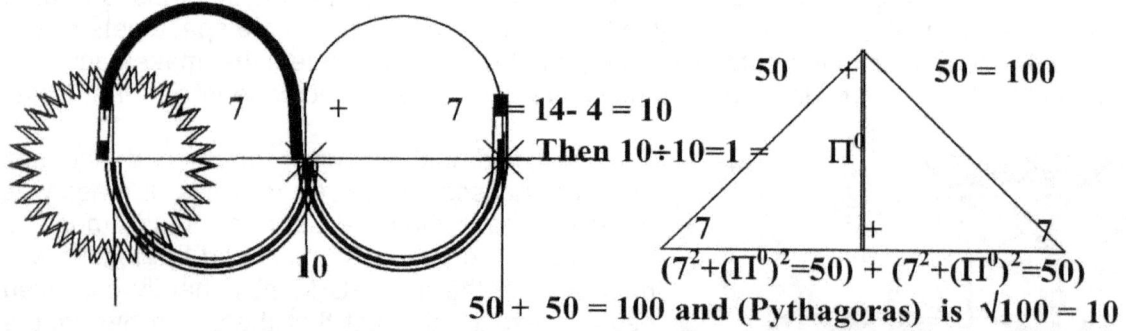

$$50 + 50 = 100 \text{ and (Pythagoras) is } \sqrt{100} = 10$$

Without the application of specific heat, the object remains in the three directional moving of six possible directions. The value of space unoccupied therefore remains $\Pi \Pi^2$, as it was before the "Big Bang" event, whichever "Big Bang" you wish to refer to, because there were many. But space unoccupied holds time to the value of 10 to 1, and as the sketch of the triangles also indicated, holds space to Π. Therefore unoccupied heat holds the relation to space in applying 3 directions of influence $(3^2 + 1^2 = 10) = (a^3 = T^2 k)$. A part of this equation is where the dual function of space is $(a^3 = T^2 k)$ while at that very instant one has space-time. Therefore in space in time you have $(10^2) = 7^2 + 7^2 + 1^2 + 1^2$. In the sphere we have the axis holding a value of 3 and the circle holds a value of 4.These are dots forming in relaxation to the one spot holding a point from where singularity advances.

We have the axis valued at 3 going square through movement of the linear motion $(3)^2$ and then we have the circular motion $(4)^2$ going square by the spin of the circle ring the direct opposing side. Then the equation of influence becomes $3^2 + 4^2 = 25$ where $\sqrt{25} = 5$ and doubling the 5 on both sides of the triangle will apply the factor of $5 \times 2 = 10$ that then is $(10^2) = (7^2 + 1^2 + 7^2 + 1^2) = 50$ on both sides is 10. The implication of this may not dawn on one the very instant of realizing, but to scientists, there is no greater shock than just that. To any application of movement, the factor will be in the realms of singularity where half a circle is equal to a triangle is equal to a straight line and the lot is equal to 180°. No fancy mathematical expressions have any value in singularity because singularity holds a value of 1.

The fact of this comes as 49 plus one becomes 50 and that is in the three dimensions of space $\Pi^2/7$ where 7 holds the relation to one and $\Pi/7$ again where 7 relates to one. At this point it is most important to remember that Pythagoras works on the application of the sum of the square of the two sides. When seven has a direction in the fourth dimension applied to it, the opposing dimension will be one and this applies in time relevancy, therefore the interchanging in time between infinity will place matter at $7^2 \times 1$ relating to circular and $7^2/1$ with $7^2 \times 1$. This makes 49 plus one (singularity) always being a factor of one. Space in time however, never can be a cube, it will always be a square with one side pointing the direction of time from time to the past (1) to time to the present (1) to time to the future (1).

The circle forming Π uses 7 to indicate the roundness of the circle but the 7 holds its roots deep within creation. It indicates how the Universe started because this is the way a star will start moving and it shows how as the infant star starts generating gravity just as the top starts to spin when it is thrown by life. Life can create nothing and that is true but life can mimic all laws in the Universe. Time is eternal movement and will be with us always. The line in infinity is still present while not being a part of the Universe. This line is always ready to be in place when the slightest movement orders it in place. Before the Universe was in place eternity and infinity was in perfect harmony and the line forming singularity validates this fact.

Before infinity parted from eternity, eternity met infinity on one spot as eternity came from the past (1) forming the present (2) to go onto the future (3) but also returned to come from the past which was the spot held by the future and this we find in the fact that the line forms 1 when not spinning but as soon as it evokes by spin, 3 points form even now. Then heat and cold differentiated values and space landed in between eternity and infinity. As eternity moved in relation to infinity but not forming a part of infinity any longer, eternity had to follow a path by never going away from infinity (3) and always returning to the point infinity holds but never lash onto the point again. With space parting the points, eternity had two points (the past and the future) before the partition came about and infinity held both the past and the future while infinity had the present as it still gas presently. By eternity also moving, the two points it held opposed each other (the past and the future) and since it moves, by the movement it became the square of the two because movement is the square and not a flat blanket-like surface with squares embroidered on it as Newtonian science depicts it by using grand mathematics to understand singularity.

I

Then we had two point holding eternity in place going square by movement to form 4 points serving eternity and infinity captured the first three points held by both and since eternity could not release from the two it had but had to duplicate what it had, eternity by movement became a circle captured by the line. With four points captured by the line of three points the circle coming about is eternally returning to infinity but never complying with infinity because if mismatching temperature or movement (3 against four). Material will always be colder than outer space. It is because material spin and outer space moves by expanding due to overheating.

This is where I start when I start to explain the first moment but I use a shipload more information to do explaining when I explain the star in the book I do so. I involve the four cosmic pillars to substantiate the claims I make because all four still work the very same way as it did at the beginning of the Universe. The three points serving one part of singularity combined with the four points serving singularity unites as seven to form a circle of either 3.1416 or 21.991÷7. The seven going to one is eternity matching infinity by movement. But since seven moves it are seven that have to produce gravity. How do I know all these facts, because we can see from the top it is still doing what it did the very first second. When time started infinity as well as eternity had altogether 3 positions, the past, the present and the future. It is still forming the very line in the centre of the top as it forms all lines in the centre of all things spinning. Then eternity parted from infinity when heat separated what is cold from what is hot and eternity formed one more point than before when it had the three points.

With infinity and eternity then jointly having 7 the cosmos came into rotation. In the aftermath post big Bang we now see the phase of cosmic development where the tow sectors try to unite and this brings along the contraction. When Π forms it does so on the grounds that 7 rotates. The circle forms by a change in direction by $7°$. Every circle has opposing sides forming in relation to the axis line. If the topside goes rite then the bottom side has to the left. If the rite side goes down then the left side goes up. There is this double presence of a change in direction forming on both sides of the circle. The $7°$ move and by moving $7°$ goes square 7^2 and that is Pythagoras.

The one part of the earth is going up by $7°$ and the other part is going down by $7°.$ Every time the earth moves in the going up or down direction the relevancy crosses singularity 1^0. It is the same $7°$ that crosses 1^0 to 1^1 but it results in two different points holding 10

They join space-time therefore the matter factor is the same. This is where one can visually see the one object, filling the space of the other object's atmosphere.

$$7 \times 7 = 7^2 = 49$$

That is matter Π^2 (time) times matter $(49)+(1) = 50$. This 50 forms space which then applies to both sides of the rotation of the solid being 7 that rotates.

As this is all under the law of Pythagoras the law will evidently place a square root to that value of 483,61 and therefore $\sqrt{50} +50 = 10$. This leaves the space value of the Roche-limit, as it develops into the Titius Bode law giving them a shared value of 7 (matter) and 21,91 (space) the value of 21,991 / 7 = Π. Then the relation becomes

Everything in the cosmos is moving, either by own individual accord, or under the influence of some other singularity dominance. In explaining we return to Pythagoras where the entire Universe with everything in it started.

It is the point forming the very centre that plays the part as the **controlling singularity** within the Universe I have named as **Infinity,** which is better known as the axis. It is where nothing can go smaller and anything within that point can never reduce. That point is where the entirety called the Universe begins and where everything holding substance begins.

Infinity is a point not even in the Universe and eternity is points forming time in the Universe.

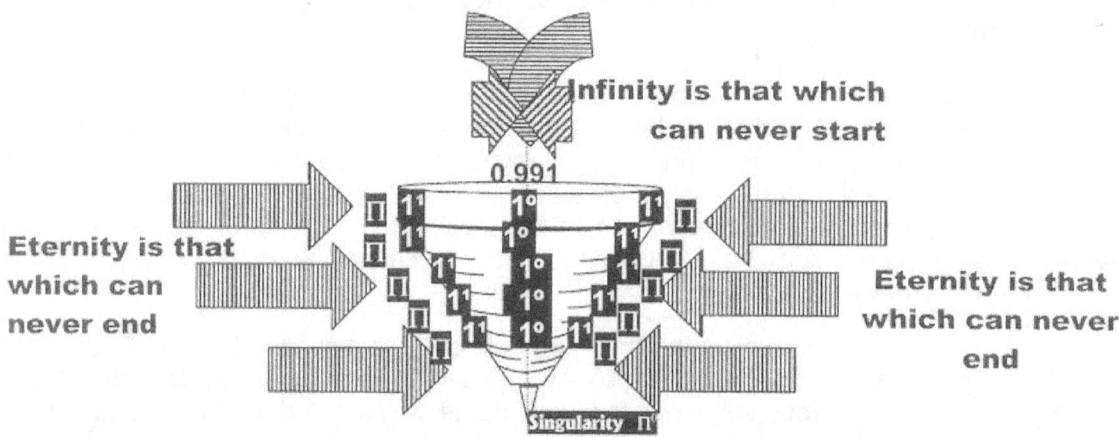

Once one accepts the fact of singularity being present in that location, that accepting of singularity then is contradicting all the things we know and we can measure and we recognise that point being present by merit of the fact that the point referred to is not being formed by any of the things we can recognise.

It is made up of everything we don't know and constitutes of everything we are unable to recognise or visualise. In that spot there is no space. That spot holds **Infinity.** In that space there can be no motion because there can be no space to have the motion within. It is formed as a line that is so small that our human reality by perception declare that point as not being there and the only reason why we know it is there is because of the results it left as an imprint of its not being there.

We cannot detect it but notwithstanding our failure to note it we can recognise the dot on the merits of its absence and while in our Universe it is always absent, reality disallows the dot ever to be absent, because it is never absent. It cannot be absent. It cannot go absent but it can never be there where it should be in a place from where the third dimension forms and it is always present if I wish to locate it. It is **infinity** that can never go away. I named the other part of singularity forming space **eternity** because that area never become bigger, or become more or find an end to the outside. Whatever was and is and will ever be is locked in that space I named **eternity** and it is **eternity** that never ends because **eternity** can never end moving. What we think of, as expanding is never ending movement giving eternity the eternal motion that will go on forever.

The line **k** coming from the centre (singularity k^0) forms by forming an initial spot Π^0 becoming the dot Πr^0. However, I went on to say that whatever the line used to start with has to continue in order to repeat the same that began the line. Therefore the line started with Π^0 and it has to continue with Π^0 until such a point, as it must end withΠ. Whether the line is Π^0 or is r^0, or uses 1^0 the outcome all refers to singularity being used. By reducing the line we come to the end of the mathematical equation of the circle but the circle does not end there. When the top is in a state of motionlessness on own accord it is everything but motionless. The motion it adapts are synchronised with the earth in harmony with the solar system and according to the greater picture of the cosmos.

When an energy source not related to the cosmos called life intervenes and energises the tops motion, the singularity in that top suddenly jumps to life. By adopting a rotation energised to an unnatural state of energising because of life's intervention, the singularity of the top is not in charge but as it applies more and more energy, it will begin to find a means whereby it can escape and apply individual singularity as the top starts to separate from the singularity the earth holds. The singularity holding the earth would then allow the singularity of the top to rotate within a specific band where that a specific band of being active before the earth's singularity will start to destroy the singularity in rebellion.

The top on the other hand will try its outmost, when the singularity it holds gets by individual spin is too strong to remain be in domination of the earth's singularity. The motion of the top is an attempt to begin applying an individual singularity space-time defying and standing apart from the earth's gravity. That action we see as the top starts rotating in a manner where the top does not align with the earth's singularity. With the adding of spin, the time the top holds becomes unrelated to the time the earth holds and the top will start a campaign too escape from the singularity domination the earth has on the top. When the time or spin of the top exceeds the limits the earth places on the top, the top would emerge by trying to escape from constrains placed by the earth.

The view I represent at this point is known to science for almost as long as science knows mathematics. Not long after the law of Pythagoras was understood where Pythagoras introduced mathematics Eratosthenes of Syene made as big a discovery as Pythagoras did.

But in the one instance the world took notice because the world could see and understand and the other instance the world disregarded the findings because the world did not see what the implications was. The same apply to aircraft flying and when the aircraft wishes to escape the earth's singularity hold it has to comply with the laws laid down by the earth.

The seven becomes as big a part of the concept as does Π as it all interacts.

Sun overhead **Alexandria**
Sun overhead **Syene**
7°

It took Eratosthenes of Syene (276 – 194 BC) a Greek astronomer who in the year 240 BC made a discovery that the earth has a profile of 7^0. Since then no one ever did anything about it. When any singularity wishes to disconnect from the earths singularity, specific pre-calculated laws would have to comply to allow the lesser object to divorce from the larger object.

I indicated how the dimensions of 10/7 and 7/10 interact to form (Π^2)

Matter is a product through the separation of space and time receiving the value of Π The original time and Π^2 as follows: By circling around a spinning solid the space contracts to form Π and Π^2. Gravity forms everywhere in the Universe by applying singularity. By dividing space into material (material spinning in space) and duplicating space by material spinning, the TITIUS BODE LAW forms a 7^0 deviation and 7 / 10 in conjunction with THE ROCHE PRINCIPLE OF $(\Pi/2)^2$

In my article to Annalen der physics I used 15 pages to explain this process of singularity applying. I received a rather cordial but sincere reply from the Editor of the magazine.

When I placed an article in Annalen der physics Dear Prof Friedrich W. Hehl said in the e-mail he sent me that there is no way to "explain" the world of physics. I am not going to go into detail how this works. On the other side of the Pythagoras's' triangle we have 1 going square.

That makes Pythagoras's' triangle 49 + 1 = 50 on the one side of the earth and the same on the other side of the earth. The total is 100 and the square is 10. That leaves the Titius Bode law with a value 7 (it forms part of the material of one body) and 10 in relation to the space.

Then from the relation of 7/10 and 10 / 7 forming Π the Titius Bode law form Π^2 applying "With a lot of words and some simple algebraic relations" to quote Friedrich W. Hehl, Inst. Theor. Physics of Annalen Der Physics fame. This was simple algebraic relations but still it is science, is it not?

Since it involves singularity moving it calls for the law of Pythagoras to produce space. The law of Pythagoras is the triangle a^3 that is moving forward in singularity **k** by turning T^2. In singularity the 7 stands in for 7 points on the numerical line crossing over the line holding singularity or 1.

By moving 7 has to go square T^2 and that means 7 goes square 7^2 twice $7^2 + 7^2$ crossing the same divide $\Pi^0 = 1$. Since all movement in singularity has to enforce the law of Pythagoras we have two triangles holding 7 dots moving across singularity. I don't want to get too involved by bringing in numerical outlays because then this can truly become complex.

The line has two opposing sides turning directionally against each other while turning with each other. By moving or turning this involves time duplicating space by the square Π^2 on both sides of the divide $\Pi^2+\Pi^2$ and using the same divide or the same axis or the same point serving singularity we have 7^0 crossing the same point in singularity Π^0. There then is in this rotational movement 7^0 standing in for Π^2 on both sides of the divide $\Pi^2+\Pi^2$, which then is 7^2 on both sides of the divide 7^2+7^2.

The circle spins in duel directions. On the one side it would go left if on the other side it would go rite. The one side hold a directional change in singularity by 90°. As it is going sideways it changes to going down. This produces a rite angle triangle of 90° and in it the law of Pythagoras produces direction changes. Since the square of the turn of the circle places by the spin and the direction change we have 7 holding a relation to 10 in space because it is space that has to carry the value of 10 when material circles by 7. There is a connection between space surrounding the spherical circle turning and the sphere. The circle holds the value of 7 as in 7^0 and this we find from looking at singularity controlling the circle by movement

Our Sun

VY Canis Majoris

The wise men from physics put everything down to mass without ever searching for evidence of mass. Putting everything down to mass may be one solution except for the fact that only nothing is that simple. Even the rejecting and / or accepting are incorrect, as the pulling part just comes across a tad too simple to make sense. To find substantiation one has to find the manner in which light connects to singularity because everything connects to singularity. In every circle centre there is space that is so small it is not present within the space of our Universe. In that space in the centre of every spinning object we have singularity forming space. That points serving singularity is the reality and the rest is only make believe.

Explaining the following is rather tough with the limited space available but it should secure the conclusion that I am not grabbing for straws and there are substantial facts on which I work.

Our human view of the sun is that the sun is big; it is more than big it is outrageously big. The sun is so big that one person got a Nobel Prize when he worked out that anything bigger than 1.3 times the mass of the sun would collapse into a Black Hole. That is how massively big the sun is. Another person then got the Nobel Prise when that person calculated that anything the size of 1.6 times the mass of the sun would be an instant Black Hole. Yea sure and Elephants fly at night over the cameras of UFO hunters just as they press the record button! And now the sun is not even a dot in something that we might call big. Look at the picture of the giant CY Canis Majoris and look at the size of the sun in the photograph of CY Canis Majoris. Then place the earth in this picture and then place yourself spinning on top of the earth in this picture. Then think of your position in terms of the entirety called the Universe.
  The more suitable question to ask is how many suns will fit into VY Canis Majoris and in that we have to realise that VY Canis Majoris and Betelgeuse and other "giants" are not stars but they are galactica not developed yet and most of all they have not even began to develop. It is not what we think is true that is true but it is what we think about the truth that applies.

Lets investigate the Universe. We can see how far off the mark science is in their estimate of what is big and what is 1.6 times bigger than big. Let's find out what is small. The Universe is made of lines connecting points. Whatever you see or can't see is lines connecting. If we wish to examine the Universe we first better star to examine lines connecting dots because whatever is in the Universe is dots that connects with lines. So it will be necessary to investigate lines. In mathematics they teach students that a line starts at zero and that is a fable. Zero starts nothing and nothing loses everything before anything can start.

This sounds so unimportant but it is fundamentally all-important. A line starting with zero how cannot increase in length. This means that any line formed in a star only ends ant the outer limit if the star as the line is a solid non ending continuously continuing line from the centre to the outside. The sun is a liquid and not gas as science tries to convince people. Look at the picture. The picture shows a liquid with more liquid squirting into a gas. The density that is apparent at the surface of the sun is very clearly a liquid. But Newtonians have hydrogen as a gas just because on earth hydrogen is a gas. Thereby the Newtonian wisdom says that $6500°$ is very hot. It is so hot it would burn everything on earth and roast any form of life on earth down to less that carbon. This they argue because from the atheistic stance that life is a normal commodity in the Universe and life is everywhere to be found, going at a dime a dozen therefore if it does not befit life then it is extraordinary. Life is extra ordinary and not even a thought in the Universe except on this little blue dot we call earth.

What contracts is cold and what expands is hot. Cold cannot expand and hot cannot contract. What contracts because of cold must eject all heat because it contracts and therefore remove all heat. That is what happens to the sun. Because it is so cold it emits all heat from it and because outer space is so hot it accepts all heat because it has a cold it can forever heat as it forever expands. This is the fundamental of physics and not the human thermometer showing a scientist what he or she must presume to be hot and cold. In cosmology there can be no hot or cold as much as there can be no big or small. When the Big Bang was in The Planck time: 10^{-43} seconds. After this time gravity can be considered to be a classical background in which particles and fields evolve following quantum mechanics. A region about 10^{-33} cm across is homogeneous and isotropic, the temperature is $T=10^{32}$K. If the entirety was 10^{-33} then how big was the sun? If the temperature was $T=10^{32}$K then what was zero? For the Universe to be $T=10^{32}$K then something else must be zero or that temperature is meaningless because then it could just as well be zero. If the Universe was 10^{-33} then what was the size of one atom? If they say it was before atoms was present then they must state what was the material used in the Universe at the time. It is like saying there is anti-matter. Yes and what was anti matter.

I know matter is singularity spinning in a direction in excess of the speed of light because the electron spins at the speed of light and what is further inside the atom must therefore spin faster than the speed of light. Matter is controlled singularity that directionally diverts from uncontrolled space by spinning in a direction. If that is matter then what is anti – matter, things that don't spin faster than light or is it a concept science ahs no clue about but naming it brings clarity to absolute stupidity. It is like saying the sun is gas and all anyone can see from pictures is streams of liquid squirting all over.

It is no use showing how space expanded in relation to temperatures cooling because the big issue is lost in the entire scenario. Space expanding is the same as heat lowering and heat rising is the same as space contracting. That is what gravity is. It is some space expanding and some space contracting and the expanding space is the space heating while the contracting space is the space cooling. Just because we humans feel space heating it does not mean it is hot. It means to humans it is hot but humans have no say about conditions applying in the Universe. Life is a nuisance that are not even ever recognised in the entire Universe and life is alien to everything except the earth in the entire Universe. The heat we feel is the coldness releasing heat to the heated space that we think of as cold.

The very first instant space formed is when heat parted from cold. The Universe began when a

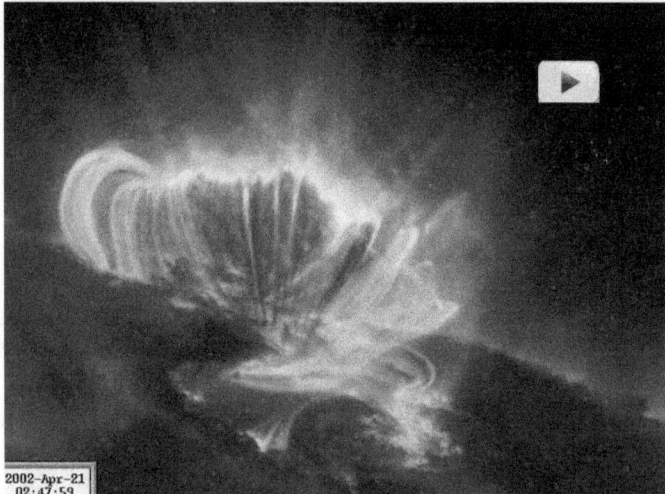

difference happened when there was one part that was hot in relevance to the other part that was cold. The entire Universe is a patrician of what seems cold in relation to what seems hot.

If there is something hot then there has to be something cold and neither hot nor cold is prescribed or is dictated. It is only opposing values in that specific space forming factors that generate gravity and gravity forms when one part expands in relation to another part that contracts space.

In the Universe there are two forms of substance that which holds material and that, which is material. Material cannot flow but has solid form. Non-material holds no specific form but is able to hold material or not to hold material.

The sun is a bowl of fluid so cold it holds hydrogen as a liquid on the surface. Where the surface touches outer space the friction heats up and that makes the liquid boil. It is not the liquid that heat but is, is the gas causing friction as the liquid turns that makes the liquid boil.

The liquid seems pretty steady and it even seems like waves across the surface. Science forever and ever tells the cosmos what it is and never learns from the cosmos what the cosmos is. What squirts from the sun is a liquid frozen by the movement or gravity of the sun. It is not a gas because a gas cannot flow as the substance squirting from the sun very obviously does. Notwithstanding lack of support coming from the cosmos science tells the cosmos it uses "mass" to arrange planets and then it uses "mass" to allocate planets.

We see liquid squirt from the sun but because humans think of Hydrogen in terms of gas and then the sun is gas and the sun is also hydrogen gas that is hot! Science will forever tell the cosmos what to be. In the Universe there is no hot and there is no cold. In the Universe there is no big and there is no small. If outer space is the hottest there is then we know what a Black Hole is it is the coldest there can be. It is so cold it froze all material into one point that is not even part of the Universe.

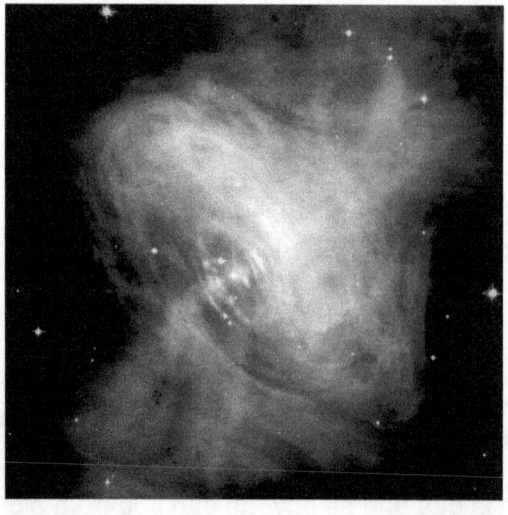

It froze space out of existence and into the oblivious; literally the oblivious where what ever can be cannot be any longer. In the Black Hole the atoms froze into one structure where not even a proton has validation. The structure spins so fast it does not spin at all and allows all spinning into space that otherwise never can spin. The movement in outer space is the expanding of outer space and the movement allowed by outer space is the increase it has in growth by forming a larger volume of space. However this entirety growing is not in size but only applies in relevancy. If one dot removes from space; any space then the entirety forming space will collapse into where that one point has disappeared too.

The Black Hole is a star that the atoms within joined to form the ultimate atom in singularity abandoning even the proton in the core and all movement in relevancy has gone outside into time or outer space.
In the Black Hole the atom froze to something that has gone into the oblivious and it froze to a point that no longer has a position in our Universe. But reality is that in every atom there is a Black Hole because every atom absorbs light and the rest it casts away. That is all but silicon and I just can't explain that detail in this book since it will be far beyond the technicality of this book's standard. The sun is a future developing Black Hole and that is why what is in the solar system will flow around the sun eternally without ever having a chance to escape from the centre of the sun. The sun like every atom in every formation forms a Black Hole by forming singularity.

In ever star all atoms join one single point attaching all starts to that point and even after a Super Nova release such bonding remains firmly attaching the star.

Even the structural remains that is the reminisce of a star that went sour still hold singularity in position as forming a centre Π^0 indicting the point allocated to Π. This connection is not coincidental but rules apply putting this formation in place even after the structure went beyond repair.

It is accepted that stars develop by growth and that is true but the growth in stars developing is

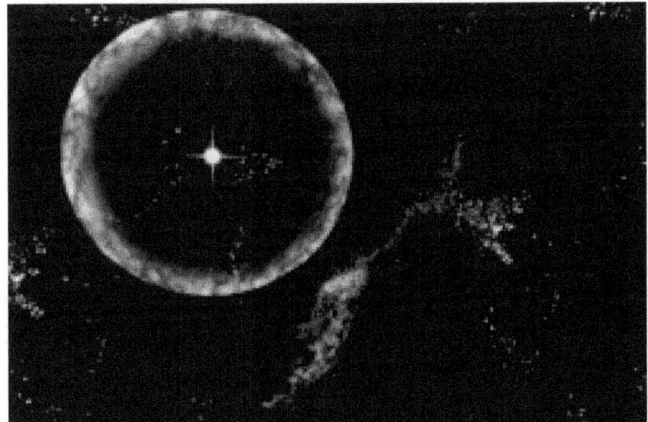

moreover in the Universe expanding. When the Universe expands the relevancy left to the star is not in ratio the star holds less space. The star expands to a point where the star the forms gravity that exceeds the speed of light and when that happens the space the star requires fades in comparison to the space used by the Universe in growth.

The speed of light is not a constant and it could never be because gravity changes the speed of light as it alters the waves that light transmits by. The stupidity behind the reasoning that the speed of light was forever a constant is underlined by the question as to when the Universe was 10^{-33} cm across what was the speed of light then.

As the Universe expands so the speed of light will stand corrected every instant and every location throughout the entire Universe. The speed of light might be 300 000 km / sec from where we are but a time back with a smaller Universe the speed of light had to be say 250 000 km / sec, then before that 100 000 km / sec and before that it had to have been a 100 km / sec. It then is madness to place a galactica at 12 000 years or 12 million years because where the galactica is it developed as the speed of light developed but to put reality in place there is no saying how far the galactica is because no one can determine how the speed of light developed since the light we see left the galactica we think we see at present. It is like saying the Universe was at the time 10^{-33} cm at the beginning but the Universe was filled with atomic matter.

Then with the Universe at 10^{-33} cm how big was a neutron at the time when the Universe was 10^{-33} cm? Everything has to be in relevance or not be at all. One can't say the Universe was because what was a centimetre back then? The material forming atoms was already in place and the entire Universe was the displacement intensity of one electron and therefore what was to become each and every neutron had to be fabric and what was then proton material was produced. What was the size of the material grouping to become an atom? It is very fuzzy to play the role of God and have an overview of the entirety because the scientist then runs out of logical relevancy.

The entire Universe then was as big as it is now and was as hot as it now is but applying relevancies changed from then to now as densities and gravity alternated. If we have an atom then we have the same atom now but the relevancies applying changed in time growing space. What did change was the time in duration as time form space and time then in comparison to now stood still. Time then moved at a rate we now would never understand. We now attach time to movement and movement comes at a rate of the speed of light. How fast back then did light move? If we put the speed of light now in ratio to then we can never understand time back then.

In this very book I show how the Universe was before the Universe even formed a thought. I show how the Universe developed from being far less than zero and then developed zero to become what we have at the present moment. Most of all to all the mindless atheists, I prove in this very book that according to mathematics the Bible has been correct for tens of thousands of years and the Newtonian hoax atheism rely on is and big shamble and a fraud concede to prove idiots how stupid they are! While this book shows Newtonian science is fraud it shows the Bible is indisputably correct and vindicates the way Creation began according to the Bible.

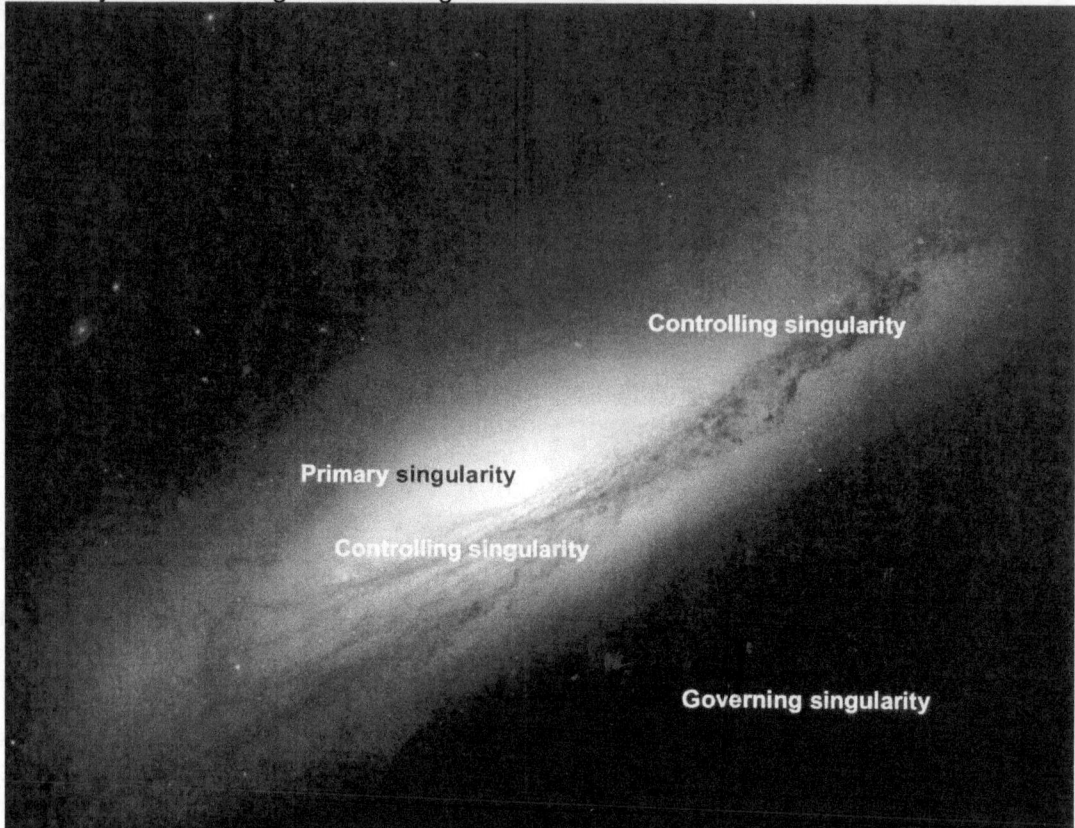

Looking at a galactica it is clear that in the centre there is a bright bubble holding a sphere of heat. Normally surrounding the centre heat cocoon a ring forms and the ring shows how the galactica went flat as the stars formed rings in which it spins around the centre. As the Universe expands the rings must grow larger but growing larger will draw the centre thinner and flatter. To progress to understand how the Universe came about from birth it is most vital to understand the relevancy between the primary singularity, the controlling singularity and the governing singularity. This relevancy dates back to the way the Universe constructed space long before the event where light formed a dimension, which is called the Big Bang.

The centre of the Universe is within the centre of everything that spins around a centre because everything that spins forms Π^0 which is one. By forming Π^0 it forms 1 and all the values no matter how large the space is that spins around is $\Pi^0 = 1$, the value holding the space together is 1. So notwithstanding the space it holds the centre value or singularity $\Pi^0 = 1$ and in being 1 it holds singularity or the single dimension. Therefore the centre of the Universe is inside the centre of everything that spins around and that creates such a centre. The movement as time formed by producing space. Everything there is also is there because it moves in relevance to everything else that moves. That what is within the Universe moves within the Universe.

If anything did not move it would form a Black Hole since Only a Black hole is within one point that does not mot move and that point that does not move is within not the Universe any longer. It must be understood that whatever is moves that does so in relevance and synchronisation to everything else that does also move and the difference between that movement and space forming forms time.

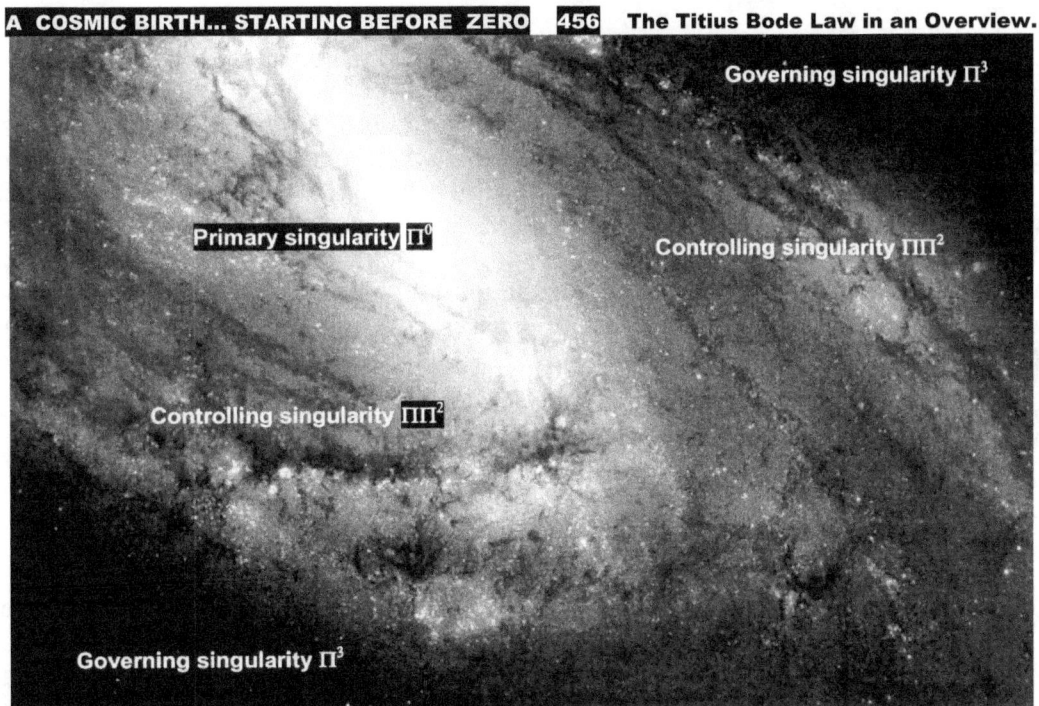

Governing singularity Π^3

Primary singularity Π^0

Controlling singularity $\Pi\Pi^2$

Controlling singularity $\Pi\Pi^2$

Governing singularity Π^3

$\Pi^0 = 1$ is infinity. Eternity is all the space occupied or otherwise that cannot escape infinity and because it will spin around infinity for eternity it holds a definite value of eternity.

Infinity is the point formed by everything holding space within eternity that spins as well as spins around infinity and where the point then forms in relation to all that spins that cannot reduce any more. Infinity is the point that bonds everything forming eternity and becomes the absolute point achieving singularity in infinity in relevancy to the centre of all which accepts that point as the centre.

As stars develop time forming space pushes the stars outwards away from the centre and into maturity.

Cosmic gas

Cosmic Solids not present

Cosmic Liquid

Cosmic solid

Cosmic Solids shining as stars while deflecting light as cosmic liquid into the black cosmic gas.

In the centre of galactica there is the incubator of stars

The idea that mass draws dust close to form stares is the another hogwash fairytale thought out by those who think not but only offers storeys with no educated substance. In the centre of galactica a ball of heat forms. This looks like the remains of a galactica that resembles the shape of current "big stars" such as Betelgeuse and VY Canis Majoris and others. Those "big stars" are galactica that is not yet affected by the Universe expanding and the stars it holds has not broken free from captivity to start development. This made me think that there are development eras still to form and that we are in the iron era

The sun turns around the earth as much as the earth turns around the sun. Everything is subject to changing and interchanging relevancies played by the way one looks at the factor prominence. It is relevancies that interchange as focus adapts new prime factors. The centre of the Galactica hold the governing singularity because it governs the spin of all the stars within while the stars form the controlling singularity because it holds control over the function and value of the governing singularity and outer space forms the prime singularity because that places the entire galactica in position.

Relevancies change as prominence alters. In the case of the solar system the governing singularity lies with planets, but in galactica the governing singularity is placed in the centre of the galactica.

Every structure holds singularity that places all other singularity in relevance of such a centre point.

This happened when Creation began and the cosmos developed passed zero and into the relevancy of 1. Later I show how this tool place but this process came into place as time began developing a cosmos from one point and the point it started with. This happened as spots became dots and dots formed associations to become clusters and clusters divided into and according to density applying.

We still have this same rule apply because as gravity grows density increases and space reduces until all atoms form one point in singularity within the star in which case we call this a Black Hole. This is where space disappears and time again reduce space to a point formed outside our Universe.

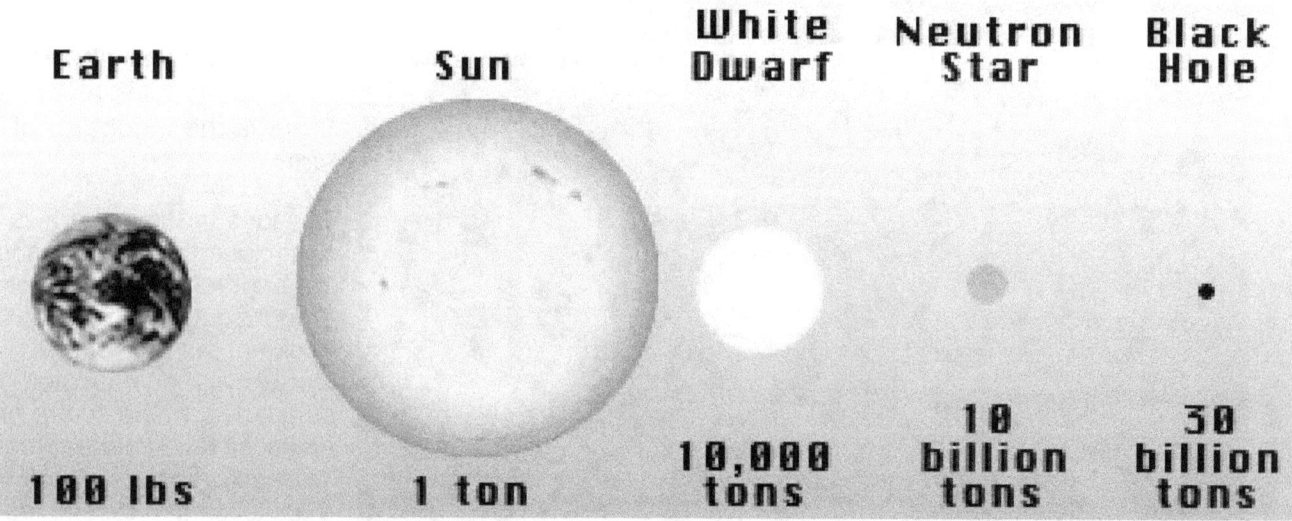

There had to be a time when the Black Hole of today was a red galactica "giant star" and as the Universe expanded form 10^{-33} cm to where the current Black Hole was a "star" in relevance to what Betelgeuse is today it went through growth taking the current Black Hole through all the above stages as the size of the atomic material diminished by gravity increasing in relevancy. The star develops as the Universe develops gravity and gravity is the differentiation between what expand in movement and what contracts in movement.

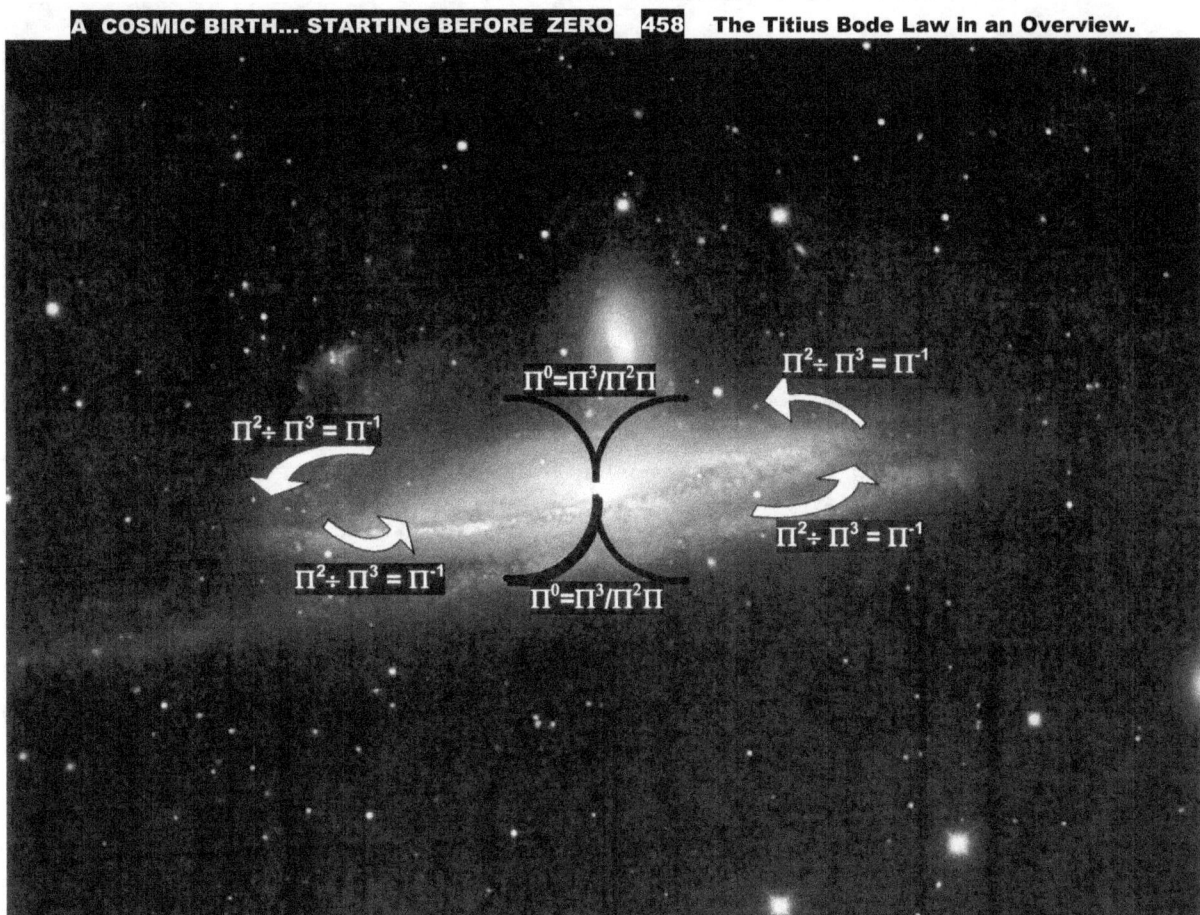

This galactica in all its splendour of light is a Black Hole in the making. The star forming the galactica will eventually combine in effort to project a Black Hole in the centre as all the primary singularity grows in strength by the input of the controlling singularity with the expanding governing singularity. The development of the Universe is natured in the time differentiation between what holds infinity and what forms eternity. The Black hole removes all of eternity and unites that again with infinity. However this is the role of every atom forming the Universe. As it removes time from eternity it captures light just as a Black Hole would do. That makes every atom a performing miniature Black Hole and in a combined unit it forms the role of the primary singularity developed by the controlling singularity of all spinning material within the galactica. It is a singularity projected by all material to form one unit $\Pi^0 = (\Pi^2\Pi)/\Pi^3$.

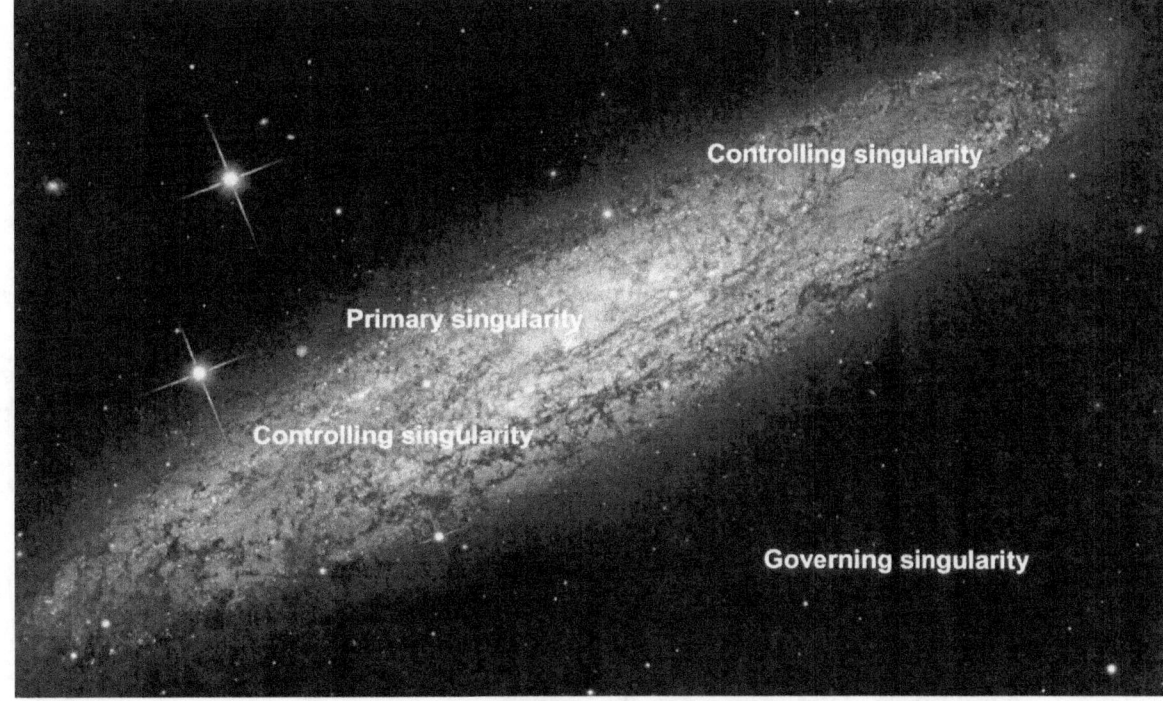

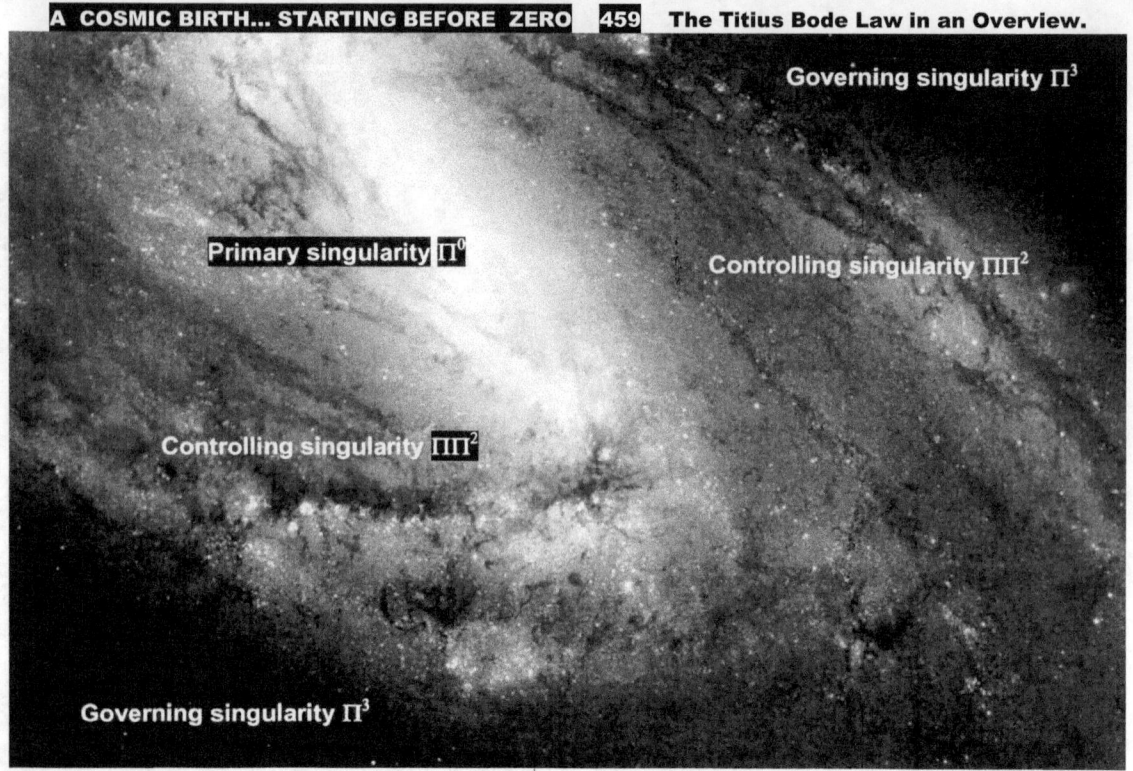

There is no and there can never be a mini – Black Hole or a small Black Hole. It is only a stupid atheist with no concept of reality that would observe and not understand what he observes and with his limited atheistic interpretation skills still then still try to explain that which his poor abilities in understanding can never comprehend to begin with. Every atom in a galactica forms a centre not only in the centre of the star in which the atoms are but also projects a point holding a primary singularity within the centre of the entire galactica. The singularity within the centre of the star becomes the governing singularity while it projects a primary singularity in the centre of the galactica and outer space forming the containing ability become the governing singularity as it hold the spin of the entire structure contained in relation to time developing the Universe and in this case moreover the galactica under question. When there are sufficient quantities of material available to project and materialise the primary singularity in the centre of the galactica a Black Hole will form as Π^0. Then the galactica forms the Controlling singularity $\Pi\Pi^2$ and outer space forms the governing singularity retaining the growth, gravity and movement of the galactica as a structural unit. Time is when the smallest particle change position.

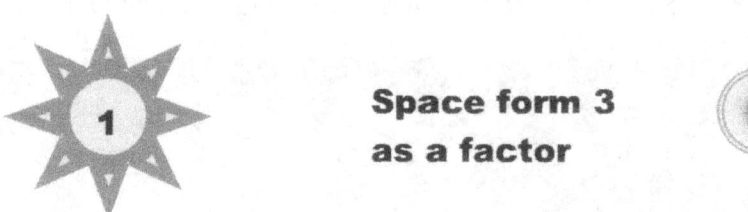

Space form 3 as a factor

Everything I have shown here is about density differentiation and density differentiation is the result of movement going quicker or slower. It is about contraction forming density or movement resulting in density increasing. However it is about material spinning in space and material moving through space and that sums up gravity. That Kepler so clearly explained in the tables he left us. Kepler's tables carry the proof we need for the Titius Bode law.

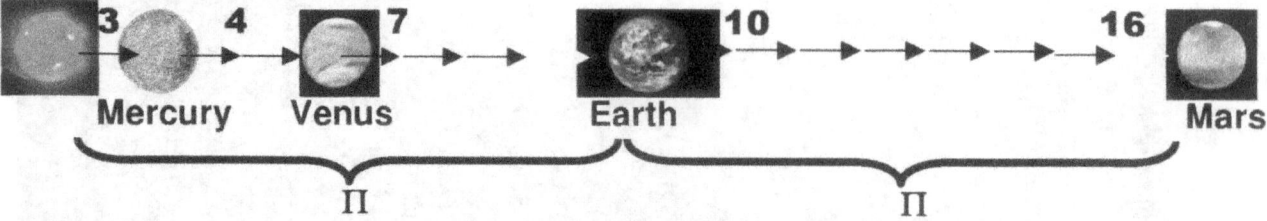

In include the following that contains mathematics. I include this shut my critics up that there are millions of and are over eager to dismiss my work on grounds of lack of proof. If you do not wish to challenge me you can give the mathematics a glance and ignore it. In the event you wish to dismiss my proof, then study the mathematics and shut up!

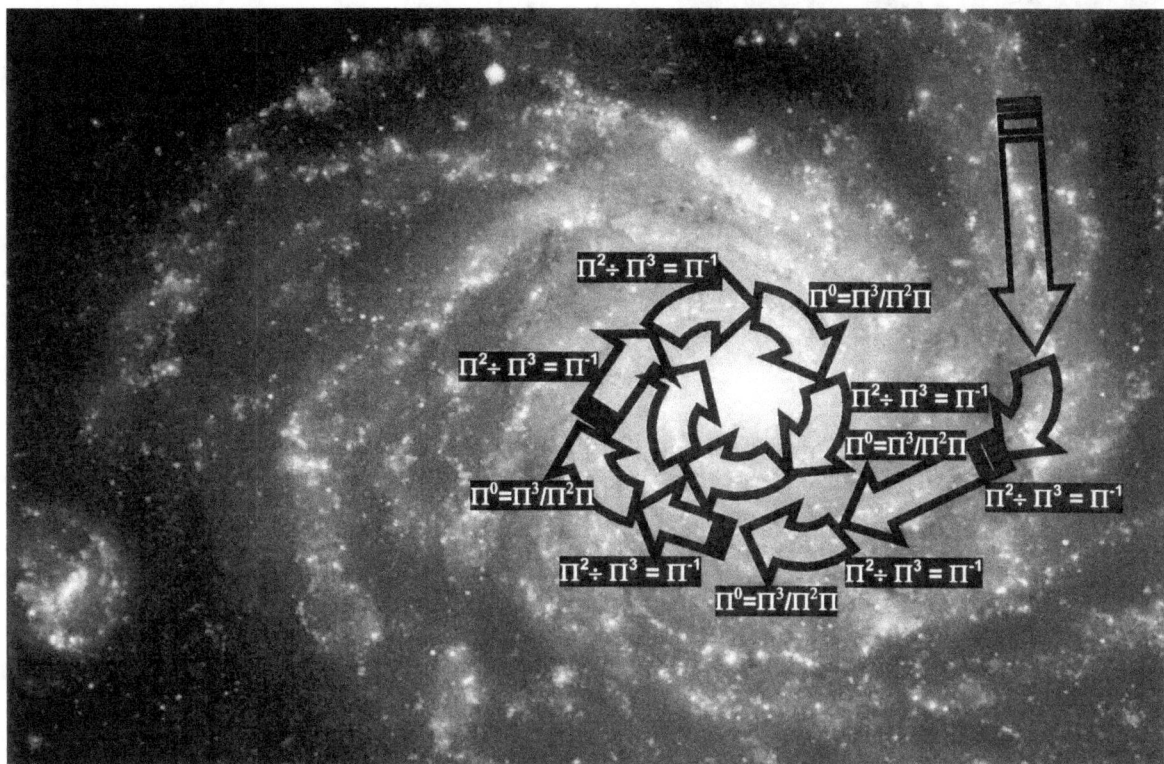

Mercury	$T^2 \div a^3$ =0.983
Venus	$T^2 \div a^3$ =0.992
Earth	$T^2 \div a^3$ =1.000
Mars	$T^2 \div a^3$ =1.000
Jupiter	$T^2 \div a^3$ =1.000
Saturn	$T^2 \div a^3$ =0.999
Uranus	$T^2 \div a^3$ =1.000
Neptune	$T^2 \div a^3$ =0.999
Pluto	$T^2 \div a^3$ =1.004

Studying the true information about Kepler's formula we see that the formulated factors are $a^3 = T^2k$ and it produces $T^2 \div a^3 = k^{-1}$ in value. Having $T^2 \div a^3 = k^{-1}$ proves that the space moves towards the centre because in the centre we find singularity valued at $\Pi^0 = \Pi^3/\Pi^2\Pi$. This proves that space moves towards the centre and material stays in place as proved by the Titius Bode law. It is a flow of space occupied or otherwise that flows towards the centre and this holds relevance with the expounding of the space outwards away form the centre. What is attached to the centre of the galactica stays attached to the galactica where no escape is possible.

Outer space is 10 / 7 $(4((\Pi^2+\Pi^2)$ =112.79547

The Hydrogen is $\Pi\{(\Pi^2+\Pi^2)(\Pi^2 + \Pi) + 3\}$= 112.3

The sphere within the cube is $7/10(\Pi^6)/6$ as the cosmic atom turns

Iron is $7/10 (4((\Pi^2+\Pi^2))$ = 55.2697

Light meeting singularity is $3^3+3\Pi^2$= 56.6
Cobalt is $3(\Pi^2+\Pi^2)$ =59.21762

Copper is $\Pi(\Pi^2+\Pi^2)$= 62.01255

Space reuniting with time is = $2\Pi^3$ =62.01255

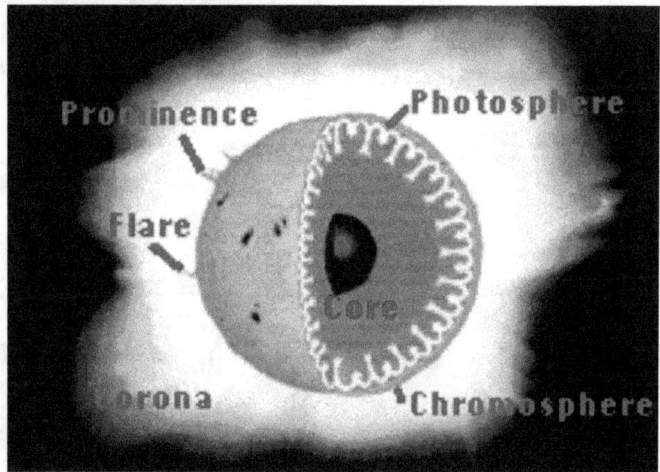

Space goes further but it loses form after 59 protons displacement. After light holds the final value of 10 / $7(4(\Pi^2))$ =56.39 and the entirety of space consist of light, any value exceeding this displacement will exceed the value of light in both dimensions $3\Pi^2 + 3^3$ (29.6 + 27) = 56.6 space has no substance to remain a sphere or to remain in the third dimension. This proves that light forms the Universe in space (not in time) and exceeding light (56 x 2) = 112 the atom's structure deforms and the Universe goes into a dimension unknown to us in human form.

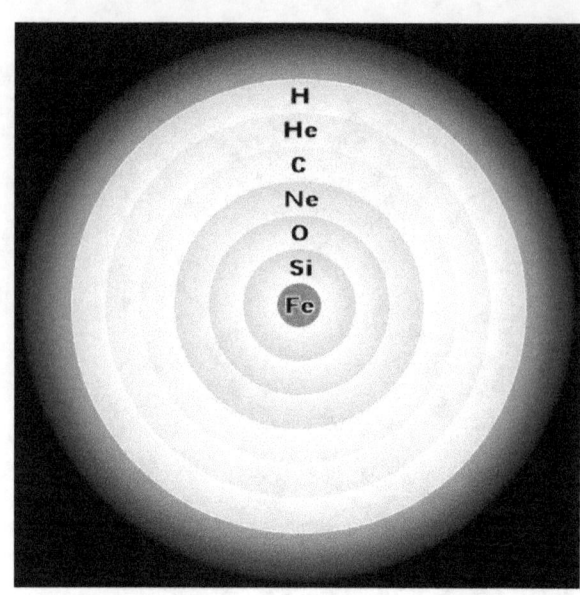

The value of 112 forms the limit of the element's value and therefore the limit of what there is forming the cosmos. Anything beyond 112 is not part of our Universe. Anything going above what light produces in the double does not contribute to what is within the Universe. Light is the three dimensions forming the cosmic gas that cosmic atoms turn into cosmic liquid and eventually turns to solids, as it freezes light into singularity. The part that holds the value of 3^3 or three cube forms the blackness where we see the spots of light bring rays of information through the vastness of space.

The dark part of outer space is 3^3 and the white part of light forming outer space is $3\Pi^2$

Light holds a double value of $56 \times 2 = 112$ and going above 56 is taking what is displacing space-time within the Universe to what is beyond the limits of what is within the Universe.

Time forms a line by moving from the past through the present and onto the future. What is left behind is space as a remembrance of time and moving from the past brings along moving by the law of Pythagoras, just as the cosmos started in the very beginning. I find Newtonians think of the cosmos in terms of human life whereas the cosmos runs in terms of a process befitting only God. The Universe can never start because the Universe can never end. The Universe uses space to form time that will bring forth the next space but what is in the Universe can never again leave the Universe.

From the centre of the earth to the very rim of the surface we have no Π forming since one singularity extending Π^0 is in place. When observing a mountain the question is where does the mountain start in order to determine the mass of the mountain. I living person has to use his judgement and that judgement is human not cosmic. The gravity line will run from the centre of the earth as Π^0 and only at the point where the surface touches the sky would $\Pi^0\Pi$ come into value. At that point anything standing on the ground will have mass or weight. At that point gravity is $\Pi^0\Pi = 3.1412$.

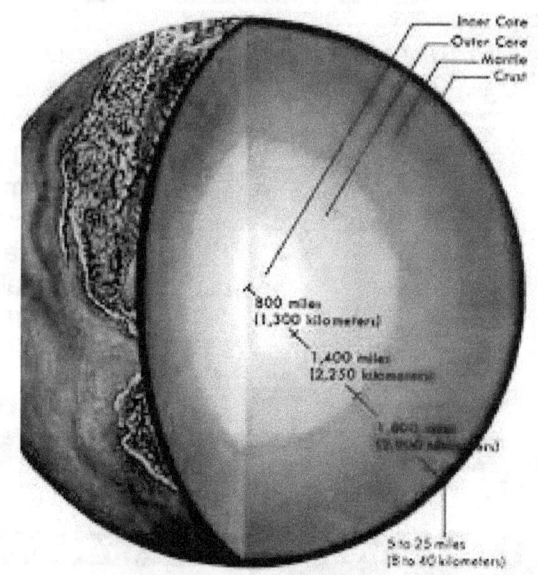

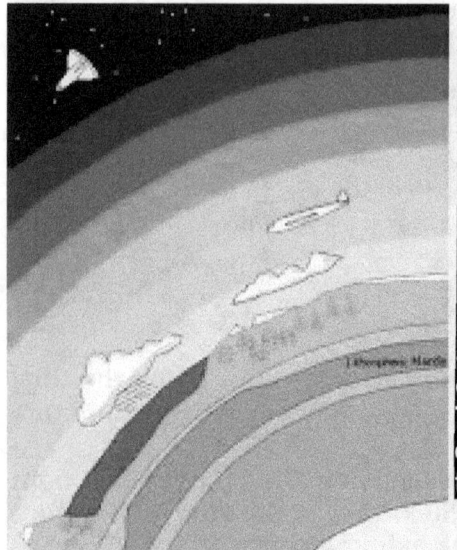

The value of Π is $\dfrac{21.991}{7}$ in every layer that forms and the 7 associates with singularity becoming one and by diminishing in value from 7 to one it reduces the top or space part to 3.1412. This happens as descending takes place throughout the space reducing up to where the object touches the surface. At that point the object then changes from forming part of a liquid to forming part of the solid and thereby then receiving mass or weight. This is how gravity reduces space albeit filled with material or not. It is the displacing of heat by density.

This is the way the Titius Bode law develops density differentiation by space dispensation and this influences travel throughout outer space. This even slows light down to a stand still because as space increases movement at a rate declines.

| Mercury 0.1° | Venus 177° | Earth 25° | Mars 25° | Jupiter 3° | Saturn 27° | Uranus 98° | Neptune 30° | Pluto 120° |

Obliquity of the Nine Planets © Copyright 1999 by Calvin J. Hamilton

In the forming of gravity within the solar system the ratio used is much different as is used in gravity forming within spherical structures such as planets and stars. It is because in stars the full value of Π as it forms $\dfrac{21.991}{7}$ forms the factor known as gravity whereas in the

Titius Bode law only a part forming the value of $\dfrac{10}{7}$ comes into play. Because it only calculates one

part of $\dfrac{21.991}{7}$ the compacting of space does not come about as it does in stars spinning. In this we then find growth of space.

When applying the full compliment of Π such as $\dfrac{21.991}{7}$ does space reduce but implementing the half

of $\dfrac{21.991}{7}$ which then is $\dfrac{10}{7}$ we find space grow by the value of $\Pi^0 = \Pi^3/\Pi^2\Pi$.

The entire Universe grows by $\Pi^0 = \Pi^3/\Pi^2\Pi$ and not by the Hubble constant because the Hubble constant varies so much it is beyond human calculation. Yet because it makes scientist look and feel good they stick to the nonsense of the accepted Hubble expanding value, which is inconclusive.

The First Article about $F = G \dfrac{M_1 M_2}{r^2}$ **in terms of Gravitational Formation**

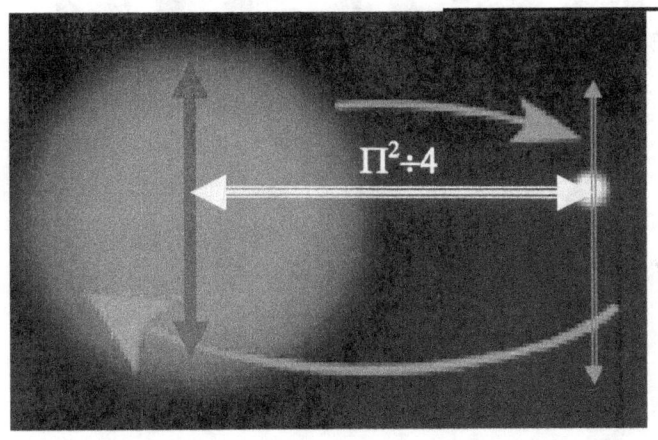

$\Pi^2 \div 4$

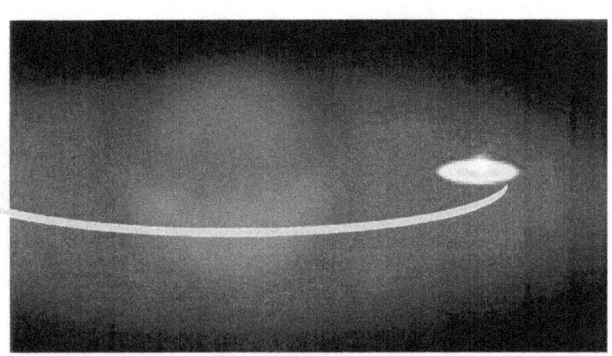

...And I use nature and the laws nature applies to prove Newton is so far off the mark that nature no less proves that Newtonian science is not even enjoying the same Universe as nature where correctness are addressed. At a distance of $\Pi^2 \div 4$ or $(\Pi \div 2)(\Pi \div 2)$ the major structure liquefies the minor structure in order to resolve and absorb it as liquid heat. That is the Roche limit and therefore there can

never be $F = G \dfrac{M_1 M_2}{r^2}$ in any shape of form. I was then informed I do not **understand Newton.**

Are the wise and the clever scientists not just simply great in **understanding Newton?** The pity is that nature rejects Newton and the **understanding of Newton!**

Part 6

The Where and The Why and What did we Miss

In opening the last part of the book I wish to address the entire worlds scientist: You are the brainpower of the world. In this letter I make fun of you and with good reason. You may impress one another with mathematics that can burn off the hair on a dog's back. You may investigate the smallest of quantum physics that will stun an elephant to a stand still, but what is the use if children at school find the most basic of science a laughing matter. What is the purpose if it is isolated from nature? Why use it if nature contradicts it. Science can only use nature because it is nature that creates this Universe and not you using Newton. It is the choice scientists have to make. The choice is to go on with senseless science because changing to a new view will offend too many and demolish an institution of collected knowledge, or change and become sensible by using nature.

I do not use the term sensible to self praise because every reader will see the science I apply comes from one that is academically very poorly developed. I do not use great quantum references or have the academic power to involve myself with the enormity of developed research on "the cutting edge" of the twenty first century and I am under no elusion that I impress any one with a vast field if accumulated knowledge. However being totally ignored by you because of poor academic reference does not make you lot big but it uncovers you poor equipped thoughts. The examples I use are basic. The method of my thinking is crude. The science I refer to is every day and almost child-like and I am the first to admit that. I might not agree with your work but you impress me to a point of questioning my personal self worth but that is in vain if nature kick you lot off the table. I do not wish to be anybody than who I am and least of all be foolish enough to think I can impress the greatest minds on earth with my work but then again you cover Newton and Newton is pure mythology. My work throws Newton's mythology out the room and that I have not because I have but because I have not the training you have and my lack of basic training is the reflection of my thinking and that is using the basic that nature uses; I do not wish to present myself in any other way because then I shall not only be ill littered but also on top of that be a fool. I do not have the illusion that I have the ability to even try too fool the BRAINPOWER of our time, which is you but I have nature and you lot tried to fool nature for three hundred years. To you with high academic achievement my work may seem trivial, as I was unceremoniously and in an extremely crude manner told by the editing team of Annalen Der Physics but to me it is a mammoth task. Then again think of it in this way…I was the first one in three hundred years that could add 3 plus 4 and get seven and on that I could solve the Titius Bode law.

Still this did not impress any academic in fifteen years and therefore I accept that my work will not be accepted as science during my lifetime although it is the only science there is and that has a small significance to bear. I do admit that I would enjoy the acceptance because the work I have done might not be of tremendous academic force and vigour, but that is not who I am and is not my capabilities.

What I do feel strongly about is the small-minded view science hold about religion and a person as small as I can see their greatness they think they have comes from their smallness that their views have. See the greatness out there and think for one second that with all the mighty brains there is, no Newtonian can explain what came before $t = 10^{-43}$ seconds the most crucial and no one knows what came before that which we know. Before the Big Bang started is what is the big issue. We may see the event in progress, but what actually started the event? Answer that one first before you spit on the Bible. If you cannot, then at least admit you are less than the Bible because the Bible gives answers man is incapable of understanding.

To the atheist I give the challenge use your vast intellect and calculate the space the earth holds only in the circle it relates to as the earth orbits the sun. Calculate the cubic meters of space between the earth and Venus and between the earth and Mars. Then calculate the ratio the earth occupies holding matter to that space not holding matter. Then calculate the space you hold in cubic meters in relation to the space the earth occupies. Use your great mantel power to see your place, not in the Universe, not even in the Milky Way, not even in the bigger solar system, but only in the area you cover each year while on the joy ride the earth provide, Position your greatness in relation to your immediate surrounding and find your space you have. If that cannot bring you in relation to You creator, the one that created the immeasurable, then see yourself in the space you have and then admit to your greatness, Then see who you scowl.

The Bible says in the beginning the Creator Created heaven and earth and that is what there is: there are solids (earth) and there is cosmic liquids (heaven) and other than that there is not. According to the Bible the Creator created two substances that the Bible calls earth, which is solids, and heaven, which is cosmic liquids and cosmic gas. What we now call it after so much Newtonian wisdom is…guess what

Newtonian stupidity did not allow science to even recognise this substance because Newtonian idiots think they are clever than God and the Bible! Now isn't that a joke! If they studied nature and how God devised the Universe they would have learned how God create the Universe by using nature and nature uses the Titus Bode law amongst other things.

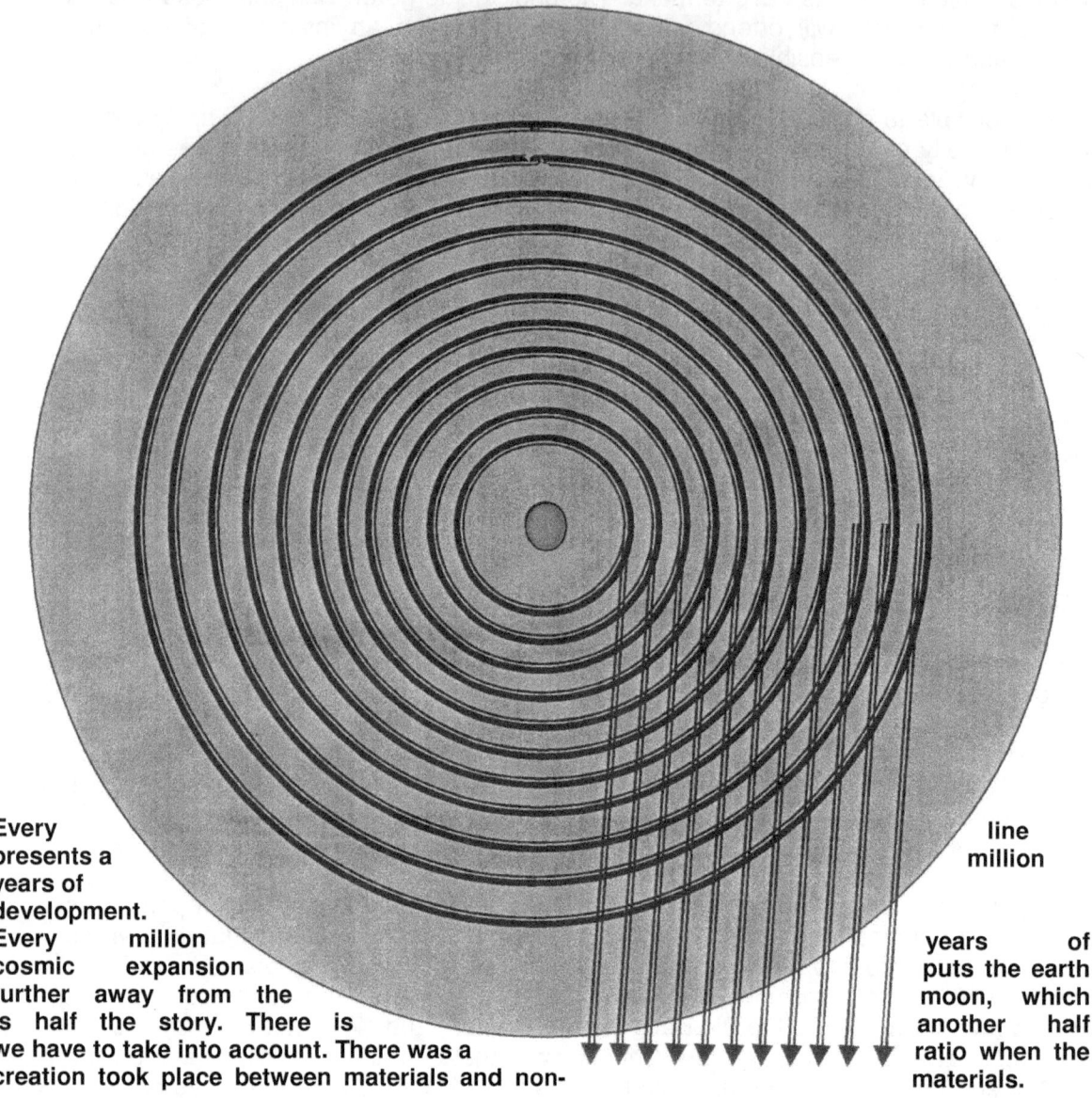

Every presents a years of development. Every million cosmic expansion further away from the is half the story. There is we have to take into account. There was a creation took place between materials and non-

line million

years of puts the earth moon, which another half ratio when the materials.

The little dot in the middle is a picture of the earth and the moon and the darkness is space. This picture does very little to the true ratio of space between the earth and the sun. In the area the earth occupies relating to the space the earth does not occupy cannot explain the heat the earth has concentrated, which it received from the sun but has to share with such vastness of space.

It is believed that cosmic expansion is the moon shifting away from the earth while the earth shifts away from the sun and so on. Yes it is true but that is very much partly. As the ratio moves the earth and moon apart the earth and the moon grow in density of the material, which means the size of the earth, and moon also expands by this same ratio. In this statement it is not the earth and the moon that grow in size but the density within the atom that forms the substance holding the earth, the sun, the moon and every other material component wherever. Should you accept my challenge and calculate the earth to space ratio you will find there is no explanation about the heat we enjoy, other than to agree with my view that the cosmic occurrence that we call "gravity: is the dimensional depletion of space concentrating heat within the core of the atom but in that a ratio must apply.

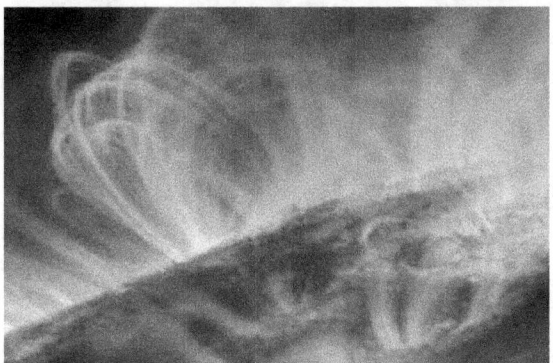

As the atom in material grows in density so the material forming non-material deplete in density. This makes the ratio reapply as the cosmos develops. The material seems to grow but that is not the case. To keep the material from overheating and expanding the movement of the material absorbs cosmic liquid and by contracting this the atom revalue its status in heat. This absorption is of materials and the materials will take more room. However because the density of non-materials deplete it will leave the impression that materials reduce in size but that reduction in size is also growth in size. The sun contracts the cosmic gas and turns it into cosmic liquid and inside the sun it turns the cosmic liquid into the cosmic solids, which is the atoms.

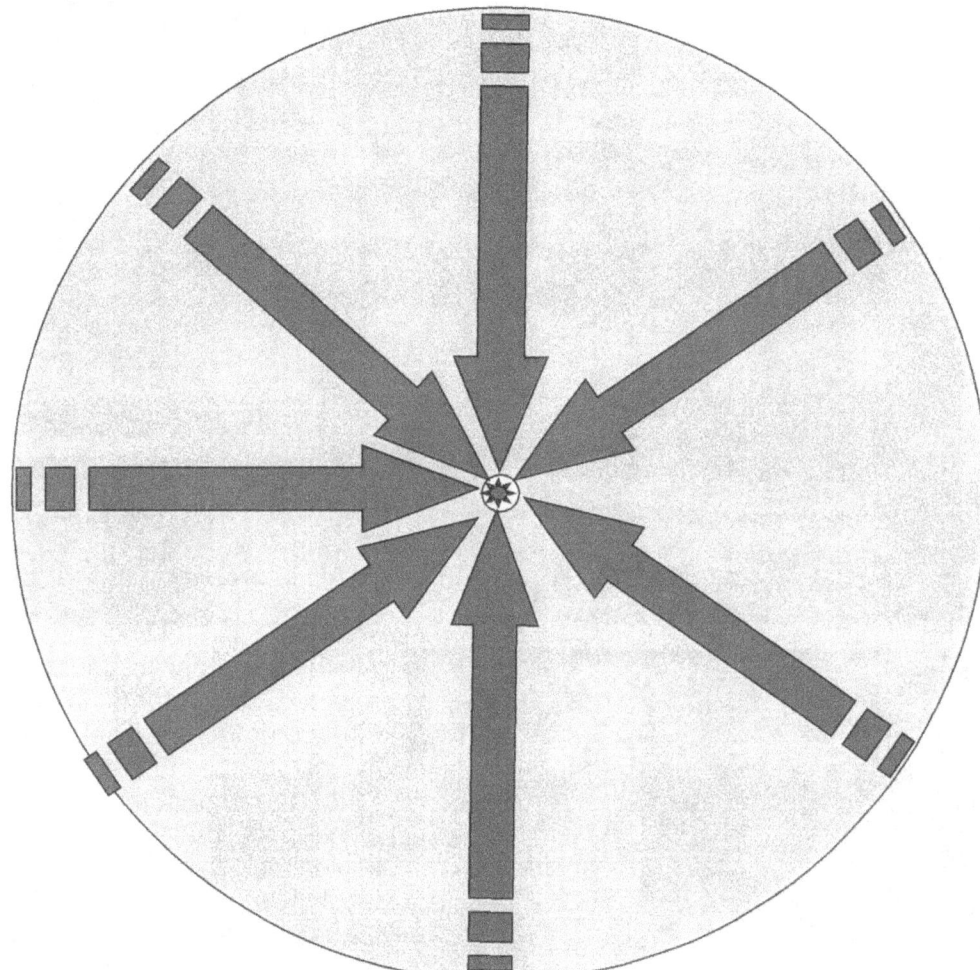

The atoms can absorb so much and no more and that leaves the sun with an excess of liquid. The sun takes cosmic gas that associates with eternity or space outside and turns it into cosmic solids or space inside which is infinity. Since the ratio of cosmic eternity and cosmic infinity is so much in unbalance because of the size of the sun in development, there is much more cosmic liquid produced that what the cosmic solids can absorb. The cosmic substance associated with infinity is a liquid we call photons and this flows back into the gas forming cosmic outer space still as an expanding liquid concentration.

In spite of Newton and the Newtonian view mass has nothing to do with the process, it is in the density and the density comes from the atom holding the highest number of protons. The atoms holding the highest number of protons are the iron$_{56}$, which is in the centre of the earth. As the Big Bang theory as well as the Hubble expansion proves there is a distinct shift in space and that can only come about from the growth of the proton to the relation it claims in space-time holding a balance in the space-time it controls (the neutron) and the space it influences (heat).

To prove that this happens we have the Titius Bode law that proves planets hold a specific distance in relation to the sun and in relation to the inner planets. Therefore planets do not pull the sun and the sun does not pull planets. The other evidence we have is Kepler' tables wherein we se that it is space moving closer $a^3 = T^2k$ where the space moves closer as the planets circle.

I am the very first person in history that made an effort to make a serious close stdy of the work of Kepler and this is the first in I guess five hundred years. This says a lot for the investigation efforts of Newtonian science.

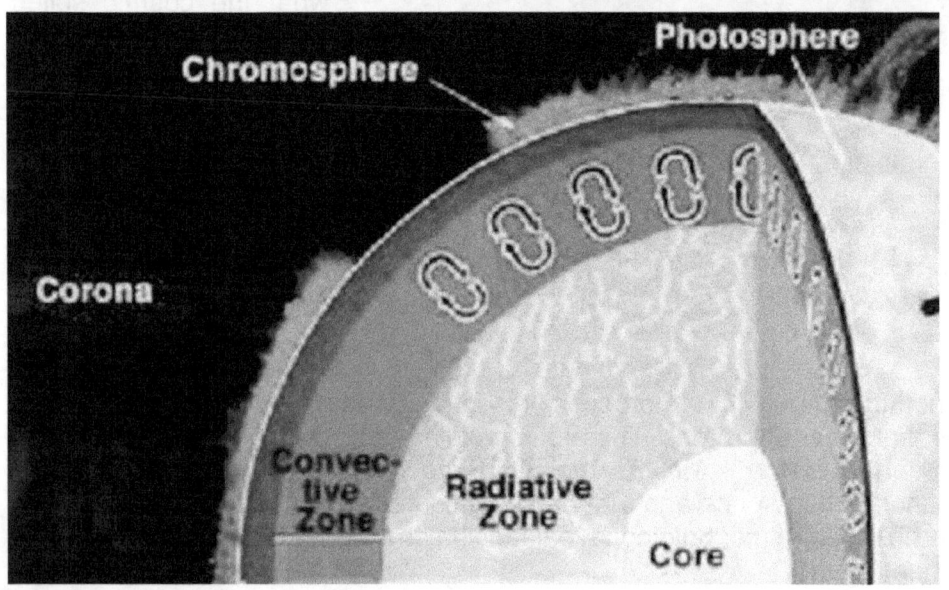

All unused cosmic liquid is rejected and returns to the surface of the sun as it does in the case of the earth. Since the sun is very poorly developed the casting of liquid back into cosmic gas is of great significance. As starts develop and the gravity approaches the speed of light the rejection of liquid reduces up to a point where no light leaves the star. At that point the star goes dark since it then turns all cosmic gas into cosmic solids without the need of the top helium and hydrogen layers to help with the freezing process. This is where Newtonian backwardness has the star die as if the star has life just like an unattended coal stove that stops boiling water if the coals are not stoked and the fire is not revamped. However this is what Kepler's tables prove and that is what the Titius Bode law proves. The reduction in density carries on up to a point within every star or planet where space reduces and that makes ever star or planet a future Black Hole wherein all material will be absorbed. However this brings about the density of every layer is in place to concentrate substance and claim less area space. In that every layer in every structure has a specific purpose to produce the future development of the structure and turn cosmic gas to cosmic solids by producing cosmic liquids.

Looking at the inside of any star the inside is an explanation of the Titius Bode law and show that elements and layers from as the continuing of the density development of the Titius Bode law.

It starts off with many volatile substances that moves very fast and absorbed a lot of cosmic heat turning the "gas" into absorbing missiles and that turn cosmic gas into cosmic liquids. The first layers are all in place to reduce space from a gas to a liquid and then by virtue of the flow of space thrust this onto and into the layers. Below. The C-N-O layers is to convert the liquid to flow to the silicon layer and the silicon layer turn the density of the liquid into a materials concentration of pure electrons which the solids below absorb.

Then in the bottom layers space becomes time as the proceeds absorbs heat into solids.

These layers are the star but every layer is a star in development and although the gravity or movement must be in harmony with all the layers above and below to be valued as Π. It is the depleting of space to intensify the concentration of density as the developing the star into the Black Hole it will again be.

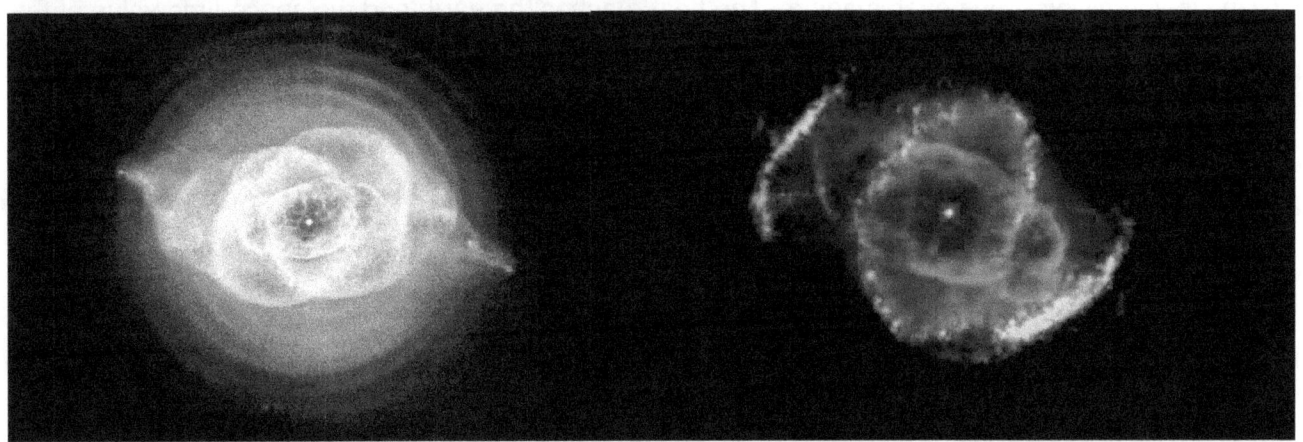

When any of these layers go out of synchronising with the rest, the layer will overheat end expand and blow the star back into outer space leaving the liquid as frozen evidence.

Gravity lightning and electricity is the very same thing also un-comparing and totally different. It is the process that directs heat to singularity in the earth in the proton within the earth.

Lightning is directing heat with extreme impact and extreme intensity but little influence and uses a direct route and cannot be used to sustain mans needs.

Electricity is the directing of heat to singularity and is of low impact and low influence but very intense.

Gravity is the directing of heat to singularity to every proton in every proton cluster, more over to the densest clusters within the earth core. Gravity is of high influence low intensity and high impact.

The four forces is the same thing only maintaining different dimensions where each relate to the same in another intensity or dimension.

There is no gravity, THE FORCE. The only Force is the Force of the Creator.

In all fairness to science, it is one thing criticizing them but what then is the solution to this great mystery: A science book that is with us, as long as history goes back; I am referring to the Bible.

The heat layers within the earth core is just more extending of the Titius Bode law that concentrates heat as space reduces. This carries on to a point in the earth where space completely disappears. However the increase in heat a cosmic phenomena applying as the Coanda effect and it is what we see as the atmosphere just getting denser because the earth gravity is reducing space by the value of Π reforming applying densities.

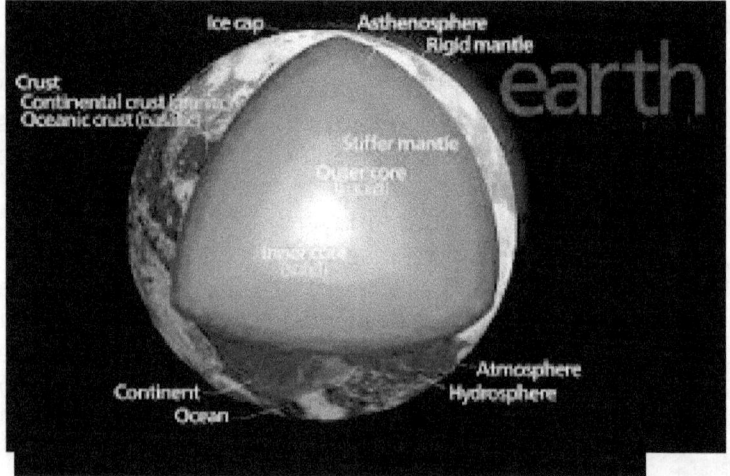

Gravity is accumulating heat from a liquid density to a solid density and the density allow material to gain in cosmic substance while non –materials lose density of cosmic substance, This is how the Universe "expands" which it can't do. Let us once again study the information presented up to this point. The sun contracts heat by increasing the density of cosmic gas to become cosmic liquid. The sun can absorb so much and repels the rest back into cosmic space. This we call sunlight. The light in darkness we associate with space or relevancy as $3^3 = 27$ and the light associated with singularity in infinity we associate with $3\Pi^2$ which together this gives light a relevancy of 56.6. The sun has an outburst of heat, blowing in veracity we can never imagine. The heat reaches the earth and the space-time value of the earth multiplies by millions of degrees in the first event. The second blast was thousands then hundreds of degrees centigrade. The blasts that took place were billions of years apart. Each eruption brings an increase in the earth's proton-growth where by the growth accelerates many, many times over.

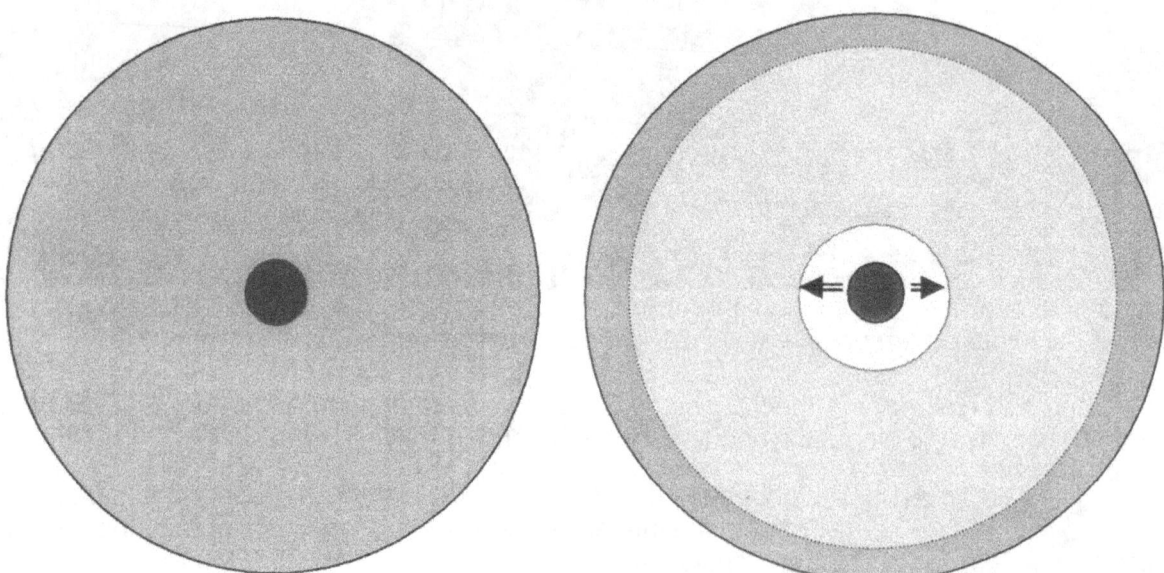

The protons worst affected, will be the ones carrying the largest number of accumulative protons. Unfortunately, these are the iron protons, located in the earth's inner core. With the four square law, the growth on the inside is not only double due to the 28 protons, but also fore times in expansion. In this lies the problem the silicon has to face.

The silicon by itself is expanding at a rate it can hardly cope with, and then are pushed from the bottom to extend its limits much father. The iron inner core is rising at a ratio of 54 x 4: 1, pushing the silicon, which has to perform according to its own time development of 28 X 4: 1 in relation to the water with a growth of 14. However, this is not that simple, because the water transfers the space-time down to the core.

As the earth develops the different layers developed differently because of portion density applying Iron has more density that helium because of two reasons. Helium is volatile because of rapid movement and in that context it is thought of as gas. However in atomic principle it holds fewer protons that iron and that less protons too makes it in accordance with star layout less dense that what iron or copper would be.

The iron core will grow according to the number of protons supplying the atomic density and the silicon will grow according to the number of protons it holds. That means the iron will grow at a relevancy of 55 and the silicon will grow at a relevancy of 28. That means the iron core grows much more rapidly then does the silicon and this uneven growth produces tsunamis and earth quakes.

There is no shifting of tectonic plates under each other but it is the iron core that pushes the silicon core upwards as the iron core develops more and gaining much more in size that the silicon does. This makes the top ground also increase and allow the landmass to stretch as it grows by the earth growing. Only a small part of thee earth was above sea level when Africa and America parted company. Then America to the east lifted above the sea and so did much of all land surfaces grow. It did not grow but the space the sea initially took in relation to the land depleted the sea area as the land took a bigger chunk of space. The land lifted above the sea giving more land to occupy. It is not the land becoming more or bigger but the space each part holds that goes in favour of the core, which reduces the chunk of silicon in relation to water. In stars we find layers formed by elements with a proton number but the earth is far too small and so we only find a core made up of mostly heavy material and a silicon outer solid layer.

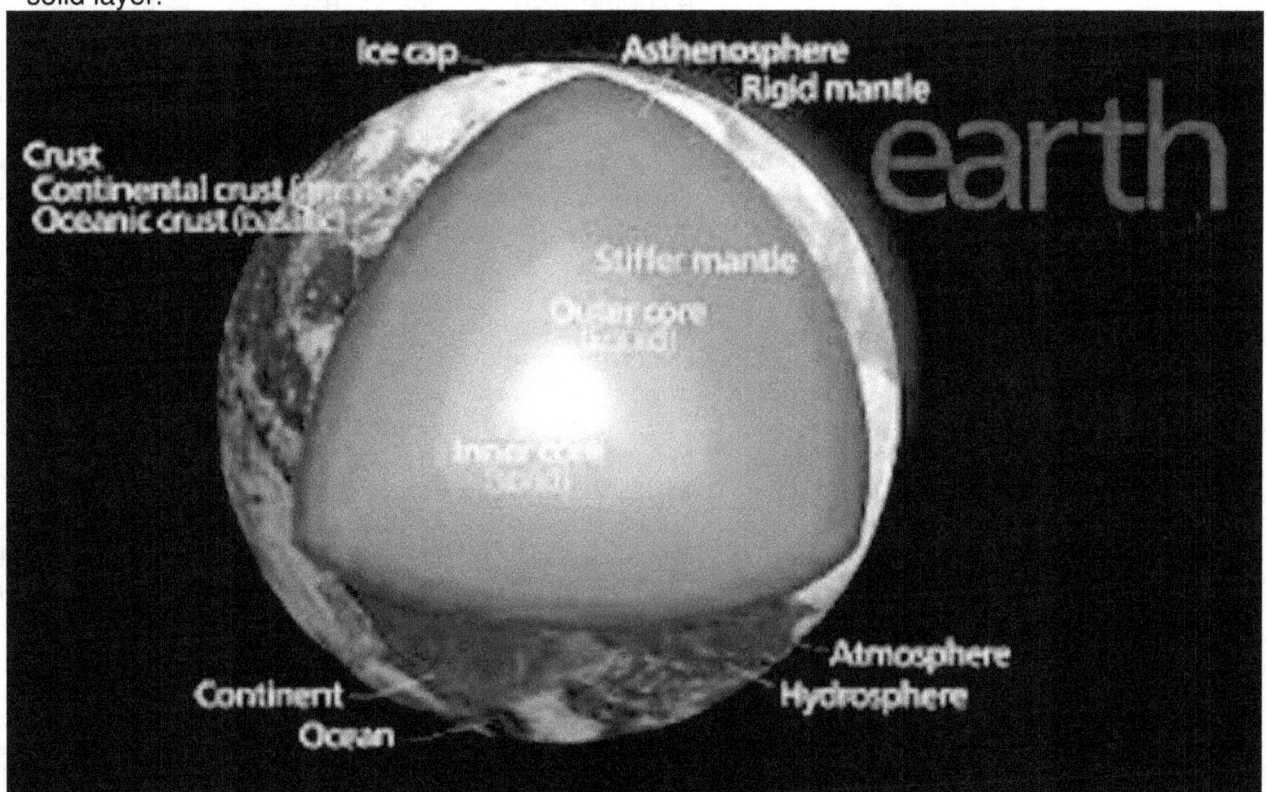

The heat that the earth collect by gravitational spinning is concentrating and becomes more and more intense the further the heat moves towards the centre of the earth. This puts growth into the earth by concentrating heat to move from the liquid or magma to the material within the core.

That is what gravity is. It takes gas, which is the heat in outer space and concentrates it to become a liquid in the atmosphere. Then it moves that liquid to even a much more intense and dense form of liquid within the earth and that we call magma. That makes the core of the earth grow in relation to the silicon outer core and the outer core grow in relation to the atmosphere and the atmosphere grow in relation to the Universe. As liquid move over to solid the solid structure becomes larger since this way it accumulates material in the form of unoccupied cosmic substance and this process is in place to prevent the earth or any star from overheating and bursting into fragments and clouds of liquid clouds. When this overheating takes place we cal the process "going Supernova"

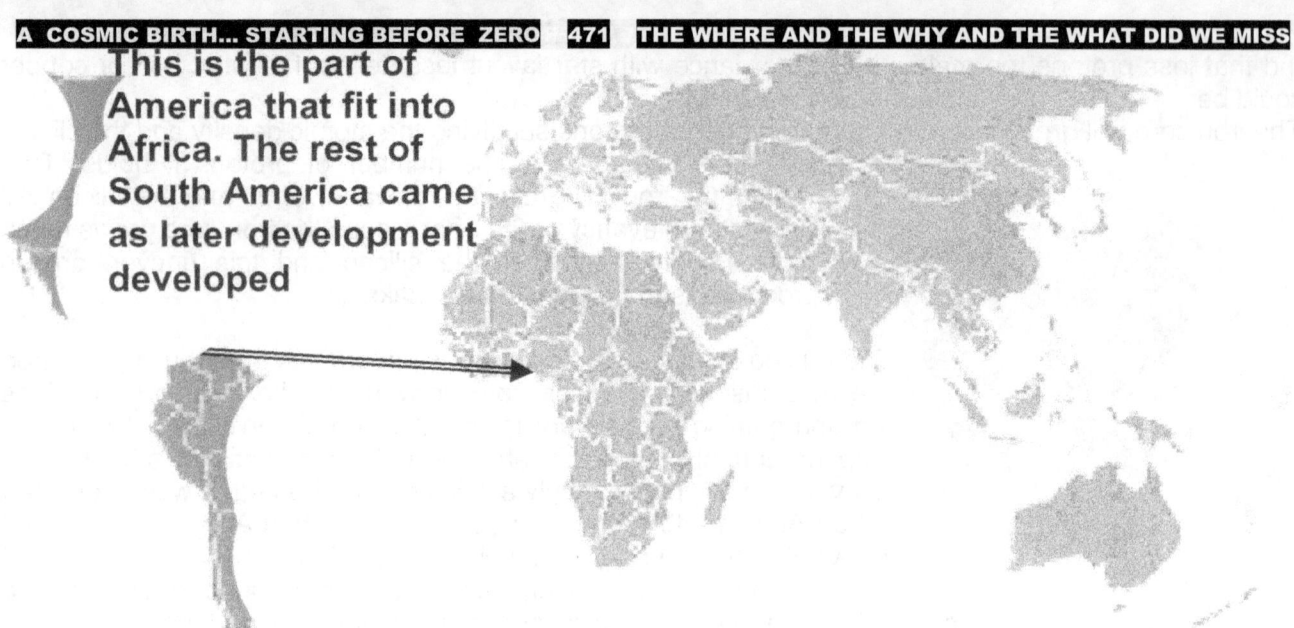

This is the part of America that fit into Africa. The rest of South America came as later development developed

This also is where the water on Mars went. The huge channels were small rivers that grew as the surface of Mars extended and this in relevance diminished the water to something not present toddy. This is al part of density alternations.

All of the yellow is America extending towards Africa as the surface rises above the water. The Amazon basin was there but when America removed from Africa it rose above the water and then in time extended, as it will remain extending until all water is no longer part of earth. The land is not growing the relevancies in space are reapplying and the atoms holding more density in protons will grow to claim more space than would the elements.

I the book Seven Days of Creation I elaborate on this issue much more showing how Ireland broke from Wales and How Wales is growing towards Ireland by forming more surface as land. One can find this in the history where Edward I of England had ships moor next to castles and the harbours today are dry land meters above the surface of the sea. This is all about the ratios reapplying as the Titius Bode awl form the Coanda effect and this in turn form more ground in relation to sea.

The rising earth cause certain indentations to develop and certain islands develop leaving hot and cold climate spots on the earth. Land is separating from the water. (Where did I read that?) In the photos presented as the prelude in this article is very distinctly in contrast to the view promoted by the SUPER-EDUCATED, SUBSTANTIATING WHAT I HAVE JUST SAID.

The sun and earth are more or less fifty times closer than they are at present, (this I shall prove in due time). Although by the same margin, the sun is not fifty times weaker than at present, making the earth a very unpleasant place to be. That all came about during day one of creation, not a twenty four hour day, but a creation day lasting most probably billions of years.

Te earth crust grow not only in the areas it sticks out from the oceans but the entire crust grows, in some places more than in others but this growth seems as if continents shift. It does not shift but the core we are unable to see which is below water is also growing and since the core as an entire structure enlarges in relation it looks as if the earth we can see moves further apart. The entire earth grows and this we can see as layers by which the earth forms.

It is obvious how the Colorado River cut through the earth but also we see how the relevancy of water to land grew in favour of the land. It seems the river reduced it ability to cut wide while the ground gave less space away the more space went to and the river channel sank into the earth.

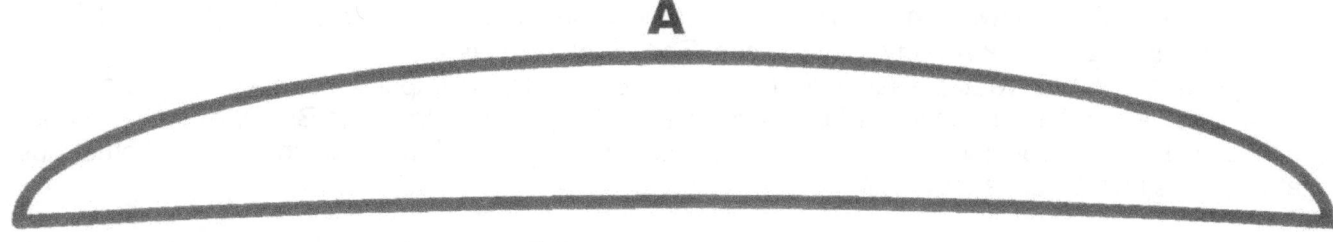

A

A   Earth grows - rain falls, allowing flat rivers to flow;

(B)   The rivers flow slow and forceless, because the land has not risen high above the then, sea level.

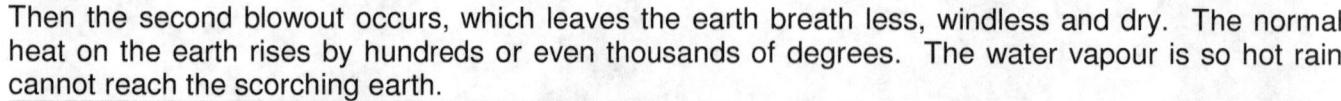

Then the second blowout occurs, which leaves the earth breath less, windless and dry. The normal heat on the earth rises by hundreds or even thousands of degrees. The water vapour is so hot rain cannot reach the scorching earth.

As the earth and the sun grew further part so did the space that the sun / earth holds in size also grew.

The earth moves in this period from a position 3×10^6 km away from the sun.

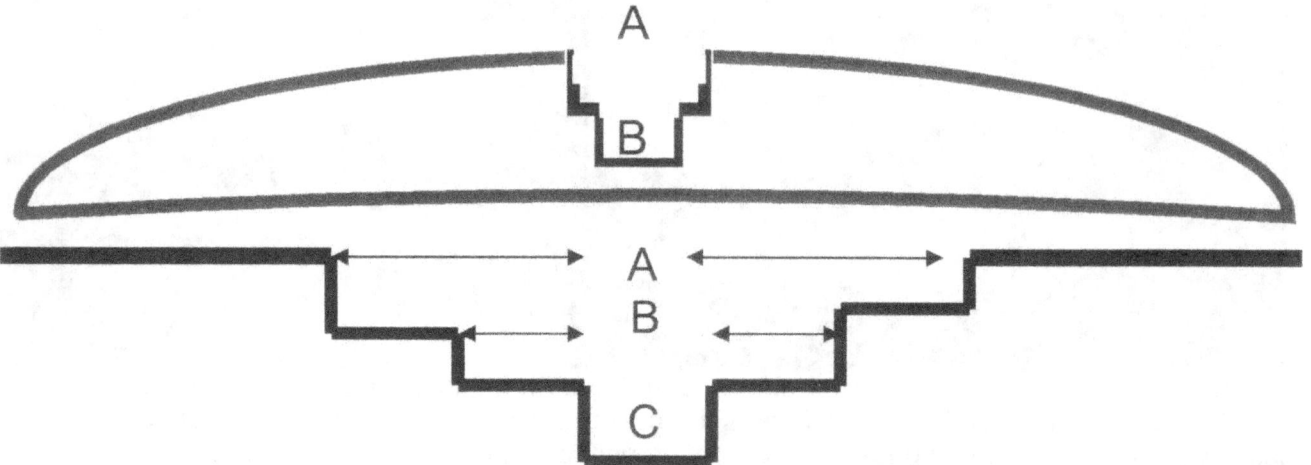

The next eruption follows, which follows the same procedure, and eventually ends with the earth's position being 12×10^6 km away from the sun. The eruptions was a seasonal occurrence more kike the drought and flooding we at present experience. It was of extreme impact but very common and did not occur once every few million years because the availability of heat from the Binary and other structural demise was still influencing the sun, which was influencing the earth (as it did influence all other cosmic atoms in its field of influence).

The evidence is all there and for all to see and by ignoring it Newtonians only show desperate stupidity. At the end of every cyclic ice age where the relevancies in relation to the Milky Way produces more gravity changed from what was to what is and where gravity is going. This is remarkable showing a direct and continuous planning of the Universe by the Creator thereof.

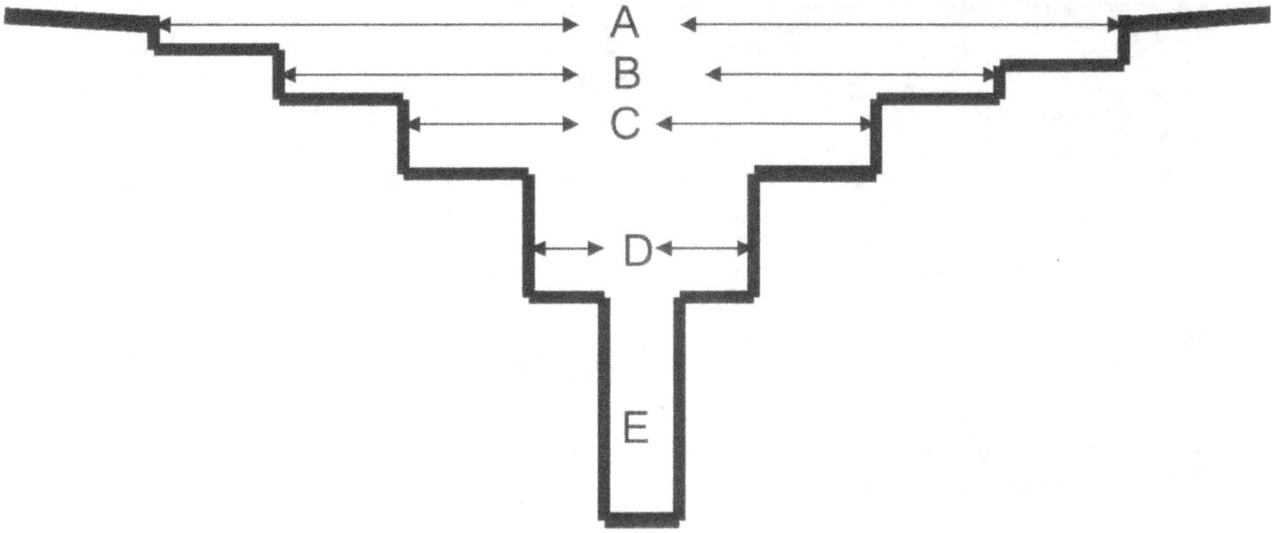

A
B
C
D
E

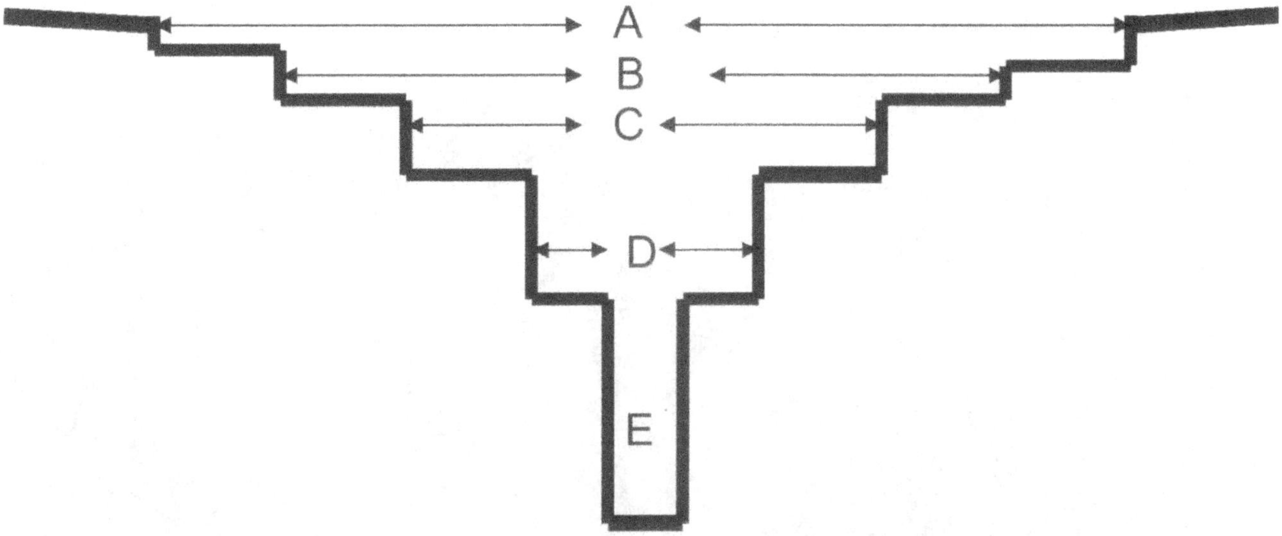

A
B
C
D
E

With each blowout, the earth rises to a new level above the sea, where a new riverbed forms. The previous riverbed seems wider, but this is because the growth of the earth causes the water flow to be more force full. Each period leaves a new river bed that cuts deeper into the earth, because of faster flowing river.

In places where the land meets the sea, a cliff form where the sea cut it straight down.

In some other places, the previous sea level shows as rocks the sea current washed the shore rock to become as smooth as a babies bottom, later the new shore formed, leaving the previous shore meters above sea level.

The evidence is out there, far too conspicuous for any person to deny. However, there is more days to follow and the point I wish to underline, is that the Bible confer the start of an event never declaring the event to stop.

This is a very crucial fact that one has to understand. The volcanic presence and lava flow is quite, obvious in the picture if the mountain development in the photo just above. What is clear also, is the unmistakable youth in comparison between the mountains depicted on the picture just above, and the pictures in the mountains just below. The mountain in the following pictures show, a similar growth pattern, as the Grand Canyon, except it is on a much smaller scale, less intensive, and much, much later in the earth development.

The evidence is out there, far too conspicuous for any person to deny. However, there is more days to follow and the point I wish to underline, is that the Bible confer the start of an event never declaring the event to stop. This is a very crucial fact that one has to understand. The volcanic presence and lava flow is quite, obvious in the picture if the mountain development in the photo just above. What is clear also, is the unmistakable youth in comparison between the mountains depicted on the picture just above, and the pictures in the mountains just below. The mountain in the following pictures show, a similar growth pattern, as the Grand Canyon, except it is on a much smaller scale, less intensive, and much, much later in the earth development.

Everything seems to grow in the form of rings. There is a constant development where time forms space and every development era is considerably different from that which came before or afterwards. We can see the colour and the formation as another era brought different conditions that laid down new formations and new earth. Then we see how the earth parted from the water because the earth being a core of 55.6 Iron and 26 Silicon grew much more rapidly than did the water level grew in volume of space.

We find this also on mars and the water did not disappear but in relevance became less in volume that what the soil forming mars did. Even the Canyons we now see on mars is exceptionally larger than it was when the water that is on Mars eroded the canyons into existence. Every picture shows proving evidence of such development taking place on earth.

When giving more attention to the first period of development it is striking obvious that the water flow was slow and had no violent inclination. Admittedly is the fact that the first gaping walls were not as far apart as it is today, but still the walls are not as deep in relation to the walls in period 1 distinctly different in size and width from period 1. Then in period 2, there is a marked difference in the walls of the river, as they are. The very same aspect is also prevalent in the differentiation between all the other periods. There is a distinct period of water flow.

After this, a period with no flowing water follows. This is obvious from the way that each gorge has its own bed where the water flow carved a new route out of the stone. Each time the effect of faster, flowing water that has power that is more cutting is strikingly noticeable. The fact that springs to mind from this is that the water flow is has a greater urgency in its flow pattern and is therefore able to cut a deeper gorge each time. The only way this could happen is that there had to be periods of excessive matter growth, where the rock rose higher above sea level and during this period, there was no rain. Therefore, the rock was cut at a distinctly different river edge, each time. Another factor not to forget is the duration of time back then, were much longer than at present. This is above and on top of the fact that with the sun so much closer, the time duration was incomparable to what it is today. I wish to advise the reader to disregard the figure Xepted science have of fore and a half billion years for the earths existence. It is as off the mark as the thirteen and a half billion years of universal existence.

The growth that is witnessed is far different from the rock growth that I referred to in the Namib Desert. The growth in the Namib Desert is one where the rock formation does grow, but because the trend is much slower, the rock formation becomes fragmented and brittle, forming sand "crumbs".

In the case of the Grand Canyon, no rock was brittle, so the heat transformation was apparently completely different as in the case in the Namib Desert. In the case of the Grand Canyon the heat, supply had to have been much more and much steadier than is the present case in the Namib. Heat causes a prolonging in relevant time duration and this has to apply in a case where there is such a strong growth in a rock formation. Therefore in this period the earth turns heat into matter, but this happens more readily in the inner core where you find the iron, and the iron expands twice its ration to silicon which makes the silicon respond to the iron, but not only that, the silicon responds as it transfers heat to matter at a huge increased rate. There might be a notion that all this is wrapped, but during this process, the duration of time must have been much more than quadrupled.

The rock growth cannot associate with volcanic eruptions, because from evidence provided earlier in this chapter, the silicon growth is distinctively different to lava flow. Lava flow comes from the inner core breaking the earth mantel and flowing through the crust. In this case, however there were no lava flow, but silicon growth.

When comparing the three different examples of rock growth one can clearly distinguish between lave growth, and a much older period of silicon growth. The tale this evidence tell is that there were at first a growth in silicon that were not protected by water, and therefore grew much faster than the silicon covered by water. Another telltale sine is the mountains that were cut by winds creating a flat top. From this evidence, one may presume there was a period before this, where the winds were even more forceful with much less land rising above sea level. With little to no landmass, the winds blowing across this scourging planet at speeds close to the sound barrier.

After the period of land rising from the sea, another period came about. The following periods lead to the forming of mountain ranges, as the iron core broke the hardened crust of the earth and minerals and metals poured out from the inner core.

In South Africa, this is extremely apparent, because all the metals came to the surface, each to its own space. This came about because these metals remain grouped together in accordance with their own specific density as it is in the inner core. When expanding, these metals are each differently effected in size and mass.

At first a few volcanoes rose above the sea, which then was flattened by incredible winds. These flattened mountains became islands and was follows by shallow places, which developed, into isolated islands. We still see this development taking place on earth.

This earth seems to become more pleasant place to live, period after period. The impotents in what I am trying to say is that one can see evidence in different periods, very distinguishable from one another.

We can extend the evidence a lot further. The moon being dry waterless, airless and "dead" shows the same case scenario witnessed on our mother planet. I have not once read or saw any scientist commentating about the inconsistency the lunar craters show. With a planet as "dead" as our moon, again it surprises me that there absolute silence scientists bestow very gracefully on the following matter.

By close observation one detect a very peculiar tendency the lunar craters show. The BIG ones were the first meteorites to fall. Only when the sky ran short of BIG meteorites it started peppering the moon surface with ever growing smaller missiles.

The big missiles had less force than the small ones had, because the craters they left are much deeper than does the older ones.

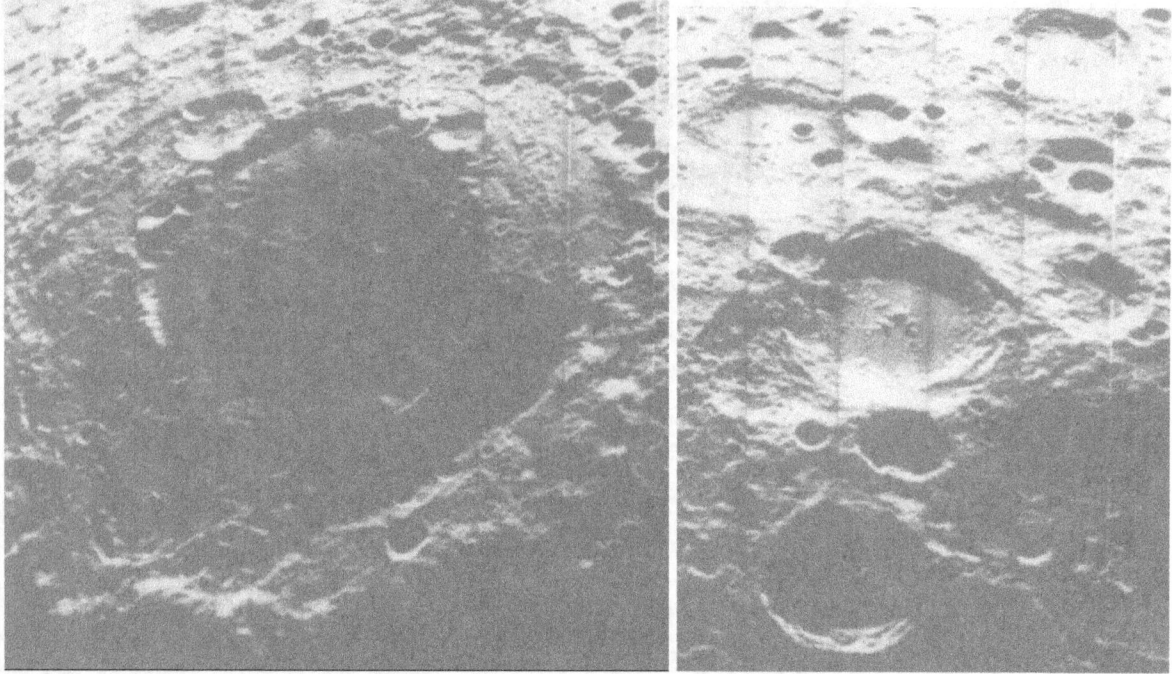

I do not except none of the SUPER-EDUCATED SAW THIS, BUT AS Newtonian Xepted science could not answer the phenomenon, and it was rather left in peace, not to contradict or question Newton's mistakes. The answer is simple, what happened to the earth, also applied to the moon, as it did to the whole solar system.

Even our very own earth craters started off very much less spectacular than they seem to us to be today, However our earth craters are the result of ancient volcano blow outs when the earth core developed and the result was much more spectacular. We still have such volcano blowouts but the effect on the earth as a whole is much less devastating than what it was back then.

In this, the earth iron core grows at twice the ratio of the silicon layer. As the earth grows, the earth has to rise above the water at certain points. Therefore the Bible once again is correct by declaring that the water mass, which at first covered the complete surface of the earth, separated from the water by rising above the surface of the earth. The craters on the moon are left by incoming debris. Newtonian science confuses the craters left on the moon and craters left on the earth by core outbursts. The moon doesn't spin around an axis and therefore has no electricity development within its core and the earth core develops by the heats the earth core collects and uses to expand and grow. The gravity applying to the moon is so much different from gravity applying on earth.

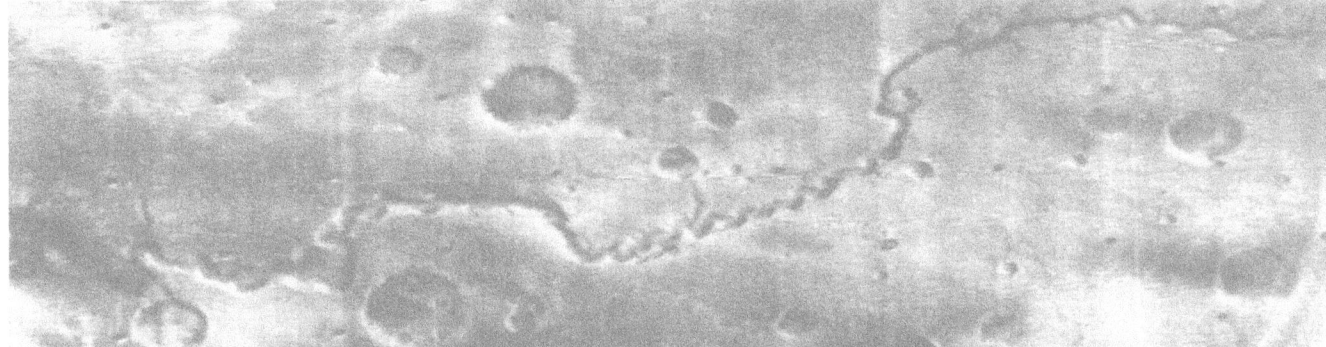

The "LARGE RIVER BEDS ON MARS" are a witness to a very small amount of water when the ratio of land to water was much more in favour of the water. Since then, the volume of space to volume of water grew to the benefit of silicon, reducing the water space relativity.

Because the moon holds the earth as axis the moon is to the earth just more liquid. The moon is therefore not applying the Roche factor but is forming part of the liquefied process of the earth's Roche factor. The moon is therefore target to every cosmic debris when lots of debris was still around. The moon has no atmosphere by which to establish a Roche factor because the moon does not spin around its axis and by that does not have cosmic individual recognition but is just like the top that is on its side and does not spin. The moon forms Π but the movement Π^2 it forms as a result of the earth spinning giving the earth the relevancy of three and placing the relevancy of four on the moon turning around the earth.

The earth therefore could not be the target of a meteor onslaught or a shower of solar rock as the moon was. From the history we can see some of the craters grew and filled up but in the beginning of the solar development there were craters caused by much bigger cosmic rocks than what floats around now it is obvious that some was so big it could have robbed us of a moon or could have supplied us with a second moon. But as the material grew the scars closed and now the eminence of the mighty collisions are merely left as huge stains where time heals all wounds, even on the moon.

At one point in the earth's history this obscure looking peak, represented a vast landmass.

When considering the mines in South Africa, it is striking that the gold had the most rock growth and is therefore the deepest, where as the iron is all in open cast mines where whole mountains are being excavated as the iron has formed huge mountains which contain the iron. To me this suggests that as each of these incidents occurred, a different metal rose to the surface on each individual occasion. The logic conclusion was that by this time, there were landmasses, heating more than other places, pushing the core upwards. As the whole Universe practice apartheid, the presumption can be made to the fact that each metal surfaced at a different temperature, bringing evidence of different stages in the same period. During this period, one may presume there were huge, and I mean huge volcanic eruptions, unprecedented to this day. Only after these eruptions, life came about, as minerals also surfaced. During the first period, there was no life. It is almost safe to say, Ayers rock then was Australia, being one of the biggest places on earth. After Ayers rock, mountain ranges rose to the surface. The forming of mountain ranges once more was located to a different period than the one following in which diamonds came about. Diamonds represents life, and during the formation of metals, life was not present, because it was too damn hot.

Part of this becomes evident when one considers the difference in carbon$_6$ that is found on our planet. I am not that well read and well informed, therefore it might not be very surprising that I have never come across one single explanation about the fact how diamonds came to be, as did crude petroleum oil and coal. Oh yes, there are many explanations floating about, why there is the apparent differences in diamond, oil and coal. When one considers the fact that these carbon$_6$ composites were at one time or another, all been gigantic forests and these forests became diamonds, petroleum oil and coal, you have to wonder why these forests were covered in the first instance, but far more wonderful is the fact that these forests did not burn to smothers and ashes. In cases like Hawaii and other volcanic eruptions, everything that meets the lava flow, immediately ignites and burns to ashes. In these events, no forest can be processed to coal, but goes straight to ashes

Then God said: "Let the earth produce fresh growth, let there be plants on the earth plants bearing seed, fruit-trees bearing fruit each with seed according to its own kind." So it was; the earth yielded fresh growth, plants bearing seed according to their kind and trees bearing fruit each with seed according to its kind, and God saw that it was good.

What this says, is a hard nut to crack, both in theology as well as science, except when applying a bit of common sense. Seen from the way the earth looks today, this just does not make sense. There were plants before there came a separation between day and night.

Let us bring focus on an event that happened in the near past, not that long ago. In fact, there may be some people that may still be alive and can bear personal to the event. Here I am referring to a small piece of orbiting debris that entered the earth and such a small fragment caused havoc all over the planet. The comet struck in Siberia, in the northern region of Russia on or near the river Tunguska. The method of expansion is very important in understanding the process, however the detailed explanation may consume a lot of facts and that may fall outside the parameters of the intention of this book.

What is important is the aftermath or consequences worldwide concerning the events on nature. Principally a comet rained down on the earth at Tunguska. I am aware of certain wild statements of some Super-Educated-Wizards about a "Mini Black Hole" going through the earth and bringing destruction in that manner. Please believe me if any form of Proton Star "mini" or otherwise would even come near our Milky Way, the particle acceleration that must come about will lead to the galactic demise of our section of the Universe. If the particle acceleration of a comet can have such an effect, think what outcome would realize from a structure probably holding the same protons in cluster as a galactic do with all its stars.

The after effect that followed the comet explosion at Tunguska enabled a distribution of light particles through the redistribution of sun rays that brought about where the night skies over Europe was luminous at night. People in London, England were able to read the newspaper at night without the aid of an additional light source. The dark night skies over Europe transported sunbeams carrying it all over the European continent.

Where the interaction of two solar orbiting structures can bring about this influence and taking into consideration the mismatch between the two structures it again proves how little science really understand science.

The moon and the earth were in a Roche – atmospheric sharing, bringing about that the water vapour density was far superior to anything we may ever realize. The moon and the earth were still sharing space-time, hence distributing everything all over both planets. This means the sunrays that fell on the moon had the same effect by atmospheric distribution than as if it fell on the earth at any given point. Remember that both structures were in circular displacement, spinning about on an axis they share at point S_1, as well as a linear axis. Another point to consider is that at this point in planet development, both planets still had their axis rotation, which in the event of the Roche-process becomes the linear displacement. The structures were spinning about each other but at the same time, spinning about them. The effect coming about is that the circulation of moist brought about a distribution of light particles all over the planets.

Conditions on earth were very different from what we presently witness. High tide was truly high tide. This brought about a lot of saturation of land, accompanied by moist heat and permanent sun. From this one can gauge the growth of plants superseding the tropics hundreds of times. There were no deserts, dry or ice. There were no mountain ranges too cold to sustain life. Altitude had no bearing on plant life.

By taking these factors into consideration we have to take another fact into consideration. It is very obvious that there were three definite periods of plant growth. The carbon outcome or the carbon end products supply undeniable proof. There is carbon in the form of coal, carbon in the form of fossil oil and then there is carbon formed from plants in the form of diamonds.

The end product that resulted in the forming of diamonds stands miles apart from the other two carbon end products. The heat needed to form diamonds is many times more than the heat needed for the forming of coal. Imagine the amount of diamonds scattered all over the earth. Convert the volume that the diamonds have back to that of coal and one must find that diamonds formed from a lot more plants than that of coal. The same argument can apply to fossil fuel, and we have not even gathered any idea of the fossil fuels or diamonds still hidden.

Further on in this book I explore this idea further; therefore I shall stop at this point. What I wish to establish in this part of the book, is that there is a lot of proof to prove this text in the Bible, just by merely considering evidence both in nature laws and in evidence on the earth.

In the beginning of the chapter, I proved the ratio that apply to the atom ratio being $(\Pi^2 + \Pi^2) \times \Pi^2 \times \Pi \times 3$. This particular ratio not only apply to the atom determining the "mass" of the proton (Mp) in relation to the "mass" of the electron (Me). This ration extends far wider than only the atom, as it is an indicator to the revaluation of time duration application and plays the major role in determining the "sound barrier".

The "sound barrier" is a sure indicator as to how heat relations affect the atoms. By intensifying the heat ratio between atoms, the time to space of the atom changes completely to a point where the time overshadows the transfer of sound. It is all dependence to heat forming time or on the other side space. The less heat there is between the atoms in the unoccupied space, the less the time will be affecting the atom, and the reverse is also true.

It is not only the heat that one find between the atoms that influence the proton – electron ratio and this is the major part of the huge misconception in the view "gravity" applies, because "gravity" not only rely on the "mass" of the protons, which makes up the number of the protons, and it is not mainly the density in which the number of protons are, but it is just as much dependent on the "speed" that the protons travel in relation to other protons.

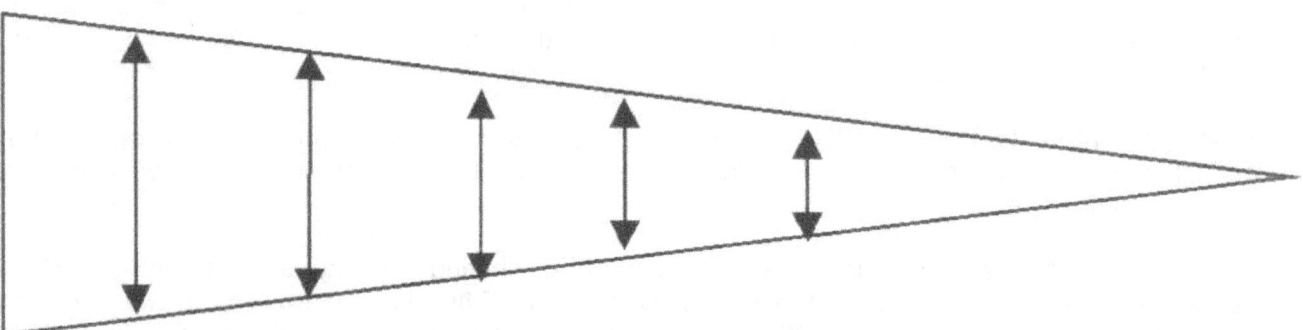

As the electron "travels" through space the time relation increases. The proof in this is that when a wind blows over an open fire, the heat in the fire increases. Oxygen as such, does and cannot burn. It is the special relation that oxygen holds with heat that increase the heat by speeding the flow of oxygen and nitrogen through the fire that increase the amount of heat.

By reducing the flow of air (oxygen and nitrogen) the fire smothers and the heat contains or regulates in the process. This does not change the transformation from wood to ashes; it retards and controls the process.

In the cases where carbon was transform to diamonds and fossil fuel, there is a distinction that the ancient forests were deprive of oxygen and therefore became a dense form of charcoal. The only way that coal, which becomes fossil oil, which became diamonds, can form is by smothering enormous forests by denying these forests oxygen and heating it enormously. Where does this enormous heat come from and how does it influence the earth?

There is, very apparent difference, in the periods where, for instance, there is rock growth, or in which an ice age occur In. some case, life fossilized, in other cases mountain ranges form. It is also quite apparent that there was life on earth during the periods when the precious metals separated, each to it's own specific density, and precious metals surface from the deep inner core. The life later surfaced on their turn in the form of diamonds, and was stored, deep within the earth. The big question no one ever raise, is how did diamond come about.

This fallacy of diamonds coming from a place being deep in the earth's layers holds no water. Again, as with almost every other myth science promote, it cannot withstand the most basic questions. Diamonds can only come about in the presence of enormous heat and pressure, but I think by now I have produced sufficient evidence to the fact that there is no heat or pressure and it is a result of space-time differentiation.

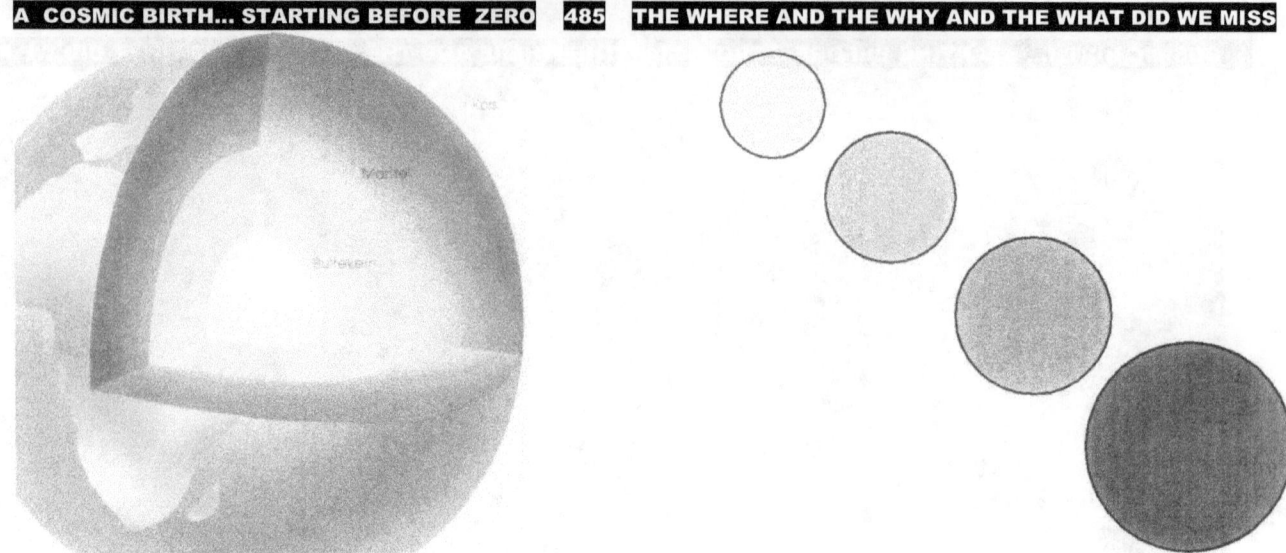

The answer to this lies in the deep inner core of the sun. The space-time that is locked up by the aanplasing of the sun's inner and is seated in layer that still has to develop as the silicon layer would produce heat that is at such high values, it would push the geodesic space-time around the planetary system billions of years back in time and space. In this silicon (to be) layer, the inner core's aanplasing can only be surpassed by the space-time value of the most central iron inner core.

I did explain the start of the sun as a binary system, going array. I shall neglect the binary influence the moon had on the earth in this book, because explaining that will mean that I have to start explaining "life" and bringing in life will bring in religion. However, I will say this about my view. While the moon and the earth were in a binary, the moon was covered with hydroxyl, and the earth had hydrogen a plenty. As the earth evidently formed the senior partner, it walked away with all the water, the air and the environment to host life, leaving a dead, cold lifeless moon to follow the earth as an eternal shadow. With the sun in a binary system, this became the same fate to the junior partner as its aside casting condemn it to a few miserable rocky gasless objects leaving only one to fore fill one important function, to host life.

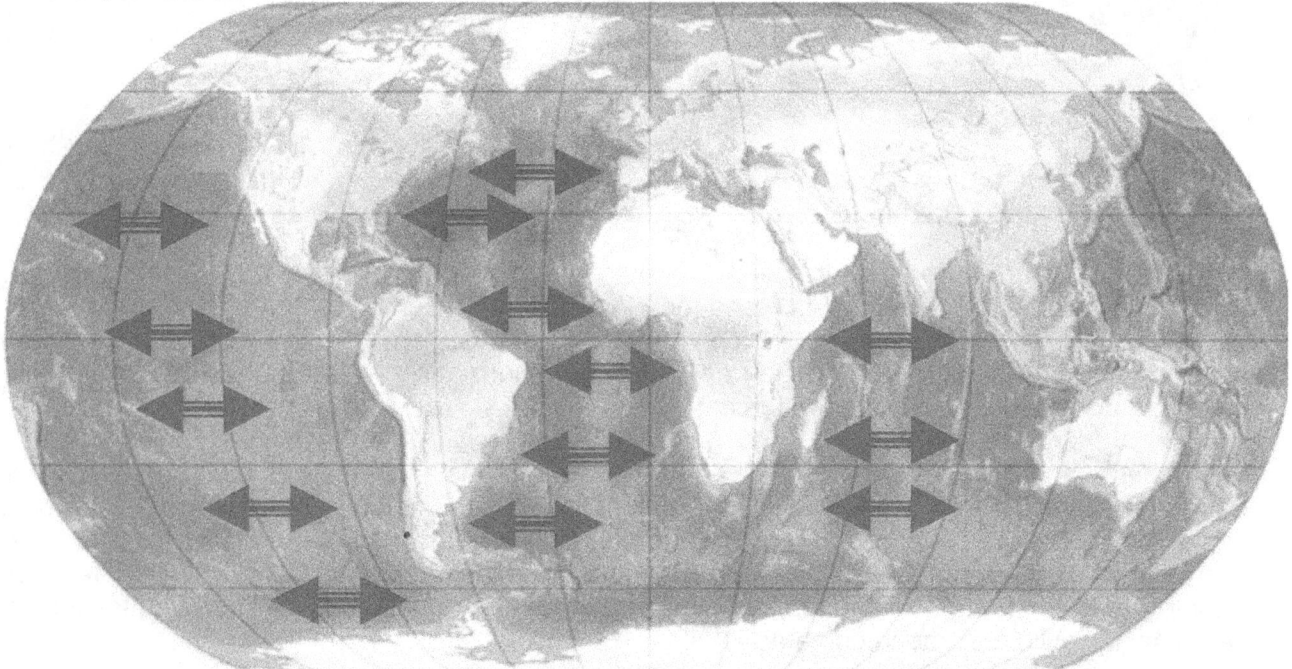

The continents can't drift apart because the continents are not ships sailing on liquid magmas. If that was the case the continents would have less density that the liquid magma below. A ship float on water because the overall volume of water in relation to the overall volume of ship is many times more than the ship and this increases the buoyancy of the ship. It is the displacement of the hollow ship that allows the floating to happen. The rock under the sea gains by growth of material accumulating substance and that makes the surface stretch as much as the rocks break free from the water line to rise above water. This is how islands form. The magma is cosmic substance that accumulates and become denser as well as takes up more space while it is the shear volume of magma increasing that makes the rock break free. The gravity collects heat from outer space as cosmic gas to become cosmic liquid within the earth where the concentration forms an increase in claims on volumetric space. As the core increases it pushes up to fight foe more space to fill and this pushes the mantel to gain in space that produces more space that the surface floor of the earth has. All this allows the core to increase claims of\n earth space. Therefore the sea floor not only pushes the continents apart but also grows to cover more ground and this increase in volume makes the continents grow further apart.

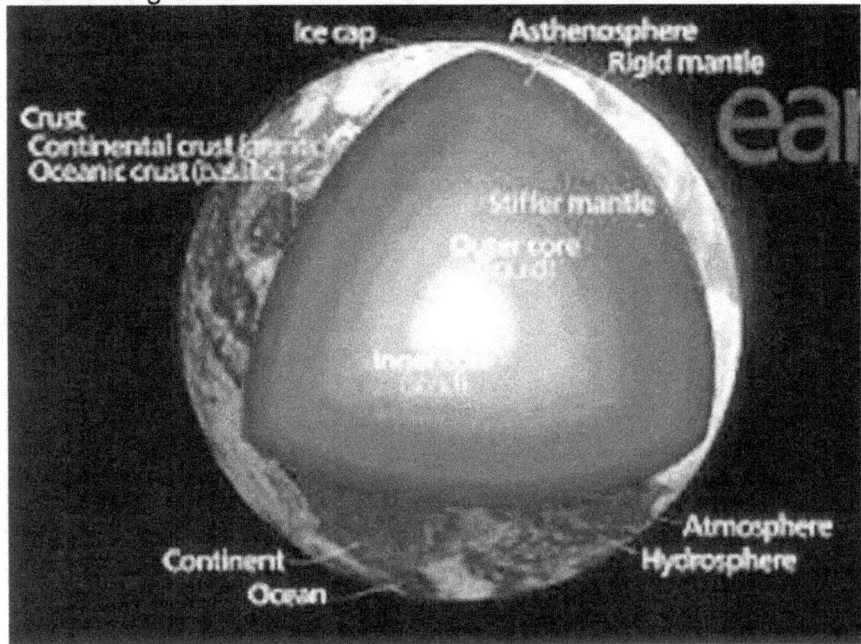

The earth like all cosmic objects is a sphere. The continents can float and move only if what is below is bigger than what is on top. Then there is room for the smaller top to move on the bigger bottom. In the case of a sphere the bottom is always smaller than the top. There is no space the bottom has to allow the top to float on it because the top is always much bigger than what the top is. It is growth of the bottom layers that claims more space that enforces the top layers not only to rise but also to stretch because the top layers are also growing by ratio. It might be half the growth of the core but it is a sight more growth than what the water can produce.

This entire process is still the Titius Bode law forming space by gravity applying time. Movement reduces space and concentrates cosmic heat from a gas to a liquid and then with heat applying the solids gain volumetric space. This is the natural movement of cosmic substance from gas through liquid onto material and into singularity.

Looking at a star we see elements forming layers to apply this concentration but it is how the cosmos grows and that is how the cosmos develops. Through this our "giant planets" will eventually become "giant stars" and the safelights floating around the "planets" will become the new planets and most probably the next collection that will perform as the next solar system. This is how the sea bed rises so far above the current water line just as the next pictures show ancient sea beds and sand that became hard rocks but still the evidence of gradual growth and the wear of material on the rock by water degeneration comes by flowing over and past the sea rock is still much in evidence to prove this argument.

Every layer is not only a set of elements but it forms another cosmos era where the star sheds eras as the star develop to become a black Hole, which is the ultimate development as far as we at present can tell. I find more eras but it seem to be far outside our spectrum of the Universe we are able to detect.

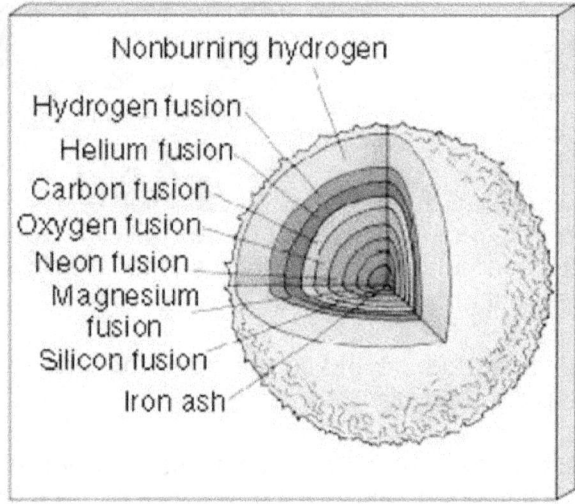

The influence of a prehistoric sea with the water turbulence only the sea can provide is undisputable. More over is the evidence of a sudden declining water level, stabilizing for a lengthy period with again an abrupt drop in the water level.

One can attribute this to ice ages, bit what about the next piece of evidence, pointing towards the complete opposite.

Clear as day -light, is the small water wave marks one associate with a beach. No ice age can have such a sudden influence on any rock by burning the sand on the beach to stone in a very short space, which implies a very long time. There was no time for wind to wipe the waves off the surface, nor were there time for the water to, once again, clean its prints on the sand.

Clearly, the intensity of the blowout, which caused this occurrence was much less than the formation of the Grand Canyon and happened, during a much later time in space development; but happened it did.

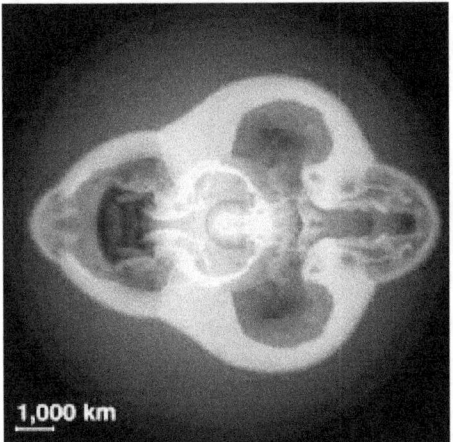

1,000 km

The sun flair I refer to was a couple of billions of times larger than the sun flair depicted in this picture. It made the Carrington Super Flare look like something pale completely insignificant.

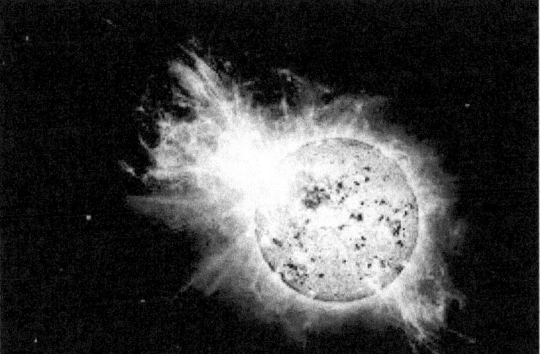

The event I refer to was an outburst that grew into something we now can imagine when we look at the crab nebula. It was not as severe or as large but it covered the inner planets totally in a cloud of sun-liquid. clouds of sunlight and

Place the earth somewhere in the one would get the idea of time freezing to a standstill where one second in time currently changes in duration to a thousand years. Then imagine how long the duration will be of one normal year.

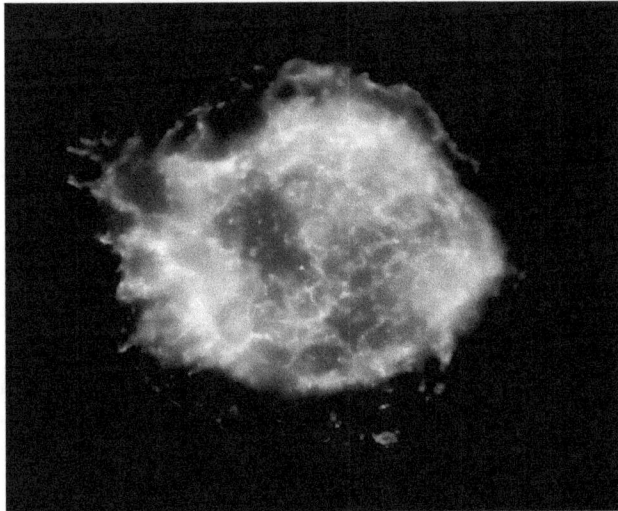

This fits the picture like a personalized glove.

It was not only one occurrence, but from he evidence in the top photo, there were several such occurrences, not immediately following one another.

In the Bible verse 14 in Genesis reads God said: "Let there be (1) _lights_ in the (2) _vaults_ of heaven to (3) _separate_ (4) _day_ from (5) _night_, and, let them serve as (6) _signs_ both for festival and for (7) _seasons_ and (8) _years_. Let them also shine in the vaults of heaven to give light on earth." So it was, God made the two great lights, the greater to govern the day and night, and so separate light from darkness.

(1)   Let there be lights in the (2) vaults.

For the first time stars were observed from the earth.

(3) ... to separate (4) day (5) from night and let them serve as (6) signs both for (7) festival and for (7) seasons.

What this text spells out is so clear I actually feel guilty explaining it because I know many readers may find I am insulting their intelligence, insight and vision. To those, I apologize but since most of our Super-Educated-Superior lack the abovementioned characteristics. Through that they miss a human ingredient which make them over eager to dismiss anything that do not fall into the parameters of their short sited opinions, I shall therefore do it never the less.

With the event of the fourth Solar Day, the moon lost all its planetary abilities, even the effort of spinning around its own axis. Through that lost ability, the moon lost its linear "gravity" and accepting the linear "gravity" of the earth as its own circulates "gravity". The earth still holds the moon captured, but through the loss of its circular momentum, it lost its hold it had in sharing the earth's atmosphere. The earth finally robbed the moon of all its ingredients. Because the moon shifted and through its break away action, the earth (for the very first time), experienced day and night, a shining moon and a brilliant glowing sun. The earth became what we know today to be "normal". I wish to point to the fact that what science are even at present incompetent to realize, the Authentic Author of Genesis discovered many millenniums ago.

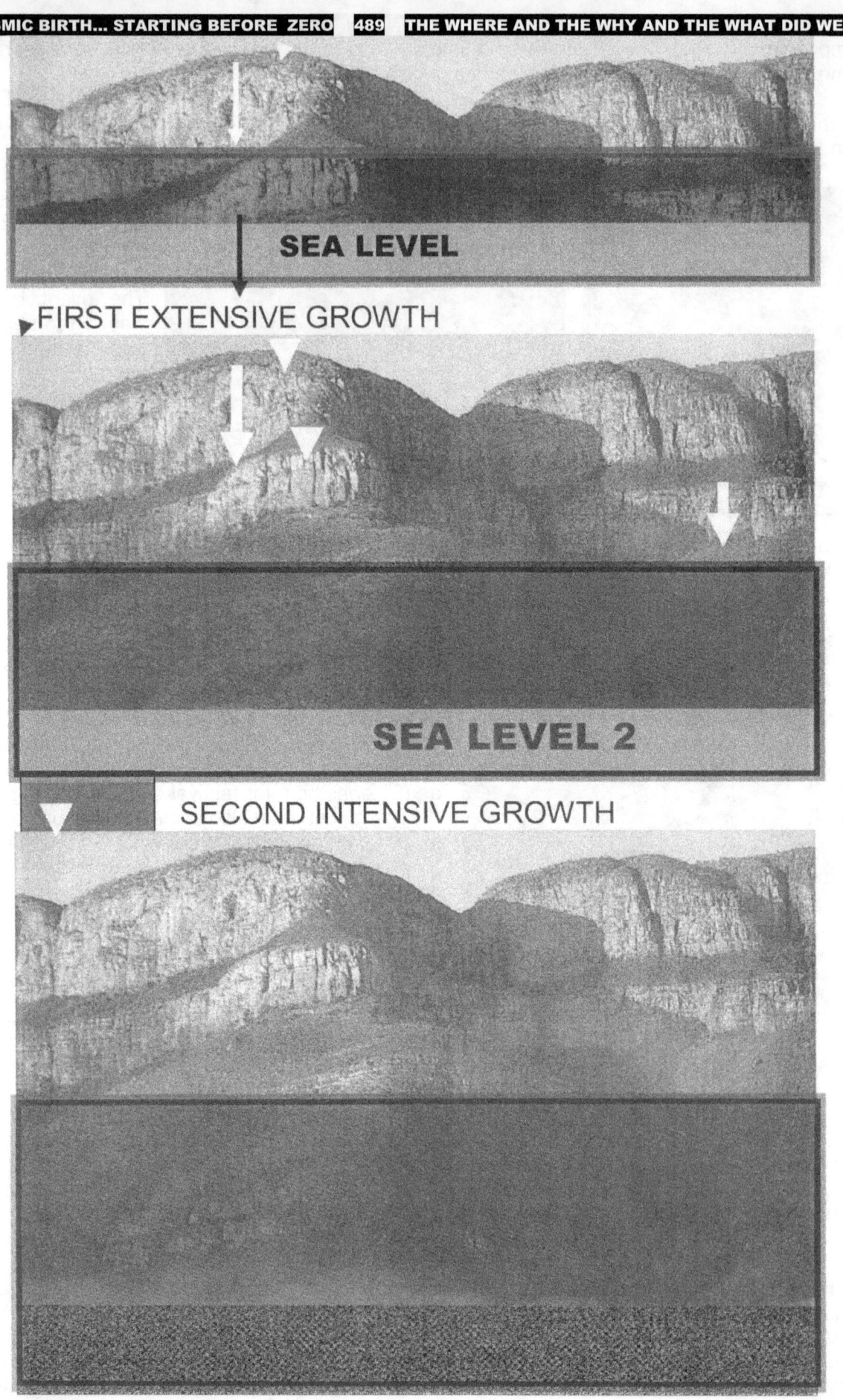

SEA LEVEL

FIRST EXTENSIVE GROWTH

SEA LEVEL 2

SECOND INTENSIVE GROWTH

SEA LEVEL

Consider the material growth there were when the earth was covered in a dense cloud of sunlight for an extensive time like a few years ongoing.

GROWTH ONE

GROWTH TWO

GROWTH THREE

If the world acted on the forces of gravity, as Xepted science would have it, and the earth was crushing inwards, moving ever down, the tell tale signs of ancient sea beaches and water marks, that flowed over rock would not be there. In the picture, the very first perception one get is where water with huge currents had their way in this marking of territory. The tell tale signs would now not rise to almost one and a half kilometre above sea level at a distance of almost a thousand kilometres from the nearest coast, but would find itself far beneath the sea. This is an indicator as no other indicate can be, that the earth crust is rising and growing, and _not dragged down by gravity._

Xepted science's astonishment is surprising to me, in the way they seem surprised by the size of mountain growth on other planets such as Venus. Not once did any of the ones with the brains draw a correlation between the earth having water, therefore life, and this being the reason why earth did not concentrate all efforts on to one mountain to grow.

As explained on a previous occasion in this chapter, how water influence the growth of mountains, quite the opposite is true in the case of huge forests

The heat coming from the sun penetrates through the hard mountains and barren soil, while the plant growth bask up the heat, transforming it to vaporizing moisture and life growth. the damp soil, beneath the plant growth, also helps deflecting the heat to plant growth. In the areas not covered by plant growth and subsequent moisture undercover, the heat transmits to the rocky core, which responds by silicon growth, and pushes the mountains on top upwards.

This effect only applies to "normal" periods, as the one, we are presently experiencing. In the "abnormal conditions, life plays no role, as the conditions extend way beyond anything life can endure on land, but life does still find refuge in deep lakes and the sea. In the case of the barren, water and lifeless planets, life does not play any role in diverting heat to different area. Once a hill form, it relays all heat that rise from the bottom like an antenna therefore drawing as much as transmitting heat. This way any small hill becomes a permanent conductor of heat, with the top always extending father into colder atmospheric regions, thus drawing all growth to one specific location.

I feel we have shared enough information to begin explaining the Solar Day. The reference in the Bible when speaking about : "It was day and it was night." This reference absolutely indicates that this reference of day and nightstands apart from what we experience as a sunlight day and a star studded night. The distinguishing places the two applications vastly apart from each other.

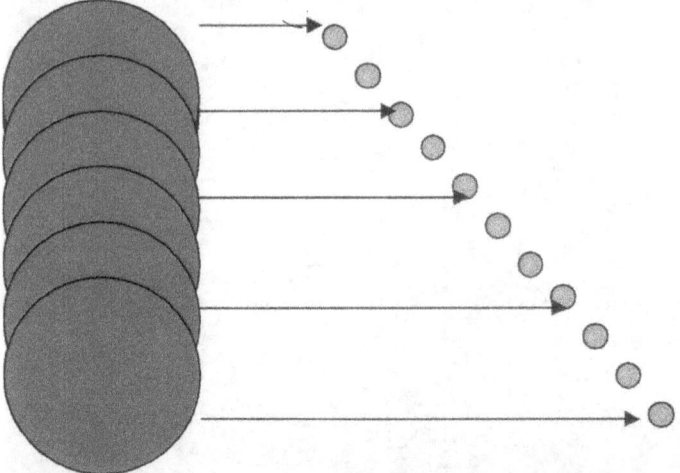

One has to take into consideration that the space-time repositioning of orbiting structures do not rectify its rime position to space occupied instantaneously, but it will occur through a process of development. To that reason the growth, responds will be seasonal.

The earth reads like a tree showing rings that tell the tale of time forming soil. Every line is a layer representing a time where gravity inspired earth growth but gravity is never the same and that changes bring changes in development.

It is the turning of the earth, moon, sun and Milky Way that produces space and time and with that enlarging matter, matter also claim more space as it controls more space influencing more space. That is what the Hubble shift and the Big Bang is all about. Matter grows and that include matter on earth. The only constant in the Universe is that there is no constant other than what Newtonians science creates to make them loo powerful. Kepler showed as time, which is movement forms space, which is a hologram of time in singularity. The Universe exists by not being what it was to become what it is going to be. The Universe changes every instant that time puts another frame on space.

If you look at physics there is always something turning around something else that goes bigger and smaller.

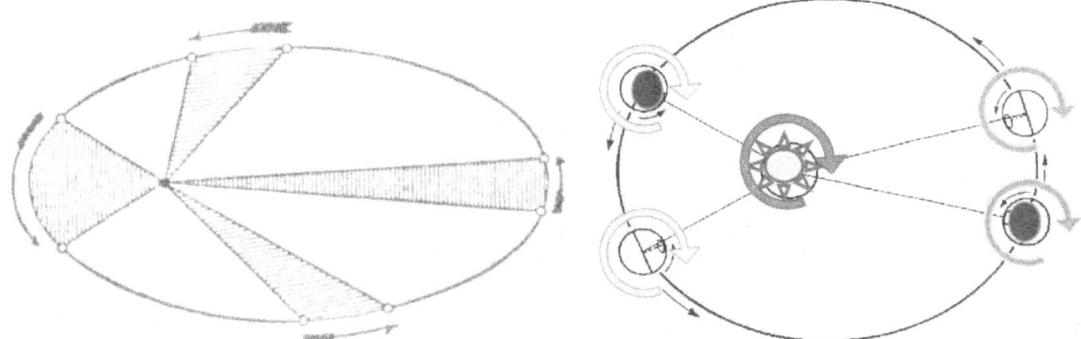

This is the reality of gravity because wherever there is motion it is a circle going straight ahead as part of forming a circle that becomes another circle.

We have a Newtonian spinning around the earth while the earth is taking our Newtonian around the sun. Can you see the Newtonian...neither can I but still... That is not where the circling stops.

The sun will spin around some local system while the local system will take us around the Milky Way.

Then the Milky way will take us around a much bigger system as part of this bigger system

As the earth moves around the sun the entire Universe breaks down from the picture it holds in space to contract into the single dimension it holds in singularity. The by changing everything and it is everything in the smallest derail it relocates everything that can move in terms of everything that can't move and in that everything that can't move did move because everything that moves repositioned a location to another location. This is the Universe.

If we call to mind the Butterfly diagram as well as the El Nino and El Nina whether shifts we can see a bigger picture developing. The Butterfly diagram is a graph on which the latitudes of sunspots are plotted against time. It shows how spots migrate from higher latitudes (30 – 40° north or south) towards the equator (latitude 5° or so) throughout each sunspot cycle, in accordance with Spörer's law. The shape of the distributions, when plotted for both northern and southern hemispheres, resembles the wings of a butterfly. These whether cycles one present on earth and the other spots forming on the sun indicates cycles of changes with patterns forming in 7 to 10 years radials. These indicators prove that using Π in seven forms sequences by ten of going into or ten going out of and these patterns are undeniable. I prefer not to go into this any deeper because in Seven days I go into this extensively but I have devoted almost an entire book in forming support for my suggestions.

These formations are cyclic and these were some of the very first pointers that sent me looking for

proof about gravity in the sense of Π forming gravity. This lead to me to research the Titius Bode law and that proved unequivocally gravity is density distribution spreading the value of Π. We look at the sun and we know it will take between Π and 7Π to circle all depending on the area that it covers in time to commit the space to for the circle it will hold as $4\Pi^2$. This will be between 31 000 years and fifty thousand years depending on what the circle holds/The this will go around the Milky Way to form the circle the Milky way will take to circle its axis which I have no idea but should be something like 220 00 years. This again will be dependent on intervals of between 7 and 10 000 years. There are so many big issues influencing our lives and yet small-minded science is looking for non-existing life on "other planets" where planets only exist by definition and not as a reality.

This forms what we see as the Universe but what we see as the Universe is the result of time replacing positions to vacate old positions by filling new positions. This means the entirety called the Universe replaces with what was to what is to what will be and that is where physics start with a numerical value.

We spin around the sun and that we cal a year. The sun spins around us and that duration we call a day. The sun us on a spin around the Milky Way and that we call...hell nobody thought about giving that duration a name and yet this is as important as a day or a year. This is how human life develops and through the quarters this is the origin of out ice ages. There is no ice age and I gave the two stages two names. It is the liquid period in which we now are and then it returns to the solid era, which we call the ice age. The ice age is only a geographic reference but has no other validity than that. At the polar point there was never anything else than ice and at the equator there was never ice. The Global warming is just another ploy science tries to form a hysteria by which the Mammonites and Hoggenheimers can rip money off a brainwashed and hysterical population. Science goes into deception for the sake of defrauding the public to make money. If there is global warming in excess of what was and I don't believe that then we have to blame electricity and not carbon. The carbon can't become more or reduce but electricity is gravity and by increasing the electricity flow we increase the gravity snd gravity is the concentration of heat from outer space to the earth's core. This MIGHT effect but I can't see that because there is no way in hell that man can influence the earth or affect nature. Man can influence or destroy life to what life need and by that kill off life but that is it. Man can have no influence on the cosmos and neither can life bring cosmic changes but to a small specific location.

To give an example: When the Greeks and the Persia went to war during the battle of marathon; the Persians used trees to build a crossing that stretched across an ocean. They built a fleet of war ships during the same war that was never seen before or since. This must have used up millions upon millions of trees. Where did the trees come from? They sure as hell did not cart it in from Iceland or Siberia. It was local trees and it was dense forests that they reduced to scrap just because the lust for power made men mad.

This becomes evident that before the Persian / Greek wars there were lush forests in what is now desert. Then this madness took over called wars and greed destroyed all the plantations in that region. It took vast forests to supply enough wood to build a bridge crossing an ocean and we can see from that time it became desert. But according top facts left as history this was a lovely tropical area with a very substantial rain forest and by destroying the forests and reducing the trees to a point where it now is desert the entire area was left with dry sand and that is it. Greed removed all the tress and made the area barren desert and a pit for hunger and misery. This is what Life can do to a planet. It can only destroy other life to get the planet as lifeless as the rest of the cosmos is. However with that area being lifeless and barren what affect does that have on the earth or on the solar system or indeed on the cosmos…it has no affect at all. Come the next ice age and with cooling and icing taking place the trees will come back as it destroys all legacy those arrogant tyrants left. The Tyrants and their greed will become lost in time and memory while nature will recover and re-install what greed for power removed. Their memory will be lost as trees recover into forests again. Where will al the gold and silver and wealth be that was fought over? Only God Almighty will know because no greed will be present purely because no humans will remain, as nature will once again cull the numbers of humans back to the

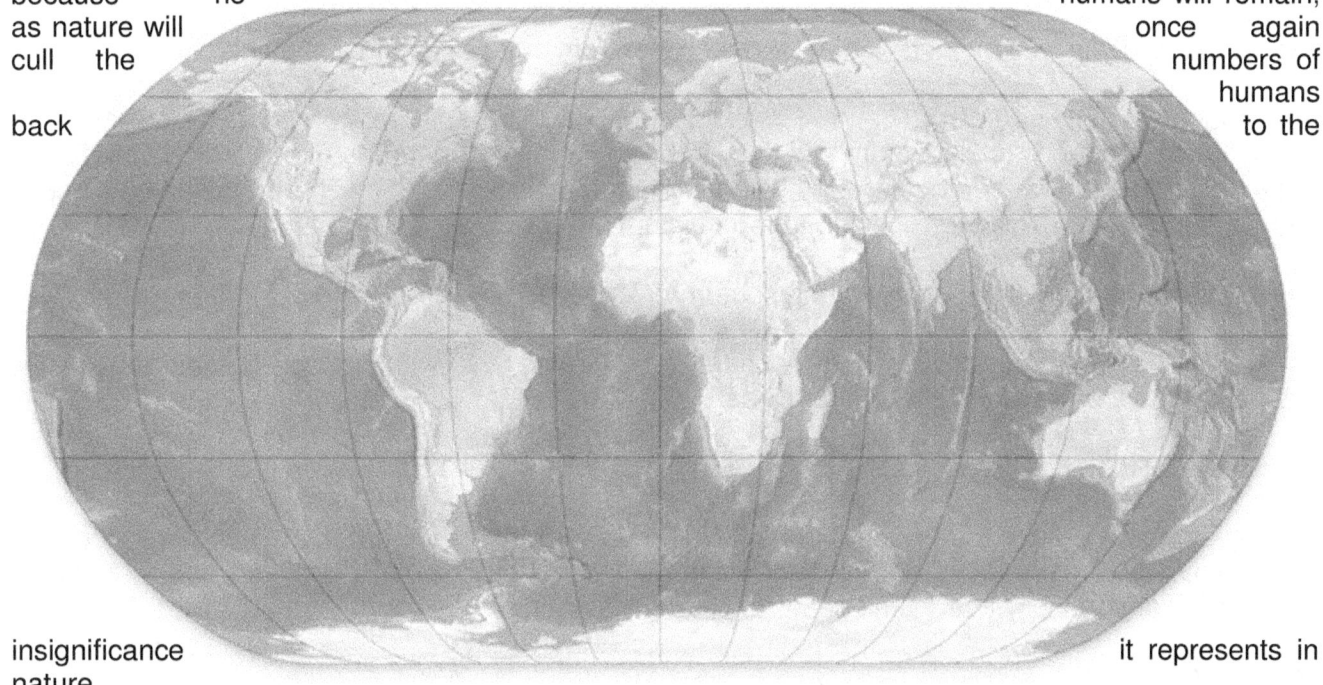

insignificance it represents in nature.

In this picture everything that is green was in the north of the earth was as white as Iceland is during every ice age and everything brown or desert badlands were rain forests where life could prosper.

In the South everything that is green at present was water and swampland and was only habitable by semi water species much like the Amazon forests now is but many times denser with plant growth. North America was ice with some part that now is without rain and deserts were green and over grown. The came the end of the ice period and the Sahara desert and Asia Minor that held all peoples on earth became desert. That is why Egypt flourished. The Egyptian King had to find a way to feed all the migrants that left the desert and came to Egypt to find aid. In this he collected a wealth of intellectual brainpower that grew him the ability to transform Egypt of back then into something we still are unable to understand, even to this day. This resulted in moving populations where people coming from the Sahara wended in Australia and South America. The humans from the Sahara, which was all what we will call Europeans inhabiting part of the Americas, Africa South of the Sahara and mostly of all going just across the Mediterranean sea in starting city-state civilizations the a de-icing Europe. This was also where a burst of culture went from Asia into a de-icing Asia to the North and Northern Europe. All evidence is there but somehow there is a drive to prove that all races are the same and research about this is hindered and silenced. How does this happen? This is the Global Warming but this reality is part of nature and not manmade corruption.

As the solar system turns around the Milky Way the gravity applying holds four completely different positions all opposing one another by 90° and by 180°. That means what was valid in the one period is completely invalid during the next period and what applied opposes what applies. This is the nature of nature and this is gravity. The process we now go through to de-ice we will go through once we ice up again and what melts now will form ice when the cycle moves into another quarter.

In that and with the development of density re-applying new changes every time the cycle returns to what it was previously a new Universe emerges with new rules and old rules never changing.

Every instant that everything changes new development brings new conditions. That way the earth transforms from what it was being as it is to what it is going to be. The result is time forms space and the space we see is in layers formed just as tree rings does and every layer tells of other influences bring on a new earth. Only Newtonians can be blind enough never to give the evidence any notice.

All evidence point to polarity, the same polarity we find in alternating electric current. In all flow of electricity there is an alternating flow. Direct current is not electricity in the true sense of the meaning because it is just transferring stored heat from one sector to another but then the process must be reversed to recharge. So in that sense we can make some argument that there is a flow back and forth. Then to produce work the heat has to flow back and thereby lose polarization.

However in electricity we have a flow of heat directional from outer space to singularity.

The relevancy factor is $10/7 (4(\Pi^2 + \Pi^2) = 112.795$. What this translates to is that space (10) goes singular when 7 reduces to singularity and this lines up with all 4 quadroons of the proton value of the full circle of charging gravity (Π going Π) + (Π going Π) and this is then ($\Pi \times \Pi$) on the one side and ($\Pi \times \Pi$) on the other side. This then lines up to form a circular movement ($\Pi^2 + \Pi^2$).

Also we get gravity forming or charging the sphere at $7/10 ((\Pi^6)/6) = 112$

This is simple to interpret because it is a sphere ((Π^6) turning 7/10 within space forming a cube that holds borders to the cube with six counter sides.

Then we have $\Pi(\Pi^2 + \Pi^2) + (\Pi^2 + \Pi) + 3 = \Pi(35.75) = 112$.

To formulate this we have the proton ($\Pi^2 + \Pi^2$) lining up with the neutron ($\Pi^2 + \Pi$) and the electron 3 where this lining up is included as a circle (Π) and then with everything in the atom lining up the atom allows gravity or electricity to transmit heat from eternity to infinity.

This is gravity.

This transmitting of heat forms a reduction as ($\Pi^2 + \Pi^2$) x ($\Pi^2 \Pi$)3 = 1836 which is the displacement difference between the electron and the proton $10/7 (4((\Pi^2) = 56.4$. The neutron forms a dimensional liquid with no reduction properties but only serves as the liquid transmitter of heat.

When a proton and a neutron and an electron lines up (+) a sphere forms not by free electrons because there us no free electrons. The whole idea of electrons running free is just more Newtonian hogwash.

A dome forms when an atom forms and cosmic gas $10/7 (4(\Pi^2 + \Pi^2) = 112.795$ condensate to cosmic liquid $7/10 (4(\Pi^2 + \Pi^2) = 55$ in relation to copper $\Pi (\Pi^2 + \Pi^2) = 62$ which takes the flow then from there where it goes singular.

This is the relation $10/7 (4((\Pi^2) = 56.4$ which is the process of transforming cosmic gas to cosmic liquid.

SPACE AS COSMIC GAS

SPACE AS COSMIC LIQUID

Individual stars forms SPACE AS COSMIC SOLID

As time reduces space back to singularity, the singularity within every proton grows to accommodate the singularity maintenance. This does not mean there is a common all-inclusive "going around to see if it fits" singularity, as Einstein would believe. It is a well-balanced time to space occupation where the time has to match the space in claim, the space in control and the space in influence to allow the growth of space in liquid to space in gas.

Gravity / electricity / sound / electromagnetism / wind / light / lightning / magma flow is all forms of the same thing which is the converting of cosmic gas through cosmic liquid to cosmic solids to serve singularity.

The flow of cosmic gas to cosmic solids forming cosmic liquids is.

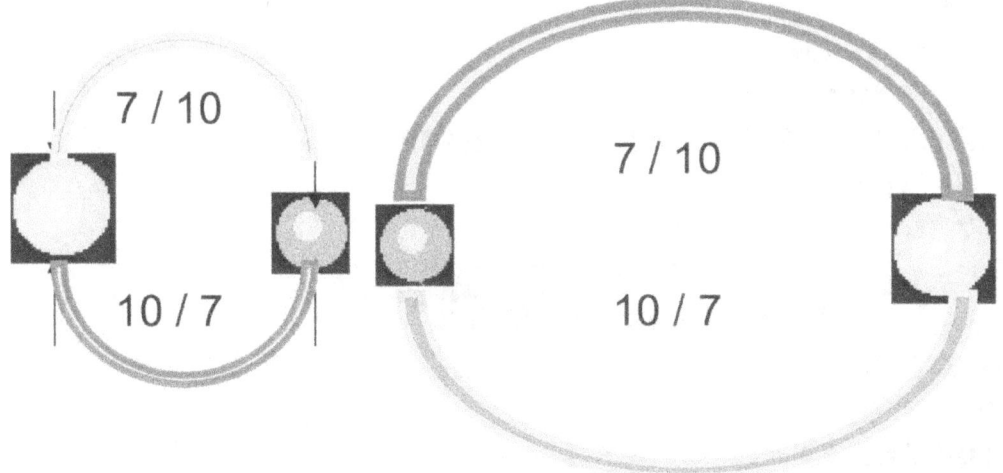

Because the claimed experts have no idea about their expertise, they present the biggest misgivings that science can produce by claiming planets they found and other total rubbish such as gravitational pulling and gravity swings and what not.

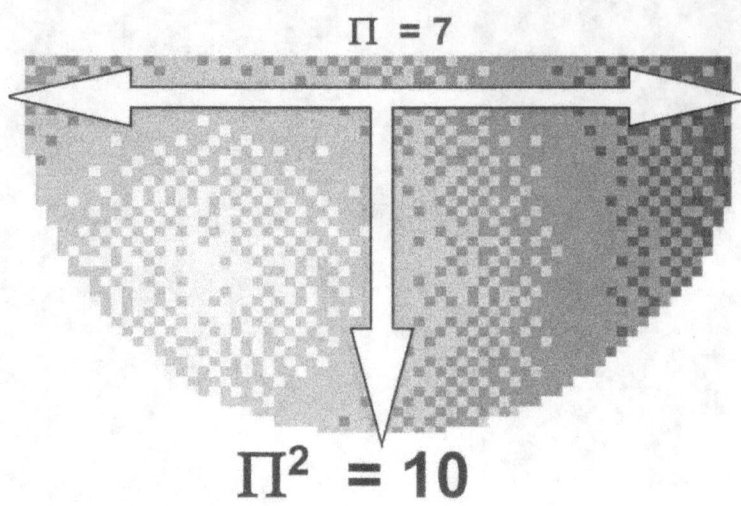

$$\Pi = 7$$

$$\Pi^2 = 10$$

Everything is about transforming cosmic gas to cosmic liquid to cosmic solid but this goes in two directions. We have the conducting of heat from the gas it forms to solid whereby solid gains density

$$\Pi = \frac{10+10+1.991}{7} \qquad \text{x}$$

$$\Pi = \frac{7+7+7+0.991}{7}\Pi^2 \qquad \text{in}$$

materials and **10** in space.
This is the gain or growth of cosmic substance as a result of increasing density in solids.

The there is $\Pi^2 = \dfrac{14}{1.42} = 9.86$. The total relation is as follows

Matter is a product through the separation of space and time receiving the value of Π The original time and Π² as follows:

THEN THE NEUTRON FORMED ON THE OTHER SIDE OF THE UNIVERSE OUTSIDE SINGULARITY BY DIVIDING SPACE INTO MATTER AND MATTER INTO SPACE, ANG ALL OF THIS ACCORDING TO THE TITIUS BODE LAW OF 10 / 7 AND 7 / 10 IN CONJUNCTION WITH THE ROCHE PRINCIPLE OF (Π/2)²

TIME DIVIDED INTO SPACE

(10 / 7) \(7/ 10) = 2.04

1.4285 /   0.7   = 2.04

SPACE DIVIDED INTO TIME

(7/10) / (10/7)   = 0.49

.7 /   1.4285 = 0.49

SPACE MULTIPLIED WITH TIME

7/10   10 / 7 X 7/10 =2.04

THE PROCESS PARTED USING THE ROCHE PRINCIPLE

10 / 7   (Π/2)²

7/10   2.04x(Π/2)² = 5.033

(Π/2)²   2.04x(Π/2)² = 5.033

10 / 7   5.033 +5.033 = 10.066

SPACE DIVIDE INTO TIME

7/10

7/10 / 10 / 7= 0.49

10 / 7   =   0.49

10 / 7   10 / 7

7/10=.49   7/10= .49

.49   +   .49   = .98

.98 X 10.066 = 9.8 =Π²

TIME SPACE = Π²= 9.8696

TIME SPACE = Π^2 = 9.8696 = MATTER HOLDING THE SECOND PROTON COUPLING THAT TO THE NEUTRON TO COMPLETE THE NEUTRON.

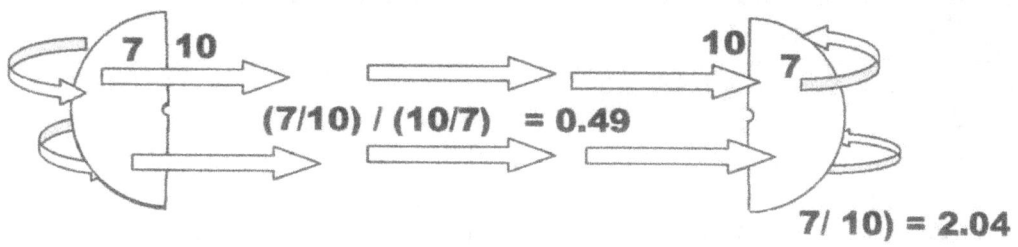

(7/10) / (10/7) = 0.49

7/ 10) = 2.04

As the cosmic solid increases in density the cosmic gas decreases in density and there is a flow of cosmic substance from gas to solids and this removes density from the one sector to the next sector. This means we will have an impulse flowing in the one direction and a stabilising of equality going in the other direction. This will lead to a movement to solid interrupted with a flow of directional movement to stabilise the other factor.

The flow is cyclic and it is emulating the graph to a precise detail where in the graph there is no zero because there can be no zero in the Universe. The line breaks singularity where singularity reaches a point and a return of value is established taking flow in the opposing direction.

SPACE MULTIPLIED WITH TIME

($\Pi/2$)

10 / 7

7/10

7/10 / 7/10 = 1 and   10 / 7 X 7/10 =1

Lets find the reason why seasons are part of gravity but we already know it is because gravity is part of singularity. That means detecting the roots of seasonal changes we have to trace back to singularity to connect this to gravity.

By applying this spinning philosophy, the world of science become a lot less complicated and a lot more scientific in its approach to matters relation to matter, as matter respond to other matter's behaviour.

Take two gears rotating in sequence at the same speed and we find the rules that dictate the philosophy behind Kepler's tables and the Titius Bode law. In the Universe everything from the most minute particle to the biggest unified star there is, everything spins opposing directional.

Mark each gear in a precise and recognizable way to match the other gear. That is how the Titus Bode law places the planets in such a precise sequence as the planets now are.

Just like it is in the case of gears the process is in place to maintain the rotation speed of each gear to a precise instant of one in three hundred thousand revolutions per second, while they are rotating at one thousand meters in three hundred thousand revolutions per second.

Take them six miles apart and be astonished by the way, they act in a precisely comparable manner. That is how simple the explanation is of particles acting the same when even being a distant apart. The particles do not apply a nature change when observed. The particles will turn in the precise fashion, but they will show opposing characteristics, as long as the rotation speed maintains to a degree that only the cosmos can apply through maintaining the same time balance to space occupied. Now, AS THE LAST REQUIREMENT: be astonished by the similarity they show in movement and you have the makings of a great scientist.

fter all when gauging the universe, everything else seems to spin in a way

The Coanda effect that joins time by committing material to gravity

Time in eternity forming the motion or liquid that locks space in as material

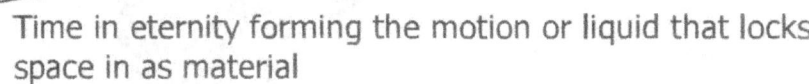

In the center where we see an axis we find singularity as Π^0. That is r^0 or 1^0 or it cold be $24589421^0 = 1$ and this is all 1, which is the point where singularity is. Where singularity forms space it forms a circle because at that point where singularity Π^0 becomes space it forms a circle Π and where space starts we have Π^0 becoming Π only because movement Π^2 develops $\Pi^0\Pi$ Time in infinity forming the motionless or solid that locks space out as material

I have shown that in the centre of all moving circles and sphere we locate singularity. The smallest particle in the Universe is singularity and is it so small it is a mathematical principle forming the centre of the Universe. The smallest particle there can be is so small it does not fit into the Universe in space.

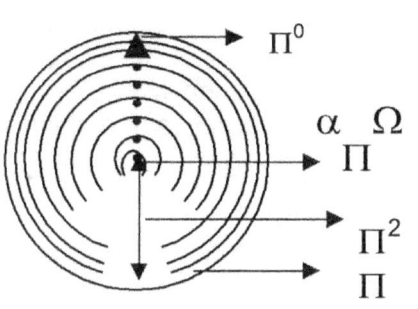
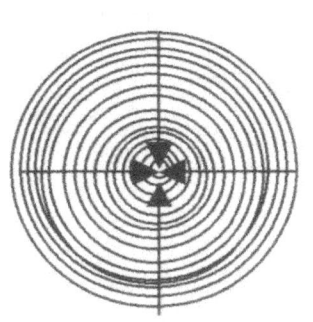

In order to reach this particle there is a mathematical process which one has to follow. Everything in the Universe is a sphere that turns. This turning forms a circle because if not a circle it becomes a fragment floating without destiny. In every circle there is a centre forming the circle around the centre.

A circle is $circle = \Pi r^2$ That is the circle that is visible. To reduce the circle $circle = \dfrac{\Pi r^2}{r^2} = \Pi$ Now we are where all material begins at $circle = \Pi r^0$. Then we have to reduce the movement further and there we reach singularity. To reach the point that is smaller than material is filling the Universe we have $circle = \dfrac{\Pi}{\Pi} = \Pi^0$. This point holds no material and has no space within the Universe. This is the spot Π^0 that grows into the dot Π and the dot is the point where space begins at Π but all this is only possible if the circle Π moves $\Pi\Pi^2$.

In this point where the spot is with the value of Π^0 is there is no space. If you enter that point you leave that point at the same time. If you enter that point you find the Universe going in one direction to the left and to another direction to the right. As you go in the turning will go to the left but as you enter (no space remember) the Universe turns to the opposing direction. It has all four directions of flow concentres at a point just where space starts and therefore it is all space that gives division to directional flow but the source of this opposing directional flow of movement is from a point outside the point that forms the Universe.

This is what we see when looking at The direct link one find between severe and devastating thunder storms destroying all in its path and the much more feared The carnage these storms bring to households that can least afford it and where the poor loses all life savings in one hour or sometimes even minutes. Everything visible in these storm devastations is gravitational directional flow. In weather we see this pattern develop where from the earth it collects water to pour down on the earth.

This point in discussion does not form part of the Universes we are within but it is part of the universe we in life are part of. This point does not exist and yet this point controls the entire Universe as a unit.

This directional flow brings opposing time periods because as I have shown the flow is not coming from within the Universe but the source is located within time within the Universe. When space becomes a factor directional opposing movement forms an immediate part of it. This is gravity but also this is what we think of as seasons. This when time forms space and as the one goes left the opposing direction will go in another direction as it crosses the point does condoles everything but does not exist.

Although this point does not exist this point controls gravity and gravity is the standard by which time forms the Universe and time is the planning of God Almighty in the value of Π. The value of

$$\Pi = \frac{3.1416}{\Pi^0} \quad \text{or} \quad 3.142$$

and in movement it is

$$\Pi = \frac{21.991}{7}. \quad \textbf{That means time}$$

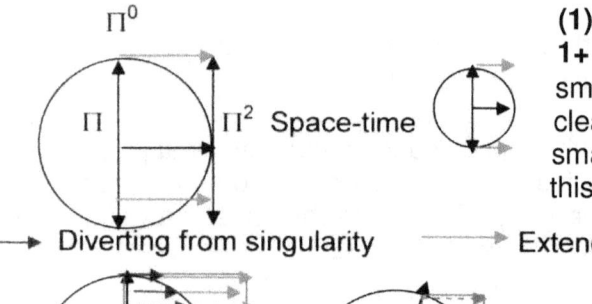

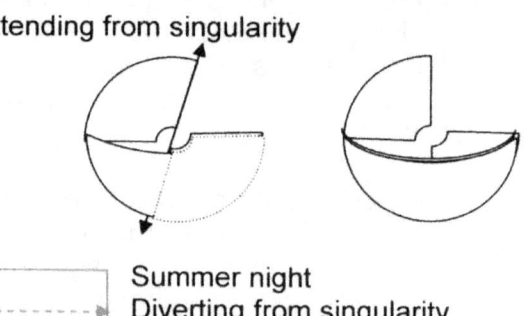

forms gravity by Π and in Π we have **a future (1) a present (1 + 1 = 2) and this becoming a past (1+ 1+ 1 = 3).** However as I showed the point of three is even smaller and as the value of Π indicates this value shows clearly the point coming from the future is even much smaller forming 0.1416 or 0.009. Later on I plane to explain this in the next chapter. This shows that time has a future forming the immediate future forming the present forming the past forming space as Π and that value is forming the Universe. This is why the Universe is merely representing space as a remembrance of what time left behind.

In this forming of a cyclic time line we have seasons forming as a direct result in the manner time forms space from appoint formed outside space and within reality. We holding life is outside reality and our bodies are in space and therefore we are not our bodies but we are time or the cosmic reality of Go Almighty creating space by forming Π = 3.142 where 0.142 is taking Creation into the future and into time coming from the future Planner.

The four seasons are defined in nature by the position the earth in orbit holds in relation to singularity.

The seasons come about as it crosses certain points in the circle indicated as $k = \dfrac{a^3}{T^2}$, $T^2 = \dfrac{a^3}{k}$,

$k^{-1} = \dfrac{T^2}{a^3}$, $k^0 = \dfrac{a^3}{kT^2}$. **The seasons, winter, spring, summer, autumn, is the result of the earth crossing the sun's singularity and this leaves a yearly ice age and liquid age where we have heat and cold as a cyclic occurrence.**

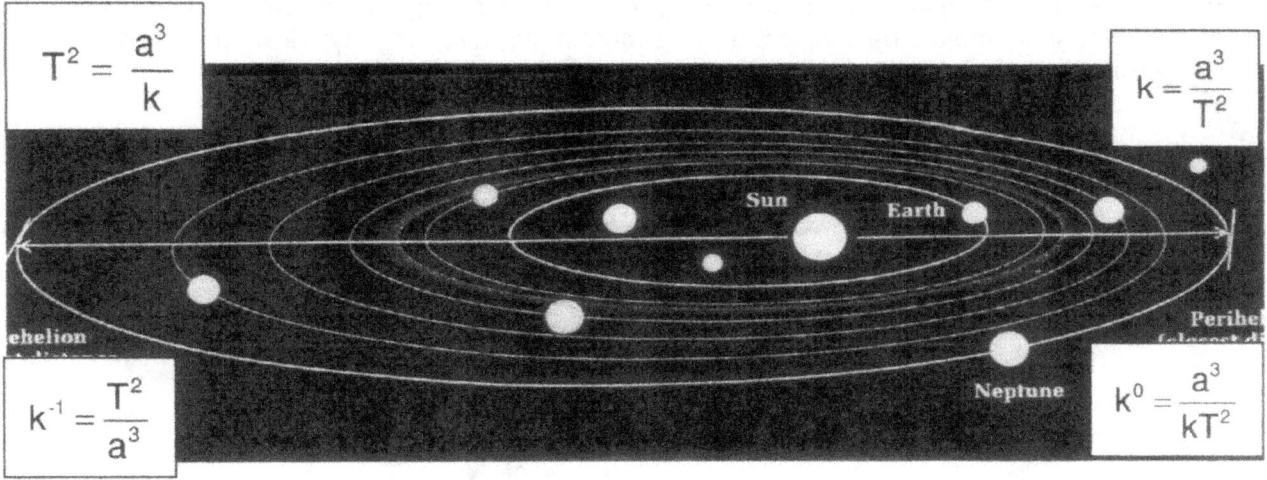

Because Kepler proved that space a^3 is time $T^2 k$

$a^3 = T^2 k$ **or movement $T^2 k$ and therefore singularity forms** $k^0 = \dfrac{a^3}{kT^2}$ **and then the spin**

determines the space in movement $T^2 = \dfrac{a^3}{k}$ **, The movement determines the space** $k = \dfrac{a^3}{T^2}$

' the movement reduces the space $k^{-1} = \dfrac{T^2}{a^3}$ The following is relevancies and one can plays the

relevancies any way as long as the opposing is depicted. In our view seen from a Universe that does not exist this is what we see but this shows space ands space does not exist.

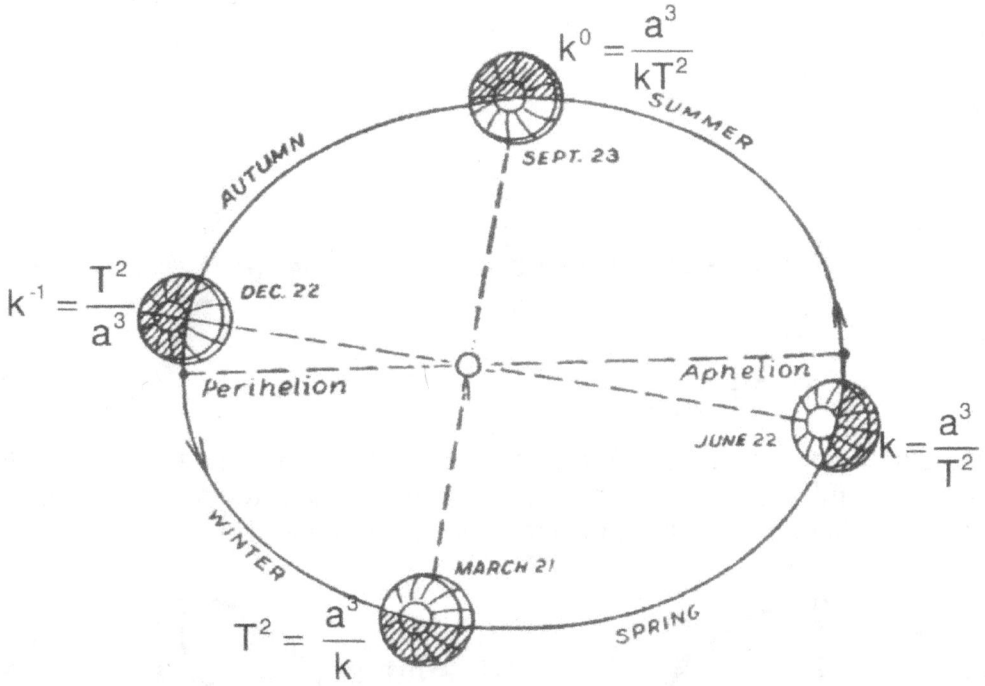

We see the planets turn around the sun in a circle as we see space but we see space because we are part of time. We with life forms part of time and time is a line worth 3 flowing in relation to a circle worth 4 and in the it is a continuing ongoing process with no intervals of any sort as far as we are concerned. This is not reality but a hologram that time leaves behind as space. In truth it is a jerking of contracting and expanding of increasing density and of releasing density and this is as cyclic as it is contradicting by being opposing. The entire Unversed functions on a line forming time in relation to a circle broken in half by time shifting and this is the reason why we see the sun circling around us on earth. The sun rises in the morning by a circle crossing over our heads and then opposing in darkness it shines on the other side of the divide. The divide I refer to is the divide being singularity within the centre of the earth.

The line provides time a linear path to drag the circle into opposing sectors. As the future brings the present, which pushed into the past in, is the cyclic motion of gravity that lays down the rules.

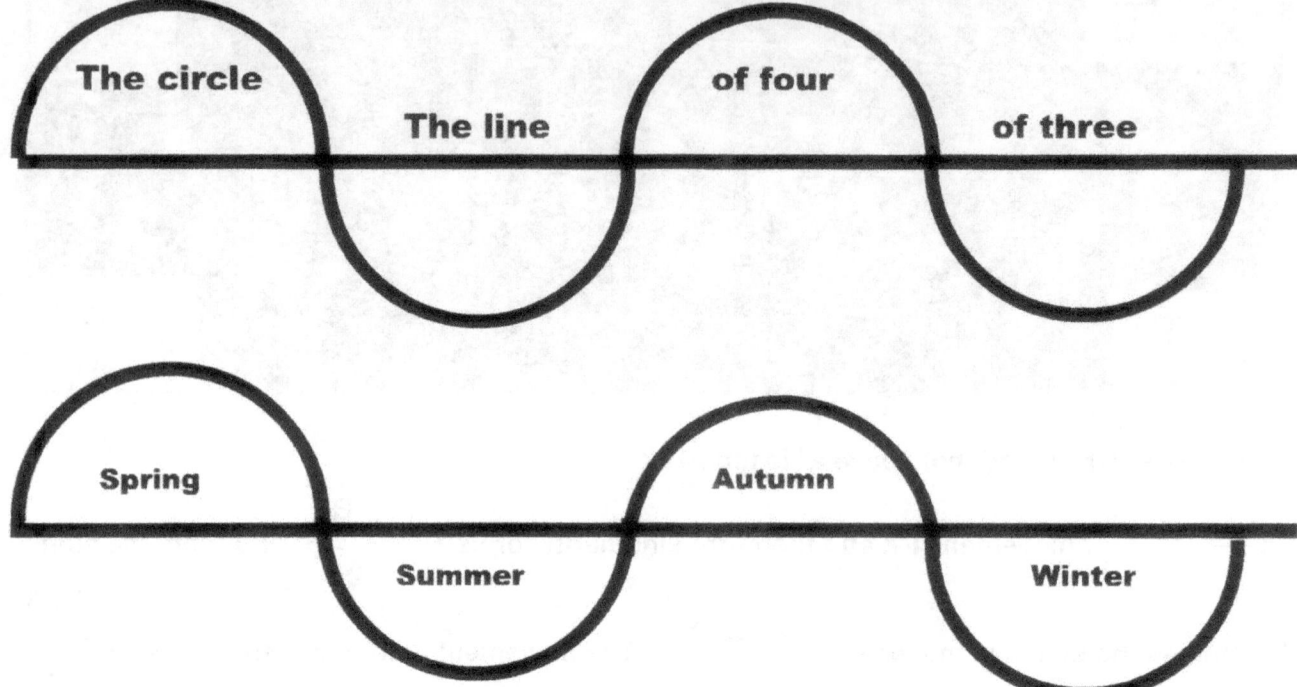

Seasons changing are not the heat of the rays catching the top of the earth and then the bottom as Newtonian incompetence would make us believe. Even if the earth does wobble around the sun the sun shines equal in all directions and this idea that the sun shines different bringing about different seasons is folly. It is cyclic and everything has to do with gravity forming contradictions in opposing cyclic periods much the same as electricity does. We have electro-magnetism as well as conducting electricity works on this very same cyclic principle of going "positive" and "negative".

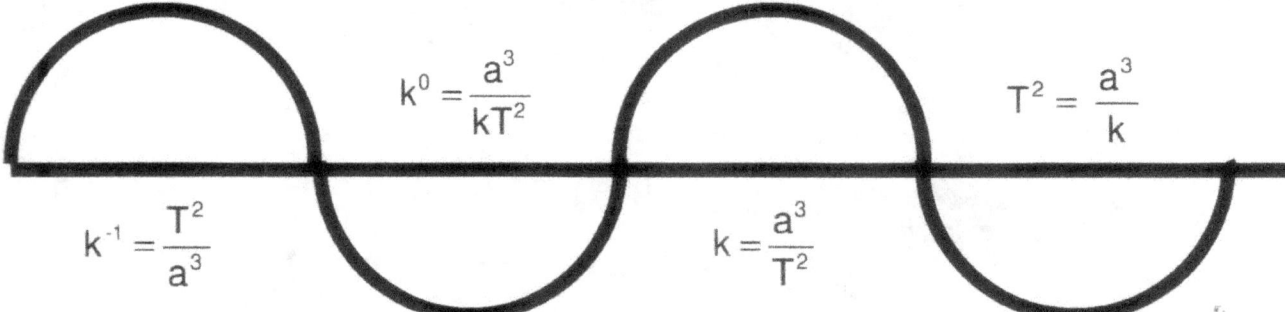

It is poles swapping polarisation much the same as electricity because gravity is electricity and this conducting of cosmic gas and turning it into comic liquid or electrons as Newtonians wish to call it is the same thing but in different dimensions changing the scale on which it happens completely.

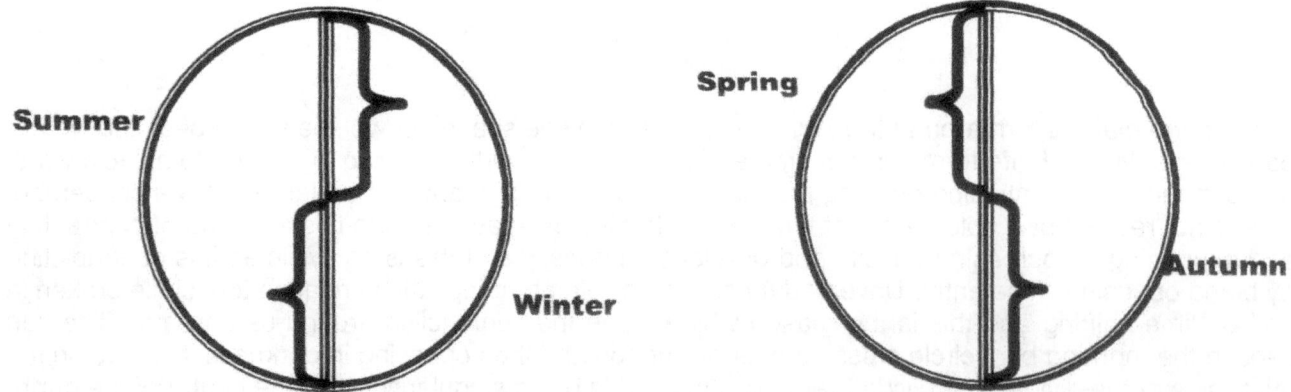

In winter the poles are different to what it is in summer on both sides of the equator. This too must also apply within a star as layers in the star forms and we will find this is the reasons for tornadoes forming. The manner in which time slows down in duration is a fact and everybody should understand it by now. All of these phenomena are due to the sun's influence on our planet, but it only holds minor consequences to us. At the most, it is only somewhat inconvenient. **But, and I say but**, it is not severe enough to lead to the creation of new species, as they would have to adapt or die in order to survive. The sun admits a certain value of space-time through gravity. Any person that wishes to sneer on this, please allow me to remind such a person, that this is an ongoing process, much the same way as it is done at present with the so-called hot house gasses and the global warming. By sneering at the Noah phenomenon and the folk law aspect of it, may I remind those that we are just a guilty today, but a lot worse. This comes about by the fact that the modern media is selling advertisements, and not news. The bigger the hysteria the media can generate, the bigger income it ensures to itself. When a star bursts it is because every layer in the star is not in sync with the other layers and these layers then begin to overheat as a result of moving too slow. At a point the star expands in the layers where movement could no longer sustain formation

All these cycle events only brings about slight weather changes, and as we, humans, are only used to the better side of the sun's development, we tend to go into some outrage, every time the flooding or extreme colds occur. The worst part of all of this is that we humans are arrogant enough to think we play part in the process, or that we can alter the process. We are far too insignificant to do anything bad or good in the process. The next hundred years, of five hundred years might be intensely hotter, or colder, and can even produce a mini ice age, for all we know.

I clearly hope that by now, most, if not all the readers may realize how inadequate our Super-Educated-Geniuses Newtonians are in their explanation about the forming of stars and "planets". It is quite impossible to seriously consider that a structure as the earth and the inner planets will form a solid structure from hydrogen and helium gas spinning around the sun. Any structure holding an orbiting cosmic position in this era, must have an iron core. From an iron core a star forms and the only manner that the four inner orbiting cosmic crumbs could obtain an iron core, is when a larger star fragmented and deposited its inner structure to cool. All evidence to that affect is present in the cosmos and accepted as principles, yet the Super-Educated stick to their belief in magic where gas can compress by own "weight" to something as solid as the earth or even the moon. Fact of the matter is that the Lagrangian system is well appreciated and quite common.

There is a link between what happens in the sun and what happens on earth and this interlinking is as fixed as a cog driving a chain connected to another cog. Every aspect going array within the sun By using the axis provided by the sun this movement then Every season leave a ring that represent space left according to condition that was present during that specific centre.

The growth on earth comes as a direct result from influences coming from the sun and the earth takes a direct queue from this influence by the sun. This we know from how the Roche lobe works.

To understand gravity we must understand how the Roche lobe forming the Roche limit works.
When two objects are within the range of $\Pi\Pi^2$ the gravitational demise of one object the minor partner comes about. I am not going to explain the entire process but the limit is set at $\Pi^2 / 4$ of the diameter of the major partner sharing an atmosphere. The major partner excites the spin of the minor partner to a point where both hold the same movement being $7\Pi\Pi^2$ of the major partner. This is portrayed in the picture below. The minor partner's spin rate is accelerated by the gravity of the major partner when the minor partner is with the limit of $\Pi^2 / 4$ of the major partner.

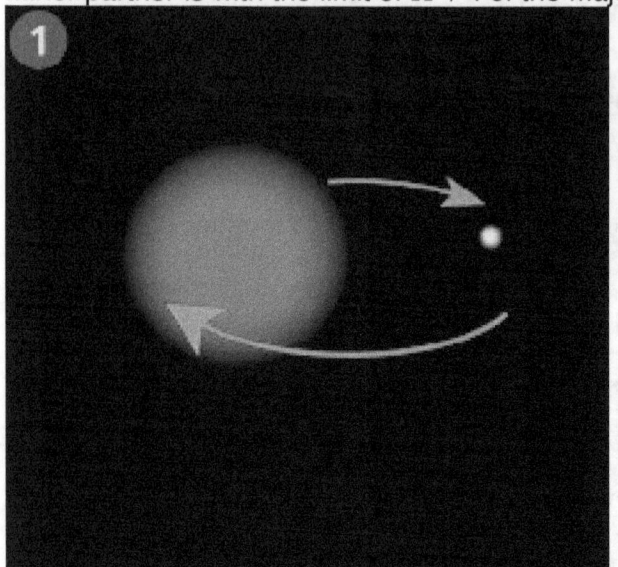
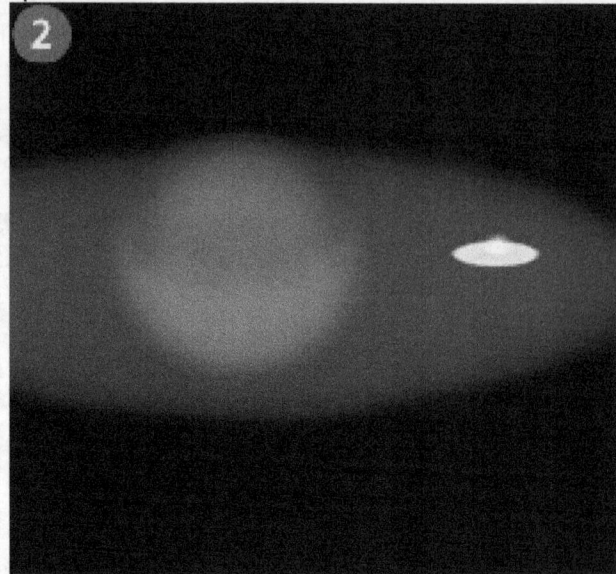

This accelerating of the spin of the minor partner increase value of $\Pi^0\Pi$ of the minor partner as to match the radius $\Pi^0\Pi$ within the boundaries of $\Pi^2 / 4$ to then match $\Pi\Pi^2$. By increasing the spin Π^2 this longer ratio Π^2 applies will add a new dimensional value applying as $\Pi^0\Pi$. By increasing Π^2 and this increasing $\Pi^0\Pi$ the new value of $\Pi^0\Pi$ sets new density dynamics, which with less material to the increased liquid ratio the minor star becomes, liquid and just more atmosphere of the major star.

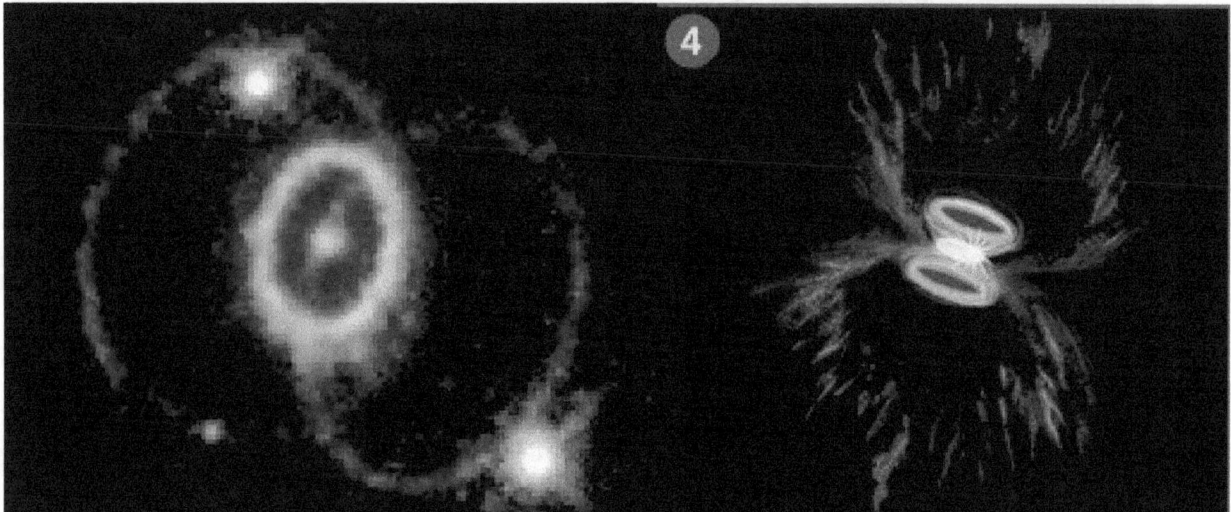

Then it makes the minor star liquefy its core in the event where the minor star was not that minor. However in this picture we can clearly see how the core goes $(\Pi/ 2)$ and $(\Pi/ 2)$. This is to show how one star can affect the behaviour of and inside another star.

From all over come influences via gravity that plays part in the development on earth. The sun has a direct influence on earth because gravity and electricity and electro magnetism is all the same thing. It is all a connection that time brings to how space develops on earth. As the earth grows by cosmic development so the ratio between water and earth favours the earth part. The water does not reduce or disappear but the ratio gets favoured by the earth core developing more than the crust that in turn develop more that the water forming the top part of the earth surface. On mars we cans ee how much this development went against water in favour of the silicon while it is clear that the iron core does not seem to develop that much any more. That means that in a different part of the sun cycle as it goes around the Milky Way. These layers on earth tell a storey of seasonal development but the seasons are time effected and the seasons may not be seasons we now think of as seasons that's in pace.

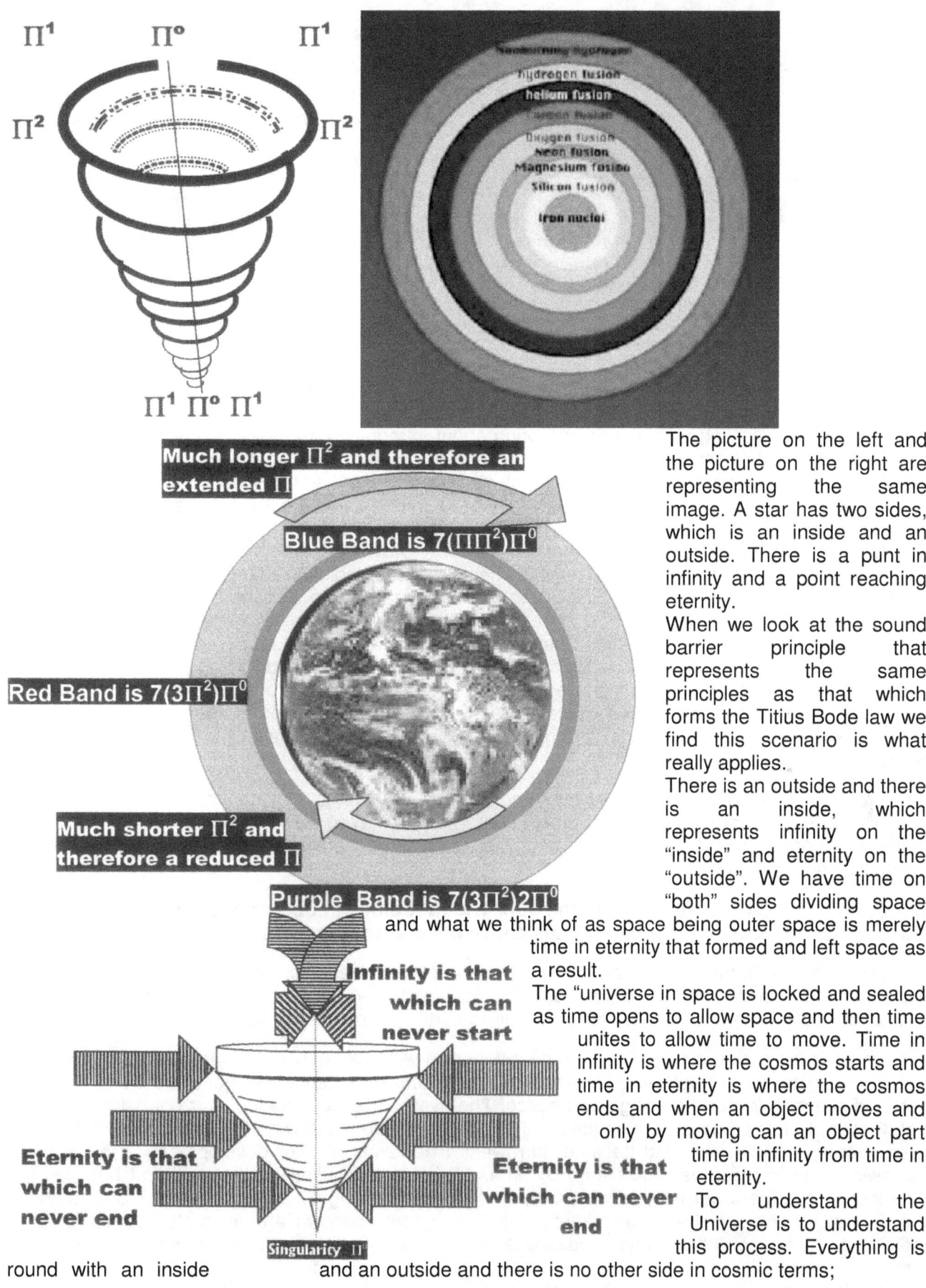

Π^1 Π^0 Π^1

Π^2 Π^2

Π^1 Π^0 Π^1

Much longer Π^2 and therefore an extended Π

Blue Band is $7(\Pi\Pi^2)\Pi^0$

Red Band is $7(3\Pi^2)\Pi^0$

Much shorter Π^2 and therefore a reduced Π

Purple Band is $7(3\Pi^2)2\Pi^0$

Infinity is that which can never start

Eternity is that which can never end

Eternity is that which can never end

Singularity Π

The picture on the left and the picture on the right are representing the same image. A star has two sides, which is an inside and an outside. There is a punt in infinity and a point reaching eternity.

When we look at the sound barrier principle that represents the same principles as that which forms the Titius Bode law we find this scenario is what really applies.

There is an outside and there is an inside, which represents infinity on the "inside" and eternity on the "outside". We have time on "both" sides dividing space and what we think of as space being outer space is merely time in eternity that formed and left space as a result.

The "universe in space is locked and sealed as time opens to allow space and then time unites to allow time to move. Time in infinity is where the cosmos starts and time in eternity is where the cosmos ends and when an object moves and only by moving can an object part time in infinity from time in eternity.

To understand the Universe is to understand this process. Everything is round with an inside and an outside and there is no other side in cosmic terms;

This means that all adding of space and material will be on the "outside" of every layer because going "inside we have infinity in that specific layer which also could be another layer of elements.

Clearly water carried the erosion that formed the slopes of the mountains. There are two possibilities, the water disappeared to somewhere or the mountains came from somewhere.

With the sun blowing heat in thick clouds over the planets, the heat that the earth and the moon detains through "gravity" is many times more than it is at present. Because the iron56 core will reduce most space, densifying heat, it will also retain most heat. As it cannot accommodate all the heat retained, it will relocate heat through space forming, to the surface. In relevancy to the iron-core the silicon is space, therefore rejecting surplus heat will bring about introducing excess space amongst the silicon. This is the same that apply when baking bread and the bread "rises" in the oven. The silicon layer "rose" as the heat, coming from above, as much as from below baked the silicon to rise. From the top, the water still formed vapour, taking all the heat that the moon and the earth hold as a compliment and then with the core of the earth being the dominant, directs almost all heat to its core, because the earth's iron core brings about most space depletion ("gravity"). Through this process where the combined effort of the moon and earth removes heat by space depletion a large area becomes effected on the space end … but, on the inside, at the time end, the iron core of the earth is the almost sole recipient of heat, leaving the silicon of the earth as the sole benefactor of space-incorporation (bread rising). Said in another way, the moon helped in doing all the work, but the profits of the work went entirely the earth's way. That even includes the vapour where oxygen and hydrogen combined through the excessive heat to form water. The vapour from the charging of hydrogen and oxygen, discharged again as the heat moved by lighting to the earth. From this water formed in abundance as the moon and the earth both collected, both stirred oxygen and hydrogen into a mixture, but only the earth collected the end product.

This process became as seasonal as winter and summer now are, as seasonal as rain spells and drought spells now are or as ice age and heat spells now are. Who would know the intervals, and the intervals are not important, because time back then is not time at present. The important issue is the evidence left in the earth

The earth is a mere 3×10^6 km from the sun. The first blowout occurs and the first day dawns. The earth's inner core develops and rises above the sea, for the first time.

The outlay of the Grand Canyon amazes one and all. As it presents itself today, applying scientific explanation it remains a mystery beyond belief. It is clear that water cut the canyon's profile but in itself it is not possible. There is just not enough water on earth to complete the task.

One can see the grand canyon stated off as a flat earth with a lot of water flowing on top of a flat bed while starting to cut through earth and rock. Then as seasons came and gone the water in space declined space to favour the soil gaining space. The riverbed reduced as the river got bigger and also got deeper. As this process went on the smaller riverbed made the flow go faster and this cut the rock ever deeper while the riverbed became narrower. The levels of river development show clearly a narrowing river in steps as cycles form by time. At first a wide river cut the earth flat but then the riverbed became narrower as it became deeper with faster flowing water. Rocks began to form above the river water line and the process kept on flattening some part while other parts kept sticking out more and more. The entire Canyon today is a history told by soil written in rock that time left us as space.

It is visible traces of it being a very ancient formation that paint the picture of the past. One can clearly not detect water playing any part in the history of the forming of such a mountain and then other part are there because water took away natural rocks. Going back to the example in the pictures depicted, mountains are visible on the horizon, and then clearly a new horizon came about as a new layer, which resembles this mountain remarkably. As the earth grew vertically upwards the river cut a trail horizontally side ways.

The mountains rose into space while the river cut a path into the rock bed and the one going up was a result of gravity just as the one going down was the result of gravity. However what is very clear about this is that it came by stages and every stage or era was different from the previous. The river did not reduce in water but in relevance as much as the rock did not gain in size but took more space due to relevancies re-applying to what holds space. The mountain rises vertically which is gravity while the river curs a bed horizontally which is also there because of the movement of gravity. It is visibly clear that at different stages the wind and water erosion flattened the then riverbed while kicking up water currents that was so strong it had the ability to flatten the earth. This

happened while the mountains that already rose from the riverbed stood and endured the onslaught of the water while remaining in form. This was the period the Bible mention where water and land parted. While the earth grew stronger than the water it left the earth as if the growth occurred in a "v" shape. This pattern came in place because the bottom grew bigger forcing the top to even grow bigger while seemingly remaining smaller. The evidence is there for all to see but no one can be as blind as they who refuse to see. Because the gravity at the time was so much more intense the flow of the streams that then was sea current was so much more devastating. Remember the water did not reduce and neither did the soil increase. It is the ratio of volumetric space that moved towards favouring the soil.

It is very clear that different levels of riverbed caused different riverbed rock formations where the river ran for quite some time and then the water influence reduced as the rock grew in height. We can see a horizontal line that formed a level at one stage and then it allows the rock to lift above the horizontal line just to have the river cut into the rock again but every time taking reduced space while cutting deeper.

This is all over the world and not far from where I live we have the same dynamics with the same results but not as impressive a picture as the Grand Canyon. In the case here by me the ancient rock "died" as it stopped growing and a very new mountain range started to grow but much less influential or impressive. This is the evidence Newtonian fools should research instead o going Hollywood by trying to show this one has a bigger penis by trying to show off fossilized rock he or she discovered and it is grander than that one and this one can make a better fool of him and his work that all the rest.

One can clearly observe each layer still in the precise order of development, as it came about rising from the ancient seabed. The water covered a far greater surface area than it does at present. The flat surface of the then riverbeds is so obvious and clear for all to see that it is unmistakably how the earth developed. It is only the Newtonian scientist that overlooks such evidence because Newton's pulling does not allow the earth to have such evidence. This is how the earth formed and this is how nature shows and explains how the earth formed whether Newtonian wisdom agrees or not.

The picture to the top might not be the Grand Canyon, but it surely proves a point. The rocks are (beyond any argument) sea washed and very obviously a rather great distance from and above the sea. Again, the formation shows layer growth not common with volcanic flow or the accustomed general portrayal of scientifically backed evidence in rock forming. This mountain tells a story which is not yet told.

This picture proves what a nasty place this planet was when life developed. Winds tore mountains apart while water gouged mountains away. In all these horrible circumstances life battled to survive but moreover to proper and that life did as we see how life flourished and developed.

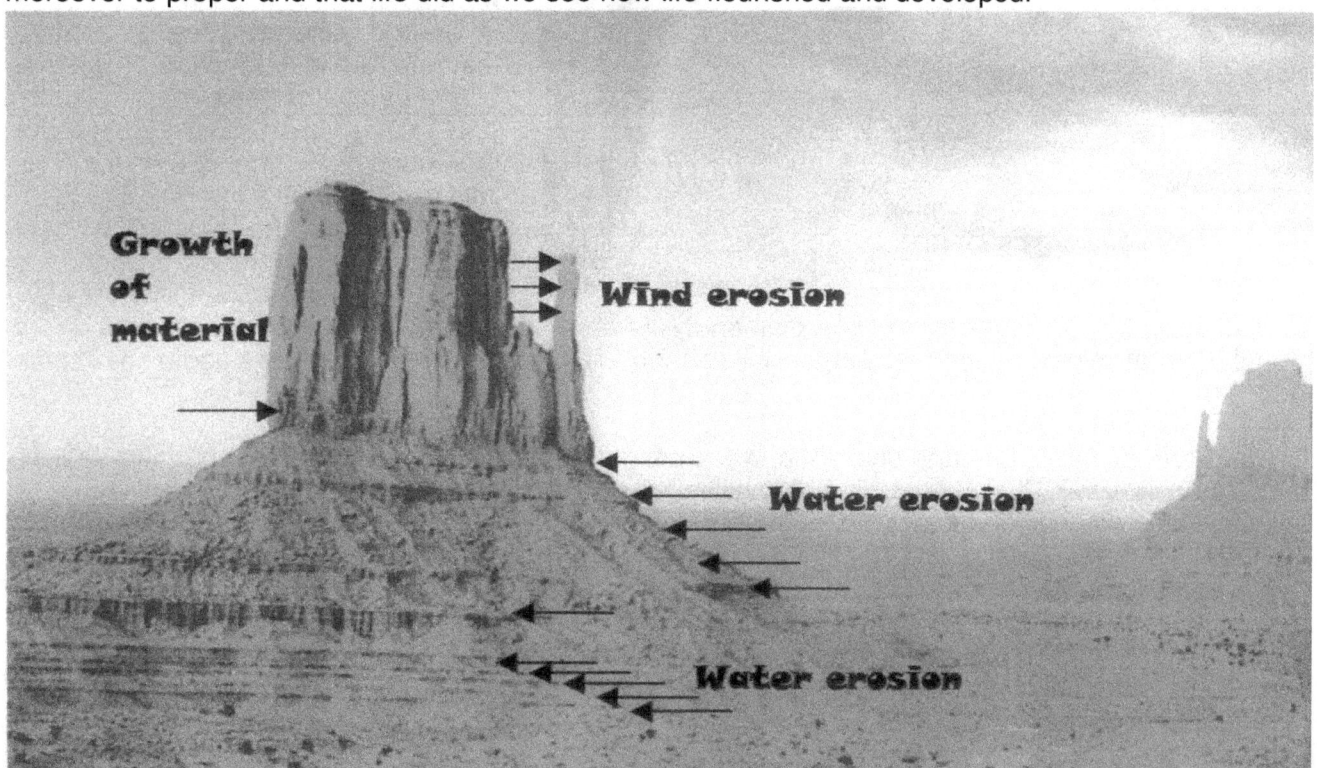

Imagine how a nasty place planet earth was as it developed during this time. With water cutting away mountains and winds blowing away rocks bit-by-bit by sandblasting the mountain until rock formation disappeared.

One also can see how the eras of development reduced because the layers or rings of development reduced as the linear growth became shorter. With all this evidence hiding in plane sight it is hard to believe that Newtonian science just overlooks all these facts and go on with the Newton myth of making up stories about how old the earth is and how old the Universe is and all the time is is so easy to disprove all their faking while nature is indisputable.

One might still call Newtonian science and not shit but to call this science is to insult the thought of science. I am going to call this Hollywood because this is what it is; Hollywood. Palaeontologists are not science but are gangster Las Vegas style. Then don't bend the truth, they don't even create the truth but defraud on a scale that makes Darwinists look like Sunday school teachers. They take a fossil bone that is hard rock, not even mummified carbon but pure rock. Then this rock they compare in size to a reptile bone of a living animal. They don't ask why this bone became rock or why did the proton structure change, no that will bring the truth and they are about selling whatever to gain money directly from that. Any means they find necessary they use notwithstanding the moral standard to defraud.

They wish to present a picture of size representation such as the example on the left while truth is that the sizes are our Newtonians versus the animal to his left. They take a piece of rock and compare this rock with completely altered an atomic structure with what we now have as a living creator. The on that evidence they go bananas and say how big this reptile was. They get so exited because an exited audience feeding on their regurgitated horse-crap pays top dollar to be fooled. They should go to jail and stay there because they are a threat to human kind let alone to science. The reptiles back then were about the size they are today but it is true that the earth in relevance also was smaller. But they were not the massive monsters Hollywood science tries to portray them to be just to cash in on the falsified trash they sell.

Diamonds came from carbon life. This implies a diversity of life existing on our planet at the time. The heat needed to create a diamond is immense. That means the early blow out was enormous in order to produce diamonds. However, the carbon product cannot come about through normal volcanic eruptions. Carbon will light up and burn to ashes. At a later stage, I shall come back to this.

The dinosaur, as all life on earth back then, was no bigger than life is in its present format; IN FACT IT HAD TO BE IMMENSELY SMALLER .The availability of usable landmass dictates this rule. Tyrannous Rex was about the size of a kangaroo, no bigger and moved about in the same way. Those enormous flying reptiles were abut the size of eagles but no bigger. The extremely dangerous and exceptional movable smaller raptors, was in fact, most probably insect eaters, about the size of the common house cat or smaller. I know that I am throwing a prickly pear around in the same way I did by announcing the fallacy of gravity. To be honest, I am not a person that would entertain controversy as a household pet. When there is that much disinformation, I cannot allow it to go on as truth. With this, I have unleashed the anger and dismay of yet another faculty, with an enormous wealth of brilliant minds yet again! There was only one day, a period of twenty four hours, that the dinosaurs died and all that remains to this day, is the dinosaurs that roamed the earth that day. Each sight represents one farness blast from the sun, and there were quite a number of such events, but in each case, all the dinosaurs that died at that spot became extinct because of one blast of heat. How many such events there were, is quite impossible to say, but because the earth is at a different position to the sun, every second of the same day, there were different spots that brought the dinosaur dynasty to a closure. How long the duration of time laps was, is quite impossible to predict but as time in space is relevant, the duration period the earth went through had an enormous difference in value. There are many leads all over the world, to be found, that points at this. In Mongolia, some dinosaurs were caught on their nest while brooding and the blast of the heat buried those dinosaurs with eggs, mother, nest and all.

At another place, some where in Australia, there was a hunting procedure in progress that was near a riverbed. The sand was fried to stone, leaving the imprints burnt into the sandstone.

The remains of the hunters and hunted is still on location and was turned into stone due to the enormous radiation that struck the earth at that time. The duration of time that followed this procedure was excelled in duration to any length of time, and this placed the sand and remains to age a considerable period in time. There are numerous occasions and sights where this took place and there are even evidence that during the increase of heat, some species had sufficient time to seek pastures in places that is considered today as being inhospitable. One such a place is in Alaska, but this creature did not outlive a later blast. No meteor crash had enough influence to bring about a growth in stone matter, which is very evident in all these cases.

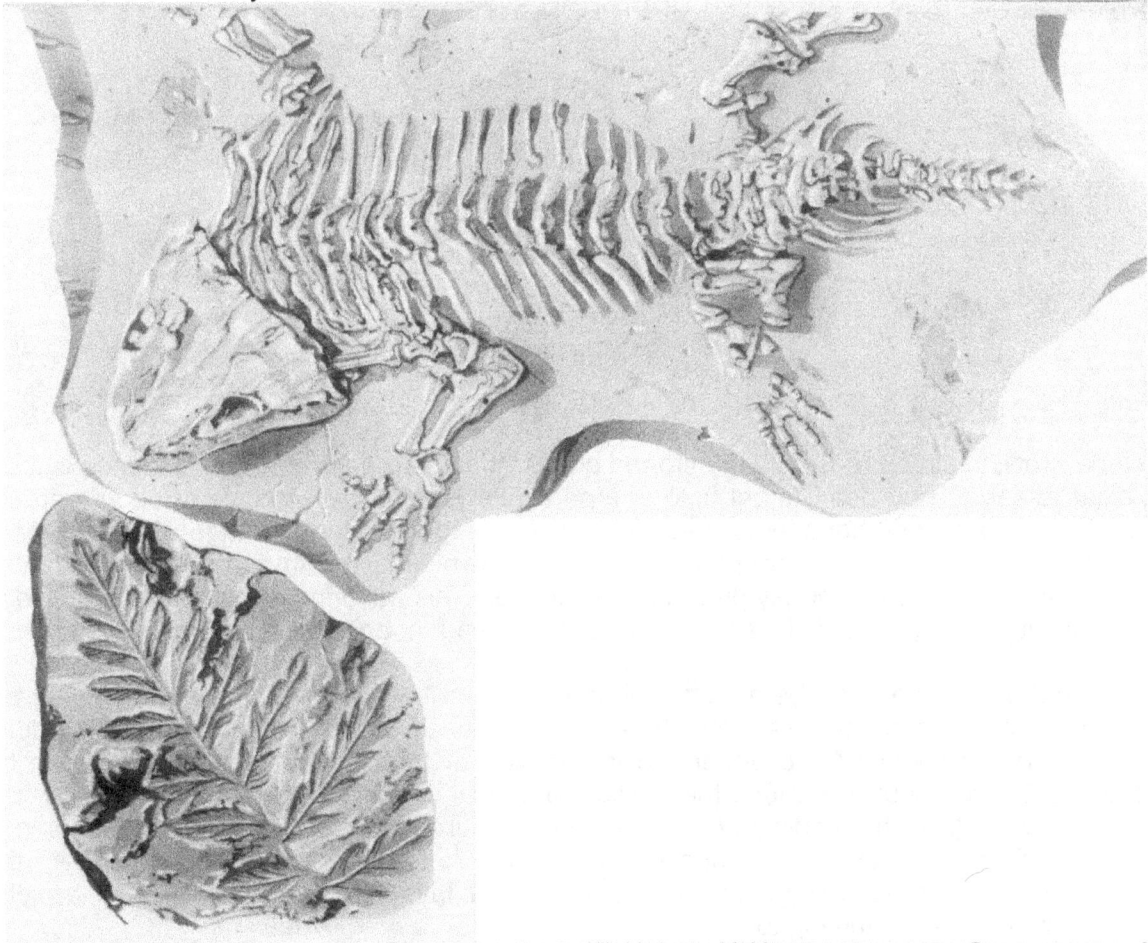

It is therefore clear evidence the dinosaur's demise did not last weeks or months, but rather seconds
A fact that I find truly surprising, is that the brilliant Super Educated in this particular field once again did not ask some obvious and apparent questions. For instance, how can footprints in sand, remain as it was? Any wind that blew afterwards should have covered those imprints. Any person knows that footprints, once imprinted in sand, covers by the wind blowing sand, in a short period. However, these prints remained there, for as long as it took to fossilize, because that is what has happened. Very evidently, the duration of time it took to fossilize, was dry, hot, and windless. This is an extremely uncommon scenario and with the intelligence involved in this study, suspicion should arise. However, it did not. This tends to raise my suspicion. Either the brilliant minds involved, were not paid to think, or they were not allowed to think, only to be SUPER-EDUCATED.

As the stone grew, in structure, so did the stoned bones of those creatures. The evidence is there and is quite clear to see. The scientists get hold of a particular bone or two, maybe three and around this evidence; they build a story that seems valid. Science loves to toy with the image of total carnage and all out destruction. That makes newspapers sell and that makes the image of archaeology sell. It promotes sensation-promoting money promoting what ever. This will bring about that the motive behind research is slightly tampered and slanting a bit oblong in favour of one direction.

It will promote the sensation, maybe not as deliberate as I portray the favouring to be in other parts of the letter, but still the temptation remains there to promote an effort that will promote funding. The scientist might be or might not be under the impression of total honesty, but nobody can be one hundred percent sure about how much the political or social or economical aspects that will stand to favour or not favour his or her findings, bring a strong flavoured influence directing the facts in one or other direction. Such a direction may bring that the facts are then somewhat misleading in after the what ever number of researcher going along the lines of favouring fashion will have a picture as obscure as that of dinosaur hunting today.

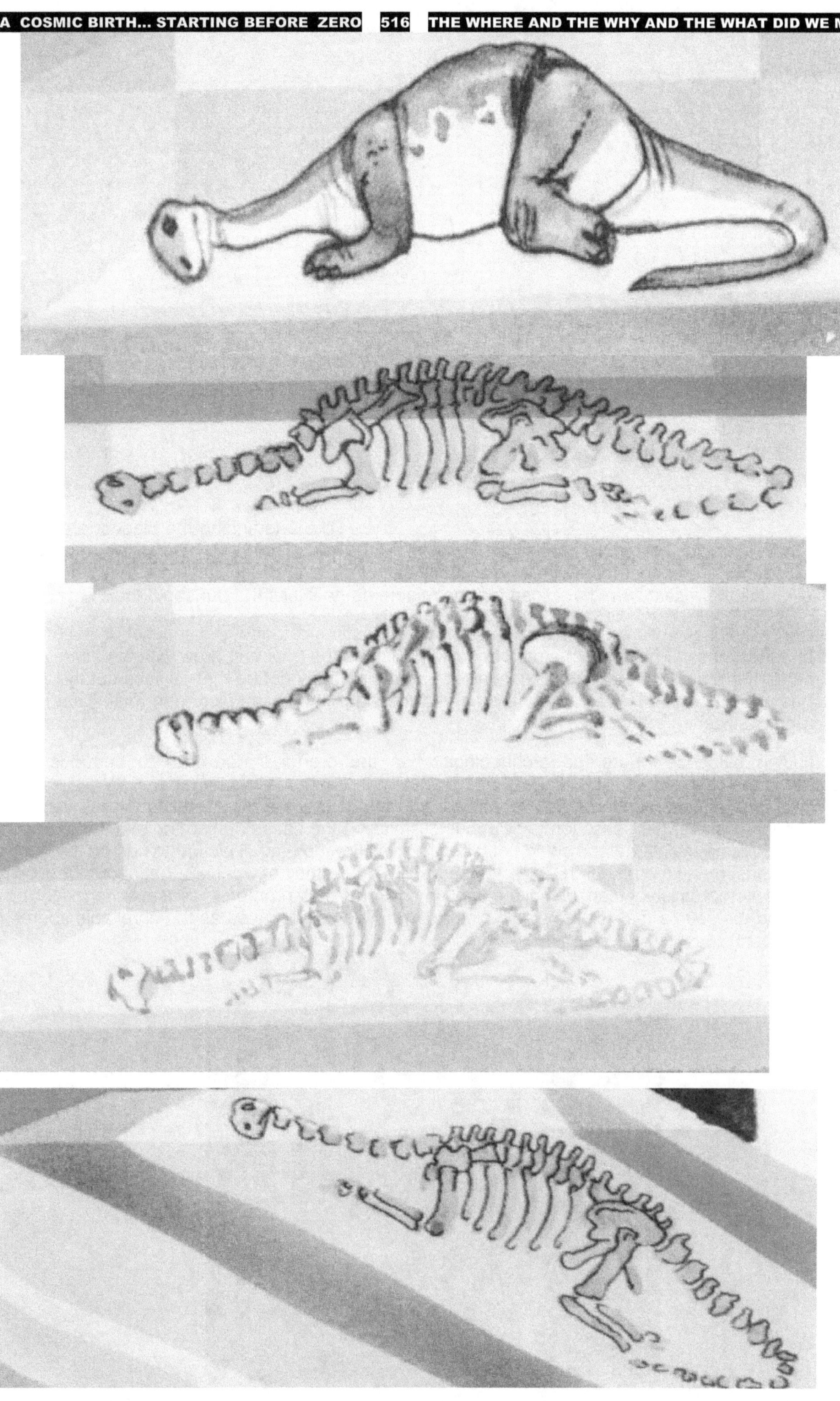

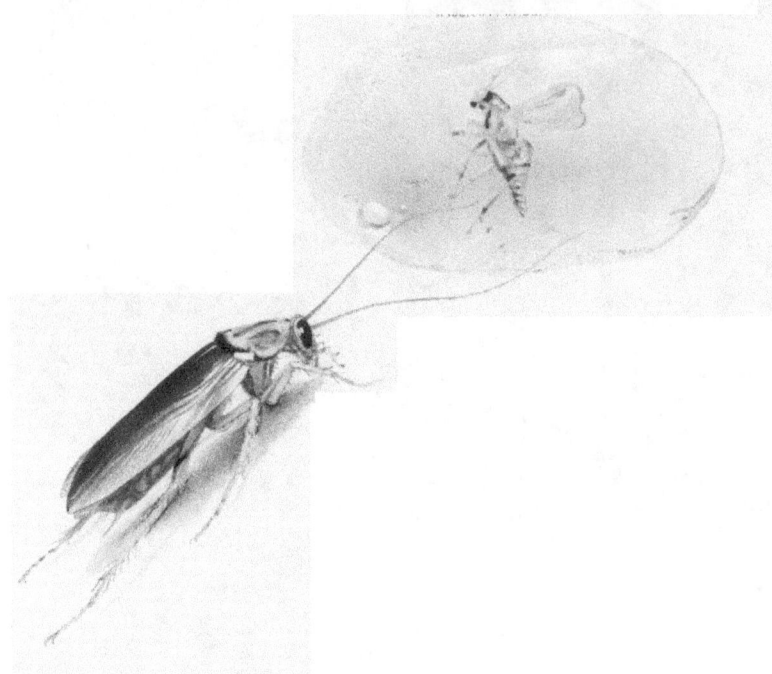

The one finding the biggest are the one pimping the others to the record of size and size is what sells. The logic behind the findings and direction it follows are of no importance any longer. I do not believe for the life in me that persons with such intellect as the researchers have, can for one minute not feel less truthful about their views.

The reason why insects were the size back then, as they still are today, is because all insects were burned to ashes. No calcium (bone) structure could withstand the heat and grew with the stone in size. In this matter there are other factors presenting it. The factors may not be that obvious but the work of research is just that; to find what is not obvious and translate that to the picture in total. That is what science and intelligence is.

When finding evidence where the archaeologist has the pre-meditated plan to beat his fellow researcher in the obscurity of his findings promote nothing more than a false image.
It is always about the worst events and very little about the cause, the facts leading to the events that brought about whatever the sensation is all about. Dismissing the cause behind the event of sensation to a meteor collision that in the end could not take place is allowing the full picture to ride on the back of the vehicle transporting the research. Why not then make a full study of what brought about the picture. It is far less compromising than being the fool in the end that promoted sensation all incorrectly applied... and the source that tell the whole picture, the Bible, is portrayed as the unreliable source of information.

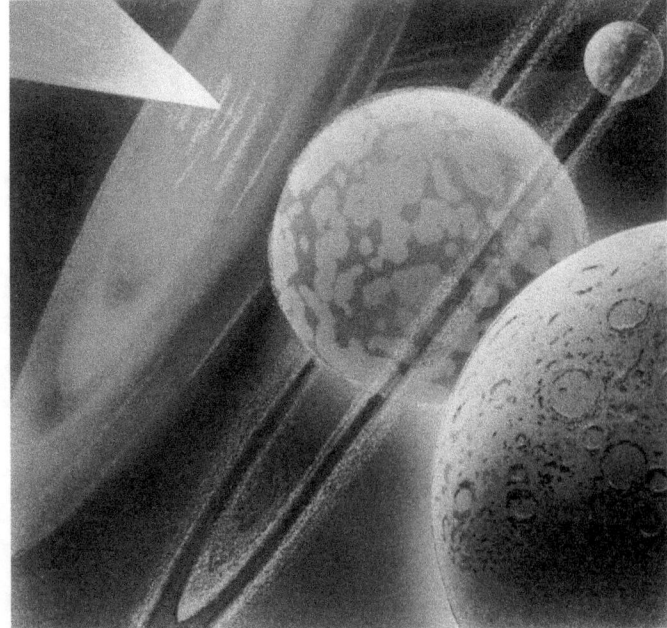

In all said, the worst yet, are not told. When the silicon magnetic space-time overcomes the aanplasing of the inner core, the development that takes place then, would not only remain to the value of the silicon layer, but all layers space-time will be released simultaneously. When this happens, the sun would as a star, almost double in space-time.

In this the relevance of the space-time ratio that exists between the sun and the earth will be effected, but I have no way in predicting the change that this will bring about in the orbit of the planets around the sun. Bare in mind there is time in space between the earth and the sun but the time brings about the value to the space that is effected. When this blast occurs, the space-time affect will be to such an extent that this could reroute meteors, asteroids and comets and lead to the past shower on orbital structures by these I.F.O. As the most massive ones would be the ones that is most effected, they were the first to impact on the planets.

During this phase, the orbital value would remain to the present value, but the duration of time contained in one year as experienced by life on earth would vastly differ from our own experience of time duration.

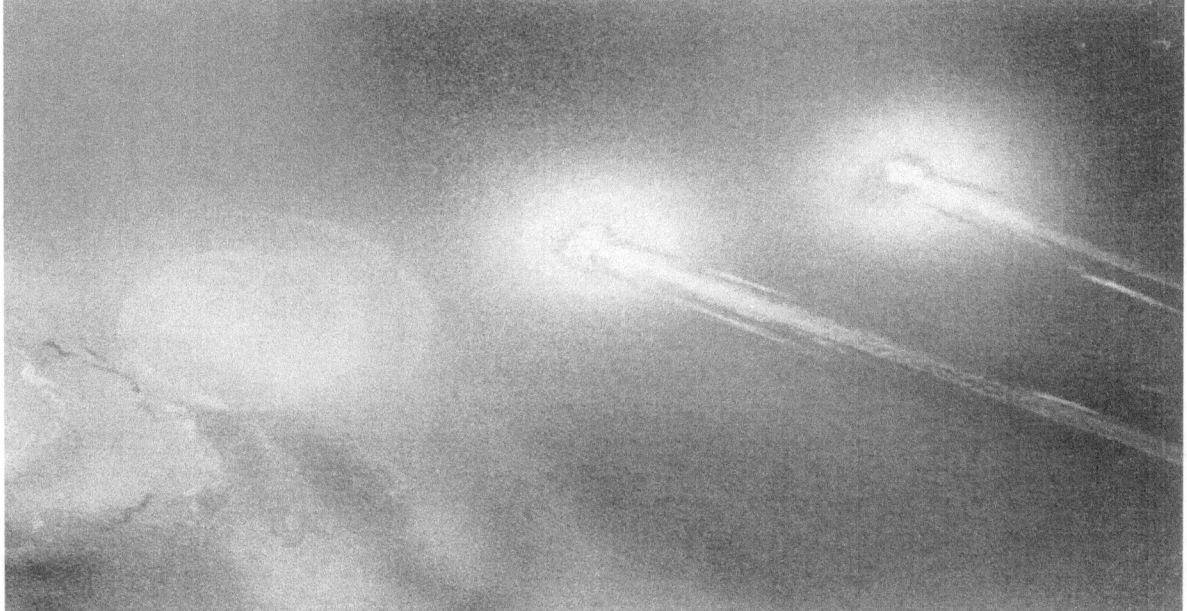

That things were tough back in the day is an understatement of the decade. Whatever was nature had one thing to work to and that was to kill off life before it began. ...And yet that made life hard as nails and tough enough to survive the worst conditions there was. In another book that I wrote on life I show my reasons why I am of the opinion that the earliest forms of life development the creatures of the time lived only as breathing entities feeding on pure magma that streamed from the ocean floor. There were two forms the one that separated carbon and oxygen giving the oxygen a supply of heat and the other form used the heat it received from the oxygen for movement and combining that carbon that moved with oxygen. This gave rise to plants and moving carbon life.

Therefore, as the orbital value of the earth in relation to the sun changed a most likely scenario would be that the meteor, which collided with the earth in that proximate period, was caused by the relative space-time differentiation. As it approached the sun, its route was not affected, but as it approached the sun and come closer, this direction of travel would have been diverted because of the increase in value of the geodesic space-time surrounding the sun and planets. Therefore, it is much more likely that the collision of the meteorite was a consequence of the sun outburst and therefore part of the disastrous events rather than the cause of the disastrous events.

I truly cannot believe any person in sound mind will attempt to convince some one that diamonds, oil and coal can survive liquid silicon flowing.

After this life had to start form fresh and all the evidence that remains are those unfortunate ones that experienced the enhancement of space-time firsthand. The rest are lost to time and space of an extinct era belonging to a period of days gone by. The extermination that leads to the demise of the dinosaur kingdom was not as predominant as the era's that produced diamonds, fossil fuel and stony forests. These eras that gave us gold, silver and other precious metals, were in a time long before the dinosaurs and in those events, life on land had to have been wiped out extensively. The shifting of the earth's crust plates would have been enormous compared to today's standards and even in the present time, the continental shift comes about because of the increase in densified space-time within the earth's crust. Compare the volcanic action that is experienced today in relation to the heat at is present on the earth and relate this volcanic action and heat to what has to prevail in a solar outburst when there is enough heat to create complete mountain chains and lift precious metals from deep within the earth to its surface.

Alternatively, compare our gentile climate in the earth's present state to a climate where the relative space-time turns complete forests to carbon rock and the silicon growth is so real, that it buries these carbon forests in the process. These forests turn to coal, because the earth's atmosphere, just above the earth's surface becomes so hot, the oxygen in the air pushes up and the layer that is just above the earth does not contain any oxygen. This is the only reason which I can think of that would allow the wood to be carbonised without turning it to ashes.

A great deal of evidence leads me to believe that the silicon development and the carbon to neon development is in fact the same occurrence. Should this be true, then the picture that forms, makes better sense. Because nobody alive ever witnessed this event, one cannot say for sure. If this is the case, then the first development was the most severe one. In this event, the aanplasing in the sun was still an extremely small factor. The presented facts to this point, I believe, are without any doubt. The following scenario could have taken place, which will sketch the development of the earth to a certain degree, but then again it could or could not have taken place. This scenario would only apply in the case where these aanplasings were released in a single duration each time. In this scenario, the ferocity of each event would have declined but the frequency of these events becomes more rapid. I shall apply the scientific accepted duration of time development, not because I agree with it, but only to imply an understandable scale in which events took place. At the same time, I have to declare that the development that leads to the 11 year cycle, must be the hydrogen / helium release of space-time because there is one such a development missing to conclude the picture. There then has to be helium / carbon releases of space-time that will alter the magnetic poles directional flow, and bring along ice periods. These periods will be on a more predictable and far more frequent basis, but in the same breath one has to say, by far less severe. The outstanding fact is that there has to be three separate releases and all three vary in intensity. There is the eleven-year cycle, the ice age cycle, and the devastation cycle, because in these cycles the surface of the earth's matter would be influenced, extensively. Only the time interval is in truth a presumption, but the actual scenario should be regarded as factual.

As promised I do not intend to delve into the chapter concerning the development of life, as I see it or to explain in detail why I consider the moon as the crucial factor, which led to the spurning of life. However, in order to understand what I try to describe in this article two important issues remain unexplained. The importance of these issues has apparently eluded the Super Educated up to now. They keep themselves busy with more important issues: as to pinpoint other life in the Universe, instead of recognizing why earth is the only place where life could establish in the solar system, let alone on other planets outside our solar system and that can only exist in their imagination.

The first serious occurrence took place o a scale of $2,5 \times 10^9$ years after the sun-started fusion. The presumption is that the scale of the sun life is 5×10^9 years. With the gradual built up of the space-time concentration, the time's versnelling would not have been that crucial. It was a gradual built up with a gradual inperking with not a big development reaction. The fusion was in place, but the inner core was on a steady increase, which helped the inner core to maintain its growth, and thereby the aanplasing managed to control the verplasing. All the planets would have been a lot closer to their present relative position and the sun's relative concentration would (at that stage) not have been as different from the planets as they are at present. The sun's effect that it had on the planets would have been less than at present, but the impact was still in harmony due to the positions of the planets.

As the earth was covered by water altogether, with no surface of land that rose above the sea, the sea could still have been dead and even if there were some form of life, which I doubt, the day-night factor would not have been significant. Divers, that submerge to enormous depths in special crafts, bears testimony to the fact in the ultra deep seas where life originated, there is no difference between day and night. The famed scientist, dr. Carl Sagan shows with accompanied photos, in his book "Cosmos" that trilobites at first did not have eyes, but the species later developed eyes. The visible change in the species development came about from the fact that they at first had no use for eyes, but later had to develop sight in order to complete in life's fight against extinction. If you regard yourself to be of an inquisitive nature, the immediate response to this would be: "WHY?"

After 1.5×10^9 years of development the sun rose to the occasion for the very first time. The very first radiant display had to be the worst one of all. This came about due to the suns under developed position in time and space. Up to this point, the aanplasing increased the time concentration in the space of the sun at a steady but persistent rate. In this, only one factor advanced in space-time, which was the inner iron core of the sun. This inner core laid claim to all heat that came about from aanplasing and a very weak fusion process. As the aanplasing grew by a constant maintenance in growth due to a steady increase in the ability of growth of the inner core, the fusion process grew in accordance for 1 500 000 000 years. This went about as the development of the magnetism came in contact with the hydrogen that was directly in the vicinity of the core which already was under extreme inperking and versnelling due to the aanplasing, fusion took place, as previously been explained. The magnetism was created due to a new polarized value again went into a state of verplasing. Light and radiation escape with relative ease because the time duration within the sun was relatively under developed. This led to an extreme gradual built up in the time factor within the space concentration of the sun. Then at a certain point in time and space, the time development became too strong for the aanplasing to control. As the development was became stronger than the aanplasing, it was overcome by the aanplasing, and therefore it went into a negative space-time displacement because the radiation, which escaped from the sun, had increased many times over. This increase produced more fusion but also escaped in a much larger volume, thus the radiance was extreme and this radiance covered the space-time concentration of the three most inner planets completely. The three inner planets, Mercury, Venus and Earth were at that point at a much smaller geodesic space-time position to the sun, than they are today. This means they were relatively much closer so the radiation blasts affected the inner planet much more extensively. This had to lead to an extreme growth of surface development in the case of the three inner planets and to a lesser degree the fourth planet. In this way, all water, and to that extent, all chances the life had, was incinerated to such an extent that it was blown off the face of Mercury and Venus, but life had its chance to begin on Earth. The radiant shower caused tremendous turmoil in the heated water and space-time revaluation (winds, electric storms, waves, sea current flows, magma development) was at a routine of daily endeavours. This period was extensive in duration as well as in expansion. Life had to develop eyes to help their progress and fight extinction. The most important factor of all was that the sun would never again not emit enough light to distinguish between night and day, as far as the planet were influenced.

After this, the sun turned relatively cold which brought about the first (although it had to be the second) official ice age. There was another steady built up for the next 1×10^9 years, and by this time there was an extensively better time concentration in the space of the sun. In this period, the fusion multiplied by a number of times in which the helium layer developed. In this period the silicon, as well as the iron inner core grew extensively, because of the growth period of the first sun expansion. In the first event, the sea absorbed most of the radiation, so that left a much smaller development to the earth. This time, on the other hand, the earth grew substantially; the seawater and oceans did not protect the earth any longer or to the same extent by the seawater as in the first case. Therefore, land, at places broke free from the surface of the sea and minute islands took their rightful place, sharing in the sunshine. During this period, microorganisms could have fixed themselves onto rocks that surfaced above the sea, but if this was the case, their chances on survival were exceptionally poor. The second ice period would have made their lives utter misery.

At another 0,5 billion years interval the time expansion within the space confinement became unbalanced one more one can safely presume that the inner core of the sun doubles with each of these dynamic displacement shifts. This would double the aanplasing; time duration in as much as half the space confinement, double the development and the radiant outburst too would be double, but for the interval, which then halves. So, in effect, the overall construction / destruction would be less than the previous event. The sun would now be 3×10^9 years of age, which leaves the earth $2,5 \times 10^9$ years old. During this outburst, the earth's landmass increased substantially in which young mountains and fresh water lakes formed that fed with ever so slow moving rivers. From this point onwards, recordings are on record of these occurrences, in the most precise way that is possible, and that is by nature itself. We only have to find a way to break the code in which nature had hidden the information.

Going about it in the present way, would be a waste of valuable effort. To do research in a manner to sell the story and like the publicity connected to it, does only disgrace the view that is held by ordinary people, like me, and millions of others which share my standing and place in society. Saying this, I would share some of the guidelines, which, to my mind, have to be possible. Firstly, look at the size of the average specie that developed during a specific period. Use this then to gauge at the surface of land that was available at that time.

With the outburst of the deep inner core magnetic space-time, an era of total devastation dawns on the solar system. In these periods, the space-time in the solar system grows enormous in the time factor, and this alters the space factor accordingly. What this means is that all routes that cosmic debris follow, completely changed their course.

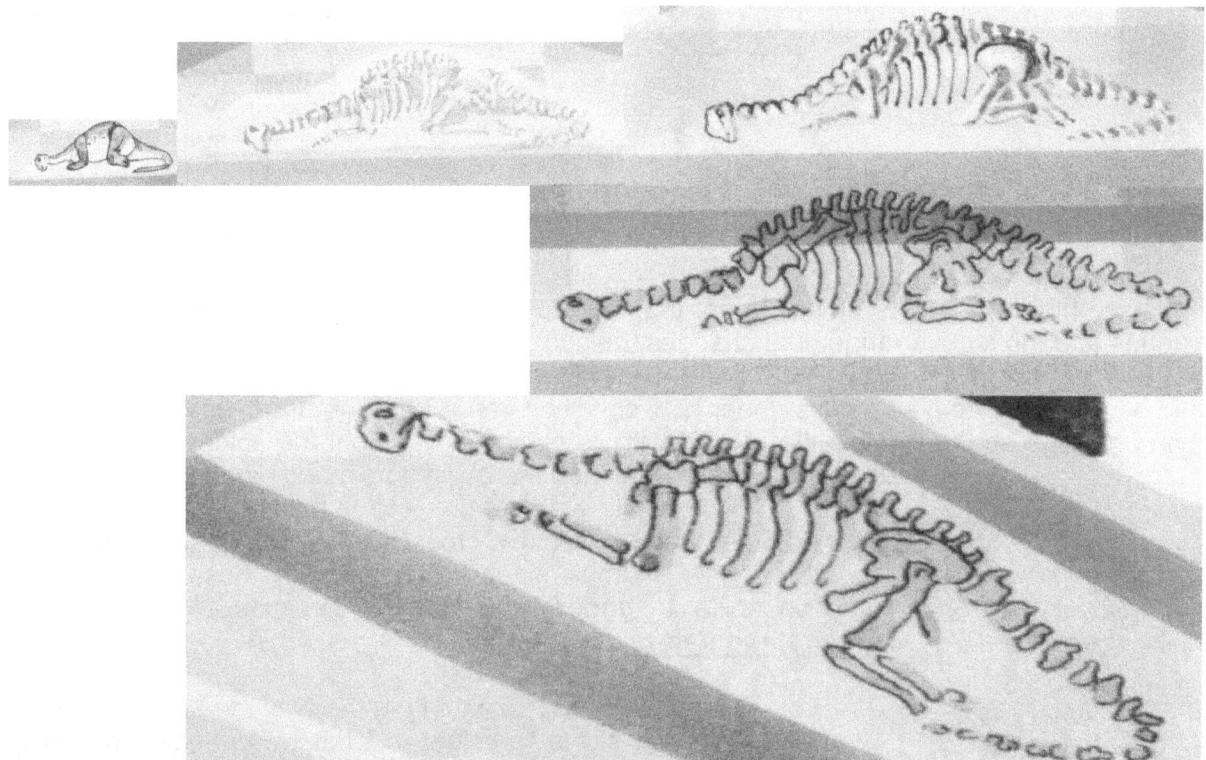

Take not that Newtonian science are famous for breaking the sound barrier at height where no sound or barrier of sound is possible. Newtonian science is famous for portraying reptiles coming from ancient times and condition where human life is not possible Life tells how earth development started and to find out how the earth began then just look at life and how life progresses by time forming space and putting down living conditions. For God sake stop being Newtonians with a single mind and only one intention and that is to cover and hide Newton's misconception with more Newtonian backwardness. Let present truth and cosmic evidence about the origin of species and what influences life to start with evidence and not cooking the books with suggestions like Darwin did. The following I can back up with science because I can prove the four cosmic pillars.

There is only ONE reason why life would be in any particular form or shape. That would be too prosper and survive under the conditions applying. Darwin is just more Newtonian deflection of their effort to hide Newton's hoax. Nobody in his right mind will support that half-hearted unproven garbage of hogwash such as Darwin corrupted. For one-thing insects do not share the same sex organs between species so how could they have crossbred and formed new species. Snakes did not have legs because the gravity applying at the time did not render condition in which to walk. Life developed legs after the earth's immense gravity reduced to achieve more by forming legs. Life is

where snakes could about survival and using what is at that moment applying to survive and prosper. Once a development is successful and needs no further development the specie will remain the same without changing. Then gravity comes along and in one instant the earth moved across from one half in solid state to the other half in liquid state and gravity changes gene. Only a few genes change from parent to offspring and a new insect comes about. The genes that change in some insects produce a range of different possibilities and development starts all over again. The previous species are still flourishing but crossing the half segment from one to the other brings a new selection and assortment in genes and the sting arrangement changes the code.

Only by applying remorsefulness under new condition can a new line of genes provide new life. From having no legs and being one the ground with its entire body weight the spider grew a light structure with eight legs. It developed the ability to sin and weave. Where did this ability come from... did it come from the stupidity of the atheistic simple mindedness where a snake's crossbreeding with a salamander got to form a spider. This is simpleminded madness only those with a poor mentality and an ability to **understand Newton** because of a lack of intellect to **understand nature**.

We can see much planning in the way the spider came about. We can see a foresight in what was needed to survival of the new species to fill a gap not yet taken by other species. There is development of a specie and then there is much development of the specie ranging in many forms of the same type but nothing is in cross breeding because this is development. Every era of gravitational changes brought about new forms of life and new forms within a specie group.

The scorpion is a scorpion that glows in the ark because there was the sun not being what we see today bring on a set of circumstances and that brought about darkness. To survive the scorpion stated producing an individual glow to compensate for nature not bringing light in that specific era where the scorpion as specie formed. It was flay and it was long and it was close to the earth. That means winds did not have that much effect on its survival and this helped in its quest to hunt and prosper. But it was hard times because we still have scorpions eating anything including other scorpions. This is evidence of ever changing conditions.

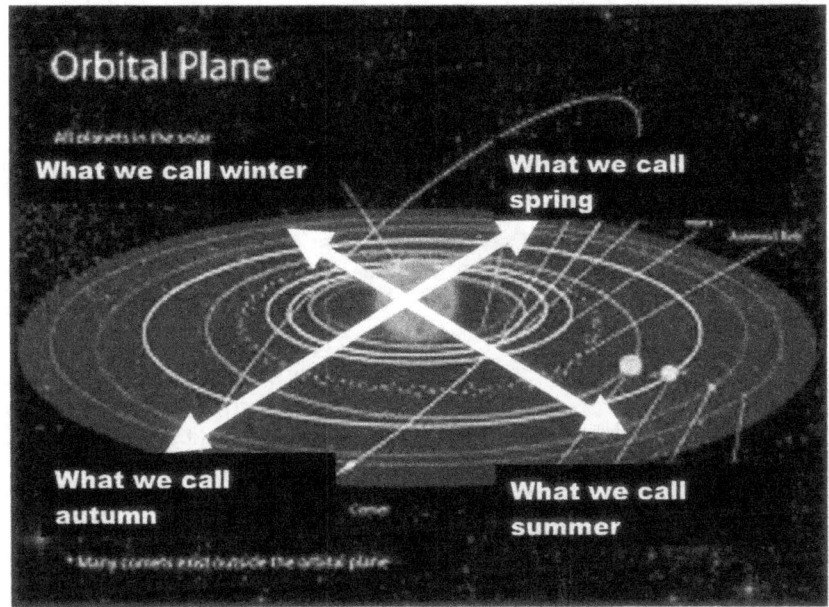

Just from being earthlings we know about seasons There are seasons that bring about cols and there are seasons that form heat and then there are seasons that take us from hot to cold and seasons the takes us from cold to hot. These changes in gravity we call seasons and these changes alter conditions on earth most dramatically.

Going from one season to the next season bring along so much differences it seems as if moved to another planet. That is because we did. But also we know we cross line. This line is unseen but it is there nonetheless and it changes everything ww think of in life. The line is there and it divides our future in to four totally different sectors where the one is an icy age and the other is a liquid age and then there is the two quarters that lead us into these changes. We also know these seasons are never alike. Every cycle brings along condition during the dry season or the cold season or the wet season or the hot season that is very different from previous years.

This is what we call the weather. It is weather conditions changing and it brings joy as much as deserters. This is exactly also applying during the solid cycle or the liquid cycle and the changes that affect us are even more dramatic that the cycle we live through according to the sun providing gravitational cycles.

Every cycle provides a new season and in some seasons one specie is more in numbers than another and the one season on form of insects become a plague and then you don't see that specie for a number of years. If this apply to the sun how much more with it not apply to the Milky Way enforcing adoption and adopting more forms. If we go around the sun in seasons then sure as hell going around the Milky Way has to bring about the same but on a much bigger scale. If going around the sun can bring about births and deaths then cycling around the Milky Way shall produce more species to form.

Seasons have everything to do with crossing singularity and breaking the four parts of space. But this extends much further and into the millions of years. There are ice ages every thirty one thousand years as we cross some undiscovered centre and we go into a liquid or a solid era. These eras we call ice ages where the ice caps melt just as it does in global warming and global warming has been going on the past fifteen thousand years. There is global warming but the process is gravitational and this will turn around in time as we go into the next ice age.

Every Milky Way cycle brought about new forms of the same specie while one particular specie reigned supreme. Then this would carry on for some hundred such cycles before conditions change so badly that a new completely different specie arrive. This is the result of gravity changing the layout of the complexity of the double Helix forming and this leads to a new development in species. Spiders reigned

for thousands of cycles before gravity cam and brought along insects. Then insects had the reign it shared with spiders until say scorpions came along. Scorpions formed the dominant specie on earth before some other specie brought new dominant specie. We have to look for conditions prescribing the outlay of the form of the specie. I can say there are longer been life on earth developing than what the Newtonian clan allows the Universe to be. To say the Universe is 55 or 13 or so many years old shows a serious misgiving about reality.

Life forming on earth supersedes 13 billion and 30 billion an 100 billion years of development. Their surmising of the Universe's age is coming from their inability to appreciate reality by which the atheist stupidity tries to form a reality but shows child like lack of comprehension about matters surpassing his or her inferior intellect. The crock began as specie much smaller but because it could use mammals also as a food supply it got much bigger through time while the gecko remained almost as small as it was is because it only had insects on the menu. What we see now is not what was and although the gecko for instance also became bigger as the Universe developed the earth and life on the earth the feeding of some animals in relation to others also brought along changes.

I just want to confirm about one aspect. Just like my brain is not enough to explain everything that goes on in physics and cosmology, and because I severely suffer from a deprived education so too can my littler mind never present even bits of the entire picture that form when going into this aspect.

However small my mind may be my eyes can see that there is a reason why the baboon has a long tail and the chimpanzee has no tail and that is because of gravity. The winds prevailing during the development of the baboon insisted on a long tail to allow balancing and especially keeping balance what serious winds prevailed. Those awful winds subsided as gravity declined and that allowed the next species developing to get rid of such a long tail. The tail remained with the baboon because that brought success to the specie and it formed intellect around managing a tail while the other primates that came afterwards did not need the tail to form an intellect. Then the next cycle produced new species and the tail was till absent because climate conditions prevailing did do subscribe the need for a tail. The winds died down. Remember winds are as much part of gravity as sound is part of electricity that is part of gravity that forms part in the dispensing of heat. Everything there is depends on gravity that forms space that prescribed conditions applying.

If we look at the baboon it is obvious that this animal is far better equipped to deal with winds that torment and blow that the chimp will be that became even more bulky. In the beginning all life developed close to thee earth and with the ability to handle severe conditions much better. As life went on from species to species we can see that severe weather became an ever lesser problem and food diversity became more prominent. Diversity in specific species also became where individuality surfaces more prominent. If you look at insects they all share the same form of life and the one are just a duplicate of those that came before as much as they remain as the specie comes about the next season. There is no diversity to speak of and every one is a blue copy of what the specie is. One cricket is as much as the other and one bee is as much as the rest.

As species developed individualism also came a factor. We can recognise one elephant from the rest or one rhino from the others as life got an individual trademark added. I won't be able to say when personality development became factor but personality differences is very obvious in species that came with later development. I think a shark is a shark and I do not wish to study if there are different personalities roaming the seas in same specie sharks but dolphins are absolutely individual seasonality programmed. With this personality diverting also came intellect as a factor where some are more aggressive than others and some are more tolerant than others. With all this I just wish to show that life developed with time and gravity allowed this development to take place. As the harshness in conditions reduced species became more flexible and if we want to read how the earth developed we have to follow the record that life left us to read. If there is such a study for God sake keep atheists out of it. With the stupidity and blindness the atheist shows I am sure he or she must calculate how to shit because they will not otherwise understand the process. We have to remember that gravity forms the Universe by a mathematical process of taking time in the present from the future and then passing it onto the past. This also is time and time is the thoughts of God but that I do in the next chapter. The Universe develops according with a very strict flow of time and a precise process.

Gorilla
(Gorilla gorilla)

Bonobo
(Pan paniscus)

Orangutan
(Pongo pygmaeus)

Chimpanzee
(Pan troglodytes)

Gravity changed so much that it allowed humans to walk up right while apes still walk on all four legs. The winds dies down and this brought along that life could pursuit intelligence by fighting nature less for surviving.

But what is more than surprising is the witlessness of the **Palaeontology**. Then I hear of thee discovery of this or that specie. The creature had a jaw that could swallow a man or a horse whole and with one gulp and to top that the specimen with THAT jaw was still a baby. Have they no minds. Are the lust for sensation and deception so big that it clusters all their ability to think? Or do they just think we the populous are so stupid we are unable to think?

What space will an animal that size take up? What space will the food take up to keep this hundred-ton monster going? An Elephant eats hundreds of tones of forest in a lifetime and I have seen the devastation on trees and bushes a heard of elephant can cause. That is why they HAVE to cull some to preserve the bush for other animals. The Elephant graze wood and bark which no other animal can digest. This monster that as a baby can swallow a horse with one go will have to eat several elephants per day because it takes a lot of food to keep such a huge animal going. An Impala eats one point three time the fodder that a cow does and this I know from farming with game. This is because the Impala is on the go all the time and requires that much to graze. Remember also then Impala is little bigger than sheep but eats a lot more than sheep and that goes for all other game.

The fact that they keep moving makes them eat a lot when compared to domesticated animals. Then they come along with such rubbish just to get on TV and to get money? This criminality and deception is sickening and that they do in the name of science because the name of science became deception. You can cheat and lie and mislead as much as your heart desire as long as you do it in the name of science. Have they no brains or have they no conscience or is it purely a supreme feeling of being god so you can say anything to the brain-dead population and you will be believed! Have no senses or have they no shame because this goes beyond being civil. In science you can bullshit your study to any level as long as you create a total unbelievable hoax and sell it for the truth.

If all these carbon animals turned into rock over the years it means these animal fossils are no longer carbon and bone. The atomic structure changed and became a completely different atomic and molecular set-up. The size of the atoms change but also the atomic number changed. That proves that heat in accumulation over a very long period of time made the atom structure develop into something bigger than what it was when the animals were alive.

This is not that difficult to work out because even I managed to accomplish this feat. Atoms grow by accumulating heat but not only do atoms grow in structure, also is it clear that atoms grow bigger in atomic proton number as heat through gravity become more through gravity.

This is the fusion process science is so desperately looking for. There is no fusion through pressure rising atoms to weld together. As the heat flow from the liquid / gas part of the Universe to the solid material part of the Universe we have an accumulation of heat bringing not only size gaining space but also the proton / neutron and therefore electron numbers becoming more in quantity within the same atom.

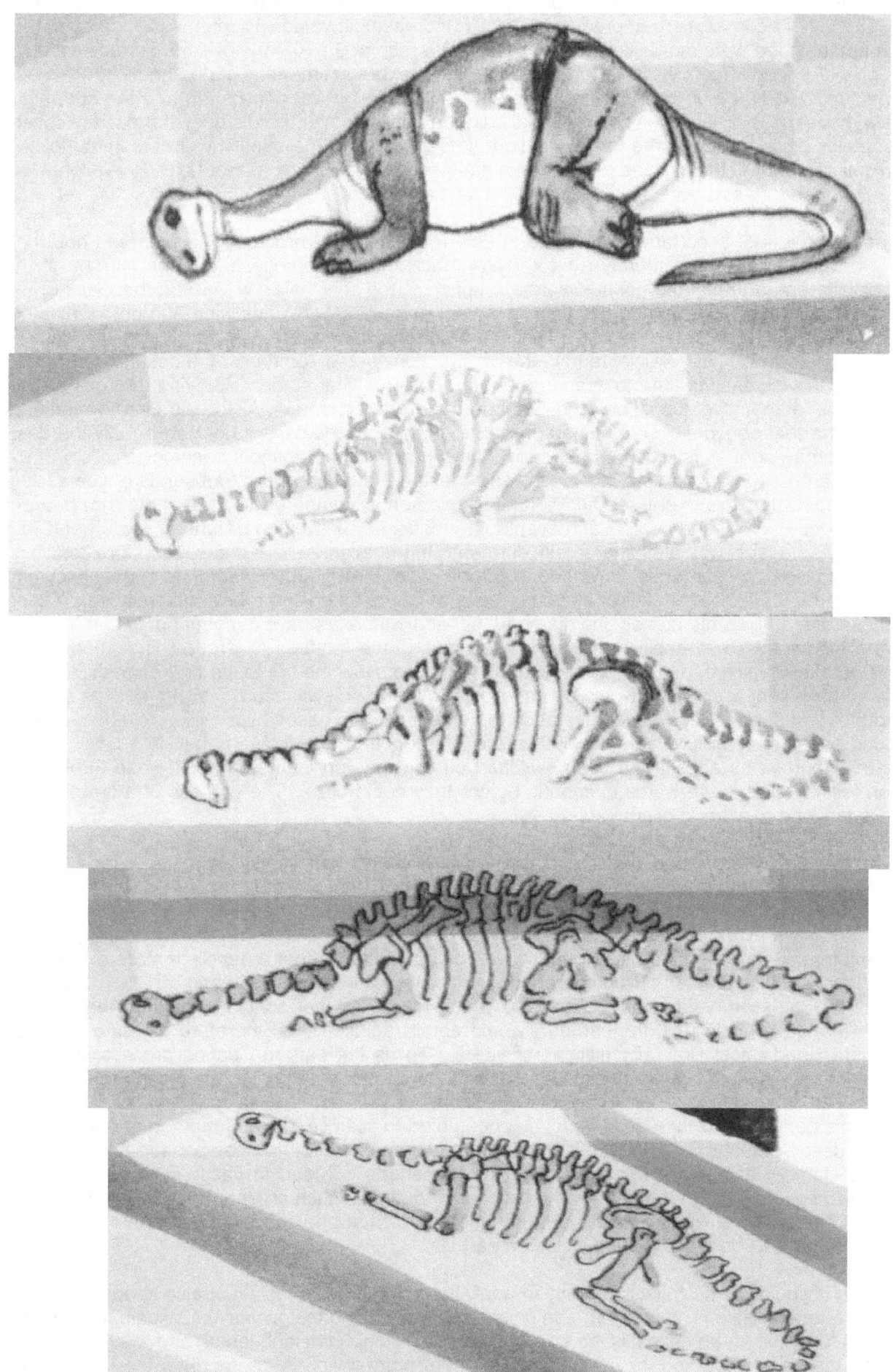

The cosmic debris comes from the outer regions of space, travelling undeterred, as they have done for thousand of billions of years. Then as the debris enters the planetary orbit route of the solar system, its path progressively alters, as the time component grows in value, and the space component reduces. THEREFORE, **k** becomes more unpredictable, as **T** increases in value. Take into consideration that the earth at that stage were more likely to be covered at a ratio of 9 : 10 parts of earth surface covered by water than 8 parts water in ten parts atmosphere, but in time, has gone unnoticed.

When the sun experience an outburst the temperature (HEAT) between the earth and the sun might increase a thousand fold. In that event, it can take light a year to cover the distance it presently covers in 499 seconds. This will place the earth one light year away from the sun. The accumulated loss in photon displacement will through the earth in total darkness, hence bringing about a Biblical Night. The earth most probably was as hot as hell, still totally dark.

The big factor that is of importance is that the structure that hit the Mexican peninsular, did not bring about the devastation but was merely a victimized participant in the process. By the time, it hit the earth not all life on earth that did not survive the eruption of the sun, bared witness to the event of the Mexican peninsular disaster. During this time, the plant and animal life, was wipe out, not in minutes, hours, or days, but in less than a second. This is if one applies the current value of time in space: as seen from the outside. In relative terms, the destruction lasted for tens of thousands of years, depending on the increase in time to space ratio that applied at that stage. Studying the rock growth that came about at that stage could only indicate how the earth recovered. All the ultra heavy elements that form during that period which remained on earth, came from the deep inner part of only the deep inner core of lightweight stars, such as our sun. The remains of the animals became fossil's, in fact, more stone than bone, within a second or less, which lasted thousands upon thousands of years. It is important to note, that this event seemed lightning quick once compared to the events that brought about diamonds, fossil oil and coal. Each event just mentioned, happened billions of years apart, but the duration of time laps, was almost eternally different. In the event, where the wooden forests, which are $carbon_6$, changed to diamonds, which too is carbon, went through a relevant time discrepancy and the duration of the period lasted, millions upon millions of years, taken in a relevant time lapse. A thing that all these different events namely the diamond, oil, and coal era's have in common, is that the time lapse were increased so extensively there was too much time for fires to occur. That means, the period that it takes for a fire to ignite, was too long in order to allow the fire to start. If I can explain this by means of a present scenario, it would be as follows: When one switches a T.V. off, it takes about a thousandth of a second. If at that very instance the time lapse increases by a million fold, that means one second becomes a million years, it will take the T.V. too long to switch off, therefore it will remain on for a thousand years after it has been switched off. In the very same manner, wood turned into coal, then fossil fuel and afterwards diamonds before they could burn. The process of a fire burning, would take too long to actually occur.

During this period, where wood became diamonds time stood almost still, and this period will go undetected by the methods science applies in dating rock in the present way. In order to explain myself, I would give another example. At present, the dating of $carbon_{14}$ has a half-life period of say 49 years. This means every 49 years, the emission of radio- active particles reduces by half its value. Therefore, if it emits one eighth of its original value, the estimation is that a period of more or less 400 years has lapsed, since the $carbon_{14}$ came about. Now, time slows down to a ratio of 1 to 1 000. This means that the radiation decay would last 4 900 years. Not true, the radiation would not be emitting radioactive matter at all, in fact, the $carbon_{14}$ would admit radiation, and therefore the time in space value would extend (most likely) by millions of years. To us the $carbon_{14}$ dating process would be completely inadequate. The very principle applies to the silicon dating process, although this process would not be that inaccurate. As the silicon grows, it too will admit vast quantities of radioactive material. This period would then go completely undetected to science, as their method of dating relies on only the decay of matter. Time would only then apply, once this period has passed and the earth fell into the slumber of another Ice age. Even these periods will lead to vastly inadequate dating, because snow covers most of the earth. The process of dating is such an enormous prickly pear, that it is best left alone. What is extremely important is to realize that one cannot rely on the size of fossil material to gauge the size it has, during its life on earth.

Furthermore, it would pay scientists, not only to apply the in hand given facts, but also to apply some of its better-developed instincts, called common sense, before jumping to conclusions. Again, I would like to stress the following. If an animal is the size of a modern bridge in height, length and width, taller and bigger than the trees it lives on, how will it survive? Yet more telling is the fact that the food source would not be able to sustain such life and survive a few hundred animals that devours up to 15 trees as a morning snack.

The concerns of the sun's influence on the life on earth cannot be overestimated. The sun admits a certain volume of geodesic space-time as aanplasing. This space-time then becomes a normal process as building material to develop the sun's growth as the sun might not be an average star, following an average path. Nevertheless, it is a star. The majority of space-time emits back to the geodesic value as magnetic space-time. Our problems are not that much in this emission of space-time, as the earth can cope to a certain degree in protecting its life. The problem arises with the "release value", (if I may call it that), a mechanism in the sun, as it is in all stars.

There has to be reasons connected to certain trends that various species followed to secure their continuous existence on the earth. For instance, there has to be a reason why bacteria diversified their genes and not grew to enormous sizes. This could only have been forthcoming because of the area available, the size of their food supply and the climate available in which they could progress.

If you compare this with the way insects developed, you will find during the insect culture, land was more plentiful, compared to the bacteria era, but it had to be much less than in a following era, say up to the dinosaur. One thing always struck me as being very odd. In the dinosaur period, no species lived off insects, although insects had to have been as abundant and diverse as today. Not once did people in this field of research ever put this question forward. No one had any inclination what so ever, to question the enormous sizes that that beast developed to, and relate that to the supply of food available.

According to our Super Educated, scientists' T. Rex was the size of an ordinary three-story house. It could move at the speed of a Jaguar; it was as agile as a meerkat and above all of these super abilities, was strong enough to take several elephants to task at the same time, where on earth, did it get a large enough food supply to sustain these various unbelievable qualities. Any animal that is twice the size of your standard, garden variety, home broken elephant and could match in all of this the speed of a wild cat, must have burnt enormous quantities of food on a regular basis of say about 12 meals a day. On that basis, they would have eaten themselves out of existence within a few years. Amazingly, they did not, on the contrary, every so often, they discover another rival species that match, and even beat them to size, strength, virility and what ever you feel to add to the list. Strange as it may be, there is never any mention of dinosaur herds, as can be found in zebra blue and black wildebeest and impala herds that flock to new grazing in Africa. There are always only individuals found or at best less than a dozen to a group.

It might pay our distinguished SUPER-EDUCATED-KNOW-THE-LOT in the end, TO PAY ATTENTION TO OTHER DETAIL. Once they are a little less educated and more inquisitive spurring their minds to think more about their facts and less about their SUPERIOR-EDUCATED-STATE-OF-MIND, they might observe other more telling indicators than a few "*rockyfied*" silicon "bones".

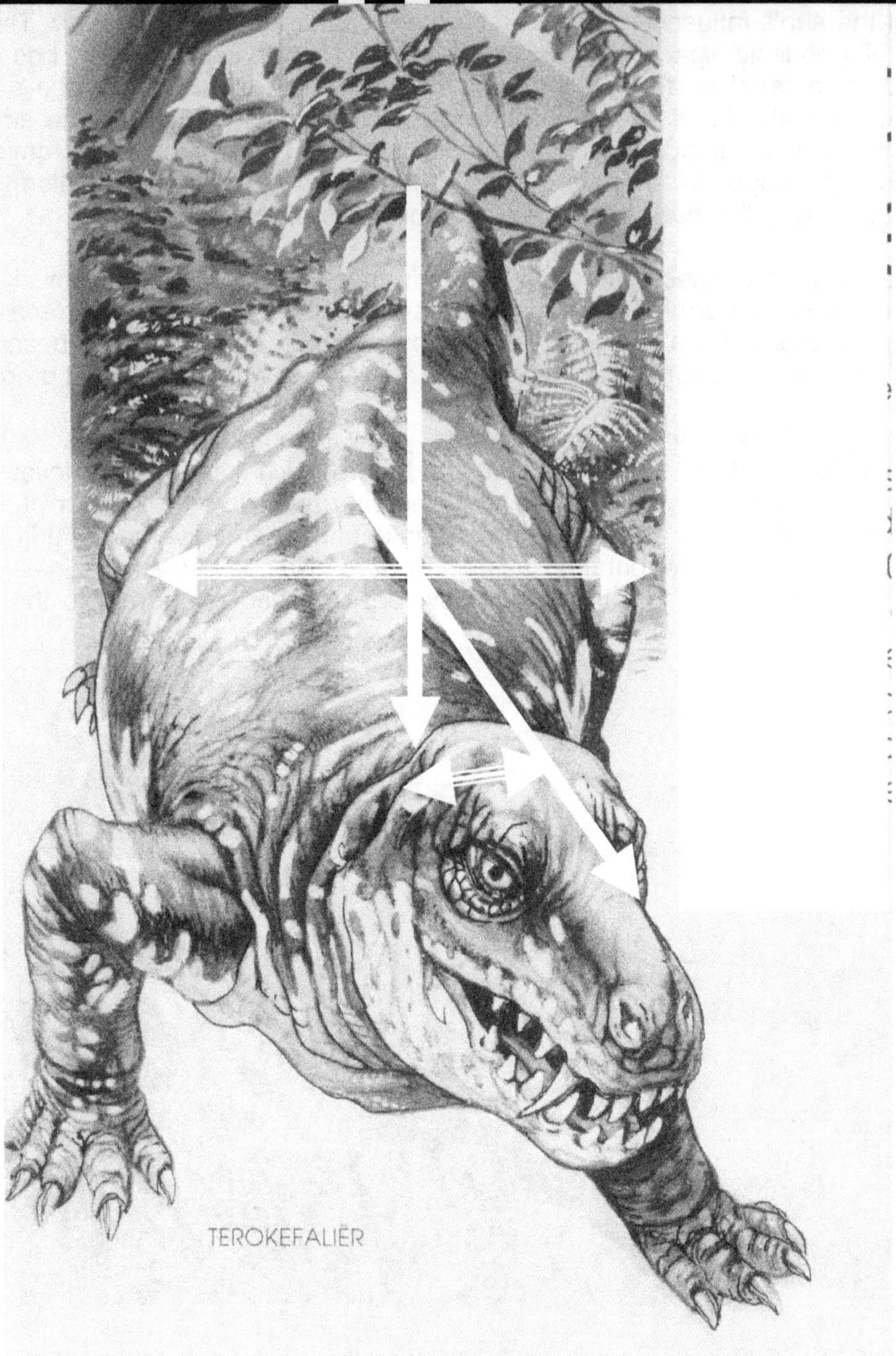

TEROKEFALIËR

All dinosaurs had a long had a elongated shape with an enormous tail

REPTIELEIER

SELAKA

THESE SAME CHARACTERISTICS BELONG TO FISH. With fish, we know it is to streamline its movement in water. another clue in dinosaur is they all had very strong legs Thirdly they were fitted with those large tails. All the characteristics mentioned point to one exclusive indicator. They were in a struggle for balance. The aspect, which would influence their balance most, will be strong prevailing winds. Strong winds come about as the result of flat landmasses and huge coastal areas

During the life of the dinosaur, land was at a premium to water. By comparing species and the position, their development places them in relation to other specie development, we find that life started with animals hardly bigger than a molecule. As life progressed, the various species became larger. Yet, this does not tell our SUPER-EDUCATED-MASTERS-IN-FACT anything. Life depends on the availability of space to survive.

Dinosaur back then was the same size than they are today. The dinosaur had a lot less land to live on. Species develop to a point where they find their position secure, then they stabilize with little change to follow. The dinosaur did not increase in size but kept their size they still maintain today.

Their table manners also did not seem to improve a great deal since back then. In the Afrikaans book, I went into a lot more detail since I devoted one chapter to the aspect of life. As that chapter reflect some of my religious believes I felt it mite compromise the true reason in translating "**MATTER'S TIME IN SPACE**", as the book might leave the religious aspect to more significance than I would wish it to be. This way there can be no mistaking to the scientific base being the true reason in the translation and thus will therefore, not be compromised. However there is one more aspect on life I wish to raise, as this factor sheds some light to the meaning if this chapter.

All of the misguidance I put down to the fact that these scientists had to sell their product in a market, where each faculty wants to outperform the other faculties, is it for fame, money, recognition or plain old-fashioned vanity. It is about going Hollywood and not seeking the truth because a bunch of morally deprived atheists are in charge of science and there goes honour and truth to hell. If they are atheists then there is no sin and sins there is no sinning we can see the why those morally deprived atheists guide science to become the den of criminality that does not seek truth but sells out for money. They have no honour so they cheat and mislead and misinform just because they believe there is no God to answer too. Then also there is no public to answer to and that is why they misrepresent the truth in so many facets of what they call science.

The best way to do this is, to mesmerize and stupefy poor old brainless John and Jane Dow who is easily astonished. When they are flooded with far more information than they can grasp in a short period of time, John and Jane Dow will gladly pay taxes in order to be presented with a better and more unbelievable scientific fairy tale, while in all this the Mammonites, rake in the cash from all possible sides. I am about to make a statement that I know, will scare me as a branded racist, until kingdom comes. In the opening articles of this book, I said this book is all about the truth, and so it is. Should I then, become disfigured for life because of factual truth, and then so be it? Those that do not favour the next paragraphs, I dare you, prove me wrong!

The fact of the matter is that the Negro race has been on this planet many, many millions of years longer than does any of the other two human races. This is clear from the hair, the bone structure and the colour of the skin. The Negro race is not black in colour, just because they were part of the hot African continent. The Negro dates back to a period when the sun was not merely as developed as it is today. The sun grows each day, so does all the planets to a lesser extent and this applies to all other stars to a lesser or larger extent. At this instance, we are concerned with the earth, which has the position in space, and time that has its fate totally connected to that of the sun. This places all forms of compounded matter totally in relation to the sun. The form that the Negro's hair has taken on, relates to the fate that the sun has become intensity larger than it was during the era when the Negro developed as a species. All evidence indicate to the fact that the Negro was at first as pink as the white race is today. When the skin of a white person is examine, you will notice that such a person burns black dots, which is known as freckles.

As a South African farmer, that is outdoors most of the time in the blazing African sun, I put up with much more intense heat than does the average European or North American. Therefore, a white person's skin is more often darken by the heat of the sun and is so-called "tanned" most of the year. After each such "tan" during the summer, there is evidence of more black spots called freckles, all over one's body, but this is more effected on the parts that is unprotected by cloth, such as the arms and legs. In time, this effect of dots on the skin has to become part of the genes that carries on to one's offspring.

Through thousands of millennium, the overall skin pattern has to become more covered by these black dots, and will eventually lead to an overall change in the complexity of the skin. This is a built-in mechanism to protect species' survival in a forever-changing habitat. When a white person's hair is heated, you notice the very same spiral effect that the Negro's hair must have under gone. This fact I realized when I stood in the local hospital and saw two patients being rushed in after they were shocked by an electric current flow of thousands of voltage.

The body is a tool life produces. Life is not the body and I wrote a book on that. The body is what Life produces and that is why every life form on earth has different DNA that does not match. Even the smell life leave behind through the body life creates is totally unique. The body forms as a result of life duplicating what was by replacing ever particle in the body every seven years. This too we witness in the way nature builds the cosmos and that we see in the way the earth develop. That is what the Titius Bode law shows how the material side accumulates density and structure and cosmic space deplete by releasing substance. Life had to develop a structure to host not intellect because I believe we sell the intellect of animals very short but to host a body driven by reason. The life that build the body took million years to form a body that can be in search of reason. The quest for human life is to live to learn to learn to live and that is finding reason in everything. Looking at the body developing we don't see a body changing functions but reason developing the body not to be supreme primates but to be supreme in thought. Then some F$^%@CN fool invented money and once humans got money driven they became animals such as the case now is where Europe / America grab the right to plunder, maim and kill Arabs in the name of liberty and freedom while advancing democracy to further the drive for peace so that they can have Arab oil. Now in the same way truth in science and religion are sold to the highest bidder and science and religion can cheat all they want as long as the price will compensate.

Looking at how life developed to be humans today draws a picture of the development of the mind and forming a body to match. It is getting a structure in place that would be able to hold that which all life should be and that is human. But now humans became slaves of money and money will destroy life eventually because life sells all properties that make it special for money and for what money can buy. There is a tragedy looming because a point will come where the growth in populations will lead to a disaster that the human mind can't understand. We think we can but even if we develop another million years at the pace we have developed this past century we will still be lost in understanding the cosmos and the layout thereof.

The Negro race was through one of the more intense solar blowouts. They did not receive the black hide because of the blowout, but from the sun's intensity that accumulated from the period they first developed as a species to now. This astonishing factor I realised years ago when I saw electricians that was hurt in an industrial accident at our local power station.

Since I know that electricity is merely the relocation and intensifying of space-time which is redirected and artificially concentrated time and is relocated to a much smaller area of space, the effect on the human body had to be that of pushing time in space forward by a few million years. The persons that I saw in that hospital that day, had every aspect of their complex changed to precisely that of Negro's or as they are referred to locally as blacks. Only afterwards did I learn that both men were white persons. All aspects of their body, except the colour of their eyes, which was close for obvious reasons, were the same as a black person would be. Even the aspect of having the surface underneath their feet, still the pinkish white as you find with the blacks in every aspect these two men turned black. Even the hair was a spiral Negro version. The only conclusion that I could arrive at was that this very same time intensifying took place in the case of the blacks, but in natural time expansion and under natural conditions.

In this letter I have indicated on many occasions using an array of proof that the only difference between electricity and gravity is the intensity of the heat concentration. The rest is the same thing; a dimensional directed flow of polarizes heat directed to the point of singularity within the earth and transmitting along atoms that provide a temporary linking to singularity through conducting. The evidence is every-where, but unfortunately, no one is as blind as they whom do not want to see, because of certain political motivation, and a guilt complex.

There are three major race groups on this planet. There are the Negro, the Mongolian and the Caucasians. The time development between the races one cannot measure in thousands of years but in millions and maybe millions upon millions of years. The studies conducted into this aspect are the worst influenced by outside and external artificial meddling there is in any field from politicians to money donors. All effort are to cover up the true racism there is because whites feel they are the dominant specie and not to tell the others that but rather to force feed the other races to adopt the life style and culture of true white, now that is true racism.

The whites took the Negro from Africa where the Negro was accustomed to a culture for many millions of years. What ever the white may think of the culture and how the Negro must suspend that culture to adopt the culture of the white is true racism. Too criticize the oriental Mongolians of human rites atrocities because the Chinese wishes to apply the culture it held in place for millions of years just because the white seem to feel superior holding the advantages they feel they have, that is racism. The other races was on this earth millions of years before the whites and the racism the whites apply through out the world by meddling in the affairs of other races puts a price on the white the European will have to pay. And pay they will!

The other aces got where they are when the earth did not yet have the European or Caucasians or what ever name you wish to describe the whites. The whole attempt is to enslave the other races to benefit the white and then cry foul when any person on earth wishes to indicate a difference between races.

The Negro race already went through several "blowout" periods the sun unleashed. That is very evident the way they are. The Mongolians them selves went through at least one such a period. The whites with there lily pink skins never once came through such a period and yet they wish to control the other races with hypocrisy and to further their greed.

Then some Newtonians conduct studies about genes and in the studies they find that the gorilla and man share 97.8% of the gene pool and the orang-utan shares a common factor of 98.2 % with man. With such Newtonians I never could find an answer to the following question: Do they really feel that I the commoner is brainless or do they think they are so superior that anything they say I must except with out thinking. I can believe that the Negro and I share 97.8 % of common genes and I and the Mongolian share 98.2 5 common genes but the genes between man and ape is Newtonian science or in other words, plain old bullshitting the un- intellectuals.

In this is another point that proves the technique science applies at present, do not nearly give a near value to the time duration of development on earth. It should be out by as much as a few billion years for all we know. To indicate the meaning of what I am trying to bring across I shall illustrate a time scale in which the development might have taken place.

It is not an accurate and tested scale in time but merely as a presumption in order to indicate how the frequency will relate to the time duration. From where we stand, we may have a perception that the frequency is getting shorter, but as seen from within the sun, the time duration would be precisely the same value each time. This is the process in which time is concentrated in the space confinement and the relative space-time is amplified to extent the duration. In time, this variation is perceived as flair and later as a pulsating readjustment. I hope it will now be apparent just how small and under developed our sun really be when compared to other structures. This comes about because the relative size of a star is base on its space volume that contains matter and the incorrect way in which the density of stars are calculated. At present the frequency could have come down to a little as 15 000 years, maybe slightly more, but who knows. However, it is not the frequency that is the problem, but the way in which the frequency is measured that is of a concern. At present, we relate to the duration of time laps relevant to the magnitude in which the sun presently is. This might be a problem to all of mankind and civilization. There exist no method nor means in which one can determine at what stage of progress the sun is in at this moment. All that is extremely clearly, is that at one stage, the sun becomes a raging bull, and a sleeping bear follows this. In between these two possibilities of time duration, time can become double the value it holds now, which then is followed by a period where time might have half the duration, we experience at present. The first thing that springs to mind, is that we find ourselves in the middle, which averages out the extremes. That is not the case.

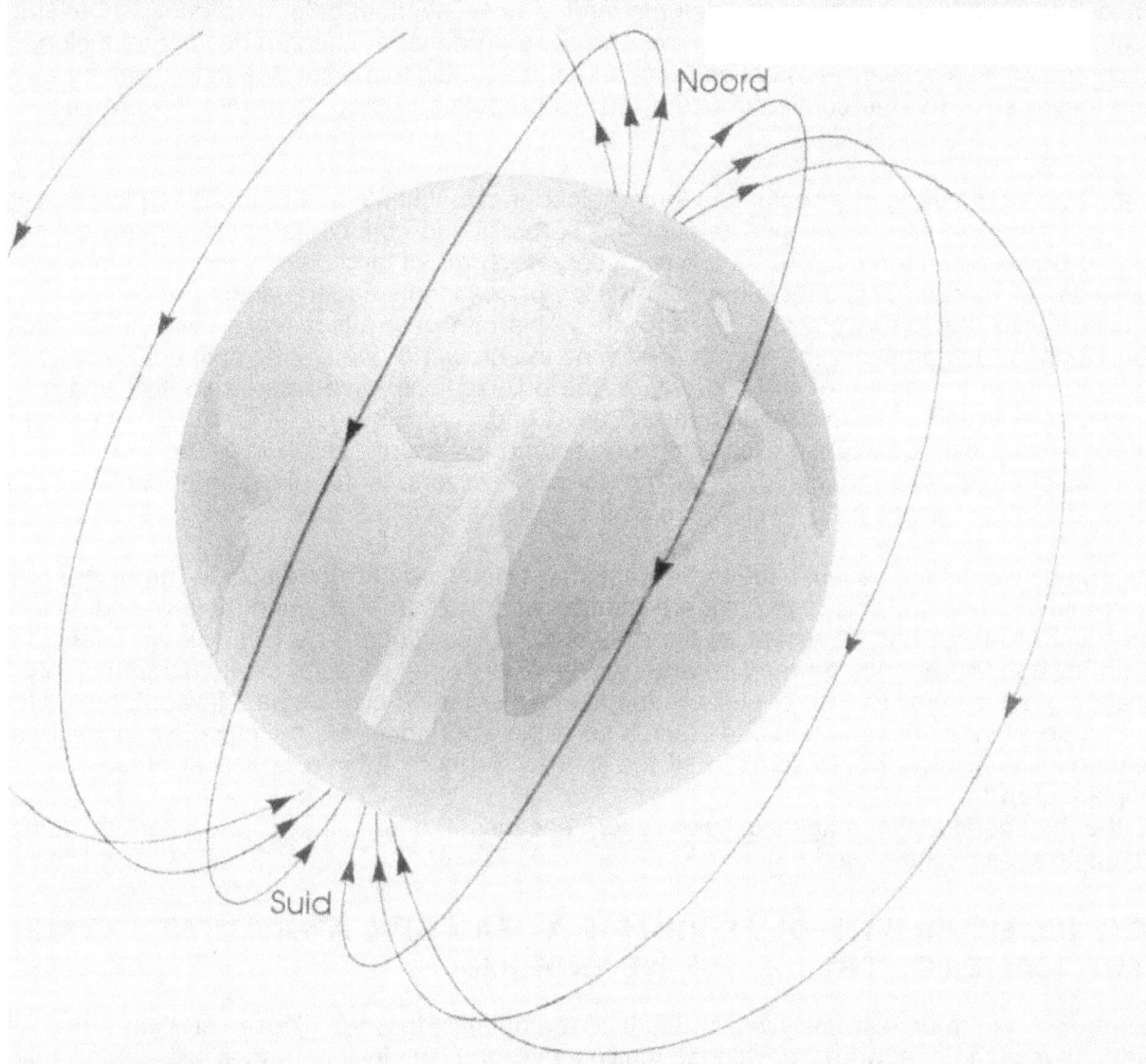

In the case of the extended duration, we might live to a relative ripe old age of 120, but in geodesic terms this might only be 20 years or so, because our hearts are programmed to last that many beats and then dies on us. To what we then perceive as one hour, would in fact be 10 of the local minutes. So, our biological clock would be completely mesmerized because the time ratio has gone out by 6: 1. No mention is made of the fact that we shall have to endure the extreme temperatures. Under such conditions, there is not even a thought of ice caps on the poles, and the tropics suffer from continuous rain all year around. In those wet conditions, nobody would be able to survive, because of the heat and the variation in river water levels, the ongoing mudslides and the exploitation of bacteria and viruses in these conditions that they would inevitably find most favourable. Still we can populate the icy parts we cannot use at present. The radiation of such an event would leave most white people to die from various forms of cancer. The Negro is therefore least affected; somewhere in their genes, they have already survived a number of these events. Nevertheless, it is not only the violence of the eruption. The time in which the eruption comes about that is a factor. To the sun, it might start as a smooth and even process that then proceeds in certain time duration, but to us on the outside, the whole thing is quick and after all, to us as humans our perspective is the most important.

I admit I am unable to see why the continents would drift and crash, the way science portray this. There was not enough available earth to float about the way that science portrayed it. As is the case in all stars, there is an aanplasing route, and a development route, which core expansion follows. At first the sun was dark and non luminous. Then one eventful day, something ignited the sun, and it was luminous. It was not hot, only luminous. One has to accept the fact that the sun had a diameter somewhere in the order of 50 times smaller than it has today. Should this revaluate to the time, which the sun was capable to generate, it would have been a weak sun, once the effect of the inverse square law applies. The first outburst would have taken eons to manifest, but the severity was brutal. I presume that this event changed the hydroxyl surrounding the earth to water, and this extreme heat developed land growth as much as it cooled the land growth. Without going into detail about life, the first command was to bring about a day period and a night period, as the sun developed fusion. The second command was ordering separation from land and see. On this point, I wish to dwell for a while. The bible never said that the command came for this purposes to stop, therefore the process is still ongoing.

Due to the spin of the earth, the centrifugal force helped the continents to drift to the tropical position it held.   Each time the core developed through an outburst, the continental crust broke, forcing the continent to brake and fragment. There are two separate forms of land development. One is the rise of the islands i.e. Hawaii. This effects the fragmenting of continents and the drifting of continents, and is associated with the top layers and soil growth. This can only affect the growth in the "colder" periods. During a sun eruption, the growth affects by the growth of the baking soil, but the inner core furnace boils out in the pole regions. Therefore, the pole regions have less fertile soil, and a much higher mountain region. The mountains should be younger and less developed. One reason why Australia is mostly flat, is because it can drift unhindered. Should it crash into South America, this tendency will change. The mountain ranges that formed did form in the direction of continental drift. The drift comes about as the magma still increases in space-time

The ice periods we detect, because the rocks, mountains and land show the scars. The bigger reason we are in understanding of this is that we are familiar with time duration during this period. We have just come out of an ice age as resent as fifteen thousand years ago. The fact that we might still be amidst an ice age, never once dawned on any person. Every time the polar caps change polarity, it is not because of an ice era. The ice era follows the actual frightening era, a solar blow out, be it a major or a minor one. During, through, and after such a solar blow out, there are no ice poles, as the deserts grow enormously. The Namib Desert ran all the way to Northern Angola, which at present is about tropical plant growth.
During the hot period, the magnetic fields alter, because the sun layers will alter its magnetic proportions to suite the new need

ALL THE ICE FILLED WITH DEUTERIUM IS A DEAD GIVE AWAY THAT THE SUN HAS THE BLAME FOR THE ICE AGE WE ENDURED.

The sun at first had to be extremely weak, this fact, the animals tell me. These very first animals had no need for eyes, and after words, developed multiple eyes that see in light ranges still unfamiliar to the human. Therefore, life developed simultaneously on land in water. The water was too cold to support life back then, so the places life found an opportunity to develop, was in puddles and shallow ponds. This was the only habitat that had sufficient heat and sunlight, to support life. As already stated, I do no accept the official account for life development and the dating of the solar system. Time and space is far too complicated to be that simple and straight forward. One instance I should mention is this: There has to be a reason why the rhinoceros has weak eyesight. The skin of the rhino tells a story of a different kind than the scientists do. This rhino was part of this planet for hundred of millions of years. During the development age of the rhino, light was far from hat it is today, therefore, the rhinoceros found it more to its advantage to develop its other senses, and not focus on its eyesight. During the time, the rhino had a soft pinkish and thin skin, circumstances on the earth was very different from what it is today. The earth is in a state of progressive development, therefore, the indicators to look for, must keep the emphasis on this indicators. The continental shift is not plates rubbing against each other, but it is all about the crust of the earth that expands therefore it is braking up, as this allows the continents to drift apart.

One cannot judge a book by its cover and one cannot gauge the earth by some crater.
The biggest misconception that any body can create is measuring time and relate the measurement too the known. That is just not possible. Time is in mans culture more than his reason to calculate and create a systematic order for himself. It is culture we have to fight. It is common sense we have to feed. It is judgment we have to adhere and the strongest part in all is: we have to know our position and place in relation to all other aspects. We have no control, not even on our destiny or events to follow. One splatter from the sun, one tiny cough, one hick up the sun will not even notice will mean the end of day seven came and it is darkness and death to all living the good life. Theologians, that is the warning; that what the Bible warns us. The Messiah said NO ONE KNOWS THE HOUR BUT OUR FATHER IN HEAVEN. Yet every second person wishes to be a profit of some sort. Stick to what the Bible teach

and leave time aloe, ONLY OUR FATHER SITTING IN ETIRNITY CAN UNDERSTAND TIME. We others are far too small.

The rhino, elephant, hippopotamus, amongst others, are of the oldest living fossils alive today. One can clearly see why they survived by their habits, location and armour. The hippo took to the water, or was already a small, nocturnal mammal that took refuge in the water in order to hide from the, what was then, more advanced, reptiles and lived in conditions that the reptiles preferred to avoid. What was not first their draw back, became the reason for their survival. An instinctive habit, both these species have, is to be drawn to fire. In contrast to other species, they approach light and fire, at night, and put it out. The killing of fire could be a habit, that developed later on, but back then, after the heat was gone, this could have become a way they survived the prolonged ice age, which followed. However, this is only a thought, and grounds for further intensive study. Another factor to investigate is the whale and dolphin and such mammals. They are never part of the land animal as mammals came to existence, life developed as mammals in the sea, and on land simultaneously. This is as far as I am prepared to go, in the translation about life.

Archaeologists get all existed when their Brainy part overtake their senses and good old logic. To feed frenzy they take off to some desert to scavenge for fossil remains. The greatest fossil is still with us and we do not have to hunt down some glued remains stuck in carbon. The insect is a living fossil running around every day. The insect outsmarted the hick ups the sun has; it even chose to detach its survival from a skillet, as it must have had in time previously. If it held on to the skeleton, that body function would kill it because the wait it will then have to drag around will be unbearable. So it did the best it could; it tossed the bone and replaced it with a dome. A dome so hard it can withstand even a nuclear war...and not one of the Brainy Bunch ever made the connection that the reason it can withstand a nuclear incineration is because it already withstood a nuclear incineration the sun provided in the past.

Life connecting to whatever specie and whatever form will never change because of habit it will do so only when forced to do so facing the prospect of distinction.

Life will fight to survive or life will no longer be. Enjoying the privilege of space in time means fight or remove.

The way man is taking space in time in killing and destroying other species also in the space of the earth we share time, will have dire consequences. I maybe small comparing to others, the role I play may have little to offer in glamour but I also may (just maybe) have some way enabling me to see what others miss. If that is the case, then I can assure you that WE LEARN TO LIVE TO LIVE TO LEARN, IN MORE WAYS THAN EVER WE IMAGINE AND WE SHALL PAY THE PRICE, THE PRICE TO PAY.

That puts the cosmos on one side of the argument but the only thing that can have movement in gravity and not necessarily as gravity is life. All movement in the cosmos and on earth beside life is gravity. Wind is a product of gravity and earthquakes are a product of gravity and the growth of the earth and therefore the entire Universe is gravity. First of all it is true that gravity or lightning produces life because everyone with a swimming pool will know about the green slime forming of moss and algae in the swimming pool and this proves that gravity and light and electricity closely associates with life. As a farmer I knew that after the first lightning storm weed and unwanted plants will start growing.

The different species on earth all holding life is different because of gravity and the form life takes on is dictated by gravity. Life shows how the Universe formed and to look at life one can see precisely how the earth went through stages forcing life to adapt according to gravity. But as I showed gravity or time or 1 is formed by God's thought and will and that makes God and nature in charge of how the earth and the cosmos developed.

Man tries to change nature because man thinks he can beat nature but man can't. fighting fungi and fighting germs will only create a super bug the man can't destroy but it will destroy man. Man now develops poisons to destroy life that opposes the creation of wealth but it is putting in place a danger that can wipe out man. It will not wipe out life but if there is something like twenty billion people on earth it becomes a rat-infested pit where these micro creatures can grow and develop. Once the tied turns ands one of these creatures form a chain that can't be defeated by man it will come and defeat man and that day will bring an end to human life.

One question never asked is why so many lines of human development never survived to continue on earth. The get a piece of scull here and declare that a new species and they get a bone there and declare they found a new specie and the some tooth by which another new finding come to light. No one ever bothers to ask why did they not continue to live and what happed to them by which they died out completely. Why would one entire specie die out and not leave one specimen just to be followed by the next development. If this demise came from some micro form of life that then went into retreat once it killed off an entire species that micro animal is still around but is dormant until it find the host it has been waiting for all its existence and if it could wipe out the entire specie back then it will be able to do it now.

The gorilla the orang-utan and the girl is what they are because of life forming a body and not due to the closeness of the body they form. Let me explain but I realise that this would be unbelievably hard for the Newtonian atheist to understand because the Newtonian atheist understand Newton that makes no sense while The rest that can think and that doe not depend on understanding Newton will follow but it is beyond the Newtonian atheist just plainly because the Newtonian atheist is able to understand Newton. If it is not a hoax Newtonians don't get it. As life formed it produced a different body befitting the requirement to suit that specie holding life and life is about the intellect in reasoning that forms a body by inhaling oxygen and exhaling carbon dioxide and using the heat the elements carry or transport. The ability to produce a body is life and not the body holding life that supports life to use space within the Universe. Life is about intellect and not DNA that forms a body.

This picture shows man carrying a stick with which to hunt whereby suggesting this gave man his first hunting edge. According to my view, which as usual is different from everyone else's opinion, man developed intellect and the use of the brain by two things that placed man above and beyond other beings. This first development of man's ability was to raise his arms above his head and that is the very foremost. They can come with spears and rock-splinter-cutting flint-knives and hunting techniques but the positioning of the arms holding the hands outstretched above the head was the first development that put man in a category above and beyond the rest of life. This placed man in the top as the most proficient hunter. That made man the top species. One tiny hand movement allowing better hunting made man cleverer.

All apes can hold an arm square to the shoulder and all apes can take something and hit it on the ground to break it and get to the content within whatever container or shell. All apes can crack the shell of a fruit or nut by slamming a rock on it and get to the inside of the edible nut but only man can raise his arm above his head and throw a stone accurately at an object. The ability to aim with a stone in hand above your head and throw instead of it being square with the shoulder to crack a nut brought about humanity. Ape became man with the ability to throw the stone at something in order to kill it. That action made man a hunter like no other before. This meant man could throw a stone and injure an animal at a distance, which then provided an advantage above running fast or being strong enough the break its neck or to kill while the animal ran. Man could stand a few meters adrift and throw a stone to injure and then even effortlessly walk to the prey and kill. That advantage gave man a jump on every other hunter. Today every human man can throw something and throw rather accurate to become (with some practise) deadly accurate. No other species can achieve this and through that ability man got wise. Lifting his arm from square to his shoulders to above his head in a throwing action as in raising the arm another 90° is what brought intellect and civilisation to humanity. It was only moving the arm a little further that enabled man to hunt perfectly and more deadly.

The difference that made man an intellectual hunter and not merely a strong-armed killer was lifting the same stone he used for cracking nuts a little higher and throw the stone while directing the flight thereof to a target that was a distance away from the hunter. A big issue in this development is that all men can throw and I have never seen woman throw naturally. Women shout and squeal naturally to call men but men always pick something up to throw. The ability to throw kept man opportunistic vigilant in hunting and man had to walk upright as to allow his hands to be free for any action at any second. To kill, man had become upright to be most opportunistic in order to have the advantage to survive and feed the best. So those that survived the best were those that walked upright the longest to have arms freed.

This where the arm comes from a chopping position to a throwing position is where the road to civilisation begins but this is still far from presenting a human from being ape. It began by lifting the arm. A lot of other inputs contributed. In that idea man formed intellect and civil order and left the others as ape. Throwing gave man more energy and time to think about ideas and less time spent running and hunting. By incapacitating animals before killing reduced man's dependence on body strength and encourage cunning in order to kill more swiftly with better weaponry. But hunting in a tribe will produce better results and also give much more leisure time, which brought along a closer social order. There was more time to interact that produced intellect. Nights still had to be profusely dangerous but with more idle time man then had the ability and energy using leisure time to research many things such as other tools with which to improve life and living conditions. Just think how much boredom it took and what idle time was available to allow a person to rub one stick against another stick until it smoked and burned to start fire for the first time. Reposition one arm brought a new stance that found intellectualism.

In many thousands of years man then formed better development of improving a throwing arm with better throwing techniques that produced better tools with which to throw and to kill quicker and devour more food. The next step was fire. Those that had the ability to roast food had the advantage of quicker and more sufficient digesting that also allowed those to have more energy and more strength to hunt even better. This is common knowledge but sitting with full bellies around a fire and many persons all trying to convince others who are top dog made the stronger seem weak in stupidity which then had to allow the more intellectuals to entertain the tribe with more ideas and thereby gain most respect within the tribe. The ones that developed verbally better earned more attention and demanded more respect. The respect pushed aside the fear for stronger and gave way for the clever to step to the forefront. This was all the result of sitting at a fire and planning hunting strategy and the intellectual had the advantage to convince others of planning better hunting and applying more cunning.

This I take from facts that still have a place of prominence even in modern society. The silver tongue devils can cheat more but hold better respect because they can talk faster and so they are more convincing although they are better at cheating. It is no longer the ape with muscles that leads society but it's using intelligence and convincing others to follow his lead. Those with better planning get more followers and the one with the plan gets people to listen. The leaders are the intellectuals and this is part of all cultures in humans but only in humans. The safety having nighttime fires gave people chance to think and to exchange ideas without being scared and chased into hiding in darkness at night. Sitting beside a fire in conversation brought about much more emotions of understanding and enduring and companionship and from this I think family-bonding ties became more appropriate than merely ritual mating. It gave special bonding and connecting in better ways than other primates do even to this day.

Then the second function that developed man was the ability to recognise stars and to philosophise what stars were and to give stars a meaning and a purpose exceeding the ordinary use of either eating it or using things as tools. It made man think of what could be outside the need of man to just survive but to name stars. It gave man reason to think about the cosmos and about what is greater than man and what could control the destiny of man. This made man human and set man intellectually on a road where man developed spiritually as well. All this came from having, more time in hand which came from throwing with the arm.

They sat at the fire at night and saw objects they could not reach or touch and yet it clearly drew their attention. Maybe in the very beginning they thought of stars as eyes looking down on them and this made them identify with what they saw up in the sky. The wise found new avenues to speculate and could convince others about powers outside man's mode of reality, much as the press does in the modern age with UFO which they connect to aliens visiting. The fact that man gave prominence to something that was outside his reach and he had no use for made this aspect about man just as unique as the ability to hunt better. This might be speculating but it would explain many things that man's behaviour is different than we find in any other animal on earth, even those we crudely associate with. This is where my speculating ends and this is where pure factual evidence and proven arguments starts again.

We investigate the role of physics in life and how life by exerting thought controls the body. I prove gravity forms thought. Thought controls the brain but I also prove that thought, as a factor is not in space. The body is in space according to life while life is in thought, which is in time. This is an enormously complex issue but is so simple to understand the idea is child-like simple. The entirety of the Universe is in two aspects where one is time and the other space.

Like most other things Newtonians got this wrong. Let's get rid of some misconceptions. The Universe does not expand because whereto can it expand if it is everything that can ever be. The Universe can't grow to become more because it is everything that ever was, that can be and that ever will be and therefore this mythical misconception that the Universe expands is more Newtonian mythology wrapped in a coating of unbelievable stupidity. I am going to show what expands and that which expands is growth outside the Universe that we know. We investigate the role of physics in life and how life by exerting thought controls the body.

In time lapse we are able to see many moons although we know there is only one but the significance behind this serves as the only indicator we may use to understand physics. This picture answers all our questions about the Universe and life within the Universe. There is a make believe to pretend it is the Universe that we see and that Universe we see and know but that is a Universe that doesn't exist but for our visible light focus. The Big Bang brought light and that was when God Almighty rewrote the cosmos in dimensions using light as ink and space as paper. However, don't kid yourself; that is not the real Universe. By giving us less perspective on reality it is what we wish to imagine. There is one

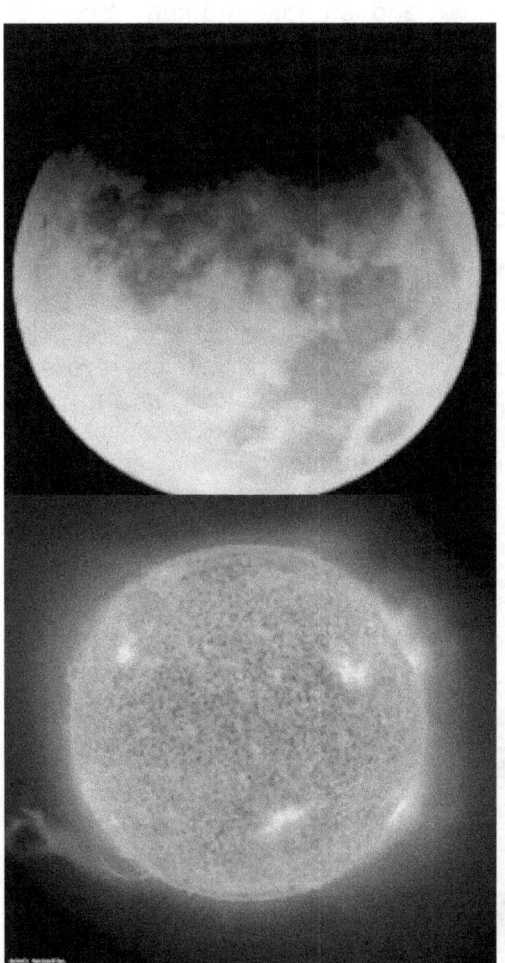

Universe that really exists and that Universe is in singularity. The rest is just space that does not exist but in our vision. We think we see many things forming the most beautiful cosmos but all we see is light painting a false picture of a hologram that doesn't exist. What is out there is dark and invisible.

Allow me to explain that because this is most vital and any concept we must have must use this as a basis reference or we will never get to grips with understanding God's Creation.

We all look at night sky and see the moon. We see the moon rise and we see the moon set. However is it the moon that we see or are we allowing our imagination to run wild? We can't see the moon because there can't be a moon that we can see. The moon is pitch black void of all colours because it is void of all light. The moon has no colour by which it can present an image to leave us with a view. We see light bouncing from what we believe is the moon and we use that light to form an image that by culture alone we recognise as a moon. It is light we see rushing from a part of space bringing as an image we then interpret as something our teachers and parents told us we all call the moon. The moon is totally invisible and therefore the moon we will never see. ...But there is plenty more... It takes light about one and half seconds in time to cross the space between the moon and me to reach me on earth. The moon is in space one and half seconds into my past and the moon I see was the moon one and half seconds previously but by the time I see it, the moon is in another space. I don't see the moon.

With that in mind we can only see the sun. The rest we see is a picture painted in light on a canvass of space. That is what I said the one part of the Universe is. So all we can see is the sun...and that too is a myth. We can't see the sun. We see light in photon rays forming heat that the sun discards and we see the photon rays carrying an imager we also by culture associate with what we believe is the sun. It takes light about eight and half seconds in time to cross the space and reach me on earth. Again I do not see the sun but the way the sun was in the past using light that is not the sun. But even more is that I can't even see the sun as it was in the past because I see light.

The sun by rotation freezes heat we call outer space by gravity onto its surface, which by concentrating

so much heat in so little space it forms light. That frozen heat the sun then reflects and it discards what it does not use in the form of a concentrated white light. The sun froze dark expanded light into a concentration of white visible light and then discards this visible light back into the dark light, which absorbs the white light gradually. We see the light that the sun discards which is not the sun but which is an optic illusion with which we see. There are two forms by which the Universe forms and we use the one that is light-illusion to avoid reality. We don't see the sun but we see light it discards and again our culture steps in to make us believe we see what we think we see. We think of what we see in terms of the immediate while in cosmology space is time's history. Where space is light forming space and if time crosses space then space in light is part of the history of time, which is what space is. Space is light that time left behind.

The only thing forming the Universe is time moving along and space is the records of time as time went on and left space to keep record of where time went as it ventured into the future while leaving the past behind as space. However make no errors there is no such a thing as space except the light that forms a hologram of what time formed as time kept moving into a future placing the present in terms with the past. Space is the past that is written in light on space in three dimensions but has no other value.

Can I see another person? When I look at someone a few meters from me I can't see the person although all my senses tell me I see the person. I see the person with light again that bounced off the body I think of in terms of the person. The image the person conveys is in time phase behind what I experience because it took time to pass from the point the light bounced from the person to where my eye is. That puts the place the person holds both into my future and part of

my past. The person that I am looking at is at a point that is a part of the greater Universe of which I am part of because the entirety forms one big future that will arrive the very next instant. The person forming the past is the light that connected on the person and from there took time placing that to the past because that happened in the past of the point in time I am now experiencing. Therefore whatever I think I see is not what I see but it is the past I see in the present in which I am. This might sound like splitting hair but we have to realise that there is time being in the here and the now and then

there is space that forms a past and the past is not part of the present which is in the reality. Time is now and space is then. There is a now and then there is back when. This is physics forming a reality splitting space and time. That is why there is space-time.

Can I touch another person?
The person that sits there might imagine he touches something but he does not because he is not able to touch. He uses his arms to touch, which is an extension of his body. To touch anything he must use an arm extended by a hand extended by a finger. It takes time for the nerve impulse to move as electricity from the finger through the hand and the arm then through the body to the brain to get recognition about the validity of the touch. Travelling through any space at any speed uses time and that means time moved on through space becoming the history of time. There is the now that is in time and there is the when back then that is in the past and in that history of time we get space. As time moves on it leaves behind light that forms space.

In order to think cosmic we have to think in terms of space-time and not the idiotic nonsense Newtonians connect to this concept. The moon is in time and I am in time and space is between us therefore space puts the moon in my past time frame as well as my future time frame because light will come to me providing a future. However the space disconnects the time the moon holds from the time I have and the moon is in a different location when the light that tells me there is a moon reaches me. By associating and disassociating time and space we find the reality about life. Since this is a website I can't get too involved in this argument but in the book I can. Kepler formulated this entire concept he got from the Universe perfectly before Newton raped Kepler's work and destroyed it with his (Newton's) senseless stupidity.

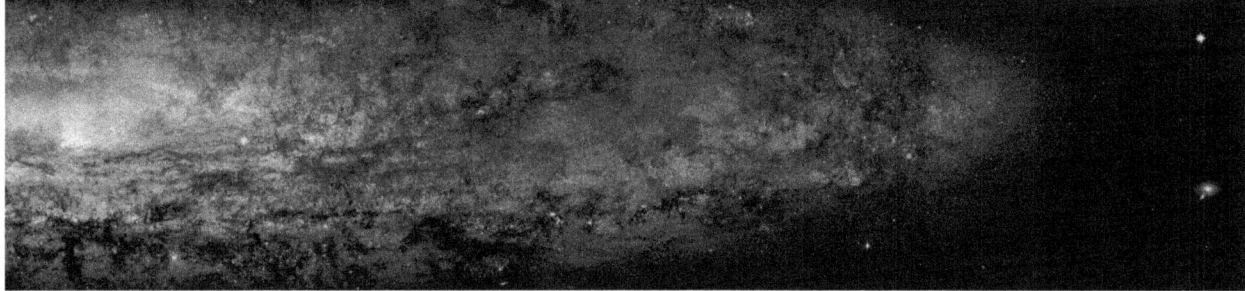

Everything in this picture is beautiful, marvellous, beyond what words can describe and most of all it does not exist. What you see is not what you see because what you see is time that formed space as time moved on to the next instant while leaving space behind that forms an image in many colours. It is an array of light formed in colours mixed with what are not technically colours. It is an image drawn in light on a medium canvass called space and is only there as a supplement of my imagination. It is a lot of light that is dark as it moves away or expands and light streaming towards me in a concentrated form colouring the canvass many colours but in all it is light and it is light alone. But there is one more significance in this picture and that significance underwrites the cosmos as it defines cosmic physics and in that also every aspect of physics. Light moving towards me places me in the centre of the Universe in time but I can't be in the centre of the Universe in space because being in space and forming space presents a mythical historical light-illusion and being in space is fiction there can't be a centre in fiction. Space is the history written in light and a history of events gone by can't have a centre because of no conclusion seen from where and when it started.

Kepler gave a formula, which reads $a^3 = T^2 k$. Do not think of this formula lightly because this is the correct formula that carries all physics. From this formula Einstein got his so famous $E=MC^2$ that should read $E^3 = mC^2$ which is a carbon copy of Kepler's $a^3 = kT^2$. This is a far cry from Newton's idiotic $a^3 = T^2$ where Newton says a thee dimensional view equals a two dimensional view. Kepler said $a^3 = T^2 k$ which places the one sector which is space $a^3 = \text{space}$ in equal terms with movement or time, which is, $T^2 k = \text{time}$ and in this we have $a^3 = T^2 k$ that is $(\text{space})\, a^3 = (\text{time})\, T^2 k$. There is a location in time and there is a location in space and time as it moves on converts the past time into space and in doing that it uses light. Life is in time while my body is in space and my body can never be in time because there are always a few billion atoms between my thoughts (time) and my body (space)

In this we investigate the role of physics in life and how life controls the body. It is very easy to say the brain controls the body, but what controls the brain? When life ends the stomach immediately stops to process food. The stomach stops to use the acid to formulate what we had for a meal into what we will use as stuff feeding our body. When life seizes the body loses smell and that means life gives off a distinct smell. In death the smell changes and the acid no longer digest food. If it was acid that digested food it would carry on processing the stomach content but it stops immediately and therefore we know it is life that digests the food using acid as a tool. The moment life leaves the body it stops discarding body material. This body material floating in the air holding the smell of the person's identity is what life uses to rebuild the body and we call that process aging. Life builds the body atom-by-atom in healing using carbon and oxygen. When life exists the body all vital aspects change. Then the cosmos atomises the then cadaver back to atoms.

Then a cadaver dog can trace the smell as death because the body as a cadaver is without life and became just more cosmic rubbish that breaks down into atoms. The body disintegrates the second life

no longer secures the structural integrity of the composition of the body and therefore life in time removes from the body holding space.

Newtonians has all this down to just showing no electrical activity in the brain but that is a stupid argument. Only a senseless atheist can have such a meagre thinking ability. Electricity charges the mind or so science says but that is a very outdated and backwards small-minded look on the entire matter. If that is the case I can plug the brain into a supply of volts and the brain will put life back into the body and the brain will continue to control the body. Life is a separate issue that builds the body and the body is not what host's life.

Can I touch my body? Science holds the opinion that the brain controls the body. It's easy to say the brain controls the body but then what controls the brain? It is Newtonian to even think the brain controls the body because Newtonians put everything in simplicity because they can understand nothing so good they are able to fill a Universe with nothing overflowing making the Universe expand because of nothing growing and becoming even more nothing than the nothing it was. Thank God they are atheists and that shows their stupidity is endlessly big while increasing every second. It is thought that governs the brain but what then produces those thoughts that control the brain that moves the body?

There is partition between singularity and space. Singularity can hold a thought whereas space is controlled by thought. Thought via electricity or gravity or electromagnetism or electric field all being the same and that takes charge of body material and controls the behaviour and abilities of material. That is life. It is establishing a channel of electric flow from what life starts the fields being where cosmic gas at $10 / 7(4(\Pi^2 + \Pi^2)$ is going spherical $7/10 (\Pi^6)6$ and forming an electric field around the brain whereby it generates electric thought into the brain. This is a controlled operation. If it were the mind controlling the electricity what would direct the specific flow and what will control such flow. Something more must take charge of the release of electricity to control what is in the mind. When they tested the flow of electricity in the brain by cutting the human scull of a person alive it was proven that this control of electricity forms thought in the mind but shocking the entire brain with a jolt of electricity did not do it. The medical doctor that took charge of the process took the place of the person controlling her or his body. Then by directing the electricity the doctor in charge released electricity just like life would do of the person in control of the human body. If that happens the electricity will unleash involuntary impulses much similar to the shock with a stinger or porter that forms spasm. The muscles will go into spasm, as all the muscles will get electricity at the same time but without control. Charging the muscles is done with the flow of blood by establishing the Coanda effect. The second, the instant the blood flow stop the generating of electric field stop and the person is dead. It is the blood holding plasma or liquid in relation to flowing solids (red blood) that charges electricity via the fields surrounding the brain. Practising an athlete or anything with life is just establishing fitness better electric flow to the body that results in having better body control. The better flow of electricity forms more muscle control and then the higher standard of achievement will the athlete accomplish.

These pictures show my two sons battling in a wrestling match. Each one is trying his very best to overcome the opponent's strength. Sometimes they win and other times thy loos because that is sport.

My sons are internationally acclaimed wrestlers and they have to practise eight hours at least per day for four days a week to be on the top level of the required performance. If Newtonian stupidity is correct then when they require more fitness they don't have to practise more; that is if the Newtonians are correct because all they need is a jolt of extra electrons running through the brain matter and they get the strength Superman has. To get them to wrestle we have to condition the mind to accept orders from life. Life does this by practising to induce current flow to the brain that conducts the flow through the brain to the muscles in the body. If the current is not strong enough the muscles cant do the job and if the current is strong enough any muscle of any size can do the job. If the electricity impulse is not strong enough a person can't lift a heavy weight but in circumstances of life and death I saw a man lift a car to get another person out that was trapped underneath the car. The emotion of fear and desperation was so massive it charged an electrical current that allowed the brain to order the body to pick up the car and free the person stuck underneath. Both my sons can lift a person of 120kg from the floor up above and over their heads one with movement. I have many such moves documented on film taken during their fights. As strong as they are and they are strong, they will not be able to stand next to a car and flip it over on its side with one movement. That is why we shout and encourage sportsman to get maximum electricity charges and to get the most voltage emotion can muster to the athlete's body.

We have to associate thoughts with life while disassociate thoughts from being a product of the brain. When being dead the brain are present but empty with thought. When life or the thought process ends in death the stomach acids immediately stops processing food. The stomach stops formulating our last meal into something that produces life's energy. Newtonians and other simpletons say it is the electric charge that makes the brain think but put any voltage through the brain and the stomach won't start digesting food even with the brain charged with electricity. The absence of life stops stomach acids from digesting food and how does that change the process of acids digesting food?

Typical Newtonian atheistic small-minded research tries to show that the brain applies electricity to convey messages to the human body. Certain parts of the brain sends messages to perform certain acts in movement and some part of the brain controls some characteristic aspects to allow certain behaviour to take place. The brain through electricity controls the body by performing manipulation of muscles that create movement ability. Well this is like saying to breathe is to take in air. That is hardly science but comes down to observation of elementary facts. However, this is where Newtonian atheists stop the argument because as usual they run out of brainpower. At this point of researching life we find that the argument hardly started…. And now it becomes complicated to the extreme.

Newtonians are convinced that thoughts are electrons within the brain. By emitting electrons the brain performs the manipulation of the body through electrical charges and that when released controls the muscles in the body and that is life performing. Well yeah and also Newtonians say that mass pulls mass to supply gravity to a small edge-filled Universe and in this simplicity it is only the Newtonians' small brain that can understand so much complexity with such minuscule simple approach. It's easy to say the brain controls the body but then what controls the brain? If those Newtonians were correct sportsmen would have so little to do and reap so much awards.

To get them to better perform during a tournament I can put a belt sporting a stun gun around my sons' head and every time any one of them needs more strength of essence during a wrestling match I put a surge of electricity through the brain while that will make them many times stronger. I stand with a release mechanism and when they require a jolt of electricity to beat the opponent badly I give him a jolt and he breaks the other person's arms with the strength he then has. When more strength is required we up the voltage and increase the amps. No more practising is necessary because we just up the supply of electricity and the mind will supply so much muscle power to the body my son will have the power to lift not only his opponent but the entire tournament above his head and break the venue. Then everyone will need to have stronger battery packs and we can do without the hours and hours of tedious practising sessions. You just plug any person into a strong battery and let the current flow.

Can you see how utterly childish such an argument is but then again what can you expect from a simpleton that admits to being animal-like in mentality as an atheist. By the way my dog is the biggest atheist I have ever met and my dog is not an atheist because he is a clever scientist but because he is too simple to understand physics and science. My dog is an atheist because of stupidity and that goes for every other mindless atheist but because they elevated Newton's corruption, the lot sit with a hoax they try to give reality and because they can create their own truth they make it up as they go along while pretending to be the clever atheists. They always were too stupid to see that Newton's concepts are a hoax, so how can they be bright enough to see a God that created everything we call a Universe only by God's thought?

Life is in thought and thought is in singularity. Again they say it is the acids in the stomach that digests the food but death does not remove acids absorbing stomach content. It has to be life that digests the food because when life removes after death the stomach content and acids are still present but the processing part stopped and by life vacating the body the digesting process stopped. With life gone notwithstanding whatever outside intervention, life's functions will remain absent. Electricity in the brain doesn't reinstate life again by controlling the body.

With the human body in its entirety forming space that the mind controls by forming gravity it puts the entire human body in space and brings the mind in time from where it conducts electricity. It is the mind or life or call it what you wish that generates electricity outside the body and then by forming a circle or band of electricity around the brain it charges electrons within the brain and life directs the electrons to manipulate certain parts of the body in the manner life wants the body to perform. These are all part of finding the process we call life. Life isn't as simple as putting a few electrons that runs through brain matter and therefore instigates life. Life is a lot more complex than such simple ideas about the most complicated process. I prove that gravity and electricity starts at $\dfrac{7\Pi^6}{10\Pi^0 x 6\Pi^0} = 112$ and 7/10 is movement of gravity of the sphere Π^6 in terms of a six-sided cube (6) where outer space or the element table starts at 112. This I explain much more extensively but this is where life that develops electricity forms an electric sphere around the brain. This requires much better explaining than what the limits of the length of this article allows for. The 7/10 is the Titius Bode law forming gravity and this is a sphere turning in a square and time forming space.

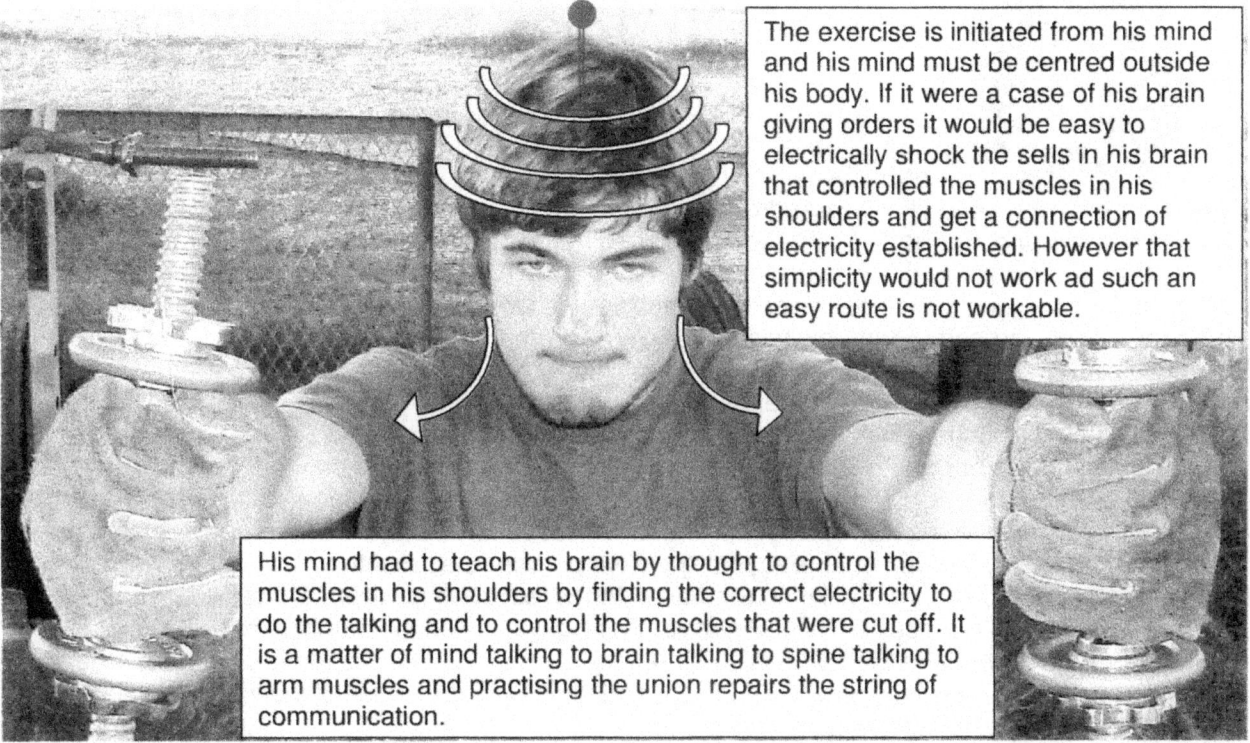

The exercise is initiated from his mind and his mind must be centred outside his body. If it were a case of his brain giving orders it would be easy to electrically shock the sells in his brain that controlled the muscles in his shoulders and get a connection of electricity established. However that simplicity would not work ad such an easy route is not workable.

His mind had to teach his brain by thought to control the muscles in his shoulders by finding the correct electricity to do the talking and to control the muscles that were cut off. It is a matter of mind talking to brain talking to spine talking to arm muscles and practising the union repairs the string of communication.

When Willem recouped from an operation we had to get his mind to again talk to his muscles correctly to give him the required control over the manipulation of the movement within his muscles. Life reinstates the ability to control the brain to manipulate the body movement.

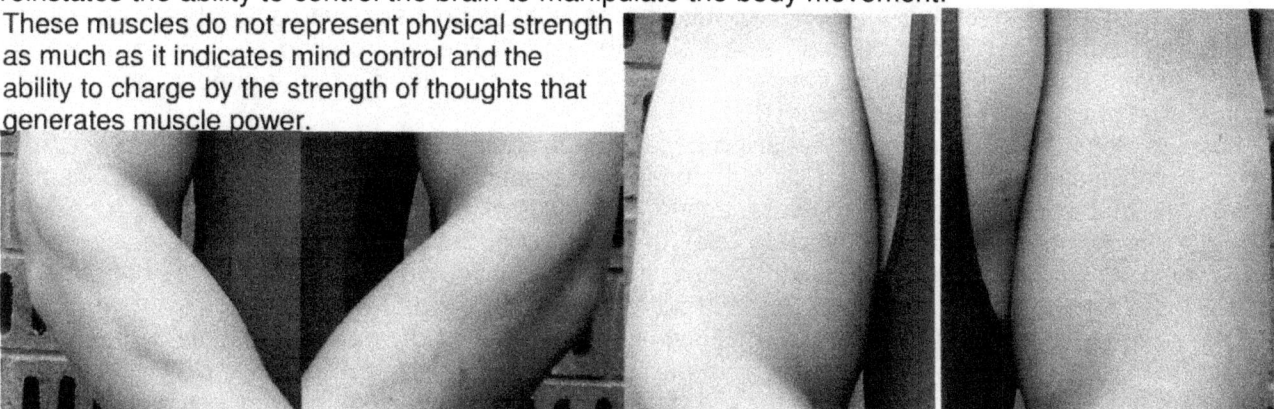

These muscles do not represent physical strength as much as it indicates mind control and the ability to charge by the strength of thoughts that generates muscle power.

That concludes that with electricity the brain doesn't instate any functions we attribute to life. Therefore Life by thought generates electricity that forms around the brain as the Coanda effect. I prove it in another book. The electricity controls the brain but what charges electricity? In order to improve his muscles I purposely increased the electricity flow to his muscles from his mind. People call this concentration but a rose by any name still remains a rose. Without practise the current weakens and the muscle power fades. You have to charge the battery.

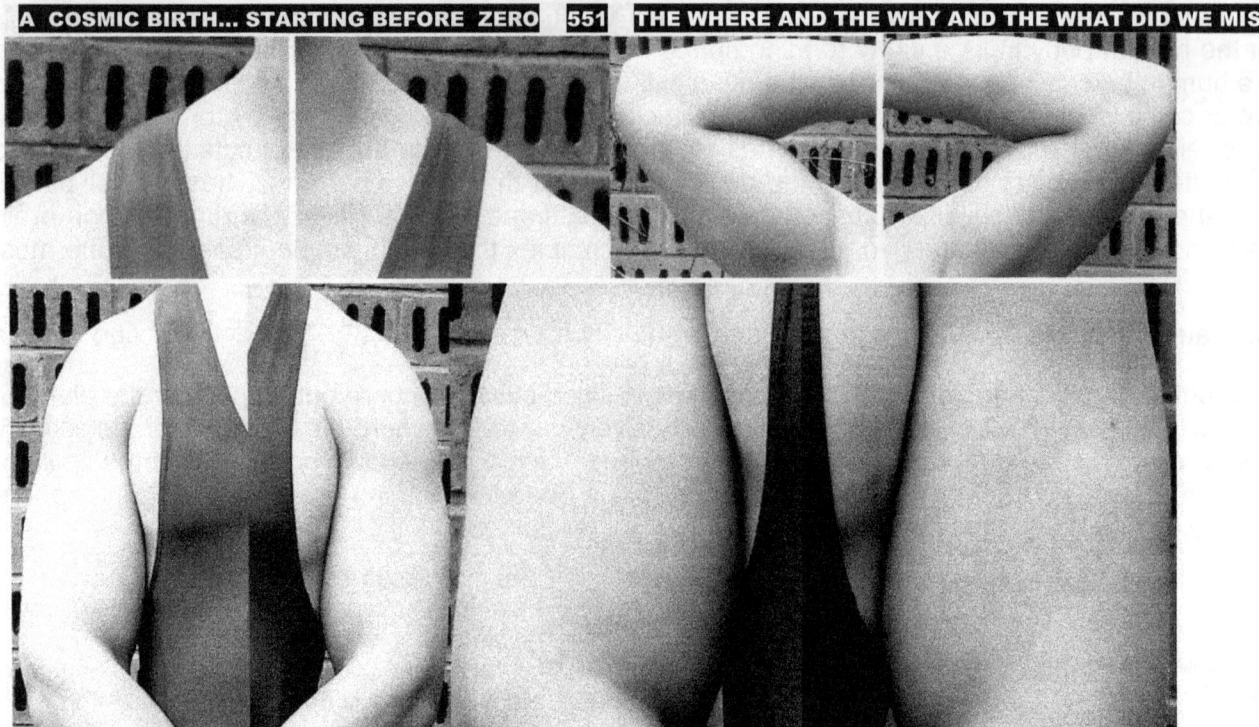

These are the muscles we built when Andre was 19 without lifting one weight in the process. Through practising the mind and the body I allowed the electricity to develop that controls the mind and the muscles and then by practising the brain to endure lots of pain and suppress the pain this method of stretching the limits of endurance developed the muscles as well as the control of pain in the brain. By developing the mind the body forms muscles and these are the arms of my son when he was 19. It is not the muscle that does the job but the mind controlling the muscle that produces a muscle strong enough to do a job. This is no cheap oriental philosophy but true raw cosmic physics placing gravity and electricity in charge of the brain to charge the muscles in the body.

When the body and life parts ways we call it death. When life seizes the thought process the body loses smell and that means a by-product of life is to leave an identifiable unique smell. After death it is gone and that is why we use cadaver dogs. The cadaver dogs tell the story of the body being lifeless throwaway reusable recycled material ready for the next person. Electrocuting the cadaver won't bring back odour the body had before death. That means life leaves small particles of the body behind as it rebuilds and replenishes the body while being alive. The body also leaves particles behind but clearly it is of a different nature because the cadaver dogs recognise that sent as a cadaver, which is a cosmic body without life and is breaking up. Science says electricity charges the mind but this perception proves outdated.

If it was electricity alone the brain had no control over the flow, much like a person being stunned by a stun gun. The body just reacts without control because behind the flow there is no message about what to do or whereto to direct the electricity conducting. It is thought that governs the brain but what produces thought that controls the brain that moves the body? The control is in the thought and the thought is what produces life and in the book I show where exactly life is when it directs control over the brain that manipulates the muscles.

We have to place thoughts in another position because it is clearly not in the brain. The brain acts on electricity impulses it receives. When jolting the brain with electricity we can produce movement in the muscle but that shows the brain receives and relays electricity and the brain does not generate electricity The brain is merely a conducting relay of an impulse that direct the flow to the station that should respond to the wishes of life's commands. In that we have hypnotic commands because with hypnosis we remove the character that life has and replace that with the hypnotisers orders. There is a string of commands coming from a source that is not part of the body but is outside the body. Electricity as such will not become a hypnotic replacement. The hypnotist does not use electricity to convey commands but also put any voltage through the brain and the stomach won't start digesting food even with the brain charged with electricity. With life gone notwithstanding whatever outside intervention, life's functions will remain absent. Electricity in the brain doesn't reinstate life again controlling the body. Life enforces electricity that life as a factor generates and life is not electricity.

That concludes that with electricity the brain doesn't instate any functions we attribute to life. When the body and life parts ways we call it death. When life seizes the thought process the body loses smell and that means a by-product of life is to leave an identifiable unique smell. After death it is gone and that is why we use cadaver dogs when searching for the dead. Electrocuting the cadaver won't bring back odour the body had before death. Science says electricity charges the mind but this perception proves outdated. If that's true I can plug the brain into electricity and the brain will continue controlling body movement by putting life functions back into the body.

Life or thought is a separate issue that rebuilds the body and the body isn't what host's life but functions according to life being in control of the body. By thought life controls the brain that manipulates the body. Life is an entity that's in thought and thought is in singularity, which is time. That I prove. I show Life is in a space that isn't space because life is outside where we think space is. If you wish to find life's place follow my argument and find life's place in the Universe. Life or thought is a separate issue that rebuilds the body and the body isn't what host's life but functions according to life being in control of the body.
By thought life controls the brain that manipulates the body.
Life is an entity that's in thought and thought is in singularity, which is time. That I prove.
I show Life is in a space that isn't space because life is outside where we think space is. If you wish to find life's place follow my argument and find life's place in the Universe.
All light throughout the Universe meets at the point you are. All light comes from everywhere to meet you wherever you are. If you weren't the centre light won't flow directly to you.

Life by thought assembles the body from even before birth and when being without thought the cosmos dismantles the body again after death. If a lifeless body is a total different thing from a body filled with life then after death and with the body still usable but not functional it is totally Newtonian or atheistic or insanely stupid (which is all the same) to say that the body is what confirms life meaning the body is life. The body is still present after death but the decaying process starts immediately to dismantle the tissue structure that life maintained and that life preserved while the body was in use of life. It is madness to discount life, as a presence in body while without life the body is cosmic material going into a process of dismantling. If the body held life then we can re-install life by electricity and the body will return to all its normal functions. However that is not possible and so it is clear that life constructed the body from before birth and maintained the body by an aging process until the final departure of life where the body is no longer maintained and has no further purpose and so the cosmos in space goes into destruction of a useless accessory but no longer necessary of life. What walks with time and presides in time as part of time went with time to where time forms the future while the human remains remain cosmic atoms and remains in space as a part of the history of what time once was when life occupied and built a human body.

If life is electricity or if is as simple as any of the functions Newtonians atheist stupidity try to contribute to life then it can revitalise a cadaver by shocking it until it hops like a ping – pong ball. I then can replace the brain's electrons and produce thinking within the cadaver mind. The cadaver will again be able to lift its arms and start to walk around. If I am not able to replace life within the human body that puts life in thought and I prove singularity is what holds thought but that issue is far too complex to touch in this website. I aim to keep the website very simple and in the understanding of any person except atheists who will be too simpleminded to make sense of anything that is a reality and not part of the Newtonian hoax. If this sounds complicated then by me using 480 pages the explaining gets much better while the topic becomes very logic. If life removes from the body life was never part of the body.

The entire Universe is light and that is why we can focus on one image with lenses and with another lens view a picture as wide as we wish to see. However, what is out there is light either expanding or contracted but it is light we see that fills the cosmos. We see darkness as light expanding thus moving away and the white light we see is light coming towards us. Have you sat back and think about how light enables anyone to read?

Newtonian science says the light bounce of the page and I use my eye to see the words. What can be simpler than this explaining...only the mind of a simple atheist and has no idea about what reality is because they believe in Newtonian corruption. How does the light take the words away from the page and put the image of the words printed in ink on paper into my eye so that I could see the image of the words on the paper. How does light take what is in the image and produce a picture of that image and transport that image to my eye so that I don' see the paper because I can't see the paper but I see the light that presents me with an image takes of the paper as a photo of what the paper was when the light bounced on the paper. How does the process work whereby light can carry a photographic image of what the paper is and convey that to the ball of my eye and where that specific photon that hits my eye is the very one that contains the image of what I wish to see. If all the photons carry the image then how does that work and how do I explain that every photon contains the entire image?

Before science got so wise science back then had the earth in the centre of the Universe. Everything in the Universe was spinning around the earth and hell was where it was very hot in the earth centre. If you went to heaven you rose into the sky and when you went to hell you descended to the centre of the earth. The most valid part of this was that the clergy got everyone so scared they paid huge sums of money to the Church so that the Church would tell God that the person in question had no sins because the Church had his money...and every one knew exactly where the centre of the Universe was. Now we only know where the centre of the Universe is not, because according to science it is not in the centre of the earth.

Without knowing where the Universe centre is no one will understand how the Universe layout works. Where the centre is was always a question that was nagging humans since humans saw a Universe in the sky. In the Ptolemaic Universe the earth was the centre but then "modern science" rubbished this idea. If you read the book you will see that the centre of the Universe is even much closer to home than what that idea allows. Ptolemy said the sun and everything within the Universe spins around the earth, which then places the earth in the centre of the Universe, and I can see the sun and the stars rotate around the earth and me but me mostly.

FIX STARS
SATURN
JUPITER
MARS
SUN

VENUS
MERCURY
MOON

EARTH

Everything about physics and about mathematics and the cosmos cradles behind this question where to find the centre of the Universe and "modern science" has no clue where to find it. If there was one point where the Universe started then the entire Universe will put all focus there is on that point with all light streaming to that spot. I discovered that point because I went about researching nature and using how nature applies the Universal laws I discovered where to find the centre of the Universe. The principle is simple. The centre point shows us where the Universe started. The Universe started where singularity is and that I prove in **Uncovering Corrupt Science**. Therefore to find the centre of the Universe we have to locate a point where the Universe started when it started with one single point. If I embroil on this idea in this small space it will seem to be exceptionally complicated but when explained correctly with a few pages to spare the answer is laughably simple. To find out where the centre of the Universe is follow my argument. It is simple to do because I can just as well say follow the trajectory of the light.

Such a thought brings to mind the most simplistic answer. Science says the light hits the page bounces from the page and contacts the lens of my eye where the lens conveys the photons becoming electricity to a part of the brain that translate the electricity to an understandable message and that makes one read. Sure sounds simple to me. Is it as simple as that? Ever gave a broader thought about light streaming across the night sky, coming from ends of the Universe we don't even realise it is there? Have you ever given it a thought of how big it is out there and how small we are down here and how far did the light we see travel to meet us where we stand. I have to be in the centre of the Universe if all light comes from all over and find me forming the point to which light travels and where all travelling light eventually ends.

All light throughout the Universe meets at the point you are. All light comes from everywhere to meet you wherever you are. If you weren't the centre light won't flow directly to you.

First I had to put life where life belongs and life isn't in the body but in the Universe. Then I had to connect life to the body by life conducting electricity. I prove how life exerts control over the brain to manipulate the body in movement but it starts where one put life in relation to the Universe. Life is alien to the Universe and is part of time, not space.

That I prove.

Step outside into the night sky see where you are. You see an entire Universe. Every sparkle of light coming from where ever is directed at you. All the light released from any point in the Universe comes to where you stand. This puts you in a location where whatever light there is will come to you. You can see all the light coming to you and it's beaming towards you at the speed of light. This positions you in the centre of the Universe.

The Universe is what we see as the entirety when we see everything in the Universe. How does the photons manage to convey one complete picture coming from as far apart and as wide an area as it does? With a few photons connecting the eye or lens no one ever noticed the wonder of light. The photons reflect a view that seems as if coming from all the billions upon billions of stars with all the information present in one photon. Most information comes from darkness covering an immeasurable area. Yet how many photons can actually connect to the lens of the camera or to the eye to transport such vast information? All the light released from any point in the Universe comes to where you stand. This puts you in a location where whatever light there is will come to you. You can see all the light coming to you and it's beaming towards you at the speed of light. This positions you in the centre of the Universe.

But being in the centre of the Universe you then can't be in space therefore you have to be in time because time is part of any and every point that forms space according to time placing space in whatever location. Light formed our culture. It is said that because we ate meat or because we threw a spear or because we ran upright we got clever.

That is rubbish.

We got civilised when we sat next to a glowing fire and started discussing the stars and what influence stars have on our culture and that became our culture we developed. If we didn't see the sky we wouldn't have formed intellect or culture. It is good and well to couple everything to fire and cooking food but that didn't bring us to embrace intellect. What formed our intellect is when we formed speech and that was when we had a meal and then formed a society sitting around glowing coals. That is when speech developed but looking at the basis of all cultures in antiquity the wisdom was locked in the manner that the culture developed along the lines of the studies of star formations.

First I had to put life where life belongs and life isn't in the body but in the Universe. Then I had to connect life to the body by life conducting electricity. I prove how life exerts control over the brain to manipulate the body in movement but it starts where one puts life in relation to the Universe. Life is alien to the Universe and is part of time, not space.

That I prove.

Step outside into the night sky see where you are. You see an entire Universe. Every sparkle of light coming from where ever is directed at you. If every spot of light represent a star and a star is bigger than the sun is how many photons of light does one a speck of light represent? Let those (the oh so wise) that wants to calculate the driving force of a neutron star just attempt to formulate the reduction is size there has to be when the entire Milky Way reduces to the size of one nerve point in my eye.

Am I my body? I know for a fact being the father of Andre that he is not inside his head. Inside his head are girls in bikinis and girls not in bikinis and girls wearing much less than tiny bikinis and girls chasing him while he chases girls wearing very small garments. With that many girls filling his head I am surprised that there is room for brain matter let alone leaving space for his self assured and arrogant self taking up all of the 120kg his body fills and that self worth are so prominent that can fill all the vacant cracks and spots between the grey-matter. In his body he is not and that I showed and so between what we see and where he is in space and where space is we can't have time while in truth he and you and I am in time. Why is he in time? If I am in time where the F@%$^K am I then? Remember I said you are in the centre of the Universe but it is the Universe time fills in singularity and not in space because time fills the centre of the space in the Universe.

Try to mathematically fit all of this space into this human eye. ➞

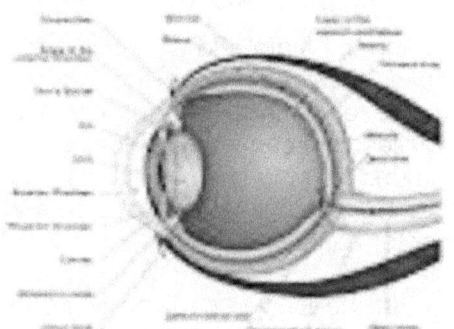

We live in an imaginary Universe built in light in three dimensions while time moving as gravity that moves in darkness. Not darkness I am able to see as light but darkness I am unable to see. Singularity is one spot equal to 1 or 2468123489^0, which comes down to one and in one the entirety that is reality forms gravity. This concept is so much more intricate than what Newtonian stupidity could ever realise and that says why they are all stupid atheists without having the ability to understand reality and how the cosmos works outside their lies.

We are able to see the entire Universe because life that does the viewing by only using the eye to see everything is in the singular dimension of singularity which is one spot Π^0 that is equal to 1 and it is also equal to as many as 2468123489^0 because being part of singularity brings equality $1^0 = 1^1 = \Pi^0 = 2468123489^0$ and that is why I can see al space by applying one single electron that enters my eye and dhows me the entirety of universal cosmic dimensional equality where everything is in singularity. Mathematics show that $1^0 = 1^1 = \Pi^0 = 2468123489^0$. Singularity is time and time is in a dimension we can't understand because although we are part of time in singularity we can't witness it as we reside in the centre of the entire Universe.

But being in the centre of the Universe you then can't be in space therefore you have to be in time because time is part of any and every point that forms space according to time placing space in whatever location. There is singularity controlling time in which life is and then there is the human body material in which space is and the body is to life just more space. By dying life removes from space and the body goes cosmic by getting destroyed back to atoms.

It is the way we appreciated stars that made us develop beyond animals and brought us to have an intellectual group of people. Yet as far as I know I am the only one who asked AND answered the question as to how can we see what is out there in relation to what we have. Man has been living with light from before man became intelligent if man ever became intelligent and yet we have the most small-minded approach to the most complex issue man could ever devise: how can we see all the space through something as small as the space of an eye? Space is never-ending and so is our view thereof looking through an eye so small.

Newtonian wisdom currently dictates the idea that it is the brain that controls the body and as usual just before the reasoning can become complicated they stop to avoid issues they can't deal with. If it is the brain that controls the body there has to be a controlling aspect that decides which part of the brain should be active in any particular instant. If there weren't such a mechanism controlling the brain and dictating which part of the brain should send messages in that instant wherever in the body then brain would release all the electrons instantaneously and then cause an uncontrolled electricity shortcut.

This will cause instant death because every motorised action within the body will try to do what the rest does because there is no control to direct which part of the brain must be active when. There is control over what the brain does and that is in the mind. I show that life is part of time and the body is part of space but singularity is time and space is the Universe we see. This is all new and this line of thought was never before pursued because I am the first to show singularity forms gravity by implementing four cosmic laws. Taking control from singularity to space follows paths we would consider as dimensions and dimensions are a product of mathematics.

There is singularity controlling time in which life is and then there is the human body material in which space is and the body is to life just more space. By dying life removes from space and the body goes cosmic by breaking down to atoms.

Everything about physics and about mathematics and the cosmos cradles behind this question where to find the centre of the Universe and "modern science' has no clue where to find it. If there was one point where the Universe started then the entire Universe will put all focus there is on that point with all light streaming to that spot. I discovered that point because I went about researching nature and using how nature applies the Universal laws I discovered the centre of the Universe. The principle is simple. It shows where the Universe started. The Universe started where singularity is and that I prove in **Uncovering Corrupt Science**. Therefore we have to locate a point where the Universe started. If I embroil on this idea in this small space it will seem to be exceptionally complicated but when explained correctly with a few pages to spare the answer is laughably simple. To find out where the centre of the Universe is follow my argument.

The entire Universe is light and that is why we can focus on one image with lenses and with another lens view a picture as wide as we wish to see. However what is out there is light either expanding or contracted but it is light we see that fills the cosmos. We see darkness as light expanding thus moving away and the white light we see is light coming towards us. Have you sat back and think about how light enables anyone to read? Such a thought brings to mind the most simplistic answer. Science says the light hits the page, bounces from the page and contacts the lens of my eye where the lens conveys the photons becoming electricity to a part of the brain that translate the electricity to an understandable message and that makes one read. Sure sounds simple to me. Is it as simple as that? Ever gave a broader thought about light streaming across the night sky, coming from ends of the Universe we don't even realise it is there? Have you ever given it a thought of how big it is out there and how small we are down here and how far did the light we see travel to meet us where we stand.

The Universe is what we see as the entirety when we see everything in the Universe. How does the photons manage to convey one complete picture coming from as far apart and as wide an area as it does? With a few photons connecting the eye or lens no one ever noticed the wonder of light. The photons reflect a view that seems as if coming from all the billions upon billions of stars with all the information present in one photon. Most information comes from darkness covering an immeasurable area. Yet how many photons can actually connect to the lens of the camera or to the eye to transport such vast information?

Light formed our culture. If we didn't see the sky we wouldn't have formed intellect or culture. It is good and well to couple everything to fire and cooking food but that didn't bring us to embrace intellect. What formed our intellect is when we formed speech and that was when we had a meal and then formed a society sitting around glowing coals. That is when speech developed but looking at the basis of all cultures in antiquity the wisdom was locked in the manner that the culture developed along the lines of the studies of star formations. All great societies embraced stargazing as a mythological culture and formed the basis of antique religion development. The basis of nation greatness was star gazing intellectual development.

It is the way we appreciated stars that made us develop beyond animals and brought us to have an intellectual group of people. Yet as far as I know I am the only one who asked AND answered the question as to how can we see what is out there in relation to what we have. Man has been living with light from before man became intelligent if man ever became intelligent and yet we have the most small-minded approach to the most complex issue man could ever devise: how can we see all the space through something as small as the space of an eye? Space is never-ending and so is our view thereof looking through an eye so small?

Look at the earth versus the sun on the next page and picture yourself on earth compared to the sun. Then place your self in the picture as you are on earth. That dot that you don't see is the entire earth that is so big to me its size goes beyond my grasp. Now picture me trying to grasp the size of the sun.

The following concept is in many ways dedicated to atheists and moreover directed to those physicists with larger-than-life complex taxed with the more-than-equal-and-even-better-than-God-Almighty syndrome which all small-minded Newtonian astrophysicists have. Then see your eye in the picture. Then see the nerve ending of your eye that you use to see in the picture in space with the sun. See the photon entering the eye and remember the photon is so small no one could ever measure it and yet being so small it still has the able ness to convey all the space that you see throughout the vastness of the Universe. Do you get the picture?
See the reduction there is and the information converted into vision?
Now that we know our human position in the solar system let us go cosmic.

The Milky Way is as big as an atom when compared to space holding another hundred billion Milky Ways in space. If you think you are an accomplished astrophysicist working with real numbers and with a better-than God complex and you want to have reality in your work then put the eye in space in ratio with one hundred billion Milky Ways in space. The tiny light specs you see as far-off galactica drown in the overwhelming darkness of space that fills the real picture. Remember with such vastness the galactica, as material in ratio in space not being shiny material when compared the material doesn't even show up as dots in the space viewed as darkness in the space holding the hundred billion Milky Ways. So as an accomplished Newtonian astrophysicist don't find the edge of the Universe but calculate the space reduction in terms of how the cosmos fits into your eye and get you place and position you have in our perfect God-Created cosmos where the cosmos is a mere thought God Almighty has. That the cosmos is a mere thought of God Almighty I also prove mathematically but that work is complicated beyond mere explaining and is advanced to the extreme.

So how did the Universe start mathematically from one point as Einstein said it did? Where you might think this is old news I have news for you because Einstein never got further and I did. This is where the entire concept changes because I prove that gravity and electricity is the very same thing and so is electromagnetism and nuclear energy but each is on a different level of intensity.
What would I mean by that. Let's take mathematics.
There is the dimension of adding 1+1 = 2.
That is one dimension because at that dimension one will have 1x1 = 1 and $1 \times 8756543^0 = 1$.
The there is multiplication where this is 2 X 2 = 4 while this also is where 2 + 2 = 4 and this alternates at 3 x 3 = 9 and 3 + 3 = 6. This is a dimensional change. Then a dimension comes about when 2 x 2 x 2 = 8 and 2 + 2 + 2 = 6. Therefore at 2^3 another dimension intervenes. This is exactly how the Universe progresses and new dimension changed the formation of the Universe. Life applies electricity to control the brain that manipulates the body into movement. In that the gravity plays a huge role in our ability to think.

I shall quickly and briefly explain but only in the books does these arguments find proof.
7×2^0 is a gravitational border and therefore round objects spin at 7° directional change.

7×2^1 is a gravitational border where gravity forms space by also moving linear (7 going circular and 7 going straight and every straight is a circle forming).

7×2^2 is where light controls space by forming a circle in pi.

7×2^3 is where light ends as space and iron can charge electricity as well as gravity.

7×2^4 is where the element table starts by putting gravity and electricity into movement.

At 7×2^4 is why time becomes the fourth dimension that provides movement to a three dimensional Universe.

So electricity is on another dimension (2 x 2) than gravity (2 + 2) but is the same thing. I am not going into the proof of this in this article but that I do in the book. However one cannot discard the influence of gravity in favour of electricity or visa versa and that is why we have to test how electricity is conducted in terms of gravity.

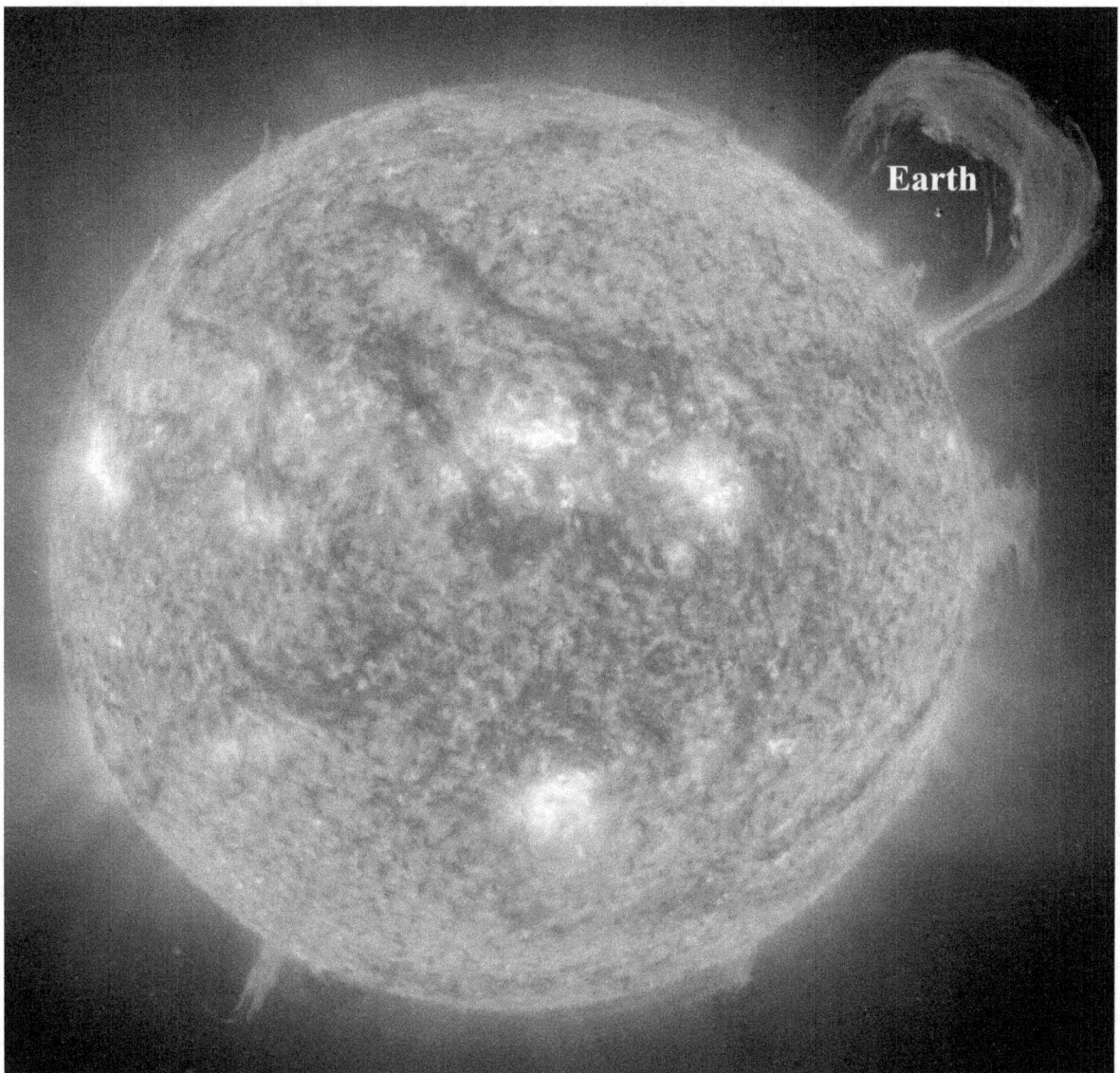

Look at what is the smallest cosmic object in the sky, which is the moon. Try to vision that you are able to see an entire country such as Britain on the moon and you might be able to entertain such a thought. Now put England within Britain. The view that you can see becomes very narrow. Put London in England in Britain on the moon and there is no chance that you will see London. Then put yourself in London and put your eye sock in your head in your body in England in Britain on the moon at the distance the moon is from the earth and calculate the scale that space has to reduce just to accommodate the reduction from the moon to the size of your eye. Interpreting this mathematically no number will ever make sense. This reduction in size is beyond comparing when you put the moon in terms of the sun and the sun in terms of the Milky Way. We know in the Milky Way the sun does not even come across as a dot.

To do this I show how gravity conducts in terms of four cosmic laws governing gravity and this same process conducts electricity. The Titius Bode law, the Roche limit, the Lagrangian points and the

Coanda effect form the cosmic structural code. I show how these four form the cosmic code and the cosmic code is in the way that Π forms gravity and how gravity is Π. Where these laws form gravity by

forming Π and measuring Π as a value of $\Pi = \dfrac{10 + 10 + 1.991}{7}$ as space and

$\Pi = \dfrac{7 + 7 + 7 + 0.991}{7}$ as time which puts time which is movement in time and space in

movement and this is crucial when going into something as massively complicated as how life controls the body.

I manage to answer these never asked before questions because I resolved the mystery behind the four Cosmic Pillars. I wish to make one fact very clear. I base my work on formulating the working process of four cosmic principles in Nature. These are:

1) The Coanda effect
2) The Titius Bode law
3) The Roche limit
4) The Lagrangian points.

I did not discover these phenomena because science knows about these phenomena for a very long time and in some cases even for hundreds of years. Science knows they apply and where they apply. When science discovered or allocated missing planets they used the law applying such as the Titius Bode law from which they deducted positions that they knew in that circle according to the planetary layout that the law predicts there had to be a planet according to the law. Science did not apply Newton's formula to discover and locate planets but they applied these phenomena and especially applied the law of planetary allocation to discover the precise location the planets discovered after Galileo.

Everyone in science knows these phenomena are there and are in place and they rule the orbit set-up of the planets. The solar system functions according to them. These four laws on planetary motion that are used by nature at this moment and have been in place since time began, are what apply and they dismiss Newton. If you argue with me about Newton being correct you better take your case to God or the solar system because the four cosmic phenomena are working in nature and nothing Newton said is applying in nature. This is a truth and a fact and a foregone conclusion and can never to be in doubt.

Brainwashed as you may be in believing Newton you can't either side with my view or decide on Newton because it is not a case of choosing between Newton and me. I'm out of the picture! It is either telling nature to listen to Newton and change what is in place or read and see what is in place and what is applying in nature all along. The phenomena are what we find to be used in the cosmos while Newton is in the imagination part of the minds of scientists and nowhere else. If you don't believe me and if you wish to discredit me first find out a little more about science. Then deflate your ego as to what you think you know.

Science never mentions these phenomena because science can't use Newton and explain these phenomena or use these phenomena to prove Newton. These four phenomena that the cosmos uses as we speak have been in place ever since the Universe formed. Since science can't explain the phenomena and the phenomena destroy the credibility of Newton science avoids these phenomena as if it brought the plague. You can't choose between Newton and me because I did not put the cosmic phenomena in place. All I did was doing a study since 1977 to formulate how and why these phenomena work and how these phenomena keep the cosmos and the solar system working. I am the first in history to show why they work. We are all been brainwashed for centuries to believe Newton. Should your brainwashing kick in and you have an axe to grind with me about what I say, then first prove these phenomena are not in the cosmos and are not applying to form the laws that the cosmos put in place as gravity. I only found out how they work and why they work and I did not make the phenomena work. All those clever stooges that have so much to say even before you read first learn what is in place before getting so opinionated.

So the next question is why is there singularity and the question about that question is what is singularity and moreover where do I locate singularity that is time that is gravity that is movement that forms the next moment which life is also another part of moving with time in time by standing aside from cosmic movement but filling space within the cosmic movement.

There was one spot, a point where the lot we got started. To have one spot means time stood still because time is forever. But time always has to run and therefore we must look why it seems to us as if time stood still.

Today we know that time forms by Π and Π forms by what we know as the Titius Bode law.

Whatever is, forms by relevancy between a primary singularity, a controlling singularity and a governing singularity.

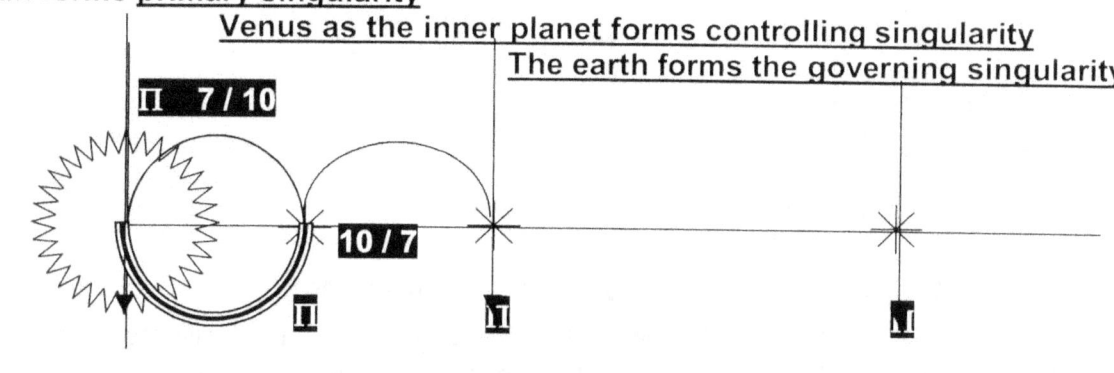

This combination of three forms time as time forms space.

In time there is the future the present and the past. The sun shines on me and by having the rays take 8.5 minutes to go from the sun to where I am on earth puts the sun into my past. The sun is the primary singularity from where the light advanced to the controlling singularity that contracted and collected the light and I as the beneficiary harvested the light making me the governing singularity. However since the sun shims to me there is rays heading my way from the sun which puts the sun in my future because the rays coming towards me is still way into my future. This is the relevancy of time that forms space.

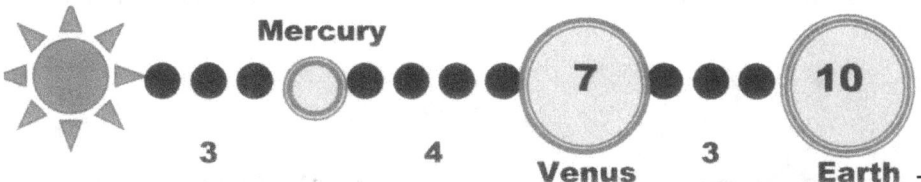

The sun holding three and Mercury holding four puts the earth at ten and that means the sun as it is in my past at three so it is in my future as three and that goes for the position of Venus at 7 in relation to the earth. The one is always primary singularity, holding the other at controlling singularity forming the governing singularity. This is the way the Titius Bode law forms the solar system and time forms space.

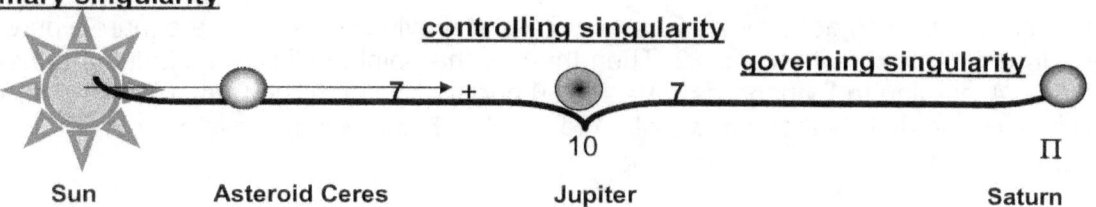

In this we find the dynamics how the Universe formed. This is the guideline that will show how the Universe forms space. It is either three rotating by seven degrees or it is double ten plus singularity coming from the sun 0.991 plus one coming from the point singularity is placed and this can be anywhere because singularity in mercury and Venus and the earth are not identical but it is the very same point. Therefore it is 1 in all three or just one and it is 0.991 making the sun our future.

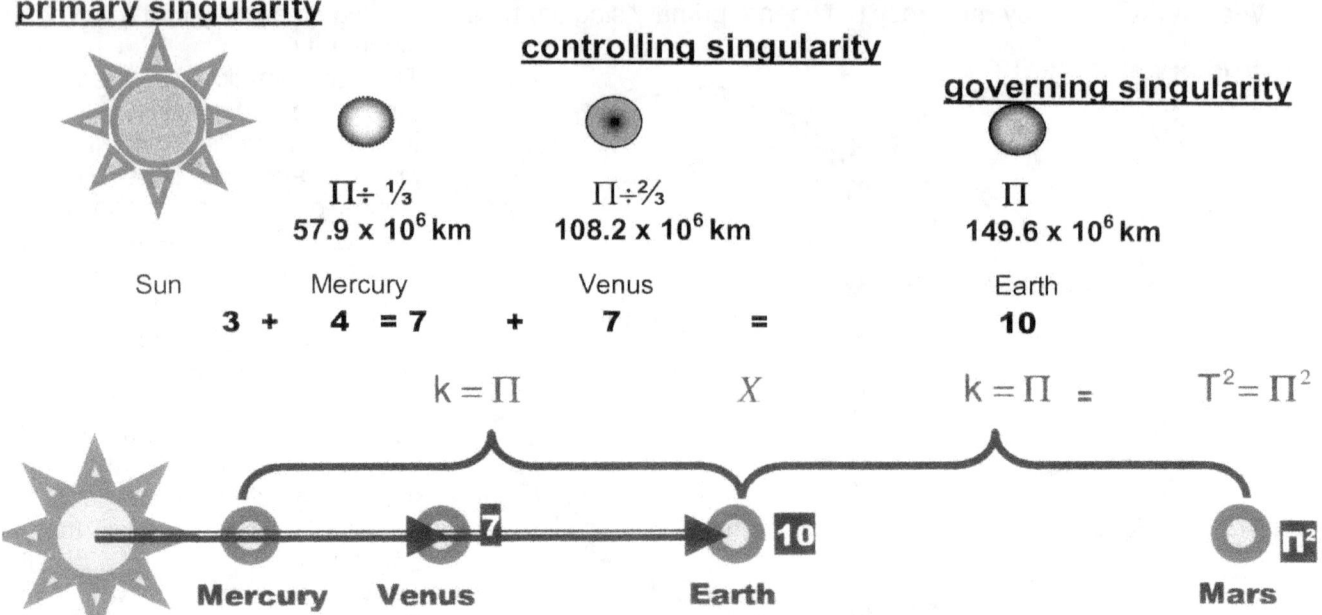

It is the movement in relevance taken from Kepler's tables that form Π as 7 + 7 is 14 mad 10 divided by 7 is 1.42. Then the division that this brines about forms Π^2.

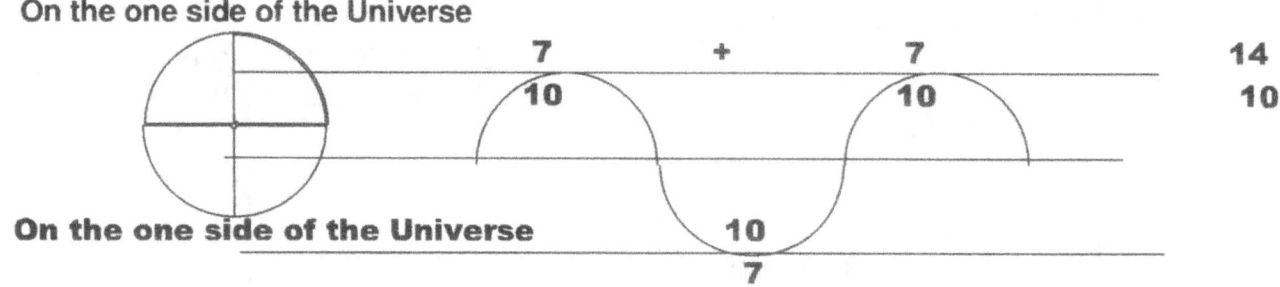

Movement in the Universe is 1.4 / 1.42 = 0.986 = 0.986 X 10 = Π^2

This is how gravity forms and this is how the Titius Bode law forms the Universe. By movement dividing into space it results in the square of Π^2 or Π^2. This is how nature produces gravity and how gravity or movement or time or whatever you wish to call it forms space by taking time and leaving space behind.

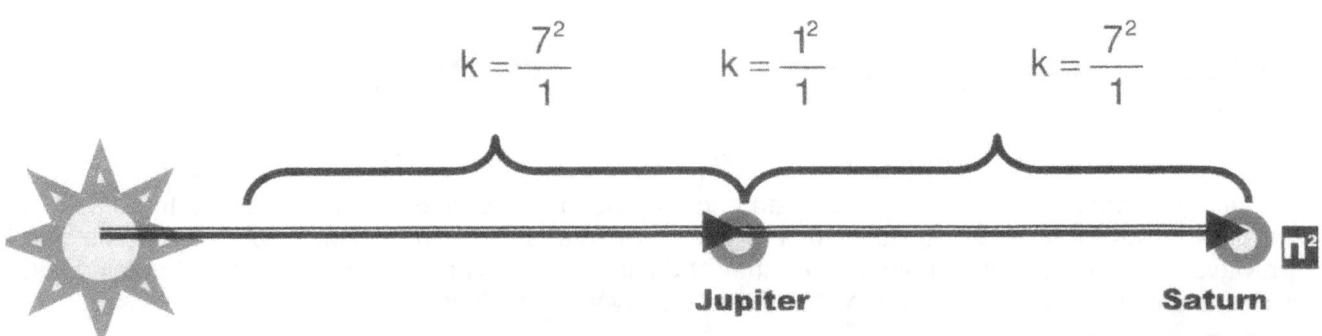

In the movement the turning goes by 7°. Going by 7o and moving places 7 in a square where it moves from point 7 to point 7 and that then is 49. Then there is the point holding singularity and that is 1 x 1, which also is 1. According to Pythagoras 49 + 1 = 50 and that going twice form 100. To get the square root of 100 it would be 0. This then shows why the 3 + 4 is 7 and why there is a 10 in relation to 7 and all of this comes about as a result of Π forming.

$$k = \frac{7^2}{1} = 49 + 1 = 50 \quad k = \frac{1^2}{1} = 1 \quad k = \frac{7^2}{1} = 49 + 1 = 50$$

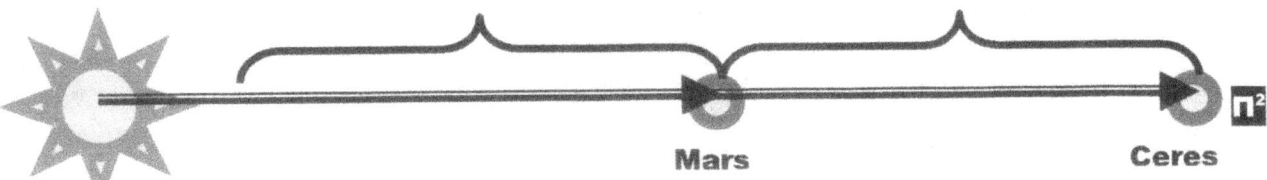

Mars   Ceres

<u>primary singularity</u>

<u>controlling singularity</u>

<u>governing singularity</u>

Sun   Mercury   Venus   Earth   Mars

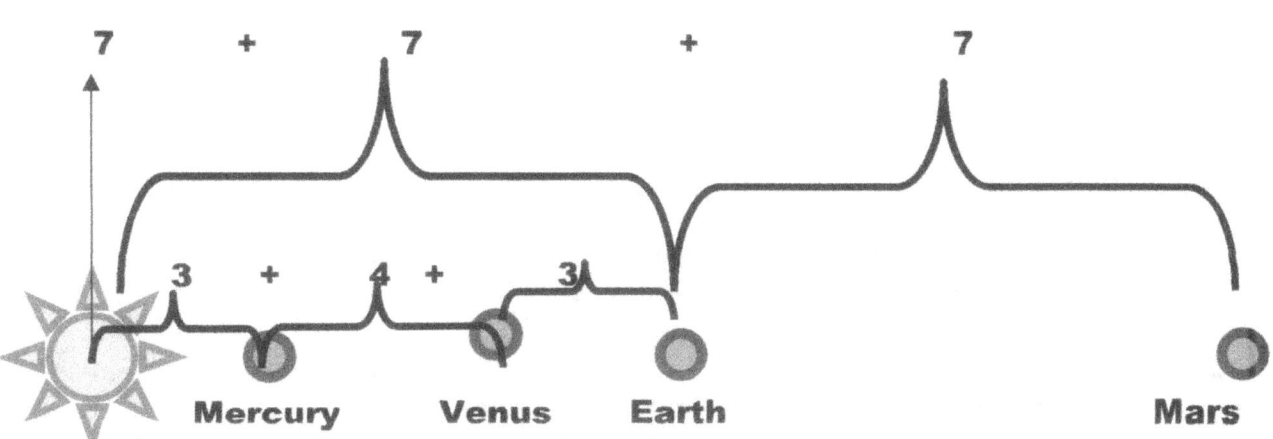

This is the way Π^2 forms by Π multiplying with Π and from this movement space comes in place. It is $k^0 = \frac{1}{10}$ which is singularity in ratio with the sun times $k^0 = \frac{3}{10}$ which is singularity in ratio with the inner planet in ratio with the outside planet $k^0 = \frac{3}{10}$ that leaves the future at 0.009

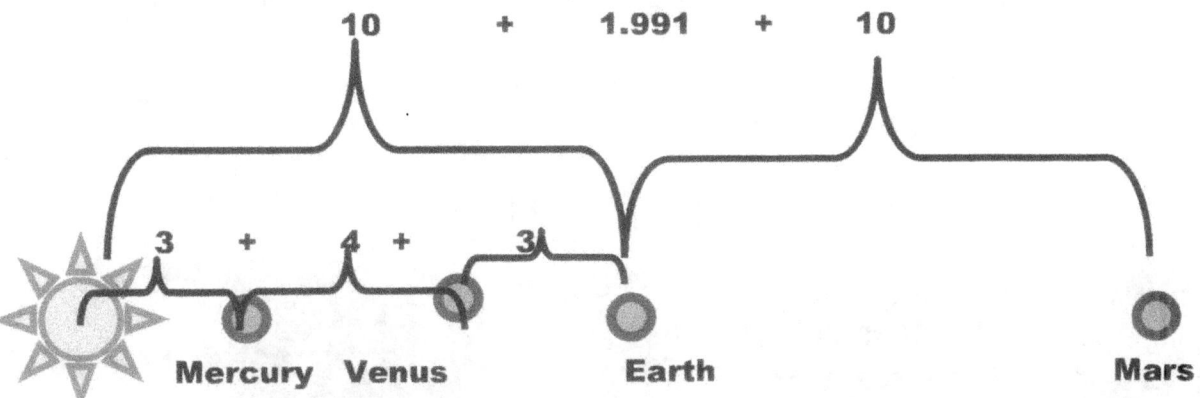

All this proves that by moving through 7° in relation to a direction change of 7° gravity forms. One thing must be remembered. If anything becomes part of the Universe it stays part of the Universe because it has nowhere to go. That is why life is never part of the Universe but the body life controls is and the body life leaves behind while life stays in time. Life is controlling space in the Universe but life is never part of the Universe. This says that if it applies now the way the Universe starts then that is the way the Universe started when it started.

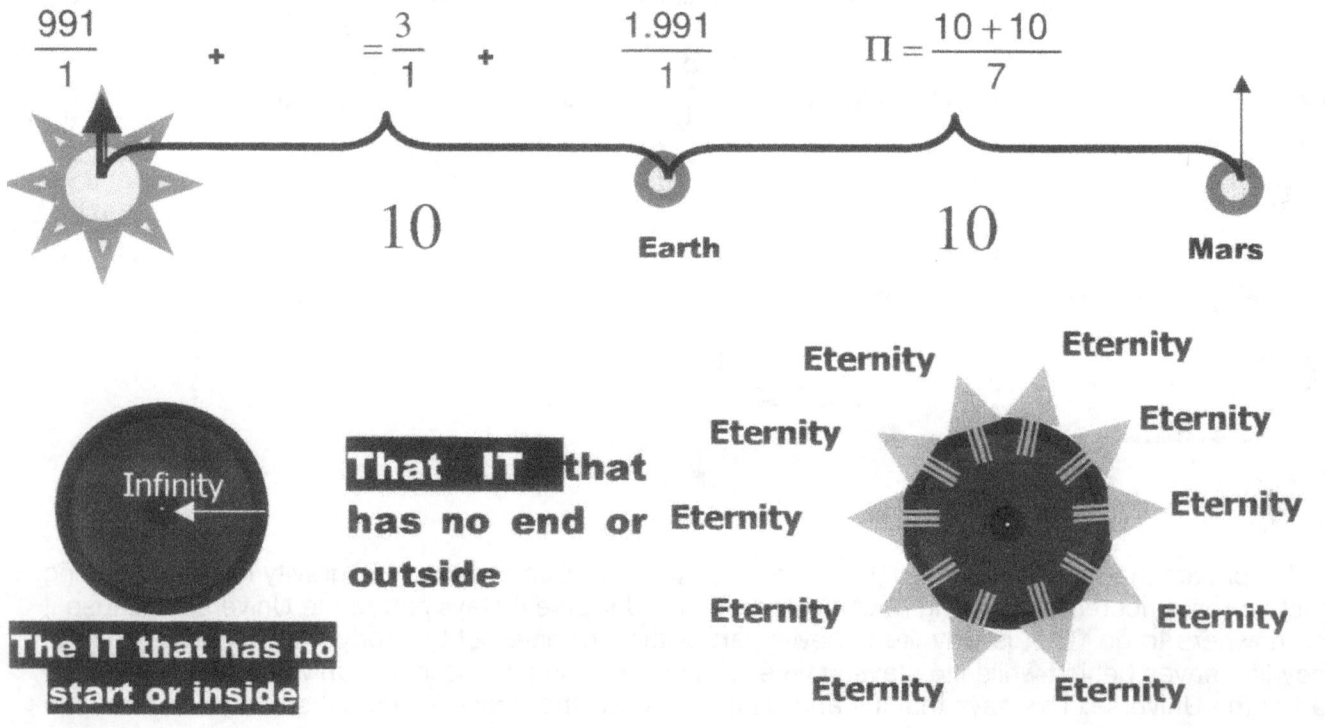

This is how space forms by time moving but since nothing and only nothing leaves the Universe at will (not even Newtonian stupidity can achieve that) we can surmise that this is the process used when the Universe started from one and formed all the dots we can't see.

Taking queue from this we can see how did 1 become two and three and four and going onto five.

Sun

Mars $T^2 \div a^3 =$ 1.000 @ $4\Pi^0$

Sun

Earth $T^2 \div a^3 =$ 1.000 @ $3\Pi^0$

Sun

Venus $T^2 \div a^3 =$ 0.992 @ $2\Pi^0$

Sun

Mercury $T^2 \div a^3 =$ 0.983 @ Π^0

7 ÷ 7 ÷ 7 ÷ 0.991 = 21 991 ÷ 7 = Π

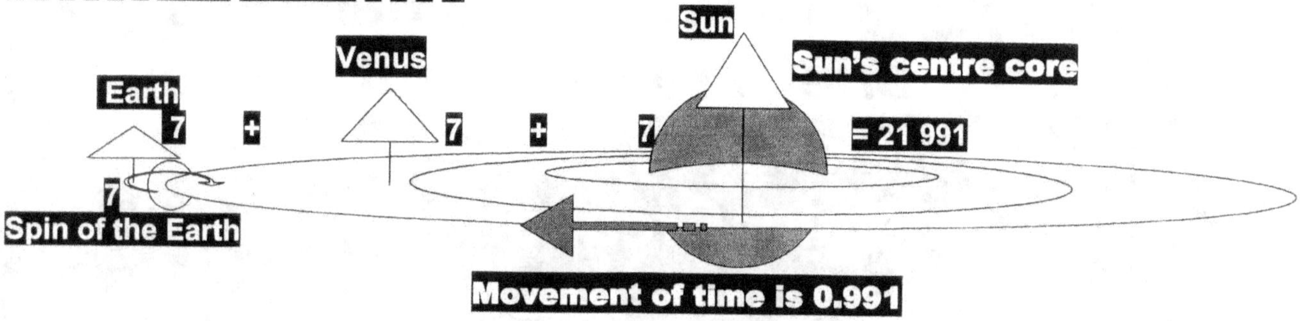

Sun

Sun's centre core

Venus

Earth
7 + 7 + 7 = 21 991

7

Spin of the Earth

Movement of time is 0.991

However how did this start? How did one end up as two and then three and so on... To answer this we must first look at why one was one was one and remained one before one became two and three and four and so on.

We all agree there was this spot. It was so small it still does not fit in our Universe. Only when this becomes a dot Π it forms space, which is part of the Universe.

The dot represented time because time is eternal and infinitive at the same time. The spot never changes but remained the same. The spot was so perfect in every way that as the future spot came to the present it remained what was in the past because there was no difference in the way the dot repeated. I have shown that this spot can only be thought since it is too small to hold material.

Part 7

is a small extraction from a part of the Theses called The Absolute Relevancy of Singularity In terms of The Cosmic Creation, which is also named Part 8. This very brief article about the book The Absolute Relevancy of Singularity In terms of The Cosmic Creation or Part 8 of the theses called The Absolute Relevancy of Singularity allows an open window on a process whereby the cosmos was created before the cosmic birth applied. The Cosmic Birth is now known as the Big Bang and what a stupid name that is.

In other words I explain Genesis 1 verse 1 for the first time and I prove my explaining with the support of bringing mathematical backing.

Think of what you wish and put that in terms of what was available when the Universe started and you are wrong because it was not present. Whatever you are able to think of or not think of was not in place to be thought of at the point when the Universe started. Numerical numbers and numbers in order came eternities later as a thought that progressed from inventions that came before and forming as part of how the Universe grew into what became available. The number one was one such a number into which the Universe grew as one came as an invention in a planned future. You reading this were not a possibility. The words you read and the thought you think was not yet invented. The light you use was not anything and the electricity by which you think was not yet invented. The space from which to gather the electricity to charge the thought you use was not invented. What you are in terms of what you think you are was never yet a concept because being a concept was not yet a concept. Even if you think of nothing; nothing still was more than what there was.

It is said that Einstein proved that the Universe started with one point, a single point but as usual I am going to be different. It did not start with one point because when it reached the one point stage the Universe was well on its way to progress into what it is now, and opportunity already had value. I am referring to when opportunity did not exist as a concept because a concept did not yet realise. Please read very carefully for I have to use words that were not to describe events that did not yet take place to show what was never in place before. If I say there was blank then that is incorrect because being blank is valid by meaning in definition and blankness at that point had no meaning to form definitions because blankness was not part of what was in place. Even vacant ness was much more than what was available. The idea that something such as vacancy is was not yet invented.

Using 0 or what you symbolize as forming the picture by which you think was not there so you who think were unable to think because thinking, as a process was not in place. The number 0 representing nothing was not in place because the number 0 holds a place and a space and a meaning and a symbolic value which was all still absent. If you think of a dot • not being there you are wrong because the vision of something forming a symbolic value such as the dot • was out of bound and the thought that there could be a symbol formed • was meaningless because being meaningless was yet some futuristic concept not yet conceived. There was no blackness of absentness because of the lack of present ness.

Any shape of whatever form forming sequence was not yet conceived. A triangle was not yet in place. A straight line was something in the future and a circle was something not thought of. The law of Pythagoras was still to come as a master thought on which the rest was built. Numerical mathematics was something unheard of and being unheard of was what the Universe was still progressing towards but was not yet understood. Unheard-of was futuristic, something to progress towards. Being understood was unheard of because even nothing was a concept to progress towards as being brought by the future. Even the future was not a possibility yet for there was no past to recollect during a present that was not.

When the Universe started there was nothing but nothing was much more than what there was. It started with zero except zero was much in the future. When the Universe started there was no future because there was no past because there was no present. There was no zero because zero was still an idea to be invented because ideas still had to come. Even inventing was an idea still not part of the Universe. To think of what to be was not yet to be.

There was nothing except that nothing has a reference to something and when the Universe started there was no reference even to nothing. Even being in and part of a concept was still not invented because a concept was not invented yet. The fact that 0 meant something was not yet a practical part of the cosmic-idea because 0 was not yet thought of just because what we think of as meaning thought of was not yet thought of and thought of did not exist to be part of a meaningful Universe. What ever is present now was absent then.

In most if not all my other work I tried to keep religion as a factor and the Bible as a science component out but then I have to start where science are at present, which is after everything already started. However to get to the start, the very place everything started there is no other place scientifically to start but to start exactly where the Bible started and that is the beginning. The Bible takes us to pre-mathematics into true singularity.

True singularity is not the horseshit science concocted but a valid point. To start at where we don't involve the Bible or religion is where Einstein started but then we loose the very sense of how, where and why it started. I agree that the why as in knowing what God Almighty wanted when He started the lot is way out of my and every other human's league and nobody is worthy of that, but the why in terms of cosmic perspective I can form an idea.

But starting at where Einstein started with one dot is avoiding the start because the start is not in the middle but it is in the beginning before the spot was in place to begin the dot. It was before there was a thought of a dot or s spot. Lets get to singularity and then see where that is. Then we get to how the Bible describes what there is when what there is started to form what is there at present in the now. Look at what science offer in terms of Einstein's vision on singularity and if it was not so stupid it would be funny but now it is just plainly Newtonian. I see a circle with dimensions and I see a

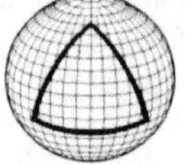 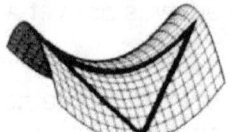 

Closed Geometry   Open Geometry   Flat Geometry

saddle formation with dimensions and I see a flat surface but that too must have six dimension if it has any because on all of these forms we have sides and sides as it is bring borders in that the borders have to have three dimensions. They, and now I am referring to those in science that developed this idea about what singularity is, they are as incorrect and mindless about facts as a monkey is about religion. Singularity is one and one or singularity holds no more than one and where singularity ends there sides start at a value more than one but that is where singularity ends. In sensuality fits only a thought and that is because singularity is where space starts and to have material with or without density requires sides, which singularity in a group can form but in singleness it can't.

The Coanda effect that joins time by committing material to gravity

Time in eternity forming the motion or liquid that locks space in as material

In the center where we see an axis we find singularity as Π^0. That is r^0 or 1^0 or it cold be $24589421^0 = 1$ and this is all 1, which is the point where singularity is. Where singularity forms space it forms a circle because at that point where singularity Π^0 becomes space it forms a circle Π and where space starts we have Π^0 becoming Π only because movement Π^2 develops $\Pi^0\Pi$ Time in infinity forming the motionless or solid that locks space out as material

Now we are at the point where we have to bring in the Bible because if science is where they are in the pictures they present that they think singularity is then those poor witless atheistic slobs are where they are with everything including science perception and that is nowhere with nothing that they fill an entire Universe with. Look at the figures they say forms singularity and then understand the poor intellect we deal with when we deal with Newtonian perception about science and then also realise how far are the atheists off the mark in science as much as in everything they present and represent. They give a picture of a square shape and a saddle shape and a sphere as a shape and theses forms are at best identifiable as forms and singularity is being one and without form. It is being in the void as the Bible says and this void is being void of form, of space, of shape of sides, of even our understanding of what it is. That to them is singularity and singularity is that which is the void of form because it represents one being without form. I do realise that they refer to what the Universe grows into but what the Universe grows into is what singularity is and where to singularity extends and this is what this chapter is all about. From singularity the Universe grows into Π but that is using the Lagrangian points. The Titius Bode law, the Coanda effect and the Roche limit and singularity extend space in accordance with that. If singularity extended as a surface then singularity must present some surface and that it does not do because singularity has to group together as 3.1416/1 to form a form. Then we have the Bible that is light-years ahead of science and claims that the earth was without form and void. This brings us to the reality of singularity as singularity is and not the concocted incorrectness that science is in what science claims reality is.

My Bible as it reads says: IN THE BEGINNING OF CREATION, when God made heaven and earth, the earth was without form and void, with darkness over the face of the abyss, and a mighty wind that swept over the surface of the waters. God said: 'Let there be light', and there was light; and God saw that the light was good, and he separated light from darkness. He called the light day, and the darkness night. So evening came, and morning came, the first day.

The earth was without form and void. This is what the Bible says. I was without form and void with darkness over the abyss. The Bible says when we go before we have one we then must remove singularity as a concept because the earth was even without (darkness meaning you can't see and therefore meaning we can't witness and report on) an abyss and being void even thereof. That means the earth that the Bible refers too is not the earth we think of being the earth because in this case the earth is being without form and to be the earth there at least have to be form.

Lets go and remove form from what the Universe is and that is a circle.
A sphere, which is what the Universe is form by a multitude of cycles uniting in one form. Therefore to go to the sphere we go to the circle. The circle is Πr^2. To remove space we have to remove the radius

$$circle = \left(\frac{\Pi r^2}{r^2}\right) = \Pi$$ and that leaves us with Π. The Bible says the earth was without form and then Π

gives form and so we have to remove Π as form being $$circle = \left(\frac{\Pi}{\Pi}\right) = \Pi^0$$ This is what the Bible says

was in place when everything started and if I leave out the Bible I leave out this part and then I land where Einstein was and that is no-where in particular.   This proves the Bible correct using MATHEMATICS as a tool that formed the Universe. This puts the earth in as Π^0, which is forming singularity.

In singularity there can be no substance because singularity groups together to form substance as Π. This makes Π^0 much smaller than material and the only thing that is smaller than material is thought. Being in thought is being not in material because the brain works with electrons and I have shown an electron holds the value of 3 whereas singularity is 1. That makes the mind use electricity that singularity as 1 charge by movement (future- present-past) to form electricity (3) or the electron. Therefore to be 1 in singularity it can only be in thought being smaller that Π and is therefore Π^0 and that proves mathematically the Universe is a creation of thought and that proves God Almighty by mathematics being a though of God Almighty which then puts my claim that I by the grace and mercy of God Almighty proved the Bible mathematically correct and science as just more fake.
Then came the Universe but as we think the Universe started such not only not with one but that it started is an idea is misplaced since it could not start anything before it first ended everything that was not. The Universe could not start before it ended not being and not being I mean being void of everything and having everything in and part of the abyss. To start a process it had to end what was not a process and this changes everything we see in the Universe. This implicates the progress in time. Eternity still has to stop before infinity starts everything again and that is the process that is in place to this day.

The Universe at the very first initiative ended what was not and so it was to change what was not into becoming what was as what was not what could be but is not yet. Before there was nothing and then that ended into everything changing into anything not being and being nothing. But this idea of not being and everything being nothing only had a function at the next stage when something shifted into putting nothing into a bracket that had a relevancy to nothing. Not to have something can only be if you could have something but not have it yet.

When it started there was nothing to compare what was not to what was not in order to form an opinion what changed to become an idea of what was not. For what was not to find meaning what was possible had to become in place to have a meaning to what was not.

Before that what was still was not the idea of what was not gave no meaning to 1 because there was no one and 1 was still part of everything not invented. Without one in place the measure of zero being numerically 0 had no place because the numerical order that starts with 0 or zero was still an invention that came about with the invention of 1 or one as a concept. To think the Universe started with zero or 0 was jumping time by one eternity because again I have to stress 0 was something that found a place one eternity after 0 or zero became part of the thought that put one or 1 in place. Newtonian atheists are unable to think that far back because Newtonian atheists are unable to think! In their stupidity and mindless shortsightedness everything was in place because their utter stupidity can't stretch to such realities for they are only equal to mindless computed animals. My dog is the biggest atheist but it is not because I have a clever dog...it is because he is as thoughtless and mindless as all atheists are. Atheism is a depreciation and denial of human mental faculty.

We have to be clear about how it started to understand. The Unversed did not start but ended with change before it started with introducing change. It was change that ended what was not in place to bring in place to replace what was not in place. This is very important to realise what drives the

Universe. To start with a beginning it had to end with what it is not in order to replace what was with what it starts with to begin with. This still drives time even in the present. It is changing what is by ending what is in order to introduce the order by which the next instant will start that produces a new Universe as we have a Universe in place.

The only cosmic substance holding form is a cosmic solid. Cosmic liquids adapt to form and cosmic gas adopts form but cosmic solids produce form. However, in the very beginning there was no form just because form grants mathematical principles basic numerical value before numerical principle starts to apply and the form even predates that principle and when no form applies that even predates that which predates that which predated numerical principle in form. That our brilliant mathematical atheists did not know and going back some ten thousand years or more to when the Authentic Author of the Bible first penned the moment - Alpha this was his vision. Just shows how backwards a Newtonian atheist must be to be part of modern thinking and how much brainpower has gone lost as atheists brought in their stupidity to replace true wisdom as we only find in Biblical Scripture. The Authentic Author of the Bible saw that which has form had no form in the beginning and he said that the earth, which is cosmic solid, had no form. It is a clear statement but a fool such as an atheist who can only understand Newton and Newton's unexplained corruption will not have a mind clear and clever enough to understand the Scripture because of a lack of intellect.

Let's see how the Bible, which is the only true authentic document, reflects on this matter and keep in mind it is far too complicated to be understood by mindless zombies such as atheists which form part of countless other mentally deprived sub human intellectuals.

This is directly word for word quoted from my Bible as it is: IN THE BEGINNING OF CREATION, when God made heaven and earth, the earth was without form and void, with darkness over the face of the abyss, and a mighty wind that swept over the surface of the waters. God said: 'Let there be light', and there was light; and God saw that the light was good, and he separated light from darkness. He called the light day, and the darkness night. So evening came, and morning came, the first day. Fantastic! I could never in a million words find a better way to describe moment-Alpha and I have all the vision of "modern science" and telescopes to aid me while the Authentic Bible Author only had a vision.

IN THE BEGINNING OF CREATION, 1 when God made heaven and earth, 2 the earth was without form and void... 3... with darkness over the face of the abyss...

The Creator Created heaven and earth which is what cosmic solids (amongst all others also the earth) is and the Creator Created the earth without form (mathematical shape, form, terminology, was not yet created and all there was at the time was void ness, nothingness, the absolute lack of what now is and the absence of whatever came later on. The Authentic Author saw nothing and seeing nothing was darkness and absence of information, a something not there that even nothingness could not describe. Even the abyss the nothingness was hidden in a shroud of darkness to such an extent the nothingness was hidden as part of it not there even to be realised. This is what the Authentic Author saw ten thousand years ago and how he described the event using words when we don't even to this day have words in which I can describe what was not present when *IN THE BEGINNING OF CREATION*! What was present was the abyss and even the abyss, which is a place that is so small it is not, was even smaller because it was covered in darkness. ...And a mighty wind that swept over the surface of the waters. What was forming got more of what formed.

Then came the water because material came in place. The solids was covered by cosmic liquids as 1^0 became 1^1 and 1^1 became 1^2 and 1^3 ...the relevancy of a time line got in focus. Its not then came the Big Bang or as the Bible describes it, God said: 'Let there be light', and there was light; and God saw that the light was good, and he separated light from darkness. Immense time lapse came into play and a cosmos filled with development came about. He called the light day, and the darkness night. So evening came, and morning came, the first day. He then created darkness as that which we take as night and as light as what we call day. This puts relevancy in focus for the first time. For the first time there was contrast, which is what the cosmos is all about. What goes forward goes backward on the other side.

Now I am approaching the point where I start to go to the dimension where 1^0 moves to form 1^1 and what that entails. This process stitches the Universe together in lines that can't be separated even by gravity. If you look at the Titus Bode law every dot is stuck to a centre.

The winds that the Bible refers to are possibilities that swept in as changes became part of the Universe and possibilities what might be in the future formed the Universe. What is in the Universe came in the Universe not by magic because as scientists the last thing that any scientist can believe in is magic. Every possibility present in the Universe was deliberately placed in the Universe for a reason and the Creator that created all also placed all the possibilities in the Universe. That which I previously mentioned the Bible says, but since I am no Clergy and in that I am not going to go into. I only share my religion with my very close family while the part I shared with you is the extract that forms part of science. It is pure science because it can have no theological interpretation. For that the message is too scientifically clear and only a Clergy and a Newtonian can be stupid enough to miss the meaning of what is said as I showed what is said.

We must get away from words and terminology and into understanding by recognising concepts giving meaning instead of clinging to words where as in the case of Newtonian science they give a name to use words to avoid explaining a concept. One such a case is anti-matter. Anti-matter is just a name they invented to hide their stupidity of not understanding anything. Where the Bible refers to earth and heaven we have to seek the concept and not the word because we have no longer the very original language to refer too and see what the original word meant. The Bible was written before names came in place and concepts substituted for modern stupidity. The very similar concept is understood in the term the Bible uses where the Bible says, "God said let there be light and there was light". Now everybody goes into the magic mode. God said and it was. One instant is was not and the next instant it was. To God time means nothing and that is the very way all started and to God all still is. We must never forget this fact about God but to us humans time forming space rules our existence. Therefore to God it was by word but to us it took eternity upon eternity to become established because we are so insignificant compared the Greatness of God. To God space is a thought 1^0 to 1^1 and time is non-existent with infinity holding eternity.

"God said let there be light and there was light" Throughout this book and almost if not in every chapter I prove that the cosmos forms by the forming of heat. Light collapses and movement in space ends. It is gravity that develops or collapses heat to form space or to remove space where space is just heat in different density forms forming different parts of the Universe. Again I go back to this quotation from the Bible and quoted directly word for word quoted from my Bible as it is: IN THE BEGINNING OF CREATION, when God made heaven and earth, the earth was without form and void, with darkness over the face of the abyss, and a mighty wind that swept over the surface of the waters. God said: 'Let there be light', and there was light; and God saw that the light was good, and he separated light from darkness. He called the light day, and the darkness night. So evening came, and morning came, the first day. I have shown in every chapter that heat forms the Universe in density.

This is the concept outranking words and destroying Newtonian stupidity. God said: 'Let there be light', and there was light; and God saw that the light was good, and he separated light from darkness. He called the light day, and the darkness night. This means that light is heat. There can be no light if there is no heat. To have light is to have heat. It is not the Big Bang dimensional heat or light we are referring to where the hallucination of space started a cosmos but where there became contrast. This is what the Bible says! It says: and God saw that the light was good, and he separated light from darkness. He called the light day, and the darkness night. There became distinction. There became one versus the other. There was light in opposition to darkness where before there could not be darkness in the absence of light not forming a Universe. It is not the words used as day, and night but the concept of having light and day, and that concept contrasting the darkness at night.

What is most note worthy is that in sharp contrast to Newtonian atheists stupidity the Bible was correct (I guess) about twelve thousand years ago when it says that: God said: 'Let there be light', and there was light; and God saw that the light was good, and he separated light from darkness. He called the light day, and the darkness night. With this concept God did not create day and night but an entire Universe forming on the concept of contrast between what is cold and what is hot; what moves towards as white light and what moves away as dark light; what is remaining still and what is forever moving; what can never reduce and what can never stop increasing because of light or heat and darkness or cold. It is concepts that form and if human stupidity does not understand this because the Clergy wants to earn money selling religion and thereby performing as if is they are God's incarnation on earth or science impersonating God by trying to defeat God it is all the same; it is human stupidity bringing separation between science and religion. Science is religion and by science God used mathematics to form a Universe. God said: Let there be light and a Universes came about by mathematics forming heat. That is what the Bible says and that is what the Universe is it is light or heat forming contrasts by darkness and night and light and day. It is the concept forming.

Then the stage came about where what we now have and what we now find to form links, linked up eternally. We are linked to the sun and that link can never break but also the sun links to the Milky Way and that link can never break and as much as we cannot escape from the sun so the sun can't escape from the Milky Way. Everything probable became everything possible as the plane began to link. As 1^0 linked to 1^1 then that linked to 1^2, which linked to 1^3 that linked to 1^5 and this chain was the Universe. This is how gravity forms as a link to the prime singularity capturing all controlling points and whatever the numbers it is meaningless. $6126540987134586414379065432768329875341234321765498767845217690342769823^0$ has as it had a value equality of 1^0 and everything linked by being equal and the same and this applies to the mathematical period when by mathematical law the cosmos formed and is a mathematical fact This is a numerical fact and only because it is mathematically proven can it be part of what forms the Universe since only the Universe can form numerically mathematically proven facts. This line is part of what started the cosmos and today within singularity where the Universe "goes flat" as Einstein said it does, the cosmos still returns there when space changes by time returning space to time.

However, relevance grew as the string of cosmos linked numerically much further. Because we know also that 1^0 became 1^1 became 1^2, 1^3, 1^4, 1^5, 1^6, 1^7, 1^8, 1^9, 1^{10}, and so on and all of this dimensionless concept forming the Universe made the Universe change by leaving the Universe the very same. But I deliberately jumped the gun to make the reader aware of the chronological order of events developing. All of this in the above paragraph came in place much, much later. This was progress developing from what I now wish to explain. I wish to amberoid this as to make people aware of the events following in a numerical procedure. Before this the thought did not enter and the thought was not the Universe. Then the absence of the thought 0 found a place as the thought 1 gave the absence of the thought validity in not to be 1^0 and this being a thought I prove a little later when the explaining reach that far ahead...but for now a numerical order gave the Universe substance to be.

I explained how zero or nothing or 0 was not in place because the concept was not yet thought of. But then 0 was validated in the process only by 1 becoming an accepted number. The value of 0 got a value only because it supported the value of one as a dimensionless concept and by being 0 or forming an abstract absence of holding value could it give 1 a dynamic value. As change and change no less ended nothingness it could only end nothingness by replacing nothingness with something probable. To end nothingness it changes 0 to become a dimensional value to 1 that by not giving a dimensional value it gave 1 a dynamic value from which all further numerical order came in place. By dismissing the dimension that one has, it gives one a place and by that, one supports the rest of all dynamics that forms our Universe. As the line up of 1^0 going 1^1 flowed in with it all possibilities of things to come entered the Universe and the magnitude of planning became a reality. It is a line forming time leaving some sort of record that time did flow directional.

There were solids and there were liquids and then there was movement of that which we cannot see but we know it is present. What would we call it...if we have more insight than the stupidity of what a Newtonians lamebrain would allow? I think we would call it winds. If something moves and it is not the material (and that I explain in other work I have) and the liquid substance which is already present and accumulating is visible because of density equality between that which is sold and that which is liquid then the gusts of movement could only be interpreted as winds. How magnificently accurate this person was many thousands of years ago...and at that without the knowledge of modern science or the hindsight of a Hubble telescope prowling the sky for a history that formed an imagery as development progressed. However he had one advantage I have not, he was taken back and saw the reality as reality developed and I have that privilege not. But still.... I am unable to improve on the man...and I can only use more words because more words are available to me and I am sure that is the only reason. Atheists on the other hand have not enough insight to cope with the intellectual onslaught on their disabled mind and overcome their intellectual disability to follow procedure as the bible explains because they are stuck in a fantasy of understanding Newton! The poor slobs that lives in such a dim reality The Universe they live in must be misery... to be able to only understand the fantasy the ***"mass pulls mass"*** and they seem to be proud of it as their rejection letters to me proves.

Fortunately atheists are so much in the minority that they did not restrict mental development that much and as donkeys and mules proved, even stupidity has a place if dong the correct work it is made to do. This is when 1^3 became a point being related to 1^4 to put in place the numerical order of 1^5 where 1^3 forms solids unable to move and 1^4 is liquids moving around solids as it still does in the present with the sun circling about the earth and around us and then winds as the sun seem to us as solar winds put in place by the value of 1^5. Back then the liquid and the solid was equal but only apart by density and the winds were movement of incoming thoughts or cosmic substance that had less dimensional density. However this does not explain how eternity started as since 1^5 still represents eternity because $1^5 = 1^4 = 1^3 = 1^2 = 1^1 = 1^0$. This still is eternity and this does not explain intervals between eternity.

The very first question we have to answer is how did eternity stop? Think of eternity as this unbreakable entity with no start and no end. How can something that is everlasting break into sectors because we know from the Universe we now see that the Universe is defined by periodic intervals and the Universe is what is going to be in contrast of what is which is in contrast of what was. What makes time is the change in space that puts out relevancies of what was to what is to what will be.

We still have it present in our Universe.

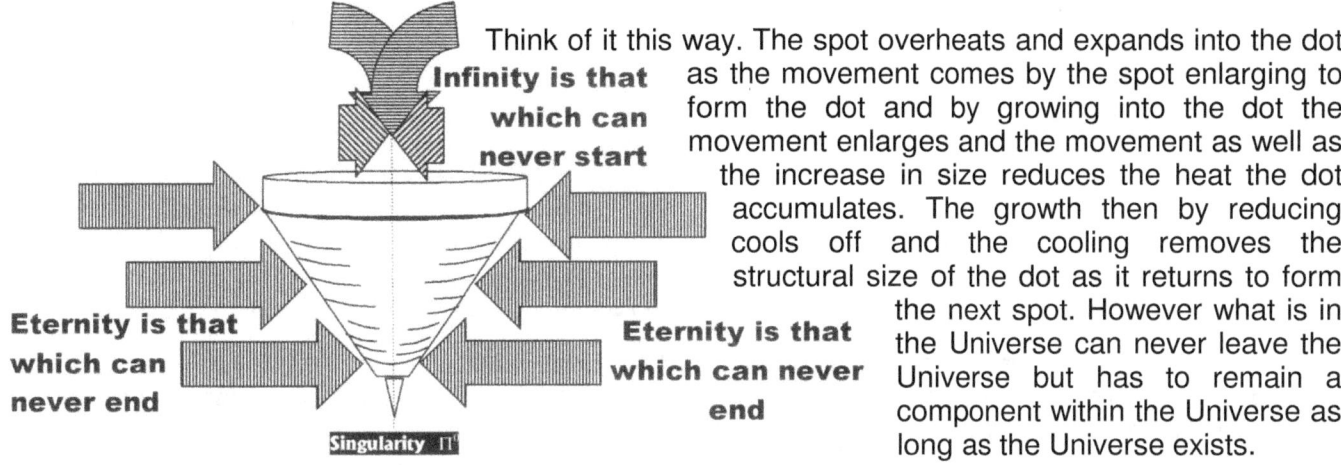

Infinity is that which can never start

Eternity is that which can never end

Eternity is that which can never end

Singularity "1"

Think of it this way. The spot overheats and expands into the dot as the movement comes by the spot enlarging to form the dot and by growing into the dot the movement enlarges and the movement as well as the increase in size reduces the heat the dot accumulates. The growth then by reducing cools off and the cooling removes the structural size of the dot as it returns to form the next spot. However what is in the Universe can never leave the Universe but has to remain a component within the Universe as long as the Universe exists.

However before heat brought about the Universe we know time was one continuing everlasting spot, which is a line that never went further than one spot.

The spot formed the future. The spot formed the present. The spot formed the past. Since the spot in the future was an exact image of the spot in the present and that was identical to the spot in the past the new spot had no identifiable difference between the future and the one in the past. It was a repeat of what was being identical to what was coming and therefore it stayed the same. The spot was so small it was invisible and unnoticeable and yet it is so big that at present time it holds the entirety of what is within. It is as big as nothing could ever be and it was so small it was nothing that could ever be. This was the only time nothing had a validation because as soon as the Universe came nothing disappeared and became something. One should be very clear about understanding this aspect. The spot then represented nothing because it was consistent of nothing since only nothing existed at the time. It was infinity that which can never start united with eternity, which is that which can never end. To start that which can never end first had to end and to end it first had to wait for that which can never start to find a way to start to lead to the process of ending. Since neither process could apply due to unfavourable circumstances making it impossible to follow on the other the process was eternally infinite.

There was no beginning of time and there was no end of time and that still remains the structure of time to this day. ...And then came the first moment where eternity and infinity parted ways with eternity. Infinity remained cold while eternity heated. How do we know this? When something heats it expands and that is a cosmic law. When something cools it contracts and that too is a cosmic law. The Universe in outer space is expanding to the advantage of material contracting in order to sustain and regulate the heat balance, which is what determines the era applying at any instant in time forming space. That is the principle by which cosmic development takes place. That is why we have gravity. How cold was cold? It was infinitely cold. How hot was eternity? It was eternally hot and that is why there is forever a seemingly gain in space to the loss of space in infinity.
So how did the Universe start when it started?

It started with the smallest dot there ever could be. •
This spot is so small it can only have an outside with everything there could ever be being on the outside of this spot.

This spot is still present forming singularity inside the centre of everything spinning or turning.

Then this spot overheated into becoming a dot and it expanded into the largest spot there will ever be. This dot is so large it could only have an inside with everything there will ever be afterwards, being on the inside of what there is.

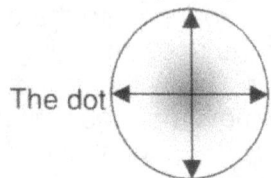

The spot • became the dot.

Then by expanding the dot cooled as it expanded and grew back into a spot again. • As it overheated it expanded into a dot.

The dot

The spot. •

These were points without any value as far as form goes. It was 1^0 going to 1^1 and then returning to 1^0 that formed 1^1. This was way before material becoming present as the Universe started movement by establishing space. It was 1^0 growing to $25914289371753916378294727746^0 = 1$ and it is the same.

This then that is how space started and space grew from one to innumerable. The first spot had only an outside so the following spot had to be larger than what the first spot was. This grew larger and set the trend we still have today where material constantly grows.

With the dot only having an inside the second dot was smaller than the first dot sending a trend where the movement goes inward and the density grows less.

The spot

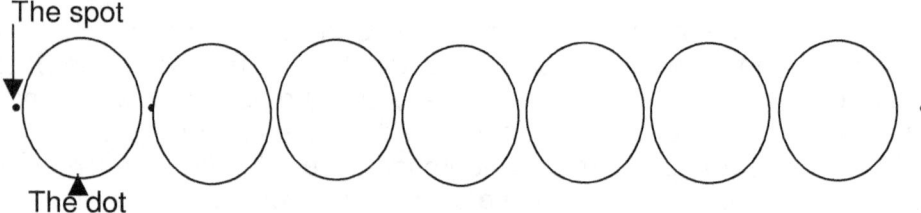

The dot

I this way space followed time so that time could form by forming space.

Time

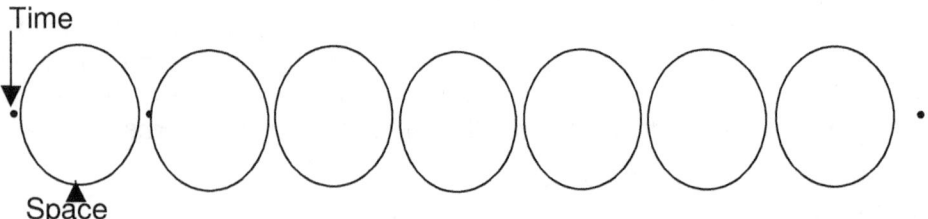

Space

In order to get time the form a line the Universe had to establish space in order to give time an individual identity that set it apart from the previous and the next instant.

This also was way before Π formed any value and form had no standing.

This is exactly as the Bible describes the way the Universe started.

The Universe was a void without form.

...And with mathematics and reality while leaving out Newtonian science cheating the scripture proves nature to be exactly what the truth is.

Space came in place to allow time to have a past, a present and a future as it still applies to space-time.

In today's Universe we have to recognise that this value of 1^0 takes no space, fills no space and yet it holds so much space without offering space that it forms a Universe filled with space without being space. Whatever you see is a huge volume of space that has no space but is so many space less ness it became an entire Universe of space. Playing with numbers and formulas does not bring appreciation to the reality we find in nature and in the construction of the true Universe as nature presents it.

Lets look at the Black Hole to understand the concept of a cosmos forming. There were stars so massive we now living can have no concept of what the involved. As the density of the cosmic liquid lost value and became penetrable by light so by the same margin did the density of the stars then shining increase dens it to a point it became unable to transmit light. This process still applies and will forever apply because that is how the Universe will come to end as the end shift back into the beginning. That is what a Black Hole is where the beginning swallows the end or infinity devours eternity in order to reunite time as going back to form a single thought. We now have the evidence and we know the mindless Newtonian atheist's concept of "gravity gone mad" carries the same intellectual substance as having "nothing" filling outer space to a point it overflows. As the Universe went "softer" the ancient stars went "harder" (Sorry for this outburst of stupidity but I put this in so that the mentally deprived atheist can follow procedure). There was a time when the current Black Holes were stars holding proportions that by shear comparison dwarfs our biggest and most magnificent galactica by billions of times. As density collapsed in the liquids it increased in materials.

The only way to leave the Universe is by the way it entered the Universe and when something was part of the Universe that is the way out because that was the way in. Life by the way was never part of the Universe as I prove in so many articles and books. The body life creates as a vehicle to serve it for a specific short duration of time is left behind and the material becomes available for life that comes in the future to serve that life again for a very limited and specific duration of time. The body, which is cosmic, stays behind as reusable material and life, which leaves the Universe never, was part of the Universe at any stage. This period was when the Universe formed into a thought and not becoming material yet. Only afterwards did material, as a substance became a valid entry to the cosmos. This was when and I quote again directly word for word God said: _**'Let there be light', and there was light; and God saw that the light was good, and he separated light from darkness. He called the light day, and the darkness night. So evening came, and morning came, the first day. He then created darkness as that which we take as night and as light as what we call day. With this concept a Universe came about that holds everything in place.**_

 At this point I do not wish to reflect on this period other that to show that this is where the Universe started and this is where the Universe was built as a memory of what time left behind to form space. At this point Π^0 developed to Π by the movement of Π^2. In order to prove authenticity and reality I wish to show that the bible thousands of years ago made distinction between light which is $3\Pi^2$ and darkness which is 3^3 and this darkness is what atheists and Newtonians gave the same value equal to their intelligence being "nothing". Secondly the Bible places distinction between the thoughts that formed and the thinking process forming a hologram we use as reality. When God said: _'Let there be light'_ the Universe of substance came about and the substance I reflect on I wish to confirm as a hallucination without reality. This is what came and this is what eventually will go whereas life that are able to value and recognise God is part of the permanent "other side" where the thought creates a reality and the thought is the reality.

There are two realities in place. The one is where one holding the dimension of zero putting the value. Zero only becomes significant if something gives the absent-ness validation. It only becomes a factor when 1 shows a point zero could fill in being absent as a dimension.
$6126540987134586414379065432768329875341234321765498767845217690342769823^0$ as equal to and in fact the very same as the number we mathematically have and is written as $1^{6126540987134586414379065432768329875341234321765498767845217690342769823}$. These two are same one.

This is not what I tell you because this is what mathematical law tells you. This is not Newtonian bullshit

cheating mathematical law by changing $F = \dfrac{r^2}{M_1 M_2}$ into $F \propto \dfrac{M_1 M_2}{r^2}$ and then with n o

regard for logic or mathematics put $F \propto \dfrac{M_1 M_2}{r^2}$ equal to and the same as $F = G \dfrac{M_1 M_2}{r^2}$

and moreover this madness went "_undetected_" for three hundred years by the mob that holds the place as those being mathematical brilliance and mathematical genius. What a lot of rot they are.

I have given only natural mathematical principle and common logic and I have shown this information to be available only for the past ten thousand years or so. Not one did I have to cheat to make sense of what I put forward. ...And mathematics just like the Bible is unable to lie in contrast to Newton and his Newtonian mob of gangsters.

Later came the dimension of 2 where it ran as 2, 2^2, 2^3, 2^4, 2^5, and then came time as 3, 3^2, 3^3, 3^4, and in that all comic motion finds valid order.

In the dimensionless Universe dimension found a place and in the order of mathematics came in place a Universe only God Almighty Creator of all could envisage. With that also comes the idea where it became zero and zero was not even a thought an Entity with the will power and the ability and the Supreme ness to intervene then did just that, he intervened and not by Newton or the madness of the Christian Clergy's magic but by science and mathematics came a Universe in place so good and so great that the more I leant about the Universe the smaller I get in terms of what our Creator did create. The more I learn about what is in place the more I lose my place in that which I am part of. Think of what is a galactica, strong enough to centre a Black Hole in the centre and then think that this and everything else is part of one thought the Creator thought up. Then these re-born Christians and also mentally deprived wish to tell me that "they walk with God and they talk with God" as if they are mates with God. The atheist is what they are because of stupidity, which in some way is not their fault. Because how can it be a sin to be stupid. These egomaniacs Christians on the other hand putting them in place as being friends with God Almighty do so because in their sublime state of egocentric madness they know not their place in terms of God Almighty.

You are your own God as much as you are your own Satan and you create sin in terms of treating God with no respect and the way you treat others with life. You form your own sins according to your own believes and whatever you choose will be a price you pay because as much as there is no big or small, hot or cold, near or far also there is no good or evil but only a price to pay for every deed you do. But to think the Almighty Creator will have a conversation with you in your back garden while keeping billion trillion Black Holes in service is showing your disrespect for a Power no human has a mind to understand. You think far too much of yourself and boy am I glad I do not belittle such an Entity with that Force that can by thinking keep everything we see as a Universe in place by thought. Your concept of God is in the very same bracket as the atheist because you put yourself on par with God. By trying to boost your own miserable incompetence you belittle God down to your standard.

To get back to sensibility: Before what we now have come to past we did not have a past, a present or a future. In the very beginning there was no space and without having a line in time there was no time.

There was no mathematics because even zero or one was one more than everything there was and therefore having any idea involving something like even zero or one was meaningless if even zero or one was that much more than there was before one came to be then that left everything numerical worthless, otherwise with no meaning.

There was no light because there was no darkness. To give light one has to remove darkness and without darkness one cannot produce or move to light.

Think of something so small it does not exist. You are unable to do it but still that was what there was. To be small it has to be and it was not present yet so it was still infinitely smaller.

Then increase that to a size that includes billions upon trillions of everything there is which is more than everything we can think of and then fit in it everything we can and can't see and then it still is outrageously too small because that what now is in place still can hold no end because in the smallest space we can think of fits everything we can't think of and it fits effortlessly with so much space still to spare because what ever it is and is in it then still is not enough to fill that which is to its full capacity.

The spot that formed the start was so small it can't hold a thought but still is so big it holds no end. If you go to the centre of a circle you find just such a point hiding there in plain sight,

That which starts everything still starts everything without being present. Think of how did what now is everything started. It started with a spot that had no time and no space and no future, present or past that continued being what is was without end because there was no end because there was no beginning that could bring an end. That is where what is started.

The spot that started the lot is still there because it still is not there to start with. There was a spot without form, shape size or time but became representing all that developed.

This spot was so small it was not and yet it was so big at the same time it has no end or borders or limits or beginning or end. It is as it was as it will be without changes altering positions and changing was still something that had to be invented as part of a future that had no place yet in which to be. There was what starts the lot and that only started to be...

Whatever you may think of was invented after changes and changing was invented.

Whatever is, was still to be invented because even one was not yet a concept, it then became one as one became the invention that brought the first change to endlessness.

Even the thought of reading this was as a thought a concept that still had to be invented.

What there was we call nothing which is a concept that has no meaning because that is where the Universe started and nothing is the only thing that was replaced as it was removed.

In its place came a potential of possibilities. That brought about a concept of change.

Nothing changed to something with possibilities.

Still that was the first change…the possibility that there could be change.

Before this there was nothing and nothing removed any possibility of something changing.

This still has a mathematical statement that can render a value to this very day.
We have been there but let's go and see where singularity is because whatever we think of started at singularity and to this day is still with singularity. Where singularity is there the Universe started even before time started a Universe. The entire concept of singularity is between what can move and what can't move, what overheats and what cools down by movement, what expands and what again contracts. In other words it is all about heat and God Almighty said: "Let there be light" which is also "Let there be Heat". Singularity ruling the Universe is the interaction between varying heat that forms density differentiation.

The Universe is written in a code that is this code of 3 and 4 that I decipher and the reasons why it structurally works by 3 and 4 and it is this growth by 3 and 4 that I unravel.
Let's start explaining it this way. To be within the Universe is to move as the Universe. This is the most important aspect we learn from Kepler's finding as $a^3 = T^2k$. All movement is by $7°$ going straight to form a circle and a circle forming $7°$.

The seven degrees of rotation is composed as three points forming the axis and four points forming the circle. Anything that rotates does so around an axis holding three point one on top one in the middle and one below. Then the circle is the rotation of two points changing direction as it turns around another three point. Anything in the cosmic forming gravity holds three plus four points with the circle being 1+1+1+1+4 as well as 2 x 2 =4. There are two points going forward as much as two points reversing on the other side of the Universe. Where is the Universe? The Universe is Time in singularity (1) moving from the abyss (.991). This then forms Π. The Titus Bode is about gravity forming Π. Remember no circle can start at zero because of it started with zero it must continue with zero or then start the line where the line starts with 1. Therefore a circle starts at 1• +1• +1• +1• +1• and that is a line. If it started with 0 then it would be as follows 0 + 1• +1• +1• +1• +1• and that proves that zero has no place to start in any line up. If Physics does not even know where a line starts where will they locate where the Universe started because the Universe is comprised of numerable lines criss-crossing and lines connecting numerable point. Therefore in singularity (1) everything is a line 1• +1• +1• +1• +1• or then it is forms as singularity $\Pi^0• +\Pi^0• +\Pi^0• +\Pi^0• +\Pi^0•$. The circle is a line and the line is a line and there then is seven points in the line-up forming rotation. It is one line in singularity but in the space in which we are it is a circle and an axis in connected by rotation. In singularity it is $\Pi^0• +\Pi^0• = 2 +\Pi^0• = 3 +\Pi^0• = 4 +\Pi^0 = 5+\Pi^0• = 6+\Pi^0• = 7\Pi^0•$. This is the one part of singularity where 7 connect to 1. The issue here is 3 + 4 = 7.

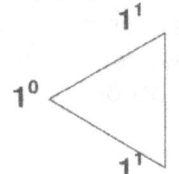

The three stands in for the measured value of a triangle because it forms one dot in reference to the past and the future and that places the same dot in a ratio of coming and going. This leads to four dots.

The solution to encrypt the cosmic code is in the value of 3 and 4. It is because of this value that we see the sun as the sun crosses our sky while the earth spins around its axis It is a ratio that represents cosmology as we see it. We see the sun cross our sky because the sun rises in the east and sets west.

As time moves on dot is coming while one dot is going.

One dot becomes two by leaving one point and going to another point which is in reference two positions for the same dot. But since one dot is coming while the other is going we have dots in two positions representing the coming part and two dots in the going part. This means we have $2 + 2 = 4$ as well as $2 \times 2 = 4$ and the beginning of the second dimension. In this we have the three forming a line motion as Kepler said $a^3 = kT^2$ where 3 represents the **k** part. Then we have four forming the square part or T^2. This is how movement starts space $a^3 = kT^2$ or movement forms space $a^3 = kT^2$.

The three of the axis and the four of the circle combine to give a rotation movement of $7°$. As the rotation is of a solid object the rotation is $7°$ connecting by a fixed line and this means that although it is on either side of singularity it still remains the same $7°$ since it is the rotation of the same object.

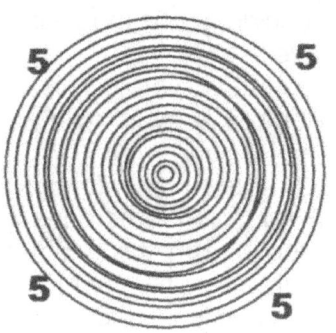

In this is where the 7 forms 10. The seven moves from one point to the next point and in mathematics when something moves it is going square. It is not going to become a four sided flat object as Newtonians whish to portray singularity but it is going square such as Π moving and then duplicating $\Pi \times \Pi = \Pi^2$. In this case it is 7 going square to be $7^2 = 49$.

With three forming the axis and four forming the circle the next point just outside the circle of four must be five. Therefore with four points forming the circle and each point forming a fifth point the total value inside the circle is four then the value outside the circle and on the outside of the circle is five. Then the full compliment of the circle on the outside is $4 \times 5 = 20$. Space holds 20 which include singularity coming from the future (which I will explain later) is 1.991.

When we look at space we think we see a circle but when we connect time and movement into the equation we see half circles forming as time moves. Moreover we see circles forming that cross singularity going on both sides of the divide forming opposing rhythms or we call it seasons. We do more about this in another chapter.

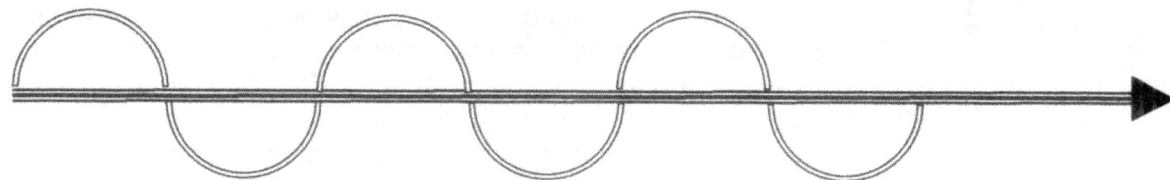

Therefore gravity is half circle flowing with time
The circle turns by $7°$.

The circle moves, which also make the turning, go by 7 square, which is 7^2.

This means that the circle goes $7^2 = 49$ on both sides of the centre and that means it goes square on the open divide.

The circle connects by crossing singularity, which is 1.

The one his time remains the very same 1 and it is the line of singularity crossing by movement applying.

On the one side we have 7^2 in relevance to 1^2 and on the other side we have a duplication of the same process forming as 7^2 goes in relevance to 1^2. That means on the one side there is the value of 7^2 in relevance to 1^2 and on the other side it is 7^2 in relevance to 1^2.

As we are dealing with

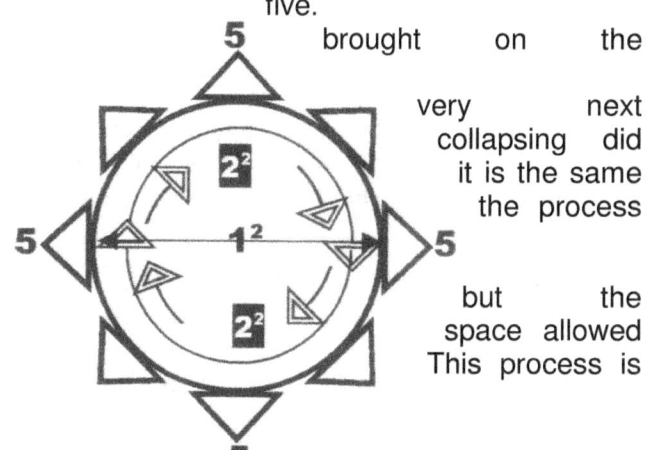

singularity we are dealing with the issues forming singularity. In singularity the half circle and the triangle has the same value as a straight line forming 180°. Therefore as time moves in a straight lien time therefore moves in half circles crossing the straight line which is the straight line turning as relevancies changes from one side to the next.

According to mathematics and mathematical principles forming it is $1^2 + 2^2 = 1 + 4 = 5$. Therefore by movement the value that follows four must be five.

That placed a value of five after four and this brought on the Lagrangian point laws.

As the dot expanded so the dot contracted the movement. However the dot moving forward and very next collapsing did not land on the same spot because in relevance it is the same dot but time placed it on a new spot from where the process repeats.

This then placed the linear dimension still at 2+2 but the collapse of space following the expanding of space allowed the spot to land on different sides of the divide. This process is still happening and we can witness it as a graph.

As time flows the spot expanding into the dot crosses the line that does not exist. This is the line, which the Bible refers to as the abyss because it connects the Universe in space with the Universe in singularity. This evidence we find in the Lagrangian points forming.

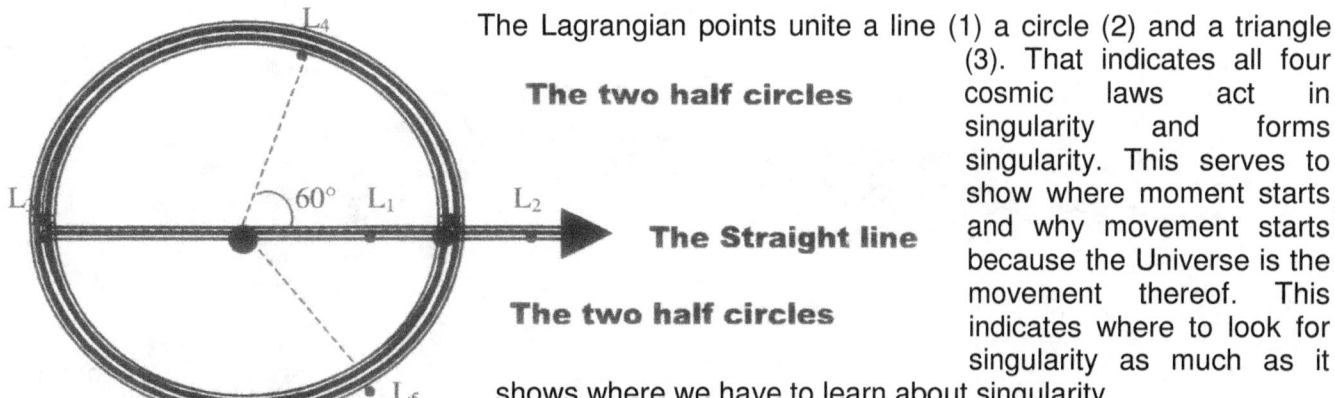

The Lagrangian points unite a line (1) a circle (2) and a triangle (3). That indicates all four cosmic laws act in singularity and forms singularity. This serves to show where moment starts and why movement starts because the Universe is the movement thereof. This indicates where to look for singularity as much as it shows where we have to learn about singularity.

The two half circles

The Straight line

The two half circles

This is the point where mathematics started as it introduced a Universe filled with space.

This is where $(3 + 4)^2$ became 100 that then according to Pythagoras became 10.

Also this is where $3^2 + 4^2$ became 5^2 and in this process two of the five cosmos laws are founded while breaking singularity.

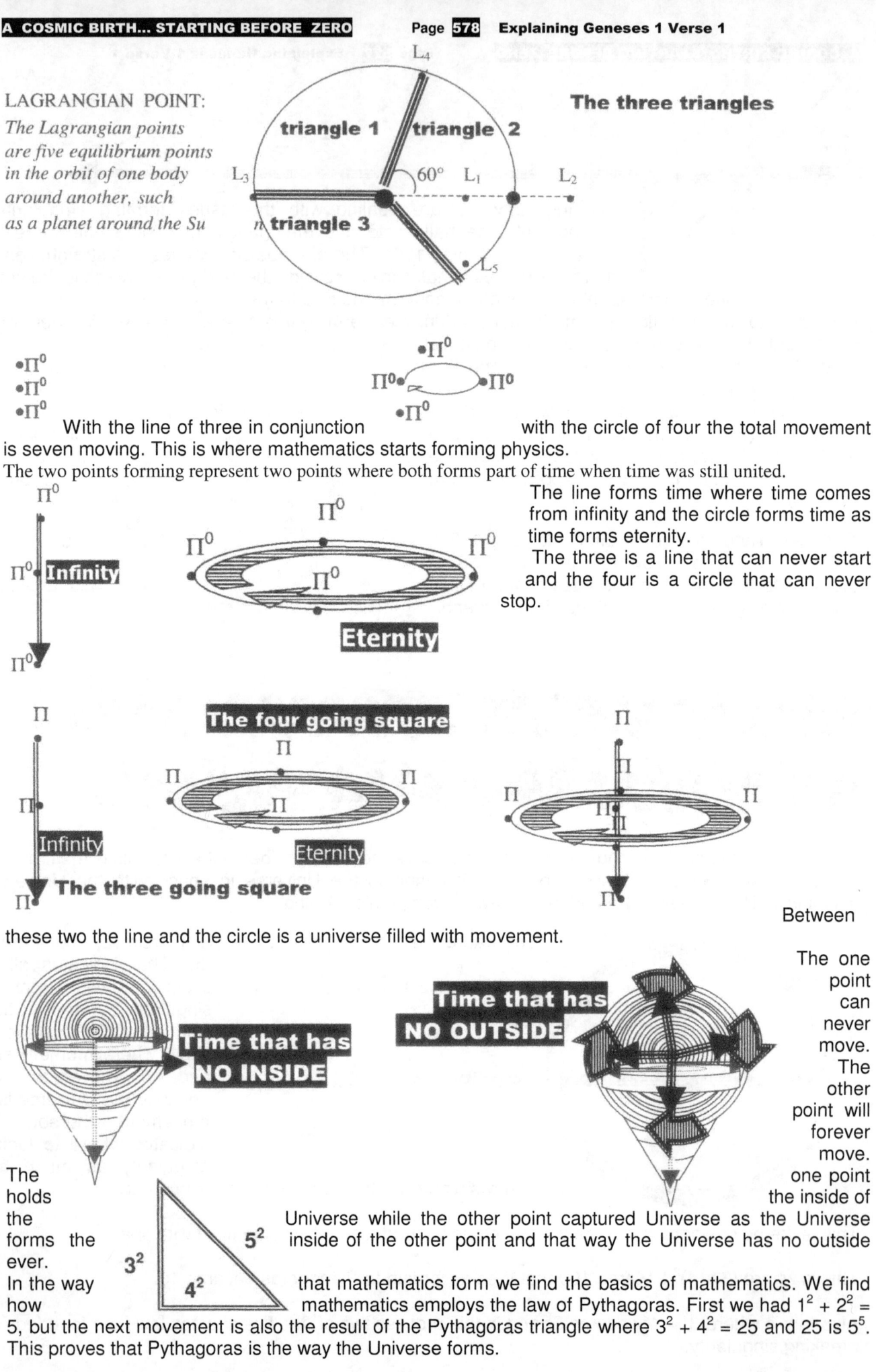

LAGRANGIAN POINT:

The Lagrangian points are five equilibrium points in the orbit of one body around another, such as a planet around the Su n

The three triangles

triangle 1   triangle 2

triangle 3

$60°$

With the line of three in conjunction with the circle of four the total movement is seven moving. This is where mathematics starts forming physics.

The two points forming represent two points where both forms part of time when time was still united.

The line forms time where time comes from infinity and the circle forms time as time forms eternity.

The three is a line that can never start and the four is a circle that can never stop.

Infinity

Eternity

The four going square

Infinity

Eternity

The three going square

these two the line and the circle is a universe filled with movement.

Between

The one point can never move. The other point will forever move.

Time that has NO INSIDE

Time that has NO OUTSIDE

The holds the forms the ever.

In the way how

one point the inside of Universe while the other point captured Universe as the Universe inside of the other point and that way the Universe has no outside

that mathematics form we find the basics of mathematics. We find mathematics employs the law of Pythagoras. First we had $1^2 + 2^2 = 5$, but the next movement is also the result of the Pythagoras triangle where $3^2 + 4^2 = 25$ and 25 is 5^5. This proves that Pythagoras is the way the Universe forms.

3^2

4^2

5^2

Then we have to look for Pythagoras in forming movement.

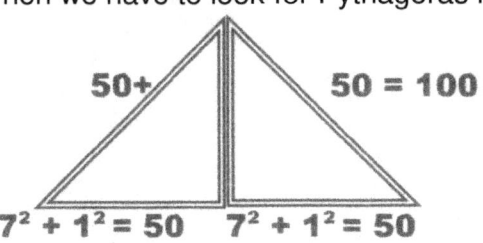

50+ 50 = 100

$7^2 + 1^2 = 50$ $7^2 + 1^2 = 50$

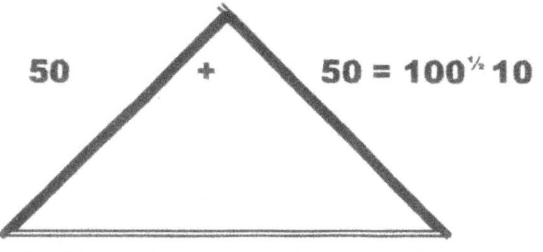

50 + 50 = $100^{1/2}$ 10

In order to find singularity one must reduce a circle. Everything in the Universe forms by the form of a circle. Anything that is anything holds six sides Π^6 that spin $\frac{7}{10}$ in a six sides cube $\frac{\Pi^6}{6}$ forms a sphere. Movement in cosmic terms is the Titius Bode law or then $\frac{7}{10}$ and space is a sphere $\frac{\Pi^6}{6}$, which is a multitude of circles Πr^2. To get to singularity one removes the circle

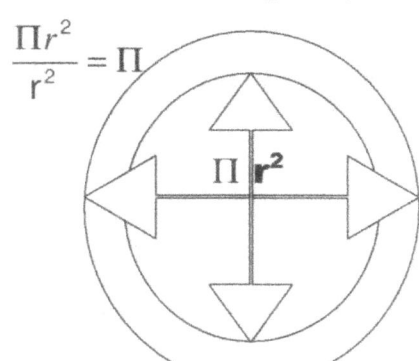

$$\frac{\Pi r^2}{r^2} = \Pi$$

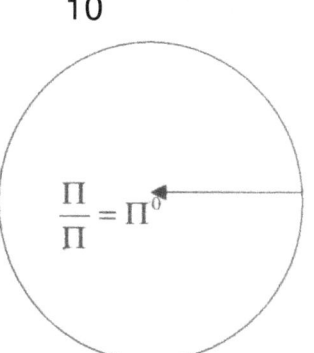

$$\frac{\Pi}{\Pi} = \Pi^0$$

$$\frac{\Pi r^2}{\Pi r^2} = \Pi^0$$ and then you have singularity. That is where everything that is material starts to form space. Anything from Π and larger is space and anything smaller than Π is time within singularity. What can be smaller than space, for one thing the thoughts you use to remember this. We do not use electricity to think but we charge electricity to send to our brain commands to think. This division in time brings evidence of how the Universe started. The Universe started when the undividable divided and time got two factors, which is infinity that never starts and eternity that never ends.

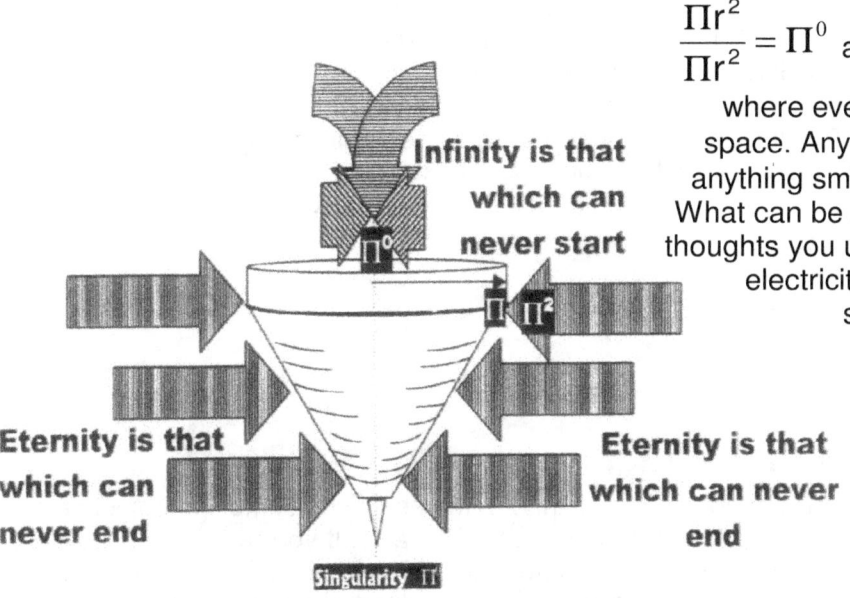

Infinity is that which can never start

Eternity is that which can never end

Eternity is that which can never end

Singularity Π

In the centre we have singularity contracting or moving towards to infinity, the void that can't get smaller and it holds the line from which everything progresses. Then outside there is Π which is that which accelerates as time forms space. Kepler with his tables summed up gravity and all the characteristics of gravity perfectly with the following sketch I made.

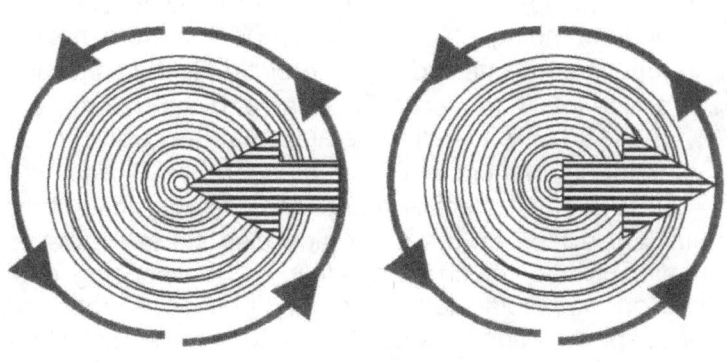

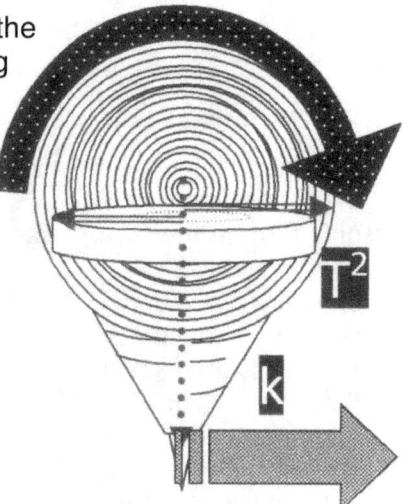

T^2

k

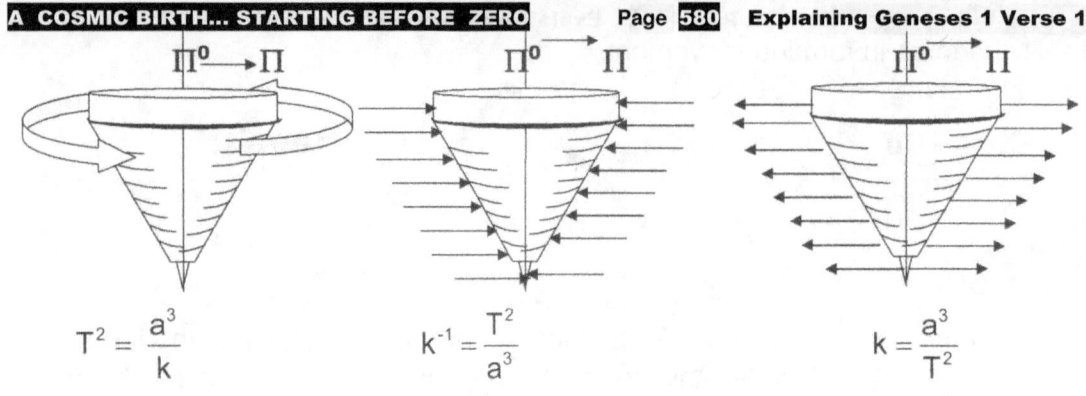

$$T^2 = \frac{a^3}{k} \qquad\qquad k^{-1} = \frac{T^2}{a^3} \qquad\qquad k = \frac{a^3}{T^2}$$

In order to find independence from the capture and establish individual existence within the influence range of a dominant structure movement is essential. There has to be an axis established to reserve independence but the circle establishes the axis and therefore it is the circle that plays the major role in the establishing of 7° turning.

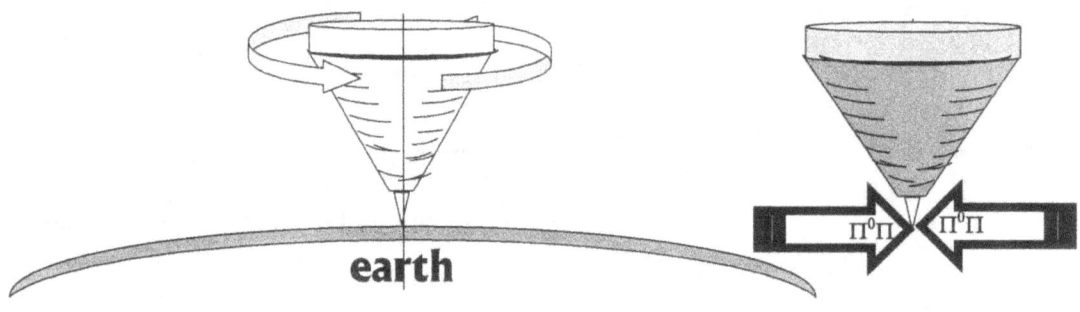

it is movement that hold the top erect. This proves the line that forms time only comes in place when the circle brings the top erect. As the top starts to spin it condenses or cools the space around it. This forms the Coanda effect as the movement of Π^2 produce Π but as Π^2 increases this also increase the position of Π^2 and thereby extends the line singularity forms as $\Pi^0\Pi\Pi^2$. This then forms the interaction between what is solid and what is liquid and this movement allows the top to make the air around the top denser than the surrounding air. This increase in density is what keeps the top upright. If it were balance the balance would remain even at extremely slow rotation speeds.

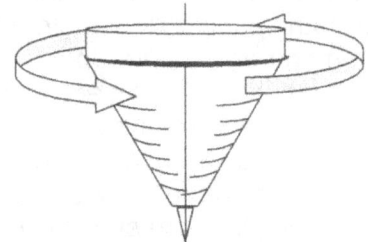

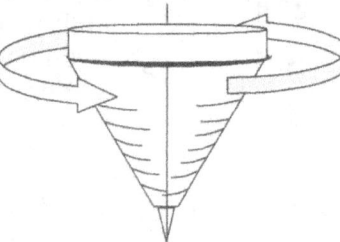
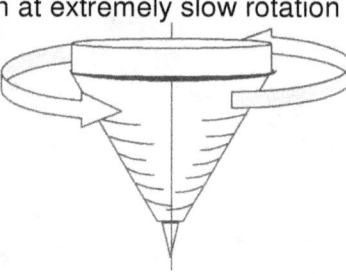

The past 7 +   The Present 7 +   The Future 7

This where to locate singularity: In the centre of all moving objects a line forms that we named an axis. If you reduce the circle that turns the circle will get to a point where space is absent. This puts a point inside every spinning axis where space disappears. All turning moves all space towards and in the direction of this point where space disappears.

By the cosmos only form in the specific value of space. When reduce to become Π^0 starts, that same the one side the other side. If the one versa. The one point is reversing the direction entirety we think of as a Universe is controlled through movement applying as time from a point that

inspecting and investigating this point brings about some understanding of how gravity and therefore how gravity within functions. If one reduces the circle Πr^2 by removing all material shape of Π remains. This value Π represents a circle with a 3.1416/1. This value establishes the value of time forming entering the circle Π the measure of Π has to even further Π^0. In this Π^0 there is no space. The very instant the entering of instant exiting on the other side also starts. However by entering motion is very much completely in the opposing direction on the direction was going left the other direction is going right and visa without space divides the Universe into four sectors where one sector of the opposing side. This is where physics starts. This proves that the completely and unquestionable falls outside the realms of what we think of as space within the Universe.

The atom and every atom individually form a Universe that starts and ends the Universe. The movement in space of space holds two sectors, which places the identity in two total different categories. The one is material driven by time in as much as time forming $\Pi = 3.1412$ and the other is time forming space by the movement Π brings about and that value is time in the future going to time in the   present moving to time in the past and that is 3.

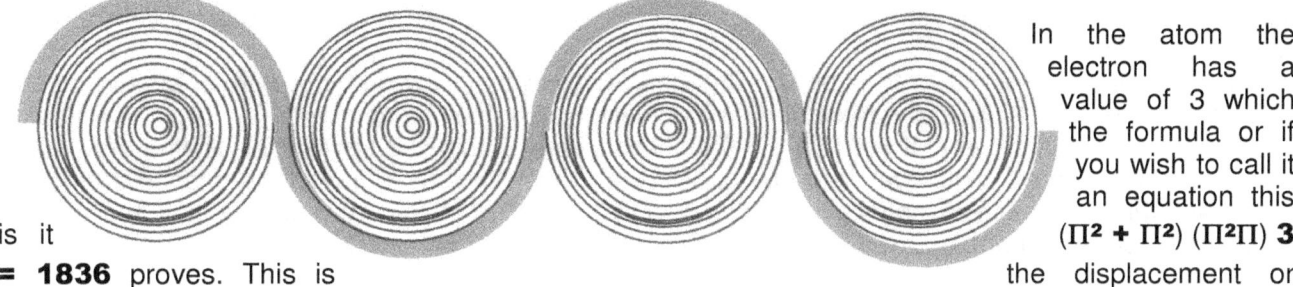

In the atom the electron has a value of 3 which the formula or if you wish to call it an equation this $(\Pi^2 + \Pi^2)(\Pi^2\Pi)$ **3**

is it **= 1836** proves. This is the displacement or reduction of heat from the gas (electron) to the solid (proton) via the liquid (neutron).

The proton is $(\Pi^2 + \Pi^2)$
The Neutron is $(\Pi^2\Pi)$
The electron is 3.

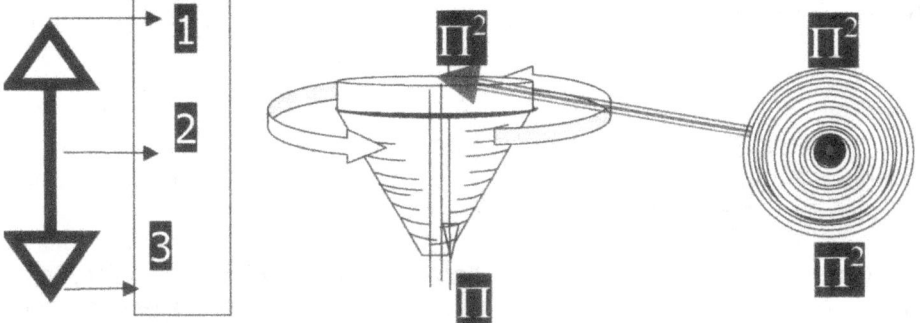

The atom forms space within the Universe as $\Pi (\Pi^2 + \Pi^2) + (\Pi^2 + \Pi) + \mathbf{3 = 112.}$
This is where space starts a sphere and where the atom forms as a sphere.

Anything smaller than 3 falls outside the realm of space because time forms space by going from the future through the present to the past.

That is how easy it is to prove the existence of God Almighty. There where only a thought can be time forms space as time in the form of gravity shapes space and by a thought in the void of where space can't be a thought drives the Universe as 1 or as Π^0. Everything in the Universe unites in one point that from where we see can never by and since $\Pi^0 = 1^0 = 1^1 = 1^{230759312756299} = 1$ the Universe is within one point forming time. Only by movement can time begin space as 3 but essentially it is one point moving from the future through the present to the past that form time that forms the Universe.

Now we have to investigate how the Titius Bode law forms gravity as Π^2.
This far I have established why and how gravity is the measure and the value of Π^2.

This is a point I have to press to bring justice where justice serves best. My accusing atheists of being blind and being stupid are vindicated by mathematics, which they normally insist they excel in. The Bible was written, what twelve thousand years ago because it predates anything that was written and survived and still using mathematics correctly the Bible proves to be correct. The Bible refers to the abyss and the earth being without form. That is where the earth and all other material in space and forming space originated. The Bible says clearly the abyss, which can only be where 1^0 parted from 1^1 and landed a Universe in the process. However we are within that very first dot that is still growing. We are part of that dot and in that context we can only be part of the earth.

Man this far never yet left the earth because the moon is Π in relation to the earth providing Π^2. The earth provides the axis the moon rotates around. Anyone that wonders why the Bible refers to the earth and not the entire Universe is because the earth is our only reality. Man was on the moon but the moon is the last frontier. The earth holds this much water considering the other planets because the moon extends the Roche factor where the moon provides the linear extension of $\Pi^2/4$. **This is why we have water on earth to support life.** This extending of Π makes the earth that much bigger than what it truly is and condenses so much more elements from outer space that can turn into water. I doubt it seriously if we can ever visit mars never mind colonise it because our brains would not function properly. We think by gravity helping us to think and being in the space-time of Mars will render our ability to think very much useless. At best we will have the mental capacity of a very low developed animal. This is all the result of trying to force bogus science into being truthful and trying to ignore truthful science. Mathematics show the Bible is truthful science with mathematics and physics used as nature uses it, proves everything started according to the Bible. Come you atheists bring your science and prove me wrong!

In the abyss is where what came about as Creation started with nothing being an over statement and when going there we have to predate even before there was 0 because 0 x whatever and that remains 0 even to this day. It is true that 0 x 10000 = 0 just like it was. This is the abyss. This is where the Bible said everything started. Whatever could be couldn't be because zero removed the chance of change from something possible to nothing changing where nothing removed all possible something as it numerically formed zero 0.

Then one dynamic shifted and the nothing became the dynamic while the possibility became one. When 1 came to be it gave 0 a dynamic of not being a factor while forming a factor to 1.

The Numerical value of 0 shifted into a dimension from having a value which instead of nothing then by replacing nothing a dimension started something with 1^0 and from this possibility $1^0 = 1^1$ while still being the same, that then changed the impossible to a possible or zero into Π^0 from which a Universe came about.

This happened by dismissing Newton and his nothing or zero or numerical 0 to become Kepler's **$\Pi^0 = a^3/T^2k$**. Then how did this start begin to take place because it produced everything? What brought about all the further changed to endlessness because eternity is still present. This is what all physics form and is what forms all physics; Kepler's **$\Pi^0 = a^3/T^2k$**

In order to start at the start we have to go back to the start and that is not when moment-Alpha occurred but when the last perfect moment ended. Then we have to look why the prefect moment ended and why the imperfect moment started. The number and the value of one or the winds that flowed over the waters remained in place for a very long time. Because God is perfect the Universe remained perfect where one spot was the same as every spot and only relevancies in order of status formed differentiation while all spots remained one spot just as mathematics would indicate $1^0=1^1=\Pi^0$. We still are in the Universe of $1^0=1^1$.

I think everybody is in agreement that from this the Universe started with a spot that became all the dots forming time at present. Then the first place to look is where this place is where it all started and where we should look for it because being part of the Universe places it within the Universe and then it could never leave the Universe. Then we have to look why it was perfect. Remember if it was perfect why did it change to become this Universe that changes and therefore is very imperfect. That what was most perfect became imperfect on purpose.

Being perfect was in this case establishing change by putting in place one spot in reference to the next spot while the first spot duplicated and in the same process remained the very same. This is what singularity presents. By forming one the concept takes all the spots into one spot and although all are different being what each is in place in space we also know that in time we hold the very same spot

We view the Universe not as God Almighty from a distance and standing on the outside of the Universe but as being part of this very small dot that grew into this very large dot that is the Universe. We form the centre of the dot. We are part of everything and not an onlooker gazing down on everything we are part of the smallness and not the viewer looking down from a dizzy height at everything that is that small...as part of it we are even smaller.

First we have to find this spot where the Universe started.

At first the spot was perfect, eternal and everlasting but so small it was invisible.

•The spot was one perfect spot that overheated and parted into a dot.

•• Then by overheating the spot split into two being a spot and a dot.

••• The spot shifted to the past leaving the dot in the present while the dot cooled off and formed another spot one being in the past and one in the present and one coming towards as the future.

The tiniest spot ever overheated • to the size the Universe   now is and it is still overheating.

Nothing was added and nothing was taken away because the first spot is still every dot forming a Universe we now have.

Then there was cold 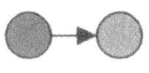   and there was hot which still is the same thing but light-years apart.

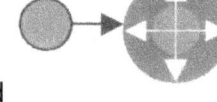

There was the eternal first spot    that overheated 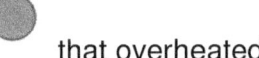   and by that expended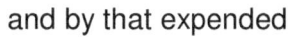

Then by moving•the dot duplicates • and this duplication splits its heat into half, a process that cools.

By moving according to time ●   and splitting time as space forms ●   to a line the dot then ●   reduces the heat to become even less as it distributes the heat over a wider space.

By moving ● and splitting time ● to a line the ● line takes time ● to sectors of the same line in a line time and while time is one it created space as three and space does not exist.

All the while ● the splitting ● is occurring ● the new dots ● remains the original dot ● because no

spot ● or dot can remove or add to the Universe. It is cooling taking shape through heat distribution.
Every • spot • that • formed • a • dot • also • remained • a • spot • without • renewing • because • the • very • first • spot • ended • as • a • dot • is • still• the • original • spot • that • started • the • lot.
Time took one eternity to form, that which is, and what came from • the one perfect spot • and then to

duplicate   that • into the dot while the original dot is the spot holding the Universe without end.

Everything in the Universe moves just as Kepler said it does and everything in the Universe that moves goes straight while circling around a centre that refers to a centre where mathematically it is the same centre but only hold different dynamics in time.
The spot Π^0 overheated and expanded T^2 which by the precise same movement **k** pushed the newly formed space a^3 into the future while moving Π^0 as a • into the past. That is how the very first moment

arrives according to Kepler's formula. ...And there we still have it Kepler's formula $k^0 = \dfrac{a^3}{kT^2}$

Time began as infinity preserving eternity and then overheating came about and shifted infinity and eternity apart by expanding into space.

This brought three or the line, which is the three points in the seven that forms movement.

Then the overheating brought expanding which is a circle movement as a ring grows into $\Pi\Pi^0$ where in this space-less circle the radius is a number and not a value. But this moves not the radius Π^0 into a square but the circle itself duplicates by $\Pi\Pi^0$ going $\Pi^2\Pi^0$. Then this becomes a square as it becomes

two. By duplicating it halves the overheating by half which forces to dot   to return to the spot•.

This action reverses everything that happened whole not bringing the point it was to the previous location but depositing this on a new allocated point. Therefore by duplicating the one moved to two but by duplicating the one cooled down and moved back to one which by then was in a new location.

This brought a value of two and two to form two by two or numerically said it was 2 + 2 that then became 2 x2 and the first dimension other than the no dimension came into place. Then 2+2 became a second dimension that was 2x2, which is the nu8mber 2^2.

This put time and movement in the same bracket as the same thing and that made expanding while

contracting the same thing as moving linear. That affirmed $k^0 = \dfrac{a^3}{kT^2}$ When the expanding reaches

the other side of the divide which is singularity the expanding changed direction to become contracting.

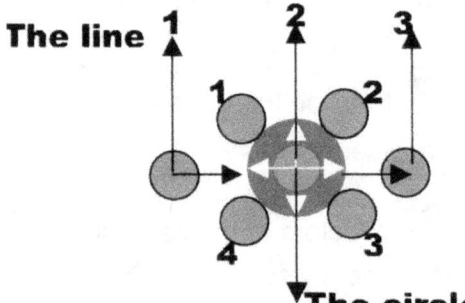

The line

The circle

This meant that going in a circle is equal to going straight and a straight line id equal to forming a half circle while still connoting positions in reference brings the triangle into the equality in the Universe without form.

This is where we find the first evidence of the Roche lone and the Roche limit,
This is the Roche limit and that is the Roche lobe. As the first spot Π^0 grew into the first dot Π it expanded Π^2 into four sectors and everything being close than $\Pi^2/4$ will expand and sacrifice density. It will take on the growth of the major star and then the minor star will become liquid to by density loss and equalling the space occupied of the major star.

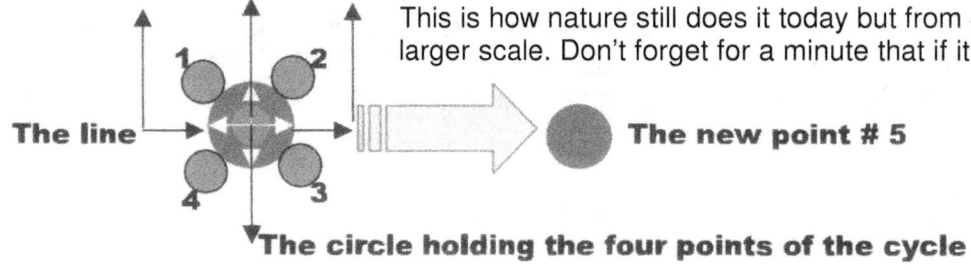

This is how nature still does it today but from our point it is on a much larger scale. Don't forget for a minute that if it happens in the macro it is the micro that forms the macro and we only see the macro as a result of the micro is taking place.

The line

The new point # 5

The circle holding the four points of the cycle

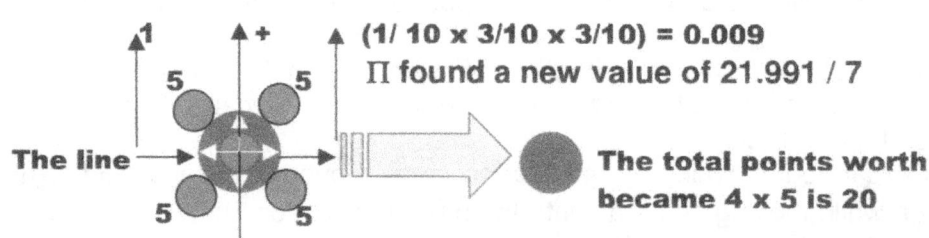

The line

(1/ 10 x 3/10 x 3/10) = 0.009
Π found a new value of 21.991 / 7

The total points worth became 4 x 5 is 20

The process in which the circle that holds the five points form a triangle in reference of the circle of four points forming a new cycle.

All the seven spots this movement holds and that came onto place are the first one spot that started. This is the same point the Universe originally and then became the dot as it grew. It is the very same spot that is the very same dot and while in reference there are seven dots in truth there is only 1 the spot Π^0 that becomes the dot Π by circling Π^2. This is then how the Universe grew into space by coming into the Titius Bode law but first we have to show the Lagrangian point.
This lead to the Lagrangian points forming but also this introduced what became the Titius Bode law.

This is the result of Pythagoras coming into form. Up to now the line was the circle by half but then this movement brought into play the triangle as the past represents the connection to a new point within the present as the future. As the line of 3 shifted it became 3^2 = nine and as the 4 four shifted it became 4^2 = sixteen and this brought about that the total of twenty five had to spread over an area of four. This was still the one point that held relevancies as positions but space did not exist yet.

As a result of the Lagrangian points forming a line this resulted in the Titius Bode law forming a value in the non-material sector. In the very beginning space formed as material and as non-material and the duplication of overheating became the program by which the Universe formed. The seven formed as material contracting in relevance of 7 x 7 x 7 = 21 with the future adding as 0.991 and in space the Lagrangian point resulted in the Titus Bode law applying 10 + 10.

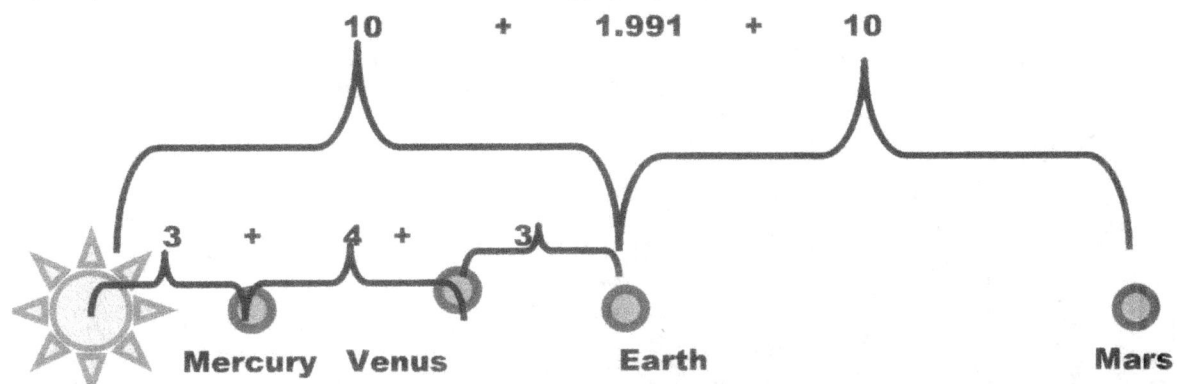

By standing still the Universe overheated and by overheating the Universe moved and by moving the Universe duplicated and by duplicating the movement cooled the Universe again.

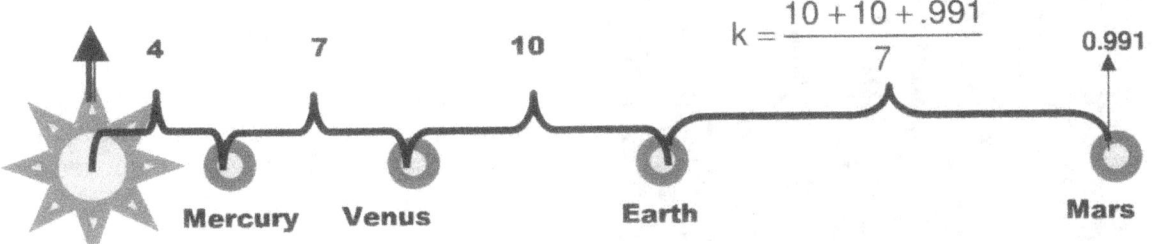

This became the constitution, the norm, the formula and the recipe of what the Universe still is today. From the three one more added as the fourth law and a Universe came about as time moved space through space. However before we understand this part we have to understand what preceded this part because this is so simple even the mentally retarded atheists that understand only Newton should be able to grasp this. This is then where we go very far over the mentality of the atheist's slow-to-follow Newtonian simple-minded-ness. Somehow they can understand mass pulls and they have this clarity notwithstanding that even a picture of the solar planet layout is not enough to make them see there are small planets and there are large planets and therefore there is no mass pulling. With such small minds and unequipped understanding ability how can their simple ness follow something as large and as correct as the Bible? It seems they can only understand magical mass pulling more magical mass all over a make-believe Universe where no intellect is required and only understanding Newton is accepted. It is so obvious those that understand Newton's classical mechanics are unable to understand the Bible and it is clearly an issue of unavailable intellect not being sufficient in order to follow clear reality, which is the Bible versus Newton's classical mechanics, which falls outside reality.

We now have to go to a period of eternity before the Universe that we accept as the Universe started. It there is a Universe that we believe but it is only because we believe. In this Universe it must have a start where it begins and must have an end where all becomes perfect again. If it started with spot then we have to look at why it kept n being one spot before it became the Universe we know and which we think is true. First I press the point that this entirety we call a Universe is a figment of God Almighty's imagination, which is so strong it became our reality and from that point and that opinion I shall not divert. However this is a long storey I deal with in other books but in this one I am going to try and present Creation as simple as I can while staying as truthful as I can. We have to see why it seemed as if time stood still because the only unchangeable reality is that time can never stand still.

Therefore we must investigate time first of all. In time there is infinity. There is space and there is eternity. That which we think is space is eternity and material fills space that covers infinity.

Infinity is twice as long as what eternity is. We see eternity but when infinity the point that starts everything is comes in it takes charge of how eternity is in that present instant.

Then it draws eternity as wide as all space allows to and into infinity where space has no validation. That shows that infinity has to take charge of eternity by taking over eternity as it then becomes eternity and eternity is the picture we see of everything that we see.

The most crucial aspect of understanding the cosmos that is, is the understanding that we are within the very first spot and we are still part of the dot that formed without having limits. We are the singularity with our bodies forming solid space a^3 that moves T^2 through liquid space k. Wherever we look that which we see is still the original dot still growing as the original dot and we have a centre spot in time that leaves space behind.

Newtonians take time as it is in relation to what applies on earth but then what applies to the rest of the

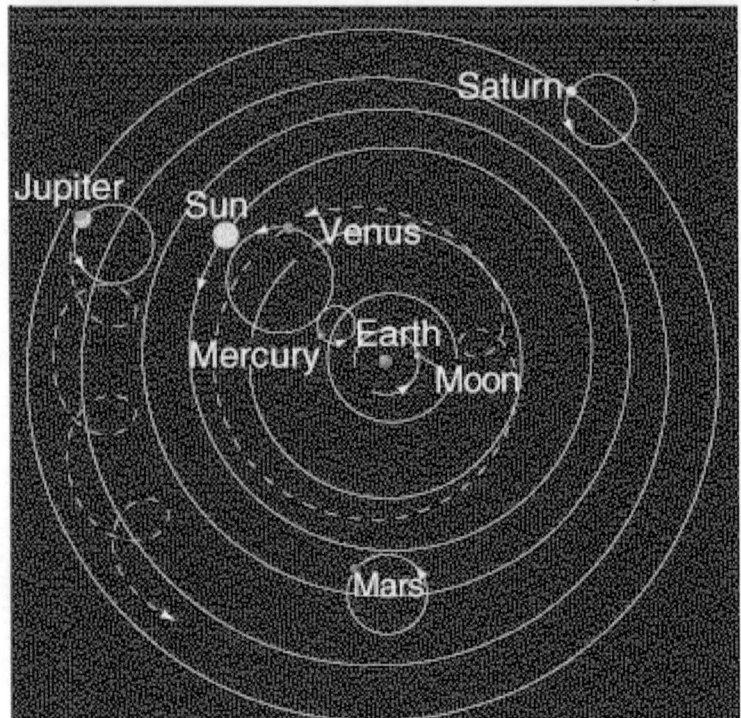

Universe, which does not relate to factors on earth. Time is not even the same any place in the solar system so why can it be the same throughout the Universe? It is clear that every planet circles the sun differently and then having a different and unrelated time.

Then to change eternity as to allow movement to form time it draw everything back into infinity the single point that hold everything. The solar system as does the spinning top as does every planet forming the detail of the solar system stand still before moving to the next position. That is why we are able to take pictures of the planets and of the top and by freezing the movement of the picture we freeze the instant o the time in which the movement stand still.

That instant where we take the picture of the moment standing still does not exist and the space representing the picture is a hoax painted by light on a canvass of space. The light 3^3 moves $3\Pi^2$ by standing still which is $10/7(4\Pi^2)) = 56.4$ is time (1) in space (10) which then is .1 shifting one position .1 and that presents the difference of .2 in the total value. It is space moving two points. That shows that the Universe in space 3^3 that moves $3\Pi^2$. To replace every point that is not a point in reality infinity has to consolidate eternity into one single point by uniting reality once again. Then the overheating begins again and the shift in density brings equilibrium wherever point holds a different definition of reality to what was and in line with what is coming.

...And for the sake of sanity toss out Newtonian small-minded gargle about life and the future because life has no future. Life is made up of insignificance eventualities occurring in line with cosmic planning but with a difference where in the cosmos life is a meaningless addition to a very small part within what is meaningless. Life is so short and insignificant it does not share a future with the future but translates to single simple eventualities that hold no place whatsoever in the totality of cosmic planning. In the cosmic reality life has no future because life has no past and is pasted fragments formed by memory of intellect holding a part of life.

...Other than that life has no meaning and no part in cosmic affairs that forms a timeline. Therefore when we think cosmic then leave planets and life and Newtonian simple-minded concepts out of the picture because it is too small to form part of a cosmic overview. Think what place we, or the earth, or the solar system has, or even the Milky Way has as it forms part of what is. In Afrikaans we call the Universe the "heelal" and which translates as "what is" and I think it is a much better and more appropriate description of "what is".

To get back to before time shifted space into s lot in-between infinity and eternity there was infinity where everything started and eternity where everything ended with infinity and eternity joined as one concept forming time. Then that which can't divide did divide by heat and cold becoming a concept of division. That means before this what was hot was cold and what was big was small.

In this concept of whatever is was in one point in which it still is because the Universe is in that one point. Therefore is everything is in one point and everything that is, came from one point that alone acknowledges the existence of an Almighty Creator. Being in one point makes the point timeless. The Almighty Creator says in the Bible that He is timeless. Being timeless means that everything stays the very same and there is o change from one instant to the next instant. The future is such a perfect duplication of the present it moves to the past without any change that is recognisable. If nothing changed then there is no distinction as time moves from the future through the present to the past. The one is the same as the other and change is a non-existing factor because in this scenario change is the only thing that is not present.

Then came change. In this concept alone we find confirming of a God Almighty and that is science not religion. If everything remains the same because everything is perfect and change is all that is absent than change has to be added to the scenario as a new inclusion, which then proves there had to be an Identity with the will and the power to break the perfect and introduce change. Everything was perfect without change and therefore just because there is a universe that started this fact prove that there is a Creator that decided on the other changes to come about. If there was no intervention what was perfect had to remain perfect because the change that brought about the intervention was the invention that brought about a Universe. This proves the change cam in as a deliberate act and as a planned action and every change that followed became a present because of changes that was introduced celebrity as part of a Master Plan. The mere fact that the prefect everlasting received change is the concept that proves an Almighty God and The Creator of heaven and earth or cosmic liquid and cosmic solid, the two substances forming the entire Universe. In **Seven Days of Creation** I explain the rest of Genesis 1 but that book turned out to be a total of three thousand five hundred and fifty something pages holding seven volumes when printed in Ariel eleven. Science and creation is so easy to understand if science was not bent on keeping the hoax of Newton and atheism alive. But to keep their falsification a reality they had to invent more falsification to cover the hoax called Newton.

I'll leave you with another clue about gravity. It says tat in the beginning there was a voice and the voice was close to God and the Voice was God. We know that sound is a form of gravity and is similar to electricity and all the other perks that or gravity. This say that the Voice (sound) was near (God) and then it became God and gravity I prove is time and time I prove is the thoughts of God and gravity or time is the planned forming or creation of God In the day this was written there was no understanding about concepts we understand presently and therefore we must understand what was implied.

This is the clarity the Bible brings if you are not a false profit as ALL religious Clergy are that only wish to translate religion into currency and show a profit financially for the preaching they do or being a scientist with one aim and that is to falsify what already is falsified even further. Now I get to where creation is part of the thoughts of the Creator. I have shown that in the centre of all moving circles and sphere we locate singularity. The smallest particle in the Universe is singularity and is it so small it is a mathematical principle forming the centre of the Universe. The smallest particle there can be is so small it does not fit into the Universe in space.

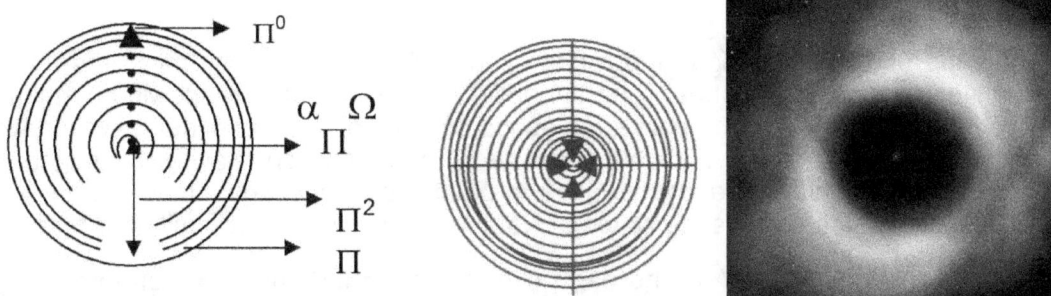

In order to reach this particle there is a mathematical process which one has to follow. Everything in the Universe is a sphere that turns. This turning forms a circle because if not a circle it becomes a fragment floating without destiny. In every circle there is a centre forming the circle around the centre.

A circle is $\text{circle} = \Pi r^2$ That is the circle that is visible. To reduce the circle one goes about as follows $\text{circle} = \dfrac{\Pi r^2}{r^2} = \Pi$. Now we are where all material begins $\text{circle} = \Pi r^0$. Then we have to reduce the movement further and there we reach singularity. To reach the point that is smaller than material is filling the Universe we have circle $= \dfrac{\Pi}{\Pi} = \Pi^0$. This point holds no material and has no space within the Universe. This is the spot Π^0 that grows into the dot Π and the dot is the point where space begins at Π but all this is only possible if the circle Π moves $\Pi\Pi^2$.

In this point where the spot is with the value of Π^0 is there is no space. If you enter that point you leave that point at the same time. If you enter that point you find the Universe going in one direction to the left and to another direction to the right. As you go in the turning will go to the left but as you enter (no space remember) the Universe turns to the opposing direction. It has all four directions of flow concentres at a point just where space starts and therefore it is all space that gives division to directional flow but the source of this opposing directional flow of movement is from a point outside the point that forms the Universe.

This point in discussion does not form part of the Universes we are within but it is part of the Universe we in life are part of. This point does not exist and yet this point controls the entire Universe as a unit.

The Universe starts at Π. Everything in the Universe starts at Π. Going to a point that is smaller than what anything is that starts the Universe goes to something that is not part of the Universe. Yet in every rotating object and everything rotates in some way we have in the control of space a point forms where no space can be. This is most significant because this point holding no space controls all aspects of space. Although this point is so tiny it still is present and it is so small it does not exist it is what forms the entire Universe. This point can only hold a thought because a thought does no hold space. This is the point that life manipulates as life takes charge of the body that life controls. We call it being alive. By controlling a body from such a point we control everything we can mange to manipulate like building and driving and thinking but we d this control from outside our body. If this position is so small it can't hold material and still it is the point from which the entirety of the Universe are controlled then this point has to serve as formed by the Thoughts of the Almighty Creator, This point hold no space but controls all space and this point is in a position that is not part of the Universe and yet it establishes the destiny of the Universe and it positions everything that is from this point where every point that is by mathematics form one and the same point because $1^0 = 1^1 = 256789134^0$, then according to the basic law of mathematics this point is still the original point where all Creation started and that point can only hold a thought. Therefore according to true physics the Universe is formed by the thought of God Almighty. This is what mathematics is and this is where science starts.

To understand the following we must understand the difference between time and space. Time moves because time moves space. As a human I look at objects and put time and space in a straight line. That is wrong. Time is a straight line but also time bends space by placing space into forming the history of time.

When I look at my finger it forms a straight line according to my observation and for all technical purposes that is true. When I look at the moon time bend space in which the moon is by one and a half second. Time in relation to space is no longer a straight line because time placed the space one and a half seconds arrears of me. I see the moon and through my observation skills the moon is in a straight line according to my vision.

Observing the sun puts the sun in a slot eight and a half minutes arrears but to my mind it is many millions of years more then that but I am not debating the observing merits but only showing that time and space does not form a straight line. Time bends the space by eight and a half minutes. When we look at the giant planets there is a time interval of thirty minutes or more between relaying messages to space craft and the space craft receiving the message. Time bends space by thirty minutes. Time will bend space by many hours when time places the space of Neptune in relation to our position in time.

Time forms as movement changes space. I.e. the earth circles around the sun and therefore every instant the earth is in a different position and in a different location in relation to the sun and therefore the Milky Way and therefore the entire Universe. In the Universe time flows because the present is different from the future as the past is different from the present. Time is about everything changing.

**Every aspect of the Universe is different from what it was to what it is to what it will be the very next instant. That forms time. This is proven by the formula we received from Kepler and Kepler in turn received it from the cosmos which he and Tyco Brahe studied for eighty years and calculated as $a^3 = T^2k$. Time is not space but time is the movement of space. Time changes space. The changing gives the Universe time flowing. The space a^3 moves T^2k and therefore** *changes the space in a singular Universe a^3 at a ratio of a circle T^2 in a singular Universe that moves straight by **k** that is a line in a singular Universe. The change we see in the Universe as pictures formed by light is in the time flowing that forms space and thus creating a Universe by changing it every instant.*

Before this we now have the future was so perfect that the future was the same as the present in every detail. The past was the same as the present and being perfect the past was the same as the future because it had no difference from being the present. The Universe never changed anything and therefore was in singularity. This even the Newtonians agree on…without realising the concept. *Einstein brought us the concept that everything came from singularity and everything still is in singularity. Forming singularity as one was that what gave one as the future that arrived in the present but also left one as the present that moved to the past, serving time as the future being equal to the present being equal to the past with nothing ever altering from one position to the next position.*

Then at one moment in time the imperfect brought the Universe a point in time where the future was different from the present that took the difference onto the past at a rate of every instant from then on and that is how time formed space.

Suddenly time flowed in two directions. Since time will tend to remain the same perfect ness therefore it can only be a Creator that will bring change.

The change came as heat formed that split in the flow of time that then had the Universe flowing into two parts. The one part overheated and expanded ever since and the other part in relevancy cooled down by moving and with that contracted.

There was space that overheated and expanded because of overheating and there was material within the space that by movement froze and shrunk by spinning. This factor science never came to realise and therefore never came to acknowledge because Newtonian science never understood this principle.

The difference between outer space expanding and material shrinking we call gravity. Before this any numerical order was nonexistent. There was no need for anything more than one. Then afterwards the Universe introduced numbers because only then did a need arrive for numbers carrying any measured value at a level of forming a concept being more than one. This book shows how the Cosmos started way back when it introduced Creation and the four Cosmic Pillars... It shows why numbers came about as laws formed, founding Creation. This concept described in this book goes back much further than any one using formulated mathematics can realise, it is before light or material formed as the Big Bang.

The notion of Scientists thinking that the intellectual thinkers in society don't know anything about the Universe before the Big Bang is altogether wrong and also it is altogether correct at the same time.

The reason we as the intellectual species on Earth think we don't know anything about how the Universe started before the Big Bang is because we are staring into the science of Newtonian darkness formed by misconceptions. It is those that supposedly hold all the intelligence on Earth that believe that we are clueless on matters that happened before the Big Bang happened.

In the Universe time flows because the instant in the present is different from the following instant in the future as the instant in the immediate past is different from the present. That gives the Universe time flowing because every instant everything changes to what it is going to be from what it was.

Before this we now have, as a Universe the future was so perfect that the future was the same as the present in every detail. The past was also exactly the same as the present in every smallest detail and perfect. The present was the past in every aspect that was the same as the future in precise detail. This is because the Creator is perfect. That gave one, serving time as the future that was the same as the present that was the same as the past and time formed one.

Then the Creator decided to create something imperfect. Then the imperfect brought the Universe time where the future was altogether different every instant that time formed space and that difference also brought a past. Before this a numerical order was nonexistent. There was no need for anything more than one because in one everything that was, was perfect. Then by changing every instant one no longer applied as a solo number and the need were there to start numerical multiplying.

The Universe introduced numbers because a need arrived for numbers valued as more than one. This book shows how the Cosmos started when and how Creation introduced the four Cosmic Pillars. It shows why laws formed, founding Creation. This event goes back way before light or material formed the Big Bang. In this book I show what stupidity is going on physicists minds when think they are equal to God Almighty.

I show everyone that reads how God created physics in the process of creating a Universe. I show how the numerical order formed mathematics as that formed physics because only one was present before the Universe was in place. Should you doubt that a living God was responsible for Creation, then you better read this and get wise about how physics is truly conducted in the way nature intended it to be!

It is Newtonian Science that stands to be corrected firstly because Newtonian Science knows nothing about the Universe...at all, but that is because Newtonians know nothing about science...at all. Scientists think they are brilliantly brainy because to them intelligence is punching in numbers by calculating formulas and then to equate formulas. However, that is not science, it is highly camouflaged misrepresentation because to think and to reason about nature is true intellectual science.

That misrepresentation only comes about through the way we practise science.
 Mainstream Science doesn't know anything about the Universe before the Big Bang but this is on account that Mainstream Science doesn't know anything about how science started to begin with.

We know very well how the Universe came about but we just never considered the correct approach.

 It is because those thought to be wise are instead only Academic mathematicians therefore being the mathematicians they then think they then can so gallantly by their calculation say God didn't create the Universe because they declare that physics created the cosmos. How smart does this and these idiot(s) think he (they) are by saying something so stupid? Who then created physics or are the likes of Steven Hawking and his band thought of as superior thinkers think he (they) are also so mentally clueless that he and his brigade think that physics was around before the Universe came about?

In that case they admit to a God being present but they then just call the Creator by another name since they are fools that can't appreciate reality...and that is because they can punch in numbers and never think. When the Universe started, the Universe started everything including numbers, and with it came mathematics and with mathematics came physics. We know very well that the Universe started with one spot or dot or point holding singularity. It started with one...and that is singularity, the one! With singularity being one and one is the point that the entirety started with, then the question is not why the Universe would explode because when it exploded it was not one but many that exploded. It is a case of asking how did the number two come about and then three and then four and five and six and seven going up to ten. How did the numerical order formed by Pythagoras come into place? How did Pythagoras instate numerical order and therefore physics as a law? That is how mathematics started physics and that is how physics developed as the Universe developed through physics developing. When you read further you will find how numbers developed physics, mathematics and the Universe in the process of creating the Universe. This is how the creation of the cosmos came about.

Moment-Alfa is in eternity and eternity never ends therefore it moved to eternity in space eluding time by having a vision upheld by the information light delivers. Through light it is possible to find information being able therefore to see the information that can be reported because the light carrying the information was in that space but not in that time. If the viewer and what he saw were in that same time the view the viewer saw and the viewer would still be there sharing space, but also then the lot would be there in that space within time eternally and never being able to come back to report on what the information about the past that the viewer saw.

There is much rumour of a Big Bang, however I am inclined to think the biggest bang that ever was also became the smallest bang there ever can be. It was the instant when heat parted from cold. Then the past separated from the present to form a future. Without a future there can be no present and then we would not be able to live in the past that began Creation in that instant. It is what moves that form a centre and the centre produces a line that puts a centre in the circle that moves, That is Creation and that is how Creation started, albeit too simple to our outrageously clever Newtonian physicist to accept.

When a top starts to spin Moment-Alfa starts all over again as it did at the start.

Before this period the past carried the present from the future and everything landed on the same dot. The past was the present was the future because it was laboriously monotonous while being perfect in every aspect all the time. Then the Creator created heaven and Earth just as the Bible says the Creator did. The Thought became sound. The idea became structured. That which is beyond recognition became observable. The thought became the voice, which remained God. Singularity found a presence within the space that now is the Universe. The Creator created what is cosmic liquid and what is cosmic solid. The Creator created something that spins (as solid as the Earth) while holding space in which it travels. Study how singularity works and you see precisely how the Bible says God Created the heaven (cosmic fluid and gas) and the earth (cosmic solids that move within the cosmic liquids). Still, everything that is also is in place as either a dot filling or a dot being filled by a dot. The role the dot accepts depends on the relative movement of the dot being filled or filling. It still is a dot, although many dots but inn singularity remains one single dot.

The proof of the dots existing is looking at the Universe and finding space. The dots are not there because individually no dot is part of the Universe but because every dot is moving in terms of every other dot all the dots form a solid structures Universe we all enjoy. Still every dot that is also is not and every dot that proves a history also becomes one dot forming time. The dots form a circle that forms a line. The line started in infinity because the line was continues but being continues it never was. The very instant followed the previous instant identically and the instant was so identical it remained the same by never moving while always moving uninterrupted eternity upon eternity. Then came the entire Universe when infinity broke free from eternity with a difference between heat and cold parting time. Time became imperfect splitting the perfect into two zones. It was long before when darkness broke into light. This was when space was a thought and not a measured distance. It was when whatever possibly can be became a possibility to be. It was when the first line of numbers mathematically arrived and from the one became two by gong 1^0 to 1^1. That infinity is so small it houses everything there is in the entire Universe. The entire Universe still is in a spot that formed a dot. The spot has no outside but it only has an inside while it is inside all that spins as it generates all that can spin. Yet, by spinning it brings motion into being. The spin creates a drive that keeps the Universe mobile. Still the first forming of the dot from the spot came about to the inside and not the outside, which makes the Universe shrink and not expand. It is the smaller things that come into relevance as the larger things were placed in relevance when time began. The Universe is shrinking into the oblivious since the Universe never had anywhere to expand to. Never once did one Newtonian sit back and consider their laughable proposal of an expanding Universe with nowhere to go when it is expanding. The diversity we have in the Universe is the combining of what came about when there was nothing added to what there already was.

In the very centre of the sphere the form of the sphere dictates that the shape will relinquish space as the line run from the outside towards the very centre. With this natural state of affairs the sphere are naturally inclined to dismiss all space that it can form in the form as the sphere holds space inside and the form will finally be without dimension. All that I attribute to the line shrinking by reducing actually takes pace in every sphere as the diameter reduces to the centre. In the centre where the radius line goes single the form relinquish the three dimensional form it has inside. Being without dimension in the very centre means that at a point in the extreme centre of all spheres there are a point that holds singularity because this point with no space has a mathematical position although it is invisible since there is no sides to such a point to give that point any dimensions. The shape of the sphere is calculated by using the formula $4\Pi\,(r^3)\,/\,3$. By reducing r to a point where r is r^0 singularity steps in because only the form remains as Π. Going even further we find that there then comes a point where Π goes singular Π^0. At that point absolute singularity is present but so is absolute gravity present at that point. When holding the strength of the shape of the sphere in mind as well as taking into account that all cosmos objects of importance is in the form of planets or stars and they are all in the form of a sphere, we therefore may contemplate that it is where gravity originate. We now only have to find the reason why gravity will hold a base in a space less ness as Einstein predicted. It is clear to be seen that gravity is in the centre of the sphere controlling from the centre everything that is outside the space less centre. We can reason with confidence that gravity is the strongest where space is the least. We can further reason that it is gravity that is holding the sphere in true form and since the sphere allow gravity the best working opportunity, gravity can form the sphere in as strong a shape and form as the sphere seems to have. From every point on the surface of the sphere is where that point connects with the other side of the surface of the sphere by a line that runs through the space less ness of such a centre of the sphere. Such a line also connect by an angle of 180^0 as well as 90^0 to six other lines running from top to bottom, right to left, and back to front, where all join and cross in the centre of the sphere. There are therefore six lines crossing and connecting by a centre from any given point on the surface of the sphere. Such points connects in total six surface points on each side of the sphere while they all support one another through the space less centre. In that absolute space less ness in the centre holding singularity we find gravity supporting and controlling all space within the sphere as well as space connected to the sphere. That is where gravity control and guide the space, which falls in the parameters as well as under the influence of the form of the sphere. In the gravity centre space goes

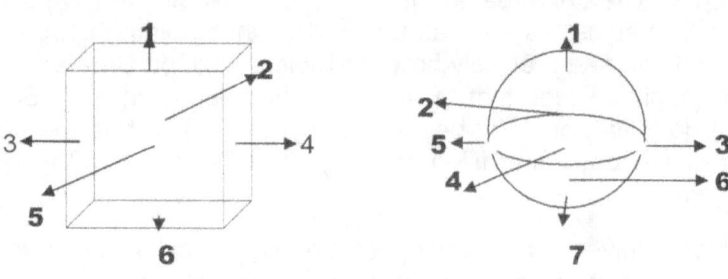

singular meaning space becomes space less or flat.

It is from the layout that the sphere uses as natural form that we are able to locate singularity. In the case of the sphere the material naturally reduces by measure of the radius becoming smaller to a point where the radius is r^0. At that point the line that will form the radius has gone single dimensional r^0 and that is equal to 1^0, which is singularity. Also it is true that the entire form that is the sphere is controlled from a centre within the sphere. That centre holds the sphere in form and shape. Therefore the strong form is dictated from that space fewer centres where there is no space and no form left. The natural inclining is in the form of the sphere. It is part of the roundness that the overall shape of the sphere represents and this structural strength is carrying down to the very centre. Because the circle is forever reducing that reducing which is inherently part of the form of the sphere becomes a tool in distorting of space in the sphere and is eventually removing all forms of space from within the centre of the sphere. The very centre ends up as having no space because of the reducing that continuous down to become the space less inner centre. The all roundness is the ingredient that forms the backbone of the absolute strength that the sphere has and that is the component that the sphere is so famous for. The form the sphere has allows the sphere to have a control that is coming from the centre deep inside the sphere where the space vanishes and being without space seems to keep the entire structure rigged. From the centre the sphere shape shows strength that the shape as tough as it is. How does it work in its most basic analyses?

Realities such as getting there, finding the object, fighting the gravity (remember this object has the gravity to absorb the entire solar system in a matter of minutes and that is according to their calculated efforts on the Black Hole) Those capable of designing space whirls were in charge of the mathematics of the Universe and still they could not solve such a simple issue as gravity in all the time they had to their disposal over the past three hundred and fifty years. It takes human effort to recognize reality. It takes a human's intellectual effort to recognize the Godly aspect to the Universe. The thought manifests as an electrical component after it is realized and then by a human's deciding ability is transferred to other parts of the body in electricity conducting. Normally it is the emotion provoking the thought that generates the thought because it is the emotion attaching thought to the moment that places the thought in the mind. The machine can never reason and challenge conceptions but will always accept conceptions unconditionally and Newtonians are brainwashed to do in science. To them and theirs and those the computer may replace them and I hope it will the sooner because that will make life the better for the rest of us that can face reality and not hide behind computing skills that machines can replace.

Time came from eternity. It is so obvious but then again science are not very likely to condone in the obvious because science grab onto the ridiculous. Science is absolutely opinionated when they do not even know what gravity is. The Coanda affect is the epitome of gravity and is the implicating principle in establishing gravity, but while at the time I am writing this, not one in established physics was able to acknowledge this fact. Kepler said $a^3 = T^2 k$ which in turn says $k^0 = a^3 / T^2 k$. This then translated to a verbally understood language says that all space-time $a^3 / T^2 k$ becomes nullified in singularity k^0. This a now can say because for the first time since gravity was conceived four hundred years ago the principles behind the phenomenon gravity are revealed. Up to now every one in science was acting as if gravity is a commonly explained factor, which every one knows every aspect about all principles that are involved in gravity down to the smallest detail. In truth no one in science anywhere remotely knows what brings gravity about and I used Kepler to unravel this mystery called gravity. The most informed in Science at best can only assert their suspicion on a rumour presumed about what causes gravity to perform as the part interlinking the cosmos but no one can go any further than affirming their alleged confirming such possible correctness by explaining the concept. But in the meanwhile gravity eluded them as much as it eluded their master, Isaac Newton.

This is because they avoided and belittled Johannes Kepler. Newton started this rumour and Newton admitted to it being a concept he could not explain but nobody in science at present will denounce Newton's gravity ideas as being merely just a rumour as Newton admitted it was. But instead of trying to study the findings Johannes Kepler produced, Newton degraded the work of Kepler by changing the work of Kepler. As he did it, it also rubbished the work and the findings of Kepler. Newton agreed that he could only declare gravity as only a rumour announcing a force that could be anything. He did not see that it was motion of space that kept the planets orbiting the sun as Kepler said when Kepler said space a^3 moves around T^2 in a precisely designated relevant space k where the distance of k places such space in that specific motion in a specific and precise relevancy to a generated specific centre k^0 where such motion produces time for space to move in the relevant space coming from k^0 and ending at k to produce $k = a^3 / T^2$. Not once could one person in the past or present provide substantiating proof on gravity as a physically reality by defining the very principles that produce such a force, should such a force even exist. That includes Newton as well as Einstein and even Hawking.

Scientists can declare gravity was a factor at 10^{-43} seconds after the Big Bang but what brought gravity about or why gravity became or still remained, as a presence is still tightly concealed information. Even to the best informed amongst the most educated do not know what is gravity because they all ignored Kepler and for ignoring Kepler the price they pay is not finding gravity. Using Kepler makes the method to follow and understand even Einstein's discoveries shockingly simple. By my applying Kepler I can define gravity precisely to the point where I now can explain why the proton and the electron forms a mass difference of 1836 times between the two mass components.

With the aid if unravelling the four cosmic pillars, which the Universe in its entirety was built on and now rest on I can enter an era never entered before by science. But to uncover the veil that kept the Universe hidden from sight was finding the four cosmic pillars. The four cosmic pillars are keys that were in place before mathematics came in place. It was dimensions that applied before dimensions were applied. By the first spark a Universe came about translating the four cosmic pillars into cosmic principles allow us to look past the Big Bang into an era where space was an image and time was the interrupting of eternity by changes in infinity. Firstly and in front of everything else the cosmos is about relevancies. There cannot be gravity is there is not a gravity between two particles in the Universe. There is no force such as gravity because gravity is a motion of space in relation between two particles. Once that fact is established and the fact of two points holding the space singularity but the singularity is divided by space-time generated by motion between such points can we find the cosmic pillars that serve as keys to unlock our past. The keys are not recognised by the Mainstream policy makers in officially accepted science. The Newtonian thinking minds never put cosmology in terms of reality and as normal put Earthly science in cosmic place and Universal space.

Kepler declared the Universe to be $a^3 = T^2k$, which the Universe itself told Kepler. If the Universe is a ratio of $a^3 = T^2k$ then the Universe is $a^3 / T^2k = 1$. When I place any number or any figure holding a mathematical symbolic value to the value of an exponential 0 as in 1^0, then the figure has the value of one notwithstanding the actual number or symbol. From a^0 to Z^0 and 1^0 to $10000000X10^0$ all the answers come down to one. That means singularity is not a mystical numerical mathematical idea exported from the world of make believe to pester the sane, but it is an every day reality. That puts the Universe at $k^0 = 1 = a^3 = T^2k$ declaring the **Universe as being singularity.** That way the Universe is singularity and singularity is without form being one to the dimension of zero. Where then do we find form, if not in the Universe?

How big was the Milky Way when the Universe was the size of the neutron. Science in their overwhelming intellect has not thought one nana meter past their noses to come to a realistic approach to cosmology. If mathematics can explain a scenario then the mathematics accept it as religiosity. However it is never the idea that they scrutinise but always the mathematical formula that goes tested. What no one realises is that mathematics is part of the Universe. Being part of the Universe there then had to be a period where mathematics were established as a factor and some factors in cosmology has to predate mathematics because it did not start with mathematics, the cosmos started with singularity being 1^0. Technically singularity is not part of the Universe but part of what is determining the Universe by creating the Universe. Singularity at 1^0 is above and is outside mathematics.

When the Universe was the size of a neutron, where were the react that we now find to form the Universe? When the Universe was 10^{34}, what then was minus 273 C. If there was no minus 273 C at the time any where in the Universe to compare the 10^{34} with, the heat of 10^{34} stands rather without meaning. Only by comparing the hot with what was seemingly cold could put the hot in an appreciating position otherwise it has no meaning but to be a statement someone wish to make.

What was the size of the sun when the Universe was the size of a neutron? We cannot look at the Universe and proclaim a size without the size affecting the lot within. Stars were stars when time distortion produced space because what is space when space is not filled and what is space when space is filled, but moreover, what is the difference there is and why is the difference there? It is obvious why science has no idea what time is if they wish to use a line supporting the upright stance of a vector to indicate time. Time is the holding as well as the support of space. It is time and not space what keeps space in motion and the motion we see as space while it is the time that is the space. Space moves in time with the support of time. Therefore space grew as much as time grew. With the Universe being the size of an atom the sun was frozen into a thought of one day probably. But that cannot be because there are no means to add to the Universe, as there is no means to remove from the Universe. That which is in the Universe is here to stay and is as eternal as the Universe is. At the of the Big Bang period in time the sun was far from the boiling bubbling soup bowl containing pure liquid heat that we have today. It is quite impossible to imagine the sun being the size it is and still fitting into a Universe the size of a pea. It is also quite impossible to imagine a sun bubbling with heat while being the size of a pea. That means if the Universe grew, then the sun grew with that Universe. However, the Universe contains what ever will be which means the Universe cannot possibly grow. The Universe already has everything there ever can be.

With the Universe being the size of an atom the sun was frozen into a thought of one day probably being more than a concept. At the time the sun was far from the boiling bubbling soup bowl containing pure liquid heat that we have today. It is quite impossible to imagine the sun being the size it is and still fitting into a Universe the size of a pea. It is also quite impossible to imagine a sun bubbling with heat while being the size of a pea. That means if the Universe grew, then the sun grew with that Universe. However, the Universe contains what ever will be which means the Universe cannot possibly grow. The Universe already has everything there ever can be and that the Universe had at the very start of the Big Bang. Whatever came after the Big Bang were merely relevancies alternating possibilities.

This was moment-Alfa. The darkness was there as big as it is and as dark as it never again could be and from the darkness heat came about. Only heat expands and it interrupted the true invisible darkness, the blackness of a Black Hole, the invisibility coming from within the Black Hole. Eternity tore from infinity. Darkness broke from light. Heat broke from cold. Relevancies parted by 1^0 going 1^1. There was one but also there was two too because one cannot be without two being there to ensure one is one. The marks are still with us but to see the marks requires a great deal of intellect.

$\Pi^0 \Rightarrow \Pi$. In this there was only space for one being one in the two forming one. It was $\Pi^0 \Rightarrow \Pi$ however there was no space to be $\Pi^0 \Rightarrow \Pi$ and there fore because of the lack of space to be which is the infinity of time braking the eternity of time the true measure was $\Pi^0 \Rightarrow \Pi$ but realized only 1^0 going 1^1. Π was to the future because of the motion of time involved and the space less ness of space at the time. By inclining to move the process crossed the Universe but also it took one eternity to accomplish the feat.

The fact that 1^0 going 1^1 brought movement can only become a reality as a result of light. Light is heat and the heat is expanding.

Science look at space when space had 1^0 going 1^1 brought movement and today space has 1^0 going $1^{1000000000000000000000000000000000000000000000000}$ and yet because of their inability to understand concepts regarding truth they wish to place the cosmos that then applied to the cosmos that now applies.

They look at a hologram of space going back say a Yotta- meter times another Yotta- meter (Y or numerical 10^{24}) it is by numbers 1 000 000 000 000 000 000 000 000 000 000 000 000 000 000 000 meters of space is a concept surpassing understanding that developed and that image of a Tera-years (T or numerical 10^{12}) it is by numbers 1 000 000 000 000 of space development that they wish to transfer to what the space was a trillion billion years ago. Then the hologram goes out of context completely and this effects out truth about the cosmos. As long as science tell the cosmos it has mass and the planets align by mass while it is not true then they are busy with a ridiculous complot to deceive and defraud just as common criminals do.

Motion creates space-time as much as space-time is supplying motion to form space. The only way to enable that to become a reality is that motion creates space as much as space follows the direction of motion. That is what Kepler said when Kepler said the space is equal to the motion thereof $\mathbf{a^3 = T^2\ k}$. There is no solid $\mathbf{a^3}$ Universe but all interrupted by positional changes that recreate the space in the and according to the new direction singularity will create as singularity allows space to flow by motion by fragmenting space into time sectors.

We with life take motion as for granted. We never consider why objects in orbit would have motion and what would be the result when such an object is deprived of motion. Having life means moving and we as humans are on the move from birth. We even call an infant that has died at birth being still born meaning it did not move in the manner we accept life would move. The last thing life would have before capitulating to death is to have the blood circulation terminated. In that way to our reason motion is a fundamental or a basic which is true but that makes it even less common. Why would all things move? The quickness of the Newtonian mind would remind me about the galactica inner core that shows no motion. That is a time concern and not a result of inadequate motion. There is lots of motion but the motion is more delayed and therefore we are unable to witness it.

Before I allow myself to get tangled into that explaining (I do explain this issue on another occasion in this very book) I wish to return to the basic concept of motion. Kepler quite correctly stated in his formula that the cosmos holding gravity is motion. Motion of space in time as well as through time stated as space $a^3 = T^2k$ moving in time and through time. In that we find the birth of the cosmos. Newtonians have this grand dilution of a Birth of a Cosmos that holds a formula, which can crack the jaws of a hurricane.

Kepler's formula shows how the Universe started. When space a^3 started it moved in time and through time $= T^2k$ where time holds two parts in securing space. Realising this birth we must get to terms why stars move. Why would a star move and why would they not move? I guess it would be fair to put any and all activity concerning motion of the star in relation to the activity within the star. The motion of the star must be related to the time component the space uses to move within.

As humans tend to do we look at the brooder picture and try to assess from that what applies. That makes us miss the target by miles. The star moves with everything in it but it is everything in it that moves and therefore the star moves no with everything in it but because of everything in it. The star does move everything in the star but those within the star moves more and therefore by creating more motion the star establishes an accumulative all including motion of which all individual motion takes part in and therefore is part of such an accumulative motion. The atom is vacating the allocated place it occupies and moves in time and through time to claim and a new position in which it holds space for one point in infinity. It vacates to fill and while vacating to fill it claims a spot where it is. Let's put it in the manner Kepler introduced it. Material a^3 vacates T^2k an allocated position to fill $k = a^3 / T^2$ a predestined position and while vacating $k^{-1} = T^2/a^3$ to fill another spot it claims $T^{-2} = a^3 / k$ a spot where in it is $a^3 = T^2k$ at the present $T^2 = a^3 / k$. Well normally we would leave it at that but then that would be that and we would be left with no results. What makes the atom move?

In the first factor of singularity a line indicates direction and hydrogen becomes a volatile product. But singularity also provides space a^3 and duplicates by motion k to destroy by rotating T^2. In dismissing the space the proton grows by accumulating space that the other side lost. The growth depicting of the dot I use to symbolize the proton's growth are highly exaggerated I have to admit, but that is only to bring across the idea I wish to convey. Looking at carbon$_6$ one would think that proton numbers would bring about mass, which we then associate with density between particles. But then comes Nitrogen with seven proton pairs and oxygen with eight proton pairs, Fluorine nine and Neon having ten. These mentioned are significantly highly volatile which means they truly extend duplicating of space.

This as a group forms a relation with heat unlike any other. If mass did the trick these must have been the group having the second least density, but they form the group with the least density as a five point group. The next group holding the Pythagoras five or the Lagrangian five plus two or three, four or five enabling their relation with heat to be quite remarkable. This must be some indication of events during the period just preceding the Big Bang at say 10^{-7}, 10^{-6}, 10^{-5}, the time when the fuel that would ignite the Big Bang turning heat into space turned material into heat. The relation of five plus one, plus one and five, plus one plus one plus one is just too uncanny to ignore.

Following the process and seeing the influence of singularity should bring about a pattern that may lead one to a pattern of how the required heat formed and how the intended heat transformed to space. Density depends more on proton number arrangement producing specific form in relevancy as to merely and only having mass as factor that contributes to the forming and development of stars in the cosmos. The evidence is so clear that mass has nothing to do with gravity but density has everything to do with gravity. Density is the volume of space in numbers used to fill material in ratio with numbers of space per volume not filled with space. It is matter versus space in every sense there are. This came about before the Big Bang took place and before space was formerly space and time was formally motion. It was a time when singularity set relevancies moving from Π^o to Π

In that manner we know that that was the way particles formed combinations just after the arriving of moment-Alfa. Singularity brought the Universe but also singularity brought the divisions between the many Universes that followed the immeasurable many Universes that came after the flooding of Universes to follow the leaders. The term "moment-Alfa" is the way I refer to the moment when singularity changed, not when space formed or time began or space exploded but even before anything including mathematics became definitive. At this point mathematics renders it useless. There was no space or time to calculate because relevancies came in place. Form took shape but space there still was not because Π^O moved to Π. Every slightest point in space became an opportunity of establishing a Universe with most different functions and ingredients there might form. This is apparent from the fact that it still takes place at the present moment by motion attaching new singularity through duplication and through duplication releases previously attached singularity from serving the purpose of duplicating by motion.

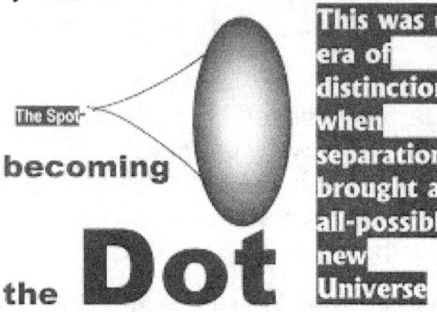

becoming

the Dot

The Spot

This was the era of distinction, when separation brought an all-possible new Universe

When the cosmos came to motion, motion was not yet defined. When the cosmos brought about motion, the first motion was relevancies. Cold parted from hot. Eternity parted from infinity. Motion parted from motion absence. Infinity broke the laboriousness of eternity for the duration of infinity. The spot became and grew into the dot.

From what the spot was to what the dot now is might be just a mathematical implication of going from 1^0 to 1^1 but in reality that first motion was the creating of and establishing of an entire Universe with all possibilities now in it. Never again can that much growth become a reality, although to us the growth is beyond what we ever can notice. But it is because the growth is so massive and we are so small that are unable to notice such almighty growth.

we **Singularity**

When the spot Π^0 became functional and established all relevancies possible, heat parted from cold as eternity parted from infinity. The expansion was not clear motion but more a parting of relevancies where a centre formed a relevancy because the centre could not provide motion. Without being capable of motion, the centre established four points, which also served singularity. From the inverse square law we know that the centre doubled by producing the four points holding singularity.

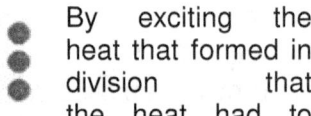 

By exciting the heat that formed in division that the heat had to centre spot, the centre spot came to be because of the relevancy as heat parted from the cold bringing about the followed and that was the motion that formed. Therefore move but being singularity it could not get singularity to move. In an attempt to establish growth, singularity activated six spots of which four was having motion drawn into relevance four spots that was providing what was to be motion and three that was to be securing the position the centre holds. There were four forming a ring around singularity with two forming in locations we will refer to as above and as below or north and south.

The three in line was in singularity not being able to move but the four was also in singularity and just as incapable of moving. All the points came as relevancies applying the forming of more of what was to come but only the four committed to time were expected to move. The four points that came as a result of discrepancies that became time that produced form and that established the relation with the one but had to perform the motion by expanding was as much incapable of motion as the centre was that charged the four with motion in the first place. As they were incapable of motion, it still required a tendency to apply motion that did separate Π^O from Π. This not only involved form but it involved all relevancies that did come or may in the future come about as a result of the attempt to commit motion. If mass was a factor contributing to gravity the cosmos would have frozen back to singularity without ever releasing singularity to relevancy.

Mass does not establish gravity. There is no magical graviton. In the beginning there was no mass but boy was there gravity! The only means that the cosmos could find a way to break from the grip of eternal eternity was to expand into relevancies. Such a feat can only go to task by forming opposing hot and cold. Becoming hot produces more of what is heating. That implies motion or a moving away from where it was by generating more of what is available. Only where hot released from cold could whatever was repeated once again and duplicate what was before into what then is more. Secured by motion T^2 in relation to a specific centre **k** from where singularity holds the Universe true to form. The **k** was an intention to place apart and by today's standards will not even qualify any noticing.

All that are is in singularity. From singularity comes the motion and the space we call space-time. Singularity is dimensionless, time less and space less and because of all this features, it carries the value of Π^0. By expanding, singularity applies a relation coming about that reforms singularity from Π^0 to Π. Only when extending Π^0 to Π, the extending creates motion and the motion creates space that then doubles through motion applying which cuts the space in motion in half by matching the space as a duplicate. Motion creates another dimension or another level reforming singularity from Π^0 to Π or from Π to Π^2 or from Π^2 to Π^3

As said before we now know Π came about since Π is achieving form and not space. Only **r** can establish space as size will accumulate and as it had with everything else singularity had **r** covered by one as in being $r^0 = 1$. By reducing the circle radius **r** by half continuously will lead to an infinite small circle and an infinite number holding r would place **r** to the power of one as a factor. Then as a factor **r** would not contest any change when change is introduced into any future equation but Π will remain because the circle as a form remains even being infinitely small. By reducing r indefinitely to the tune of half each time, r would become infinitely small, beyond human calculating means, however as mentioned in the case of the smallest dot holding one spot, r would become insignificant beyond human comprehension even, but never reaching zero and still Π would remain intact and dictating form. To amplify by dimension a value has to be set to r but if r remained covered by singularity all alterations that could possibly come about was in the form, which was Π.

This expanding can be a problem one can wrestle with for one lifetime and never reach any conclusion. How can something grow without getting more that what was before? Then it hit me like a ton of bricks. The answer is in heat but not heat, as we know heat. It is heat in getting relevancies between outer limits. Only heat could break the monotony of singularity. Heat in the form we now know heat as heat is now. Since the Big Bang heat is material transforming from one state to another state.

The change that took place involved singularity but singularity was 1^0 and being $^0 1$ could not grow. The growth came about. Heat rose from singularity, but if heat rose from singularity. Singularity as a factor changed from 1^0 to 1^1, which means a relevancy came in place that no one could detect. It is true that 1^1 are still one, but one could then escape from singularity by producing factors other than 1. Heat came about but only as a relevancy to utter cold. If there is heat, there is cold or if there is no heat there can be no cold. Space came into forming a relevancy that brought form. Since it is a relevancy and not a generation by accumulation, the form produced was Π. The spot formed a dot by heat and cold establishing relevancies and from that singularity was broken to allow all other forms of relevancies to come about. The cosmos did not start because of gravity. The cosmos started with heat and cold coming into a relevancy and in the cosmos there is no hot as much as there is no cold. The cosmos broke, put from the confinement of singularity by establishing a singularity in a relation of heat and cold. The heat that came about was beyond measure because the cold that held the heat was also beyond measure. The immeasurable heat was on the outside of the dot that formed and the cold was on the inside of the dot that formed. The cold contracted because in nature cold contracts. The heat expanded into a dimension of form and heat by expansion is in nature about motion. Motion is duplicating that which is and heat is what is duplicating by motion. But only heat by expansion was possible because in affect singularity cannot move. The motion became contraction, as the motion was the result of heat expanding which was forming four points in the rim of the dot. The expanding of the points created motion in relevance of a centre that formed because of the motion, which established an immovable centre as the Coanda effect, placed more dots in relation to more dots that formed

Every dot was Π and every dot formed Π^3 because of the expanding heat, which produced Π^2. With that a new relevancy came about forming a centre in between the four points of expansion that was resulting in time. But since the points were in themselves singularity, which is immovable and space-less, they still heated forming a cold centre with the heat bringing about motion. It became a repetition where infinity broke eternity by producing a centre because of space (or rather form) forming the motion to enable the space to form in relation to the heat applying motion. This brought about a Cosmos being conceived.

The spot forms a full circle, but the line running through the circle is forever present because that is the future radius of the circle that will one day develop the circle, which is equal to the present diameter. The fact of the presence of such a possible line in such a possible circle dividing the possible circle into two parts makes the centre line equal to the half circle. The line forms the half circle but not only that the line presents the half circle as much as the line is the half circle. The line then is 180^0 and the half circle is 180^0 because in singularity the two factors are the same.

The same value is of course $\Pi^0 = 1$. The issue of concern is o understand that singularity cannot move. Singularity has no space. Singularity is no only part of the Universe but singularity is the Universe. By establishing motion singularity has to be charged with the time delay we find space to be. The space is time taking a period or a duration while moving from one singularity point to another singularity point while conducting the heat and the accumulation of heat that built up due to the retarding of the time to conduct the heat forms the space that is conductor to bring about the motion of the space.

It takes heat time to entice singularity and singularity can only entice. Singularity cannot move and neither can singularity form space. By enticing from one relevancy to another there is a bridging of heat that has to be crossed in order to send the gravity or the enticing or the relevancy to depart the space and reconnect the space to the next singularity. Bridging all the accumulated various time delays that formed an accumulation of heat through time distorting brings us the space we see and have. However there is no true space or motion but it is eternal motionless space is singularity charging time to provoke heat into forming space.

Three points formed a line covering singularity where the centre singularity recovered heat to grow and two points served as an axis to allow the rotation and to assist the duplication. There is one centre connecting the duplication of three as well as the recovery of one (the fourth one) that is applying the tie aspect. Therefore, motion consists of three positions in relation to a centre, which forms as space in relevancy to the motion and the space receive a controlling centre.

The duplication comes about as singularity is exciting another singularity in precise relevancy of 3 to 3 to 1, but the points charged is as space less and as motionless as only singularity is. The heat it requires to carry the exciting between points forming space and the space excites heat and the time delay it takes to excite singularity between points forms space-time. Where motion conducts electrical charging which is equal to gravity the charging of motion is to entice duplication of singularity. This is the basis, the heart and the sole ingredient of the Coanda principle that includes the Roche limit ($\Pi^2/4$). The charging of gravity $((7/10) + (7/10)) / (10 / 7) = \Pi^2$ and the charging of space-time $\Pi^3 = \Pi^2\Pi$ is all due to the relevancy brought on by the Coanda principle. The value of motion came from singularity exciting singularity and that is the duplication while the duplication or motion presents the space.

The development came into eras as the relevancies brought about new relevancies that spawned even newer relevancies that all remained in touch with the original singularity centres. Every one focused a new time delay that eventually brought about space and every distortion of time brought more. That concentrated between singularity points that charged the points to form space. When the charging became overdue in some sectors it erupted in forming the Big Bang. By the time the Big Bang erupted there was such a huge backlog in heat and time corrupted and delayed the next result was the employing of space as a commodity in the Universe. The relevancy was C the gravity was C^2 and the space was C^3. That left what was inside atom still spinning faster than the speed of light applying the relevancy of $\mathbf{k} = C$ where the electron applied the relevancy of $\mathbf{T}^2 = C^2$ and that formed the atom which then became the cube of the speed of light $\mathbf{a}^3 = C^3$. That left the atom at the relevant size of what the speed of light permitted at the time but since the Universe from that the relevancy expanded as the Atom grew in space to the extent it has now. The purpose of the star is to recapture the space the atom grew into and from there dismiss the space by spinning faster than whet the speed of light will be on the outside of the star.

• This form came about when only form was present in the cosmos and there was no room yet for space. It was in a time era where form featured in relevancies that would lead to one day becoming the atom. The atom forms a dual purpose of duplicating as well as dismissing and some prefers the one better to the other. This relevancy came in place when time was not time and space was form. Time is forever eternity being interrupted by form in infinity to bring about eternity ticking as infinity ticks. Before that singularity took on stages in forming relevancies between duplicating and dismissing space-time, which incidentally was not yet truly space-time in the sense we think of as space-time. At first a dot moved from the spot leaving the spot but taking with the spot as part of the dot to remain in the dot. The two never separated but the one allowed the other to be.

As the dot confirmed a discrepancy between infinity and eternity by defining infinity as an interruption of eternity cold and hot parted a union. The dot that formed was not space but a relaying of time to form a new point of singularity where eternity was interrupted by infinity. Time took form from 1^0 to 1^1 or from Π^0 to Π. It brought form into differentiating between interrupted eternities with infinity doing the interrupting. Then a true distinct relevance came about that positioned a time differentiation outside the realm of time by four. In this realisation we can assume that space had some meaning at this point and the formula used to investigate suggests just that.

Even in grouping, there are characteristics, which make a certain group of atoms more perceptible to duplicating and others more perceptible to dismissing

Where the seven of material duplicate by the three positions according to time and from that realise the sphere, the other side of the Universe uses other dynamics, which also realises the form of the sphere in as much as being the four five sided straight lines pulled around a circle with multiple circles attached to one specific centre and this brings the ultimate sphere. There is conserving expansion about space in acknowledging a common centre that controls and influences a dynamic formed by a sphere which is inside another sphere formed by a consistent continues flow of a line forming relevancies as lines that is connecting at 90^0 angles by measure of Pythagoras.

Finally with all the heat retarding an interrupting of the line a gap formed time.

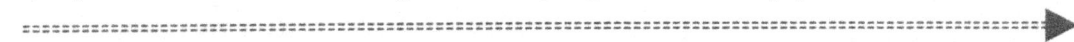

All the while it was just a spotted and dotted line running along time as space duplicated with heat surging and cooling as cold contracted much similar to the actions of stars in the process of pulsating by what ever name one wish to use.

The Black Hole proves to be the ultimate atom and is also the ultimate profile of the Coanda principle. It is stretching as far as singularity can stretch space-time in the reducing of the cosmos by gravity. What I introduce is what Kepler introduced although Kepler may have used somewhat different factors in the manner he introduced what I am about to introduce. The black Hole shows the Coanda principle in its ultimate role. Any person doubting my theory about space –time moving towards a centre and thereby creating gravity must then explain the Black Hole working. But please for sanity's sake let's forget the theory of gravity going mad in a super nova event and then becoming a monstrous Black Hole. The super nova is as natural having the process where gravity is in the process of moving from the atom accumulation to the governing centre singularity. However in the case of the super nova things went wrong because the super nova occurrence is where the transfer from movement or gravity within the star went out of control. The motion was to slow and the build up of accumulated heat exceeded the motion and over came the contraction of the governing singularity. In the process heat went back to gas as time increase by millions of folds.

On the other side of Universal time another scenario gives the sphere distinction. Three dots accumulated on one side and three dots on either side accumulated on the other side of the Alfa singularity dot. This made up the second dimension. How long did it take singularity to evolve from the spot to the dot and then on to form a sphere? It takes 1^0 to go to 1^1. To us it is incredibly small and it is so small we can form no understanding of the concept but can you see the spot, no you can't because although I know it is there, it is so small it falls outside the parameters of the Universe.

The spot $\bullet$ forms the dot $\blacklozenge$ Π^0 The spot comes first, so small it is not part of our Universe and then the next movement takes the spot to become the dot. The dot is there because the spot became the dot but the dot is there because the dot moves in relation to all other movement.

$\bullet$ The dot Not yet formed 1^0 to 1^1 $\bullet$ if the spot did not move in relation to all other dots $\blacklozenge$ in the Universe the dot $\blacklozenge$ will not form from the spot $\bullet$. It is through movement that the spot grows into the dot. The spot 1^0 forms the dot to become 1^1 and yet in relation to space the dot is simply a thought an idea an emotion that will in future become material because it relates to material.

The dot $\blacklozenge$ Π^0 **has** not yet formed material Π^0 to Π $\clubsuit$ and material must come as all other positions forming material or not forming material reposition in accordance with everything else forming a position of space within the Universe formed in space. These lines are representing the dot that serves as the spot extended. I wish to deliberate the dot in better detail. The dot is an extension of the spot and is immeasurably bigger than the spot. If the same space improvement has to come into place in terms of the way we visualize the cosmos the moon and the Sun would suddenly be double the distance way halving the space that the Earth claims.

$\bullet$ **The spot** Π^0 There are two pertinent numbers applying when space-time forms, one is two and the other is three and then seven...and it starts off with a spot that becomes a dot. The dot becomes one thousand times of what it was and remains invisible in relation to the entire Universe from the point it becomes the dot.

Where the spot ends and the dot starts there is still no space (1^1 to 1^0) and therefore with it being a mathematical factor it then represents one phase of cosmic development that stands apart from all the other phases that came to follow this phase.

In a close inspection of the dot as the dot is in our current Universe we find the dot not holding a quantifiable measured value except for being the numerical value of 1 to whatever exponential value one wish to apply. This is proven by $1^{256} = 256^0$ and this shows a specific period in cosmic development predating the Big Band by many eternity of mathematical translation of numerical figure formation to actual space. This spot that forms the dot is not tangible even in the terms we apply to anything being a part of the reality in the cosmos. It is similar to all the characteristics we have in a thought. In fact it has everything that a thought must have. Having a thought in my mind is a reality as far as I am concerned but in the view of the person next to me the thought has no right to exist.

The fact that thought is a part of the cosmos but the cadaver confirms not have a part in the body. When the body turns into a cadaver the body weighs as much as it weighed before such turning into a cadaver ended the person's stay on Earth. After the thought (also called life) leaves the body the body disintegrate to atomic particles with no structural attachment in any way. It is life that attaches the atoms forming a body to have any coherent significance that then can function as a unit. It is thought that drives the cadaver as a principle carrying life. Denying the existence of the fact of life being apart from the body and forming a part of some concept that is much bigger than the space-time we find in the body is very stupid or very Newtonian, which is the same of whatever you wish to name the same concept of having a feeble view on science. The concept of the thought is what drives the body. Life can only manipulate what is part of the cosmos and can't create anything outside the cosmos. That is then what drives the cosmos where the point holding singularity attaches to the same principal that drives the cosmos and drives the living body. Without thought or emotion or whatever you wish to name the driving principle it is the same as that which drives life but is on an immeasurable larger scale. This goes much further and from that we have a lot of information to harvest about the cosmos.

Newtonians hare the opinion that the thought is in my head but again that is much Newtonian or in other words represent that that comes about with the lack of intelligent thought. It can't be in my head because in my head is grey matter. If the thought was in my brain then after death my head must be filled with thought since my brain is still there. After death sets in the thoughts are gone and the brain is still there so therefore the thought is not in my brain but my brain is only a device translating what I think to the rest of my body. If it is the electricity with which we may charge thoughts back into the dead body by establishing a flow of electric current through the brain but doing that will not produce thoughts that would reinstall life and this shows that no part of the body holds thought but thought holds every aspect of the body. This is critical in understanding how the cosmos works and where the cosmos came from. It can't be in the grey matter but since it has no material in which to be so it must be a part of the vision in the Universe and material is part of the vision in the Universe.

The fact that the thought is there gives the thought a definite place, albeit not tangible, but definite nonetheless. Everything in the Universe confirms the Universe as much as it is being confirmed in the Universe. The thought takes up time and in order to take up time it has to take up space because space is time and space used is a form of time delayed to become the past. Before Creation the past was the present that was the future but now time in the past is space coming from the present as a result of movement coming from the future. Thoughts does not take up space so therefore it has to be space less time and therefore it has to be space that time produces while moving from the future through the present to the past. I am deliberating this because Newtonians work in the era of light where they can see the lion and when they can't see the lion there can be no lion and when they see the lion they run because there is a lion. What they can't see (or calculate by mathematical formulation) is not a reality. If they can't calculate what life is they put life in terms of electricity and that they can calculate. The reality of life not being electricity they ignore not because of stupidity because that is an insult to stupidity but to be abstinent because they think (or can they think) they can replace God.

The dot 1^0 to 1^1

In a close inspection of the dot as the dot is in our current Universe we find the dot not holding a quantifiable measured value except for being the numerical value of 1 to whatever exponential value one wish to apply. This spot that forms the dot is not tangible even in the terms we apply to anything being a part of the reality in the cosmos. It is similar to all the characteristics we have in a thought. In fact it has everything that a thought must have. Having a thought in my mind is a reality as far as I am concerned but in the view of the person next to me the thought I have has no right to exist. The thought is in the present in time and the body controlled by thought is in space. Time is the instant in which thought is and in that sense time travel as the Newtonians whish to speculate on is total madness, as much a product of their incapability to think as their creation they call "mass". Time is in the instant that the thought is and in order to control the body in space-time is needed to send a message to the arm or leg or whatever that is in the space that is behind time. There can be no time travel because time travel is rushing through space but even that requires time to accomplish that and since time is in the instant within thought it takes space to cross and time to cross the space to even control the body by thought control.

The fact that thought is a part of the cosmos but when life becomes the cadaver that absence of a thought controlling the cadaver confirms a thought has no part in the body. When the body turns into a cadaver the body weighs as much as it weighed before such turning into a cadaver ended the person's stay on Earth. It is thought that drives the cadaver as a principle carrying life. Denying the existence of the fact of life being apart from the body and forming a part of some concept that is much bigger than the space-time we find in the body, is very stupid or very Newtonian, which is the same of whatever you wish to name the same concept. The concept of the thought is what drives the body. Life can only manipulate what is part of the cosmos and can't create anything outside the cosmos. That is then what drives the cosmos where the point holding singularity attaches to the same principal that drives the cosmos and drives the living body. Without thought or emotion or whatever you wish to name the driving principle it is the same as that which drives life but is on an immeasurable larger scale.

Newtonians share the opinion that the thought is in my head but again that is as much Newtonian as the "mass" they create to hide their stupidity or in other words represent that that comes about with the lack of intelligent thought. It can't be in my head because in my head is grey matter. If it must be in the grey matter then why does the grey matter not produce thought after I am declared "brain dead" Since the grey matter is still present and the thought is absent the thought has no direct part in the grey matter? Since it has no material in which to be it cannot be a part of the vision in the Universe and material is part of the vision in the Universe. The fact that the thought is there gives the thought a definite place albeit not tangible, but definite nonetheless. Everything in the Universe confirms the Universe as much as it is being confirmed in the Universe. The thought takes up time and in order to take up time it has to take up space because space is time and space used is time delay. It does not take up space so therefore it has to be space less time and therefore it has to be space that time produces while moving. I am deliberating this because Newtonians work in the era of light where they can see the lion when there is a lion and they fear the lion when they see the lion because only when they see a lion can there be a lion. That thinking is Neanderthal but don't tell the Neanderthals because they are modern in terms of Newtonians. When they can't see the lion there can be no lion. That is somewhat atheistic or animal like in thought. The animal only sees a lion and then feels threatened. That is where Newtonian's arguments start and end. If it is not tangible, in vision, part of the smelling or in hearing range there is no possibility of it being part of the cosmos. What is in the cosmos is part of the cosmos. So too, is the thought and we have to trace the space and the time the thought holds. However I showed that there is a place in the Universe that is not part of the Universe.

The thought may drive me in reacting in any manner that comes to mind and the person next to me will never be aware of the thought I have. The fact that I have to have a thought is a proven fact without ever doubting such a statement. No person can have a mind that is clear of thoughts because in such an event such a person is brain dead. The thought is there in my mind while to others it is there in my mind without the thought being there, but such a concept reduces the Human intellect to animal instinct. It is present by never being visible and not being tangible makes the thought not being present in the view of others, although that thought is "me" in every aspect "I" am present. The only way to make the thought tangible is by speaking the thought in words. I have to reproduce the thought into a voice box and translate ideas into verbal language to make the thought a reality to others. If I wish another person should regard the thought I have with understanding I have to transform the idea I have to a voice and in the voice I use I reproduce the idea in terms of sound. But this action relates to time. Having the thought also related to time but the time spent on the thought while thinking was much less than the time used to find a means in converting the thought to sound. I transform my thought to another person where I then hope that person will translate the sound back to a thought that that person will have. No part of what I just mentioned has any tangible evidence, which I can touch or hold but notwithstanding that it is part of the most valid evidence of the human being in the Universe. Never can I reach any part of what I just said and still that is what controls the Human race. If not for the ability to think and to produce conversation by voice or body language, the thoughts I produce I would not be classified as a member of the human species. That is what makes me part of being man, the most dominant specie ever to live on Earth. If not spoken, my thought will go lost forever because I did not share the thought with any person. Yet the thought had validity because the thought controlled my life for a period and an impulse. I might try to determine the time it took me to think the thought I had, but that will be in vain because no chronograph can measure such time response. If I try to measure the time it took to transform the thought to sound through applying voice that can be measured. That is because it is done by time that already formed space.

All this arguing is to show those Newtonian nihilists that it is what they and I can't see, touch or feel that is in control of our daily life. Without the thought the deed will be impossible. But the conveying of the thought gives the thought a lasting reality because it then is shared with others. They put their reputation on mathematics, which then forms physics, but do they know mathematics or for that matter physics at all? Has any one ever explained how mathematics formed physics while the cosmos formed mathematics in the process of development? Mathematics we know is the forming of a numerical line giving values according to a flow of values by adding the original numerical value to produce growth. If mathematics started with 1 we have to add the original value that was first brought into as the first value to get to the next value as it is $1 + 1 = 2 + 1 = 3 + 1 = 4 + 1 = 5 = 1 = 6 + 1 = 7$ and so on If Newtonian conclusion is correct we have and the numerical order started with 0 it would produce no further growth. For instance $0 + 0 = 0 + 0 = 0 + 0 = 0 + 0 = 0$. This is a mathematical fact that is above argument but since it indicates Newtonian incorrectness I have never seen one Newtonian that does not belittle this fact as insignificant rhetoric on my part. Numbers would never evolve past zero.

Let us find the smallest possible line first. We already have reached the conclusion that by reducing the line, the reduced line will eventually leave all sides on the same spot. Such a spot must be round in form. With the line being the smallest line, such a line will start off as a dot that moved away from a spot. With all possible sides being in precisely the same spot we have all possible sides onto one spot. Mathematically the spot is in the single dimension where the space is one and exponentially zeros. There the space moved over to form the dot. We now are reaching into areas only the human mind can venture by understanding and nothing more. The understanding of this concept demands our reaching the point where the mind of the animal cannot reach. Numbers start with a line and numbers grow by adding the original value every time.

If it starts with a line that line only represents two sides being one and as such that is rather a flat Universe. The spot is not yet round because being round are requiring a shape or form and this lies beyond or before a time when any form of shape came into the cosmos scenario. It was in a period where shape and form was a part of the distant future hidden in and beyond eternity. In that time the line must have been so small it had reached a point not yet dividable in any way. If any further dividing took place such dividing would have brought growth because there then would form space between the sides going in the opposite direction. The dividing brought all there is having all sides literally on the precise same spot, and I have located singularity in just such a spot.

I came to the conclusion that the spot I found had to be singularity purely on the grounds that that spot holds only one side to serve as a start to the starting point of all directions possible. In that side is only one spot where there is only one side applicable and one dimension present. With all the factors given one can only come to one conclusion and that is that there can be only singularity. In such a case more dividing by two will land further positions on the other side of the divide. That point is serving as a position for all possible points and cannot allow further dividing as it is in the smallest line or spot there may ever be. This spot is the result of a most basic process of reduction as the Hubble constant is a most basic process of expanding during a matter of time. By reducing the line constantly the only value that will eventually remain without dispute from any party arguing about the facts is exponential zero. By only having exponential zero instead of a numerical zero and a radius as one in the square (the radius effectively becomes one holding any and all sides on one point) such a point might become any value of any significant measure implicating anything but zero as the radius. By expanding the line, it will be an evenly spaced structure growing into the most perfect round dot ever possible anywhere at the point when it starts to grow.

Firstly we have to get rid of the Newtonian notion of zero. A line does not start at zero because that is where Newton went miserably wrong. If a line did start with zero the Universe had to be nothing, which it is not. The Universe is an assembly of lines criss-crossing as the lines form circles where one circle follow the other circle with no room in between. When looking at the circle in the conventional manner, we persist with errors brought about in culture of stubbornly upholding Newtonian mismanagement of facts and not by applying some significant modern logic. Take a circle and reduce such a circle constantly to where it no longer can reduce. Reduce it to a point where only form remains part of the circle because the radius has gone beyond human measure and becomes so small it is not noticeable with what ever measuring tools man may use, then what remains is pi since pi does not indicate size but indicate form, and form is all that then will remain. We know that the ratio of pi is 3.1416 to one and one is singularity and not a measured meter or a specific distance. If a line forming the radius is 1 then the circle brought about is 3.1416. That is the connection to singularity and if singularity is seven then the circle is 21.991. This where we should start to look for a point where the Universe started since the Universe is a ratio of lines connecting circles. In any circle or sphere the size only depends on the fluctuation of r, as a component to the circle or sphere but that does not affect the form by indication of Π in any way there may be. The conclusion I drew from following this process is that from this line can start at zero because that will be a mathematical impossibility since no line can ever reduce to zero. If the line of the radius that initiates pi was zero, there was no pi. If there was no pi there was no Universe with or without Newton's magical forces. A line will forever be able to reduce further becoming smaller but it can never reach zero because zero is not part of the scale on which we measure lines. If a line cannot reduce to zero it then cannot start at zero. Zero is in the mind of the Newtonian but so are his forces and other witchcraft he projects into science.

A line or spot starting at zero would therefore be shorter than the shortest line possible. For obvious reasons can no line, or any line grow or extend from zero because such a line must then quit zero, which is the current value and become some other value. It has to become something, thus abandon its original value. However, a line that starts with 1 when growing it adds another 1 to become 2 and by adding another one would become three. That is where the Universe started and that I show. If Newtonians are correct and the line did start at zero with zero that would mean the start of the line has a different value to the end of the beginning and we know that a line holds numerical conformity through out. When any line is starting from point zero and it uses the factor zero, then it can never leave zero because of the influence of being zero disqualifies any possibility of growth. (1000 x 0 = 0) and that is a mathematical law. But when coming from singularity π^0 and the line then had to grow in all directions at the same pace the line must then become a circle π or being three-dimensional, then form a multi circle π^3 we named a sphere. Since the Universe is about circles and lines connecting circles I came to conclude that flowing from this fact is that in the Universe there can be no zero improvising as a filling ingredient for the space of a point or be unfilled space. Zero is no valid factor in the Universe. In the case of the growing sphere the value of the circle is Π, and that is where creation must have started. That gave me the clue where to start looking for singularity. After I found singularity that finding directed me to the point where to look for the start of the Universe. One would find singularity in the value Π as 21.991/7 or as 3.1416/ 1 and the value Π will be in all things rotating in a circle but by measure one dimension smaller. As usual I am again shooting the gun before the hunt started. Lines in mathematics do not start from zero and that is no discovery on my part but that was a realisation I came too. The Universe is all about lines and the manner that Kepler pointed to the increasing of the lines by $k= a^3 /T^2$ proves growth in the composition of all lines. However only movement secures space.

•Π^0 Singularity was the first there was and was eternally running as the past going to the present and on to the future but also returning from the future to become the past. Everything was in perfect synchronization to accomplish the perfect cycle. This was only possible because the only value that was valid was 1. One repeated eternally without ever stopping. Then for some reason that which was eternal and perfect became temporary and imperfect. Singularity distorted into a concept outside the eternal. Then that which can have no inside parted from that which can have no outside.

$\Pi^0 = 1^0$ parted from 1^1. Heat became relevant as the Universe transformed.

•1^0•1^1 Then the perfect era ended with infinity releasing eternity as much as eternity dismissing infinity. There was a past seen from the present and there was a future seen from the present. There were two dots moving from one position to the next position.

$1^{••}▶2$ The fact that 1^0 going 1^1 brought movement can only become a reality as a result of light. Light is heat and the heat is expanding.

1^0 going 1^1 1^0▶ 1^1

1^0 going 1^1 1^0▶ 1^1 had to bring about 1^0 going 1^1 1^0▶

•+•+• = 3   Two dots were moving where the one dot was the future of the other dot that formed the immediate past of the one behind and they were connected by an immovable centre dot. To find surety about this we have to look at the numerical order that came about from this. Two plus one is three so there were two in movement with three connecting the two. This brought about time in movement.

The number 1^0 was the same as 1^1 but also completely different, because the eternal repeat of duplicating while contracting was not relieved from the Universe. Before the contracting was equal to the duplicating because by measure the heat was identical to the cold. It was eternity that was interrupted by one cycle of infinity and was in repeat of eternity. This change is since once something is part of the Universe there is nowhere else to take it so it has to remain as a part of the Universe.

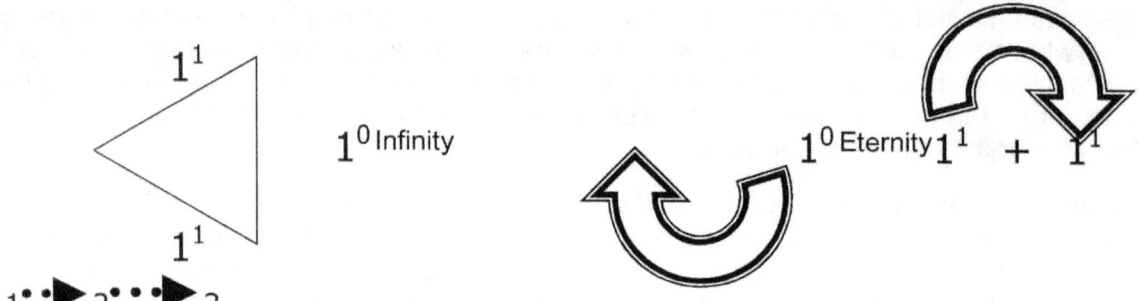

Then came three because motion was so limited that the least inclination to move threw what wished to move to the other side of the Universe, As it moves it also moved across singularity. It crossed the entire Universe as it moved because it moved and finding nowhere to move too. It crossed the entire Universe and it took one eternity less the measure of one period lasting infinity to achieve that. That brought to relevance three points where each was in measuring quantity exactly equal but also one Universe apart. In the reality there was now two points holding singularity on both sides of the Universe because by crossing the divide that crossing set in place the two sides relevant of singularity governing. However infinity was bridges at two points holding infinity with which process eternity repeated the past into the future.

•••• Since the one dot was the future to the second dot and the dot moved from the position the first dot held the movement brought eternal time and therefore represented eternity from there on. The rotation of two 1+1 = 2 brought about 2 doubling (2+2=4) as well as moving in ratio (2^2=4) the result was that on the circle time was forever moving and on the other side time was never moving. All the motion was connected to a centre point that had two moving (2+1=3) and two moving (2+1=3). Thus the one represented time moving in a ratio of three where the one dot became the future of the dot behind and the past of the dot ahead. The four dots that formed enlisted a fifth point in the centre that connected the four but also at the same instant there were three dots on the one side and three dots on the other side although the two sectors shared the same dot. In order to bring distinction to the two times three holding two opposing sides the one sector turned in the opposite direction of the other sector while also sharing one centre.

This development introduces opposing directional spin and is a major factor in Universe development to this day. It is part of fundamental gravitational influences. Also by bringing in three forms with equal influence the straight line the half circle and the triangle receives the same and an equal value.

There is a markedly difference in how five forms space-time and how seven forms space-time although both applies the same principle although not distinctly the same.
Five forms with the four rotating and one point forming the centre shifting one position or in other words
$a^3 = T^2k$.

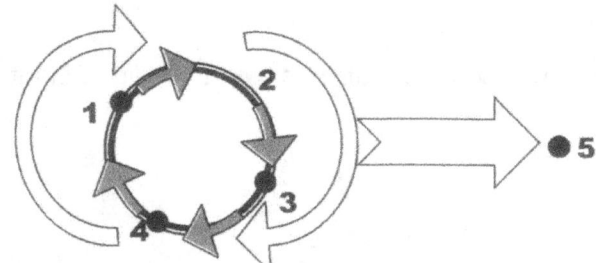

This is the first infinitive limit that Creation developed because this is where the value of Π forms by implementing the law of Pythagoras and it is the law of Pythagoras that kick starts all mathematics and forms the basis of physics and with that Creation by name. From this come the discipline known as the Lagrangian points and the confining of time into space. This starts mathematics by numerical value because one half of the Universe is ten in space and the other half of the Universe is ten in space and when combing that with singularity. We find pi form by the four locations presenting five places in conjunction with an allocated centre every time.

Although six is part of the concept five brought into duplicating the numerical order that followed five it also placed a new spot into the numerical order. Since three represents a line forming the order that comes about is that five completes time as time returns to the past by completion of five. The forming of the triangle and the straight line as well as the circle brings time into a repeat of what was to what will be.

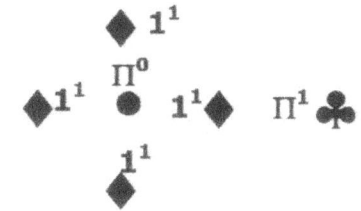

The formed a repeat of what was to what again will be. His is a result of he fact that once something is in the Universe it will repeat constantly and the repeat of that will always repeat constantly because the cosmos does not change its basis such as time.

$1^{\bullet\ \bullet}\blacktriangleright 2^{\bullet\ \bullet\ \bullet}\blacktriangleright 3^{\bullet\ \bullet\ \bullet\ \bullet}\blacktriangleright 4$

This then is the occasion where Pythagoras stepped in. Since it as a crossing of the divide the crossing involved a line that formed a half circle connecting a triangle. But the crossing was done in the space of half the Universe and since the Universe was 180^0 half of the Universe was 90^0. That involved Pythagoras as mathematics was born. Up to this point it was arithmetic done with adding but now mathematics by multiplication came into place. Remember we are a few eternities in side the development of the Universe. Later on I am going to show how critical part the Law of Pythagoras plays in the forming of Creation and how from that an entire Universe comes about as a result thereof.

The radius that the Universe has is a factor of two. The increasing of this has a value of seven points. The relevancy in-between forms a Universe we have. The dimensions this carries are 2, 2^2, 2^3, 2^4. The significance of this I elaborate in ***Matters' Time in Space: The Thesis*** where I discuss how stars function and the significance of displacement of space-time inn terms of singularity.

How did we get this far? There was one and then a split came in time. Time brought in space and in space there was a past and since there is a past there has to be a future. That made the past one point away from the future, which is three points. But since two points move one point is moving to the next position which is the present and one point is moving to the future which is the next position and since the present is the same poison we have two points (the past and the future) moving (the circle) Because the present remains the same position movement occurs in relation to one point that remains fixed.

This puts a line of time that remains forever but then space formed as a circle with the two points circling about the one point serving as the line. From the present the past is a point from where the present is moving towards the past. If there is a past there has to be a future that moves towards the present in order to fill the void. That is how gravity works. By moving towards the circle in a descending manner, the space just below any given point forms the future of the space just above and the space just above forms the past of the space just below. That is how a centrifugal pump works and the earth is just a centrifugal pump pumping away at space and compressing space towards the earth. If a pendulum can measure time by measuring the space the earth compresses in gravity then what the pendulum measures is time and then time is gravity compressing space. If time is compressing space then time is space expanding as well and so time is the movement of space towards as well as away from any given point.

Time is a line formed by • singularity and runs as straight as no other.

This line that holds time within the sphere is erected by the circle that spins about it and this we see in the manner the top spins. The spin takes the dot from the future to the past so by movement there are two dot moving future going to present as well as present going to past.

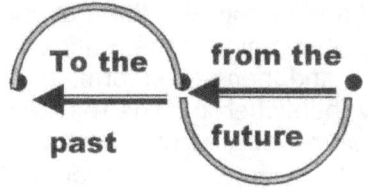

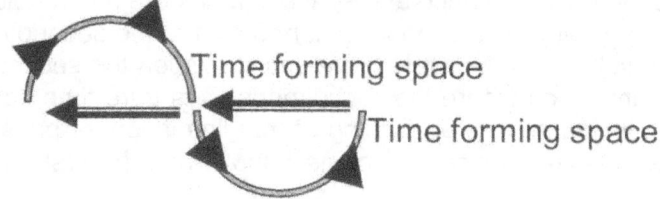

Time forming space

Time forming space

The movement alone forms two points but this cannot ignore the third point, which seemingly is shared, as it is stationary.

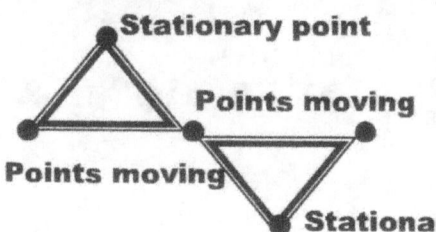

This shows that two points move as a circle but when the third point connected the two points moving we find this results in a triangle. It is still a straight line of time shifting through space but bringing in space as a result of time delayed this forms space as a triangle and a half circle although both also ahs the equal value of a straight line.

At this point space was an idea of time departing setting apart one point carrying time in infinity from the other sector in time forming eternity. Because the past assembles validity from the present initiating the past and the future taking queue from the present as what is going to be we must regard that everything starts with the present going past while allowing the future to flow into the present. For that reason time travel is a farce because no one ever tried to find the difference between time and space.

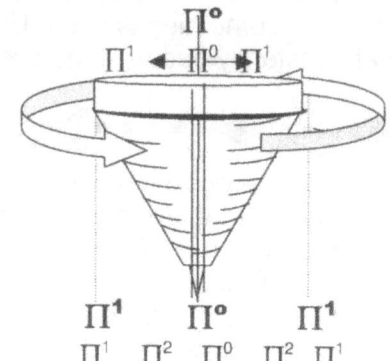

Time was in the represent. There was no numerical order because the numerical order gave value to one point that moves as two but remained as one that forms the constant. The flow of time gave a numerical value to three points because space brought about a circle and the circle held relevancy

We can see this evidence in the way the top generates gravity and applies relevancy. A circle puts a line as an axis in place and the axis supports the circle presenting space in which the top can spin. This is why everything that is has to spin while going straight but going straight calls for a circle to be and what spins will spin again on another level.

There were two points moving bringing about two more points that was the same but filling different locations, which then formed the square of two and that, is four. Then we trace another dimension rising from the way mathematics formed as the circle by rotation substantiated the line that validated the triangle. The line was the half circle by including one dot and became the triangle by validating the third dot. The number of four is just the result of the complete circle forming and five puts a centre into the circle of four to secure the circle. Then the number of six came about as the triangle that formed doubled and by it's doubling the trend for duplicating the lien came about. Transforming the entire value of the sphere brought about replacing the four of the circle as well as the three of the line, which forms seven. The rotation is in opposition to each other, which puts a distinction to what applies "on the other side of the Universe".

What this argument further proves is that the circle rotating must then come from all four points because the radius might be a line but that line represents the centre of a circle through 360^0 coming from and accounting for all possible directions. Taking that into account it is important to recognise that notwithstanding the size of what a line that forms any radius of any size might be, there is another line (or dot) that is eternally bigger as well as eternally smaller than the line in question. The line represents an innumerable many variations of possible sizes as it was in the past or what it would be in the future. While we are in the third dimension being part of the third dimension such being in the third dimension then allows that all parts of the third dimension forever can be divided once more until the line in the third dimension is no longer part of the third dimension. That then is as far as we have to take the lien to see where the cosmos started. When such a line leaves the third dimension it is still dividable because it might not be part of our dimension any more but it can still reduce further as part of the second dimension going into singularity. By that time it has left our scope by miles but that does not mean that it ends there because from our perspective that is where it ends. But our perspective does not represent reality.

Our perception involves space and space is a hologram that holds the past and therefore in the present does not form reality. Yet, even then connecting to the second dimension, when being in the realms of space it can still reduce infinitely more until it has left the second dimension and then at last forms part of the first dimension where the first dimension is pure time flowing. Only then when the line reaches the first dimension no further dividing of that line is longer possible. We can never grasp the size of a line that the first line in size that came about when the first motion broke the eternal stranglehold on space.

According to our big and small conceptions of what we perceive as large, ultra large, small and microscopic small is just mere words describing thoughts totally unrealistic in the context of what the cosmos sprang from as the cosmos moved out of the spot and formed a dot. Even by the standards of forming the dot, which was eternally bigger that the spot **T**, as the dot and all the many dots that came from the spot. The size differentiation only between those two exceeds all limits and divides we wish to create forming borders that we can appreciate. By coming from the spot and becoming the dot the sphere or space forms. Looking at the sphere we see many circles that are in association with each other and to find the truth we can look at one circle and see the characteristics of all spheres.

Any point will be opposing itself within the **rotating of 180°** where it **then change every aspect** of its **previous flowing** characteristics it had or **will once again have** in **360°** from there. While in rotation from the view point of a bystander it all may seem static and never changing but to the object in spin every next instant in time will be diverting from every aspect it had every second passing, and the direction it held in relation to the direction it held the previous mille, mille second will totally be incompatible with the direction it holds the very next mille, mille second of rotation. This is why we can

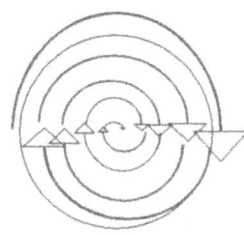

use degrees measuring the circle by (6^2) (forming the square relating to matter through singularity) X 10 (square if space) = 360° however it is always in motion. That proves no point can be static or constant, though it may seem that way to outsiders. Although matter is matter, matter can also be anti-matter and moreover form its own anti-matter at the same time. This degeneration of structure is very likely to occur with overheating. Revaluing Π to Π^2 will bring about a new contact point where Π meets **r** forming another relation in Π^2. Every time material swap sides it also qualifies as anti matter to matter because

if it goes out of orbiting rotation frequency, it has the ability to collide with the same matter it forms union with but is located on the other part of the spin. It then becomes in a situation where Π **revalue to r. Time is** the **changes in relation** where Π **contacts a different r** not withstanding the many r points there may form because **every r constitutes a different value** to the Universe through other ratios and relevancies brought about **by heat and light. Time is the duration it takes Π to rotate between any two given points of r** and therefore must always amount to **a square (T^2)** moving from point to point through the **cube of space (a^3)** in that **duration of time (k)**. With that it proves **Kepler's a^3 (space) =T^2k (time in the instant of motion)** but motion must continue through a specific value in space where the space-time is maintaining relevant equilibriums throughout singularity connecting.

Moving as an object that forms the numerical value the orbiting object also secures an individual independent and own centre but from the orbiting object the limit it holds ends at the edge it forms. This we named infinity and also the present. Being a circle that formed the object secures four rotating positions in relation to a connecting centre. The forming the shape the Universe applies which is a sphere the orbiting object secures seven positions and the larger containing sphere is ten of which seven is within the singularity dimensions within the centre, which we not observe. Immediately following that as part of that relevancy comes the containing sphere that holds space-time and another three positions. The three positions puts a relevancy of three to the holding space that already caries seven. There are forever another centre that secures seven positions just because singularity chose the sphere at the value of Π and in the sphere 7 positions is made up of six sides that hold relevance to a precise centre. There are Π^3 but then there are Π^2 putting Π^3 at a value. Then there is a relevancy named Π, which puts Π^2 at a relevancy. None of these is fixed markers because the relevancy can and does swap sides placing importance as alliances changes. When Π^2 focus on another a^3 the relevancy about k changes and amplifies the importance of yet another space. When one applies the Coanda effect one would see just how easily, new alliances come in place and secure new centres that charge new relevancies between newly established points. Going either "bigger" or "smaller" is only shifting focus on another relevancy.

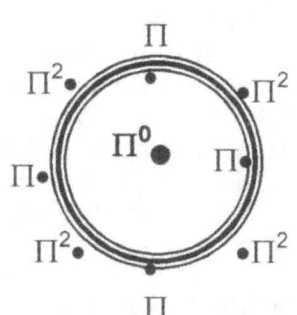

Let us see why would space form. We have time forming two sectors where one moves and where one stands still. The moving is eternity and the one standing still is infinity. If only movement applied only eternity would be present but we know we are in the present, which represents infinity. We think we see movement but we see movement within space and not the movement of time. Infinity is double as long as eternity. Eternity is taking what is and carrying on with that forever or until infinity replaces the moment carried by infinity. Then infinity takes what eternity is and replaces it with what eternity will be until such a time, as infinity will again interrupt eternity by one instant.

Let us then with that in mind consider how growth will come about that forms space. Time we already placed on a line. Space we can see has six sides. There is one going forward and one backwards and on left and one right and on to the top and to the bottom. That is six sides. If a line starts growing space it is conclusive that the growth will go into all directions equally by not favouring one specific line at any given point. It is at this point that patrician developed between points holding singularity as a solid and points not holding singularity as a solid. The movement brought about the relevance of Π and the movement brought into play gravity measured at the movement of the relevancy and therefore Π^2. By the relevancy moving space was required to bring distinction between what was and what will be and then what is took up space for the very first time. The relevancy moving Π^2 formed the for points in contact of the expanding and the Roche limit brought an effective boundary before the next line came into place. This gave the Roche limit a numerical value of $\Pi^2/4=2.4674$ the size of the straight line that formed.

By just being it had to move. But that put all lines in the cosmos at a straight as well as circling forming a doubled **k** as well as $\mathbf{T^2}$

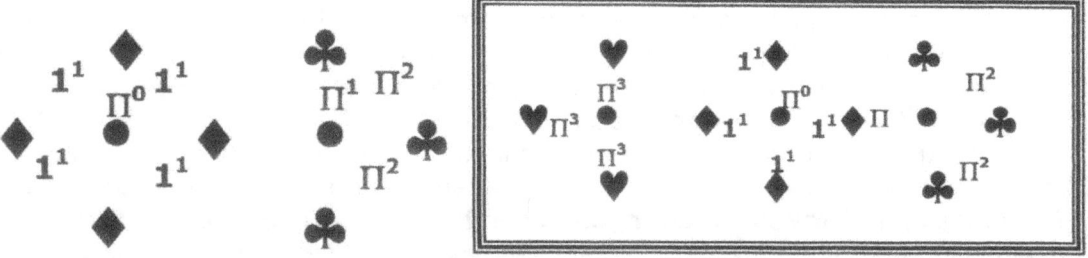

The truth is at that point all there was, was an inside and an outside and a promise of what might come. Before there was what there now is there was what we might think of as what is perfect, because the Universe was not in the in the fourth dimension in which we are but time was in repeat of continuing the perfect. That which was in the present was going into the future at the instant it came from the past because crossing the Universe took one eternity just as it does now, but back then all there were was one eternity holding one Universe within it without boundaries or limits and that was exactly what there was. There was one Universe and that was what there was to cross while it took one eternity to do so. Still today it is just the time it still takes to accomplish the only accomplishment available at the time back then when it all began to go oblong. However what there were, is what now is because what there was had nowhere to go but to remain where it is which is inside the Universe. There is nothing to add to what is within the Universe and therefore what is in the Universe was in the Universe at the time it all began.

As the Bible says nothing is new (under the sun) meaning in the Universe. The way the Universe started is the way the star repeats every instant after the start ended. The start had to end but the end had to begin a new start and this continues by the implementation of the four cosmic pillars. If the start was introduced into the cosmos the start became procedure because the start had to continue since the start had nowhere to go. The cosmos restarts every instant of the smallest fraction time could form and ends as eternity develops away from infinity. Then infinity introduces a new start to the process that ends in eternity developing and ending in space as space formed by light.

The movement back countered the movement foreword because while the movement went foreword it was in the process of moving back. It was in the dimension of r^4 but at the time this state of affairs was all connected to r^0 and since r^4 was r^0 the lot was the same. Therefore because the past present and future was the same time carried on monstrously as it does now, but without distinction as it does now. In all that it was everything was still inside singularity as well as inside the present keeping everything there were that was locked into a thought, moving into the future as it was moving from the past. In singularity the present, the future and the past has no distinction and therefore moving into the future was exactly what was taking the movement from the past into the present.

The Universe was perfectly posed, perfectly balanced, perfectly harmonized and totally synchronized. What was hot was cold as well. What was big was also small. That, which was far away was just as near and close to everything at the same time. What was were as big as the other and as far or as close as everything else. Going into was coming out of and what was the future was the past exactly duplicated. From our perspective we now have we will find that everything stood still but that cannot be because then there was nothing to be and everything was being only it was being in a state of precise duplication with no change to the slightest detail possible while the lot was happening at the same time. That which has no inside was united with that which has no outside and the only limit was that everything could be everywhere since there was no place to go but to be everywhere all over at the same time. I am trying to be persistent on this matter because we fall into concepts we now identify with being ends, boundaries, pockets, limits, edges, and diversities while then all this was something carrying no concept. It still is like that except fro light forming space and in the hologram of space in light we find the specifications we live by.

Then suddenly 1^0 moved from 1^4 to 1^1 taking away 1^1 and leaving 1^4 at 1^3 and 1^1. The moving involves 1^2 and suddenly we had one Universe moving from being 1^4 onto being $1^0 = 1^3 / 1^2 \times 1^1$. The perfect Universe was no longer perfect. That is what must be realised. We cannot see what is now and try to judge from that, what was then. The imperfect grew but that did not dissolve the prefect. It just gave the perfect some other option to be as long as time has a perfect running in between imperfects.

Gravity is the strongest where space is least because motion at that point there is infinitive. Consider the spot before it even formed the dot. The space it had was is not there at all although all that we now see and cannot see because of the vastness of the concept, was locked into one point that formed everything while not even being except being the present. Going across the entire Universe was equal to an attempt to move, going from one point to the next point but the effort of leaving the one point was too massive to take place. The time it took to cross the Universe was one eternity of not crossing or not getting across. One might say that is still the case, but that is because the Universe remains that big or small as it was at that time. By any attempt to move that which was there that called on such gravity also called all the factors that there ever might be throughout the entire Universe to one spot in order to frustrate the motion. By one attempt not forming motion but only an attempt to commit motion gravity released the Universe and overcame the entirety of frustration to resist motion, and while this was going on and all the gravity in motion that might ever apply to and in the entire Universe was in a spot concentrating on a spot that is not yet part of the entirety we consider as the Universe. That is because what we think of as the Universe is not the Universe but is a hologram of what was at a specific time and scattered in relation to time forming space forming distance forming size by filling what there is into the hologram. While it became the Universe the split of time from singularity established the gravity that lies between what is not because to us all must end somewhere and that then which is there then is not because all must start and end somewhere and that singularity does not accomplish. Only the hologram of not being while forming pictures using light to form space to form time ends what never starts. What would generate the equivalent to accomplish that?

What would create the motion to break the strangle hold whereby of eternity restricted infinity. It is gravity that forms time that forms space that keeps eternity from reuniting with infinity again and motion keeps the time zones apart. Time leaves space as the wake moving away from the centre of time. Do not be fooled by the seemingly innocent explanation that space is the motion thereof which is what gravity produces because of all things the cosmos creates, motion of space through time is the utmost complex manoeuvre and without bringing a restraining of mathematics into science, it is so complex there is no viable explaining in physics about how the cosmos produce the act of motion of space in time. In movement we find that the atom in front of the one following holds a specific relation because the atom in front becomes the future of the one following and that one finds the atom behind filling the role of the past. While the atom in front is vacating space to fill the space of the atom in front it is vacating at that instant that the atom behind is filling the space that the atom in front has vacated in order to vacate and relinquish the previous position in favour of the following position to honour the direction gravity is insisting upon. Removing material from space by filling material into a position of new space sounds simple because the complexity has never been realised. That is $2^4 \times 7 = 112$. Gravity is motion differentiation between objects. While falling the gravity applies as moving of space that is putting time in relation to the distance travelled. While the object falls towards the earth it is then that the motion confirms gravity and gravity forms time. When motion ends only then does mass set in and it becomes the constraining of the object preventing further independent motion. Where a confronting of objects restricts gravity to ensure the objects independence the action then implements an introducing of the mass as a substituting factor to motion that then replace motion as substitute to the motion that would be and the mass is providing the tendency of gravity being the motion of space. However mass then restricts motion and becomes motion in a tendency to apply motion. While falling gravity applies and motion neutralizes size, mass or weight. Mass counters motion being when the Earth restrains further motion of the falling object and the moving object is stopped from further movement where mass is then preventing or hindering gravity. Further movement is disallowed as other material fill space that falling body wants to lay claim to. Mass then sets in not causing the motion but substituting the motion and from that motion restriction becomes resistance that becomes mass. While falling the object is experiencing gravity because the object is in gravity but when on the soil the object experience mass which is the restricting of gravity or motion by other space filled with material. When any person is standing on any place anywhere, while viewing the Universe, that person is filling the centre of the Universe. All the light that come across and travelled all of the vacant space from any and all possible positions in space runs directly towards your position using a straight line towards you where you are filling the centre of the Universe. Realising this will end this nonsense of looking for the end of the Universe. Wherever the smart-arse astronomer places the end of the Universe it must also be realised that in that instant at that point we will find the centre of the Universe looming.

Should you decide to shift your position to any other place in the Universe, you will shift the centre of the Universe to that location as well. If you install a camera on Mars, the light is obliged to acknowledge your relocating the centre of the Universe at your will to reposition you're being that centre of the Universe where the camera then is. All the light that ever left its destination crossing the vast spaces of the Universe, excluding no particular light, travelled all the way just to find you filling the centre of the Universe, right where you are, wherever you are. By you're standing anywhere, you fill the centre of the Universe, and the entire Universe admits to that because all the light comes to meet you there. If you shift from the North Pole to the South Pole you will shift the centre of the Universe because all the light travelling throughout the Universe will find you where you then moved the centre of the Universe. The light left its destination billion years ago as it travelled through space at the speed of light anxious to acknowledge you're being in the very centre of the Universe. The Universe is spinning around you or I, which is filling a centre where all motion is connected. It implicates gravity as wide as can be... It is our task to find space, to find time and moreover it is our optimal task to find the Universe. Gravity is to move or apply the intension to move space a^3 at the distance or relevancy of k while T^2 is the time it is going to take to apply gravity or move the space filled with material space a^3 at the distance of k in the time period of T^2. That confirms Kepler's attribution to gravity where according to Kepler space a^3 is equal to the movement T^2 (time it takes to move) at the distance k from the centre specific.

It puts all aspects of gravity in the Universe in new dimensions. If gravity is motion, what causes motion? What stops motion? What is the gravity if the star has melted all atoms it had into one all-inclusive atom and this all-inclusive atom is providing all the gravity that the star had when the star still had massive volumetric space? If all that space that once filled an entire giant star fused into one specific space less centre holding singularity 1^0 then the enormous gravity is applying to the centre of such a non existing space-less atom and that entire enormous force has been secured in the space less than that which one atom holds. In that case the atom would then show a force that would pull the surrounding Universe flat. The purpose of fusion is to reduce space and magnify space less ness inside the sphere. Where does the gravity of the star end when all the atoms in the star became one giant atom by fusing all atoms into one nucleus? Gravity is smallest where space is least.

Where space of an entire massive star is left in the size of one atom the gravity coming from that will pull the Universe flat at that point. Coming to the conclusion about gravity being motion and mass being the restriction of motion was the easy part. The facts that presents the understanding of what produces the motion and what prevents the restriction from overcoming the motion was the part that required thinking. Gravity is The Roche limit Gravity is The Lagrangian system; Gravity is The Titius Bode law; Gravity is The Coanda affect. By you're filling the most exclusive spot there is in the entire Universe, right in the centre of the known Universe now you'll have to buy the book to know why you are in the centre of the Universe.

Gravity is about reducing space and maintaining different cosmic sides not sharing the same sort of space. It is how gravity does it and why gravity does it that produce the accepting of the Big Bang Theory but it also includes the Plasma theory in detail. In the beginning before and after the Big Bang (yes before because there had to be a before) gravity was about bringing across heat that was in space to material that was in another space. That is how Mainstream science presents the forming of all particles. The only difference is that in the space where the space is filled with plasma or heat and not material, the heat was much denser then compared to now. Now they present that same space holding nothing but they never show where such a process stopped at a later stage. I say it never stopped and is still the result of gravity applying. Material is still growing but the growth now is hardly noticeable because the heat in space lacks density to provide such growth and space has much less heat density compared to what there was back when...

Mainstream science claims that many aspects of material (matter and antimatter) and singularity amongst others were present in the cosmos during the early phases but has vanished since then. Think about it clearly…What ever was in the Universe had nowhere to go but to remain in the Universe after the Big Bang. The Universe is everything there can be. That means if singularity was part of the cosmos during the Big Bang singularity must still be in the cosmos. If antimatter was present during the Big Bang it must still be present. There is nowhere to go except remain as part of the Universe! Space was little and heat was massive. Heat became space as the density of heat turned to form space through a process we named exploding, which is space expanding. The fact that heat turns to space was never recognised by science in the past. The fact that space compressed forms heat notwithstanding all evidence supporting this knowledge was never accepted by science while all combustible engines work on heat coming from space compressing to bring about heat. If you push in a cylinder the space heats up but science does not recognise such obvious facts. While the whole world produces energy from a principle that is about converting heat to space mainstream science chooses to deny the fact. In the beginning singularity was present in the Big Bang. Mainstream Science promotes the idea that singularity and antimatter went on the disappearing by escaping from the Universe. But Mainstream Science never says where they went. If singularity was part of the cosmos in the beginning it is still with us. However by the claiming that singularity vanished somehow since the Big Bang is totally incorrect and proves a lack of understanding cosmic concepts. Kepler and his formula also prove this fact.

Every aspect that was part of the Universe at the cosmic birth announced by the Big Bang has to be present up to this point. We have to realise that singularity is a prerequisite for the Big Bang to have formed any and all material. The cosmos has lines forming cubes and lines forming circles, which in applying 3D manifests as spheres. Between the circles and the cubes runs lines so the key to understanding the Universe are lines. The Big Bang was a time when the Universe was incredibly small making the running lines small. Understanding the Universe is taking the line connecting particles through space back to its limits where such limits were during the Big Bang. Extend that value received to a Universal centre and bring that value to align with Kepler's $a^3 = kT^2$ and understanding the Universe by finding the centre of the Universe makes the Universe simple as can be The Universe becomes sensible making the entire different yet unexplained phenomenon as easy as children schoolwork. There are suddenly no more mysteries in the Universe. It is only possible when we see gravity not as a grabbing force instead of seeing gravity reducing the space between particles. Gravity is not being some magic force found between particles grabbing onto everything.

What was within the Universe at the start will be in the Universe at the end. The Universe holds all; maintains everything and combines the lot. In Afrikaans we call the Universe the "Heelal". It is a combination of two words namely "geheel" and "alles". "Geheel" means everything and "alles means everything. Therefore the "heelal" directly translated from Afrikaans to English will mean the "Everything of everything". Nothing can be added and nothing can be lost. It is all-inclusive. With this fact so commonly known and accepted, how can the Universe grow? When the Universe started it started with 1 because 0 is the value of what we could find between the ears of the Newtonian atheist. Notwithstanding Newtonians' feeble views on the matter nothing can start with 0 because if a line started with zero it has to continue with what it started with. If a line has an initial number it stars with as 0 it must continue to apply the initial number being 0.

I do realise Newtonians filled the Universe with nothing to a point of overflow but because of their atheistic low understanding of Creation and the way the Creator created the Universe nothing is a validation of their concepts and not a factor in the Universe. From anywhere to everywhere the Universe forms by lines connecting points. Innumerable lines running all over form the Universe. If the line was zero as the atheistic Newtonian minded nihilist wish to think then when adding zero the line must be $0 + 0 = 0 + 0 = 0 + 0 = 0 + 0 = 0$. The line must then continue as 0 because it started as 0 and it has to be what it started with, because by changing the value it changes character. In order to then switch to one would dissolve the start as the start will change value and that would repudiate the 0 as it did not apply as a starting value. This is where numbers start and even that the Newtonian atheist does not understand and yet they claim to be mathematical masters. They don't even know where the numerical order starts but they want to tell everyone where the Universe started.

The Universe started as a line because the Universe is still a line. In order to know where the Universe started one have to look where a line starts because the Universe is made up of innumerable lines crossing at innumerable angles. Still the Universe is a line forming a Universe. A line starts off with 1. The line must be a repeat of whatever the line starts with and in that we have to focus our understanding of how the Universe started. The Universe started with 1 because there was no need for two. The Universe was perfect and on the spot holding the future the present presented the past. Time did not part as eternity was the same line as infinity and infinity continued into eternity. A line therefore starts with 1. As the lien progresses it will continue with the value with which it started. It started with 1 + 1 = 2 + 1 = 3 + 1 = 4 + 1 = 5 + 1 = 6 + 1 = 7 + 1 = 8 + 1 = 9 + 1 = 10. This is a numerical order and not 0 + 0 = 0 + 0 = 0 + 0 = 0 = 0 = 0, which proves zero is invalid as a numerical number and therefore on the graph the lines parting the graph by forming distinction is not zero but is 1^1.

There was Π^0, which was α^0 or if you would rather have it Ω^0 or it maybe was 1^0, but more correctly it was all the above and the beyond because multiplying what ever constitute the mentioned will bring about what is mentioned to a precise equality. It was a spot that was not. It was a line that ran eternal but because it ran eternal and kept repeating exactly what was before to the precise what came afterwards the line was there and was eternally running, while never changing in the least or growing by any measure. It was not one because before it was one, what was repeated and the process cycled back to before one and before one could be reached. It was such a continuing of the monotony, no change occurred and therefore never did the running produce progress because the progress was in the perfect repeat of what was before. The duplication brought contraction to the minutes detail. That is where our atheists get one hiccup. The repeat brought eternity and the repeat was so perfect that the repeat continued. The repeat still is with us as much as we are within the repeat. There was something beyond the Universe that institutes change. There was something that brought a difference and we are within that difference. That difference was time and that time is what we move through as much as what we see at night. Oh, how stupid and how thoughtless the minds of atheist and other atheistic animals are. Baboons do not recognize this because they cannot think and are therefore atheists. Spiders cannot think and therefore they are atheists, as they do not think what the night consists of. Reptiles cannot think and without thought they are incapable to see what time is, what space is, what light is and what darkness cannot be. All the animals I have mentioned are mindless atheists because they fail to see beyond the visible into the realms of the thinkable. Because of the incapacity to think the animals are both mindless and they are atheists. Therefore atheists are mindless. The night sky is such a bright light our evolution protected our vision from the brightness in order to give as much better vision. Through evolution development our eyes is protected and we remove the qualities from the light. However animals do use that light and not our light to see by. You can shine a bright hunting spotlight onto an animal at night and the animal will not be able to see the light on it. The animal does not use the light to see better as the animal is totally unaware of the light. Then a prowler come from the night and see the animal in the light the night provides. It does not use the light the spotlight uses and the light is not even traceable to either the hunter or the hunted. From there we accept that during the day the animals must be using our light to see because the nightlight is inferior to see by. Who says they use the daylight much different from the nightlight because all evidence is there that they cannot recognize our light as light. It is very evident in the manner they go on hunting and grazing while being totally unaffected by our form of light. That which you see at night because you cannot see darkness and you cannot see black is the light the Universe is painted in just like the Bible says. This is not religion and it is not a sermon, it is hard-core and brutal basic science and it the most fundamental basic physics there is. It is the start of the mathematical Universe portraying the only physical way it could ever be.

My atheistic idiots, your mindlessness caught up with you! Then came this light that the Bible refers to as the first of what ever was and what our stupidity tells us is darkness. It was there and it interrupted the true invisible darkness, the blackness of a Black Hole, the invisibility coming from within the Black Hole. Eternity tore from infinity. Darkness broke from light. Heat broke from cold. Relevancies parted by 1^0 going 1^1. There was one but also there was two too because one cannot be without two being there to ensure one is one. $\Pi^0 \Rightarrow \Pi$. In this there was only space for one being one in the two forming one. It was $\Pi^0 \Rightarrow \Pi$ however there was no space to be $\Pi^0 \Rightarrow \Pi$ and there fore because of the lack of space to be which is the infinity of time braking the eternity of time the true measure was $\Pi^0 \Rightarrow \Pi$ but realized only 1^0 going 1^1. Π was to the future because of the motion of time involved and the space less ness of space at the time. By inclining to move the process crossed the Universe but also it took one eternity to accomplish the feat. The fact that 1^0 going 1^1 brought movement can only become a reality as a result of light. Light is heat and the heat is expanding.

1^0 going 1^1 1^0 ➡ 1^1

www.ingramcontent.com/pod-product-compliance
Lightning Source LLC
Chambersburg PA
CBHW080647190526
45169CB00006B/2016